The River of Golden Sand

The Narrative of a Journey through China and Eastern Tibet to Burmah

VOLUME 1

WILLIAM JOHN GILL

CAMBRIDGE UNIVERSITY PRESS

Cambridge, New York, Melbourne, Madrid, Cape Town, Singapore,
São Paolo, Delhi, Dubai, Tokyo

Published in the United States of America by Cambridge University Press, New York

www.cambridge.org
Information on this title: www.cambridge.org/9781108019538

© in this compilation Cambridge University Press 2010

This edition first published 1880
This digitally printed version 2010

ISBN 978-1-108-01953-8 Paperback

This book reproduces the text of the original edition. The content and language reflect
the beliefs, practices and terminology of their time, and have not been updated.

Cambridge University Press wishes to make clear that the book, unless originally published
by Cambridge, is not being republished by, in association or collaboration with, or
with the endorsement or approval of, the original publisher or its successors in title.

CAMBRIDGE LIBRARY COLLECTION
Books of enduring scholarly value

Travel and Exploration

The history of travel writing dates back to the Bible, Caesar, the Vikings and the Crusaders, and its many themes include war, trade, science and recreation. Explorers from Columbus to Cook charted lands not previously visited by Western travellers, and were followed by merchants, missionaries, and colonists, who wrote accounts of their experiences. The development of steam power in the nineteenth century provided opportunities for increasing numbers of 'ordinary' people to travel further, more economically, and more safely, and resulted in great enthusiasm for travel writing among the reading public. Works included in this series range from first-hand descriptions of previously unrecorded places, to literary accounts of the strange habits of foreigners, to examples of the burgeoning numbers of guidebooks produced to satisfy the needs of a new kind of traveller - the tourist.

The River of Golden Sand

William Gill (1843-1883) was an explorer and commissioned officer in the Royal Engineers. After inheriting a fortune from a distant relative in 1871, Gill decided to remain in Army and use his inheritance to finance explorations of remote countries, satisfying his love of travel and gathering intelligence for the British government. He was awarded a gold medal by the Royal Geographical Society in 1879 for his scientific observations on his expeditions. This two volume work, first published in 1880, is Gill's account of his expedition from Chengdu, China through Sichuan, along the eastern edge of Tibet via Litang, to Bhamo in Burma, a region little explored by westerners before him. Gill describes the cultures, societies, settlements and economic and political situation in the region in vivid detail. Volume 1 covers the area around Chengdu and includes an introductory chapter by the eminent orientalist Henry Yule (1820-1889).

Cambridge University Press has long been a pioneer in the reissuing of out-of-print titles from its own backlist, producing digital reprints of books that are still sought after by scholars and students but could not be reprinted economically using traditional technology. The Cambridge Library Collection extends this activity to a wider range of books which are still of importance to researchers and professionals, either for the source material they contain, or as landmarks in the history of their academic discipline.

Drawing from the world-renowned collections in the Cambridge University Library, and guided by the advice of experts in each subject area, Cambridge University Press is using state-of-the-art scanning machines in its own Printing House to capture the content of each book selected for inclusion. The files are processed to give a consistently clear, crisp image, and the books finished to the high quality standard for which the Press is recognised around the world. The latest print-on-demand technology ensures that the books will remain available indefinitely, and that orders for single or multiple copies can quickly be supplied.

The Cambridge Library Collection will bring back to life books of enduring scholarly value (including out-of-copyright works originally issued by other publishers) across a wide range of disciplines in the humanities and social sciences and in science and technology.

THE
RIVER OF GOLDEN SAND

FIRST VOLUME

MOUNT LIEH SHAN-LIANG IN MONGOLIA.

THE
RIVER OF GOLDEN SAND

*THE NARRATIVE OF A JOURNEY THROUGH CHINA
AND EASTERN TIBET TO BURMAH*

WITH ILLUSTRATIONS
AND TEN MAPS FROM ORIGINAL SURVEYS

By Capt. WILLIAM GILL, Royal Engineers

With an Introductory Essay

By Col. HENRY YULE, C.B., R.E.

THE FERRY

IN TWO VOLUMES—VOL. I.

LONDON
JOHN MURRAY, ALBEMARLE STREET
1880

All rights reserved

CONTENTS

OF

THE FIRST VOLUME.

INTRODUCTORY ESSAY.

§ 1. Captain Gill's Communication with the Writer before going to China. § 2. Introduced to Baron von Richthofen and Mr. T. T. Cooper. § 3. Title of the Book Explained; and Scope of the Introduction. § 4. The Great Parallel Rivers issuing from Eastern Tibet: (i.) The Subanshiri. § 5. (ii.) The Dihong, and the Identity of the Great River of Tibet with the Great River of Assam. § 6. Comparative Discharges of Rivers involved in the Question. § 7. Theory of Identity of Tsanpu and Irawadi. § 8. General Grounds for Adhering to Former Conclusion. § 9. (iii.) The Dibong; does not come from Tibet. § 10. The Lohit, or True Brahmaputra; (iv.) its Identity with the Gak-bo or Ken-pu of Tibetan Geography; Murder of the Missionaries Krick and Boury. § 11. The (v.) Tchitom-chu or Ku-ts' Kiang; almost certainly a Source of the Irawadi. § 12. The (vi.) Lung-Ch'uan Kiang or Shwéli River; The Lu-Kiang (vii.) or Salwen. § 13. The Lant'sang (viii.) or Mekong. § 14. (ix.) The KIN-SHA, RIVER OF GOLDEN SAND, or Upper Stream of the Great Yang-Tzu Kiang. § 15. The Ya-lung, or Yar-lung (x.); its Confluence with the Kinsha. § 16. The Min-Kiang (xi.), the Great River of Ssŭ-Ch'uan. § 17. Its Ramifications in the Plain of Ch'êng-tu. § 18. History of the Problem of Direct Communication between India and China; First Chinese Knowledge of India. § 19. Indications in Greek Writers. § 20. Alleged Chinese Invasion of India in Seventh Century. § 21. Mediæval Counter Attempts from India. § 22. Marco Polo's Journey in Western Yun-nan. § 23. Ta-li-fu and the *Panthés*; Bhamò. § 24. The Treaty of Tien-tsin; Previous Journeys of Roman Catholic Missionaries. § 25. French Missions on the Tibetan Frontier. § 26. Curious Episode in Connection with these Missions. § 27. The Journey of Huc and Gabet. § 28. Klaproth's *Description du Tübet*. § 29. Blakiston's Exploration of the Upper Yang-Tzŭ.

CONTENTS OF

PAGE

§ 30. The French Camboja Expedition, and Excursion to Ta-li under Garnier. § 31. Garnier's Later Efforts and Projects. § 32. T. T. Cooper's Attempts to Reach India from China; and to Reach China from India. § 33. Major Sladen's Expedition from Bhamò into Yun-nan; its Consequences. § 34. Baron Richthofen's Attempt to Reach Ta-li by Ning-yuan. § 35. The Abbé Desgodins. § 36. The Deputation of Augustus Margary, who successfully reaches Bhamo from China. § 37. Murder of Margary on the Return Journey. § 38. Negotiations that followed; and Grosvenor's Mission to Yun-nan. § 39. The Agreement of Chefoo; and the Margary Proclamation. § 40. Reports from the Grosvenor Mission—Mr. Baber's; and his subsequent Journey to Ning-yuan and the Lolo Mountains. § 41. A more recent Journey by Mr. Baber by a New Route to Ta-chien-lu. § 42. Journeys of Protestant Missionaries; Mr. McCarthy, the First Non-official Traveller to Bhamò from China; Mr. Cameron's Journey in Captain Gill's Traces. § 43. Captain Gill's own Journeys: that in Pe-chih-li; that to the Northern Mountains of Ssŭ-ch'uan; the Non-Chinese Races of the Western Frontiers. § 44. Man-tzŭ and Si-fan; Comparison of Numerals. § 45. Captain Gill joined by Mr. Mesny, and his Debt to that Gentleman. § 46. The Journey by Ta-chien-lu; the Tea Trade with Tibet; some recent Information by Mr. Baber. § 47. Currency of English Rupees in Tibet; and of Tea-bricks, which they are superseding; Singular Kinds of Tea Discovered by Mr. Baber; § 48. Ta-chien-lu; its Tibetan Names; its Position in regard to the Great Plateau. § 49. The Mu-sus and Li-sus. § 50. The remarkable MS. brought by Captain Gill from the Mu-su Country. § 51. Nature of Work achieved by Captain Gill; Baron Richthofen's Opinion. § 52. Concluding Remarks . . . . [15]

CHAPTER I.

'OVER THE SEAS AND FAR AWAY.'

China resolved on—Preliminary Visit to Berlin—Baron Ferdinand v. Richthofen—Marseilles, and Voyage in the 'Ava'—Sea-flames —The Straits—Chinese Practices first realised—Approach to Saigon—The City—Hong-Kong reached and quitted—Shanghai and its peculiar Conveyances—The Chinaman's Plait—Voyage to Chi-Fu—The Chi-Fu Convention—The Minister Li-Hung-Chang—Ceremonial of his Departure—Voyage to the Pei-Ho— Difficulties of Navigation up that River—Scenery—Arrival at Tien-Tsin—Choice of Conveyances—Carts of Northern China— Mongol Ponies—'Boy' found—Horse-buying—Scenes on the Tien-Tsin Bund—Risky Building and Chinese Devotion—Tien-Tsin Hotel—Preparation for Travel—Suspicious Wares—Arrangements for a Start completed . . . . . .

THE FIRST VOLUME. [7]

CHAPTER II.

'CHINA'S STUPENDOUS MOUND.'

PAGE

Departure from Tien-Tsin—Rural Characteristics—Pictures by the Way—Chinese Hostelry—The 'Kang'—Long First Day's Ride—Early astir—Approach to Peking—A Fair going on—The British Legation—Visit to the Temple of Heaven—Money Arrangements and Currency Difficulties—Carts or Mules?—The Latter and their Packs—Bread in China—Chinese Lamps and Candles—Items of Travelling Stock—Visit to Prince Kung—Chinese Whim of picking Melon-seeds—The Journey commenced—The Difficulties of a Name—The Pei-Ho crossed—Picturesque Villages—Incidents by the way—City of Chi-Chou—Tombs of the Manchu Emperors—The Great Wall first seen—Profusion of Fruit—The Day's Routine—Frank Curiosity of People—Roadside Pictures—Pass of Hsi-Feng-K'ou—Village Inn at Po-Lo-Tai—Bean-curd, what it is—Oak-leaf Silk—Millet—Pa-K'ou Ying—Exaggerated Ideas of Chinese Population—Foreigners as seen by Chinese Eyes—Journey to Jehol abandoned—Chinese Encroachment on the Mongols—Halt to Rest—Ta-Tzŭ-K'ou—Ideas of a Lion—Splendid Autumn Climate—Palisade Barrier of Maps non-existent—Stentorian Commandant—Smoking in China—Lovely Sunset Scene—The Inn at San-Cha—Impressions of the Great Wall—The Sea in sight—Shan-Hai-Kuan 35

CHAPTER III.

'ATHWART THE FLATS AND ROUNDING GRAY.'

Halt at Shan-Hai-Kuan—Deserted City of Ning-Hai at Sea Terminus of Great Wall—Ancient Inscriptions and Vandal Treatment—Dirty Habits of Chinese—Salt Manufacture described—Itinerant Pieman's Wares—Chinese Love of Hot Water—Mouth of the Lan—Mud-flats by the Sea—Rich Monotonous Plains—Kublai Kaan's Shooting Grounds—Chinese House Fittings—Swampy Plains—Tumultuous Curiosity at Fêng-T'ai—River of Pei-T'ang—Coal expected to be found to Order—The Pei-T'ang Commandant—The Ferry—Mud-flats and Fragrant Artemisia—Dutch-like Landscape—Water-logged Country—Weary Swamps—Rest at Urh-Chuang—Hawking Ground of the Great Kaan—Chinese Wrangles—Chinese Agriculture and Use of Sewage—The Traveller's Sympathy with Marco Polo—Swamp-Lands again—Lin-Ting-Chen—Marshlands—Hsin-An-Chen—Figures bound for the Fair—Roadside Pictures—Pleasures of Travel—Peking reached again—Recreations there—'Bis cocta Recambe'—Chinese Dinners—Sir T. Wade's re-

CONTENTS OF

gretted Departure—The Inscribed Stone Drums—Visit to the Summer Palace—Needless Havoc—A Great Bell—Modern Cambaluc—Public Examinations—Departure from Peking—The Hilarious Serving-Man—The Kang Mismanaged—Tien-Tsin, a Land of Fat Things—Shanghai once more . . . 87

CHAPTER IV.
'A CYCLE OF CATHAY.'

Cradle of the Chinese Nation—Their Settlement in Shan-Si—Characteristics of Chinese History—Manifold Invasions of China, but Chinese Individuality always predominates—Imagination essential to Advancement, but the Chinese have it not—Inventions ascribed to them—Stoppage of Development—China for the Chinese—Foreign Help inevitable—The Woo-Sung Railway—Defects of Chinese Character—Shanghai in Winter—Pigeon-English—Perverse Employment of it—The word 'Pagoda'—Preparations again for Travel—Detail of Packages—The Shanghai Theatre—Chinese Treatment of Chinese—The Chinese Performers—Sport at Shanghai—A Shooting Excursion on the Yang-Tzŭ—Names of the Great River—The so-called 'Grand Canal,' and its Humours—Shooting Scenes—The Mixt Court at Shanghai—The Yang-Tzŭ Steamers—Additions to our Following 139

CHAPTER V.
THE OCEAN RIVER.

Start up the Yang-Tzŭ—The Dog Tib, and his ethnological perspicacity—On Board the 'Hankow'—Transhipment and Arrival at Hankow—Manufacture of Brick-Tea—Tea in Chinese Inns—H.M.S. 'Kestrel'—Boat engaged for Upper Yang-Tzŭ—The Lady Skipper and her Craft—Our Departure—Chinese Fuel—Our Eccentricities in Chinese Eyes—The New Year Festival—Chinese View of the Wind-points—Hsin-Ti—Entrance of the Tung-Ting Lake—Aspects of the River—Camel Reach—Vicissitudes of Tracking—Chinese Duck-Shooters—Great Bend—Wild Geese and Porpoises—Rice as Food—Huo-Hsüeh and Sha-Shih—River Embankments—Tung-Shih—Meeting with H.M.S. 'Kestrel'—The Hills Entered—Walks ashore—Population dense only on the River—Passage in a Cotton Boat—The Telescope Puzzles—Arrival at I-Ch'ang—The Chinese Gun-boat . . 172

CHAPTER VI.
THE GORGES OF THE GREAT RIVER.

The I-Ch'ang Mob—Chinese Unreason—Departure from I-Ch'ang, and Adieu to English Faces—The Gorges Entered—Vicissi-

THE FIRST VOLUME. [9]

PAGE

tudes of Ascent—Scenery of the Gorges—Preparation for the Rapids—Ascent of Rapids—Difficulties of Nomenclature— Puzzles of Orthography—Yang-Tzŭ Life-boats—Dangers of the Rapids—View at Ch'ing-Tan—Ch'ing-Tan Rapid—Waiting our Turn—Pilots of the Rapid—Successful Ascent—Kuei-Chou— Its Means of Support Obscure—Mode of Tracking—Coal Districts Entered—The She-Skipper's History—Cave of the Cow's Mouth—Expedition in the Hills—Position of Pa-Tung—Long Gorge of Wu-Shan—Weird and Gloomy Aspect—Oracular Rock Discs—Cave Habitations—Rough Scrambling—Opium hostile to Bees—Town of Wu-Shan—Perplexing Eddies—Kuei-Chou-Fu—A Visitor, and his Monstrous Torrent of Speech— Method of Local Salt Manufacture—The Kuei-Chou Aqueduct —Credulity and Untruth—Civility of Ssŭ-Ch'uan Folk— Troubles in the Rapids—River Wall of Miao-Chi—Great Haul of Fish—City of Yün-Yang—Extraordinary Richness of Soil— Importance of Rape Cultivation—Chinese Skill in Agriculture Overrated—Their True Excellence is in Industry—Ample Farmsteads—Rock of Shih-Pao-Chai—Story of Miraculous Rice-Supply—Luxuriant Vegetation—'Boat Ahoy'—The Ch'ung-Shu Flowers—St. George's Island—Cataplasm for a Stove-in Junk—' The River of Golden Sand ' at last—Fu-Chou —A Long Chase—Effort to Reach Ch'ung-Ch'ing—The Use of the Great Sculls—Arrival at Ch'ung-Ch'ing 201

CHAPTER VII.

CH'UNG-CH'ING TO CH'ÊNG-TU-FU.

Arrival at Ch'ung-Ch'ing—M. Provôt, Monsgr. Desflêches, and the French Missionaries; their Cordiality—Official Visits—The Last of our Lady-Skipper—We are satirised in verse, and enabled to see ourselves as others see us—Mê, the Christian— Persecution of Christians—News of Tibet—Unfavourable Change of Feeling in Tibet as to Admission of Europeans—The Consulate at Ch'ung-Ch'ing Proclaimed—Consular Relations to Missionaries—Cheapness of Ch'ung-Ch'ing—General ' Fields-within-Fields '—Difficulties of Mr. Baber's Photographer— Exchange Troubles—Preparations to Start for Ch'êng-Tu— Elaborate Coolie Contract—Chinese Commercial Probity— Baggage Arrangements—Adieu to Baber—Joss-Houses and Paper Offerings—Characteristics of the Rice-Culture—Wayside Tea-shop—Pleasing Country—Poppy-fields—Yung-Ch'uan-Hsien—Everything Roofed in—T'ai-P'ing-Chên—Grand Thunderstorm—Paved Roads—Inns of Ssŭ-Ch'uan—Chinese Parsimony—Good Manners of Ssŭ-Ch'uan Folk—Jung-Ch'ang-Hsien—Characteristics of a Restaurant—Coolies at their Meal

CONTENTS OF

—Cooper's Dragon Bridge—Realistic Art—Want of Ideality in Chinese Character—Lung-Ch'ang-Hsien—Charming Inn—Wang-Chia-Ch'ang — Niu-Fu-Tu — Hsien-Tang — The Brine Town of Tzŭ-Liu-Ching—The Christian Landlord—The Brine-wells and Fire-wells—Mode of Boring and of Drawing the Brine—Further Details and Out-turn of Salt—Politeness of the People—River Ch'ung-Chiang — Tzŭ-Chou—Gratuitous Diversion for a Village School—A Chinese Regiment—The School System—Sanitary Paradox of China—Red Basin of Ssŭ-Ch'uan—Transfers of Property—Yang-Hsien—Safflower Crop—Thickly Peopled Plain—Still Ascending Valley of the Ch'ung-Chiang—Thrift in Excess—Arrival at Ch'êng-Tu-Fu—Public Examinations in Progress 260

CHAPTER VIII.

A LOOP-CAST TOWARDS THE NORTHERN ALPS.

I. CH ÊNG-TU TO SUNG-P'AN-T'ING.

Kindness of French Missionaries at Ch'eng-Tu—Arrangements with Mr. Mesny—Endeavour to take a House—Mystifications on the Subject—Pleasures of French Society—Proposed Excursion to the North—The Man-Tzŭ, or Barbarian Tribes—Preliminaries of Departure—Leave Ch'êng-Tu-Fu—Pi-Hsien—Engaging Official there—The Escort—Irrigated and Wooded Country—Halt at Kuan-Hsien—Scope of Excursion Extended—Frantic Curiosity of People, but no Incivility—Irrigation Works—Rope Suspension Bridge—Coal-beds—Yu-Ch'i Charming Inn—Yin-Hsiu-Wan, and Water-Mills—Hsin-Wên-P'ing—The 'Min River'—Macaroni-making—Wên-Ch'uan-Hsien—First Man-Tzŭ Village—Pan-Ch'iao—Traces of War—Relentless Advance of Chinese—Miraculous Sand Ridge—Hsin-P'u-Kuan—Rapid Spread of the Potato—Excursion to Li-Fan-Fu in the Man-Tzŭ Hills—Scenes that recall the Elburz—Carefully-made Hill Road—The 'Sanga' of the Himalayas—Angling—Village of Ku-Ch'êng—Peat Streams—Musk Deer—Arrival at Li-Fan-Fu—The Margary Proclamation—Tales of Local Wonders—The Traveller fain would see—The Lions of Li-Fan-Fu—Search for a Man-Tzŭ Village—Man-Tzŭ here a term of reproach—The I-Ran Tribes and their Language—Ku-Ch'êng—Local Wonders again—Return to Hsin-P'u-Kuan—Resume Valley of Hsi-Ho (or 'Min River')—Wên-Ch'êng—The Himalayan Haul-Bridge in Use—Polite Curiosity at Mao-Chou—Grandeur of the 'Nine Nails' Mountain—Precipitous Gorges—Wei-Mên-Kuan—Difficulties of the Road—The Su-Mu, or White Barbarians—Alpine Scenery—Ta-Ting—Tieh-Chi-Ying—War with the Su-Mu—The Yak seen at last—Travellers'

THE FIRST VOLUME. [11]

PAGE

Disappointments—Glorious Mountain View (Mount Shih-Pan-Fang) — P'ing-Ting-Kuan — Expulsion of Man-Tzŭ — Maize Loaves—Wood Pigeons—Ngan-Hua-Kuan—Delicious Tea— Smoking in Ssŭ-Ch'uan—Country of the Si-Fan—Sung-P'an-T'ing 317

CHAPTER IX.

A LOOP-CAST TOWARDS THE NORTHERN ALPS—*continued.*

II. SUNG-P'AN-T'ING BACK TO CH'ÊNG-TU.

Si-Fan Ponies—Reports of Game—Musk Deer—Si-Fan Lama—Language—Crops—Butter, Fish, Yak-Beef — Bitter Alpine Winds—Foreign Remedies Appreciated—The Traveller Quits the Valley ('Min River') and Turns Eastward—Si-Fan Lamassery—Herds of Cattle and Yaks—Desolate Hospice at Fêng-Tung-Kuan—Tibetan Dogs—Reported Terrors of the Snow-Passes—Summit of the Plateau of Hsüeh-Shan—Descent Begins —Forest Destruction—Verdure of Eastern Slopes—Splendid Azaleas—Slaughter by the Si-Fan—Luxuriant Gorges—Inhabited Country Regained—Chinese Dislike of Light—Shih-Chia-P'u—Miracle-Cave — Hsiao-Ho-Ying—Iron Suspension Bridges—Peculiarities of Chinese Building—Road to Shui-Ching-Chan—Gold-Washing—' Ladder Village '—*Mauvais Pas* —Lung-An-Fu—Approach to the Plains—Ku-Ch'êng—Fine Chain Bridge—Road to Shen-Si—Chiu-Chou—Chinese Fête Day—Moderate Drinking of Coolies—Preserved Ducks' Eggs, a Nasty Delicacy—P'ing-I-P'u—Boat Descent of River (Ta-Ho) —Great Irrigation Wheels and Bunds—Populous Country Again—Chang-Ming-Hsien—City of Mien-Chou—Road Resumed—Ill-smelling Fields—The Red Basin Again—Lo-Chiang-Hsien—Tomb of Pong-Tung—His History—Pai-Ma-Kuan, or Pass of the White Horse—Summer Clothing of the People— Method of Irrigation—Fine Vegetables—Roadside Pictures— Official Civilities at Te-Yang-Hsien—Progress to Han-Chou— Marco Polo's Covered Bridge—Hsin-Tu-Hsien—Rich Country —Suburbs of Ch'êng-Tu—End of Excursion 376

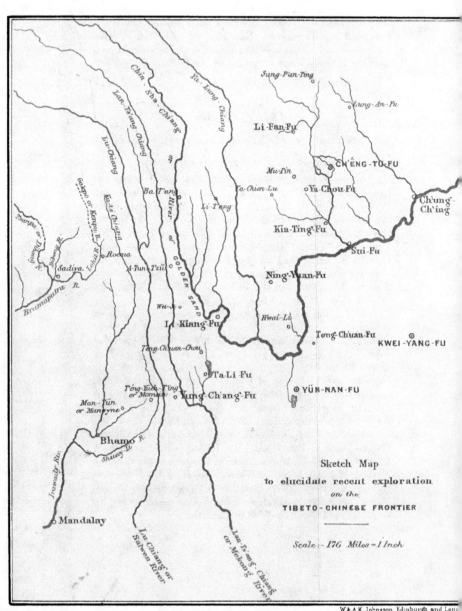

Sketch Map to elucidate recent exploration on the TIBETO-CHINESE FRONTIER

Scale.— 176 Miles = 1 Inch

London: John Murray, Albemarle Street, 1879.

ILLUSTRATIONS *and* MAPS *to* VOL. I.

MOUNT LIEH-SHAN-LIANG, IN MONGOLIA . . . *Frontispiece*
THE FERRY . . . . *Title-page*
SKETCH MAP TO ELUCIDATE RECENT EXPLORATION ON
   THE TIBETO-CHINESE FRONTIER . . . *To face page* [13]
ROUTE MAP. SECTION I. FROM CH'ÊNG-TU TO SUNG-
   P'AN-T'ING, AND RETURN . . . . „ 317
GENERAL MAP OF CHINA . . . . . . *In pocket at end*

Errata in Vol. 1.

Page 10 line 15 } *for* 'Yang-Tzŏ' *read* 'Yang-Tzŭ.'
„ 13 „ 12 }
„ 22 line 25, *for* 'was not a very good walker' *read* 'was a good walker.'
„ 24 note, *for* 'two mulecarts,' *read* 'two-mule carts.'
„ 46 „ 14, *for* ' was ' *read* 'were.'
„ 50 „ 23, *for* 'plain hot water' *read* 'plain water almost boiling.'
„ 58 „ 3 *for* ' Tan-Ho ' *read* ' San-Ho.'
„ 58 „ 3 from foot. The date '*September* 28' belongs to the preceding paragraph.
„ 67 „ 3 from foot, *for* ' Po-lo-Tai' *read* ' Po-Lo-Tai.'
„ 74 „ 9, *for* 'N.N.E.' *read* 'N.N.W.'
„ 74 „ 10, *for* 'N.W.' *read* 'N.E.'
„ 75 „ 11, *for* 'gevas' *read* 'jivas.'
„ 76 „ 18, *for* 'was it' *read* 'it was.'
„ 120 The cut belongs to the preceding page.
„ 127 line 2 from foot to page 129 line 8. This matter has been misplaced. It should be a note to page 263.
„ 149 note, *for* ' May 6 ' *read* 'page 306.'
„ 165 lines 12 and 15, *for* ' Bath'ang ' *read* ' Bat'ang.'
„ 176 line 18, *for* ' Ch'ang-Ch'ing ' *read* ' Ch'ung-Ch'ing.'
„ 178 „ 15. The semicolon should be after the word 'China,' not after the word 'Ssu-Ch'uan.'
„ 192 „ 4, *for* 'birds' *read* 'buds.'
„ 252 „ 7 from foot, *for* ' Wu-Shan ' *read* ' Wu-Yang.'
„ 255 „ 5, *for* ' hole ' *read* ' hold.'
„ 267 heading, *for* ' Bishop Provôt ' *read* ' Monsieur Provôt.'
„ 316 „ 23, *for* ' Sieh T'ais ' *read* ' Hsieh-T'ais.'
„ 351 „ 14, *for* ' Sieh-T'ai ' *read* ' Hsieh-T'ai.'
„ 367 „ 9 from foot, *for* ' twinkling ' *read* ' tinkling.'
„ 368 „ 2 from foot, the date, June 2, should be inserted here.
„ 381 „ 2 from foot, *for* ' Wing Cave Pass ' *read* ' Wind Cave Pass.'
„ 385 line 27, *for* ' has ' *read* ' had.'
„ 386 „ 21, *for* ' planted ' *read* ' flaunted.'
„ 393 „ 16. The passage should read thus, ' Running between partly cultivated mountains, the peaks of which were about 2,000 feet high.'
„ 394 „ 7, *for* ' them ' *read* ' it.'
„ 398 „ 6 from foot, *for* ' fostering ' *read* ' festering.'
„ 410 „ 18. The sentence ' A little further we came to,' &c., should be 'A little further we came to a temple,' &c.

INTRODUCTORY ESSAY.

§ 1. My friends, the author and the publisher of this work, have called on me to write a preface to it. I confess to a strong interest in the book, which I have seen through the press from first to last, during Captain Gill's absence on duty in the Levant. That is, indeed, an office which nature does not easily permit to be done by deputy; and I am told that I have left some of his flowers only planted, when they ought to have flaunted, and his banners to flatter when they ought to flutter, whilst I have made his bells to twinkle when they only tinkled.

But my interest in the journey which the book relates began long before the book was even an embryo, and with the first hour of my acquaintance with its author. Three years and a half ago, he was indeed well known to me by name as a brother officer who had been an enterprising traveller on the Turkoman frontier of Persia, and a still more enterprising candidate for a metropolitan borough. But we had never met when, in the end of May 1876, Captain Gill visited me at the India Office, and announced that he was meditating an expedition, by way of Western China, into either Eastern Turkestan or Tibet.

Though I had during many years past travelled much in those regions, my journeys had been accomplished in the spirit only, not in the body—those of the latter character never having extended into regions more remote than Ava in one direction, and Java in another; hence I was gratified by the motive of Captain Gill's visit, and did my best to justify it. This was not so much by any information that I could furnish, or suggestions that I could venture,—except, indeed, that of making Marco Polo his bosom friend, a hint that he has cherished and acted upon throughout his travels,—as by introducing him to two men who could advise him from singular practical experience, I mean Baron Ferdinand v. Richthofen, and the late Mr. T. T. Cooper.

§ 2. Of Baron Richthofen I will venture to quote words written on another occasion :

'It is true that the announcement of his presence at the evening meetings (of the Royal Geographical Society) would draw no crowds to the doors; no extra police would be required to keep the access; no great nobles would interest themselves about engaging St. James's Hall for his reception . . . but it is a fact that in his person are combined the great traveller, the great physical geographer, and the accomplished writer, in a degree unknown since Humboldt's best days. In the actual extent of his journeys in China, he has covered more ground than any other traveller of note, and he has mapped as he went. His faculty of applying his geological knowledge to the physical geography of the country he traverses is very remarkable, but not more so than his power of lucid and interesting exposition.'[1] Baron Richthofen's advice and information were communicated with a fulness and cordiality which Captain Gill has recorded near the beginning of his book.

---

[1] Letter to Sir Rutherford Alcock, then President of the Royal Geographical Society, of which an extract was read by him at the annual meeting, May 27, 1878.

INTRODUCTORY ESSAY. [17]

Mr. Cooper, though far from any pretension to be classed as a traveller with the one just spoken of, has been justly characterised as one of the most adventurous explorers of modern times; and had himself made two bold attempts to force that Tibetan barrier, which remains yet unpierced, between India and China: once from the side of Ssŭ-ch'uan, and once again from the side of Assam. And it is a circumstance worthy to be noted here, that whilst Mr. Cooper (it was in my room in the India Office) was one of the last persons with whom Captain Gill took counsel regarding his journey before quitting England, it was the same Mr. Cooper who received the traveller with open arms and hearty hospitality at Bhamo, when he emerged from the wilds of the Chinese frontier in November 1877. A few months later (April 24, 1878), poor Cooper, in his solitary residence at Bhamò, was murdered by a soldier of his native guard.

§ 3. The 'general reader,' whose eye may be caught by the title of this work, will not, we trust, be misled by the familiar melody of Bishop Heber to suppose that the traveller will conduct him to 'Afric's sunny fountains.' The 'River of Golden Sand' is a translation of the name *Kin-sha-Kiang*, or (in the new orthography in which I find it hard to follow my author) *Chin-Sha-Chiang* (Gold-Sand-River), by which the Chinese, or at least Chinese geographers, style the great Tibetan branch of the Yang-tzŭ, down at least to its junction, at Sü-chau (or Swi-fu, as it is now called), with the Wen or Min River, descending from Ssŭ-ch'uan. Of other names we shall speak a little below.

It is proposed now to indicate some of the points of geographical interest in the little known region of which the River of Golden Sand is as it were the axis,— that region of Eastern Tibet which intervenes between the two great historic continents of India and China,—and to sketch the history of explorations in this tract previous to that of Captain Gill. If in this task I sometimes use words that I have used before, on one or other of the somewhat frequent occasions that this dark region, from which the veil lifts but slowly, has attracted me,[2] let me be forgiven. And all the more one may overcome scruples at such repetition in seeing how persistent error is. Within the last few months I have read of 'an able argument' (I certainly did not read the argument itself) to prove the identity of the Tibetan Tsanpu and the Irawadi. Life seems too short for the study of able demonstrations that the moon is made of green cheese, but, if these are still to be proffered, there can be no harm in stating the facts again.

I do not forget the pungent words with which Abbe Huc concludes his sparkling *Souvenirs d'un Voyage*: 'Quoiqu'il soit arrivé au savant Orientaliste, J. Klaproth, de trouver *l'Archipel Potocki*, sans sortir de son cabinet, il est en général assez difficile de faire des découvertes dans un pays sans y avoir pénétré.'[3] But as regards a large part of the country of which I am going to speak we are all on a level, for no one has seen it, not even the clever Abbé himself and his companion; and of *geographical* information regarding the region in question, they can hardly be said to have brought anything back.

[2] E.g., in a review of Huc and Gabet in *Blackwood*, 1852; in connexion with the *Narrative of Major Phayre's Mission to Ava* (Calcutta 1856, London 1858); in the *Journal of the Asiatic Society of Bengal for* 1861, p. 367; in the notes to *Marco Polo*; and in various papers in *Ocean Highways* and the *Geographical Magazine*, and discussions in the *Proceedings of the Royal Geographical Society*.

[3] The name of Potocki Islands was given by Klaproth in honour of Count Potocki, under whom he had served on a Russian mission to Peking, to a group of eighteen islands in the Gulf of Corea. This sheet of the Jesuit map of China had been mislaid or omitted when D'Anville engraved it. Klaproth afterwards became owner of the missing tracing, and on it, *sans sortir de son cabinet*, found these islands, and claimed their discovery.

INTRODUCTORY ESSAY. [19]

§ 4. Everyone who has looked at a map of Asia with his eyes open must have been struck by the remarkable aspect of the country between Assam and China, as represented, where a number of great rivers rush southward in parallel courses, within a very narrow span of longitude, their delineation on the map recalling the fascis of thunderbolts in the clutch of Jove, or (let us say, less poetically) the aggregation of parallel railway lines at Clapham Junction.

Reckoning these rivers from the westward, the first of importance (i.) is the Subanshiri, which breaks through the Himalya, and enters the valley of Assam in long. 94° 9'. This is a great river, and undoubtedly comes from Tibet, i.e. from Lhassa territory. Some good geographers have started the hypothesis that the Subanshiri, rather than the Dihong, is the outflow of the Tsanpu; but recent information shows this to be impossible.

§ 5. The next of these great rivers (ii.) is the Dihong, which enters Assam in long. 95° 17', and joins the Lohit—or proper Brahmaputra—near Sadiya. Though the identity of this river with the great river of Central Tibet, the Yaru Tsanpu, has never yet been continuously traced as a fact of experience, every new piece of evidence brings us nearer to assurance of the identity, and one might be justified in saying that no reasonable person now doubts it. This was the belief of Rennell, who first recognised the magnitude of the Brahmaputra, long before we had any knowledge of the Dihong, or of the manner and volume of its emergence from the Mishmi Hills.[4] Many years,

[4] 'On tracing this river in 1765, I was no less surprised at finding it rather larger than the Ganges than at its course previous to its entering Bengal. This I found to be from the east; although all the former accounts represented it as from the north; and this unexpected discovery soon led to inquiries, which furnished me with an account of its general course to within 100 miles of the place where Du Halde left the Sanpoo. I could no longer doubt that the Burrampooter and Sanpoo were one and the same river, and to this was added the positive assurances of the Assamers "that *their* river came from the north-west, through

a 2

however, before Rennell's work was published, in fact twelve years before Rennell was born, P. Orazio della Penna, writing in Tibet (1730), had stated that the river was then believed to join the Ganges, explaining (from such maps as were available to him in those days) 'towards Rangamatti and Chittagong.' A conjecture to the same effect occurs in the memoir on the map of Tibet, by Pere Regis, at the end of du Halde. Giorgi, in his *Alphabetum Tibetanum* (Rome, 1762), says the like.[5] The same view is distinctly set forth in the geography of Tibet which is translated in the 14th volume of the great French collection of *Mémoires concernant les Chinois*, a document compiled by order of the Emperor K'ang-hi, and issued, with others of like character, in 1696. This represents the Yaru Tsanpu as rising to the west of Tsang (West Central Tibet), passing to the north-east of Jigar-Kungkar (south of Lhassa), flowing south-east some 400 miles, and then issuing at the south of Wei (or U, East Central Tibet) into the region of the *Lokh'aptra*, 'tattooed people' (*i.e.* Mishmis *et hoc genus omne*); then turning south-west it enters India, and discharges into the southern sea (pp. 177-178).

The Pundit Nain Singh, on the journey to Lhassa which first made him famous (1865), was told by Nepalese, Newars, and Kashmiris at that city, that the great river of Tibet was the Brahmaputra; whilst all the natives who were questioned also declared that, after flowing east for a considerable distance, it

the Bootan Mountains."'—*Mem. of a Map of Hindoostan*, 3rd edition, pp. 356-7, see also p. 259. Rennell's actual knowledge of the Brahmaputra extended only to long. 91°, a few miles above Goalpara, but his sketch of the probable entrance of the river from Tibet is very like the truth. On the other hand, it is curious how he was misled as to the source of the Ganges, which he identified with what are really the upper waters of the Indus and Sutlej. The importance of the Dihong was first pointed out by Lieutenant Wilcox in 1826 in the *Calcutta Gazette*. (See *As. Res.* xvii.)

[5] '*Sese tandem in Gangem exonerat.*' But Giorgi's information was derived from della Penna and the other Capuchins.

flowed down into India. The Pundit's information on his last great journey, when he crossed the river somewhat further to the eastward, before striking south into Assam, did not add much, but it was all in corroboration of the same view.[6] And this is still further confirmed by the latest report of exploration from the Chief of the Indian surveys. We have only a sketch of this exploration, and await the details with great interest. But we learn that the explorer (N—m—g) took up the examination of the Tsanpu at Chetang, where it was crossed by Nain Singh on his way from Lhassa to Assam (in about long. 91° 43′, lat. 29° 15′), and followed it a long way to the eastward. He found that the river, before turning south, flows much further east than had been supposed, and even north-east. It reaches its most northerly point in about long. 94°, and lat. 30°, some 12m. to the north-east of Chamkar. The river then turns due south-east, but the explorer was not able to follow it beyond a place, 15 miles from the great bend, called *Gya-la Sindong*. There, however, he saw that it flowed on for a great distance, passing through a considerable opening in the mountain ranges, to the west of a high peak called *Jung-la*. Chamkar appears in D'Anville's map as Tchamka, and in one of Klaproth's[7] as *Temple Djamga*, in a similar position with regard to the river. And Gya-la Sindong seems to be the *Temp. Sengdam* of the latter map, standing just at the head of the '*defilé Sing-ghian Khial*,' by which Klaproth carried off the waters of the Tsanpu into the Irawadi. If the position of Gya-la Sindong as determined by the explorer is correct, its direct distance from the highest

---

[6] See *Journal of Royal Geographical Society*, xlvii. p. 116. It is remarkable that the information collected by the Pundit on his first journey was most accurate as to the position where the river turns to the south, which he placed in about long. 94°. (See Montgomerie, in *J. R. G. S.*, xxxviii. p. 218, *note*.) His later conclusion was less accurate.

[7] In vol. iii. of his *Mémoires relatifs à l'Asie*.

point hitherto fixed on the Dihong river from the Assam side is only about 100 miles.[8]

§ 6. We have mentioned above that some have supposed the Subanshiri to be the real continuation of the Tsanpu. The idea seems to have been grounded in part on an exaggerated estimate of the volume of the Subanshiri, and partly on Nain Singh's indications (in 1874) of the course of the Tsanpu, which seemed to bring it in such close juxtaposition to the Subanshiri as to allow no room for the development of another river of such volume as was attributed to the latter. The last of these foundations for the theory has been removed by the new explorer (N—m—g)'s extended journey, carrying the south-eastern bend of the Tsanpu so much further to the East; and the first also was erroneous. Careful and detailed observations by Lieut. Harman in 1877-78 give the comparative volumes of the Assam rivers with which we have to do, at their mean low level, as follows:—

	Cubic feet per second.		Cubic feet per second.
Dihong	55,400	Dihong and Dibong before union with Bramahputra ('Lohit')	82,652
Brahmaputra ('Lohit') above Sadiya	33,832		
Ditto at the Brahma-Kund	25,000	The combined (Brahmaputra) river at Dibrúgarh	116,115
Dibong	27,202	The Subanshiri	16,945

We see here how the Dihong vastly surpasses in discharge not only the Subanshiri, but also the Lohit Brahmaputra and the Dibong, while both greatly exceed the Subanshiri.[9]

[8] This is just the space at which Rennell, 100 years ago, estimated the unknown gap. (See p. [19] above.)

[9] It is of some interest to compare these measurements with those made by Bedford and Wilcox in 1825-26. They were as follows (see *Asiatic Researches*, vol. xvii., but I take them from J.A.S.B. xxix. p. 182):—

	December 26, 1825.	March 29, 1826.
Dihong (after a correction)	(*a*) 56,000 ft.	
Brahmaputra at Sadiya	(*b*) 19,058 ft.	(*a*) 33,965 ft.
Dibong	(*b*) 13,100 ft.	
Dihong and Dibong	69,664 ft.	(*a*) 86,211 ft.
Subanshiri, 'in dry season'	(*a*) 16,000 ft.	

§ 7. Very eminent geographers have, however, not been content to accept the view of the identity of the Tsanpu and the Brahmaputra, and several have contended that the Irawadi of Burma was the true continuation of the great Tibetan River. D'Anville, I believe, was the first to start this idea.[1] It was repeated by our countryman Alexander Dalrymple, the compiler of the 'Oriental Repertory' and much else, the founder of the Hydrographic Department of the Admiralty, and a very able geographer, in a map on a small scale which he put together for the illustration of Symes's 'Mission to Ava' (1800). The idea was maintained at a later date with great force and insistence by that remarkable and erratic genius Julius Klaproth, who in demonstration played fast and loose on a great scale with latitudes and longitudes, and produced Chinese documents from the days of the T'ang dynasty to those of K'ang-hi in corroboration. His dissertation in its latest form[2] is, like almost everything that Klaproth wrote, of high interest. We need not, as some other things in his career suggest, doubt the genuineness of the Chinese documents. Some of them at least are to be found translated in independent works before his time. But everything is not necessarily true that is written in Chinese, any more than everything that is written in Persian —or even in Pushtu! Chinese writers find leisure to speculate on geographical questions as well as Europeans. And some of them, finding, on the one

---

The close approximation in those marked (*a*) to Lieutenant Harman's recent measurements is remarkable ; whilst in (*b*) the discrepancy is great. All Lieut. Harman's measurements were taken in March. In some the rivers had risen, and the low level discharge was arrived at by calculation. But it is a pity that no notice is taken of the older measurements in the publication of the recent ones. The suggestion of the facts on the surface is that the recent observations do not represent the lowest level, or that the rivers in December 1825 were unusually low.

[1] *Éclaircissements Géographiques sur la Carte de l'Inde*, Paris, 1753, p. 146.
[2] *Mémoires relatifs à l'Asie*, vol. iii.

hand, the Tsanpu flowing through Tibet, and disappearing they knew not whither, and finding, on the other, the Irawadi coming down into Burma from the north, issuing they knew not whence, adopted a practice well known to geographers (to Ptolemy, be it said, *pace tanti viri*, not least) long before Dickens humorously attributed it to one of the characters in Pickwick,—they 'combined the information,' and concluded that the Tsanpu and the Irawadi were one. Klaproth's view that this was so, and that the actual influx took place near Bhamò, was adopted by many Continental geographers, and staggered even the judicious Ritter. Maps were published in accordance with the theory, some bringing the waters of Tibet into the Irawadi by the Bhamò River (down which Captain Gill floated in Mr. Cooper's boat on the last day's journey which he has recorded), and others through the Shwéli, which enters the Irawadi some eighty miles below Bhamò.

§ 8. It seems hardly worth while now to slay this hypothesis, which was moribund before, but must be quite dead since the report of N—m—g's exploration. Its existence was somewhat prolonged, especially in France, by the fact that some of the missionaries in Eastern Tibet, of whom we shall speak presently, had carried out with them elaborate maps, compiled under the influence of Klaproth's theory; and the ideas derived from these had so impregnated their minds that in communicating geographical information which they had collected on the scene of their labours it was confused and tinged by the errors of Klaproth.

The main bases for what we may style the orthodox theory of the Irawadi are found in the constant belief of natives above and below the Tibetan passes, and in the evidence of direction and volume. The lamented Col. T. G. Montgomerie, in his most able analysis of the Pundit Nain Singh's first journey, deduced from the particulars recorded by the latter, and a careful oral catechisation, that the discharge of the Tsanpu, where crossed below Jigatze (or Jigarchi), could hardly be less than 35,000 feet per second. We see that the discharge of the Dihong, on its emergence from the hills of Assam into the plains of Assam is 55,400 feet. These are in reasonable ratio. Now the discharge of the Irawadi, so far down as the head of the Delta, is not more than 75,000 feet, and at Amarapura it cannot, on the best data available, be much more than the 35,000 feet attributed to the Tsanpu on the table-land of Tibet, at a point which would be at least 1,200 miles above Ava along the banks, if the theory of identity were true.[3]

§ 9. The third river (iii.) is the Dibong, which joins the Dihong before its confluence with the Brahmaputra. This has, on Mr. Saunders's map of Tibet accompanying Mr. Markham's book, been identified with the Ken-pu, one of the rivers of Tibet delineated on D'Anville's map. The Ken-pu, however, we shall see strong evidence for identifying with a different river, whilst there is positive reason to believe that the Dibong, in spite of its large discharge, does *not* come from Tibet. At a meeting of the Asiatic Society of Bengal in 1861, at which I read a paper connected with this subject, Major (now Major-General) Dalton stated that the people of Upper Assam admitted only two of their rivers to come from Tibet, viz. the Dihong and the (Lohita) Brahmaputra. An attempt was made in 1878 by Captain Woodthorpe, R.E., who has done much excellent work in the survey of the Eastern Frontier, to explore the sources of the Dibong. He was not successful in penetrating far up the river, but he

---

[3] See Appendix to *Narrative of Mission to the Court of Ava* (Major Phayre's), pp. 356 *seq.*; and a paper by Major-General A. Cunningham in the *Journal of the Asiatic Society of Bengal*, vol. xxix., pp. 175 *seq.*

considered himself to have derived, from extensive views, and native information in connection with them, 'a fairly accurate knowledge of the sources of 'the Dibong, and the course of its main stream in the 'hills;'[4] and in the map representing this knowledge the river is indicated as having no source further north than about 28° 52′.

§ 10. We next come to the (iv.) true Brahmaputra, or Lohit, which enters Assam at the Brahmakund, or Sacred Pool of Brahma. This I believe to be identical with the Gak-bo of the Tibetan geographies, and the Ken-pu, or Kang-pu, of D'Anville and the Chinese.

Granted, as we may now assume, that the Tsanpu is the Dihong, the Ken-pu can hardly be other than either the Dibong or the Lohit. We have seen that the Dibong does not come from Tibet. But there is a very curious piece of evidence that the Ken-pu is the Lohit.

I have just alluded to a paper connected with our present subject which was read at Calcutta in 1861. This was a letter from Monseigneur Thomine des Mazures, 'Vicar Apostolic of Tibet,' and then actually residing in Eastern Tibet, to Bishop Bigandet of Rangoon (himself well known for his works on Burmese Buddhism, &c., and who had been very desirous to establish direct communication with his brethren in the north), and which contained some interesting geographical notices, though they were, as has been already indicated, impaired in value by the erroneous ideas as to the Tsanpu, gathered from Klaproth, with which French maps were then affected.[5] The paper was read with a comment by the present writer.[6]

[4] Letter of Captain Woodthorpe, dated Shillong, August 10, 1878, forwarded by the Government of India, in their letter of October 31, *id.*

[5] Particularly the map, on which Bishop Thomine relied, of *Andriveau Goujon*, Paris, 1841.

[6] See *Journal of the Asiatic Society of Bengal*, vol. xxx. pp. 367 *seq.*

Now in this letter Bishop Thomine spoke of the series of rivers in question, beginning with the Lant'sang, or Mekong, and travelling westward. Next to the Lantsang was the Lu-ts' Kiang (Lu-Kiang or Salwen). Beyond that the Ku-ts' Kiang, of which we shall speak presently, and then the *Gak-bo Tsanpu*, 'called by the Chinese *Kan-pu-tsangbo*.' The Bishop, influenced by his Klaprothian map, stated this to join the Irawadi. And this would only have made confusion double but for a circumstance which he proceeded to mention. 'In that district,' he wrote, 'according to the Tibetans, is the village of Sâmé, where our two priests, MM. Krick and Boury, were murdered.' Here was a fact that no theories could affect. These two gentlemen were, in the autumn of 1854, endeavouring to make their way to Tibet from Upper Assam, by the route up the Lohit, attempted fourteen years later by Mr. T. T. Cooper, when they were attacked and murdered by a Mishmi chief called Kaïısa. On the receipt of this intelligence, and after a detailed account of the circumstances had been obtained from the servant of the priests, a party was despatched by the Assam authorities into the Mishmi country to capture the criminal chief. This was very dexterously and successfully effected by Lieutenant Eden, who was in command. In the beginning of March Kaïisa and some of his party were taken, and were tried and convicted by Major Dalton. Dr. Carew, the Roman Catholic archbishop, interceded with the Governor-General for a mitigation. But Kaïisa was hanged. It is an old story, but so creditable to several concerned that it has seemed well worth being briefly told here.

Now the place at which these two travellers were

The Bishop's letter as sent to the Society had been done into English, and not always lucid English. In my present quotations I have corrected this.

murdered was *Sime, on the banks of the* (Lohit) *Brahmaputra*, a place entered from native information in Wilcox's map some thirty years before, and some fifteen or sixteen miles above the place where Cooper was turned back in 1869.

I can hardly conceive of better evidence than this regarding a country unexplored by European travellers, and I have repeatedly adduced it in proof that the Gak-bo or Ken-pu is identical with the Lohit, and that the latter comes from Tibet. This, too, being established, there remains no possibility of communication between the Tsanpu and the Irawadi, unless the Tsanpu pass athwart the basin of the Brahmaputra.[7]

Thus, singular to say, from the blood of those two missionary priests, spilt on the banks of the Lohita (the 'Blood-red'), is moulded the one firm link that we as yet possess, binding together the Indian and the Chinese geography of those obscure regions.

§ 11. (v.) In the Chinese maps, and in Bishop Thomine des Mazures' list of rivers, there comes next a river variously called *Tchitom* (D'Anville), *Tchod-teng*, or *Schété* (Des Mazures) Chu, all probably variations of the same name, and also *Ku-ts' Kiang* (Des Mazures), and in Klaproth's map the *Khiu-shi-Ho*. This river, which he calls 'rather inconsiderable,' the Bishop identifies with the Lung-Kiang or Lung-ch'wan Kiang of the Chinese, or Shwe-li of the Burmese, which flows a little east of Momien (called by the Chinese Teng-yueh-chau), and which eventually joins the Irawadi 80 or 90 miles

---

[7] The only possible doubt is that of the identity of the Gak-bo and the Kan-pu or Kang-pu, but I think there is no room for this It is asserted by Bishop des Mazures, and a comparison of the course of the *Ken-pou* of D'Anville's map with the *Kakbo Dzanbotsiou* of the Chinese map given by Klaproth in his edition of the *Description du Tübet*, entirely corroborates this.

below Bhamo. The Shwe-li does, according to Captain Gill's report, appear to bring down when in flood a vast body of water,[8] but it has not been seen by any European north of where he crossed it. Dr. Anderson, however, who accompanied Major Sladen's expedition, states that he was positively informed that its sources were only 40 or 50 miles north-east of Momien.[9] Bishop T. des Mazures, in his identification of the Schété or Ku-ts' with the Shwé-li, was perhaps again unduly biassed by maps founded on Klaproth's theories, and thus we cannot feel confidence that his statement on this point was derived from native information. Chinese geographical speculators have identified more than one river of Tibet with the Shwé-li, some of them supposing it to be the same with the Gak-bo or Ken-pu.[1] I have long been inclined to conclude that the Ku-ts' Kiang of the Bishop, the Tchitom-chu of D'Anville, represents the unseen eastern source of the Irawadi, which has been the subject of so much controversy. Dr. Anderson's Shan informants gave the unvisited eastern branch of the Irawadi the name of *Kew* (Kiu) *Hom*, a name possibly identical both with the *Khiu-shi* of Klaproth and with the Ku-ts' of Bishop Thomine des Mazures. In any case, judging from D'Anville's map the best authority we as yet have, the sources of this river, and therefore under my present hypothesis the remotest sources of the Irawadi, will not lie further north than 30° at the most. If so, the extreme length of the Irawadi's course will still fall far short of that assigned to the Lu-Kiang, or Salwen, and to the Lant'sang, or Mekong, to say nothing of our 'River of Golden Sand.' And this will be consistent enough with the calculations regarding the discharge of the

---

[8] See the present work, vol. ii. p. 357.
[9] *Report on Expedition to Western Yunan*, Calcutta, 1871, p. 188.
[1] See Ritter, iv. 225.

Irawadi, which will be found in the places quoted at p. [25] above.

§ 12. (vi.) The Lung-ch'uan Kiang, Lung-Ch'iang of Captain Gill, and Shwe-li of the Burmese. Of this we have spoken under No. v.

The next of the parallel rivers (vii.) is the Lu-Kiang or Nu-Kiang of Chinese maps, the Lu-ts' Kiang of Bishop des Mazures, the Salwen of Burma, under which name it enters the Gulf of Martaban. Rennell thought that the Nou-Kian (or Lu-Kiang) of the Jesuit maps must be the Upper Irawadi. And since then doubts have been thrown on the identity of the Salwen and the Lu-Kiang of Tibetan geography, by myself many years ago, and more recently by Dr. Anderson; but I am satisfied that the evidence had not been duly considered. The chief ground for discrediting its length of course and its Tibetan origin was its comparatively small body of water as reported. This may, however, be due mainly to a restricted basin,—and as far as we know the river from Yunnan downwards, the basin is very restricted;—but also we see not only how various the relations between the length and the discharge of considerable rivers may be, but how deceptive, as in the case of the Subanshiri, comparative impressions of discharge are apt to be, in the absence of measurements. The French missionaries who were for some years stationed near the Lu-Kiang, about lat. 28° 20', speak of it as a great river. Abbé Durand, June 1863, describing a society of heretical lamas who had invited his instructions, and who were willing to consign the paraphernalia of their worship to the waters, writes, 'What will become of it all? The *Great River*, whose waves roll to Martaban, is not more than 200 or 300 paces distant.'[2] . . . A river so

---

[2] *Ann. de la Prop. de la Foi*, Tom. xxxvii.

spoken of in lat. 28° 20', or thereabouts, may easily have come from a remote Tibetan source. It is hard to say more as yet, among the uncertainties of the geography of Tibetan steppes, and the difficulty of discerning between the tributaries of this river and that of the next; but the Lu-Kiang, or a main branch of it, under the name of Suk-chu, appears to be crossed by a bridge on the high road between Ssŭ-ch'uan and Lhassa,[3] four stations west of Tsiamdo on the Lant'sang. We may hope for more light if Colonel Prejevalsky's present journey is attended with the success that it deserves.

13. (viii.) The Lan-t'sang, or Mekong, the great river of Camboja, which rivals the Yangtzŭ itself in length, has its sources far north in Tibet, but attended with the uncertainties that we have spoken of under No. vii. Its lower course has long been known in a general way, but only accurately since the French expedition, from its mouth up into Yun-nan, in 1866-67. The town of Tsiamdo, capital of the province of Kham, which stands between the two main branches that form the Mekong, in about lat. 30° 45', was visited by Huc and Gabet, on their return under arrest from Lhassa; but whatever quasi-geographical particulars Huc gives seem to have been taken, after the manner of travellers of his sort, from the Chinese itineraries published in Klaproth's 'Description du Tübet.' Kiepert, in his great map of Asia of 1864, had apparently so little faith in Huc's statements of this kind, that he makes the two branch rivers of Tsiamdo, after their union, form the source of the (Lohit) Brahmaputra. This was a somewhat wild idea even then; but now, when Tsiamdo has been visited by later missionaries (as by Bishop des

---

[3] See *Description du Tübet*, translated by Klaproth, p. 222, and compare Ritter iv. 252, and 225-6; also *Huc*, ii. 445. The bridge is his Kia-yu-Kiao, and had fallen just before his arrival.

Mazures and Abbé Desgodins in 1866[4]), travelling from and returning to the Chinese frontier, and following at no great distance the course of the Lant'sang, there can hardly be a reasonable doubt as to the course of this river as far north as Tsiamdo ; and this is shown roughly in M. Desgodins' map.

§ 14. (ix.) The Kin-sha (or Chin-Sha), is that which gives a title to Captain Gill's book, a title justified by the fact that he followed its banks, with occasional deviations, during four-and-twenty marches on his way from Bat'ang to Ta-li-fu. This river is probably the greatest in Asia, as it is certainly the longest,[5] and one of the most famous ; but it would be excelled even in length were the Klaprothian view of the identity of the Tsanpu and the Irawadi correct ; and far excelled by the Hoang-Ho if we could view that river with the eyes of a puzzle-headed ecclesiastical traveller of the middle ages, who traversed all Asia, from Astrakhan to Peking, and who seems to have regarded as one river, which was constantly 'turning up' on his route (and *that* identical with the Phison of Paradise), the Volga, the Oxus, the Hoang-Ho and the Yangtzŭ. Well might he say with pride : ' I believe it to be the biggest river of fresh water in the world, and I have crossed it myself!'[6]

The sources of the Kin-sha are really, according to the best of our knowledge, in or about long. 90°, —*i.e.* almost as far west as Calcutta. Its upper course, though far below the source, was crossed by Huc and Gabet in the winter of 1845 ; and reached, though not crossed, by Colonel Prejevalsky in

---

[4] Desgodins, *La Mission du Thibet*, pp. 80–83. The missionaries call the place *Tcha-mou-to*.
[5] In length the order of the rivers of the world seems to be: (1) Mississippi (including Missouri), (2) Nile, (3) Amazon, (4) Yangtsze Kiang (or Kin-Sha-K.), (5) Yenesei. But probably the Congo ought, as now known, to take a high place in this list.
[6] John Marignolli, in *Cathay*, &c. p. 350.

January 1873, about long. 90° 40′, lat. 35° 50′. Huc crossed the river on the ice, and says nothing of dimensions, though he leaves on our memories that famous picture of the frozen herd of yaks. But from Prejevalsky we have information as to the great size of the river even in this remote portion of its course: the channel, when seen, 750 feet wide, and flowing with a rapid current, but the whole river-bed from bank to bank upwards of a mile wide, and, in the summer floods, entirely covered to the banks, and sometimes beyond.[7] It must have been in this flooded state that it was crossed by a Dutch traveller, Samuel Van de Putte (who has left singularly little trace of his extraordinary journeys), sometime about the year 1730.[8]

The name given to the river in this part of its course is (Mong.) *Murui-ussu*, or *Murus-ussu*, the 'Winding Water,' and (Tib.) *Di-chu*, or *Bhri chu*, the 'River of the (tame) Yak-Cow,'[9] from one or other of which Marco Polo seems to have taken the name Brius which he gives to the river in Yun-nan.

In leaving the steppes, and approaching the jurisdiction of the Chinese, it seems to receive from them the name of Kin-sha Kiang, and this name is applied, at least as far as Swi-fu, where it is joined by the Min River coming down from Ssŭ-ch'uan. Here the Great River becomes navigable to the sea, though the navigation is impeded, as Captain Gill's narrative forcibly

---

[7] *Prejevalsky*, ii. 221.

[8] 'After traversing this country one reaches a very large river called Bi-chu, which, as Signor Samuel Van der, a native of Fleshinghe, in the province of Zeland, in Holland, has written of it, is so large that to cross it in boats of skins he embarked in the morning, and landed on an island in the evening, and could not complete the passage across till the middle of the following day.'—P. *Horace della Penna*, in Appendix to Markham's *Tibet*, p. 312.

[9] These are Klaproth's interpretations, in his notes to *Horace della Penna*. See also Prejevalsky, u.s.

depicts, by numerous rapids and gorges hard to pass.[1]

Of all the Tibetan and Yun-nan part of the river, excepting in D'Anville's maps, of which the value in this part has always been a little doubtful, we have had, previous to Captain Gill's journey, nothing of actual survey.

§ 15. The next great river (x.) belonging to this series is the Ya-lung Kiang of the Chinese, a corruption of the Tibetan Jar-lung, or Yar-lung.[2] It rises in the mountains called Baian-Kara, on the south of the Koko-nur basin, about lat. 34°, and flows with a course generally southerly, and parallel to the Kin-sha, till it joins that river in the middle of its great southerly elbow, about lat. 26° 30'. In its upper course it is called, according to Klaproth's authority, Gnia-mtso, which seems to be the same as the Nia-chu of Captain Gill (II. 135). The Jar-lung valley was the traditional cradle of the Tibetan monarchy,[3] which only at a later time moved into the western highlands of Lhassa. The river was passed some 260 miles north of the mouth, by Captain Gill on his way from Ta-chien-lu to Lit'ang, by a coracle ferry (II. 139); near this the width varied from 50 to 120 yards, with a rapid broken current. Baggage animals had to be swum across.

---

[1] Geographical names are largely names given, or at least defined in their application, by geographers, and one should always speak cautiously as to how a river or mountain-chain in Asia is called by natives on the spot. Blakiston, at the furthest point of the river ascended by him, found it only known as the 'River of Yun-nan.' So streams are, or used to be, locally known in Scotland only as '*the* watter,' or perhaps the 'watter of —' such a place. In one part, Capt. Gill tells, the great river is known as 'the River of Dregs and Lees.'

[2] Ritter gives the meaning of this as 'White River' (iv. 190); Klaproth as 'Vaste Rivière' (*Description du Tübet*, 190). I can find neither in the *Tibetan Dictionary* of Jaeschke. The Tibetan vocabulary in Klaproth gives *ghiar* 'ample, vaste' (p. 145). *Kar-po* is white; and it will be seen that in its lower course the Chinese do call it Pe-shui, or 'White Water.'

[3] See Sanang Setzen in Schmidt's *Ost-Mongolen*, p. 23 and *passim*.

INTRODUCTORY ESSAY.   [35]

The confluence of the two great rivers Yar-lung and Kin-sha was visited by Lieut. Garnier and his party in 1868. Garnier thus describes the junction:

'The Kin-sha is here by no means shut in as it is at Mong-kou' (where they had crossed the eastern limb of the great bend); 'and it is reached by a hardly sensible declivity. Little naked hills line the banks. The river comes from the south-west, then describes a curve inclining to 10° south of east; and it is at the apex of this curve that it receives the Ya-long Kiang. The latter arrives from the north, shut in closely by two walls of rock absolutely perpendicular, so that no passage along the banks is possible. Its breadth is nearly equal to that of the Blue River;[4] and its current, at least when we saw it, was somewhat stronger. I could not measure the depth of either, but it seemed considerable. As at Mong-kou the flood-rise was 10 mètres. I was surprised to learn that the country people here gave the name of Kin-sha Kiang to the Ya-long—i.e. to the tributary—and that of Pe-shui Kiang, 'White-Water River,' to the principal stream. If, as regards volume, there was, at first sight, some room for doubt between the two, the aspect of the two valleys showed at once which was entitled to keep the name of Kin-sha Kiang. The mouth of the Ya-long is a sort of accidental gap in the chain of hills that lines the Blue River, and the orographic configuration of the country indicates clearly that the latter river comes from the west and not from the north. . . . This anomaly in their nomenclature will seem less surprising if we remember that in China river-names are always local, and change every 60 miles. About Li-kiang you again find that the Kin-sha has got its proper name, and it is the Ya-long that is there called Pe-shui Kiang.'[5]

§ 16. The last of these great parallel rivers with which we have to deal is that great branch (xi.), called on our maps Wen and Min Kiang, which we regard geographically as a tributary of the Kin-sha or Yangtzŭ, but which the Chinese hydrographers have been accustomed to regard rather as the principal stream. We find this view distinctly indicated in that oldest of Chinese documents, the Yu-Kung.[6] It comes out again prominently in Marco Polo's account of

[4] So the French term the Yang-tzŭ.
[5] *Voyage d'Exploration,* i. 503. Garnier gives a view of the confluence.
[6] See Richthofen's *China,* i. 325: 'On the Min-shan begins the course of the Kiang. Branching eastward it forms the To . . . . &c.' The Min-shan is the mountain country north-west of Ssŭ-ch'uan.

INTRODUCTORY ESSAY.

Sin-da-fu (or Ch'êng-tu-Fu), which is quoted by Captain Gill at the beginning of his second volume. ... 'The name of the river is *Kian-Suy*,' i.e. as the late M. Pauthier explains *Kiang-Shui*, 'Waters of the Kiang' (or River Kiang, see *He-Shui*, a little below). The same view appears in Padre Martini's 'Atlas Sinensis' (1655);[7] and very distinctly in a paper professedly (and probably in reality) indited in 1721 by the great Emperor K'ang-hi, which Klaproth has translated in that dissertation of his already spoken of regarding the course of the river of Tibet:

> 'From my youth up,' says the Emperor, 'I have been greatly interested in geography; and for such purposes I sent officers to the Kuen-Luen mountains, and into Si-fan. All the great rivers, such as the Great Kiang, the Hwang-Ho (Yellow River), the He-Shui (Black River, the Kara-Ussu of the Mongols), the Kin-sha Kiang, and the Lan-t'sang Kiang, have their sources in those regions. My emissaries examined everything with their own eyes; they made accurate inquiries, and have embodied their observations in a map. From this it is clear that all the great rivers of China issue from south-eastern slopes of the great chain of *Nom-Khûn-ubashi*, which separates the interior from the exterior system of waters. The Hwang-Ho has its source beyond the frontier of Sining, on the east of the Kulkun mountains.... The Min-Kiang has its origin to the west of the Hwang-Ho, on the mountains of *Baian-Kara-tsit-sir-khana*, which is called in Tibetan *Miniak-thsuo*, and in the Chinese books *Min-Shan*; it is outside of the western frontier of China; the waters of the Kiang issue from it.... According to the Yu-Kung the Kiang comes from the Min-Shan. This is not correct; it only passes through that range; this is ascertained. This river runs to Kuon-hien,[8] and there divides into half a score of branches, which reunite again on reaching Sin-tsin-hien; thence it flows south-east to Siu-chau-fu, where it joins the Kin-sha Kiang.'[9]

Captain Gill, so far as we are aware, is the only traveller who has traced this river above Ch'êng-tu, to the alpine highlands, doubtless the Min-Shan of the Yü-Kung, from which it emerges. This he did

---

[7] To this remarkable work I have tried to do some justice in an article in the *Geographical Magazine* for 1874, pp. 147–8
[8] The Kwan-hsien of Capt. Gill, vol. i. p. 330.
[9] Klaproth, *Mémoires relatifs à l'Asie*, iii. 392.

on that excursion from Ch'êng-tu to the north, in the months of May and June, 1877, which is described in the last two chapters of the first volume, entitled, ' A Loop-cast towards the Northern Alps.'

§ 17. Captain Gill has pointed out that, of the many branches of the river which ramify through the plain of Ch'êng-tu, no one now passes through the city at all corresponding in magnitude to that which Marco Polo describes, about 1283, as running through the midst of Sin-da-fu, ' a good half-mile wide, and very deep withal.' The largest branch adjoining the city now runs on the south side, but does not exceed a hundred yards in width; and though it is crossed by a covered bridge with huxters' booths, more or less in the style described by Polo, it necessarily falls far short of his great bridge of half a mile in length. Captain Gill suggests that a change may have taken place in the last five (this should be *six*) centuries, owing to the deepening of the river-bed at its exit from the plain, and consequent draining of the latter. But I should think it more probable that the ramification of channels round Ch'eng-tu, which is so conspicuous even on a small general map of China, like that which accompanies this work, is in great part due to art; that the mass of the river has been drawn off to irrigate the plain; and that thus the wide river, which in the thirteenth century may have passed through the city, no unworthy representative of the mighty Kiang, has long since ceased, on that scale, to flow. And I have pointed out briefly (II. 6) that the fact, which Baron Richthofen attests, of an actual bifurcation of waters on a large scale taking place in the plain of Ch'êng-tu—one arm ' branching east to form the To '—(as in the terse indication of the Yü-Kung)—viz. the To-Kiang or Chung-Kiang flowing south-east to join the great river at Lu-chau, whilst another flows south to Sü-chau or

Swi-fu, does render change in the distribution of the waters about the city highly credible.[1]

The various branches, except those that diverge, as just said, to the Ch'ung-Kiang, reunite above Hsin-chin-hsien (Sin-tsin-hien of Richthofen, Sing-chin of the general map), which was Captain Gill's second station in leaving Ch'êng-tu for Tibet. Up to this point the main stream of the Min is navigable, whilst boats also ascend the easternmost branch to the capital. Indeed, vessels with 100 tons of freight reach Ch'êng-tu by this channel when the river is high.[2] At Kia-ting-fu the Min receives a large river from the mountains on the west, the Tung-Ho, which brings with it both the waters of the Ya-Ho, from Ya-chau (see Vol. II. p. 47), and those of the river of Ta-chien-lu. Kia-ting is an important trading place, the centre of the produce in silk and white-wax, and situated in a lovely and fertile country. Below this the Min-Kiang is a fine, broad, and deep stream, with a swift but regular current,[3] and obstructed by only one rapid, at Kien-wei, but that a dangerous one. It joins the Kin-sha, as so often mentioned, at Sü-chau or Swi-fu.

§ 18. We have spoken, perhaps at too great length, of the great parallel rivers which form the most striking physical characteristic of the region between India and China. Let us now say something of the history of a problem that many attempts have been made to solve : that of opening direct communication between these two great countries.

How difficult a problem this is will be, perhaps,

---

[1] A short but interesting notice of the irrigation and drainage of the plain of Ch'êng-tu is given by Richthofen in his 7th letter to the Shanghai Chambers, p. 64. He mentions that the existing channels, though not those close to the city, reach in some instances to a width of 1,000 feet.
[2] Richthofen, p. 71.
[3] Cooper says,' often a mile wide;' but the river was unusually high, for he says, 'unbroken by a single rapid.' Richthofen specifies the frequent wrecks in the rapid at Kien-wei.

most forcibly expressed by the circumstance that in all the multiple history of Asiatic conquest,—and in spite of the fact that you can hardly lay your finger on an ordinary atlas-map of Asia without covering a spot that has at one time or other been the focus of a power whose conquests have spread far and wide,—at no time did a conqueror from India ever pass to China, nor (unless with one obscure and transient exception, which will be noticed below) a conqueror from China to India, nor at any time, omitting the brief passage of Chinghiz, who barely touched the Punjab, did the conquests of any conqueror embrace any part of both countries.

Moreover, Chinese history seems to establish the fact that India first became known to China, not across these lofty highlands and the vast fissures in which the rivers flow of which we have spoken, but by the huge circuit of Bactria and Kabul. The idea that there was a more ancient intercourse between the two great countries, and that the Chinas of the Laws of Menu and of the Mahabharat were Chinese, must, I now believe, be abandoned. The *Chinas*, as Vivien de St. Martin and Sir H. Rawlinson have indicated, are to be regarded as a hill-race of the Himalya, probably identical with the *Shinas* of Dardistan. The first report of India was brought to China in the year B.C. 127, in the reign of Hsia-wu-ti of the Han dynasty, when Chang-kien, a military leader who had been exploring the country about the Oxus, returned after an absence of twelve years, and, among many other notices of Western Asia, reported of a land called *Shin-tu*—*i.e.* Sindu, Hindu, India—of which he had heard in Tahia, or Bactria, a land lying to the south-east. moist and flat and very hot, the people civilised, and accustomed to train elephants. From its position, and from the fact that stuffs of *Shu* (*i.e.* Ssŭ-ch'uan, see Vol. II. pp

17, 35) arrived in the bazaars of Bactria through Shin-tu, Chang-kien deduced that this country must lie not far from the western provinces of China. Several efforts were in consequence made to penetrate by the Ssŭ-ch'uan frontier to India ; one got as far as Tien-Yuĕ (Burma or Pegu), but others not even so far. When communication opened with India some 200 years later it was by the circuitous route of Bactria, and so it continued for centuries.

§ 19. If the acute general of the Han was right about the stuffs of Shu, the trade that brought these stuffs must have been of that obscure hand-to-hand kind, probably through Tibet, analogous in character to the trade which in prehistoric Europe brought amber, tin, or jade from vast distances. But it is curious to set alongside of these Chinese notices of obscure trade reaching to India that remarkable passage in the Periplus, a work of the first century A.D., which speaks of *Thin*, and of its great city *Thinae*, 'from which raw silk, and silk thread, and silk stuffs were brought overland through Bactria to Barygaza (Bhrōch), 'as they were on the *other hand by the Ganges River to Limyrike*' (*Dimyrike*, the Tamul country, Malabar). Ptolemy, too, a century later, says that there was not only a road from the countries of the Seres and of the Sinae to Bactriana by the Stone-Tower (*i.e.* by Kashgar and Pamir), but also a road to India which came through Palimbothra (or Patna). It is probable that this traffic was still only of that second and third hand kind of which we have spoken, and the mention of Palibothra recalls the fact that Patna is the Indian terminus at which the Fathers Grueber and D'Orville arrived after their unique journey from Northern China by Tibet.

Returning to the Periplus, the passage that we have referred to is followed by another speaking of a rude mongoloïd people (it is the shortest abridgment of the description) who frequented the frontier of *Thin*, bringing *malabathrum* or cassia leaves. These, I think, may undoubtedly be regarded as some one or other of the hill tribes on the Assam frontier, and I should in this case regard the mention of *Thin* as vaguely indicating the knowledge, as already popular in India, that there was a great land bearing a name like that beyond the vast barrier of mountains. In a like way we find the name of Maháchín applied in the 15th century by Nicolo Conti, and in the 16th century by Abu'l-Fazl, to the countries on the Irawadi ; and I remember, many years ago, seeing a Tibetan pilgrim at Hardwár, whose only intelligible indication of where he came from was ' Maháchín.'

§ 20. As our subject is the history not of communication generally between China and India, but only of that communication across their common highland barrier, we are bound, so far as our knowledge goes, to stride at once from pseudo-Arrian to Marco Polo. There is in the interval, indeed, an obscure record of a Chinese invasion of India, which should perhaps constitute an exception.

In 641, the King of Magadha (Behar, &c.) sent an ambassador with a letter to the Chinese court. The Emperor, who was then Tai-tsung of the T'ang dynasty, probably the greatest monarch in Chinese history, in return sent one of his officers to go to the King with an imperial patent, and to invite his submission. The King *Shiloyto* (Siladitya) [4] was all astonishment. ' Since time immemorial,' he asked his courtiers, ' did ever an ambassador come from Mahachina ? ' 'Never,' they replied. The Chinese author

[4] This Siladitya is a king of whom much mention is made in the Memoirs of Hwen-T'sang. He was a devout Buddhist, and a great conqueror, having his capital at Kanauj, and a dominion extending over the whole of the present Bengal Presidency, from the sea to the frontier of Kashmir.

remarks here, that in the tongue of the barbarians, the Middle Kingdom is called 'Mohochintan' (*Mahá-chínasthāná*). A further exchange of civilities continued for some years. But the usurping successor of Siladitya did not maintain these amicable relations, and war ensued, in the course of which the Chinese, assisted by the kings of Tibet and Nepal, invaded India. Other Indian kings lent aid and sent supplies; and after the capture of the usurper *Alanashun* (?), and the defeat of the army commanded by his queen on the banks of the *Khien-to-wei*, 580 cities surrendered to the arms of China, and the king himself was carried prisoner to that country.

Chinese annals colour things, but they are not given to invention, and one can hardly reject this story.[5] It is probable, however, even from the story as it is told, that this was rather a Nepalese and Tibetan invasion, promoted and perhaps led by Chinese, than a Chinese invasion of India. Lassen, as far as I can discover, does not deal with the subject at all. The name of the river on which the Indian defeat took place, *Khien-to-wei*, would according to the usual system of metamorphosis represent *Gandhava*; qu. the Gandhak?

§ 21. The story told by Firishta and others, of an invasion of Bengal by the Mongols, 'by way of Cathay and Tibet,' during the reign of 'Alá-ud-dín Musa'úd, King of Delhi (A.D. 1244), has been shown

---

[5] The account is found in Stanis. Julien's papers from Mat-wan-lin, in the *Jour. Asiat.* ser. iv. tom. X. See also *Cathay, and the Way Thither*, p. lxviii., and Richthofen's *China*, pp. 523, 536-7. It is stated that Wang-hwen-tse, the envoy who went on the mission that resulted in this war, wrote a history of all the transactions in twelve books, but it is unfortunately lost. The Life of Hwen T'sang states that that worthy, when in India, prophesied that, after the death of Siladitya, India would be a prey to dreadful calamities, and that perverse men would stir up a desperate war. The same work mentions as the fulfilment that Siladitya died towards the end of the period *Yung-hwei* (A.D. 650-655), and that in conformity with the prediction, India 'became a prey to the horrors of famine,' of which the envoy Wang-hwen-tse, just mentioned, was an eye-witness. But no mention is made of the Chinese invasion.

by Mr. Edward Thomas to have arisen out of a clerical error in MSS. of the contemporary history called *Tabakát-i-Násirı*.[6] But two preposterous attempts were made in the 13th and 14th centuries, at the counter-project, the invasion of the countries above the Himalya from Gangetic India.

The first of these (A.D. 1304), was the adventure of Mahommed Bakhtiyár Khilji, the first Mussulman conqueror of Bengal, and ruler of Gaur, of whom the historian just quoted says, that ' the ambition of seizing the country of Turkestan and Tibet began to torment his brain.' The route taken is very obscure ; the older interpretations carried it up into Assam, but Major Raverty's conclusion that it ascended the Tista valley is perhaps preferable. The Khilji leader is stated to have reached the open country of Tibet, a tract entirely under cultivation, and garnished with tribes of people and populous villages. The strenuous resistance met with, the loss in battle with the natives, and the distress of the troops from such a march, compelled a retreat ; they were sorely harassed by the men of the Raja of Kamrud (apparently *Kámrúp*, of which Assam was the heart), and Mahommed Bakhtiyár finally escaped with but a hundred horsemen or thereabouts, and soon after fell ill and died.

The second attempt was one of the insane projects of Mahommed Tughlak, which took place in 1337. It was, according to Firishta, directed against China, but it must be said that there is no mention of China as the object in the earlier accounts. The account given by the historian Ziá-ud-din Barní, who wrote in the next generation, is as follows :—

'The sixth project, which inflicted a heavy loss upon the army, was the design which he formed of capturing the mountain of Kará-jal. His conception was that, as he had undertaken the conquest of Khurasan, he would (first) bring under the dominion of

[6] See Thomas's *Pathán Kings of Dehli*, p. 121.

Islam this mountain, which lies between the territories of Hind and those of China, so that the passage for horses and soldiers, and the march of the army, might be rendered easy. To effect this, a large force, under distinguished *amirs* and generals, was sent to the mountains of Kará-jal, with orders to subdue the whole mountain. In obedience to orders it marched into the mountains, and encamped in various places; but the Hindus closed the passes, and cut off its retreat. The whole force was thus destroyed at one stroke, and out of all this chosen body of men only ten horsemen returned to Dehli to spread the news of its discomfiture.'[7]

The account given by the traveller Ibn Batuta, who was then at the court of Mahommed Tughlak, is to the same effect ; and though he mentions the names of two places that were taken by the troops, *Jidiya* before entering the mountains, and *Warangal* in the hill-country, Ibn Batuta does not aid us by these (the last of which is altogether anomalous) in fixing the locality, any more than he helps us to understand the object, of the enterprise.

§ 22. Coming now to Marco Polo, whose steps it would be hard for any traveller in a little known region of Asia altogether to avoid, we may briefly say that on the first important mission to which he was designated by the Great Khan Kublai, in making his way to the frontier of Burma (*Mien*), he travelled from Ch'êng-tu (*Sin-da-fu*), by the route which Captain Gill followed, as far probably as Ch'ing-chi-hsien . This was Captain Gill's ninth march from Ch'êng-tu. We do not know the length of Marco's daily journeys, but after five such from Ch'êng-tu, he was already in Tibet. Probably the country which was counted as Tibet, in those days, began immediately on passing Ya-chau and entering the mountains. From Ching-chi-hsien the routes diverge. Captain Gill, bound for Ta-chien-lu and Bat'ang, strikes north-west ; Marco Polo's route continued to bear south-south-east, towards the city of Ning-yuan-fu, the exist-

---

[7] Elliot's *History of India*, &c. (by Dowson) iii. 241-2.

ing capital of the beautiful valley of Kien-chang, the Caindu or Ghiendu of the Venetian. This is the route on which Baron Richthofen's journey met with an unfortunate interruption (see p. [64]), and which has recently been travelled by Mr. Baber. It is the road by which the greater part of the goods for Bhamò and Ava used to travel from Ch'êng-tu, before the Mahommedan troubles in Western Yun-nan. Those goods went on by a direct road from Kien-chang to Ta-li-fu. But Marco Polo's road led him south, and across the great elbow of the Kin-sha to the city of Yun-nan Fu (his *Yachi*). From this he travelled to Ta-li-fu (*Carajan*), and thence to Yung-chang-fu (*Vochan* or *Unchan*). Beyond this there are difficulties as to the exact extent and direction of his travels, concerning which some discussion occurs in Vol. II. Chap. VIII. of Captain Gill's book, as well as in my own commentary on the book of Marco. It would hardly profit to enter here on a detailed recapitulation of a discussion which as yet has confessedly received no satisfactory determination.

§ 23. Ta-li-fu, which is so often spoken of in these pages, and is so prominent a point in Captain Gill's narrative, is indeed a focal point on this frontier at which many routes converge; and for ages it has been the base of all operations, military or commercial, from the side of China towards Burma. It may still be regarded as the capital of Western Yun-nan, as it was in the days of Marco Polo. Ta-li-fu, for some centuries before the master whom Marco served, Kublai Khan, conquered it (A.D. 1253), had been the seat of a considerable Shān Kingdom, called by the Chinese Nan (or Southern)-Chao: this latter term being a Shan word for 'prince,' which still figures among the titles of the kings of Siam, and of all the other states of that wide-spread race. During the recent brief independence of the Mahommedans or *Panthés* (probably themselves as much Shan as Chinese in blood), Ta-li again became a seat of royalty, and here reigned Tu-wen-hsiu, alias Sultan Sulimán, from about 1860 to 1873, when the city was captured by the Imperialists, and the Mahommedans were massacred. The king himself took poison, but his head was sent in honey to Peking.[8]

Mr. Baber, quoted at p. 303 of Vol. II., says that the terms *Sultan* and *Suliman* were quite unknown on the spot. The fact is that in Indo-Chinese countries Islam has never assimilated the nationality of those who profess it, as in Western Asia. This is the case in some degree in Java, as it is in greater degree in Burma, and no doubt more than all in China. The people, in these countries, professing Islam, are to be compared with Abyssinian professors of Christianity. At the court of the Mussulman Sultan of Djokjokarta, in Java, I have had the honour of being introduced to half a dozen comely sultanas, and of shaking hands with them; whilst I have seen the Sultan and his Court taking part in a banquet at the Dutch Residency, and in drinking a number of toasts, of which a printed programme in Dutch and Javanese was distributed. In the capital of Burma, where professing Mahommedans are much less secluded from the influence of more orthodox Moslems than those of Yun-nan are, they have been characterised in passages of which I extract the following: 'As might be expected, they are very ignorant sons of the Faith, and in the indiscriminating character of their diet, are said to be no better than their neighbours; so that our strict Mussulmans from India were not willing to partake of their hospitalities.' And as regards names: 'Every indigenous Mussulman has two names. Like the Irishman's dog, though his true name is *Turk*,

[8] Tu-wen-hsiu, or, as Cooper calls him phonetically, Dow-win-sheow, had been a wealthy merchant in Tali.

he is always called *Toby*. As a son of Islam, he is probably Abdul Kureem; but as a native of Burma, and for all practical purposes, he is Moung-yo, or Shwé-po.'[9] The style of 'Sultan Suliman' &c. was no doubt confined to the few Hajjis or Mollahs that were at Ta-li. That there were such is proved by the Arabic circular which was issued, and which reached the Government of India in the way mentioned at page [52] below. The following is an extract from that document: 'O Followers of Mahommed! in telling you how it fared with us, we offer grateful thanks to the Almighty. It behoves you to rejoice in the grace that God hath shown to us. . . . God gave us courage and created fear in the hearts of the Idolaters, so we, by the decree of God, did defeat them. . . . Therefore we have set up a Mahommedan Sultan; he is prudent, just, and generous. . . . His name is *Sádik*, otherwise called *Sulimán*. He has now established Mahommedan law. . . . Since we have made him our Imam we have been, by the decree of God, very victorious. . . . The metropolis of infidelity has become a city of Islam.'

Bhamò, again, a small stockaded town, in lat. 24° 16', stands on a high bank over the Irawadi, on its eastern side, about two miles below the entrance of a considerable stream, which we have been used to call, from the Burmese side, the Ta-peng River, but which Captain Gill, who followed its course almost the whole way from Têng-yuëh-chau (or Momien) to its confluence with the Irawadi, calls the Ta-ying Ho, or T'êng-yuëh River. Here, or hereabouts, has long been the terminus of the land-commerce from China; and as early as the middle of the fifteenth century we find at Venice, on the famous world-map of Frà Mauro (who no doubt got his information from Nicolo Conti, who had wandered to Burma earlier in

---

[9] *Mission to the Court of Ava* in 1855, pp. 151-152.

that century), on the upper part of the river of Ava, a rubric which runs : *Qui le marchatantie se translata da fiume a fiume per andar in Chataio.* 'Here goods are transferred from river to river, and so pass on to Cathay.' And in the first half of the seventeenth century there is some evidence of the maintenance here of an English factory for the East India Company.

§ 24. The right to travel in the interior of China was first conceded by Article IX. of the Treaty of Tien-tsin,[1] which conferred it on all Englishmen. And this treaty undoubtedly constitutes a land-mark from which we are to date the commencement of modern exploration, and of a more exact knowledge, only now being slowly built up, of the physical geography of the country, of its natural resources, and of the true characteristics of the cities and populations of China. But here it is necessary to interpose a caveat. When we speak of the commencement of modern exploration in China and Tibet, or allude to any modern traveller as being the first to visit this or that secluded locality in those regions, it must always be understood that we begin by assuming a large exception in favour of the missionaries of the Roman Church : for those regions have to a great extent, and for many years past, been habitually traversed by the devoted labourers who have been extending the cords of their Church in the interior, and on the inland frontier of China. Geographical research is not their object, and for a long period publicity was only adverse to their

---

[1] 'Art. IX.—British subjects are hereby authorised to travel, for their pleasure or for purposes of trade, to all parts of the interior, under passports which will be issued by their Consuls, and countersigned by the local authorities. These passports, if demanded, must be produced for examination in the localities passed through. . . . . If he (the traveller) be without a passport, or if he commit an offence against the law, he shall be handed over to the nearest Consul for punishment, but he must not be subjected to any ill usage in excess of personal restraint. . . .'

purpose ; and thus their labours and their journeys in those remote regions, which long preceded the treaty of Tien-tsin, though often recorded in the *Annales de la Propagation de la Foi* and similar journals for those who seek them there, have only occasionally come before the notice of geographical societies, or of the public in Europe. There are, indeed, notable exceptions, of which we shall presently take account ; but apart from these, in hardly any instance has a traveller penetrated in this region to a point where he has not found a member of these Roman Catholic missions to have been before him.

§ 25. We have already alluded to the letter written from Tibetan territory by an eminent member of these missions, which reached the Asiatic Society of Bengal, to their no small surprise, in 1861. When Lieut. Garnier and his party made their rapid and venturesome visit to Ta-li-fu, in 1868, their guide and helper was their countryman M. Leguilcher, of the same mission, whom they found in his seclusion near the north end of the Lake of Ta-li-fu, and with whom Captain Gill made acquaintance nine years later at the city itself. Not only at Ch'ung-ch'ing and at Ch'êng-tu did Captain Gill find kindly aid among the members of these missions, but at Ta-chien-lu, on the acclivity of the great Tibetan plateau, like Mr. Cooper before him, he found cordial welcome from the venerable Bishop Chauveau, an old man whose noble presence and benign character seem to have equally impressed both travellers.[2]

[2] See Captain Gill, Vol. II. pp. 111-112. Mr. Cooper says: 'I perceived a venerable old man, dressed in Chinese costume, with a long snow-white beard. I shall never forget him as long as I live. He was sixty years of age, forty of which he had spent in China as a missionary. But long illness made him look older. His countenance was very beautiful in its benignity ; his eye, undimmed by age and suffering, lighted on me with a kindly expression ; and he bade me welcome in English, which he had learned from his mother, an English lady, with

Members of the same body were found by both travellers also at Bat'ang, in the basin of the Kin-sha, and on both occasions, at nine years' interval, the Abbé Desgodins was one of their number. Bat'ang appears to be at present the furthest station of the missionaries towards Tibet; nor have they any now within the actual Lhassa dominions. But at one time they had for some years establishments within the political, as well as the ethnical, boundary of Tibet. Abbé Renou, the first of the body to make an advance in this direction, obtained in 1854 a perpetual lease of Bonga, a small valley in the hills adjoining the Lu-Kiang on its eastern bank, for a rent of 16 or 17 taels. This is under the Government of Kiang-ka, where officials both Chinese and Tibetan reside. The missionaries of Bonga cleared a good deal of land, erected buildings, and began to have considerable success in making converts, both among the wilder tribes of the hills, and among the Tibetan villages around them. But in 1858 they were violently ejected by the person who had given the lease, aided by an armed party. No redress was got till 1862, when the Treaty of Tien-tsin began to take actual effect; the suit of the missionaries was heard in the Court at Kiang-ka, and they were reinstated at Bonga. Three years later, however, the neighbouring Lamas, who, as Captain Gill several times explains, are very unpopular themselves, and who were all the more disposed to view with jealousy whatever success the missionaries had among the people, took advantage of disorders in the Province, and expelled the missionaries from Bonga and other settlements outside the Chinese political frontier.

a tremulous but musical voice.' (Page 181.) And again: 'The kindness of the people of Ta-tsian-loo had made a deep impression on me, and in taking leave of the kind old Bishop, who, with tears in his eyes, invoked a blessing upon me, my emotion checked all utterance.' (Page 222.)

MM. Desgodins and F. Biet, who were at Bonga, after a good deal of violence on one side, and some administration of presents on the other, were allowed to carry off their flock into Chinese territory, but their establishment was sacked and burnt (29th September, 1865). MM. Durand and A. Biet, who directed an out-station at a place called Kie-na-tong (among the Lu-tse), on the Salwen, just within the Yun-nan boundary, were driven away, and the former was shot in crossing a swing bridge.

Monseigneur Chauveau, who had at this time succeeded to the government of the mission, established his head-quarters at Ta-chien-lu, on the borders of what we should in India call the Regulation and the Non-Regulation Provinces, and outstations were still maintained at Tseku and Yerkalo on the Lan-t'sang; the former under Yun-nan, the latter in the Bat'ang territory, but none in Tibet proper.

§ 26. In January 1867, the Kaji Jagat Sher, an envoy from Maharaja Jung Bahadur to the Court of China,[3] was passing through Bat'ang, and made the acquaintance of the missionaries there. Their communications were in English, which was probably indifferent on both sides; but what the Nepalese envoy said led the French fathers to suppose that the British Government in India had heard of their sufferings at the hands of the Tibetans, and had requested the Nepal Government to make inquiry.[4] M. Desgodins accordingly sent by the hands of Jagat Sher a very interesting letter, written in very imperfect English, and addressed to the Resident at Katmandu (then Colonel George Ram-

---

[3] Cooper met Jagat Sher both at Ch'êng-tu, and near Bat'ang in returning. The Envoy had met with very bad treatment from the Chinese, and was not allowed to proceed beyond Ch'êng-tu. (See *Cooper*, pp. 158 *seq.*, 398 *seq.*

[4] It does not seem to have been the fact that any news of the kind had reached India.

say), with a full account of their circumstances, of the violent treatment they had met with, and of the murder of M. Durand. The Governor-General, in replying to Colonel Ramsay's communication of this letter, expressed the deep interest with which he had read it, but intimated that the only intervention in their favour possible, would be through the Maharaja of Nepal, and through our Minister at Peking. The Government letter went on :

> 'You will, at the same time, however, observe that if the Government may be permitted to offer an opinion to men animated by higher considerations than those of mere personal security or success, these reverend gentlemen would do well to abandon the country in which their sufferings have been so great, and settle in British India, where there are extensive and peaceful tracts, such as Lahoul, Spiti, and Kulu, containing a semi-Thibetan population, likely to receive Christianity with favour.'

Copies of the correspondence were sent to our Minister at Peking, and of the letter intended for the missionaries, not only thither, and to Nepal, but to Ladak and Upper Assam. This shows how difficult any communication is across the iron wall that separates British India from the Chinese frontier;[5] and it is greatly to be questioned if any one of the four copies ever reached its destination. That sent by Nepal was suppressed by the Chinese Amban at Lhassa; the messenger *viâ* Assam failed in making his way, and after going fifteen days' journey from Sadiya, returned; the copy from Ladák was forwarded by Dr. Cayley through the inauspicious medium of a monsignore of the Tibetan Curia, who

---

[5] There are but three cases in our time that I can recall in which the iron wall was pierced by a piece of intelligence. The first was the murder of MM. Krick and Boury, of which we have spoken above. The second was this communication from the priests at Bat'ang to the Resident at Katmandu. The third was the Arabic proclamation or circular, issued in the name of the Panthé rulers at Ta-li-fu, for the information of the Mahommedan world, which also reached Col. Ramsay at Katmandu. A copy of it was given me by the lamented Mr. J. W. Wyllie, and it was printed by my late friend Lieut. Fr. Garnier (to whom I gave it) in the appendix to his *Voyage d'Exploration*, vol. i., p. 564.

was returning to Lhassa. Of that sent by Peking the fate has not reached us ; it is doubtful, from the allusion to the subject in a collection of notices on Tibet by M. Desgodins, whether it ever was received.[6]

§ 27. This is, however, anticipating in chronological order. The first picture of Eastern Tibet in modern times was that set forth by the Abbé Huc in the famous narrative of his journey with Gabet, which astonished the world in 1850. It is true that occasional letters from both Huc and Gabet had appeared in various numbers of the *Annales de la Propagation de la Foi* in 1847–1850, but the circle to which that publication speaks was probably more limited and exclusive then than it is even now ; and I cannot find that practically anything was known to the public of their remarkable journey prior to the publication of the work. Sir John Davis, indeed, has told us how he furnished Lord Palmerston, as early as 1847, with some particulars of the journey, which his secretary, Mr. Johnstone, had obtained from Gabet, who was his fellow-passenger to Europe, and these appear to have been printed, for there are most curiously confused allusions to them in the article 'Asia,' in the eighth edition of the 'Encyclopædia Britannica,' published in 1853.[7] And up to 1855 there is absolutely, so far as I can discover, no notice of Huc or his companion in the Journals of the Royal Geographical Society, or in the annual dis-

---

[6] See that work (*La Mission du Thibet de* 1855 à 1870, Verdun, 1872) pp. 115–116. The facts in the text are gathered from a correspondence in the India Office. I lately read a Roman Catholic paper on Lord Lawrence, which, while doing him noble justice in most respects, spoke regretfully of the narrow Ulster type of religion in which he had been educated, or words to that effect. I will only say that the Viceroy who despatched the letter quoted above, and took all this trouble for these remote French Roman Catholic priests, was Sir John Lawrence, whilst the signature of the letter is that of Sir William Muir.

[7] 'Our scanty knowledge of Tibet has lately received a valuable addition in the journal of the *Revd. Mr. Puch*, a French missionary, who

courses of its Presidents, except a singularly meagre one in Captain (afterwards Admiral) W. H. Smyth's address of 1851, a reference which is certainly a notable example of scientific puritanism, true though it be that Huc does not belong in any sense or measure to the scientific category.[8] Just as little was he entitled to be ranked, as he is by a late pretentious French writer on Chinese matters, with Pauthier (who with all his faults was a genuine and enthusiastic student), and with that modest and indefatigable scholar Mr. Alexander Wylie, lumping all three together, as this writer does, as 'excellents sinologues.'[9] That Huc was, as a sinologue, next door to an impostor, and that his brilliant and, in the main, truthful sketches of travel in Tartary and Tibet were followed by later works of a greatly degenerated character, is undeniable. But it is equally undeniable that Huc was a daring and distinguished traveller, and the author of one of the most delightful books of travel ever written.[1]

§ 28. Many years before Huc's book appeared, we had, indeed, in the immortal work of Carl Ritter,—at once a quarry and an edifice,—a full, and, as far as all

proceeded from Peking, through Mongolia and Tangut, to L'Hassa, the capital of Tibet, which he left for China by the road through Kham. An English translation of his MS. journal was recently published under the auspices of Lord Palmerston.' The final redactor of the article was evidently unable to make anything of the 'Rev. Mr. Puch,' and unwilling to disturb the references of his predecessor, so he tells us that 'the travels of Huc, Gabet, and *Puch* have made some additions to our knowledge of Tartary and Tibet.' (8th edit. vol. i. p. 754.)

[8] 'The Narrative of a Residence in the Capital of Thibet, by M. Huc, a Lazarist missionary, contains some corroborative details respecting a country imperfectly known to Europeans.'—*Jour. R. Geog. Soc.* XX., p. lxx.

[9] See the *Athenæum*, August 18, 1877, in which there is a review by the present writer, of the work referred to.

[1] I have spoken more fully regarding Huc in the Introductory Essay to my friend Mr. Delmar Morgan's translation of Col. Prejevalsky's travels, and have there defended the substantial truth of his 'Souvenirs' against the ussian traveller's charges. That Huc embellished, and especially in his dramatic reports of conversations, no one can question.

our subsequent information goes, an accurate account of the great road from Ch'êng-tu to Lhassa, by Ta-chien-lu, Bat'ang, Tsiamdo, &c., with the detail of its daily stages. This is taken from Klaproth's French edition of the Chinese *Description du Tübet*, as rendered into Russian by the priest Hyacinth Bichurin (Paris, 1831). Huc makes a good deal of use of this itinerary, which describes the road which he followed on his return from Lhassa, in the very scanty contributions to geography which his narrative contains; but had it been printed as an appendix to his book, we should have followed his journey with more intelligence. In judging of his work from a geographer's point of view, however, it is fair to remember that, on this half of the journey at least, he and Gabet were travelling under arrest.

At the time of Huc's return the Roman Catholic missions had apparently no outpost beyond Ch'êng-tu. It was, as we have seen, about eight years afterwards that they began to establish themselves on the Tibetan frontier and beyond it. And apart from their little known movements, it was not till 1861 that any new endeavour occurred to penetrate those regions.

§ 29. The first attempt to act in this direction upon the concessions of the treaty of Tien-tsin, was the voyage of Captain Blakiston, Lieutenant-Colonel Sarel, and Dr. Barton, accompanied by Mr. (now Bishop) Schereschewsky of an American mission, up the Yang-tzŭ. Their object was to penetrate by Tibet, and across the Himalya, into India. That was a bold aim, which even at this date, eighteen years later, has never been accomplished. But they were the first to ascend the Great River above Hankow, and penetrated to some fifty miles above the confluence of the Min River at Sü-chau (Swi-fu), reaching the town of Ping-shan. Here it was found impossible to go on, for their boatmen refused to advance any further on the river, and a land attempt was impracticable in the then disturbed state of the country. Captain Blakiston was a diligent surveyor, and brought back a detailed chart of the river for 840 miles.[2] Blakiston and Sarel left Hankow in March 1861, and reached it again at the end of June. The work which Captain Blakiston published on the subject of this voyage[3] contains much of interest, and has excellent woodcuts from Dr. Barton's sketches. Turning to another side of the geographical territory of which we are speaking, we should mention here an attempt made by two members of the Government service in Pegu (Captain C. E. Watson, and Mr. Fedden of the Geological Survey) to penetrate northward to Thein-ni, on the direct road between the Burmese capital and Ta-li-fu.[4] They reached a point within little more than a march of Thein-ni, but the place was then in the hands of an insurgent chief, and they were obliged to turn back. The road is thus one which remains unexplored. It runs through the secluded Shan principality of Kaingma, in about latitude 23° 32′, and thence to the Chinese city of Shun-ning-fu, called by the Burmese Shwen-li, and by the Shans Muangchan. At one part of this road, between Theinni and Shun-ning, it enters a tract partaking of the excessively unhealthy character ascribed by Marco Polo and by Captain Gill (II. 345-6) to the same region a little further north, and the road then crosses the Mekong by an iron suspension bridge.

§ 30. In 1868, no less than three attempts from

---

[2] A comparison of Blakiston's chart with the old Jesuit representation of the river as given in D'Anville's maps is very favourable to the general correctness of the latter. Captain Gill, who made the comparison at my request, says: 'Generally the agreement is very remarkable. The greatest difference in general conformation is between I-tu and the entrance to the Tung-ting Lake.'

[3] *Five Months on the Yangtsze*, &c. London, J. Murray, 1862.

[4] *Selections from the Records of the Government of India in Foreign Department.* No. xlix. 1865.

three different points were made to penetrate the obscurities of the region of which we are treating: one by the French expedition which started from Saigon ; a second by Mr. Cooper, from Ssŭ-ch'uan ; the third by an English expedition from Bhamò on the Irawadi.

The great effort of the French party under Captain Doudart de la Grée of the navy, had been the exploration of the Mekong, which they ascended and surveyed from the delta, as far as Kiang-Hung, in lat 22° 0' (a place that had been reached by Lieutenant, now General, W. C. McLeod of the Madras army, on his solitary journey from Maulmain in 1837).[5] From this point they travelled through Southern Yun-nan, to the provincial capital, Yun-nan-fu, which they reached at the end of 1867, the first time in our knowledge that any European traveller (not being a missionary priest) had seen the Yachi of Marco Polo, since he himself was there, *circa* 1283.

In view to examining the upper waters of the Mekong, and to other objects not very clear, but of which one perhaps was merely that of penetrating to a place which had been the subject of so much speculation, and the scene of such a singular revolution, the leaders of the party were very desirous to reach Ta-li-fu, then the capital of the chosen sovereign of the Mahommedan, or quasi-Mahommedan, rebels of Yun-nan, whom we, after the Burmese, call *Panthés*. The Chinese imperialist authorities at Yun-nan-fu received with laughter and amazement the proposal of the Frenchmen that they should be allowed to pass direct from the capital to the rebel outposts ; but they were bent on success, and achieved it at a later date, starting from Tong-ch'uan-fu, in the northern part of the province (lat. 26° 25½'). Captain

[5] The latitude of McLeod agrees perfectly with that of the French; there is a difference of 9' in their longitudes.

Doudart was too ill to take part in the expedition, though his danger was not then suspected; and the conduct of this digression fell to Lieutenant Francis Garnier. Starting from Tong-ch'uan, January 30, 1868, they crossed and recrossed the River of Golden Sand on the eastern and southern limbs of the great southern curve, passing near Hwai-li, and crossing on the second occasion near the confluence of the Yar-lung with the Kin-sha. In the advance nearer Ta-li the party owed much (as has been already noticed) to the patriotic aid of M. Leguilcher. The meeting of the party with this gentleman in his remote parsonage at Tu-tui-tse, near the northern end of the Lake of Ta-li, is not unlike the famous meeting of Stanley and Livingstone:

"One of our guides pointed out to me, some hundred mètres below, a little platform, hung as it were in mid-air against the flank of the mountain; there were a few trees planted in rows, and a group of houses surmounted by a cross. I began running down the break-neck winding path, and before long I came in sight of a man with a long beard standing on the edge of the platform, who was attentively regarding me. In a few minutes more I was by his side: 'Are not you Père Leguilcher?' I said. 'Yes, sir,' he answered with a little hesitation, 'and no doubt you are come to announce Lieut. Garnier, from whom I have just had a letter?' My dress, my unkempt look, my rifle and revolver, no doubt gave me in the Father's eyes the look of a buccaneer; it was evidently not at all what he expected in an officer of the Navy!— 'I am the man who wrote the letter, mon père,' I said, laughing, 'and I see you take me for my own servant. . . . . . .' We exchanged a cordial grasp of the hand, and I introduced the members of the expedition as they came up in succession." [6]

Accompanied by M. Leguilcher the party reached Ta-li-fu, but they had to leave it in hot haste (March 4) within thirty-six hours of their arrival. The success of their retreat was due to the tact and boldness of Garnier. They returned to Tong-ch'uan by the route they had come, and on their arrival found

[6] *Voyage d'Exploration*, p. 510.

that their gallant leader, Captain Doudart de la Gree, had died in their absence.

§ 31. Some years later, after having completed a splendid and valuable book, and after taking an active part in the defence of Paris in 1871, Garnier returned to China, bent on fresh exploration. What he accomplished before he was called away to another field, on which he fell, was chiefly in the detailed examination of the navigation of the Upper Yang-tzŭ, and of some of the scarcely known tributaries of the great river in Kwei-chau and Hu-nan.

But the object which he had made specially his own aim was the exploration of the virgin field of Tibet. Indeed, in this direction he had, like my friend Captain Gill, aimed very high :

'I am come to China,' he wrote, 'as you conjecture, to endeavour to penetrate Tibet. My object is to reconnoitre that part of the Yárú-tsang-pu which lies between Lassa and Sadiya. If I am able—but I doubt it sorely—I should wish to return by the west, i.e., by Turkestan. I have just returned from Peking, where I have been to ask for passports, and letters of recommendation to the Chinese ambassador at Lhassa. I have seen reason to think, however, that these passports will have no great value, and that the difficulties to be encountered in penetrating Tibet will be very great. And they will be enhanced by this, that instead of aiming at Lhassa by the usual road, I wish to adopt a more southerly line (about the 29th degree of latitude) so as to cross the sources of the Camboja and the Salwen, and to make an attempt to explore the sources of the Irawadi. The Brahmaputra-Irawadi question is, in my judgment, far from being absolutely settled ; and you have yourself, in the maps attached to Marco Polo, prolonged the Irawadi hypothetically beyond the limit assigned to it in your map of 1855. . . . .'[7]

In another letter, one of the last received from him, he recurred to the subject :

'I thank you much for the paper you sent me on the hydrography of Eastern Tibet. I must have said more than I intended, if in my last letter I led you to suppose that I inclined to the identity of the Irawadi and Tsang-pu. All chances and probabilities seem to me the other way, and in favour of the Brahma-

[7] Letter, dated April 17, 1873, to the present writer.

putra, and my general map expresses this sufficiently. But we have to do with a country so singular, and so little like any other, that what would elsewhere amount to proof positive, leaves us here still in doubt. Like you I have no doubt that the continuation of the Irawadi is to be sought in some river of Tibet. The reasons which you assign for identifying this river with the Kuts' Kiang or Chete Kiang of Monsgr. des Mazures, are very forcible. Did I tell you that we were informed in Burmese Laos that the Irawadi continued northward as a great river, which the Laotians call the Nam-mao, and which they distinguish from the Nam-Búm and the Nam Kiu (Myit-ngè and Myit-gyi).[8] The Nam-mao appears to be the Kuts' Kiang. . . . . I desire to avoid forming a theory, even in my own mind, for nothing hoodwinks a traveller like the adoption of a preconceived idea, ... but I repeat as regards the Brahmaputra the probabilities require to be corroborated by material demonstration.

'The south-eastern region of Tibet, as far as we could judge on our approach to Li-kiang-fu and Tali, is a country full of surprises. The rivers vanish and appear again. A stream will bifurcate, and, by help of the caverns which abound in that limestone formation, the two branches will sometimes change from one basin into another, discharging into two different rivers. My *impression* —you will think it a strange one—is that, as regards the Brahmaputra and the Irawadi, or, in more general terms, at some point of the connection of the fluvial system of Tibet with that of India and Indo-China, there is a *perte du fleuve*—a phenomenon in fact analogous to that of the Rhone, but on a larger scale. We have seen this happen in Yun-nan with small rivers. And I am just returned from a journey to the frontiers of Szechuan and Kwei-chau, where I have been eye-witness of some ten varieties of this very phenomenon,—rivers passing over one another, splitting in two, and changing from one basin to another. Nothing could be more curious, or more difficult to determine geographically, than the hydrographic network in the basin of the U-Kiang (the river of Kwei-yang—that river which some have assigned as the line of Marco Polo's return to Szechuan). Now there is a striking analogy of geological formation and orographical character between this tract and the south-east of Tibet. It is altogether on a much smaller scale, that is all. Might not we expect to find in the course of the great rivers, of which we have been speaking, some such solution of continuity, which would explain the obscurity which actually hangs over them? This, I repeat, is no more than impression; I take good care to keep from making it into a theory. . . . Pray make me useful in every way that can help your work. I read it carefully whenever I pass over any fraction of Marco Polo's itinerary. As yet I have found nothing of interest to say, unless it be that it seems to me the most exact and faithful impression of all that can be known at this day of the acts and deeds of the traveller, and of the state of the countries which he

---

[8] These are the Burmese terms for 'Little River' and 'Great River.'

traversed. . . . . As soon as I shall have conferred with Admiral Dupré, and have definitively settled my plans, I will write again. I should of course be very glad of the support of the English authorities, should I succeed in emerging by Assam or Nepal.' [9]

§ 32. The second enterprise of 1868 to which we have made reference was that of Mr. Cooper. He left Hankow on January 4, 1868, Ch'êng-tu on March 7, and Ta-chien-lu on April 30, following, to Lit'ang and Bat'ang, the road over the high plateau, afterwards traversed by Captain Gill. Mr. Cooper's hopes were raised at Bat'ang by the information he received that the town or village of Roemah (on the Lohit Brahmaputra), from which Assam was not far, could be reached from that point in eighteen days. These hopes were, however, speedily extinguished by the prohibition of the Chinese authorities. Mr. Cooper then decided on travelling to Ta-li-fu and Bhamò. His route beyond Bat'ang diverged from that followed by Captain Gill. Instead of following the River of Golden Sand he chiefly followed the valley of the Lan-t'sang. He spent a night at Tse-ku, within the Yun-nan boundary, on the western bank of that river, where the French missionaries had an out-station among the aboriginal tribes, and an estate which they had purchased from one of the chiefs, occupied chiefly by converts from those tribes, Lu-tse (from whom the name Lu-ts'-Kiang, by which the river Salwen is known on this frontier, is taken), Lu-sus or Lissus, Mossos or Mus-us, and what not. This is the most westerly point that has been reached by any traveller from China in the region of the great rivers north of Bhamo. And Mr. Cooper appears to be almost justified in stating that he was here within 80 miles of Manché (on the Upper Irawadi), in the Khamti country, which was visited by Wilcox from India in 1827. The distance is,

[9] Letter dated Saigon, August 28, 1873.

however, apparently nearly 100 miles. South of this Mr. Cooper reached the Chinese town of Wei-si-fu, nearly due west of Li-Kiang-fu, and there obtained passports from the military commandant to go on to Ta-li. He advanced three days further, but a local chief of a tribe whom Mr. Cooper calls Tzefan, on the border of the Ta-li territory (then under the 'Panthé' Sultan), refused to let him proceed, and on his return to Wei-si he was imprisoned and threatened with death by the civil officer in charge, who apparently believed him to be in communication with the Ta-li rebels. After five weeks' imprisonment he was allowed to depart (August 6), and returned by the way he had come as far as Ya-chau. Thence he diverged to the south, travelling through a beautiful country of tea-gardens, and of the white-wax cultivation, to Kia-ting-fu, a famous river-port and *entrepôt* upon the Min river. This he descended to Swi-fu, where the two great contributaries of the Yang-tzŭ unite. Thence he descended the Great River to Hankow, which he reached November 11, 1868.[1]

In the following year Mr. Cooper made an attempt from the side of Assam to penetrate to Bat'ang. He started from Sadiya October, 1869, and passing up the line of the (Lohit) Brahmaputra, through the Mishmi country, reached Prun, a village about 20 miles from Roemah, the first Tibetan post, and half that distance from Samé, where MM. Krick and Boury were murdered. From this he was turned back.

§ 33. Major Sladen's expedition, sent under the authority of the Government of India, left Bhamò February 26, 1868. After long detentions on the way, by want of carriage and other obstacles, placed in the way of the party, it was supposed, by the influence of Chinese merchants afraid of injury to their commercial monopoly, they reached Momien

[1] This is called 1869 in Mr. Cooper's book, p. 450.

(Teng-yueh-chau of the Chinese), then the frontier city towards the west of the Mahommedan Government of Western Yun-nan. The Governor received and entertained the party with great courtesy and hospitality, but entirely objected to their proceeding further, on the professed ground of danger to themselves from the disturbed state of the country. They reached Momien on May 25, left it July 13, and arrived again at Bhamò on September 5, 1868.

Major Sladen gave an account of the journey before the Royal Geographical Society, June 26, 1871,[2] and Dr. Anderson, the medical attendant of the party, and a good naturalist, has recorded all the proceedings and observations of the expedition in a work which contains much of interest. But there was not much geographical information collected, and an officer who had been specially attached to the party as surveyor was allowed to quit it and return to Burma, for reasons which it is not easy to understand, when they were about half-way to Momien.[3]

Sir R. Alcock has pointed out how inevitably the friendly intercourse into which we entered, on this occasion, with the representatives of a body in revolt against China, must have created distrust in the Imperial Government and its partisans in Yun-nan, and not improbably led, more or less directly, to a tragical catastrophe, when the attempt to explore the trade routes of the Yun-nan frontier was renewed six years later. The suspicion of foreign interference had perhaps another effect, in stimulating the Chinese Government to effective measures for the extinction of the Mahommedan revolt.

[2] *Proceedings of Royal Geographical Society*, xv. pp. 363 *seq.*
[3] Dr. Anderson's account was printed by Government at Calcutta, 1871, *Report of an Expedition to Western Yunan*, large 8vo. In another work, published in London, 1876, *Mandalay to Momien*, he gives an account both of this and of Col. Browne's expedition, of which also he was a member. And his scientific collections have been separately published in 4to.

§ 34. We pass now to 1872, in the March of which year Baron Richthofen was at Ch'êng-tu, engaged on the last of those important journeys which formed the basis of his work on China. Art is long, and life is short! We see with pain months passing into years without the appearance of any second volume of that great work. The expedition which he projected and commenced from Ch'êng-tu brings him within the category of explorers in the region which is our subject, though it came to an untimely end. His project will be best explained in his own words :

'Although my journey. . . as originally contemplated ended at Ching-tu-fu, I could not resist the temptation of trying to add to it a trip through the south-westernmost portions of China, and to explore the mountains of Western Sz'chwan, as well as the provinces of Yün-nan and Kwei-chau. Besides hoping to contribute to the general knowledge of the geography, geology, and resources of these unknown regions, I wished to examine the metalliferous deposits that are widely spread through them, and to gather some information respecting the many independent tribes inhabiting South-Western China, and their languages. My final object, however, was to explore the road from Ta-li-fu to Burma. I had some difficulty in collecting the necessary information, but finally settled upon the plan to travel by way of Ning-yuen-fu to Ta-li-fu, a journey of about five days, and thence to go to Teng-yuè-chau [Momien], the last place reached by Major Sladen on his way from Bamo to Yün-nan. From that city I intended to go again eastward, by Yün-nan-fu and Kwei-yang-fu, the capitals of the provinces of Yün-nan and Kwei-chan, to Chung-king-fu on the Yangtze.' [4]

The traveller had accomplished half his journey to Ning-yuan-fu when, on the high Siang-ling pass, he was involved in a collision with a body of Chinese troops, whose outrageous aggression on his party, and its consequences, compelled him to retrace his steps, and to give up a journey from which a richer harvest might perhaps have been expected than even from any that had preceded it.

The journey has since been made, and Ning-yuan

[4] *Letters to the Shanghai Chambers*, No. VII. p. 3.

has been visited by Mr. Baber, as we shall see ; but we remain without any details of his journey. These details would be of great interest, for the country is secluded, and otherwise entirely unexplored ; and to me and some others the interest would be of a still more special kind, because Ning-yuan is the capital of the valley and district of Kien-chang, which has been demonstrated (as I think), by Richthofen, to be the Gheindu or Caindu of Marco Polo, a country of which, with its cassia-buds and other spices, its strange Massagetic customs, its currency of gold rods and salt-loaves, the old traveller gives so remarkable an account.[5]

§ 35. In speaking of the labours and incidental journeys of the Roman Catholic missionaries, we have mentioned Abbé Desgodins, a gentleman of great intelligence, and who takes much interest in geography. A book was published at Verdun in 1872, professedly based upon his letters to his family. It contains a good deal of information for those who bring to its perusal some previous knowledge, to serve as amalgam in the process of extracting what is valuable ; but it has been compiled by a relative of the missionary without much clear acquaintance with the subject, and contains a good deal of matter of a kind which appears to be due to this circumstance. The history of the Abbé Desgodins is not a little remarkable, and shows the persistent character of the man.

When first he quitted France as a recruit for the missions, in 1855, he was directed to proceed by way of British India, and to attempt to make his way to the mission establishments across the Tibetan highlands, in order to avoid the great détour and expense of the usual journey by the ports and broad interior of China. His first attempt was made by Darjeeling,

[5] See book ii. chap. 47, and the notes to the 2nd edition (vol. ii. p. 57).

where, as might have been expected, he had kindly relations with Mr. Bryan Hodgson, who was then living there. After various endeavours to negotiate admission to Tibet by the Sikkim frontier, he was obliged to give it up, and, accompanied by M. Bernard, an older member of the fraternity, proceeded to the North-West Provinces, in order to attempt an entrance by Simla and the Sutlej. The priests were at Agra when the mutiny of 1857 broke out, and spent the summer in the fort there, with the rest of the 'sahib-log.' After the relief, they were able to proceed to Simla, and went on by Rampúr to Chini on the Upper Sutlej. Here M. Desgodins was summoned back, and ordered to proceed by the more usual route to join his mission. We find him again at Agra in the hot weather of 1858, and then doing duty as Roman Catholic chaplain to a British force at Jhansi. From this he writes to his parents:

'You will think I am going to become a regular Crœsus when I tell you that the Government of John Bull gives me for my services as Military Chaplain 800 francs a month, or, as they say here, 320 rupees. . . . . . . However, when you know the state of things in India, and the prices, it is no small matter to make both ends meet; so my dear nephew must not count on a fortune from my savings. Moreover, I hope not to be long in John Bull's service, but soon to be able to join my mission; I shall feel richer there with next to nothing, than here with my 800 francs.'—*La Mission du Thibet*, p. 36.

Receiving a fresh summons from Bishop des Mazures he took his departure (after drawing at Agra a sum of about 1,000 rupees for his services with the army). During his journey to the interior he was arrested, imprisoned, and sent back to Canton. Starting again under a new disguise, he finally reached the residence of the Bishop, near the frontier of Tibet, in June 1860, five years after his departure from France.

§ 36. We now come to the journey of the gallant young traveller who, after being the first to open the way from China to the Irawadi, had hardly taken the first step on his return when his blood was left upon the path.

In the spring of 1873 the Imperial Government in Yun-nan succeeded, as has already been noticed, in finally crushing the insurgents who had maintained their independence for some seventeen years. The Government of India decided on now renewing the attempt to explore the road, and the facilities for trade between the Irawadi and China, which Major Sladen had been unable to carry out, owing to the state of political affairs when he visited Momien. Colonel Horace Browne, of the Pegu Commission, was appointed to lead the mission; and it was settled that an officer of the consular service should be sent across China to Bhamò to meet the mission there, and to accompany them back to China as interpreter and Chinese adviser.

The officer appointed to this duty was Augustus Raymond Margary, a young man of high character and promise. It is needless to detail a story still fresh in the public mind. His journey led him from Hankow across the Tung-ting Lake, and by the regions, hardly known to Europeans, of Western Hunan and Kwei-chau to Yun-nan-fu, and thence to Ta-li and Bhamò,—the first of Englishmen to accomplish the feat that had been the object of so many ambitions, and to pass from the Yang-tzŭ to the Irawadi.

Margary reached Yun-nan-fu on November 27, 1874, and writing home from this point he says: 'I quite enjoyed the journey; everywhere the people were charming, and the mandarins extremely civil, so that I had quite a triumphal progress.' The same good treatment was continued through Yun-nan. He started again on December 2, and on the 14th or 15th reached Chao-chau, 20 miles from Ta-li (which, as the map will show, lies about ten miles off the direct road from Yun-nan-fu to the Burmese frontier). There was some unwillingness to let him visit that city, from a dread, probably real, of popular turbulence; but this was overcome; and he writes home, on returning to his quarters at Chao-chau:

> 'I visited the mandarins in turn, and had a most successful interview with all, but especially with the Tartar General, who treated me with extreme civility, very much in the style of a polished English gentleman receiving a younger man. I was perfectly delighted with his reception. He complimented me over and over again on my knowledge of Chinese, and . . . . said he hoped on my return I would spend a few days with him. . . . . . "I should naturally wish to see everything, if I visited your country," said he, " and I shall have a house ready for you and your honoured officials when you return."'[6]

The General gave Margary the place of honour beside him. The Tao-tai, a young man, had omitted this courtesy.

He reached Momien on January 4, 1875, and Manwain, the place where he met his death seven weeks later, on the 11th. Here he was visited by 'a furious ex-brigand called Li-hsieh-tai, who attacked our last expedition in 1867, and has been rewarded lately for his services against the rebels with a military command all over the country.' This is the man who was afterwards loudly charged with the murder of Margary. On this occasion, to the traveller's great surprise, he prostrated himself, and paid him the highest honour.

On January 17 Margary reached Bhamo, safe and triumphant. 'You may imagine,' he writes, 'how full of delight I am at the happy results of my journey, and the glowing prospect ahead."[7]

§ 37. After an unsuccessful attempt to proceed by a more southerly line from Bhamô, through Sawadi, Colonel Browne had to revert to the route by

[6] *Margary's Journals*, pp. 236, 278.   [7] Page 308.

which Margary had come, and a start was made from Tsit-kau on the Bhamô River (Ma-mou or Sicaw of Captain Gill, II. p. 384) on February 16. The rest is best told in the words of the editor of his journals :

'Early on the morning of February 19 Margary crossed the frontier with no escort but his Chinese secretary and servants, who had been with him through his whole journey, and a few Burmese muleteers. The next morning brought letters from him, reporting all safe up to Seray. He had been well received there, and had passed on to Manwyne. The mission followed slowly, reaching Seray on the 21st. .... On the 22nd, in the early morning, the storm broke. The mission camp was almost surrounded by armed bands, while letters from the Burmese agent at Manwyne to the chief in command of their escort told that Margary had been brutally murdered at Manwyne on the previous day. But for the staunchness of the Burmese escort—who resisted all offers of their assailants of heavy bribes if they would draw off and allow them only to kill the 'foreign devils,'—and the gallantry of the fifteen Sikhs who formed their body-guard, the whole mission must have shared the fate of their comrade. .... At Bhamô they eagerly sought for all particulars of the murder, but without much success. The most trustworthy account was that of a Burmese who had seen Margary walking about Manwyne, sometimes with Chinese, sometimes alone, on the morning of the 21st. This man reported that he had left the town on his pony, to visit a hot spring at the invitation of some Chinese, who, as soon as they were outside the town, had knocked him off his pony and speared him.'

§ 38. Then followed Sir T. Wade's unwearied negotiations with the Chinese Ministers, and the deputation of the Hon. T. G. Grosvenor, accompanied by Messrs. Baber and Davenport, to be present at the Chinese investigation at Yun-nan-fu.

The Chinese Government had given the strongest assurances that the investigation should be conducted with a view to the production of trustworthy witnesses, and the punishment of the real offenders. But the fact was far otherwise. No witness of the murder was allowed to be produced. The story which Mr. Grosvenor was pressed to accept was that Margary had been murdered by savages ; that Li-hsieh-tai (or Li-chên-kou, as he was officially designated in China) had organised the attack on Colonel Browne; that the Momien train-bands had not been moved out of Momien, but had stood there only on the defensive. The manner in which the affair had been dealt with showed that what had happened in Yun-nan had been done, if not by the direct order, at least with the approval after the fact of the Central Government, and our Minister could only express his entire disbelief in the case put forward, and decline to agree to the execution of any of the persons whom the Chinese investigation professed to incriminate.

§ 39. The termination of the affair was one of the matters embraced in the 'Agreement of Chefoo,' signed September 13, 1876. This provided, among other things (Sect. I. ii.), that a proclamation should be issued by the Chinese Government, embodying a memorial of the Grand Secretary Li with an imperial decree in reply. These documents embraced a statement of the facts of the deputation and murder of Mr. Margary, a recognition of the gravity of the outrage, of the necessity of observing treaties, of the anxiety of the Imperial Court to maintain friendly relations with foreign powers, and of its regret for what had occurred, with an injunction on local authorities to give protection to foreign travellers, and to study the treaty of Tien-tsin. It was also agreed that for two years to come officers should be sent by the British Minister to different places in the provinces to see that this proclamation was posted.

This is the *Margary Proclamation*, so often referred to by Captain Gill in all the remoter part of his travels.

The agreement also provided (*ib.* iii.) that an imperial decree should be issued directing that whenever the British Government should send officers to Yun-nan the authorities of that province should select an officer of rank to confer with them, and to conclude a satisfactory arrangement regarding trade.

The British Government was also (*ib.* iv.) to be at liberty for five years to station officers at Ta-li-fu, or other suitable place in Yun-nan, to observe the conditions of trade.

Passports having been obtained the preceding year for a mission from India to Yun-nan (Colonel Browne's), it would be open to the Viceroy of India to send such mission when he should see fit.

An indemnity (*ib.* v.) was to be paid on account of the families of those killed in Yun-nan, on account of the expenses occasioned by the Yun-nan affair, and on account of claims of British merchants arising out of the action of officers of the Chinese Government; and this indemnity was fixed at 200,000 taels.

When the case should be closed, an imperial letter of regret was to be carried by a mission to England (vi)

Under Sect. III. i., several free ports, including I-chang, on the Upper Yang-tzŭ, were added to those already constituted, and the British Government were authorised to establish a consular officer at Ch'ung-ch'ing, to watch the trade in Ssŭ-ch'uan.

Also by a separate article it was provided that the Tsung-li Yamen should, at the proper time, issue passports for a British mission of exploration, either by way of Peking through Kan-su and Koko Nor, or by way of Ssŭ-ch'uan to Tibet. Or if the mission should proceed by the Indian frontier to Tibet, the Yamen should write to the Chinese resident in Tibet, who should send officers to take due care of the mission, whilst passports also should be issued for the latter.

It is hardly necessary to say that no residents in Yun-nan have been appointed under this agreement;

nor has any mission again entered Yun-nan, nor any official mission of exploration been sent to Tibet.

§ 40. Going back a little, I may record that Mr. Grosvenor's mission to Yun-nan left Hankow November 5, 1875 ; reached Yun-nan-fu on March 6, 1876 ; Ta-li-fu on April 11 ; Momien on May 3 ; and Bhamò I don't know when, for I have searched the reports, as published, of all the members of the mission without being able to find the date.

Mr. Arthur Davenport, one of the members, has made an interesting report on the trading capabilities of the country traversed by it, forwarded by Sir T. Wade to the Foreign Office, October 9, 1876.

Another of the officers attached to Mr. Grosvenor's mission was fortunately Mr. E. Colborne Baber, a gentleman who seems thoroughly imbued with the true genius of travel, a spirit which leads him apparently to spend his holidays in exploring fresh fields and gathering fresh stores of knowledge, though we cannot say that he is as diligent in communicating it to the world. His notes on the latter part of the route followed by the Grosvenor mission (that between Ta-li and Momien) have been published by Her Majesty's command. These notes, and the maps which accompany them, give Mr. Baber *per saltum* a very high place among travellers capable of seeing, of surveying, and of describing with extraordinary vivacity and force. In fact, we are seldom happy enough to meet with a traveller who combines so many valuable characteristics. His report of the journey has as yet been given to the public only in the unattractive form of a Parliamentary paper, and its circulation even in that form has perhaps been restricted still more by the unusual price put upon the 30 pages which contain it. But I fervently hope that Mr. Baber will yet give to the world, whether officially or non-officially, a narrative not of this only, but of other journeys as well.

On one of these, accomplished in the autumn of 1877 from his consulate at Ch'ung-ch'ing, he succeeded in completing the journey which Richthofen was compelled to abandon, making his way from Ya-chau to Ning-yuan-fu. The fact of this journey being made was duly reported to the Foreign Office, and a detailed narrative and map were promised. But these have never reached the Foreign Office, so I borrow from a letter of Mr. Baber's to Captain Gill the only particulars available, hoping that the hint thus given may help at least to obtain something more than these crumbs for hungry geographers :

'Since I reached Ch'êng-tu and received your memorable note accounting for your sudden flight, I have never ceased to wonder why you selected the roundabout route by the well-beaten Lit'ang track, instead of taking the direct road to Tali by Ning-yuan-fu. The latter is absolutely unknown to Europeans, is very easy, and less than half the distance *viâ* Lit'ang. Even if you had waited ten days for me at Ch'êng-tu, you would still have saved a month, and would have travelled easily through an entirely new country all the way from Ch'ing-ch'i-hsien (at the foot of the first high pass called Ta-hsiang-ling) to Tali.

'. . . . . Often regretting you I trudged down to Ning-yuan-fu, through a glorious hill and valley region, inhabited by Chinese soldier colonists, and those interesting mountainers, the independent Lolos. Ning-yuan lies on the east side of a rich valley about four miles wide, which extends, with unimportant narrows and low passes, as far doubtless as the Yangtzŭ ; but on reaching Hui-li-chou, I turned east and struck into the poorest conceivable region of bare sandstone hills, among which we had some difficulty in procuring food. At Hui-li the rains began, and continued with few intervals all the rest of the way. . . . . I struck the Great River in lat. 26°54', and at this point obtained a good series of lunar distances, the only chance I met with on the whole journey.

'Crossing the stream, I determined to follow its course, as nearly as possible, to Ping-shan.[8] But we had to quit its banks immediately on account of a flood, and plunge into a desperate maze of mountains, leading to a high plateau, in which most of my coolies, unable to procure provisions, deserted me. I pressed on with diminished forces, through rain and dense fogs, and soon found myself in clover. The skies cleared, range after range of snowy mountains shot up. I bought a sheep for 200 cash, drank more buckwheat whisky than I fear you would deem reputable, and one

[8] The highest point reached by Blakiston. See above, p. [55].

glorious day scrambled down a portentous gorge on to the Yangtzŭ, running fifty miles east of the position assigned to it by geographers. You will be able to form an opinion of the errors of the map when I tell you that Chao-t'ung, Yung-shan, and Lui-po-ling[9] are nearly on the same meridian.

'I had still many a stiff climb before me, but I reached Ping-shan in good order, had an interview with the Lolo hostages detained there, most of whom are the same that I saw in January, 1875, and then dropped comfortably down stream to the old quarters in Ch'ung-ch'ing. I am now working at report and map, which latter comes out most satisfactorily.'[1]

We wish it *would* come out!

In another letter of Mr. Baber's published in a Shanghai newspaper, he affords a few words more on the subject of this interesting journey:

'..... Passing Ning-yuan-fu I went to Hui-li-chow; then turned east and crossed the Yangtzŭ into Yünnan not far from Tung-ch'uan.[2] Thence through the wildest and poorest country imaginable, the great slave-hunting ground from which the Lolos carry off their Chinese bondsmen—a country of shepherds, potatoes, poisonous honey, lonely downs, great snowy mountains, silver-mines, and almost incessant rains. . . : No European has ever been in that region before myself, not even the Jesuit surveyors; the course of the Yangtzŭ, there called the Golden River, as laid down in their maps, is a bold assumption, and altogether incorrect. A line drawn S.W. from a mile or two above Ping-shan will indicate its general direction, but it winds about among those grand gorges with the most haughty contempt for the Jesuits' maps.'

§ 41. Mr. Baber in 1868 made another important journey, of which he speaks thus[3]:

'I returned to this super-heated, and for the moment typhus-stricken city on the 25th. I have no time to write at length to you. Sufficient to say that my journey, begun as a holiday, and speedily eventuating in very serious hard work, has been very interesting. I have collected reams of information, and my chart, depending for longitude on lunars, D.R., and chronometric differences, comes out in a manner which astonishes me beyond measure. Assuming as the true longitude deduced from lunars E. and W., which

---

[9] Chao-t'ung and Lui-po-ling will be found in the general map attached to this book as *Chow-toong* and *Looi-po*.

[1] Dated Ch'ung-ch'ing, November 28, 1877.

[2] Between Hui-li-chau and Tong-ch'uan Mr. Baber's route must have been the same, or nearly the same, as that of Lieutenant Garnier on the way to and from Tali (see p. [58] above).

[3] Letter to Capt. Gill, dated Ch'ung-ch'ing, June 28, 1878.

agree mutually to a fraction of a minute (i.e., 0° 1′ 0″) and which were taken somewhere near the longitude of Ta-chien-lu, I plotted out the route-chart on a Mercator's projection, and you are capable of judging of my delighted surprise when I found that it brought the position of Swi-fu (Blakiston's Sü-chow, on the Yangtzŭ) into precisely his position. I could not separate the points with a divider. It is beautiful!'

This last journey has been reported, but alas! only in abstract, in a second Parliamentary paper. Mr. Baber started on his holiday with the intention of making a rough survey of the river (Min, Wen, or what not) between Kia-ting and Swi-fu (Sü-chow, Sioo-choo of maps), and of crossing the mountains westward from Kia-ting to Fu-liu (not in the maps) in long. 103°. Near the last place Mr. Baber was robbed of his travelling funds and other property, a misfortune which he turned to good account. Though detained for some time, whilst a communication was made to the governor-general of the province (of Yünnan), and to an old friend who occupied the office of Tao-tai, after eleven days the messenger returned with a very considerate letter, and a loan of money from the Tao-tai, and with orders from the governor-general for the apprehension of the burglars.

'The magistrate had received such stringent orders to make good my losses, that a scheme I had formed of deriving advantage from the misadventure, by refusing reimbursement, and insisting that I had nothing for it but to go on to Ta-chien-lu and obtain funds, would not even bear proposal. Very conveniently, however, he could not pay me on the spot, but wished me to wait a few weeks until the money arrived from Yueh-hsi-Ting. This I altogether declined to do, and the end of the negotiation was, that I offered to travel on to Ta-chien-lu, and to receive payment on my return.'

Mr. Baber accordingly travelled north by a mountain-path till he struck the high road between Ta-chien-lu and Lit'ang (that followed by Captain Gill), and he walked into the former town on April 23, 1878, staying there three weeks, and returning by the high road to Fu-liu, where the magistrate duly paid over the amount of his loss—viz. 170 taels. He returned to Kia-ting by the way he had come, and the meeting with a Lolo chief afforded an opportunity of making notes of the customs and language of his tribe. The following passage, describing the first transition from a Chinese to a Tibetan atmosphere, is a good specimen of the style which makes Mr. Baber's reports, whilst abounding in valuable information, almost as unique among blue-books as the autobiography of his illustrious namesake—I suppose we cannot say ancestor—is among Asiatic volumes :

'The remainder of the journey was impeded by nothing worse than natural difficulties, such as fevers and the extreme ruggedness of the mountain ranges. We quitted cultivation at the foot of a pine-forest, through which we travelled three days, ascending continually until we came to a snowy pass—the only pass in the country which, as the natives say, 'hang-jên,' stops people's breathing. Descending its northern slope, we soon found that we had left China behind. There were no Chinese to be seen. The valley was nearly all pasture-land, on which were grazing herds of hairy animals, resembling immense goats. These I rightly conjectured to be yaks. On entering a hut, I found it impossible to communicate with the family, even a Sifan, whom I had brought with me, being unintelligible to them ; but they were polite enough to rescue me from the attack of the largest dogs I have ever seen, and to regale me with barleymeal in a wooden bowl, which I had to wash down with a broth made of butter, salt, and tea-twigs. Further on we met a company of cavaliers, armed with matchlock and sabre, and decorated with profuse ornaments in silver, coral, and turquoise; a troop of women followed on foot, making merry at my expense. A mile or two further, and I came to a great heap of slates, inscribed with Sanskrit characters, whereupon I began to understand that we were in Thibet; for although Thibet proper is many hundred miles west of this point, yet traces of Thibetan race and language extend right up to the bank of the Tatu River —a fact which I had not been led to expect.'

§ 42. In this review we have had occasion to speak frequently and largely of the enterprising devotion of the Roman Catholic missionary priests in the obscure regions with which we have had to do. It has been the fortune of the present writer to spend many years in a Roman Catholic country without feeling in the least degree that attraction to the Roman Church which influences some,—indeed, he might speak much more strongly the other way. But it is with pleasure and reverence that one contemplates their labour and devotion in fields where these are exercised so much to the side of good, and where there is no provocation to intolerance or to controversy except with the heathen; no room for that odious spirit which in other regions has led the priests of this Church to take advantage of openings made by others to step in and mar results to the best of their power. The recognition of the labours and devotion of which we spoke just now has often led to sarcastic contrast of their work with that of Protestant missionaries, to the disparagement of the latter,—such as occurs not unfrequently in the narrative of Mr. Cooper; in this I have no sympathy. There may be much which the members of Protestant missions should carefully study (and which some of them probably have often studied) in the results that provoke such comparisons, but it is a shallow judgment that condemns them on a superficial view of those results. In any case, the discussion would here be out of place, and I have no intention of entering on it. Though it is only of late years that Protestant missionaries in China have contributed to our geographical knowledge of the western frontier, we must not overlook what they have done. Mr. Williamson's excellent work[4] does not reach our limits, as he was not nearer than Si-ngan-fu. But my valued friend Mr. Alexander Wylie, long agent at Shanghai of the Bible Society, was one of the earliest in our day to visit Sŭ-ch'uan, and to give us an account of its highly civilised capital, Ch'êng-tu. His visit occurred in 1868.[5] More recently, some of

---

[4] *Journeys in North China, Manchuria, &c.*, London, 1870.
[5] See *Proceedings of the Royal Geographical Society*, vol. xiv. p. 168 *seq.*

the numerous agents of the society called the China Inland Mission have been active in the reconnaissance of these outlying regions.

Mr. McCarthy, one of the agents of this society, was the first non-official traveller to accomplish the journey to Bhamò. This he did from Ch'ung-ch'ing on foot, travelling south to Kwei-yang-fu, and then onwards to Yün-nan-fu and Ta-li, and so forth, reaching Bhamo on August 26, 1877, a little more than two months before Captain Gill's arrival at that place. Mr. McCarthy wore the Chinese dress, as the members of his mission appear frequently to do, but made the character and object of his journey generally known. He was nearly everywhere treated with civility, often with kindness. 'Throughout the whole journey,' he says, 'I have not once had to appeal to an officer for help of any kind, and in no case has any officer put an obstacle in my way.'[6]

Mr. Cameron, another agent of the same society, followed Captain Gill not long after that officer, leaving Ch'êng-tu, on September 13, 1877, and after an unsuccessful attempt to make the directer road to Ta-chien-lu, had to adopt the usual and more circuitous line by Ya-chau, taken by Captain Gill. He also followed in Captain Gill's traces to Lit'ang, Bat'ang, and A-tun-tzŭ. He was kindly and courteously received by the French priest at Bat'ang (M. Desgodins). At A-tun-tzŭ the solitary traveller was laid up for many days with a bad attack of fever. On his recovery his further route deviated from Captain Gill's, as he went further to the west, by Wei-si, where Cooper was imprisoned in 1868. He

---

[6] Letter from the traveller to Mr. T. T. Cooper, British Agent at Bhamò, dated September 4, 1877, in *China's Millions*, the periodical of Mr. McCarthy's Society, for 1878, p. 61. Mr. McCarthy also read an account of his journey before the Royal Geographical Society; see the *Proceedings* (August), 1879, pp. 489 *seqq.*

reached Ta-li-fu on December 23, and Bhamò at the end of January, 1878. Mr. Cameron's journal is that of a simple and zealous man, and from his being without a companion, and thus seeing the more of the people, has many interesting passages. But there is hardly any recognition of geography in it; less a good deal than in Huc's narrative. For example, the passage of the famous Yar-lung Kiang is only noticed as that of 'a small river' below a place called Hok'eo.[7]

§ 43. The long passage through which we have conducted our readers—or some of them at least, we trust—in this introductory essay, must not close without a brief section devoted to my friend Captain Gill's own journeys.

His first journey, in the north of Pe-chih-li, to the borders of Liao-tong, and the sea-terminus of the Great Wall, was but a trial of his powers. His ascent of the Yang-tzŭ, though full of interesting detail, is on a line that has been described by several predecessors since Blakiston. The more important and novel itinerary begins with his excursion from Ching-tu to the Northern Alps, to those Min mountains of the ancient Yü-Kung, from which the Kiang of the Chinese—'The River' *par excellence*—flows down into Ssŭ-ch'uan. I am not aware of any traveller who has preceded him in this part of China.

Captain Gill on this occasion came into the land of the highland races whom the Chinese call Man-tzŭ and Si-fan. It is difficult to grasp the Chinese ethnological distinctions, though doubtless there is some principle at the bottom of those distinctions. The races generally along the western frontier are, as

---

[7] See Capt. Gill, Vol. II. p. 137. Mr. Cameron's journal is published in *China's Millions* for 1879, pp. 65 *seq.*, 97 *seq.*, 109 *seq.*

[80] INTRODUCTORY ESSAY.

Richthofen tells us,[8] classed by the Chinese as *Lolo, Man-tzŭ, Si-fan,* and *Tibetan.*

The Lolo are furthest to the south, and occupy the mountains west of the Min, and west of the north-running section of the Kin-sha—fiercely independent caterans, a barrier to all direct intercourse across their hills, and frequent in their raids on the Chinese population below. Further south in Yün-nan the term *Lolo* seems to be generally applied by the Chinese to tribes of Shan blood; but whether this is true in the present case may be doubted. We await with great interest the result of Mr. Baber's inquiries about these people. Captain Gill did not come in contact with them.

The *Man-tzŭ* are regarded by the Chinese as the descendants of the ancient occupants of the province of Ssŭ-ch'uan, and Mr. Wylie has drawn attention to the numerous cave dwellings which are ascribed to them in the valley of the Min River. The name is applied to the tribes which occupy the high mountains on the west of the province up to about 32° lat. North of that parallel, beginning a little south of Sung-pan-ting, the extreme point of Captain Gill's excursion in this direction, are the Si-fan ('western aliens'), who extend into the Koko-nur basin, through an alpine country which remains virgin as regards all European exploration.

§ 44. Both terms, Man-tzŭ and Si-fan, seem, however, to be used somewhat loosely or ambiguously.

Thus, Man-tzŭ is applied to some tribes which are not Tibetan, whilst it is also applied to people, like those on the Ta-chien-lu road, who are distinctly Tibetan.[9]

---

[8] *Letters to Shanghai Chamber of Commerce,* No. vii. p. 67. Shanghai, 1872.

[9] The *Description du Tübet,* translated by Klaproth, says expressly

INTRODUCTORY ESSAY.   [81]

Thus, also, Si-fan appears to be sometimes applied to the whole body of tribes, of differing languages, who occupy the alpine country between Koko-nur and the Lolo mountain country, and sometimes distinctively to a Tibetan-speaking race who form a large part of the occupants of that country on the north-east of Tibet, and in the Koko-nur basin, the *Tangutans* of Colonel Prejevalsky.[1] And in this sense it is used in Captain Gill's book; for the Si-fan of whom he speaks use a Tibetan dialect, as will presently be manifest, and also (from specimens that he brought away with him) use the Tibetan character. They seem to correspond to the *Amdoans* of Mr. Bryan Hodgson, in the passage which I am about to quote.

This passage exemplifies the wider sense of the term Si-fan :[2]

'From Khokho-núr to Yúnnán, the conterminous frontier of China and Tibet is successively and continuously occupied (going from north to south) by the Sókpa. . . . . ; by the Amdoans, who for the most part now speak Tibetan; by the Thochu; by the Gyárúng; and by the Mányák. . . . . The people of Sokyúl, of Amdo, of Thóchú, of Gyarung, and of Manyak, who are under chiefs of their own, styled *Gyábo* or King, *sinice* 'Wang,' bear among the Chinese the common designation of Sifan, or Western aliens; and the Tibetans frequently denominate the whole of them *Gyárúng-bo*, from the superior importance of the special tribe of Gyarung. . . . The word *Gyá*, in the language of Tibet, is equivalent to that of *Fan* (*alienus, barbaros*) in the language of China.'[3]

The fact mentioned in the last lines of the extract, if correct (and no one's statements are more full

---

that the people about Ta-chien-lu belong to the same *souche* as the Tibetans, and have the same manners (p. 266). Cooper, on this road, uses *Man-tzŭ* as the Chinese synonym of Tibetan (see p. 174, *et passim*). But ethnologically, *Tibetan* is analogous in value to *Latin*.

[1] *Prejevalsky's Travels*, translated by Mr. Delmar Morgan, vol. ii. *passim*, and note at p. 301.

[2] Mr. Baber again, in his printed letter, quoted from in § 39, calls the tribal chief with whom he had to do, a long way south of Ta-chien-lu, a *Si-fan*.

[3] *Hodgson's Essays*, 1874, part ii., pp. 66–67.

VOL. I.   e

of knowledge or more carefully weighed in general than Mr. Hodgson's), would imply that the Tibetans proper do not regard these Si-fan tribes as of their own blood, even those of them who now speak Tibetan; and possibly we may have to apply this to the Man-tzŭ also adjoining the Ta-chien-lu road. Mr. Hodgson, in speaking of some of the authorities for the vocabularies which he gives of the Si-fan languages, tells us that his *Gyárung* came from Tazar, north of Tachindo (i.e. of Ta-chien-lu), whilst his *Mányaker* was a mendicant friar (of the heretical Bonpa sect), a native of Ra'kho, six days south of Tachindo. These are the only data I find as to the position of the two tribes named. We shall presently find a third as to the position of the Tho-chu, which also will fall into its proper place in Hodgson's series, and confirm his accuracy.

I proceed now to insert the numerals of three of the tribes as collected orally by Captain Gill (A, B, C);[4] to which I add for comparison the spoken Tibetan (D), and the Tho-chu (E), from Hodgson's comparative vocabularies:

	A	B	C	D	E
1	chek	ár-gú	kí	chik	ári
2	nyĭ	ner-gú	nyĕ	nyi	gnárí
3	sĕ	ksir-gú	song	sum	khshíri
4	zhĕ	sáir-gú	hgherh	zhyi	gzháré
5	knā	wár-gú	hná	gná	wáré
6	trŭ	shtúr-gú	dru	thú (druk)	khatáré
7	dăn	shner-gú	ten	dún	stáré
8	gyot	kshár-gú	gyĕ	gyé	khráré
9	guh	rber-gú	kăr	gúh	rgúré
10	pchĕ	khád-gú	chĭ-thomba	{ chúh *or* { chú-thámbá	hadúré
11	pchĕ-chek	khát-yi	ki-tze		
12	pchĕ-nyĕ	khá-ner	chu-nye		
20	nyĕ-shĕ	ner-sá *or* ne-sá	nye-ka-thomba	nyi-shú	

[4] A is the language of the 'Man-tzŭ' at Li-fan-fu; B that of the 'Outer Man-tzŭ' there—or people further west; C that of the 'Si-fan' about Sung-pan-ting.

Now the first thing apparent here is that A and C—i.e. the so-called ' Man-tzŭ ' of Li-fan-fu, and the ' Si-fan,' are both Tibetan dialects.

Next, a comparison with E shows that the 'outer Man-tzŭ' of Li-fan-fu are the race which Hodgson calls Tho-chu, and that their language is not Tibetan. They will be near Li-fan-fu, in their place according to Hodgson's series from north to south, the ' Si-fan ' being assumed to be his *Amdoans*, whilst his *Gyarung*, north of Ta-chien-lu, are probably the Man-tzŭ of Abbé David at Mou-pin ; and his *Manyak* are probably Mr. Baber's ' Si-fan,' south of Ta-chien-lu.

Again, we observe that though the essential parts of the numerals in B and E are identical, the persistent affixes (or, as Hodgson calls them, 'servile' affixes) are different—*gú* in the one, *re* or *ri* in the other. In his comparative table we find the servile affix *ku* in the numerals of another language—a Chinese dialect which he called *Gyami* ; and in the *Manyak* we find a similar affix *bí*.[5]

§ 45. On his return to Ch'êng-tu Captain Gill was joined by Mr. Mesny, a gentleman from Jersey, who has passed a good many years in the interior of China, and particularly at Kwei-yang-fu, in the service of the Chinese Government.

Captain Gill had intended in his preface to render his thanks and a tribute of praise to his companion for the assistance which was derived from him during the journey from Ch'êng-tu to Bhamò. And now that circumstances have caused this prefatory essay to be written by another hand, he still desires that

---

[5] Thus 1, tábí; 2, nabí; 3, síbi; 4, rêbi; 5, gnábi; 6, trúbi; 7, skwibí; 8, zibi; 9, gubi; 10, chêchibi. Here, comparing with D, the essential part of 2, 3, 5, 6, 9 and 10 is evidently Tibetan ; the others diverge. These 'servile' affixes perhaps correspond to the numeral affixes or co-efficients which are necessary to the use of numerals in Chinese, Burmese, Malay, Mexican, &c., and which change with the class of objects indicated. This would account for the variation between B and E. China ' Pigeon English ' replaces the whole of these co-efficients by the universal ' piecey.'

the following words of his own may be introduced here :

'If Mr. Mesny's name occurs but rarely in my book, it is but because he was so thoroughly and completely identified with myself that it seldom occurred to me to refer to my companion otherwise than as included in the pronoun ' we.' But I should be loth to let slip this opportunity of thanking the companion of so many long and weary marches for the persistence with which he seconded my efforts to achieve a rapid and successful journey; for his patience under difficulties and some real trials, and for the courage he showed when it was called for. Above all I desire to say how much I feel that, in our dealings with the Chinese officials, the friendly relations we were able to maintain with them, and the aid we were able to obtain from them, were in large measure due to Mr. Mesny. Especially in the negotiation for our passage between Yün-Nan and Burma, was Mr. Mesny's help invaluable. And I feel that whatever credit may attach to the successful accomplishment of the journey, a very large share of it is due to Mr. Mesny, who, for the love of travel alone, gave up a remunerative employment under the Chinese Government to become my companion. As long as the events of those sixteen weeks shall have a place in my memory, so long will the kindly support of my companion be among the freshest and pleasantest of them all.'

Hitherto, Captain Gill's aspirations had been directed to a journey through Kan-suh to Kashgaria, and thence through the Russian dominions to Europe. But the troubled aspect of affairs between Russia and England, which had become more imminent, now threatened to render this issue impracticable ; whilst at a time of possible war, when duty might be calling him to quite another field in the west, he felt especially unwilling to risk being shut up in some Central-Asiatic *cul-de-sac*. Thus, though all preparations had been made for the long journey, he was forced to the conclusion that his steps must be directed homewards. Fortunately, however, this homeward journey might be made by a route which had never yet been successfully achieved : that, namely, which Cooper had attempted nine years before, by Lit'ang, Bat'ang, and Ta-li. So the start from Ch'êng-tu for England, *viâ* the Irawadi, was made July 10, 1877.

§ 46. The first place of importance reached was Ya-chau, the *entrepôt* and starting point of the trade with Tibet. The staple of this trade is the brick-tea, or rather cake-tea (afterwards broken up into brick-tea). Captain Gill has given some interesting particulars of this (II. 47); as he has in a previous part of his book (I. 176 *seq.*) regarding a similar manufacture carried on by the Russians established at Hankow, for the market of Mongolia.

Whilst I was writing these paragraphs a report was put into my hands, in which Mr. Baber gives most curious details respecting this Tibetan tea-trade.[6] The tea grown for it is peculiar. It is not derived from the carefully manipulated leaves of carefully tended gardens, but from scrubby, straggling, and uncared-for trees, allowed to attain a height of nine or ten feet and more. Even of these plants only the inferior produce is devoted to the use of the barbarian: in fact, what is mere refuse. 'I saw great quantities of this,' writes Mr. Baber, 'being brought in from the country on the backs of coolies, in bundles eight feet long by nearly a yard broad, and supposed it to be fuel; it looks like brushwood, and is, in fact, merely branches broken off the trees and dried in the sun, without any pretence at picking. It sells in Yung-ching for 2000 cash a pecul at the outside, and its quality may be judged from a comparison of this price with that of the common tea drunk by the poorer classes in the neighbourhood, which is about 20,000 cash a pecul.'

Mr. Baber then describes the process of pressing this stuff into the cakes or *pao* spoken of by Captain Gill. At Ta-chien-lu these cakes are cut into the portions—about nine inches by seven by three—which the Chinese call *ch'uan*, or 'bricks,' 'containing a good deal more stick than leaf.' Mr. Baber cor-

[6] In supplement to *Calcutta Gazette*, November 8, 1879.

roborates Captain Gill's estimate of the extraordinary weights carried by the porters of these *pao* up to Ta-chien-lu, mentioning a case in which he overtook a somewhat slenderly built carrier freighted with 22 of the Ya-chau packages, which must at the lowest computation have exceeded 400lbs. in weight![7]

The quantity which annually paid duty at Ta-chien-lu he calculated on good comparative data at about 10,000,000 lbs., worth at that place £160,000. A good deal besides is smuggled in by Chinese officials, for it is by means of this tea that those gentlemen feather their nests. Of these administrators and their gains the Tibetans say, ' They come to our country without breeks, and go away with a thousand baggage-yaks.'

§ 47. Mr. Baber, like Captain Gill, speaks of the remarkable manner in which the British-Indian rupee has become the currency of Tibet—a circumstance of which my friend General Hyde was probably not aware in his endeavours to estimate the existing amount of current rupees for the Silver Committee of 1876. ' Those (rupees) which bear a crowned presentment of Her Majesty are named *Lama tob-du*, or ' vagabond Lama,' the crown having been mistaken for the head-gear of a religious mendicant.'

Before the introduction of the rupee, tea-bricks were used as currency (just as Marco Polo tells us that in an adjoining region loaves of salt were used in his time), and ' even now in Bat'ang a brick of ordinary tea is not merely worth a rupee, but in a certain sense *is* a rupee, being accepted without minute regard to weight, just like the silver coin, as a legal tender. Since the influx of rupees this tea-coinage has been very seriously debased, having now lost 25

---

[7] The *pao* purport to weigh each 18 catties, or 24 lbs., as Captain Gill states. But this, according to Mr. Baber, is when saturated. The theoretical weight is a good deal reduced when they are dry.

per cent. of its original weight. The system of double monetary standard is approaching its end, at any rate in Tibet ; for in May last the Lamas of the Bat'ang monastery, having hoarded a great treasure of bricks, found it impossible to exchange them at par, and had to put up with a loss of 30 per cent.'

Mr. Baber has some judicious remarks as to the outlet for Indian tea into Western Tibet. The obstacle to this, as well as to the admission of European travellers, is the jealous hostility of the Lamas, jealous of power, jealous of enlightenment, jealous, above all, of their monopoly of trade. It is evidently a mistake to suppose that the main difficulty lies in Chinese aversion to open the landward frontier, real as that probably is. The feeling among the Lama hierarchy is evidently very different from what it was in the days of Turner and Bogle ; and judging from the reports of both Captain Gill and Mr. Cooper, their rule over the people is now become intolerably oppressive.

We must not lengthen this too long discourse, but the temptation is great to draw upon Mr. Baber, whose reports, whilst they convey a remarkable amount of information, are full of good sense, and as diverting as any story-book !

One fact more, however, we must borrow, before bidding him a reluctant adieu ; and that is his discovery (Fortuna favet fortibus !) upon his last journey—see § 39 above—of two singular local qualities of tea, one of which is naturally provided with sugar, and the other with a flavour of milk or, more exactly, of butter !

§ 48. Ta-chien-lu, Captain Gill's first place of halt after leaving Ch'êng-tu, is a name that is becoming familiar to the public ear, as the Chinese gate of Tibet, on the Ssŭ-ch'uan frontier. Politically speaking it is more correctly the gate between the 'regulation

[88] INTRODUCTORY ESSAY.

Province, of Ssŭ-ch'uan, and the Chinese 'non-regulation Province' of the Tibetan marches. Captain Gill has told the story of the Chinese etymology of the name (II. 76-77), probably fanciful, like many other Chinese (and many other non-Chinese) etymologies that find currency. The name appears from the Tibetan side as *Tarchenton, Tazedo* or *Tazedeu, Darchando,* and *Tachindo,* and is probably purely Tibetan.[8]

The place stands itself at a height of 8,340 feet above the sea-level, but the second march westwards carries the traveller to the summit-level of the great Tibetan table-land, on which, with the exception of one or two early dips into the gorges of great rivers, he might continue his way, did Lamas and others withdraw their opposition, without ever descending materially below 11,000 feet, until he should hail the Russian outposts on the northern skirts of Pamir, 1,800 miles away. This great plateau here droops southward as far as lat. 29°, and below that sends out a great buttress or lower terrace, still ranging 6,000 feet and upwards above the sea, which embraces, roughly speaking, nearly the whole of Yün-nan.[9] In the descent from the higher to the lower terrace, and for a long distance both above and below the zone of most sudden declivity, this region of the earth's crust seems in a remote age to have been cracked and split by huge rents or fissures, all running parallel to

[8] The termination *do* is common in Tibetan names—as Ghiamdo, Tsiamdo—and means a confluence. For the forms above see P. Horace della Penna in *Markham,* 2nd edit. p. 314; Pundit Nain Singh in *J. R. Geog. Soc.,* vol. xxxviii. p. 172; the Nepalese itineraries given by Mr. Hodgson in the *J.A.S. Bengal,* vol. xxv. pp. 488 and 495; and another itinerary from Katmandu, given by him at an earlier date in the *Asiatic Researches,* vol. xvii. p. 513 *seq.* This last itinerary is obviously not genuine beyond Lhassa, from which it makes 'Tázédó' only thirteen stages distant, in a beautifully cultivated plain, producing not only peas and potatoes, but rice and mangoes! But it gives us the Tibetan *name.*

[9] Height of Ta-li-fu, 6,666 feet; height of Yün-nan-fu, 6,630 feet; height of Tong-ch'uan, 6,740 feet; and height of Hui-li, 5,898 feet.

one another from north to south: for not only the valleys of those great rivers, of which we have said so much, but the gorges of their tributary streams exhibit this parallelism.[1]

§ 49. The ethnography of the manifold tribes on the mountain frontier of China, Burma, and Tibet, is a subject of great interest, and respecting which very little is yet known. We have touched it already in a loose way in a preceding paragraph regarding the tribes that look down upon Ssŭ-ch'uan, and we should be tempted to do so again in the region of the great rivers descending from Tibet into Yun-nan and Burma, but for the great scarcity of material. Two of these tribes—the Mossos (or Mu-sus) and the Li-sus—are most prominent, and are not without claims to civilisation. The Mu-sus, who call themselves *Náshi*, are said to have formerly possessed a kingdom, the capital of which was Li-kiang-fu, which the Tibetans, and hill-people generally, call Sadam. Their King bore the Chinese style of Mu-tien-Wang, and M. Desgodins, from whose authority these facts are derived, says that frequently, during his journeys on the banks of the Lan-t'sang and the Lu-Kiang, he has come upon the ruins of Mu-su forts and dwellings, 'as far north as Yerkalo, and much further,' therefore as far north as Kiangka, or nearly so. It is possible that they are the same as the barbarian *Mo*, who are mentioned in Pauthier's extracts from the annals of the Mongol dynasty as being occupants of the Li-kiang territory in the thirteenth century. The men seem to have adopted the Chinese dress and pigtail, and Cooper says they are 'quite Chinese in appearance;' but the women retain a picturesque and graceful costume, which from his description seems, like many of the other female costumes of the non-Chinese races of the Yun-nan frontier as depicted in

[1] Gill, Vol. II. p. 228.

[90]　　　INTRODUCTORY ESSAY.

Garnier's work, to have a strong analogy to the old fashions of Swiss and Pyrenean valleys, popular types for fancy-balls. Captain Gill met with some Mu-sus at and near Kudeu, on the Kin-sha, and he was struck by the European aspect of a lama (or quasi-lama) who visited him—'more like a Frenchman than a Tibetan.'[2] This recalled to him what Mr. Baber says of two women, called 'of Kutung,' whom he met outside Ta-li.[3] But the data will carry us no further.

The Li-sus, or Lissaus, are described by Dr. Anderson as 'a small hill-people, with fair, round, flat faces, high cheek-bones, and some little obliquity of the eye.' The men adopt the ordinary Shan dress, and the women, like those of the Mu-sus, a picturesque costume of their own.[4] In the upper parts of the great valleys the Li-sus seem intermixed with the Mu-sus, but they have a wide and sparse distribution further to the west, and further to the south.

§ 50. Vocabularies of their languages have been sent home by M. Desgodins, and, though I have not seen these, M. Terrien de la Couperie, who has paid much attention to the philology of the Chinese and bordering tribes, tells me that the two vocabularies have 70 per cent. of words common to both, and show a manifest connection both with some of the Miao-tzŭ tribes and with the

---

[2] Vol. ii. p. 270.

[3] 'Their oval and intelligent faces instantly reminded us of the so-called Caucasian type; and in every step and movement there was a decision and exactness widely different from the sluggish inaccentuation of the Chinese physique. The younger was particularly remarkable for a peculiarity of her long hair, which was naturally wavy, or "crimped," a feature which is never met with among the Chinese. While watching these people I felt in the presence of my own race.'—*Baber's Report*, 1878, p. 5.

It may not be inappropriate to add here that I have several times seen among Burmese women a perfectly *Roman* type of countenance, a thing I have never seen in a Burmese man.

[4] Anderson, *Exped. to Yunan*, Calcutta, p. 136.

Burmese. The last point is corroborated by the statement of Dr. Anderson regarding the Li-sus, that the similarity of the Li-su and Burmese languages is so great that it is hardly possible to avoid the conclusion that the two people have sprung from one stock.[5]

Captain Gill, when at Kudeu, obtained a remarkable manuscript, which he has presented to the British Museum.[6] I have seen the manuscript, but I derive the following account of it from the greater knowledge of M. Terrien de la Couperie, who is engaged in systematic study of the origin and relations of the Chinese characters, and is deeply interested in this document. It is written in an unknown hieroglyphic character, and consists of 18 pages, measuring about $9\frac{1}{2}$ inches by $3\frac{1}{2}$. The characters read from left to right; there are three lines on a page; the successive phrases or groups of characters being divided by vertical lines. Among the characters are many of an ideographic kind, which have a strong resemblance to the ancient Chinese characters called *chuen-tzŭ*. With these are mixed numerous Buddhistic emblems.

M. Terrien possesses another document in similar character, but less mixed with Buddhistic symbols, which was traced by M. Desgodins from the book of a *tomba*, or sorcerer, among the Náshi or Mu-su, a kind of writing which that missionary states to have become obsolete.[7] He considers Captain Gill's manuscript to be probably much older. It is not possible to say whence it came, because it may have been an object plundered in the long disorders of the Yun-nan frontier. But M. Terrien is inclined to regard it as a survival

[5] Anderson, u.s.
[6] Additional MSS. No. 2162.
[7] There is a bare allusion to the subject in the book *La Mission du Thibet*, where M. Desgodins speaks 'des livres de sorciers que j'ai eus entre les mains, mais dont je n'ai pu avoir la traduction' (p. 333).

of a very ancient ideographic system, perhaps connected with that of the Chinese in very remote times. The late Francis Garnier, during one of his later journeys in Hu-nan, was assured[8] that in certain caves in that province there were found chests containing books written in *European* characters, and judiciously suggests that these may have been books of the extinct aborigines, in some phonetic character. M. Terrien recalls this passage in connection with Captain Gill's manuscript. And he observes that a thorough study of the character, and of the dialects, for which we have as yet very little material, may be most important in its bearing on the ethnographic and linguistic history of ancient China. Very ancient Chinese traditions speak of these races as possessing written documents.[9]

§ 51. There must be an end to this commentary. I have become through circumstances, and especially through the traveller's friendly confidence in me, too closely associated with his work to put myself forward as a judge of its merits. But I am bound to call attention to some facts.

Captain Gill was weighted with serious disadvantage as a traveller in China by his unacquaintance with the language. No one could be more sensible of what he lost by this than he is. Yet he was singularly fortunate, during two large sections of his travels, in his interpreters,—having the aid of Mr. Baber in the voyage up the Yang-tzŭ, and that of Mr. Mesny across the Tibetan and Burmese frontier. And his success on a journey in which he has had no forerunner, and had no companion,—that from Ch'êng-tu to the north,—shows that he carried in his own person the elements of that success,—patience, temper, tact, and sympathy.

[8] *Bull. de la Soc. de Géog.*, January, 1874, p. 19.
[9] One recalls the tradition of the Karens, that they too once had a book, but a dog ate it!

INTRODUCTORY ESSAY. [93]

It is needful to inform or remind readers,—at least the more serious portion of them,—that the bright personal narrative contained in these two volumes does not represent Captain Gill's scientific results. Let anyone who desires to appreciate the real character of his labours look at the report of his journey published in the 'Journal of the Royal Geographical Society' (vol. xlviii. pp. 57 *seq.*) The detailed maps which illustrate that and the present work have been constructed from the route-survey which Captain Gill kept up unbroken from his first departure from Ch'êng-tu towards the Northern Alps till his arrival at Ta-li-fu. From Ta-li-fu the weary task was abandoned, as the route-survey thence to Bhamò had already been accomplished by Mr. Baber, from whose work that portion of the survey is borrowed.

Observations for altitude were made with Casella's hypsometric thermometer, and with two aneroids: observations of the latter being taken three times daily when halting, and ten or twelve times on each day's march. From the readings so taken, after needful corrections, the altitudes of 330 places have been computed, affording data for the extensive sections exhibited.

The itinerary appended to the report in the 'Geographical Society's Journal' will be found to contain a mass of minute detail of the road travelled, and its natural features, filling, between Ch'êng-tu and Momien, forty-six pages of very close print. Those only who have tried can judge how much resolution is required to keep up the labour of such a record throughout a fatiguing journey of several months, in protracting the day's journey, and writing up the diary and itinerary every night, as was done by Captain Gill regularly, with two or three exceptions at most. Here I am happy to be able to introduce some remarks from the high authority of Baron Richthofen :[1]

'Captain Gill's results have been of the highest interest to me, particularly those of his journey north of Chengtu, and of his route between Ta-tsien-lu and Atentze. He is an acute observer of men and nature, and stands very high indeed by the accuracy and persistency with which he has carried through his surveying work. It is not quite easy to do this, and it requires the mind to be firmly set upon this one purpose. Many a famous traveller might learn in this respect from Captain Gill. The determination of so many altitudes is, too, a very important part of his work, and it has brought about results of extreme interest. I regret, however, that he did not put down on the map all that he was able to see. . . . I presume that Captain Gill . . . . wished by the tendency to the utmost possible exactness to abstain from laying down on his map whatever was lying at some distance from his road. I think it would be well if he could be induced to supply this want. This would be of great interest in the valley of the Upper Min, and the passage eastward of Sung-pan-ting, where indeed a few very elevated summits are marked on his maps.

'Altogether his journey is one of the most successful and useful which has been performed in Western China, and it is to be hoped that he will have a successor who will extend his explorations into the unknown regions west of the Min and north of the road to Bat'ang.'

Captain Gill himself would desire to put in the caution that his map must not be regarded as absolutely accurate. Allowance has to be made for the fact that much of his journey was accomplished in rain and fog ; some small part of it in the dark. But these are small drawbacks to the accuracy, none to the merits of such a performance. That such a work should have been kept up openly and continuously over so extensive a journey, speaks volumes for the tact as well as for the perseverance of the traveller.

§ 52. The anonymous writer who edited the journals of Augustus Margary, with so much judgment and good feeling, concludes his biographical sketch of the young man in words from which I extract the following :

'Whether, and how soon, his countrymen will be able to travel in honour and safety the route which he was the first to explore,

[1] Letter to the present writer, dated Bonn, November 15, 1879.

will depend upon the faithfulness with which they copy his example. As soon as Englishmen shall be able, as he did, to find 'the people everywhere charming, and the mandarins extremely civil' (p. 134)—in spite of all the serious and petty vexations, discomforts, and discourtesies which met him day after day, and which he had to brush aside with a firm hand, but without losing temper —the route will open out and become as safe to them as it proved to him on his lonely westward journey. For his short story, if read aright, and in spite of its violent ending, adds yet another testimony that a little genuine liking and sympathy for them, combined with firmness, will go further and do more with races of a different civilisation from our own, than treaties, gunboats, and grapeshot, without it. If the route is ever to be a durable and worthy monument of the man, it must be opened and used in his spirit, by fair means, and for beneficent ends.'

These are just and admirable words, and I think all candid readers of this narrative will recognise that my friend its author has been not unworthy, tested as those words would have him tested, to do his part in keeping open the track which Margary first explored. He has done that, and more. And I am happy to think that he also is still young, and thus, as this has not been his first adventure in the conquest of knowledge in distant regions, neither will it, I trust, be his last.

H. YULE.

December 18, 1879.

THE
RIVER OF GOLDEN SAND.

CHAPTER I.

OVER THE SEAS AND FAR AWAY.

China resolved on—Preliminary Visit to Berlin—Baron Ferdinand v. Richthofen—Marseilles, and Voyage in the 'Ava'—Sea-flames—The Straits—Chinese Practices first realised—Approach to Saigon—The City—Hong-Kong reached and quitted—Shanghai and its peculiar Conveyances—The Chinaman's Plait—Voyage to Chi-Fu—The Chi-Fu Convention—The Minister Li-Hung-Chang—Ceremonial of his Departure—Voyage to the Pei-Ho—Difficulties of Navigation up that River—Scenery—Arrival at Tien-Tsin—Choice of Conveyances—Carts of Northern China—Mongol Ponies—'Boy' found—Horse-buying—Scenes on the Tien-Tsin Bund—Risky Building and Chinese Devotion—Tien-Tsin Hotel—Preparation for Travel—Suspicious Wares—Arrangements for a Start completed.

WHY not China?

Such were the words addressed to me by a friend I met in Trafalgar Square early in May 1876.

Up to this moment I had never thought of China. My attention had never been directed to it, and my notions regarding it were crude in the extreme: dim ideas of pigtails, eternal plains, and willow trees; vague conceptions of bird's-nest soup and puppy pies. I had never been particularly attracted to the country, and naturally replied, 'Why should I go to China?'

At the time I gave the matter no further consideration, and it was with some surprise that, a fortnight later, I was met with the same question ; this time, however, my friend had some reasons to adduce, the result of which was that, on June 26, a fine breezy morning, I stood on the deck of the Ostend steamer lying in Dover harbour.

A fresh north-easterly breeze just crisped the tops of the waves, and a bright sun lighted up the Dover cliffs as they gradually merged into the mist. For the first time for many days, I had time to think, and when at last the cliffs were lost to view, I seemed to have launched into a new and unknown sea; for whither fate would lead my steps I could not say : all that was definite was, that I was going to Peking.

Through the kindness of Colonel Yule I was furnished with a letter of introduction to Baron von Richthofen, the greatest of modern explorers and geographers, whose long travels in China had made him the first authority on the country ; and it was to make his acquaintance that now I bent my steps to Berlin.

It was a lovely summer day, and the haymakers were busily at work as we dashed past them, and past smiling villages, and lazy Belgian streams : here a quaint steamer with its paddles scarcely in the water, making more splashing and noise than a man-of-war ; there a barge drifting slowly onwards towards the sea ; now a country château with its trim lawn and bright flowers from which we could almost catch a breath of fragrance ; through many a village where the stout Flemish horses drew the quaint long-backed country carts ; by waving corn-fields where the blue corn-flowers seemed to nestle lovingly in the shadow of the wheat and barley, on through busy Liège to Verviers and Cologne, where I spent a few hours wandering about the quaint old streets, and then in the evening continued my journey to Berlin.

Some Japanese were my travelling companions, dressed in black coats, tall shiny hats, and white shirts, of which they seemed remarkably proud; they took off their boots, and placing their feet on the seat next me, we soon all fell asleep. In the morning they were pitiable objects to look upon, their black coats and neck-ties covered with dust, their faces shining with a greasy glow, their collars and wristbands without any visible signs of starch, and their thick black hair that had been carefully parted the night before now standing on end 'like quills upon the fretful porcupine;' and even then I thought that it was not all gain to the Japanese, when they abandoned their national dress and their ancient customs, and threw themselves recklessly into the arms of western civilisation.

I was fortunate in finding Baron von Richthofen in Berlin, and the week that I spent in his society passed only too quickly. Hour after hour he gave up his valuable time to me, and opened volumes from his rich store of information; day by day I grew wiser, and little by little true pictures of China and Chinese life formed themselves in my mind. Baron von Richthofen possesses in a remarkable manner the faculty of gathering up the details presented to his view; putting them together and generalising on them with rare judgment; forming, out of what would be, to a lesser genius, but scattered and unintelligible fragments, a uniform and comprehensive whole. During all my conversation with Baron von Richthofen, not one word passed his lips that was not gold seven times refined, not one hint was given me that did not subsequently prove its value; his kind thoughts for my comfort or amusement were never ceasing, and his refined and cultivated intellect and genial manner rendered the recollections of my stay in the German capital some of the most pleasant of my life.

Leaving Berlin, I journeyed leisurely to Marseilles, awaiting a telegram without which I was unable to start. The delay was rather troublesome, as by this time I knew enough of China to be well aware that early in the winter the province of Pechili is entirely frozen up; and it was in that province that I intended to make my first journey.

At last the welcome telegram arrived, and on July 27 I found myself at Marseilles. I had never been there before, and although it was rather warm, I was pleasantly surprised with the town; for instead of a picturesque place, I had formed in my mind ideas of nothing but dirty streets and busy quays. The shop fronts are screened with the gayest of gay awnings, striped with blue, red, grey, and all sorts of colours; on the sides of some of the streets magnificent plane trees, as high as the four-storied houses, meet overhead and afford delightful shelter from the southern sun; and at a little distance there are some charming gardens, where now some glorious masses of geraniums were in their full beauty.

On July 30 the ship 'Ava,' of the Messagéries Maritimes, steamed out of Marseilles, having on board but a small number of first-class passengers, as this, the hottest time of the year in the Red Sea, is not a favourite one for travellers.

On a sea like glass we glided through the Straits of Bonifacio, steamed into the Bay of Naples, and left it again before the town was well awake. That morning's sun set like a ball of fire behind Stromboli. Scylla frowned, and Charybdis hissed, as if in impotent rage that coal and iron had robbed them of their terrors, and the lights of Messina shone awhile over the summer sea; but one by one even these faded, and the last glimpse of Europe was gone from our view.

A voyage is always rather tedious, and during August the Red Sea can hardly be considered pleasant; the days went by, however, although there was but little incident to vary their monotony.

Before arriving at Galle, the sea was one night so phosphorescent, that none of the old sailors on board could recollect a similar scene. The vessel left a trail of fire far behind her in her wake ; as her bows pierced the water she seemed to dash up liquid flames that danced about her sides as if by magic, every wave that broke illuminated itself and lit up a sheet of phosphorescent light, and all around, in every direction, as far as could be seen, fires innumerable seemed to sparkle in the ocean.

Passing through the Straits of Malacca, we steamed into Singapore on the morning of August 26, and I was rather disappointed with its scenery, of which I had heard so much. The entrance to the harbour is certainly exceedingly pretty; there is a wonderful richness in the verdure, and the trees at the water's edge contrast beautifully with the deep red of the soil. Perhaps it is, that after some days at sea people are always in a frame of mind to exaggerate the charms of the first land they see, or perhaps it is, that the ships being able to come within twenty or thirty yards of the shore, the beauties are more apparent than in other places.

Here I passed a delightful day, enjoying the hospitalities of the Governor, Sir William Jervois.

Government House is a fine building, on the top of a little hill looking over rich green trees and green grass, to the blue sea, the town of Singapore stretching out on one side along the edge of the harbour, where there is a great deal of shipping and many boats. In the town there is an enormous Chinese population, and here for the first time I understood the mystery of using chopsticks. Up till now I had cherished the fond delusion that it was customary to take the rice up grain by grain; I had sorely exercised my mind on the consideration of the length of time that a Chinaman would occupy in consuming a hearty meal. I was therefore much interested in watching the process. The bowl, something like a large teacup without a handle, is held in the left hand close underneath the chin, the chopsticks being used as a shovel, by which the rice is pushed into the mouth, an extraordinary gobbling noise accompanying the proceeding. The grains of rice, moreover, even when cooked by a Chinaman, are not invariably all separate, and it is easy for a skilful performer to take a good deal of rice between his two chopsticks. The method of holding the chopsticks is almost impossible of explanation, but the art is acquired with a very little practice, and, once learnt, it is not difficult to pick up the smallest grain.

In the afternoon I rode with the Governor to the Botanical Gardens, on a pony which upset the popular theory that all horses tremble in the presence of lions and tigers; for he could with difficulty be kept away from the bars of a cage in which there was a tiger that had been presented to Sir William Jervois by a neighbouring rajah. In the evening I was obliged to take my leave, and steaming out of Singapore early the next morning, we arrived off Saigon on August 29.

The mouth of the river is rather pretty; as the steamer runs up, on the starboard hand are hills about one hundred or two hundred feet high, covered with forest, in which there are here and there open patches of beautiful green grass; the trees come down to the water's edge, the coast is broken into innumerable little creeks and bays, native villages are scattered about, and on the other side the low coast is seen two miles away. In a very short distance the hills disappear; the river, about half a mile wide, is very tortuous, and winds through a flat, swampy, uninteresting country, covered with low jungle, where I was told there were a great many tigers; but as Frenchmen seldom hunt savage beasts for sport, they probably exaggerate the number of them.

The town of Saigon lies fifty miles up the river, and is close to a very large and important Chinese town, the seat of ancient trade; it was for commercial purposes necessary to establish the colony here rather than at the mouth of the river, where there would have been a more picturesque, more convenient, and far more healthy site; strategically, too, there were good reasons for choosing this rather than Point St. Jacques at the entrance to the river, for with torpedoes, the navigation of the tortuous channel would be almost impossible to a hostile fleet, while an attack on the point from the open sea would be comparatively easy. The Messagéries Company wished to avoid the waste of time consequent on the navigation of this troublesome fifty miles, and applied to the French Government for permission to establish a station at Cape St. Jacques, and to perform the inland service in small steamers. There would have been no difficulty about this, for the roadstead is always safe, and small vessels can in any weather ascend to Saigon. The French Government, however, refused permission, and the mail steamers thus lose forty-eight hours on their passages, without any apparent compensating advantage.

As soon as we had anchored, a fellow-passenger accompanied me ashore, and we hired a carriage, that would in our India be called a shigram, drawn by the tiniest of tiny ponies, which, notwithstanding their diminutive size, galloped along at a rapid pace. Taking a drive round the town we saw Government House, a fine building, but not so imposing as ours at Singapore; this is, however, partly owing to the natural beauty of the Singapore situation. Here we noticed the marines on guard, in the stewy heat of this climate, dressed in dark blue cloth coats.

With regard to the town itself, the French have certainly made more of the little that nature has provided them with, than we have at Singapore of a much better site. The principal street of the town is a fine broad boulevard, with trees on both sides, where there are a few French shops amongst those of the Chinese. The public buildings are plain, and do not deserve much notice; there are of course cafes and restaurants, in as close imitation as circumstances permit, of the gay French capital. There is no gas at Saigon, as there is at Singapore, but the streets and houses are well lighted with petroleum. This is said to be a very unhealthy place, residents being liable to a form of dysentery that nothing appears to cure; the governors, whose salary is 8,000*l*., are rarely able to remain more than two years. We found that, with an admirable idea of how most to inconvenience the public, the Post-Office was closed till 4.30 P.M., the officials being busy preparing their mails; so we took another drive, and when we returned we found that the *poste restante* business, the selling of stamps, and the receipt of valuable articles, were all conducted by one official at one little pigeon-hole.

People had been dropping in one by one during

the past hour, and the street now presented something the appearance of one of our west-end thoroughfares on the night of an entertainment, with a long string of carriages on each side of the road. When at length the pigeon-hole was opened, a crowd of Chinamen, French soldiers, sailors, officials, and people of all sorts fought for the services of the man inside; we also engaged in the conflict, and at length succeeded in posting our letters. Before returning to the ship we had to listen to the most doleful jeremiads of a sleepless night in store for us, from the size and virulence of the mosquitos, with which the river was said to swarm; visions of large dragon-flies, with the stings of scorpions, presented themselves to me as I turned in, but happily the reports were exaggerations, and we none of us suffered much.

Leaving Saigon we steamed on again to the East, passing the Ladrone Islands, famous in the days of yore, where the old Portuguese navigators first entered these waters, and where, finding themselves the unfortunate victims of the numerous pirates and murderers that cruised about among these narrow channels, they called this beautiful archipelago the Ladrone or Robber Islands.

The times have changed, but the nature of the people is not much altered; and though at a distance the fleet of junks, with their red sails bellying in the freshening breeze, might be mistaken for mackerel boats on our own English shores, and though by profession the people follow the peaceful avocation of fishing, they are still on occasions robbers, pirates, or buccaneers.

It was a delightful change at Hong Kong to pass a couple of days amongst kind friends; it was refreshing too, once more to see English soldiers looking as smart as only English soldiers do; and after so many weeks of walking up and down the deck of a ship, a real hill was quite a treat. But our time was soon up, Hong Kong gradually disappeared, and we sailed away again over the blue waters, where the extraordinary number of fishing junks formed a marvellous sight. All day and all night the steamer passed through a swarm of these vessels that seemed to fringe the whole coast; at one time I counted 150 in sight in one quarter of the compass, and we were obliged to stop our engines two or three times to avoid the nets.

My journey in the 'Ava' was drawing to a close, and on the morning of September 8 we entered the Yang-Tzŏ-Chiang, or Ocean River, which here flows majestically through a perfectly flat country, cut up by innumerable small canals, where the vegetation appears wonderfully rich, and where there seem to be plenty of fine trees. No hedges or walls were to be seen dividing the fields, and on the river there were a great number of fishing and trading vessels, all of one shape, but of various sizes, with two, three, four, or five masts, stuck in without any regard to the angle at which they were stepped, and all the more picturesque on account of their irregularity.

At Shanghai I presented a letter of introduction to that most hospitable of firms, Gibb, Livingstone, and Co. Here I enjoyed a dinner on shore, and afterwards went on board the steamer that was to convey me to Chi-Fu.

A machine called a jinnyrickshaw is the usual public conveyance of Shanghai. This is an importation from Japan, and is admirably adapted for the flat country, where the roads are good, and coolie hire cheap. In Japan, I have been told, they are also used on hilly ground. In shape they are like a buggy, but very much smaller, with room inside for one person only. One coolie gets into the shafts, and runs along at the rate of about six miles an hour; if the distance is long, he is usually accompanied by a companion who runs behind, and they take it turn about to draw the vehicle.

The jinnyrickshaw is, however, only for the rich; for poor people there is another description of conveyance. This is the wheelbarrow, so well known in all the plains of China, with a seat at each side of one high wheel, on which the people sit sideways as on an Irish car.

Except in Shanghai, the Chinese contrive that the wheels of these shall creak, for a Chinese coolie always seems to require some noise to assist him in his work; when carrying a load in the usual way, by means of a split bamboo over his shoulder, he gives a peculiar grunt at each step, and chair-coolies almost always do the same thing. I was told that in the early days of Shanghai, the noises made by coolies and creaking wheels became so great as to be at last utterly unendurable to European nerves, and a regulation was made, which was at first enforced with much difficulty, forbidding coolies to groan, or wheels to creak, within the boundaries of the Concession, and imposing fines for a breach of the rule. Inside the settlement both jinnyrickshaws and wheelbarrows abound; these are licensed, just as hackney carriages are in London; the tariff is fixed by law, and licences suspended for misconduct or breach of regulations. On my way to the steamer, in the cool of a glorious starlight night, the reverie into which I had been gently soothed by a fragrant Manilla, such as is rarely to be met with in England, was suddenly broken by a violent bump, and I awoke to the fact that one of the wheels had suddenly come off the jinnyrickshaw. The driver, if such an appellation is permissible, did not seem at all disconcerted; he picked up his wheel, put it on, took a new linch-pin from some mysterious fold in his garment, whilst with a smart shake of his head he whipped the end of his plait into his hand. It was the work of a moment to unplait a little of it, break off a lock of his hair, and by the light of the paper lantern always carried, put the tie thus improvised through the hole in the linch-pin. In five minutes we were off again as if nothing had happened, and I learnt that a Chinaman can find a use for anything, even for his plait.

The plait was first imposed upon the Chinese as a badge of servitude by the Manchus when they took the country; but the origin of the appendage has been long forgotten—it is now valued almost as dearly as life, and to be without one is considered the sign of a rebel.

I was told that once a Chinese gentleman was riding in the settlement of Shanghai in a jinnyrickshaw, when he allowed his plait to fall over the side; it was a long one, and the end was soon caught in the axle, which gradually wound it up. The poor fellow shouted to the man drawing him to stop, but the coolie imagining that he was being urged to greater efforts, only went the faster, until the unfortunate occupant, with his plait nearly wound up to the end, and himself nearly dragged out of his carriage, was in a pitiable plight. A British sailor at this moment happened to pass that way, and observing the desperate predicament, with the readiness of resource for which nautical people are famed, he drew his knife and in an instant severed the plait from the Chinaman's head. He thought he had done a kindly act, but instead of thanks he received little more than curses, and his life was not considered safe until his ship was well beyond the limits of the Shanghai river.

There were at this time three companies that owned steamers running between Tien-Tsin and Shanghai—one English, one American, and one Chinese. The 'Zin-Nan-Zing,' belonging to the English company, was the first to leave Shanghai after my arrival. I had engaged my passage by it, and we sailed at 4 A.M. on September 10. All the steamers here, including the magnificent vessels that ply on the Yang-Tző between Han-Kow and Shanghai, are built on the American plan, with the first-class accommodation forward, where the passengers are free from smells of cookery, oil, or engines, but where there is this disadvantage, that if there is any pitching motion it is sure to make itself felt.

The coasts of Shantung are generally breezy, and soon we found ourselves in rather a heavy head sea that sent the spray flying over the deck, and reduced our speed to four or five knots ; thanks to the pleasant captain, I was able to take shelter in the wheel-house, and read in comfort, until the thermometer suddenly descending to 74° F., the temperature felt bitterly cold after the steamy heat of Shanghai.

At about nine o'clock on the evening of September 12 we dropped our anchor in the quiet harbour of Chi-Fu. The wind had dropped, the clouds had cleared off, and the stars were shining brilliantly over the smooth water that reflected the riding lights of numerous merchant vessels lying here.

Chi-Fu is the watering-place of Shanghai, charmingly situated on a deep bay, sheltered on the north by a long low spit of land ending in some low hills ; it is open to the N.E., and when the wind is from that quarter a heavy sea comes rolling in, and prevents communication with the shore. To the E.N.E. are some rocky islands which protect the harbour from that quarter; at the head of the bay is about a mile of flat country closely cultivated and very green; and at the back a range of hills, which run down to the coast on either side, end in picturesque bluffs. To the west is the large and important Chinese town, where a fleet of quaint-looking junks were lying at anchor. The European quarter is small, containing not much more than the consulates, three hotels, and a few stores where European goods are sold at rather startling prices. Here, when the heat of Shanghai is at its worst, the wearied merchants find a pleasant and invigorating change in the fresh air and sea bathing.

The now celebrated Chi-Fu Convention was at this time being arranged, and Sir Thomas Wade, H.B.M. minister, Li-Hung-Chang, the celebrated Chinese minister, and some members of the other foreign legations were here, with three English, two French, and one German man-of-war in the harbour, besides Admiral Ryder's despatch boat the 'Vigilant,' and numerous Chinese war vessels. I found two very fair rooms in an hotel close to the European town; my quarters faced the sea, and I could look out upon the British flag floating proudly from the mast of the 'Audacious.'

I was furnished with letters of introduction to Sir Thomas Wade, whose reputation for hospitality has become a proverb in Peking. Though pressed with business, he found time to talk over my plans, and I can never be sufficiently grateful to him for all his kindness and cordiality. Here also I made the acquaintance of Mr. Carles, a consular officer, who subsequently became my companion in my first trip in the province

THE MINISTER LI-HUNG-CHANG.

of Pe-chi-li, a trip that turned out to be but an introduction to Chinese travel, and the precursor of a much longer and more serious enterprise.

At length the convention was signed; the whole party broke up; ministers, European and Chinese, were to return to Peking, and Chi-Fu was to be left desolate and deserted.

It was admitted by all to be a great concession on the part of the Chinese, that Li-Hung-Chang had come to Chi-Fu, instead of waiting at Tien-Tsin for Sir Thomas Wade to come to him; much wordy warfare had been waged over this first point, and report said that on more than one occasion negociations were very near being broken off. The ministers left Chi-Fu together, but on the voyage the wily Li-Hung-Chang managed to get his boat ahead of the 'Vigilant' carrying Sir Thomas Wade, and so saved much of his reputation in the minds of his countrymen, as he was the first to land in Tien-Tsin.

September 15.—The steamers here have no regular hours of departure, but discharge or take in cargo immediately on their arrival, and start again as soon as ready. I was told that the vessel that was to take me to Tien-Tsin would probably come in during the night, and get away again very early in the morning. I was therefore ready soon after 6 A.M., and spent the morning watching for the steamer. Her smoke at last appeared on the horizon at about 1.30 P.M., and she dropped anchor at about 3 P.M. Taking a boat from the beach in front of the hotel, I went on board the 'Chih-li,' an American vessel of about 1,200 tons, with the saloon and first-class sleeping accommodation forward.

I now had an excellent view of all the ceremonies and displays attendant on the departure of the great

Li-Hung-Chang, one of the most powerful men in China.

Li rode in a covered sedan chair, preceded by a man carrying an immense red umbrella; his escort appeared to number about forty men, picturesque fellows in blue coats and red trousers, armed with rifles, and besides these there were some wonderful-looking men with cutlasses. The commander of the escort was a a most unsoldier-like and ragged-looking person, perched on a Chinese saddle, high above the back of an exceedingly small and abject pony.

A battalion of infantry was drawn up near the landing-jetty, and about forty war-junks were anchored in a triple line close by; these most picturesque and old-fashioned vessels were armed with one gun each, and gaily decorated with an immense red flag, some of them having a second banner striped red and white.

The Chinese steam-gunboats in the harbour were all 'dressed,' as was the Chinese merchant steamer by which Li-Hung-Chang travelled.

When Li-Hung-Chang arrived at the quay, the battalion fired a *feu de joie*, the Chinese steam-gunboats saluted, and the war-junks all let off their pieces somewhat promiscuously.

Li stepped into a cutter which was towed by a very small steam-launch in command of Europeans, and was soon alongside his vessel. The soldiers then on board fired a *feu de joie*, the whistle gave a few screeches, the anchor was up, and away went Li, escorted by the steam-gunboats.

The 'Vigilant' followed almost immediately, the soldiers marched home, the booming of the cannon ceased, the smoke cleared off, and as the sun descended in the western horizon, Chi-Fu, so lately the scene of such busy and hot arguments, so nearly the site of diplomatic rupture between England and China, seemed to throw off the garb of war, and smiling pleasantly after the departing grandees, to wrap itself in the mantle of that peace that it had just given to the world.

September 15.—At half-past six our anchor was weighed, and as the stars came out we steamed across the Gulf of Pe-chi-li.

This was a very comfortable steamer. The captain, two officers, and two engineers were American, and, with the exception of two Malay quartermasters, the crew were all Chinese. The captain said that he preferred the Chinese as hands to Europeans or Americans: they never give any trouble, never drink or quarrel, and although in cases of danger he admitted that at first they sometimes slightly lost their heads, yet he declared that, with proper leaders, this lasted a very short time, that then they really had no fear, and would work as quietly and as well as under the most ordinary circumstances. The captain is not without experience, as on one occasion he ran on to a rock in this vessel, and the ship was in so critical a position, that at one time they almost lost all hope of saving her.

It is not gratifying to our western pride to find that, in almost all walks of life, the Chinaman can compete with and beat the European, surpassing him in industry, sobriety, and carefulness of living. The problem of the future intercourse of Europe and China is a difficult one, and must furnish much food for reflection to thoughtful minds.

The party was a very pleasant one, when we sat down to dinner at seven o'clock. The American minister and his wife were on board, and perhaps it was in their honour that a remarkably good table was kept during the short voyage. The Americans are almost as celebrated as the Scotch for their cakes and bread, and in the morning the table groaned beneath the weight of the different descriptions made of wheat and Indian corn.

As we approached the Taku Bar, it became an exciting question whether we should be able to cross it or not. When we left Chi-Fu we drew fourteen feet, but the captain had shifted the cargo so that now we drew only thirteen. It was, however, questionable whether we should find a pilot, as so many ships had preceded us; we fortunately secured the services of the last, and between 1 and 2 P.M., crossed the bar, with many a bump on the soft mud.

The entrance to the Pei-Ho, or River of the North, with its wide expanse of mud flats, would certainly come up to any preconceived expectations of dreariness; but as Tien-Tsin is approached, although the country is still perfectly flat, the life, activity, and close cultivation around render the scenery, to say the least, cheerful.

Of any possible combination of annoying circumstances, the navigation of the Pei-Ho must be the most trying to the temper of a ship captain. The river bends and winds about in the most exasperating manner with the sharpest turns; after a straight run of perhaps a little less than a quarter of a mile, it becomes necessary to round a sharp bend of at least a semicircle; if the bend is to the left, the bow of the ship is aimed straight at the bank on the starboard hand. All may seem to be going well, when the current probably catches the vessel, and with the helm hard a-starboard, she runs hard and fast aground on the bank, in such a way that a pebble could be dropped ashore from the deck. The ship then sticks, and will not move; a warp is laid out to the bank on the other side of the river, and the donkey engine set to work. Perhaps the strain is too great, and the warp parts; this has to be replaced, the engines then are backed, the helm put amidships, the donkey engine set to work again, the helm put hard a-starboard, and at last her head is got round; she moves again and reaches the next bend, when just at the critical moment a junk steers between the steamer and the shore. The engines must be backed to prevent the junk being jammed between the ship and the bank, and in three minutes as much ground is lost as has been gained in the last half-hour. Now the steamer touches a bank in the middle of the river: the current running like a mill-race slews her round, right across the stream, and stops all navigation. Under these circumstances the captain seemed to me to exhaust the whole of his nautical vocabulary. Once we pulled the warping-post out of the bank; once, in passing a great junk, whose anchor was laid out in a millet field, our wash was so strong that, taking her broadside on, she tore her anchor adrift and went afloat on her own account. Under similar circumstances the swearing of English sailors would have been terrible, but the worthy Chinese seemed to take it in the day's work, and, laughing all the time, quietly laid their anchor out afresh, although I must admit that I subsequently found the swearing powers of the Chinese sailors to be in no way inferior to the capabilities of our troops in Flanders. Until seven o'clock in the evening our captain struggled manfully with the twists and turns, when at last we ran so hard aground that with a falling tide no more could be done that night.

The captain must have been possessed of an angelic temper: he never said a single word except to give his orders in a quiet voice, but at the most aggravating moments, when most people would have used bad language, he would violently chew the end of his cigar, and by this means relieve his feelings.

The yellow Pei-Ho winds its tortuous course through a perfectly flat plain, and, as far as eye can see, there is not the smallest elevation. The whole country is closely cultivated, chiefly with millet, which now nearly ripe, stands about five feet high; villages are close together, the huts of mud with tiled roofs, and the streets as narrow as possible, whilst round the houses a few green willow trees look homely and pleasant. In the gardens a peculiar kind of yam grows abundantly; the root, which is the esculent portion, is like a large horse-radish in appearance, it has a leaf like a convolvulus, and is trained up on crossed sticks to a height of about six feet. The leaves twine over these in a thick mat of dense foliage that contrasts pleasantly with the yellowish tinge of the ripe millet.

Every now and then passing a village close to the banks, where little brown children with their incipient plaits on each side of the head, and no clothing to speak of, would be playing in the dirt with the family pigs, our wash rolling up would give them all a muddy and unexpected bath.

September 17.—We all retired early, and it was well for us that we did so; for at about four o'clock next morning the donkey engine began to work. There was no more sleep for any one, and as we lay awake we could hear the captain's continued commands—starboard, port a little, &c. &c., and the same heart-breaking process was continued as we worked slowly up.

The morning broke, giving hopes of a lovely day, that were by no means belied, and at eight o'clock we thought that we should breakfast at Tien-Tsin. There was only one more bend in the river, but that a very difficult one, and it seemed as if the vessel's head never would come round. No sooner had she come up half a point than she would viciously shoot forward a few yards, an eddy would suddenly take hold of her bow, and she would fly right off; at last, a tug coming down the river gave us a friendly pull, and we were safely round the last point. The command was given, full speed ahead—Tien-Tsin was but two miles off. The captain threw away the end of his cigar, and for the first time did not light another. We all began to prepare for going ashore, as the ship sped gaily on up the straight reach, when suddenly she ran on to a bank in the middle of the river, and as the tide had now fallen too low, all the captain's efforts to get her off were unavailing. We descended to breakfast at nine o'clock, and afterwards, as the distance was so short, most of us went off in a boat to the bank, where landing in the mud was a matter of some difficulty. It was accomplished, however, with nothing worse than muddy shoes, and we walked to the British Consulate.

The journey from Tien-Tsin to Peking, of a minister who is taking as his guests two admirals with their suites, is a very serious matter; and I thought to myself that the British Legation must be a very elastic building, to accommodate so many; but where a minister is of such a royally hospitable nature as Sir Thomas Wade, difficulties soon disappear.

Sir Thomas and some of his guests were going by boat to Tung-Chou, whence a short ride would land them in the Legation. These river boats are long, flat-bottomed affairs, with houses on the stern, which a good travelling servant knows how to make fairly comfortable in a very short time. In cold weather the chinks must be covered with paper, but at this season it was unnecessary. One boat is usually kept as kitchen and dining-room, and at stated hours the different boats come together for meals. The vessels are mostly tracked against the stream by ropes made fast to the head of the mast which is right in the bows, but if there is a fresh fair wind, they sail. In this manner the journey to Tung-Chou occupies from three to four days.

As the river winds and twists about in the flat alluvial plain, and the boats, especially when tracking, do not travel very fast, it is easy to get out and walk along the bank, and by cutting off corners, keep up with the fleet. Thus the tedium of being confined in a very limited area is relieved, and as in September the weather is neither hot nor cold, the journey this way is far from unpleasant.

Another method of travelling is with carts, which perform the journey from Tien-Tsin to Peking in two days, unless the traveller prefers making three shorter stages; but the jolting and bumping of these springless carts over the rough tracks cannot be imagined by those who have never travelled but in carriages with springs over the made roads in England, and is really so unpleasant, that this system would hardly commend itself to any one who was not a very good walker, and by using his legs, could save his bones from being sorely bruised. The Chinese travel a great deal in this manner, and the Chinese ladies sit cramped and cooped up all day long with wonderful patience and endurance. European ladies, too, sometimes make long journeys in these carts; and though, perhaps, accustomed to all the luxuries of Western civilisation, put up with the discomfort attendant on a journey of this kind with a pluck that is delightful to witness.

The Peking carts are without exception the most admirably suited to their work of any I have ever seen. Springs, such as those made in Vienna to do duty over the Roumanian cross-roads, might possibly last over one or two journeys from Peking to Tien-Tsin; but it would be a rash experiment, for once broken it would be difficult, if not impossible, to get them repaired. The carts of Northern China therefore are made without springs; considering their very great strength, they are marvels of lightness, and the workmanship in them is really excellent. A hood, provided with a little window at each side, covers them, and sometimes in hot weather there is an awning in front to protect the driver, or keep the morning and evening sun from penetrating into the inside.

One mule is generally put into the shafts and another as leader; the traces of the latter are both attached to the offside of the body of the cart, passing through a steel ring six inches in diameter fastened near the end of the off shaft. This ring is always polished up in a way that would refresh the heart of a captain of field artillery, and the carters keep their equipment altogether in first-rate order; the reins are generally of rope, very light—indeed, in China the lightness of the harness, in which strength and durability are quite sufficiently considered, is a remarkable contrast to the heavy and useless leather-work with which we in England load our horses.

One hundred li, or thirty-three miles, is considered an average day's journey, and when sufficient inducement is held out to the carters, the way in which their carts will day after day complete these long stages over the most trying roads,—sometimes deep in mud, at others through heavy sand, or in the mountains up and down severe and rocky gradients, where the ground is often strewn with huge stones and boulders,—is very startling to anyone who has been accustomed to the slow and short marches of carts in India.[1] The mules of Northern China are excellent, and are well looked after and fed by their owners.. On one occasion, when we were travelling in Mongolia, Carles went out to ask our carter some question, when he turned round reproachfully and said, 'Don't you see that I am now attending to my mules? That is a very serious matter and I cannot be interrupted.' It must not be supposed that this was impertinence. This carter was one of the very best Chinamen I ever had anything to do with,—always cheery, contented, and respectful.

But by far the most pleasant way of travelling in China is on horseback,—one pony will do the journey from Tien-Tsin to Peking in two days,—and I at once made up my mind that I would ride.

The ponies in Northern China, stout, hardy little animals, come from the Mongolian plateau. Much has been written and said of the excellence and endurance of these animals, but while not denying their many good qualities, I must admit that I was somewhat disappointed with them, and in no way do they come up to the wiry little creatures that are sometimes found in Persia. Of these last I remember one especially that was bought out of the stable of a post-station not far from Teheran; that pony carried a rather heavy servant for many days in succession, and for very long marches, the last three of which were forty miles, forty miles, and seventy miles. For the last ten miles of the last

---

[1] Richthofen states that the journey from Si-Ngan-Fu to Ili (Kuldja), 2,673 miles, is performed as a matter of course by two mulecarts, carrying three and a half tons, in eighty stages, though practically more than eighty days are required for the journey.

march we cantered nearly the whole distance, and when about a mile from our destination, recognising the place where he had been some months before, the pony took the bit into his teeth and fairly ran away with our man, who was unable to stop him until he arrived close to the old camping-ground. The Mongols until quite lately have never taken the least care in breeding their ponies, but since pony-racing has become so universal at all the treaty ports, and such very large sums of money have been given, especially at Shanghai, for good animals, they have begun to make a certain selection in their ponies for the stud, and the breed is already showing signs of improvement.

The Mongol ponies are generally very vicious, the result, in all probability, of ill-treatment when they are young; they will nearly always try to kick or bite anything or anybody that comes near them, and in this are a very remarkable contrast to the Tibetan ponies, which are the most perfectly docile creatures imaginable.

It was a long business, getting everything ready for the large party of the minister, admirals, and suites. All the luggage was in the Chih-Li, hard and fast on a mud bank, two miles down the river. Somebody had to find a steam-launch and go down after it; boats were to be hired, provisions bought, and all sorts of arrangements to be made; but nevertheless, some of my newly made friends found time to come and help me in my affairs. I had now to discover a servant and to buy ponies.

The word 'boy,' as applied to a servant, has been transplanted with curry and rice, punkahs, compounds, godowns, and tiffins into China, and the word 'servant' is scarcely ever used amongst Europeans at the Treaty Ports.

A 'boy' had been sent down from Peking for Mr. ——, who had just arrived from England to join the British Legation ; but at Chi-Fu Mr. —— had picked up a treasure, and now Chin-Tai, for such was the boy's name, found that there was no master for him. I learned long afterwards that he already had an English master at Peking, and that he had come down here on his own account, thinking that service in the Legation would pay better than any engagement beyond its walls. I did not know this at the time, and at once proposed that he should be my boy ; he was however very loth to give up the idea of joining the Legation, and at first would have nothing to say to me. At last I told him that if a vacancy occurred, I would at any time give him leave to step into it, and so, with a wistful glance at Mr. ——, he eventually smiled, and with a nod consented to become my property.

This matter being satisfactorily accomplished, I found that there were several besides myself who wanted to buy ponies, so we made up a party to visit the dealers' yards.

Carles had come up in the 'Vigilant' the night before, and had prevailed upon a pony dealer to send a large assortment of what he considered magnificent animals to the compound of a certain European doctor of sporting proclivities. Thither we wended our way, and on arrival found two most sorry-looking steeds. One especially excited our commiseration : a grey pony with a shoulder rather worse than straight, and a huge and inexplicable lump on his withers, whilst his hind quarters sloped away like an alpine hill-side. Standing in what was to him a natural position, he seemed to get his hind feet somewhere under the middle of his back, and a very hairy Roman nose completed the sum of his beauties. Horse-dealing is for some reason or another a mysterious process all over the world, and the exhibition of animals that combine in a remarkable degree every bad point seems to be the invariable prelude, in Eastern countries, at all events, to more serious business; so, regarding as a necessary part of the performance the examination of this extraordinary animal, which could have been kept for no other purpose than to serve as a foil for other ponies, we went to another yard, where there were seven or eight passable animals. The day was yet young, and there was a third dealer in Tien-Tsin; so promising to call again, we walked on to the last place. Here we found one really very good pony, and three or four shocking bad ones. The proprietor of the place said that the good one had just been sold for forty dollars, but he thought he could get it back for forty-five dollars if we could wait till to-morrow, and that he also had another much finer and more beautiful animal.

I was the only one of the party who was not leaving that evening; so I said I would look in the next day, and returning to the second dealer, two ponies were eventually bought for forty dollars each, after the amount of mysterious bargaining usual in all countries.

Horse-dealing was thus over for the day, and on coming back to the Consulate we found that everything was settled for the boat party, who departed almost immediately, and left the consulate in its ordinary quiet state.

There were two hotels in Tien-Tsin, one kept by a European, and the other by a Chinaman in the European style, where everything was fairly comfortable. I had been advised to choose the latter; so taking Chin-Tai with me, I walked out to see if the

'Chih-Li' had yet got off the mud bank, and if so to get my things up to the hotel. This happy consummation had not yet arrived, and so I took a stroll on Tien-Tsin bund,—for, as in India, the wharf is called the bund.[2] Tien-Tsin is a very lively place at this time of year. There are always some half-dozen steamers lying alongside the wharf, taking in or discharging cargo. Underneath the trees, which are planted in a row along it, sit numerous vendors of eatables, fruits, cakes, bits of meat, &c. &c. Of these the piemen seem to play the most important rôle; they have in their baskets all sorts of pastry, cakes, and sweet things, and in their hands a cylindrical wooden box, seven inches long, open at one end, in which there are some twenty or thirty sticks. All day long these fellows are here, shouting out 'Pies for sale! Who will buy delicious tarts? Come and buy! buy! buy!' and at every shout they rattle the sticks in the box, until up comes Simple Simon. The box is then shaken again, and he draws a stick: if he draws a lucky one, he gets a pie for nothing; if he is unfortunate in his choice, he has to pay his penny and go empty away.

The fruitsellers sell grapes, apples, pears, peaches, and melons cut up in slices. The grapes in the north of China are delicious, are bought for almost nothing, and are in season for nine months in the year. The Chinese have some method unknown to Europeans, of keeping grapes, by which they will retain their bloom for months after they have been gathered. It seems that they bury them in the ground; but whether they wish to keep the method a secret, or whether it is so simple that no one has taken the trouble to find it out, I cannot say; and notwithstanding constant inquiries that I made of Europeans and of my boy

[2] In India the wharf would be, not *bund*, but *bunder*. (Y.)

Chin-Tai, I never succeeded in satisfying myself about it.

The meat-sellers have a small portable stove, and sell little bits of cooked beef, mutton, sausage, pork, soups, and all sorts of food : delicious and savoury to a Chinaman, but revolting to a foreigner fresh from Europe.

Then there are the fish-sellers, who seem to do a thriving trade in fish, some the size of whitebait, others weighing ten or a dozen pounds.

Hundreds of coolies are aways bustling about with a stick, generally a split bamboo, six feet long, over their shoulder ; from each end of this is suspended by cords or chains, a bucket or basket, that comes down to within a couple of feet from the ground, and in which they carry their loads.

There are numbers of ponies to be hired on the wharf, and on these the British sailors gallop wildly up and down the streets in the English settlement. Furious riding is as strictly prohibited here as it is in Rotten Row, but the prohibition is not quite so severely enforced. A couple of tars, just in harbour after a long sea voyage, will step ashore, and hiring each a pony, without stopping to critically examine the animals or their saddlery, will jump up and go off at full gallop, the proprietor sometimes running behind. Jack has probably no socks, and only a pair of shoes, so that the stirrup-iron catches his bare instep ; but of this he takes little notice, nor of his trousers, which ruck up a long way above his knees. All goes well until the pony comes to a familiar corner, where, notwithstanding that Jack puts his helm hard a-port, the pony turns sharp round to the left, Jack falls overboard, the pony gives one kick of its heels and gallops off to its home. Not in the least disconcerted, Jack jumps up behind his mate, who, on seeing the accident, has brought up all standing, and away they go again until the second pony manages to relieve itself of its double burden.

I was told that before a winter at Tien-Tsin was over, the sailors who had been here all the time became wonderful riders, and would go gallantly across country, taking the ditches with wild delight; and one of the features in the Tien-Tsin races is a race for sailors. Wherever you find him, the Englishman of course is nothing without his club, and at most of the treaty ports of China a club of some sort has been established. Tien-Tsin is no exception; and here the merchants and consular officers usually meet of an evening to play a game of billiards, have a chat, or read the paper before dinner, and in connection with this club a story is told very creditable to the character of the Chinese.

The Chinese burn really excellent bricks, but at Tien-Tsin they appear to build their walls without any 'bond,' using for mortar nothing but mud with just a little patch of lime on one or two points of each brick; not because lime is expensive, but because, they say, more lime would spoil the mortar.

The Tien-Tsin club was built on this remarkable system, and it can hardly be a matter for surprise that one rainy day it completely collapsed. At the time there were no foreigners about. The headman in charge, a Chinaman, saw the first crack appear in the ceiling, and although fully comprehending the catastrophe that was about to follow, boldly led the way for the other servants, and with them removed as much as they could of the furniture, notwithstanding the pieces of plaster from the ceiling that were falling about them all the time. It was not until the walls began to crack that they finally retreated from the building, which in its collapse crushed four of these servants, killing one on the spot. It must be admitted that this is a very remarkable instance of courage and devotion to duty.

At last, at about 6.30 in the evening, the 'Chih-Li' succeeded in getting off the mud bank and reaching the wharf; so taking Chin-Tai on board, I pointed out my innumerable packages to him, and let him bring them to my rooms.

After a pleasant dinner with the acting consul, I returned to the hotel and prepared for bed. I remember reading in the 'Times' some bitter complaints from travellers in Switzerland of the noises made in Alpine hotels by British tourists starting early in the morning on a mountaineering expedition. I also have suffered somewhat from that cause; but of all awful disturbances I ever heard, the worst was made here by a nautical person in a room divided from mine by only a very thin partition. Just as I was getting to bed—it was about midnight—he began shouting for his boy in the tone of voice he would use to his maintopman in a gale of wind. The boy at first took no notice; but the sonorous tones of that seafaring man grew louder and louder, until it seemed as if the vibrations must bring down the house, and even the boy was unable any longer to pretend he did not hear it. He then gave an order to be called at four o'clock, and immediately afterwards began to snore almost as loudly as he had previously shouted.

I seemed scarcely to have closed my eyes, when a terrible clattering in the passage was followed by the invasion of my room by a being, who from the depth of his stomach evolved some fearful sounds, and made me painfully aware that the coolie whose business it was to awaken the sleepers had mistaken my room for that of my neighbour. On hearing a growl from me he fled precipitately, and immediately afterwards the skipper began making as much noise in getting under way, as he had in bringing himself to moorings; and as almost at the same time the people on the bund outside the windows were getting astir, there was no more sleep to be had; so jumping up, I commenced to rearrange my portmanteaus, which, before starting on a fresh journey, required a thorough overhauling.

The floor of the room was soon strewn with a medley of revolvers, Worcester sauce, prismatic compasses, books, clothes, Liebig's extract, musical boxes, pen-knives, carbolic acid, candles, compressed vegetables, lucifer matches, hats, and a collection of articles from which it was necessary to make a selection suited to the campaign immediately before me. Fortunately for myself, I had given the subject some consideration during the last few days, and when Chin-Tai appeared with my early tea, I was able to sit in a chair and direct the operations, making at the time careful lists of where everything was stowed: a method I strongly recommend to any one who is going to undertake a long journey; for I know of nothing more heart-breaking than the search through perhaps half a-dozen or more boxes for some small article that seems always to escape into the very last corner of the very last package that has to be examined.

This being finished, I went out to get some money. I found that the letter of credit I had provided myself with was more useful than circular notes would have been. It is not only in China that I have found this to be the case, and I mention it for the benefit of any who may be contemplating an expedition into out-of-the-way places. The money current here, as at

Shanghai, is the American dollar; it is somewhat surprising that the use of a coin of fixed value has as yet penetrated so short a distance beyond the treaty ports, more especially as bank-notes are an ancient institution in China. A very few miles from the main road between Peking and Tien-Tsin, the dollar is of no use whatever, and recourse must be had to the cumbersome method of weighing out lumps of silver. For small change, the brass cash are universal: these are round coins with a square hole in the middle; there are some Chinese characters on them, and they vary in value from about one-tenth to one-fifteenth of an English penny, according to the exchange.

The next thing I had to do was to discover, and secure if possible, my guns and cartridges. Before leaving England I had been led to believe that almost wherever I went in China I should find birds and beasts of every description only waiting to be shot at, and I had provided myself with cartridges and firearms in proportion. These had been despatched by an agent in London direct to Tien-Tsin, but where they were I had as yet no conception; so I made the tour of all the foreign 'Hongs,' as the Europeans call their business establishments in China, and eventually found that my artillery was in the Custom House, where it had caused much speculation.

At all the Treaty Ports the higher Custom House officials are foreigners (mostly Englishmen) in the pay of the Chinese Government, and thus, as a rule, a European traveller has no difficulty about clearing his goods. In this case, however, a number of cases, contents unknown, and consigned to nobody in particular, had suddenly arrived for an unknown person. They naturally drifted to the Custom House, where, as naturally, they were opened by inquisitive

Chinese, who suddenly discovered a very remarkable amount of gunpowder. This at once conjured up in the minds of the Chinese officials all sorts of fearful plots against the Imperial Government; an embargo was laid on the goods, and when at last I appeared to claim my property, I was introduced to a very polite French gentleman, who lectured me severely on the wickedness of which I had been guilty in sending out guns and cartridges without consigning them to some proper person; but who, at the same time, comforted me with the assurance that they would in all probability be handed over to me in the course of a few months.

September 19.—I was early awakened by the awful noises of the steamers in the Tien-Tsin river, and spent the day in making the final preparations for my first journey in China. Thanks to the acting English consul, I rescued my cartridges from the Customs without much difficulty, and then went to find out if the pony that I had already seen was to be purchased. As the owner refused to part with him, I had to look about again, and eventually I bought a strong, white, rather coarse, underbred-looking animal, thirteen hands high, with a tail reaching to the ground, a thick hogged mane, and a very long coat.

I had not as yet provided myself with a Ma-Fu (or horse-boy), so the pony, turning up at the hotel in the course of the afternoon, was casually tied to a clothes-line, until some one could be found to look after him. Chin-Tai was now called upon to produce that necessary article, and he persuaded one of the men who let out ponies on the bund to come with me, and to bring an animal from his own stud with him.

So at last all arrangements were complete, and after a final dinner at the Consulate, I turned into bed ready for my first experiment in Chinese travelling.

CHAPTER II.

'CHINA'S STUPENDOUS MOUND.'

Departure from Tien-Tsin—Rural Characteristics—Pictures by the Way—Chinese Hostelry—The 'Kang'—Long First Day's Ride—Early astir—Approach to Peking—A Fair going on—The British Legation—Visit to the Temple of Heaven—Money Arrangements and Currency Difficulties—Carts or Mules?—The Latter and their Packs—Bread in China—Chinese Lamps and Candles—Items of Travelling Stock—Visit to Prince Kung—Chinese Whim of picking Melon-seeds—The Journey commenced—The Difficulties of a Name—The Pei-Ho crossed—Picturesque Villages—Incidents by the Way—City of Chi-Chou—Tombs of the Manchu Emperors—The Great Wall first seen—Profusion of Fruit—The Day's Routine—Frank Curiosity of People—Roadside Pictures—Pass of Hsi-Feng K'ou—Village Inn at Po-Lo-Tai—Beancurd, what it is—Oak-leaf Silk—Millet—Pa-K'ou Ting—Exaggerated Ideas of Chinese Population—Foreigners as seen by Chinese Eyes—Journey to Jehol abandoned—Chinese Encroachment on the Mongols—Halt to Rest—Ta-Tzŭ-K'ou—Ideas of a Lion—Splendid Autumn Climate—Palisade Barrier of Maps non-existent—Stentorian Commandant—Smoking in China—Lovely Sunset Scene—The Inn at Sun-Cha—Impressions of the Great Wall—The Sea in Sight—Shan-Hai Kuan.

September 20.—After an early cup of tea we started at six o'clock. The Ma-Fu rode in front on a very good iron-grey pony, in shape and size something like my own. The Ma-Fu had nothing on his head but his plait; he wore a loose blue coat padded with cotton wool, and loose blue cotton trousers, and he rode on a Chinese-made English saddle. I rode next on a saddle that I had brought with me from England, with large flax-cloth saddle-bags and leather wallets. These saddle-bags proved excellent, and if my experience is worth anything, good flax-cloth saddle-bags will last quite as long as any traveller can need; they are much more convenient and far lighter than leather ones, which latter become very awkward in rainy weather, but the seams should be lined inside with a strip of leather half an inch wide.

At this season of the year in Northern China the sun has lost its power, and a helmet is not necessary. A white English felt hat, Norfolk jacket, breeches and gaiters, completed my costume.

My three baggage-carts came next, in one of which Chin-Tai reposed as comfortably as circumstances would permit.

It was a dull, grey morning as we started from the hotel, and marched through the Chinese city of Tien-Tsin. Here the roads are of clay without any paving, and about fifteen feet wide; the houses are also built of clay, and in the main street, through which we rode, nearly all of them were shops. These have no upper story, are always quite open in front, and there is an occasional peep through them into a back yard. There is generally hanging outside the shops a gaily painted sign-board, on which the nature of goods on sale within is written. A bit of matting is sometimes stretched half across the street, as an awning for a shop front, and the street is here and there entirely roofed in with matting supported on poles stretching from side to side.

It was a matter of some difficulty to force our way through the crowds of people. Coolies were carrying fish in buckets or baskets, others with baskets of fruit, or huge bundles of long reed grass, millet, or Indian corn stalks; everybody was shouting, pushing, and in a hurry, and carts lumbering along often blocked the way entirely, but the people seemed rather to like this, as it gave them an opportunity of stopping for a gossip. The most unpleasant people to meet were the coolies carrying buckets of liquid manure, nor did they assist to sweeten the air, which in Tien-Tsin, as generally in Chinese towns, reeks with abominable smells of every description. In all the bustle and hubbub pariah dogs ran about doing scavengers' work, assisted by the pigs, which persistently placed themselves under the legs of the ponies or mules.

At length we were clear of the town, and breathed the fresh country air. The Ma-Fu, who knew nearly twenty words of English, took me under his care, and leaving the carts to find their slow way behind us, we rode on ahead.

The country here is quite flat, without an elevation of the smallest description, except the houses and river embankment. Behind the latter, masts and sails of hundreds of junks can be seen. Every inch of the ground is cultivated with millet or Indian corn, and in the fields there is often an undercrop of sweet potato or a small bean.

There are often cotton and castor oil plants bordering the edges of the fields, but the great feature is always the millet, standing about eight feet high, with reddish brown or yellow stalks.

In the immediate neighbourhood of Tien-Tsin there are not many trees, but a little further into the country the villages have more trees about them, almost entirely willows and Chinese date trees. These latter (in reality the *Rhamnus Theezans*, a kind of buckthorn or jujube, in no way whatever allied to the date palm) bear a fruit in appearance and taste very like a small date ; the tree itself is more like an olive than anything else, and is very common in Northern Persia about the neighbourhood of Sharood.

A few miles on, the road skirts large plantations

of willows, and the landscape is very like the scenes in some of the pictures of Karl du Jardin. In the Dresden Gallery there is rather a stiff picture by this artist of a grove of trees, with a herd of swine underneath. Now, not far from Tien-Tsin, this landscape is reproduced almost exactly; there is the identical row of willow-trees in a perfectly straight line, and all of precisely the same height; and as I passed, the very same herd of swine was feeding underneath: the only thing wanting to make it complete was the gay cavalier out hunting.

Round the villages there are always gardens with little square patches of lettuces, cabbages, turnips, and yams trained on sticks like convolvuli, all models of neatness and regularity.

Another great feature is the threshing-ground, where at this season men and women are busy threshing out the corn on a flat floor of puddled mud, in appearance and size very like an open skating rink. The brown mud colour of the houses against a background of willow trees, the rich brown stalks of the Indian corn, merging into a madder red, and mingled with the green and yellow undercrop of beans, and the sober blue of the people's clothes, as they sit round the floor, combine to give a charm even to this generally uninteresting plain.

The women often pretend to be afraid of a foreigner, and run away when they see one coming; and notwithstanding their deformed feet, they seem to waddle about very comfortably, though their gait is remarkably awkward.

After a ride of about twenty miles I arrived with the Ma-Fu at Yang-Tsun, the first halting place, and here for the first time I made acquaintance with the luxuries of a Chinese inn.

CH. II. THE HOSTELRY ; THE KANG.

Riding through an archway, with a room on each side used as a sort of restaurant, there is an open court-yard ; on one side of it there is what in England would be called a long, low hut, divided into several rooms : these are the sleeping apartments of the guests at the hotel ; on the other side a large open shed is the stable or feeding-place for the horses and mules.

At the farther end of the yard is a grand room, with a smaller one leading from it on each side : this is only awarded to guests of distinction, or in other words to those who can afford to pay.

Knowing nothing of the arrangements, I went, where I was shown, into one of the little rooms at the side, about ten or eleven feet square, and the same in height, the floor of brick and the walls of mud. Dirty paper, with many holes in it, pasted over the rafters formed the ceiling, and some wooden lattice-work, covered with dirty paper, full of holes, did duty for a window.

The great feature in every room in every inn in Northern China is the *kang*. This is a hollow raised dais, about eighteen inches high, covering half the floor, over which there is usually laid a bit of thin straw matting, the home of innumerable fleas ; in the winter a fire is lighted under this, and through the bricks or mud of which it is built a pleasing warmth is imparted to the traveller, who, rolled up in his blanket, lies on it to sleep.

During the daytime a little table about nine inches high stands on the kang ; a person sitting on the latter can just make use of this by twisting himself round into an impossible attitude, which after any length of time eventuates in aches all over the back. There may be in addition a broken-down and exceedingly filthy table and arm chair, about the height of ordinary European articles. The chair very clumsy, heavy, stiff, straight-backed, and uncomfortable, with legs which, thrust out in a sprawling fashion, seem to have the most unhappy knack of being always in the way; and the table with a ledge underneath just where an ordinary person wants to put his knees, and a bar below to interfere with the free movements of his feet. Such is the accommodation and such the furniture a traveller invariably meets with in the inns of China. In the course of an hour my carts appeared; Chin-Tai was sorely indignant with the innkeeper for not having put me into the place of honour, and his contempt for a Ma-Fu who could care so little for his master's dignity was delightful to witness.

After six weeks on board ship a ride of forty miles appeared rather long, and as the evening drew in I began to make inquiries about the distance. The Ma-Fu, holding up six fingers, said it was now only six li, or two miles,—but that mile and a bittock! no matter whether it is a li as in China, a cos as in India, or the abominable farsakh of Persia, the weary traveller always finds the 'bittock' much longer than the mile. On this occasion the distance lengthened out, and the Ma-Fu, in answer to my numerous inquiries, sometimes said there were only four li more, sometimes five, once the number was reduced to one, but immediately rose to three. The fact was he did not know the road, and in the dark was wandering about in a state of hopeless confusion. Presently a light appeared in front which turned out to be a lantern hanging from a cart, whose destination was the same as our own; as the light thrown on the road made the work easier for the ponies we kept close behind, and at last at nine o'clock rode into the yard of the inn at Ho-Se-Wu. With my saddle-bags for a pillow I was soon sound asleep, and did not wake till Chin-Tai appeared with the carts, and said that it was time for dinner. Chin-Tai early discovered a weakness for cookery that subsequently proved very troublesome ; he never could be brought to understand that something to eat as soon as possible after arrival was better than an elaborate meal in the middle of the night. Once produced, however, my dinner was soon dispatched, the mattress was laid on the kang, and at about midnight I was fairly in bed.

September 21.—The carts were hired only for the journey to Peking, and it was therefore the interest of the driver to get there as soon as possible. The gates of the city are always closed at sundown, and as no power on earth can then get them open till the next morning, there was no fear of the carters starting late. The people of Northern China are all, however, very early, and when after a cup of tea a start was effected at 3.45 a.m. the town was all astir, many of the shops were open, and the furnace of a blacksmith cast a bright glare across the street as the sound of his hammer resounded in the clear morning air.

As long as it was dark it was advisable to ride in the light of the paper lantern dangling behind one of the carts ; but when the dawn appeared the Ma-Fu took me on at a huntsman's jog to the halting place at Chang-Chia-Wan, where, as I sat at my breakfast, a cockroach came out of his residence in a crack in the filthy table to share my repast.

Leaving the carts to follow, we started as soon as the ponies were fed. Riding still over the flat plains the distant blue mountains presently came in sight, and soon afterwards the unmistakable walls of Peking,[3] with the great high three-storied building over the gate.

The road from Tien-Tsin to Peking runs over a sandy, clayey soil; there is no attempt at a made road, but in dry weather it is very easy for mules and horses, and the latter may be galloped the whole distance (eighty miles); but after rain the track becomes very heavy, the mud is deep, and the work, even for horses with only a light load on their back, is very severe, while for carts it becomes a continual struggle.

To-day there was some sort of fair going on in Peking, and the scene was very remarkable,—quite unlike anything to be seen elsewhere. The street was very wide, and on each side were the same wretched houses that so soon become familiar to the traveller in China. Between them the space was closely covered by the wares that the sellers of goods had spread out on the ground: old clothes, old rags, brushes, baskets, string, rope, eatables, drinks, fruit, crockery, and almost every conceivable article of household equipment, were exhibited for sale; each seller was surrounded by a mob of buyers, their friends, and lookers on. The streets were absolutely thronged with people walking, riding, or in carts; the hubbub and confusion were appalling, and progress at times seemed almost impossible. Pigs and dogs took their usual share in the proceedings, and evil smells were not absent. The inhabitants of Peking, and of all the towns and villages along the road from Tien-Tsin, have seen so many foreigners that a European causes little remark; here they were

[3] In Chinese, Pei-Ching, i.e., the northern capital. So also Nan-Ching (commonly called Nanking), the southern capital.

mostly too busy with their buying and selling to pay much attention to anything else, and with the exception of a few people who must have come in from the country, and who could not help laughing at the comical sight, no one took much heed of the Englishman moving slowly in the motley crowd. We threaded our intricate way through the mazes of this fair for very nearly a mile, when turning out of it into a bye street, a smart canter brought us at 4.45 p.m. to the gate of the British Legation.

The British Legation in Peking stands in grounds sufficiently extensive to contain the Minister's private residence and state reception rooms, chancery, houses for three secretaries, a doctor, and an accountant, quarters for ten students, a church, fives-court, bowling alley, reading-room, and billiard-room.

Two large stone lions guard the entrance to the Minister's house, and passing between these the first building is reached. This is nothing more than an empty antechamber with a garden beyond, where there are a few trees; at the other side of this there is a second antechamber, with a suite of two or three rooms on each side; and, finally, traversing another garden, the door of the Minister's residence is gained.

This was built by a former emperor for his son. There is no upper story, but the rooms are lofty, and beautifully decorated in the Chinese style. This is very different to anything European, and the harmony with which, in the deep dark shadows, a brilliant lapis-lazuli blue will mingle with an emerald green is at first rather startling to an eye educated in the principles of modern high art.

September 22.—During my stay a large party made an excursion to the Temple of Heaven, one of the sights of Peking. After riding through the filthy streets, in which the smells and the dust impressed one most, we reached the Temple. The grounds are square, and enclosed by walls about half a mile long, where the fresh mown grass is shaded by long straight rows of yews and laburnums. It is one of those places almost impossible to describe, and leaves upon the mind confused ideas of grandeur and utter ruin,—recollections of wonderful blue encaustic tiles, and marble stairs, with rank weeds growing between the slabs,—visions of elegant bridges and rich but broken carvings,—vivid impressions of a general covering of dirt and filth, and the surprise of a patch of kitchen garden in an unexpected corner.

The Emperor comes here at certain times to pray, and on these occasions, after a bullock has been made a burnt-offering, he should pass the night sitting upright in a stiff and straight-backed chair; but the attendants naively exhibited the luxurious bed for which his Imperial Majesty vacates the uncomfortable arm-chair, and they had no hesitation in admitting that economy was now strictly carried out, that the flesh of the animal was sold, and nothing burnt but the skin and bones. Familiarity with celestial affairs seems to have bred contempt in the minds of the servants about the place, for they were liberal in their offers of bricks, tiles, or bits of glass, of which tourists are generally so fond. I did not load myself very heavily, and trusted to my memory rather than my pockets to carry away souvenirs of the Temple of Heaven.

Money arrangements had now to be made for the journey we were about to undertake, for dollars do not pass current far from the walls of Peking, or the great high road to Tien-Tsin. In the city of Peking itself the private banks issue notes, but these are worthless half a dozen miles from the capital. Over

CURRENCY AND ITS TROUBLES.

nearly the whole of China payment is made by means of a lump of silver weighed in a balance.

The silver is cast into ingots of various sizes, and of two shapes; the largest are something like a shoe in form, and weigh about thirty or forty ounces, the smaller ingots are cast into pieces almost hemispherical, and weigh from one to ten ounces. The silver is of various degrees of purity, but a Chinese banker or merchant, accustomed to transactions in bullion, knows almost instinctively the quality of the metal, and rarely makes a mistake.

Provided with these ingots, the traveller finds his troubles now begin. To make small payments, pieces of silver of a less size are necessary. The ingot is therefore carefully weighed, and sent out to be chopped up by anyone who will undertake the task. The village blacksmith is the usual operator, and when he returns the silver it has to be weighed again, for the owner to satisfy himself that the full amount has been returned. This is, however, but the first of many vexations. Every time a purchase is made, when the price of an article has been finally agreed upon after the amount of bargaining always necessary to complete a transaction, the vendor will generally manage to find some fault with the quality of the silver, and will want an extra payment in consequence.

In travelling about from one city to another there is a further difficulty to be overcome, for every place has its own scale, and what is an ounce in one town will perhaps be less than an ounce in the next, so that the weary traveller, after having, as he thought, finally concluded the tiresome transaction, is quietly told that his scale is not a good one and the silver must all be weighed afresh in a balance of the place. For weighing silver a Roman steelyard, with a bar of ivory neatly marked, is usually employed; but bankers or others who have extensive transactions generally use a large pair of scales.

It was some time before I thoroughly understood the mysteries of taels and balances; I was fortunate in having Carles to initiate me, and by his advice Chin-Tai was sent off with some 500 dollars to buy lumps of silver, as well as to look after mules, and get the many odds and ends necessary for a two months' journey. In the meantime another pony was brought for sale, a strong, rough, meek-looking bay with black points, about thirteen hands high.

The owner asked seventy dollars, and thirty-five was promptly offered; he refused that sum and went away, but presently returned with his demand reduced to forty dollars, and ultimately the pony was bought for thirty-eight.

There was a knotty question to be settled whether we should carry our baggage on baggage-mules or in carts. Chin-Tai was very strongly in favour of carts because they were so much less trouble, but Carles was of opinion (an opinion fully justified by subsequent experience) that carts would be unable to move over the roads by which we intended to travel; ultimately, much to the sorrow of Chin-Tai, we decided for mules. The usual way of hiring mules is definitely from place to place, and as we wanted to depart from the fixed custom and intended to travel about, first in one direction then in another, according to our fancy, our difficulties were considerably increased. There were plenty of muleteers who would have been willing enough to engage themselves to go to any definite point and return, but to go wandering over the country, no one knew whither, was an idea the novelty of which was so startling that very few muleteers would venture to hazard themselves in the uncertain undertaking; at last, however, one was found who consented to let out his mules for $\frac{6}{10}$ of a tael daily (that is, about 3s. 5d.). That sum covered all expenses connected with the animals, and provided for a sufficient number of muleteers. The only objection to this arrangement would be that the muleteers might refuse to make the full day's march; this, however, as a fact, they seldom did, indeed not more than muleteers or coolies would have done if hired by distance instead of by day.

The packsaddles of the mules in the north of China are very well adapted for baggage that is to be carried many days in succession, but they are not convenient for travellers who continually want to open their boxes. The saddle is composed of two parts, the saddle or pad, and a framework to which the load is tightly lashed. When all the luggage has been made fast, and everything is ready for a start, then the framework is lifted up by two men, one on each side. The mules, accustomed to the operation, stoop their heads and walk underneath almost of themselves, and the framework is dropped down on to the pad, no other lashing or fastening being required.

The advantages of this system are, first, that if the goods are not unpacked the operation of loading and unloading the mules is a very short one, and secondly, that the burdens need never be put on any of the animals until the moment of starting. The disadvantages are that if the things are required every night the whole business of lashing and unlashing has to be gone through, which is a much longer process than unfastening a package from a mule whose packsaddle is of the ordinary description, all in one piece; and worse than all, these packsaddles very frequently give the mules sore flanks, for the framework not being fastened to the pad, and being kept in its place chiefly by its shape and balance, is never really steady, and so, in rough countries especially, galls the animal. It is true that with the other system, as the loads are lashed to the saddles themselves, and as it is of course impossible for all the animals to be ready at the same moment, some mules must always be standing with their burdens on their backs some time before the start. Still under all circumstances, and taking everything into consideration, the system, pursued in the north, of the saddle and framework, does not seem to be the best.

We were not going away for more than six weeks, and so our luggage was limited in amount. In the north of China there is always the kang on which to lay a mattress; a bedstead is therefore never required; and as the weather at this season is very temperate, and no rain is to be expected, a quantity of clothes is unnecessary. Chin-Tai at this time was unequal to the task of making bread, and we took a good supply of biscuits as a substitute,—for bread to suit the European taste is not to be found in China.

The Chinese eat very little bread with their food, although in towns and villages there are always a great number of shops and stalls devoted to the sale of various descriptions of bread and cakes; but it is a mistake to say, as has been stated, that the Cl inese do not make bread of wheaten flour.

The round dumplings, the sight of which is so familiar to anyone who has been further into China than the European Concession at Shanghai, are made of wheaten flour, and are leavened, but, instead of baking, the Chinese steam them. They are very heavy, and somewhat indigestible, but, when cut into slices and toasted, are a very fair substitute for European bread.

Light is one of the chief difficulties in China, as the rooms are always carefully arranged to exclude every glimmering of sun. The Chinese themselves seem able to work without any light, the miserable glimmer of a bit of wick hanging over the edge of a bowl of oil being hardly worthy of the name. A dirty bowl is the form of lamp in general use, which is as disgusting as it is inefficient. No one who has not seen a Chinese lamp of this kind can form an idea of the unutterable state of filth in which these lamps invariably are. I never met any person who had ever seen a new one, and these articles of household equipment are apparently handed down as heirlooms from generation to generation, no one venturing to remove the dirt consecrated by antiquity and sacred from ancestral associations, or it may be that the most discerning person would fail to recognise in a new and clean lamp any representative of the extraordinary accumulation of filth to which he had been accustomed.

The Chinese have also what are called wax candles, in which there is a certain amount of wax and a good deal of fat. These candles are moulded with sticks at the bottom, by which they are held or stuck into any convenient crack in the table, but excepting for use in lanterns these are luxuries only enjoyed by the rich. The wicks are very thick, and of course not plaited; the operation of snuffing is always performed by flipping off the wick with the fingers on to the ground.

To a traveller in China, candles are almost as much a necessity as food or drink, for if writing is put off for twenty-four hours, one quarter of the information obtained during the day is forgotten; in a week half of it has escaped the memory, and after any longer time very little is remembered. To do much writing in a dark and filthy inn by the glow-worm glimmer of a Chinese light is almost hopeless; for after a hard day's work it is always more or less an effort to set to work in the evening, and, when darkness is piled upon discomfort, it requires almost more than human determination to resist the temptation of leaving the writing and turning into bed.

We therefore carried with us an allowance of two good English candles a night, a luxury that can only be appreciated by those who have attempted to do much writing in the dark.

Besides these, we were obliged to take with us tea, salt, sugar, and many other small articles.

Although tea is held to be the universal drink of the Chinese, it is often impossible to procure it in Northern China, and, in the few out-of-the-way places where it is to be bought, it is always very bad. At this long distance from the tea-growing districts, it is a great deal too expensive for any but rich people, and the poorer classes in the north either make a decoction of the leaves of some tree, or drink plain hot water, for, in all the length and breadth of this vast empire, the Chinese universally hold cold water in abhorrence for either external or internal use.

The salt made by the Chinese varies considerably, but in small places, to which the worst articles always seem to gravitate, the salt to be bought in the shops is generally so full of dirt as to be really uneatable.

I had bought in Marseilles a quantity of compressed vegetables in packets, which, with Liebig's extract of meat, make really delicious soup. The space occupied by these is very small, and if enough vegetables for the day's consumption are in the morning put into a pickle-bottle full of water, and corked up tightly, soup can be prepared in the evening as soon as boiling water can be procured. During all my wanderings I almost invariably had this soup, which is wholesome and nourishing. Attention to small details of this kind makes the greatest difference in the comfort of a traveller, and it was not until after much consultation and deliberation that it was finally agreed that all necessaries had been provided.

September 25.—The pleasant ring of the mule-bells sounded in the morning, and eight mules arrived to be loaded with our goods. Each load was covered with a square of oiled cotton waterproofing, and our two servants arranged a quantity of wadded quilts on two half-laden mules, thus forming very comfortable seats for themselves. At eleven o'clock they started to await us at Tung-Chou, while we joined the suite of the Minister who went to pay a visit to the Prince of Kung. The Prince received his guests in what in Europe would be called a very poor room, in which there were two round tables and a number of cushioned and uncushioned chairs. Amongst the Chinese the left hand takes precedence of the right, and on the left of the Prince of Kung our Minister of course took his place. Tea was first served round, in small cups, without milk or sugar, then plates and dishes of fresh and preserved fruits, innumerable little cakes, apricot kernels, and water-melon seeds. The taste of the Chinese for water-melon seeds is one of the most extraordinary imaginable. Huc, though not always a safe leader, has made some remarks on this predilection, in his usual humorous style, worth quoting :

'The water-melon is in China a fruit of great importance, above all on account of its seeds, for which the Chinese are possessed of a veritable passion, or rather of an insatiable appetite. . . . .

'In certain localities, when the harvest of water-melons is abundant, the fruit is without worth, and the proprietor sets no value on it, except for the seeds. Sometimes whole cargoes of them are taken to the most frequented highways, and are given away gratuitously to travellers, on the condition that they will carefully collect the grains for the proprietor. . . . .

'These water-melon seeds are, in fact, a veritable treasure, to amuse at a cheap rate the three hundred millions of inhabitants of the Celestial Empire. In the eighteen provinces these deplorable futilities daily furnish delicacies to the whole population.

'There can be nothing more amusing than to see these astonishing Chinese before their meals dallying with these seeds, trying as it were the good disposition of their stomachs, and gently sharpening their appetites. Their long and pointed nails are, under these circumstances, invaluable. The address and celerity with which they crack the hard and tough shell of the seed in order to extract therefrom one atom of the kernel, or perhaps none at all, ought to be seen;—a troop of squirrels or apes could not be more skilful.

'It has always been our opinion that the natural propensity of the Chinese for all that is factious and deceitful inspires them with this frantic taste for water-melon seeds, for if in the whole world there exists a deceitful food or a fantastic diet, it is incontestably a water-melon seed.

'Moreover, the Chinese give you them in all places and at all times. If a few friends meet together to drink a cup of tea or rice-wine, the necessary accompaniment is invariably a plate of water-melon seeds. They crack them during journeys, and as they walk about the streets, engaged in business matters. If a child or a workman has a few cash to dispose of, it is on this kind of gluttony that they spend it. They are to be bought everywhere,—in the towns, in the villages, and on every highroad and byway.

'Should one arrive in the most deserted country, absolutely wanting in provisions of any description, one may rest assured of not being reduced so low as to lack water-melon seeds.

'Throughout the whole empire there is an inconceivable consumption of these seeds, and one that exceeds the wildest flight of imagination. On the rivers huge junks may be seen loaded entirely with this precious commodity.

'In truth, one can hardly help fancying oneself amidst a nation of rodents. It would be a curious inquiry, and one well worthy of our great statisticians, to examine how many water-melon seeds must be consumed daily, monthly, or yearly, in a country whose population amounts to three hundred millions.'[4]

Thus Huc; and every step I took in this singular country, and every wayside restaurant at which I stopped, convinced me not only that in this case Huc is guilty of no exaggeration, but that it is utterly impossible to convey to the most imaginative of Europeans the extent of this preposterous fancy.

On this occasion our hosts were not beneath their reputation, and soon the floor was strewn with hundreds of the shells. There were a number of high Chinese officials present, with finger nails of the most astonishing length, some of them at least an inch beyond the

[4] Huc, *L'Empire Chinois*, ii. 47 (ed. 1862).

tip of the finger. They asked all sorts of curious questions about the guests : who they were, what they did when they were at home, whether they were rich or poor, what were their honourable ages, and made many other inquiries that appear inquisitive to Europeans, but are amongst Chinese little more than the polite formulæ of familiar intercourse. These ceremonies occupied so much time that it was five o'clock in the evening before Carles and I were able finally to get away. Putting our horses into a smart canter Peking was soon left behind, and I was fairly embarked in Chinese travel.

The shades of the autumn evening were already beginning to fall as our ponies' hoofs clattered over the great imperial highway from Peking to Shan-Hai-Kuan. This is one of the paved roads of China,—a relic of departed grandeur,—like all else, in a sad state of ruin, with its stones displaced, and full of huge and dangerous holes. The fourteen miles to the gate of Tung-Chou were accomplished in a couple of hours, and though it was long after sunset, Carles had not much difficulty in persuading the warders to unlock and withdraw the heavy bolts. We entered a dark and gloomy street, bounded on each side by high walls, with the narrow garden doors closely shut and fastened. For about a mile we followed the deserted thoroughfare, and then suddenly found ourselves in the busy business quarter. Here the shops were dimly lighted with a bit of flaring candle, or a miserable oil lamp, that cast an uncertain glare on the faces of the dealers and purchasers, and formed many a study that would have delighted Dow or Schalken, or others of those old Dutch painters. Night and day seem all one to these busy Chinese people of the north, and as we rode out of the north gate of the city we left them still chattering over their bargains, without any thoughts of closing for the night.

After leaving the city we rode through the suburb known as the Barrier Inn to the hotel of 'Virtue and Prosperity,' where we found a very good room with a small chamber at each side; in both of these there was of course the usual kang, on which our beds were already made. The large room was adorned with sundry paintings, in some of which that omnipresent bird, the magpie, was conspicuous.

In this part of China the magpie is a very important feature in a landscape. Fat and tame, these birds walk about the roads, jump into trees, or fly over the houses in parties of eight or ten. Scarcely deigning to get out of the way of the horses' feet, they hop off to one side of the road, and stand looking on with the same impudent air that they have all over the Eastern world.

Another picture represented a willow-tree, with a bridge; and besides these there were scrolls, with inscrutable passages from obscure but revered authors of antiquity.

A cold fresh breeze had set in from the north as we finished our ride, and we were both quite ready to do justice to the excellent dinner that was ready for us.

On counting up our party before going to bed, we found that we numbered thirteen people, viz. Carles and myself, our two servants, one Ma-Fu, and eight muleteers. There were besides four ponies, eight mules, and last, but not least, Carles' liver-and-white spaniel 'Spot.'

September 26.—We were obliged to wait here for our passports, as they were not ready when we left Peking. Before getting a passport in China it is necessary to be provided with a name,—not a very simple matter there.

The Chinese having no alphabet it is necessary to choose for each syllable some Chinese word that sounds as much like it as possible. As a rule the attempt only ends in failure, and old residents in China usually have a Chinese name of one word, that may or may not approximate in sound to the first syllable of their European name. Whilst waiting, news was brought that the Emperor's brother, who had just returned from a visit to the imperial tombs, wanted to come to the rooms we were occupying; but when he learned that they were in the occupation of foreigners he turned away, so that his eye should not light on any of the hated race. For this man has the reputation of detesting foreigners to such an extent that he would exterminate them if he could; and report has it that the present boy was made emperor, in part for the sake of rendering his ferocious relative innocuous, the near connections of an emperor being unable to hold office.

The hotel-keeper was very thankful to have been spared the visit of the exalted personage, who would have taken much and given little; and when at last the courier from Peking galloped in with our papers, when the mules were loaded and we took our leave, our host was none the less cordial in his adieux, that we had been the means of saving him from the infliction.

Outside the inn a long string of camels laden with tea was on its way to Mongolia. They went off towards the north, and as we rode to the west, the deep sound of the bell these animals wear round their necks was soon lost in the distance. We soon arrived at the Pei-Ho, and as there was no bridge we passed it in a ferry, a somewhat lengthy operation, as it was necessary to unload all the mules.

This morning I rode the innocent-looking bay pony, who promptly kicked me directly I approached him; and as he shied at everything by the roadside, and ran away when I took a map out of my pocket, I began to fear that his mild and inoffensive appearance might after all be deceptive.

The road was rather heavy in places, and showed signs of recent floods as we travelled over the same flat country as before. The large millet, another smaller millet (the only crop it is said that the locusts spare), and Indian corn were the chief crops, until we arrived at the 'Harmonious and Benevolent Inn' at Yen-Ch'iao.

September 27.—We very soon found out that although the people, when about their own business, are always up and away at all sorts of hours in the morning, they are by no means so anxious for an early start when they are engaged for others, and notwithstanding that we were up at four o'clock, we were unable to get our caravan under way before a quarter to six.

The long and straggling villages in this part of the country are very picturesque; the footpath on either side of the wide main street is well raised, for the roadway evidently becomes a roaring torrent after much rain; the paths are lined with a row of trees, and the cottages behind are built of mud, with cucumbers or pumpkins trailing over them. There are not many people about, either in the fields or in the villages, indeed some of the latter seem almost deserted, until an opening discloses the threshing-floor where nearly all the inhabitants are collected, either threshing the corn or assisting as spectators.

A fresh north-westerly wind came down from the

mountains, and after the long period of tropical heat it felt quite cold in the early morning.

We crossed the Tan-Ho at a ferry, where, although some of the mules objected strongly to entering the boat, we were delayed only twenty minutes. After this 'Spot' disappeared, Carles galloped back after him, and on passing his boy, mounted on the red pony, that amiable animal could not resist a good kick, by which he broke the reins, deposited the boy on the ground, and gained his freedom. By-and-by we met a military official, who, to conceal his curiosity, pretended that it was his duty to stop Chin-Tai and make inquiries about us. After halting for breakfast at the 'Hotel of the Law-makers' in the village of Tsao-Lin, we spent the night in the small, but fairly clean, 'Inn of the Everlasting Harmonies' at Pang-Chün.

In the little country inns large tables and chairs were very rare, and there was usually nothing but the little low table on the kang. Some of these, made of very beautiful wood highly polished, and black with age, were, although very dirty, really handsome pieces of furniture; but sitting to write at them in an unnatural twisted position is not very comfortable after a long day's ride, and there was generally enough work to occupy several hours of an evening.

As we marched through the village before the sun was up, our fingers tingled in the keen air, and seeing a man selling hot sweet potatoes, our party nearly cleared out his stock. After this for some time the hot potato became a regular event in the morning's ride.

September 28.—'Spot' was a source of keen controversy and much speculation amongst the villagers we passed: none of them had ever seen other dogs than the villanous pariahs that frequent the streets, and do the scavenger's work, the same snarling curs that may be seen in every village in Turkey, Egypt, India, and Persia,—so at the sudden appearance of 'Spot' the wise people of the town would lay their heads together, some saying he was a sheep, others a sort of cow; none ever suspected that he was a dog, and in the universal interest we attracted but a very minor share of notice.

The plain was thickly populated, and the villages and towns were scarcely more than a mile apart; as we proceeded the road gradually approached a fine range of mountains; now stone entered into the composition of the buildings; large boulders, lying about the main streets of the villages, were sure indications of sudden floods in the rainy season; and soon we were near enough to see the Buddhist temples perched on the tops of almost inaccessible crags.

The city of Chi-Chou lay on our road to-day; its walls are massive, in good repair, and flanked with strong towers. The gate is protected by a double barbican, so that an invader, before he can enter, must pass three strong doors. It was market-day as we marched through, and all the sellers were sitting down at the sides of the street with great sacks of grain, chiefly wheat and millet; there was scarcely any rice.

During these marches the red pony was a never failing source of amusement or excitement. Chin-Tai, thinking the back of a pony more comfortable, or more dignified, than a seat on a mule, turned the Ma-Fu off his grey, and took it for himself. The Ma-Fu, a long-suffering person, walked for some time, but getting tired, he rashly attempted to ride the spare animal without a saddle. This was a liberty the red pony would not permit for a moment, and he disposed of the Ma-Fu no less than three times early in the day. At last, as the pony was more difficult to catch each time that he got loose, and the delays were becoming serious, authority was obliged to interfere: Chin-Tai was replaced on the mule, and the Ma-Fu restored to the saddle on his iron grey.

At Shih-Men the inn was small, but the landlord very civil and attentive; he took an immense fancy to my trousers, and wanted to buy them on the spot. Then all the people of the village came to have a look at us and our things. Very few Chinamen can resist the temptation to feel the texture of European cloth, and as the Chinese are quite ignorant of the method of curing leather, the smoothness of our belts and instrument cases used to excite universal admiration. Fortunately the villages were not large, so that nearly all the inhabitants could come in turn, and feel our clothes and belts.

September 29.—As we were starting this morning, our head muleteer dilated with enthusiasm on the magnificence and comfort of the inns at a place called Ma-Lan-Yu, and to gratify him we consented to make it our halting-place for the night.

Our road still lay over an alluvial plain, and passing through a gap in the hills we entered a basin bounded on the south by a long ridge called the Dragon Hill, and on the north by a fine range whose peaks were about 1,000 feet above us. Away to the back, amongst these mountains, are the imperial tombs, which, according to an informant here, cover a tract of country extending over seventeen mountains. The sacred ground is not enclosed by a wall, but being covered with forests abounding with game and wild beasts, and being entirely devoid of roads, the sanctity of the place is never invaded. Very little information could be obtained about the country or the position of the tombs. The emperors of the present dynasty have all been buried near the village of Ma-Lan-Yu, but some of the older graves are many miles distant amongst the mountain fastnesses. When an emperor is crowned, one of his first duties is to come here and select the site for his grave. At the time of our visit the tomb of the last emperor was not yet finished, and many hundreds of workmen living in wretched tumble-down shanties outside the walls were employed on it. The village, partaking of the miserable aspect of the workmen's dwellings, was very dirty, and though large, contained only one filthy inn, which we found on our arrival already crowded with visitors. We ordered the mules to march to the next village, and then rode off to see the Great Wall, which was a short distance to the north. The road to it was wide and level, and ran up a pretty valley through fine pasture-land; there was a row of tall willows and acacias on each side, and for the first time sheep and oxen were feeding in the meadow. The Chinese eat very little beef or mutton; they do not think it grateful to kill the useful animal that draws the plough; they consider mutton very poor food, and the butchers' shops are always kept by Mongols. In these, however, both beef and mutton can be bought for 3d. or 4d. a lb., while pork, which is considered by the Chinese as the greatest delicacy, sells for double the price.[5]

At Ma-Lan-Chen the great wall of China comes down from the hill-sides and runs across the valley, and here for the first time I saw this extraordinary work.

[5] Marco Polo, vol. ii., book ii., p. 204: 'I should tell you that in all the country of the Manzi they have no sheep.'

At this point a stream runs through the wall, which has here been broken down, leaving a gap fifty yards wide which existed before the memory of the oldest inhabitant of this place. An old gentleman who lived here invited us to take tea, and led us into a little room about twelve feet square. All the inhabitants of Ma-Lan-Chen tried to follow, but as there was not space for more than about fifty, and as the population must have been four times that number, three-fourths were sadly disappointed. Carles endeavoured to extract some information from the inquisitive but well-disposed crowd, but as each of his queries elicited nothing but questions in return, we soon said good-bye, and rode away, followed some distance beyond the houses by many of the villagers, a large proportion of whom had very severe goître.

Returning through Ma-Lan-Yu we continued in a level plain well cultivated, closely populated, and bounded on the north by a high range of mountains. The road was very bad, with ruts nearly a foot deep, and in places under water.

We halted for the night at Tsun-Hua-Chou, a city surrounded by good walls and a wet ditch.

September 30.—As we marched through the town in the morning, there were great quantities of very beautiful fruit and vegetables for sale : rosy apples, and others of a russet-brown ; pears, green, yellow and red ; peaches of all sizes, from little things no larger than an apricot, to others finer than the finest grown in England ; walnuts and chestnuts were in abundance ; and above all great bunches of purple grapes with a fresh bloom on them that would have caused an envious pang in the heart of a Covent Garden Jew. There were yams, brinjalls, great fat pumpkins, and a kind of hawthorn-berry, in size and

THE DAY'S ROUTINE.

appearance like a white-heart cherry. Where the fruit comes from it is difficult to say; for fruit-trees are rarely noticed, such a thing as a vineyard is utterly unknown, and a vine is scarcely ever seen except when occasionally an open back-door gives a peep into a garden, where one may be seen trailing behind the house, with its yellow leaves all glowing in the sunshine. Who it is that eats so much fruit is a mystery, for neither meat, bread, nor grain are to be seen in anything like a proportionate quantity.

Our days here used to begin with a slow and tedious march of six hours, generally broken by a good deal of walking. Then came the mid-day halt for breakfast, and for resting and feeding the cattle. To prepare our breakfast it always seemed necessary to unpack the loads of half the mules, for Chin-Tai had the most inveterate habit of packing the knife in one box, a fork in another, and a plate in a third; the cooking utensils he would invariably distribute over all the animals, and the food was never in a convenient place; ultimately this became such a nuisance that, as all our remonstrances were in vain, we gave up a hot meal and contented ourselves with whatever we could carry in our saddle-bags. During the halt I used to write up my notes, so that the time was fully occupied. The march in the afternoon would generally be of about the same length as that of the morning, and on arrival in the evening we used both to set to work at our writing, which, with an interlude for dinner, kept us well employed till it was time for bed.

On the road, at almost every hundred yards or so, a string of donkeys is seen, laden with merchandise, jogging along at a wonderful pace, and the traveller in these Eastern countries cannot fail to be struck with the value of these well-abused animals. They are made use of by the poorest classes, for they cost little to buy and hardly anything to keep, and from the numbers that are employed in every way, whether as beasts of burden, as draught animals, for riding purposes or for grinding corn, it is difficult to understand how the people would live without them.

Few of the villagers here had ever seen a foreigner before; but they always treated us with great civility, though, like all Chinese, they were curious and inquisitive in the highest degree, and we were objects of the deepest interest to them.

On arrival at an inn they would crowd round and make remarks on our clothing and appearance.

'What a curious-looking fellow that is,' says one, 'he has no plait, and does not shave his head.' 'No,' replies another, 'and look at his tight clothes, why it is absolutely indecent.' 'So it is; and do look at their hats, what queer things, and not even alike. What ugly eyes they have too. Their boots, however, are excellent; do not you think so?' 'Oh, yes, indeed; and I am told they never wear out.' 'Really, is it so?' 'Oh, yes, that is a fact, and water cannot get through them.' Then to Carles, 'And pray, sir, what is your honourable name? What is your honourable country; and what do you want here? Do not you find your clothes very cold, and what is your honourable age?' 'Why, indeed, you look double that, and that gentleman with you, he does not talk our language?' and so on.

Whenever we sat down to write, crowding round, they would all try to see at once; a lead pencil invariably exciting many comments.

If ever the curious people were kept out of the room they would then collect round the window; the more fortunate ones in front would poke peepholes in the paper with their fingers, and for hours would gaze stolidly, wondering at the way we ate, endangering our eyes and mouth with a pronged fork, would marvel at the odd and illogical mixture of cold drinks and hot viands, and above all fail to comprehend for what possible reason we could have left our homes to wander about in their country.

The hilly country we pass through is extremely pretty, and the villages generally very picturesque. Here is one with a little patch of tobacco of the richest green ; here a careful cottager has built a little trellis-work in front of his house, where a gourd trailing overhead casts a pleasant shade on the road, and inside he is sure to have a lark in a cage, with the feathers burnt off its throat to make it sing.

Leaving the village, a charming valley is ascended, with a stream of water clear as crystal brawling in the bottom, and shut in at each side by high hills or mountains, where on the top of the most inaccessible pinnacle a body of Lamas have built a temple.

The valley is green with groves of willows and poplars, and just outside the next village there is a plantation of walnuts and chestnuts. As we enter it a man passes selling potatoes. Our head muleteer inquires the price, and says they are too dear, and passes on ; but the seller, anxious for some business, calls out a lower sum, and in a little time a bargain is struck. Then two or three great hot potatoes are put in the scale (for nothing is ever sold in China that is not weighed or measured), weighed, and paid for, and the muleteer makes his morning meal.

Next we meet a fruit-seller, and the muleteer, after making as close a bargain as possible, selects the best pears he can see, taking care to offer one to Carles or myself with a respectful bow. He cannot afford grapes, which are still to be bought in abundance.

At this time of the year, when the leaves are all turning yellow, the villages with the great millet or Indian corn stalks piled against the houses for firewood look very homely.

As the day wears on every one gets rather tired, especially the muleteers, who, being a little footsore, go, like Agag, delicately. One of the boys, perched on the top of his laden mule, goes to sleep; presently he gives an extra heavy nod, and nearly falls off. Riding behind him I tried to sketch his figure; but the white pony seeing a blade of grass on the other side of the road, goes off with a jerk to get it, and spoils the picture. Returning to his place in the caravan, my greedy steed cannons against Bacchus, as the mule is called that carries the wine. This is a fearful animal to have anything to do with, his load projects on each side to a prodigious distance, and the corners of his boxes are most uncompromising. The bay was called Tom Bowling, for 'his virtues were so rare,' indeed, no one ever found out that he had any at all; every man's hand was against him, and his hoofs were against every man, and he was altogether such an unpleasant animal that even the other ponies would have nothing to say to him.

October 1.—There was only one ferry-boat on the river Lai, which we had to cross in the morning, so it was rather a long business, and to while away the time, after examining the heaps of anthracite coal lying on the banks, we bought hot sweet potatoes and fed 'Spot,' who was particularly fond of these luxuries. We passed the walled town of Nan-Yang-Cheng, where the walls and all the houses inside it had tumbled down, and the population now was entirely outside. Then, ascending a valley where the slopes of the hills were covered with beautiful long grass and dotted with yews, walnuts, chestnuts, and willows, we crossed a saddle about 600 feet above the sea, and descended to Hsi-Feng-K'ou, a pass in the Great Wall, never, I believe, before visited by a foreigner. Here officials, who were not particularly polite, demanded our passports, and took an extraordinary amount of trouble to copy them. At this place the Great Wall is about thirty feet high, and in very fair repair; it is built up, for a height of about seven or eight feet, of great granite blocks, above which are fifty-five courses of bricks, each about four inches thick. None of our party were ambitious to emulate the example of some of the naval people at Peking, who made an expedition to that modern wing of the Great Wall, which runs down not far from the capital. They carried off some bricks as trophies, and stowed them amongst their liquor; when the box was opened to assuage the thirst of these adventurous mariners. it was found to contain little but broken bits of glass and brickbats.

After passing the Great Wall the scenery changed completely. Instead of a smiling valley, with green hills and trees, we entered an almost savage country. The great mountains, on either hand, were rising up nearly bare, and even in the bottom, by the side of the stream, the trees were very few. The road was bad and rocky, and seemed almost impassable for wheeled traffic; but the mules made light of the difficulties, and we arrived in good time at a very poor inn in the mountain village of Po-lo-Tai. Here a great barn-like apartment, where the hotel master and his family lived, where passing travellers were accommodated, and which served as a kitchen for the whole establishment, led by an indifferent and dilapidated door to our room; this was about twelve feet square. The kang was used as the store for the family supply of millet seed, over which our beds were spread. The table was so filthy that the tablecloth I carried about was more a necessity than a luxury, and the place was soon filled with the pungent smoke of the millet stalks used as fuel in the cooking-stove close by. But these minor evils appear trifling to the hardened traveller, and we passed our night without discomfort, if not in luxury.

October 2.—Whilst waiting for the mules to get under way we watched the process of making bean-curd cakes.

The use that the Chinese make of beans is very remarkable; they cook them in all sorts of ways, eat them pickled, put them into potato patties, and convert immense quantities into bean-curd cakes.

The ordinary black and white beans are ground between two circular blocks of granite about two feet in diameter; there is a small hole in the upper stone, through which the beans are swept, water being poured on at the same time.

As the upper stone is turned a thick white cream runs out from between the stones, and is caught in a receptacle. This thick cream is then boiled with water, a very little rock-salt being added. After a time quantities of froth rise to the surface; this is skimmed off and thrown away, the remainder being tied up in a cotton cloth and squeezed tightly, after which it is put into a flat pan to set. It is finally cut up into squares, and is ready for use.

The Chinese are particularly fond of this preparation, and in the smallest village even, if nothing else is to be procured, one or two people will be certain to be found selling the bean-curd cakes.

We had a delightful march amongst lovely scenery, the foliage of every hue, from the rich green of the young leaves to the deep yellows and reds of autumn. In the early mornings the distant outlines in the narrow valleys were lost in a deep blue haze, while the hill-tops just caught a glow from the rising sun. After some miles we emerged from some low hills into as charming a view as it is possible to conceive. The Pao-Ho was winding through a broad and fertile valley, where, in one or two places, the bends ran below small precipices; numerous villages were dotted along the banks, and in the background the fine granite mountains closed the view. We halted for breakfast at a place called Kuan-Ching, where a good deal of silk is made from the silk of worms that are fed on a kind of oak (*Quercus obovata*). The workshop consisted of two rooms fourteen feet square; in one of them the silk from the cocoons was spun by a man who said he could do twelve hundred cocoons a day, and in the other a machine, the pattern of which has been in use for five hundred years, is used for weaving the silk. When woven the material is very strong, but very coarse, and were it not for the frayed edge it would be difficult to recognise its identity. Nearly all of it is sent to Peking, where it sells at one tael (about 5s. 9d.) for four yards of stuff about two feet wide.

This silk weaver appeared to combine his trade with that of 'patissier,' for in the room with his antiquated loom his whole family were engaged in making pies. They had a bucket of green vegetables chopped up, and about six of them, men, women, and children, were seated on the kang before a low table, with a quantity of flour and water. One of them kneaded the dough, another made it into flat pancakes, a third handed up the green stuff, which a fourth rolled up in the pastry, while a fifth arranged the patties on a large, flat, circular tray, ready for the oven.

We bid our good old host adieu, and leaving the room, which was now crowded with about twenty people, marched on to our night's halting place.

October 3.—The next morning we descended a valley where nature had draped the landscape in such gorgeous autumn tints, that she seemed, in some wanton mood, to be challenging the feeble hand of man to imitate her wealth of colouring. The mountain sides that rose up on either hand almost precipitously glowed in golden yellow or red; down by the rill, which leapt merrily from stone to stone, the young willows had the fresh green foliage of early spring; and the very weeds growing by the roadside vied with the trees in the richness of their hues.

The villages were very neat, and more tidy than usual; we stopped in one of them to have the ponies shod. In Mongolia, horse-shoes can be obtained at almost any blacksmith's; they are like our own in shape, but are not turned down at the heel. While the operation was going on we watched some people threshing out the millet; no flail is used, but a blindfold donkey is driven round in a circle, drawing a light stone roller over the corn. After this, the winnowing is done in the old-fashioned way, by throwing up the grain into the air, and allowing the wind to blow away the chaff. The quantity of millet grown in this part of Mongolia is very great; every day we saw long strings of donkeys carrying sacks of millet south, and others bringing cotton goods in return.

We arrived in the evening at the large military

station of Pa-K'ou-Ying. Pa-K'ou is thus called by all the people in the neighbourhood, and this name generally appears in our maps ; but according to Dr. Bushell, in the year 1778, when the system of government in this province was remodelled, Pa-K'ou-Ying was elevated from the rank of Ying to that of a city of the second order; its name was changed and it was called Ping-Chuan-Chou. It is still a town of great importance, and the bank notes issued by the bankers of this place pass current anywhere within a radius of one hundred miles. This is a very remarkable and almost exceptional phenomenon, no other city in China, not excepting Peking itself, possessing bankers whose credit is known half a dozen miles from its gates.

In the main street of Pa-K'ou, the houses stand about twenty feet back from an open sewer, six feet wide, that runs down the centre, and is full of a foul black slime. There are few private residences in this street, nearly all the houses being shops, of which the greater part are kept by pawnbrokers. These have an ornamental pole in front of their houses, and from their appearance seem to drive a thriving trade. They will accept anything as a pledge. Our head muleteer had some time previously pledged a donkey, and the chance of being able to redeem it with the cash advanced by us before starting, was probably his chief inducement to enter our service, as the rate of interest charged by a Chinese pawnbroker would startle a London Jew.

As there is no glass in China, except where it has been introduced by Europeans,[6] the shops are always

---

[6] When glass is used, it is curious to note the conservative instinct of the Chinese, who generally fasten it over the thick latticework, which is necessary when the windows are covered with paper, but of course not only is unnecessary when glass is made use of, but blocks out half the light.

quite open to the front, so that the passers-by may see the goods for sale. Private residences are usually withdrawn from a thoroughfare by a court, through the wall of which there is but a very narrow door, almost always kept shut; and riding into a town after nightfall, it is a very striking contrast to turn from a street of private houses, enclosed on both sides by a wall, pierced with nothing but the closed doors, where the very few people met with seem like spectres stalking through a city of the dead, and suddenly to enter a street of shops full of men and animals, pushing, scrambling, shouting, bargaining, or gossiping, where the mass of people collected would give a very false idea of the population of the place.

It is no doubt owing to the fact that the Chinese seem to pass their lives in the chief thoroughfares of the towns they inhabit, and leave the by-ways always deserted, that the early foreigners, who saw little but the crowded parts of the great towns, formed such exaggerated and erroneous ideas of the population of the empire. They knew that the towns covered a great area, they forgot that the houses, unlike European ones, were no more than one story high; they saw the main thoroughfares thronged with a mass of humanity, they were probably unaware that the private residences were nearly all empty during the daytime, and they seldom turned into the by-streets. Under these circumstances it is not surprising that, in the last century, the few Europeans who had landed in China came away with the impression that the whole surface of the country was covered with teeming millions of people, scarcely able to find space even in an area so immense.

Further, when they first made incursions from the great trading ports, they passed only through the fertile and alluvial plains where the population is enormous; they ascended rivers which are the great highways of China, where fleets of junks of all sizes crowd the water. Again they argued from the analogy before their eyes; again they omitted from their calculation the vast expanse of mountainous regions, where in the narrow valleys but little room is left for population, where vast forests cover acres upon acres of ground, and where along the wild by-paths for miles there is scarcely a habitation.

Who can wonder that they estimated the population by hundreds of millions?

Pa-K'ou, though a dirty town, contained an excellent hotel, from which, after a little patience, we succeeded in excluding the throng of people that had followed us. In the large towns, the curiosity and inquisitiveness inherent in the nature of the Chinese becomes rather tiresome, and often very inconvenient; but their intense desire to look at foreigners is not to be wondered at. A foreigner in European clothes had never been here before. To a Chinaman's eyes a Western is as hideous and strange as a Chinaman at first is to ours; to his mind our clothes are not only uncouth and uncomfortable, but indecent; and to his ideas a light-haired being is diabolic; indeed the very animals seem to share this belief. A story is told of a red-haired, red-bearded Englishman who one day was walking in a country place; meeting a cart, the animals were so frightened by the extraordinary apparition, that they started, and upset the vehicle into a ditch. The Anglo-Saxon good-naturedly went to assist in setting matters straight, when the carter entreated him to get out of sight as soon as he could, as his awful appearance only terrified the animals the more.

We had a long consultation here with the innkeeper as to roads and distances. Our objects on leaving Peking were first to see the road to Hsi-Fêng-K'ou, and thence to Jehol; and afterwards to travel, as much as time would permit, in the low ground south of the mountains, and between them and the sea.

As far as Hsi-Fêng-K'ou all had been simple enough, but although Jehol lies NNE. of Hsi-Fêng-K'ou, up to this moment we had been travelling NW., our guides assuring us each day that to-morrow we should bear away towards Jehol. That city lay now to the south of west, and seemed as far off as ever, and as time was drawing on, it became a matter for consideration whether we should not abandon our intention of visiting it.

Jehol is the summer hunting-palace of the Emperor; it was here that Earl Macartney was admitted to an audience by the Emperor Chien-Lung in 1793; and it was here that when Peking was attacked by the allied armies of England and France, the court took refuge.

Our innkeeper at Pa-K'ou was an intelligent man, and he gave us a good deal of information; we ultimately with much regret abandoned the idea of finishing our journey to Jehol, and determined now to get to Shan-Hai-Kuan by the shortest route. There appeared to be several roads, and on the recommendation of our host we adopted the most northern. The maps we had with us were not of the slightest use; and it was not until we began to approach the sea, that we fully appreciated the extraordinary and circuitous route, by which we had been despatched many miles away to the north-east, before finally descending in a southerly direction to Shan-Hai-Kuan.

The lengthy discussion with our host being concluded, we sat down to enjoy our dinner with appetites sharpened by the keen mountain air. The weather was now getting very cold up here, the nights especially, we felt chilly when sitting to write with our feet on the cold stone or plaster floors, until Chin-Tai suggested that a pair of Mongol shoes would be much better to sit in than anything we had. These are made of felt, with soles of a kind of *papier maché* about three-quarters of an inch thick, not unlike those of the shoes called *gevas* in Persia. When once we had invested in them, we found that we could sit for hours before our feet became chilled.

October 4.—Our march the next day was in a much less interesting country: among great bare mountains, and through undulating broken plains with no trees except round the villages, where the willows, like the magpies, seemed omnipresent. The villages here are all clean and thriving, the people seem happy and contented, growing little but their millet and Indian corn; no fruit is ever seen here, and none is to be bought; and as the road penetrates further to the north into the great Mongolian plateau, large herds of sheep and cattle are seen feeding on the rich pasture.

As on all the borders of China, the Chinese are here gradually pushing back the aboriginal Mongols; wherever corn will grow the agricultural Chinaman is by degrees superseding the pastoral Mongol. The Chinese never take to pastoral pursuits, but in some of the valleys between Pa-K'ou and Ta-Tzŭ-K'ou, Mongol encampments may be seen, where the Mongols are beginning to cultivate the ground, and abandon the nomad life of shepherds. Whether by this means the relentless advance of the Chinese will be stopped, time alone can determine.

Our Ma-Fu used to give us a good deal of amusement. He generally had a couple of animals to lead, and now seemed to have given up riding altogether; he was usually to be seen tailing off at the rear of our caravan, with two, or sometimes three ponies dragging at his arms, and trying to stop and nibble at every blade of grass. As he walked, he used to shuffle in some strange manner altogether peculiar to himself, and at last we felt that it was really an act of charity to put Chin-Tai on his pony, and let him go on in advance to get the inn ready for us before we arrived.

October 5.—The muleteers to-day were all foot-sore, and the mules knocked up, especially Judas, so called because he carried the money-box. He with two others had already fallen down twice, and poor 'Spot' could hardly walk, so to refresh everybody was it advisable to halt for a day. Carles and I went out for a stroll with our guns, after warning Chin-Tai that we depended on him rather than on our weapons for our dinner: a fortunate precaution, as the only bird we obtained was a tame duck that our guide presented to us to compensate us for our want of luck.

October 6.—The next morning, although our party were not as much refreshed by the rest as we could have wished, we again started, after putting 'Spot' into a basket and covering him with a net. Our head muleteer had brought one man with him so old and decrepit that he appeared to have no vitality left; our continuous long marches had quite knocked him up; the head muleteer was now obliged to put him on a donkey, but the poor old fellow was so feeble that he could hardly keep himself from falling off; his appearance was most melancholy, and he ultimately received the sobriquet of Lazarus, for he looked like one raised from the dead. We afterwards found out that his age was only forty-five, though he looked eighty at least.

On the road we met a poor family who had been driven by famine from their own village in Shan-Tung, and had come here, a distance of 700 miles, on foot in search of a living. Poor creatures, they had not found much to do, and were at sore straits to keep from absolute starvation.

In the evening we arrived at the gates of the walled town of Ta-Tzŭ-K'ou, a military station of some importance. We entered a street where there were no shops but pawnbrokers, all the other houses being solidly-built private residences. The street was about fifty or sixty feet wide, and the absence of shops gave it a deserted appearance, but presently our nostrils were met with the flavour of the familiar drain, and we knew that we were approaching the crowded part of the town.

In the size and bouquet of its open sewer, the main street bore a strong resemblance to that of Pa-K'ou; we soon found ourselves in the busy thoroughfare, surrounded by the usual crowd of wondering but good-natured people, who followed us to the inn. Our room here was fortunately in a courtyard of its own, the walls of which kept out all but the most inquisitive of the people; these, however, forced an entrance, and poking holes in the paper windows with their fingers, were able to examine us and our singular doings.

Fearful and strange stories were told us before going to bed of the awful nature of the country we were about to traverse, and of the ferocious brigands that infest it. How a Corean gentleman of high rank and importance had been stripped a few days back and left naked; how a Roman Catholic priest who was travelling here had been robbed of everything he possessed; how the terror of the lawless bands that roved about the country was so constantly before the eyes of the inhabitants, that yesterday when Chin-Tai was sent forward to arrange the inn accommodation, he spread consternation amongst the peaceable townsfolk, who, because he was a stranger, thought he must also be a robber.

October 7.—We were not much impressed with fear however, but as our servants would insist on all being armed, sundry mule-loads were overhauled, and our artillery prepared. Chin-Tai trusting to size chose to carry a seven-bore duck-gun, and nothing less than fifty cartridges would satisfy his bloodthirsty mind. Carles' servant having the next choice, selected a heavy double rifle, and thus equipped they swelled in importance. Descending the river, which here was coursing through an undulating plain with rounded hills on both sides, we presently entered a little gorge, where the stream ran between high cliffs; beyond there was another wide, flat, and richly cultivated valley, where many little villages, clustering by the edge of the water, were surrounded by homely clumps of elms, on some of which a great quantity of mistletoe was growing that brought fond recollections of our island home away in the Western seas. A little further on, on either side, a very remarkable mountain, with high pinnacles 1,000 feet above the plain, stood sentinel over the pass.

At the village of Chi-Chien-Fang, where we made our halt, the people all declared that 'Spot' had ears like a lion. It is not surprising that they did not know what a lion was like, but it is difficult to understand why they should all have declared that 'Spot,' was like a lion. They had never seen one, nor had they ever seen pictures of one, nor is a lion an animal frequently mentioned in Chinese books ; but as the people with one accord compared 'Spot' with a lion, of which by all sound reasoning they ought to know nothing, the only conclusion to be drawn is that if a man had to evolve a lion out of his own inner consciousness, he would adorn it with long ears like those of a spaniel.

A man asked Carles to-day if it was true that a queen reigned over us ; and when he received an affirmative reply, said that he supposed that Providence arranged for all the queen's children to be girls, so that there should never be any danger of a rupture of the Salic succession.

Leaving Chi-Chien-Fang, we marched over an undulating downy country, something like Salisbury Plain in appearance. The crops had nearly all been gathered ; the poverty of the soil was very apparent, and accounted for the sparse population and the absence of traffic. Passing the village of Ha-Go-Ta, where some few Mongols have settled down as agriculturists, we arrived in the evening at San-Tai.

October 8.—Riding to the southward, and passing here and there some small village surrounded by a wall to protect it from the troops of wolves that in the desolate winter scour the plain of San-Tai, we presently saw the mountain K'ou-Lung-Shan raising its strange head from behind the low rounded hills that enclose the valley ; K'ou-Lung-Shan, or the 'Hole and Dragon mountain,' is so called because some wanton freak of nature has driven a tunnel through its summit, and even at the distance of ten miles we could see the sky beyond like some monstrous Cyclopean eye watching over the fortunes of the plain.

The climate of October in this part of China is simply perfect; it is never hot, nor very cold, it scarcely ever rains, strong winds are of rare occurrence, and the fresh crispness of the morning air is wonderfully exhilarating. The atmosphere is exceedingly dry, inducing the most prodigious appetite, and Carles, adapting what Falstaff said of borrowing, remarked that eating only 'lingered and lingered it out, but the disease was incurable'; altogether there can be nothing more enjoyable or health-giving than an October in Mongolia.

In the summer the heat is oppressive, the land dried up, and the sun scorching. In the winter the cold is bitter, the soil is frozen many feet below the surface, and icy gales sweep across the dreary plateaux; but for just these few weeks nature seems to combine all her charms of scenery and climate, and in one short month to make amends for the alternations of excessive heat and cold.

Our march to the south must have taken us across the supposed position of the line of palisades shown on all maps, including Williamson's.

We saw nothing of them, not even one of the gates mentioned in the extract below. In all probability they have long since been used as firewood, and, notwithstanding his map, Williamson thus writes:—

'Kirin, or Central Manchuria, is bounded on the north by the Soongaris, on the east by the Usuri or Russian territory, on the south by Corea and Liau-Tung, and on the west by the Soongari, and a line of palisades, which exist only on the map, and in the imagination of H.I.M. the Emperor of China: though there is a sort of gate at the passes, and a ditch or shadow of a fence for a few yards on each side.'[7]

[7] *Williamson's Journeys in North China*, vol. ii. p. 52.

October 9.—From Ku-Ch'iao-Tzu ascending a narrow and steep valley, the road reaches a saddle, and a glorious view over mountain-tops is disclosed ; descending again, in a narrow gorge, the road skirts the mountain of Yang-Shan, where there are some precipices 200 or 300 feet high, and sometimes following a stream in a wooded plain, sometimes shut in by precipitous hills on both sides, the road at length reaches Kang-K'ou.

This place is a great dépôt for millet, and is a military station for 150 soldiers. The commandant who paid us a visit was accompanied by half the people in the place, who sat down beside us, felt our clothes, joined in the conversation, and asked all sorts of extraordinary questions. This officer had the longest and loudest tongue of any man I ever met ; he shouted at and harangued the populace in a voice of thunder, and talked to Carles as if he were as deaf as a stone. He accepted our invitation to take some tea, and sitting down, smoked, with the assistance of his servant, innumerable pipes.

There are two kinds of pipe in use in China. The upper classes affect the water pipe. This is an elaborate contrivance in which highly scented tobacco, ground into a powder like snuff, is smoked. To use it with any effect, the constant assistance of a servant is necessary, who fills the minute bowl, and offers it with a lighted match to his master ; the small quantity of tobacco only permits of two or three whiffs, after which the pipe is returned to the servant, who, lifting the bowl from its place, blows out the ashes, refills it, and again hands it to the smoker. The poorer people smoke the coarse rank tobacco of the country in pipes the bowls of which, no larger than a thimble, are made of white metal ; reeds are used for the stems, which vary from one to four feet in length. These pipes are their constant companions, being converted into walking-sticks when smoking is not an immediate necessity.

Whilst our military friend was haranguing us we could watch the muleteers in a corner preparing their frugal repast of Indian corn meal.

There is a popular idea that all Chinese live entirely on rice; like many other popular ideas, it is very far from the truth.

Rice is well adapted to the necessities of the warm climate of Central and Southern China; yielding an enormous harvest to the agriculturist, it is the cheapest food that can be procured, and has for these reasons become the national food of the people.

But in the colder climate of the north, or in the higher mountain regions, rice is not so suitable as Indian corn or oats, nor can it be so easily cultivated. One or the other of these grains consequently takes its place.

Here our muleteers lived almost entirely on Indian corn; they ground it into a meal, mixed it with a little water, and steam-baked the preparation.

October 10.—We marched from Kang-K'ou up a narrow valley, where the mules continually stumbled and fell over the great rocks scattered pell-mell over the path, to a ridge from which we looked down upon the most extraordinary and confused mass of mountain-tops. Descending again and crossing successive ridges, where, in one of the valleys, we found some thirty men washing the sand for the very minute quantity of gold in it, we gained the summit of the Ku-Ling Pass, and, when we put our feet upon the crest, a view so lovely burst upon us that an exclamation of wonder involuntarily escaped our lips.

Great mountains in front of us, lit up by the setting sun, shone in all the glory of golden hues, which changed to purple on the lower slopes, and deepened into blue in the valleys below.

This picture was, as it were, set in a frame by the narrow gorge that ran down from where we stood, and whose rocky and almost perpendicular sides, rising high above our heads, presented every variety of colouring. The bluish grey of the beetling cliffs mingled with the mellow glow of autumn on the trees, and here and there a touch of delicate green showed some young sapling just starting into life.

We neither of us could leave the spot, though the hour was late, but stood and watched the changing tints on the mountain-tops. The rosy blush faded from them one by one as the sun fell in the west, little by little the transparent blue in the valley deepened, and the darkening shade stole up the mountain sides; at length the last flush melted from the highest peak, one star shot its first faint ray of light, and heralding the night reminded us that we must tear ourselves away from the glorious scene.

A little temple, perched on the very summit of the northern face, bowered in trees and shrubs, and overlooking the quiet valley, stands in a spot at this season of the year as fair as any on the earth.

By the light of the stars that shone with marvellous brilliancy in that cold clear air, we followed the stream by a rocky path to San-Cha, where we halted after a march of thirty miles. The inn here was wretched. A long low shed like a barn led through a door that would hardly close into a small place beyond, where there was no floor but the natural soil, through which great stones jutted up; a few dirty shreds only remained to remind the traveller that the window had once been covered with paper, and the dust lay inches thick on the furniture. The kang was lighted, but the pungent smoke came pouring out upon us through yawning cracks in the dilapidated structure. These trifles, however, made but little impression on us, and we slept as soundly as on the most luxurious of couches.

October 11.—From San-Cha the road descends a rocky valley to I-Yuan-K'ou, a pass in the Great Wall not previously visited by foreigners, where our passports were again examined by some very civil officials, who invited us to stop and take tea with them. Except just at the crossing of the stream, the Great Wall is here in the same good state of preservation as at the other points visited by us; it is the same marvellously massive structure, and there is something very imposing in its appearance, as, with towers at regular intervals, it ascends the steepest mountain sides, is carried over the tops of the highest hills, plunges deep into the valleys, and crosses many a mountain torrent. There is something very impressive in the thought that for centuries this wall has looked upon the same country, has seen emperors come and dynasties go, and that for hundreds of miles, away towards the west, it stretches across the boundary of many provinces, till it is lost in the distant deserts of Gobi.

The running stream at this point appears to have washed away part of the foundations, for the wall had fallen away, leaving a gap through which anyone might ride; the farce, however, is still kept up of soldiers at the gate, which might be knocked down by a strong man's kick.

We halted for breakfast at a garrison town, where we excited more than the usual amount of interest. The commandant came to pay us a visit, bringing with him a powerfully smelling rabble, who soon blocked up every available corner in the apartment.

This was a foolish old man who had no information to impart, and could ask no more intelligent questions than our honourable names and country. After having watched him smoke many pipes, we found a want of excitement in the amusement; and leaving him, we threaded our difficult way through the dense but good-humoured crowd that thronged the innyard, and continued our journey over a very stony road, through an undulating country, gradually leaving the mountains on the left. Crossing a good many small streams, where slippery stepping-stones were provided for pedestrians, we gained the crest of a small spur thrown out from the mountain, and now the calm blue sea lay stretched before us, flecked here and there with the red sails of a fishing-junk.

Seven miles away lay Shan-Hai-Kuan, and behind it rose the steep mountain up which Fleming climbed on a blazing summer day, and where he so nearly lost his life.[8]

We sent Chin-Tai forward on the red pony to get things ready, but about a quarter of an hour afterwards an animal was seen approaching that bore an uncommonly strong resemblance to 'Tom Bowling'; and in a few minutes the familiar form of that virtuous creature, riderless and without a saddle, kicking up his heels here, and nibbling a blade of grass there, was unmistakable.

In his playful way he had set to work kicking, broken his saddle girths, and disposed of Chin-Tai; and as he now declined to be caught, we deployed our force across the plain and drove him before us, detaching parties to the right or left whenever the

[8] Fleming's *Travels on Horseback in Manchu Tartary.*

cunning beast made long flank movements. At length his foot became entangled in his bridle, he found himself unable to move, and surrendered, but with a very bad grace.

We took up our quarters in a rather small inn, in a suburb outside the west gate of Shan-Hai-Kuan, where we were very quiet. It was quite a new sensation to be able to wash, eat, drink, read, and write without being surrounded by a gaping crowd.

October 12.—Shan-Hai-Kuan means 'mountain and sea barrier,' and is so called because the mountains here run down to the sea.

Old Lazarus had lately proved such a serious charge, that we determined to send him to Peking with letters; and, as we expected henceforth to be always on the level plain, we hired a cart to make up for the loss of the animal that the old muleteer was to take with him.

CHAPTER III.

'ATHWART THE FLATS AND ROUNDING GRAY.'

Halt at Shan-Hai-Kuan—Deserted City of Ning-Hai at Sea Terminus of Great Wall—Ancient Inscriptions and Vandal Treatment—Dirty Habits of Chinese—Salt Manufacture described—Itinerant Pieman's Wares—Chinese Love of Hot Water—Mouth of the Lan—Mud Flats by the Sea—Rich Monotonous Plains—Kublai Kaan's Shooting Grounds—Chinese House Fittings—Swampy Plains—Tumultuous Curiosity at Fêng-T'ai—River of Pei-T'ang—Coal expected to be found to Order—The Pei-T'ang Commandant—The Ferry—Mud Flats and Fragrant Artemisia—Dutch-like Landscape—Water-logged Country—Weary Swamps—Rest at Urh-Chuang—Hawking Ground of the Great Kaan—Chinese Wrangles—Chinese Agriculture and Use of Sewage—The Traveller's Sympathy with Marco Polo—Swamp-Lands again — Ling-Ting-Chen — Marshlands — Hsin-An-Chen — Figures bound for the Fair—Roadside Pictures—Pleasures of Travel —Peking reached again—Recreations there—'Bis cocta Recambe'— Chinese Dinners—Sir T. Wade's regretted Departure—The Inscribed Stone Drums—Visit to the Summer Palace—Needless Havock—A Great Bell—Modern Cambaluc—Public Examinations—Departure from Peking—The Hilarious Serving-Man—The Kang Mismanaged— Tien-Tsin, a Land of Fat Things—Shanghai once more.

October 13.—We halted a day at Shan-Hai-Kuan, and the next morning, just as we were starting, we heard a very loud altercation in the courtyard, and found old Lazarus wrangling with the head muleteer, who owed him some wages. The old man declined to start for Peking without payment; he had laid an embargo on the headman's donkey, and was sitting down in the gateway, stolidly holding the reins, which he declined to give up. The head muleteer, having no money, thus found the animal that he had only lately redeemed from pawn again seized as a pledge, and matters seemed fairly to have arrived at a dead lock. They wrangled thus for upwards of an hour, anyone passing by, of course, stopped, not so much to listen as to join in the dispute ; two soldiers, who had been sent to escort us, became very energetic, and looked as if they were quite capable of solving the problem on Solomon's system, even if they did not conclude by cutting off the heads of a few of the bystanders by way of encouraging the others ; the little boys ran about in high glee, or crawled between the legs of their elders, thoroughly enjoying the sport ; until at last we ended the dispute by promising to see the old man righted.

Matters thus being settled without any bloodshed, we mounted our ponies, and with our escort rode away a couple of miles to Ning-Hai, an ancient and deserted city, built a few yards from the edge of the sea.

The Great Wall forms the eastern face of this city, and originally must have been built out some distance below high-water mark. Time, however, has laid its hand on the venerable building, and now at this point it is little better than a ruin.

Cantering round the south-west angle of Ning-Hai, we suddenly saw some huge English characters painted up in white letters on the ancient city wall. For a few seconds we were both so surprised that we were unable to read the words. We very soon realised the situation, and understood that ships cruising in these waters were accustomed to land their crews and desecrate the noble old walls of the venerable city, which for centuries has looked upon the lapse of time, and has watched the varied fortunes of the empire, from the days when its civilisation led the march of the world, through its declining fortunes, to its present decay.

Close by stand two very ancient slabs; no man now can say when first they were planted there, or in memory of what, or whom; on one was written the words 'God divides the land from the sea,' and on the other, an obscure quotation from some ancient classic which might be interpreted by the Scotch proverb, 'every little makes a mickle,' and might refer to the Great Wall, which enormous structure has been made by the accumulation of many bricks; but whatever might be its meaning, as we stood in the solitude, made more solemn by the deserted appearance of the city, whose walls enclosed little but crumbling ruins, the scene was very impressive; the thoughts of all the events that must have happened, in the long roll of ages that had elapsed since first the hand of man commenced to raise this pile, crowded through our minds, and we could not but regret that Europeans should thus desecrate what must almost be a sacred spot, should destroy the day dreams in which fancy here would love to indulge, and force upon the mind of the passing traveller that he was living in an unromantic age of steam and iron.

British sailors were not alone in the ruthless act; crews of all nations had daubed their vessels' names, and worst of all, on the very face of the old slab that bore testimony to the builder's belief in an Omnipotent Power, the commonplace vulgarity of huge sprawling letters made us ask what manner of men can these be, who take ashore great pails of whitewash, and in the futile attempt to perpetuate the memory of a modern ship carry to these distant lands nothing better than the spirit of the scribbling tourist!

Time, however, will avenge itself; nature's elements will in due course remove the hideous records, and long after the bodies of the authors have turned to dust, still, for years, may these old buildings look out upon that same sea that bore them here.

Turning away, we spurred our ponies into a gallop over the wide plain.

The country here is thickly populated and closely cultivated. At every half-mile we passed a village surrounded by trees, and in the fields the second crop was now nearly ready for the sickle; but as we rode westwards the amount of cultivation gradually diminished, and we presently came to wet and heavy roads. Wastes of swamp and mud stretched between us and the sea, from which creeks here and there ran up, receiving rivers from the now fast receding mountains. As we cantered on, our escort fell further and further behind, and as the day waned, they turned away to a neighbouring village, tired of following the mad foreigners in their long long ride.

At half-past six o'clock, after a march of more than forty miles, we met the head muleteer by the side of a stream.

He had come out with the innkeeper's son to welcome us, and lead the way to the hotel. This was a pleasant sign of hospitality and goodwill, and, with the excellent dinner that Chin-Tai gave us, helped us to forget some abominably dirty tricks we had discovered in the investigation we had had time to make amongst our baggage, during our halt at Shan-Hai-Kuan.

Foreigners, when they return to their own country, often lament the good servants they have left behind in China, daily yearn for their service, and deplore that they cannot have all Chinese servants at home.

These are undoubtedly excellent, but the dirt and filth that would accumulate in an English kitchen, after about a fortnight of Chinese management, would make the rash housewife repent her of her importation. The Chinese people are dirty beyond description in all their habits.

Their ablutions are usually limited to passing a wet rag dipped in hot water over their faces. All through the winter they wear the same clothes night and day; and as the cold weather advances, it is positively ludicrous to see the people gradually looking fatter and fatter as wadded garment is added to wadded garment. The children especially, in the depth of winter, look like dumplings rolling about the street. As the ice thaws again, and summer approaches, one after another the extra clothes are abandoned, until the people resume their natural and normal size.

In their mode of eating they are not more cleanly than in their persons; even amongst the richest classes the table, after a dinner, is covered with pieces of food, and quantities of grease that have been spilt on it from the overflowing bowls, whilst a débris of bones, kernels of fruit, and lumps of gristle are collected on the floor around the feasters. As might be supposed, their dwellings are as dirty as everything else. Their rooms are never cleaned; dust, dirt, and rubbish of all kinds may sometimes be swept up underneath the bed, or behind some lumbering piece of furniture, but there it lies for years, unheeded and untouched, except when some active-minded person chooses to increase its volume.

October 14.—The road from Niu-T'ou-Yai to Sha-Ho passes over a flat plain, where large stretches of cultivation are interspersed with marshy tracts, and the villages, tolerably numerous though small, are embosomed in fine clumps of trees.

At P'u-Ho-Ying, a miserable place, situated in the middle of a desolate salt marsh, we saw the operations in the manufacture of salt.

A ridge of mud, six inches high, encloses a space twelve feet by four feet. At one end a little drain is formed by piercing the ridge, and a hollow is scooped out in the ground below the drain. The earth in the neighbourhood is all strongly impregnated with salt, and lumps of it are put into the tank formed by the mud enclosure.

Fresh water is then poured over the earth; this

FIG. 1.

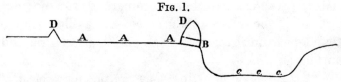

A A A. Tank where the earth is placed.
B. Small drain.
c c c. The receiver or hollow scooped out to receive the liquor draining from A.
D. Earthen ridge.

drains slowly into the hollow, and in its passage becomes a strong solution of salts. The water is then boiled three times in flat circular dishes.

By this successive evaporation, the different salts are thrown down at different temperatures, or by the varying strength of the solution. Common salt, or chloride of sodium, being the most easily held in solution, is not deposited until the final operation, and thus salt of more or less purity is obtained.

The process of evaporation is carried on in little circular enclosures of straw, to prevent the wind disturbing the surface of the liquor; and in the neighbourhood of this town the whole plain is dotted with these queer-looking erections.

From P'u-Ho-Ying a heavy sandy road leads to Sha-Ho, an unimportant place, where we were lodged in a room that smelt like a sepulchre.

A mist hung over the plain as we left the village of Sha-Ho, and little could be seen of the flat country, until, as the day advanced, we reached the town of T'uan-Lin, surrounded by vast quantities of graves, and situated at the edge of a mud flat half a mile wide, beyond which the stagnant waters of a broad lagoon could be seen through the haze. A sandy road led westwards through a busy, thriving country, where in the distant north the dim outline of a mountain could be seen, where the only crop left standing was the cotton, and where the people were busy in the fields pulling up the roots of the millet.

After the crops are cut, the stubble is not ploughed up immediately, but the people first go over the field with hoes, digging up the roots one by one. Afterwards they are followed by others with wooden mallets, who beat all the earth from the roots, and on a dry and windy day make a great dust, very unpleasant to the travellers' eyes. This being done a third party collects the roots in large baskets, takes them into the villages, and stacks them for fuel. The field is, however, not yet ready for the plough, for now a rake is taken all over it, and the few straggling stalks or bits of grass carefully collected. These also serve as fuel.

Whilst sitting at breakfast, an itinerant pieman came in with a tray, and we investigated his wares. Vegetable patties, such as we had seen made by the silk-spinners at K'uan-Cheng, are very favourite delicacies amongst the Chinese, and the pieman had plenty; also puddings of broad beans, enveloped in a covering of mashed sweet potato, and fried in a superabundance of oil. Cakes of wheaten flour, with treacle in the middle, and steam-baked, and others of ground millet with scorched millet sprinkled on the top, figured amongst the more simple wheaten cakes that the pieman had for sale, and which serve to vary the monotony of the Indian corn diet of the people.

Fond as the Chinese are of variety, and dearly as they generally love a bit of fish, strangely enough there was none to be bought in any of these coast villages. In the south of China vast quantities of fish are caught and salted, and form a very large proportion of the food of the people ; but here, notwithstanding all our inquiries, we could never hear of fish to be had, fresh or salt.

Some people with whom we were conversing offered us hot water to drink, and regretted they had no tea, which they said they could not afford. It is a very common thing in this part of the country for people to drink hot water, as they, like all Chinese, dislike cold drinks; milk, as everywhere else in China, is unknown, and of course butter also.

We continued our march over the plain to the river Lan, and thence to Lê-Ting-Hsien, which all maps show on the left bank of the river, although it is really seven miles distant from it, and on its western or right bank.

It is difficult to understand how European geographers have contrived to introduce such an error, but a Chinese map in my possession is the only one in which this town is shown in its proper position.

October 16.—From Lê-Ting-Hsien the country did not vary in appearance : still as flat as a billiard-table, closely cultivated, and with many people in the fields ; the villages, enclosed in the same thick clumps of trees, were dotted about at almost every quarter of a mile.

The absence of watercourses in this country is very remarkable. From one large river to the next, the road, though running across the line of drainage, will often pass over scarcely the smallest rill ; and to this is owing, in a great measure, the serious floods that constantly occur, and do so much damage.

We arrived in good time at Tang-Chia-Ho, and attempted to go out duck-shooting ; but, being followed by people enough to beat for a tiger, we very soon gave it up in despair.

October 17.—The room we slept in was exceedingly dirty and stuffy, and even the carbolic acid that had been thrown all over it, utterly failed in overcoming the very ancient and fishlike smell that pervaded the apartment, arising from the manufacture of shrimp sauce, in which the innkeeper had been engaged some weeks previously. The same fire that cooked the dinner passed through the kang, so that, as the night was very close and oppressive, we were nearly baked ; and as it was the first day of the new moon, the jovial spirits of Tang-Chia-Ho spent the night just outside the house in letting off crackers that exploded with a report like a rifle.

Long before our usual hour, we were fairly driven out of bed by the fleas, and calling Chin-Tai, we bade him shake out our blankets ; and then we overheard the innkeeper, who felt his reputation injured, declare that they were not fleas, but only bugs.

We rode down to Lao-Mu-K'ou on one of the mouths of the Lan, where there were a few people about who seemed to have nothing in particular to do. We sat down on some logs amongst the sailors, who told us that the navigable entrance to the river was some miles away, and that junks entering there supplied the Tien-Tsin arsenal with coal, which was brought down by the river from Jehol.

From Lao-Mu-K'ou to Ma-T'ou-Ying the road passes to the north of the border of waste lands which fringe the coast, and there is no variation in the aspect of the country.

October 18.—From Ma-T'ou-Ying we rode down towards the sea, the soil getting poorer and villages more scarce as the coast was approached. In many places the fields were enclosed with mud walls to keep out the floods; and after a few miles we entered a dreary mud flat, on which nothing grows but a miserable weed; the monotony is not varied even by the wild fowl generally found in the most desolate places; for even they appear to think this country too fearful, and abandon it to the few miserable human beings who live here in wretched and tumble-down shanties.

Everything is salt, the earth, the air, and the water; not a tree breaks the dismal outline; so doleful is the place that it would seem to have been created to teach humanity contentment; for however little a man might have to be thankful for on coming hither, he would at least go away blessing Providence that the lines had fallen to him in less unpleasant places.

This stretch of mud-flat extends for about five miles in-shore; we were glad to leave it and ride back to Ma-T'ou-Ying, where a fine bunch of grapes was very refreshing to our palates, dried up as they were by the saltness of the air.

Leaving Ma-T'ou-Ying behind us, as we approached Ho-Chuang, the country again gradually resumed its thriving aspect. All around us stretched a wide expanse of perfectly flat and cultivated plain, where not a square yard was lost. No land here can be wasted on hedges or walls, and nothing marks the divisions between the fields save the change of crops, or an alteration in the direction of the furrows.

Towards evening, the sight of an unusual crowd outside a village made us aware that we had arrived at our halting place. We rode through a double rank of people, who closed in on us as we passed, and followed us to the inn.

They soon began to fight for front places at the show, poked holes in the windows, and nearly broke down the doors in the frantic excitement caused by the production of the mysterious pen and paper. They were a good-natured crowd, however, and, after indulging for a few hours in the simple pastime of looking on, gradually retired, and all became still.

October 19.—We found the road to the north from Ho-Chuang very heavy, running through deep sand, in which our cart-wheels sometimes sank nearly up to the axles; and although the mules struggled manfully, encouraged by the cheery voice of the carter, the journey proved a long one.

Though the plain is densely populated, there did not seem to be many labourers in the fields; but the villages were proportionately busy, and the people were engaged about their houses, renewing the fences of millet-stalk with which they surround their little patches of cabbage garden.

The monotony of the scenery, as Kai-Ping is approached, is varied by the mountains, which rise abruptly from the plain, immediately behind the little village, to a height of some 3,000 or 4,000 feet. They do not appear to run in regular chains, but are almost entirely detached from one another, or joined by very low saddles.

This great alluvial plain of Pe-Chi-Li, here forty miles in width from Kai-Ping to the sea, where not the smallest irregularity breaks the horizontal outline, has evidently in former ages been submerged, and the salt waves must then have washed the feet of the Kai-Ping Hills.

It must have been here that the great Kaan went out with elephants and hawks, and enjoyed the royal sport described by Marco Polo.

And now, the times again are changed; the extensive fields of very good coal, that have lain for ages untouched amongst the mountains behind Kai-Ping, are to be worked; a railway is to be laid from that village to Peh-T'ang; and over the fields, where the great Kaan was used to fly his hawks at cranes, steam and iron will lead the van of civilisation.

At Kai-Ping we slept at an inn, which was admirably characteristic of Chinese architecture and carpentry.

It is said that in the remote ages, when the people, who were originally nomads, settled down to an agricultural life, they built their first houses as nearly as possible like their tents. Hence the wavy outline of their gables, and their eccentric custom of building a heavy roof on strong upright beams, filling in the walls afterwards.

Although the chairs and tables are neatly put together, the people care nothing for finished carpentry about the buildings of an ordinary house.

The beam that supported the roof of our room at Kai-Ping was nothing more than a crooked tree; no attempt had been made to straighten it, and, except that the rough outside had been just taken off with an adze, it had scarcely been touched with a tool. The door-frames, if frames they can be called, are always made of young crooked trees, left unplaned, so that when the doors are closed, gaps are often left two or three inches wide.

In some of the larger inns of the important towns things are better done, doors and window-frames being to a certain extent fitted ; but even in the best there is generally a big hole under the door, where the mud and bricks have been gradually kicked away. A window that will open is very rare in Chinese houses, and the doors are invariably fastened with a sliding latch. I do not recollect ever to have seen a door

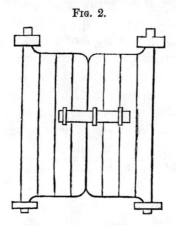

Fig. 2.

fastened on any other system, or hung in any other way, than with a couple of pivots, one above and one below, each fitted into a socket ; sometimes a hole in the floor is substituted for the lower socket.

October 20.—At Kai-Ping we again turned our backs upon the mountains, and marched over the usual plain, where in some of the fields the young winter crops were now about two inches high. When the winter sets in they become smaller rather than larger, and remain, as it were, dormant through the snow and frost until the spring, when they shoot up with great rapidity.

The people were now very busy threshing out the corn, stacking the fuel, and making things ready for the winter; in the little enclosures before the cottages of every village we used to see the whole family sitting down husking the Indian corn.

A huge mat would be spread on the ground before them, into which they would throw the grain; the husks would be put aside to feed the pigs; and the leaves thrown outside the enclosure into the streets, to dry for fuel.

The charm of variety was usually lent to the most tedious day, by some fresh freak on the part of the red pony. At Han-Ch'êng, whilst we were calmly discussing our breakfast, we heard a great uproar, and looking out, found that very original animal dashing about the yard with a light horse-trough tied to his halter. The simple Ma-Fu had fancied that, like the other ponies, he could be cheated into the belief that he was secured, no matter how trivial the object to which the rope was fastened.

This red pony was evidently looked upon as a sort of outcast by the other animals, who if separated by a few yards were always unhappy, and would keep up a lively conversation. But 'Tom' never joined in this, and used to walk on, with his ears back, ever on the look-out for an opportunity of kicking somebody or something.

Between Han-Ch'êng and Fêng-T'ai the country gradually changes in appearance; patches of uncultivated land are seen, with here and there a large extent of swamp; and were it not for the wide ditches at both sides of the roads, travelling would be difficult at any time, and impossible after heavy rain.

Fêng-T'ai is a large military station on the left bank of an important river that falls into the Pei-

T'ang, and is here forty yards wide, and navigable for boats drawing two to three feet of water, and carrying about eighteen tons.

Next to Shan-Hai-Kuan, it was the largest town we had stopped in, and being a very busy thriving commercial place, the crowds that came to see us exceeded anything we had yet encountered. The hotel fortunately was very large, and there was an inner apartment having no direct communication with the outside.

The news of the foreign arrivals soon spread abroad, and every man and child in this populous place helped to crowd the streets in the neighbourhood, and to fill the inn-yard.

Our quarters were not long sacred, and the mob in them at length became so unpleasant that we were compelled to turn them out and bar the door. The peace that now reigned inside for a few moments was a striking contrast to the wild hubbub and tumult that prevailed in the court.

But soon the surging mass of humanity burst the entrance, and poured into the outer apartment. The crash of the splintered door, for the moment frightened all but one of the boldest spirits, who advanced a couple of paces; but the moment one of us rose, his courage vanished, and he fled precipitately. This thoroughly alarmed the crowd, who, seized by a sudden panic, stampeded towards the street, tumbling one over the other in the most ludicrous fashion; and 'Spot,' hearing a rushing of many feet, dashed valiantly at the smallest boy he could see, and caught him by the skirt of his garment, before we could recover from the laughter into which we had been plunged at the ridiculous spectacle.

But curiosity soon overcame timidity, and the

crowd advanced afresh. We succeeded, however, at length, in fastening the door securely; and after a couple of hours, the people, tired of looking at nothing, gradually dispersed and allowed us to spend a quiet evening.

October 21.—The river Pei-T'ang, which flows near Fêng-T'ai, is fringed with beds of tall reeds, and carefully embanked; it winds and twists through the flat plain which gradually assumes a poorer aspect as Lu-T'ai is approached, and at Pei-T'ang, where the river empties itself into the gulf of Pe-Chi-Li, there is nothing but a waste of mud-flats in every direction.

From Fêng-T'ai to Lu-T'ai there is a road on each side of the river.

At first, as we advance towards the south, though the marshy tracts are larger and more frequent, and the land is intersected with ditches varying from ten to twenty feet in width, yet the few villages that dot the plain seem thriving enough, and the people appear happy and contented, carefully cultivating their neat gardens, in which they raise magnificent vegetables.

In China all riverside commercial towns seem squalid and straggling in proportion to their prosperity. Lu-T'ai is no exception, and being a thriving place is as dirty as any in the province of Pe-Chi-Li. Sea-going junks, of about ninety tons burden, come up here from Pei-T'ang, and discharge the grain with which they are laden; this is then stored in the town, reshipped into smaller boats, taken up the river, and distributed over the country.

The tide in the gulf of Pe-Chi-Li is felt as high up as Lu-T'ai; the river is from eighty to a hundred yards wide, and at low water there is a depth of ten feet off the town. When the tide is high the banks are no more than a foot above the surface, and during the rainy season the river overflows, and floods the neighbouring country.

On arriving here we were much astonished to learn that there was another foreigner in the place. This turned out to be an English mining engineer, who had been employed by the Chinese Government to examine the coal districts above Kai-Ping.

He was travelling with a Chinese gentleman who could talk English, and was his interpreter. Of course we all dined together, and passed a festive evening in that strange out-of-the-way place. He gave us the first items of news we had heard since leaving Peking; rumours of war between England and Russia, and another appearance, at or near Shanghai, of the great sea-serpent.

He was congratulating himself upon the success of his researches in the Kai-Ping Hills; for he had been previously sent to examine the coal-beds in another province, and when he had reported unfavourably on them, the Chinese Government had intimated that they had a very mean opinion of a mining engineer who could not find coal when ordered to do so!

October 22.—From Lu-T'ai we found the country more and more wretched as we again approached the coast. A few miles to the south we passed through a town on the banks of the river busy with some salt works; and then, until we reached Pei-T'ang, with the exception of one solitary hut in the middle of the fearful swamp, not a building, nor a tree, nor one spot of cultivation broke the monotony of the mud flat, over which a strong and cold north-east wind drove clouds of dust, that every now and then completely shut out the view.

The river was sometimes close to us on the right, at others, taking a sweep, it bent away till in the mist we could hardly distinguish the masts and sails of the junks that crowded the watery thoroughfare.

As we approached the mouth of the river, the North Pei-T'ang fort came into sight, and turning aside from the direct road we galloped off to visit the commandant.

He invited us into his house to take tea, and we were at once struck with the unusual cleanliness of his room. The regular warming apparatus was wanting, but its place was occupied by the raised dais, also called a kang, which, though common in Southern China, is rarely seen in the North. Two mattresses covered with chintz were spread on the dais, with one of the usual little low tables between them. At the centre of the end wall behind, there was a kind of altar with some yellow dragons painted on it, and a small table and a couple of chairs were arranged on the other sides of the room, in perfectly symmetrical, but unpicturesque order.

Instead of pictures, mottoes or verses from the ancient writings, in very large Chinese characters, were hung up, completing the furniture and ornamentation of the apartment.

Fresh clean paper, in which a pane of glass about six inches square had been let in, covered the window, and for a Chinese house the room was marvellously clean, and a striking contrast to the generality of the dwellings of even the richer classes, where dirty floors seem never to be swept, dirty paper never removed from cobwebby window-frames, and dirty furniture never cleaned.

This officer's servants were more respectful than servants usually are; the people did not crowd into the room, or join in the conversation unasked; and altogether our host was rather a favourable specimen of a Chinese official.

The pipe was, of course, handed round as a matter of civility, and then tea was brought in, which, made of salt and muddy water, was neither refreshing nor palatable.

After a little polite conversation we took our leave, and rode on to the wretched hamlet that is built on the left bank of the river opposite Pei-T'ang.

The river here has opened out to a breadth of two hundred yards, and is crowded with many junks of all sizes. When the wind is strong it is a dangerous matter to cross it in the crazy ferry-boats of the place, and sometimes impossible.

We had much difficulty in persuading the ferryman to venture, and it was only by the promise of great largesse that we ultimately overcame his scruples, and this not until we had spent about an hour in a little shanty filled with a filthy and ragged crowd of the poorest of even poor Chinese.

They brought all sorts of fearful odours with them, but by the aid of tobacco, we succeeded not only in tolerating their presence, but in extracting various scraps of information.

At length the heart of the obdurate boatman was softened, and leaving our friends on the bank we embarked on our perilous journey.

The first boat-load was composed of our two selves, our two servants, two muleteers, the cart, and five mules. We fully expected an accident, as the crowd of men and animals was so great that there was hardly room for the men rowing; and as the boat had no bulwarks, and was only just wide enough to let a horse stand across it, it was a marvel that one of the animals did not put its hind legs into the water. We crossed, however, in safety, and reached Pei-T'ang, in appearance the dirtiest, the most squalid, and utterly poverty-stricken place that can well be conceived.

The inn was in complete accord with the foulness of its surroundings ; its yard, deep with rotting mire, was barely large enough to accommodate all our animals, who whisked their tails in at the door of the room where we sat, pervaded by a reeking odour from the accumulation of abomination outside.

October 23.—When the dismal town of Pei-T'ang had been left behind we rode over the most miserable mud-flat, where there was not a bird, and where not even a blade of grass would grow. After two miles there was a little weed on the damp mud, but we saw not a living thing until we reached the banks of a canal where there were a few people about.

Here immense piles of long dry grass were stacked, and sweet-smelling artemisia, that filled the air with its perfume. We marched another mile without much change of scenery, but after crossing another canal the aspect of the country suddenly and completely changed.

The ground was covered with the same long dry grass, and sweet artemisia, which the people were cutting, stacking, and carting to the banks of the canal. Just at this moment the clouds which had threatened rain all the morning dispersed, and the sun shone out cheerfully on a landscape that was very pleasant after the miserable country we had now happily left behind.

There were no villages to be seen, for during the summer rains all the neighbourhood is under water.

An excellent firm road followed near to the northern bank of the canal, and after a march of eighteen miles, we arrived at a small quiet village, where the people did not attempt to crowd about the inn ; and where the landlord, a most respectable old man, quite won our hearts by his kindness to poor 'Spot,' who was completely knocked up by the incessant and lengthy marches.

The pleasant inhabitants of Huai-Tien differed in opinion as to the distance of Urh-Chuang, where we hoped to pass the night. Some said it was thirty, and others only fifteen miles ; it was quite clear, however, that the march was a long one.

We started as soon as possible, and a ride of eight miles over the same sweet-smelling country brought us to Fêng-T'ai-Tzŭ, where the canal was crossed by a ferry. Sitting on the banks waiting for the boat we were able to enjoy the charms of a quiet evening landscape.

The canal opened out here into a broad fleet, and a quarter of a mile away across the water there was a town that called to mind many of the old Dutch towns at the mouth of the Scheldt. The sun was setting behind it, and threw its nearly horizontal rays athwart the wide expanse of marsh, where great lazy herons stood with their feet in the mud, watching for the fish. On the water there were many boats, laden with the sweet artemisia, or carrying a couple of peasants who were taking home in their big baskets the purchases they had been making in some neighbouring market town.

On the bank beside us a few people and donkeys were waiting for their turn in the ferry-boat which was being slowly punted from the other side.

Our cart and mules went over first, and this was a most comical sight. The boat was flat-bottomed, flat-sided, square stern, and square-bowed. Two feet broad at one end, it widened to five feet at the other. The cart was run backwards over the narrow end, its two wheels outside, and its axle resting across the boat; the shafts were tilted up in the air, and a man squatted behind to keep the back down; the narrow end of the boat was not more than a couple of inches out of the water, and the body of this unstable-looking craft was loaded with three mules and three or four men.

In this manner the transit was safely effected, and eventually horses, mules, cart, servants, and ourselves arrived at the further side.

Fresh inquiries as to roads and distances did not tend to reassure our minds; we were told that the roads were bad, that they twisted and turned like a river in a plain, and that we yet had a long march before us. Chin-Tai was therefore ordered to get a guide, at any price, from one of the villages, which were tolerably numerous in this locality.

The villages were all exactly like one another; built of mud, with mud roofs, they were surrounded by the most luxurious cabbage-gardens, each raised about a couple of feet above the level of the country, to save them from the floods that occur regularly in the summer; every villager seemed to have his own little plot, separated from the next by a miniature canal.

The roads near the villages were also raised, and at each side there were wide ditches, sometimes opening out into little ponds, where men and boys were busy in the black mud arranging the fish traps that seemed to give them plenty of occupation, for the time that they could spare from threshing corn, making bean-curd cakes, or grinding flour.

As we left the houses behind, the road, no longer higher than the swampy ground, became very difficult, full of great holes and soft places, where the cart could hardly pass; and as night closed in, we came to a forced halt at a regular morass.

Our hunchbacked guide now made a cast to the left to find a crossing; away he plunged into the darkness, and following the sound of the cracked voice that we heard at intervals, we floundered about in the quagmire, now to the north, then south, round to the east, and back again to the west.

The cries of the carter, as he urged his tired mules to fresh efforts in the deepening bog, scarcely overpowered the croak of the bull-frogs.

At length, finding ourselves close to a village that we had left half an hour before, we began to have serious doubts of the local knowledge of our guide, and we tried to catch him, that we might question him as to where we were, but he was like a will-o'-the-wisp, and when anyone came near him he danced off into the darkness and disappeared: there was nothing for it but to follow our quasimodo as best we could—now knee-deep in mud and water; now, our feet resting for a moment on a bit of hard ground, hopes were raised only to be speedily dispelled by a sudden plunge into a fresh hole; now the shrill voice of the guide was lost behind a wall of tall reeds; and now again we found ourselves almost at his heels. So we went on, until at length perseverance was rewarded, and the fen was left behind.

Very slow, however, was our progress over the still soft roads, until the hunchback comforted us with the assurance that we had but another three miles before us.

With raised hopes we strained our eyes to catch the first glimpse of the dark irregular line of the houses against the sky. At length a light for a moment flashed a sickly glare, and we all pushed forward with lightened hearts, when our pilot, suddenly turning sharply away again, plunged us into the depths of despair.

On again, riding behind Chin-Tai, I could just make out his pyramidal shape swaying from side to side as he nodded asleep on his tired steed, which turning to nibble a blade of grass nearly upset the rider into the mire. Again we fancied for a moment that we saw the loom of a village, only again to be disappointed.

As those three weary miles lengthened out, I heard Carles sorrowfully call out that our moon was nearly finished, and so it was, for it was setting.

Forward again, till we thought the burden of the Wandering Jew had fallen on us, and that we were condemned to struggle till doomsday.

At length another village; but we dared not lay any flattering unction to our souls, until our hunchback, having turned from the road towards the houses, and from the street into a courtyard, we found ourselves, after a march of thirty-eight miles, at the inn in Urh-Chuang.

There was no news of our mules, although we had despatched them from Pei-T'ang long before we had started ourselves, but the feeling of difficulties overcome, and our haven gained, enabled us to bear the intelligence with equanimity; and though we had no bedding or blankets, we lay down on the kang, and, hard as that was, enjoyed a sound and well-earned sleep.

October 24.—After our adventures of the day before, we were not altogether sorry that the non-arrival of our mules compelled us to halt a day. A messenger whom we had sent to Tien-Tsin arrived with supplies of all kinds; but even fresh bread, and butter in a lordly tin, could not compensate for the absence of the letters and newspapers we had hoped for; and as I took my gun and strolled off to a great swamp close by in the hopes of a duck, Carles pointedly observed,

'Blessed is he who expecteth nothing, for he shall not be disappointed.'

There were plenty of duck and wild-fowl of every kind, but as they remained in the very centre of a shallow lake, even wire cartridges were of no avail, so I turned my steps homewards down-hearted; but as I was wading through the slush I put up a snipe, then another, and another, and I hastened back to get some more suitable ammunition and convey the joyful intelligence to Carles, who quickly joined me, and we had a couple of hours of excellent sport.

Marco Polo says that the Emperor 'starts off southward towards the Ocean Sea, a journey of two days. . . .

'But they are always fowling as they advance. . .

'And all that time he does nothing but go hawking round about among the canebrakes along the lakes and rivers that abound in that region. . . .'[9]

This country around Urh-Chuang is admirably described in the above passage, and I should almost imagine that the Kaan must have set off south-east from Peking, and enjoyed some of his hawking not far from here, before he travelled to Cachar Modun, wherever that may have been.

The mules turned up in the course of the day, and as each successive muleteer narrated the fearful events of the previous night, we should have felt the most profound pity for their sufferings, had not their varied tales brought Falstaff and his men in buckram so forcibly to mind that we said to one another, 'These lies are like the father that begets them, gross as a mountain, open, palpable.'

We found out afterwards that, owing to their vile pronunciation, they had strayed to a wrong village

[9] M rco Polo, vol. i., book ii., chap. xx., 2nd edit., pp. 389, 392.

for the midday halt, where, as we did not appear, they spent the night.

October 25.—Chin-Tai had a violent wrangle with the innkeeper this morning.

Chinese servants as a rule are marvellously quiet in their ways. During meals they move about without a sound, and never rattle the plates, knives, and forks in the manner so dear to the hearts of English waiters; but for their quiet movements they make up by their noisy tongues. Once they begin to wrangle, they shout as if they were drilling a battalion, and if (as usually is the case) the quarrel happens to be about money, Babel itself would be comparatively a tower of silence. But though, from the sounds, it would seem as if they were ready to kill one another, as a rule they mean no harm, and are merely indulging in the pastime that all Chinamen thoroughly enjoy, a good wrangle about money. So great, in fact, is the pleasure they take in this amusement that even the one who gets the worst of the dispute thinks it better than no wrangle at all.

As I strolled about the village before starting, I could hear the shrill voices from the furthest end of the street, and when I returned the discussion was more violent than ever. Ultimately, as Chin-Tai seemed inclined to spend the remainder of the day in this fashion, it became necessary to interpose, and tear him away.

Turning into the innyard to mount our ponies, we found the Ma-Fu apparently chopping at Tom Bowling's head with a gigantic chopper.

The Ma-Fu was a constant source of amusement to us. Nothing seemed to please him better than to have four loose ponies running about, entangling their legs in their bridles, and kicking at everybody and everything. On coming into the yard of an inn, if he could, he would always leave them loose, and when we did prevail upon him to fasten up one or two, he would always make use of some twine that would scarcely hold a puppy. This was all very well for the quiet ponies that only cared to go to sleep in peace; but it was never a success with Tom Bowling, who was always on the look-out for a chance of playing some trick. On this occasion the Ma-Fu had discovered a piece of rotten cord which he thought would make an excellent headrope, and having fastened it very insecurely to the headstall, we found him cutting off the end of the knot with the gigantic chopper. Of course he was only holding the end at which he was chopping, so that if he had succeeded he would have been like the man who sitting on the bough of a tree sawed it off between himself and the stem.

Once fairly off we heard Chin-Tai recounting aloud to himself all the details of his dispute; and by the way in which he chuckled, and from his evident enjoyment of the recollections, it was tolerably clear that the innkeeper had not worked much to windward of our close sailing attendant.

We gradually left the swampy ground behind, and entered a slightly undulating country, where there were no signs of floods, where the villages were again frequent, and a few people were out ploughing in the fields.

The ploughing of the Chinese is very poor and unscientific. They scarcely do more than scratch the surface of the ground; and, instead of the straight lines so dear to the eye of an English farmer, the ridges and furrows in China are as crooked as serpents.[1]

We struck the Pei-Ho river at Pei-Tsai-T'sun, where there was no room for us at the inn, and where, as we were obliged to wait about whilst inquiries were being made, a considerable crowd soon collected around us.

The Ma-Fu, for the first time during our acquaintance, showed a faint spark of intelligence, and warned the bystanders of Tom Bowling's queer temper. One man, however, in his eager curiosity, took no heed, and was kicked in consequence. In similar cases the bystanders often attribute the misfortune to the

---

[1] It is difficult to understand how the Chinese have acquired such a high reputation amongst Europeans for scientific farming. The real secret of their success lies in the care they take that nothing is wasted. They use no other manure than the sewage of the towns, and not one particle of this is lost. Householders sell the produce of their latrines to agriculturists. The sewage is collected from the houses in buckets every few days and carried to the fields. The stench that pervades the courtyard of a Chinese house, and penetrates the inmost recesses of the dwelling is of course abominable.

It is a recognised fact amongst modern sanitarians that sewage is most dangerous when confined in unventilated drains. It is perhaps owing to the non-existence of covered drains that the horrid smells of China seem innocuous. At all events the Chinese do not appear to suffer any ill-effects from their system, which at least has the advantage of simplicity.

The removal of sewage is carried on at all hours, and coolies with buckets of sewage may constantly be seen in the most narrow and crowded thoroughfares during the busiest time of the day.

There is a fable that the Chinese barbers never let the cuttings, or shavings of the hair of their customers fall to the ground, because—so the story runs—they consider anything sacred that comes from the human head. This is pure fiction, and although in some cases the scraps are preserved, they are kept solely that they may be used as manure on some pet plant or patch of cabbage garden.

On the high-roads in China, and especially in the neighbourhood of the great riverside towns, it is no uncommon thing to find dozens of latrines by the roadside, each belonging to some individual who has erected it, not for the sake of decency, but that he may obtain the manure.

foreigner, but here the victim found no sympathy, and was told that he was served quite right.

In this town the Ma-Fu bought a new rope for the red pony, a stiff cable big enough to hold a man-of-war. He made it fast to the headstall with a knot about the size of a bird's nest, and as the poor animal had been endowed by nature with a very large head, he looked extraordinary with this hawser wound round his neck. It was not, however, of the slightest use, for the Ma-Fu still pursued his system of choosing some particularly fragile article to fasten him to.

The Ma-Fu was one of those people who are always trying to do right, but ever succeed in doing wrong. If it happened that he was wanted on the road, he was certain to be a mile behind, and just when his services could not possibly be required, he would dash past us at a gallop, or if there happened to be a particularly dusty spot, would ride in front or to windward of us.

October 26.—There was at first not much change in the aspect of the country, and as I sat at my diary in the evening, after having for about the ninety-ninth time conscientiously recorded the fact that 'the country was well populated and closely cultivated,' the spirit of my old friend Marco seemed sometimes to enter into me, and I almost found myself writing that 'the people were all idolaters and used paper money.'

We were now in the neighbourhood of the Pei-Ho, and the habits and customs of foreigners were not matters of so much interest to the people, but still, as we approached some village, a man at work in his cabbage-garden in the outskirts would spy us, and call out to a friend that some of those queer foreigners were coming. Then, one after another, all the people would come out, followed by the children, the little ones clinging to the skirts of their fathers' clothes, and the bolder pushing about amongst their elders' legs to get a front place. They would all laugh at our queer eyes; but many would wish us good day, and ask us whence we came, and whither we were going, and a rude word was scarcely ever spoken.

As we marched eastwards we again approached the inundated country, where miles and miles of spoilt crops were still standing in wide expanses of swamp. As the sun was setting, a mist rose from the marshes and wet fields, with promises of miasma and ague for the inhabitants, who nevertheless seemed to be a fine race of men.

The smoke from numerous little fires in the fields, where useless weeds were being burnt, blew across the road, bringing a fragrant perfume from some sweet herb or grass in the flames. As it gradually grew dark, the road became deserted, and we met scarcely anyone but here and there a peasant, who had been later in his field than usual, and was now wending his weary way homewards.

The bull-frogs began to croak in the swamps, and except the hoarse quack of the wild geese as they flew over in long strings, there was no other sound to break the stillness of the quiet night.

Passing several little villages, where great piles of millet-stalks were stacked for the winter fuel, and the sweet scent of the artemisia was borne to us on the evening air, we arrived at the quiet little hamlet of Huang-Chuang, where we halted for the night.

October 27 —A grassy country lay to the east of Huang-Chuang, but in the vicinity of Lin-Ting-Chen the signs of recent inundation were very apparent, and some patches of spoiled millet were still standing. The villages here were all raised above the country, and surrounded by wet ditches lined with banks of high reeds, and with a punt tied up in some quiet corner. Neat stone bridges spanned these dykes, that often opened into ponds, where some fish-traps were sure to be seen, and perhaps a man groping in the mud.

Lin-Ting-Chen is a large straggling town on the right bank of the river, which is here crossed by two bridges, one a very good stone bridge, with five arches, hardly large enough for the rush of water that must come during the rainy season ; the other, about a quarter of a mile higher up, a small wooden bridge which is probably carried away in floods.

Here we saw a number of the quaint jointed boats in use on this river, of which the bends and twists are so sharp that a long boat cannot be navigated on it.

The people find the most convenient plan is to join two boats together stern to stern. If the wind is fair the boats will sail, but if not they are tracked up by coolies, who tow the boats with a rope fastened to the mast-head.

For a mile or two beyond Lin-Ting-Chen the road was good ; but then we found that, owing to the inundation, it was necessary to ride on the top of the river embankment.

For the last five years the whole of the country in this neighbourhood has been more or less under water, and has not dried up even in spring, which is the driest season. It looked to us as if a great sea lay spread out before us, from which the villages rose like islands. Except on the top of the river embankment communication was altogether stopped, for this year the floods were not deep enough for boat traffic, but still the ground was too soft for carts or animals.

It seems surprising that the people can live, but the Chinese will get something even out of an inundation; and here the swamps are turned into gigantic fish-ponds, where great fat carp are bred, where fish-traps are set at every twenty yards, and where men are seen fishing with nets wherever it is deep enough. Quantities of wild fowl congregate also, and the hoarse qua-qua of the wild geese, as they flapped heavily overhead, became one of the sounds so frequent that we almost missed it when it stopped.

A steady jog-trot of seventeen miles from Lin-Ting-Chen brought us again to Fêng-T'ai, where we stayed to feed our ponies and inquire about the road.

We tried to get information about the distance to Hsin-An-Chen, but the accounts were most contradictory.

We found out the reason of this afterwards. The fact was that the floods were so extensive that the ordinary roads had ceased to exist, and the people all had different ideas of the extent of the marshes, and of the distance that it was necessary to go round to encompass them. Of one thing they all seemed sure, however, viz. that we must ride along the river embankment to Wo-La-Ku, and then make further inquiries. Our ponies had already travelled for seven hours at a steady jog-trot, but after an hour's rest we were obliged to mount again and keep up the pace.

Although our destination lay nearly due north we were sent off nearly east, and in this direction followed up the river for about eight miles.

There was certainly something mysterious about Hsin-An-Chen; half the people we met had never heard of it; those who had were most wild in their estimate of its distance. They seemed to know nothing of its position, and in answer to our inquiries would stare vacantly, and waving their hands vaguely in a northerly direction, tell us to go due west, when the road would immediately lead us to the east, and after a twist or two finally settle down into a southerly zig-zag, dodging about amongst the ditches and swamps.

At Chang-Yai-Chuang, we were ferried across another branch of the river, where the boatman comforted us with the assurance that Hsin-An-Chen really had an existence, though of the distance he was not quite sure. Soon afterwards, as it grew dark, we found a guide, and we arrived in safety after having trotted the ponies steadily for upwards of eleven hours.

October 28.—At Hsin-An-Chen, the river is about forty yards wide, and there is a floating bridge across it. From this we sent the mules by a direct road to Pao-Ti, and went round ourselves back to Lin-Ting-Chen, where we halted, and after breakfast strolled down to the waterside to see the boatmen, and have a chat over our cigars. But the men were far more anxious to ask us questions about ourselves than to answer any of our interrogations; so we went off again up the left bank of the river, under a fine grove of willow trees, where the road was raised six feet above the country, with ditches at each side. Presently we crossed back to the right bank by one of the bridges so common in this part of the country.

The piers are built of stones laid horizontally one over the other, and the roadway is formed of long stones laid across; some of these stones are twenty feet in length, whilst spans of fifteen feet are by no means uncommon.

Riding through the villages in the middle of the day, when the people were mostly busy in the fields, the children seemed to take little or no notice of us, until some of the bigger boys or men began to make comments, then the children, thus early discovering their imitative faculties, would follow the

Fig. 3.

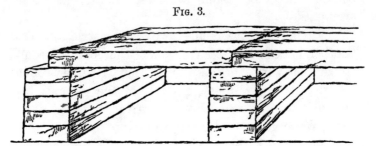

example of their elders, and run after us gazing until we were out of sight.

October 29.—There was a great fair going on somewhere in the neighbourhood, and the people we met were dressed in their best clothes; the men with their heads freshly shaved and oiled, and the women shuffling along on their hideous and deformed feet, with a little bit of colouring somewhere about their usually sober dresses.

The coiffure of the women was extravagant, and fastened with pins as big as skewers, whose elaborate and fantastic tops mingled with the artificial flowers that served as decorations for the hair.

They were a hideous set of old hags. But, not devoid of a share of that coquetry which all women have inherited from Eve, they tried to eke out the scant measure of their beauty with some little bit of finery about their clothes.

Then there were the mothers leading their children; the latter in gay-coloured coats, with their queer rudimentary plaits at the sides of the head.

When the boys are very young, the hair on each side of the head behind the temples is formed into two plaits, which stand out like a pair of horns.

When sufficiently long all the hair is combined into one large plait.

The richer people were on mules, or donkeys, or in carts, all with harness polished up, buckles brightened, and little bells round the necks of the animals, looking as smart as possible.

Here and there was a man who had no time for frivolities, and as he rested a minute or two in his work-a-day clothes, with his hand on his plough or hoe, he seemed to throw an envious glance at the happier folk hurrying along to amuse themselves.

Presently we came upon a whole family sitting in their own little patch of cotton, about thirty yards square, busily engaged in picking the cotton from the pods. It would seem impossible that it could pay to grow cotton in such small quantities, but the Chinaman likes doing everything for himself, and if it be but on a trivial scale he thinks that better than joining with others.

We soon left the fair behind us, and in a very unpleasant dust-storm rode on to the west, the road improving as we left the wet country behind. We came again upon the accustomed villages ensconced in groves of willows, where the pert magpies chased one another from branch to branch; further on, the country was quite dry, no crops were left standing, and all the stubble was dug up; plantations of young poplars and willows skirted the road, which took us to within a quarter of a mile of the Pei-Ho river, where thousands of masts could be seen above the river embankment.

As I closed my diary that evening at Tai-Tzŭ-Fu, I could not help regretting that our journey was so near its end. After the weary monotony of a board-ship life, the activity of travel had been very delightful; and when I thought over the glorious weather, and the keen crisp mountain air of Mongolia, and called to mind the many beautiful scenes we had passed through, all the petty and trivial discomforts of dirt and inquisitive people began to fade away, and I recollected little but the fine free life, and the pleasant society of the best of companions. But the days were slipping by, and in a short time what a change would come o'er the spirit of the scene! all this country would be bound in an icy grasp; bitter winds would sweep across the unprotected plain; and in thinking of them I could understand why the people planted such thick groves of trees round their villages. It was time our journey came to a close, and if I did not heartily share, I could at least appreciate the high spirits of the servants and muleteers.

October 30.—The distance to Peking was thirty miles, and Carles and I determined that in that city we would breakfast.

There was no difficulty about getting people up this morning; everyone was on the alert, and ready for a start. No need for us to wake ourselves at some fearful hour in the night; long before we wanted to stir, the muleteers and Ma-Fu were busy at work in the courtyard.

No need for us to urge the lazy muleteers to waste no more time in dawdling about before saddling their animals; long before we were ready to give up our boxes and our beds, the men came into the room with restless glances at our open trunks.

The morning was dark, and we had to pick our way carefully for half an hour; but presently the rosy glow of the rising sun lighted up the eastern sky, and we were able to push on to Tung-Chou, where we crossed the Pei-Ho by a ferry, and rode through the busy town.

We had to look about us in the crowded streets, and to take care of the little strings stretched across the road about eight feet from the ground, to which cotton cloths are tied to protect the shop-fronts from the sun.

Here we were greeted with the familiar smell of the filthy sewer, a smell foul enough to sicken the strongest stomach. I said to Carles, ' How can people live in this fearful stench ? ' ' How can a man live there ? ' he replied, pointing to an old-clothes shop, where the merchant was standing sniffing up the fetid odours of the sewer, and the horrid aroma of the foul rags, old clothes, and tattered sheepskins with which his shop was crowded to the very ceiling.

The merchant mistook Carles's motion for a note of admiration, and stepped forward with a bland smile to offer for sale any or all of the contents of his shop ; he seemed sorely disappointed when, taking no heed of his offer, we pushed on as well as we could through the teeming crowd in the narrow streets.

We made short work of the eleven and a half miles between the western gate of Tung-Chou and the gate of Peking, and, covering the distance in eighty-nine minutes, we were in time to realise the anticipated enjoyments of a civilised breakfast.

Although the sun was still warm, the weather had suddenly become very cold, and in the evenings, as I sat over the roaring fires kept up in the cheerful European houses, I soon abandoned any regrets at the termination of our pleasant trip.

In Peking, almost the only fuel in use is wood, the cheapest of which comes from America to Shanghai, is transhipped thence to Tien-Tsin, and brought up to Peking in boats. Even then it is cheaper than coal, which costs from 3*l*. to 4*l*. a ton, although there is enough amongst the mountains to supply the world, and some of it only thirty to forty miles away. But if fuel is expensive, this is compensated for by the cheapness of provisions. Beef is 3*d*., and mutton 4*d*. a pound, one partridge costs 2*d*., and a pheasant 3*d*.

The days slipped by very pleasantly in Peking. Men from the bazaars used to bring in great piles of embroidery to tempt the unwary ; costly furs of every description used to cover the floor of my room ; old curios, and modern shams, bits of bronze worth almost their weight in gold, and marvels of ancient porcelain were displayed in lavish profusion. But better than all were the newspapers and letters. I had not received a letter since leaving Europe, as I had travelled from Marseilles to Peking with one mail, and had left the northern capital before the arrival of the next.

A morning was spent in those quaint dark shops in the by-streets of Peking, and an afternoon in one of the regular fairs.

This was an amusing sight, and very like a European fair. There are stalls where every description of cheap trifles is sold, and nothing expensive is to be found. Children's toys, dolls, clay models of spiders, grasshoppers, and all sorts of insects ; groups of men, women, and children, cleverly modelled in clay, and highly characteristic ; ribbons and bits of finery for the women, pipes and chopsticks for the men. Then there are the eating-stalls, where divers savoury dishes are prepared, and the hot-potato men and the sweetmeat-sellers offer their attractive goods. Pigeons too are sold in great numbers, for the Chinese are great pigeon-fanciers. And in every corner there is a surging crowd of people, laughing, pushing, buying, or selling; the sellers calling out the virtue of their wares, and begging people to come and buy; the purchasers bargaining, and chaffering, and all enjoying themselves thoroughly.

One evening we had a Chinese dinner in the most famous of the Peking restaurants, the 'Restaurant of Virtue and Prosperity.'

I shall not attempt to describe a Chinese dinner, for although the subject may be of a nature to present some amusing details for a European, yet, as the humorous Abbé observes: ' These details are so well known that we should fear to abuse the patience of the reader. We have besides remarked in the *Mélanges Posthumes* of Abel-Rémusat, the following passage, which would quite suffice to dissipate the idea, if ever it possessed us, of giving a nomenclature of the dishes which were served to us :

' " Some years ago, on the return of a European embassy from China, where the officers composing it had not found much to boast of in the success of their mission, it came into their heads to offer to the readers of the Gazette an account of a dinner that had been given them, they said, by the officials of some frontier town. According to their account never had guests been more sumptuously regaled; the quality of the dishes, the number of courses, the play-acting during the intervals, all had been carefully arranged, and furnished a magnificent example.

' " To those who were in the habit of reading old books there seemed something familiar in the account of that dinner. More than one hundred years before the time of these officers, certain Jesuit missionaries had partaken of precisely the same repast, composed of exactly the same dishes, and served in the same style.

But there are many people for whom everything is new, and although it is certain, '*qu'un dîner réchauffé ne valut jamais rien,*' this *réchauffé* at all events was found excellent, and the public, always greedy for peculiarities of customs, and even for the details of cookery, did not trouble itself as to who had been the real diners. It was pleased with the singularities of the Chinese service, as well as with the gravity with which the guests, in eating rice, executed manœuvres and evolutions which would have done honour to the best drilled regiment of infantry." '[2]

If now I should present our bill of fare I should be suspected of having dined with, or of plagiarising, Mrs. Brassey!

Gelatine is the foundation of every delicacy that forms part of a high-class Chinese dinner. Swallow's-nest soup, shark's fins, sea-slugs, and sea-weed are nearly pure gelatine. For flavour, the Chinese seem to know but duck and pork; and the succession of gelatinous foods, flavoured the first with duck, and the next with pork, is tedious in the extreme.

European wines are utterly out of place with a Chinese dinner, and even the most conservative Englishman will find that hot rice-wine, with a bouquet of rose-water, sipped from cups not much larger than thimbles, is preferable to the driest vintage of Heidsick, or the rarest *cuvée* of Lafitte.

A European generally finds the first Chinese dinner he eats very good, the second indifferent, and the third nasty.

The restaurant-keepers at Peking, Shanghai, and Macao doubtless invent fantastic dishes utterly unknown to an ordinary Chinaman, in order to satisfy the well-known English love of the marvellous. At

[2] Huc, *L'Empire Chinois*, vol. i. chap. 5.

all events, it would be as fair to judge of an English household dinner from a Greenwich feast, as it would be to consider one of these made-up and elaborate entertainments a type of a Chinese gentleman's usual meal.

But even of that, as well as of the diversified and lengthy repasts served up in these restaurants, visited at intervals by curious Europeans, it may be said with tenfold the force with which the remarks may be applied to a Greenwich banquet, that 'the appetite is distracted by the variety of objects, and tantalised by the restlessness of perpetual solicitation, not a moment of repose, no pause for enjoyment ; eventually a feeling of satiety without satisfaction, and of repletion without sustenance ; till at night, gradually recovering from the whirl of the anomalous repast, famished yet incapable of flavour, the tortured memory can only recall with an effort that it has dined off'[3] gelatine and grease!

I have heard it said that the Chinese use paper pocket-handkerchiefs. This, however, is not the case, but the idea may have originated in the little squares of paper that are laid beside each diner, and are used for wiping the chopsticks after partaking of any dish; for one pair of chopsticks must serve for the whole dinner.

November 6.—Sir Thomas Wade left Peking in the afternoon, and to say that everyone regretted him is to convey but a faint idea of the blank his absence caused in Peking society. His hospitality and liberality were unbounded, and the regrets that his departure gave rise to will only cease on his return.

At the gateway of the Confucian Temple at Peking there are some stone drums, as they are called, for the

[3] Coningsby.

Chinese have no other word for cylinder. These drums were discovered somewhere about the year 600 A.D. lying half buried in the ground in the department Fêng-Hsiang-Fu, in the province of Shensi; they are supposed to have been inscribed between B.C. 827 and B.C. 782. The locality in which they were discovered was a portion of the ancestral territory of the founder of the Chou dynasty. Tan-Fu (B.C. 1325), afterwards styled T'ai-Wang in the sacrificial ritual of the dynasty, removed to the foot of Mount Ch'i in the present district of Ch'i-Shan in the department now called Fêng-Hsiang. Subsequently, after the establishment of the Chou dynasty by his descendants, the south of Mount Ch'i would appear to have been a favourite resort of the imperial hunting expeditions; and it is supposed that these stones were erected in commemoration of one of them. Originally large water-worn boulders, they were roughly chiselled into their present cylindrical shape, and were removed to the Confucian Temple of Fêng-Hsiang-Fu, where they remained till the end of the T'ang dynasty (A.D. 937), but were again dispersed and lost from sight during the wars and troubles of the five dynasties. Under the Sung dynasty literature again flourished, and Ssŭ-Ma-Ch'ih, prefect of Fêng-Hsiang-Fu, collected and found nine out of the ten drums, and placed them in the gateway of the Imperial College. The missing one was discovered A.D. 1052, and thus they were again complete.

When the Khitan, or Liao Tartars, invaded Northern China, the Sung Court fled south, taking with them the drums, which were set up in Pien-Ching (now K'ai-Fung-Fu, in Honan) their new capital (A.D. 1108). A decree was passed at this period ordering the characters to be filled with gold to preserve them. In 1126 the Kin, or Niuchih Tartars, captured the city and took the drums to Peking, the gold was dug out from the characters, and the drums were more or less neglected until 1307, when they were placed in the gateway of the Temple of Confucius, where they now are. ('Journal of the North China Branch of the Royal Asiatic Society,' new series, No. viii. Shanghai, 1874.)

November 7. — At about half-past eight this morning, a riding party assembled for early breakfast before a visit to the Summer Palace. A cart with luncheon had been already sent off very early, and starting ourselves at half-past nine we reached the building by eleven.

The ruins of the Summer Palace, though very beautiful, are very sad. One seems to be brought here face to face with the wreck of an empire. The builders of this palace seem to have been imbued with something of the spirit of those who in the middle ages raised in Europe such noble monuments of their devotion and piety. The whole soul of a man must have been in the work; no part was neglected, no money, time, or labour spared; infinite care was bestowed on every detail, and notwithstanding the desolations and ruin, there still seems to breathe over all the spirit of a master mind. Roaming about the palaces now overgrown with weeds, or looking out on that still lake whose mirror-like surface must have reflected so many and such curious sights, one cannot help feeling that the architect must have had a faith in something, even if it were only in the possibility of complete human happiness.

In the Wang-Tua-Shan enclosure there are now only two buildings left standing; one a beautiful little pagoda of red, yellow, green, and blue tiles;

the other a temple in the same style at the top of the hill. Both were originally covered with porcelain figures of Buddha; but now the heads have been chipped off from all within reach, and in some places there are great cavities where people have been trying to extract whole tiles. It is very humiliating to see the greedy way in which Europeans chip off the figures that in their mutilated state can be of no possible utility, and are not by themselves in any way ornamental.

Surely the Chinaman cracking his water-melon seeds is at least as dignified as the wandering European desecrating shrines with his vulgar name, or destroying beautiful monuments, for the sake of glorifying himself in the eyes of his gaping country cousins, by the exhibition of a tile, or the head of a Buddha!

Here, too, it would seem to be unnecessary to carry any further the cruel work of demolition, for, by groping in the heaps of rubbish that litter the place, amongst dust and stones and broken tiles, our party found plenty of relics, some of which, terra-cotta tiles with raised figures of Buddha on one side and an inscription in three languages on the other, were at least as valuable curios as bits knocked off a building.

The parks, in which the palaces stand, inclose many acres, interspersed with hills, some real and some artificial, looking over lovely lakes, where there are inlets spanned by elegant arched bridges. Standing on the crest of the highest of these hills, the barrier of the mountains that buttress the Mongolian plateau is seen to the north, whilst to the south the eye roams over the wide and rich alluvial plain, dotted with villages and trees, the walls of Peking in the distance showing sharp and clear through that crisp, dry, frosty air.

Here there were no noisy tourists to disturb reflection; no gabbling cicerone with his automatic tongue; and mournful though it must be to think of what has been and what now is, it is with difficulty that at length one tears oneself away from the scene, at once so fair and sad.

The last emperor ordered the palaces to be rebuilt. The ministers scraped together a small sum of money, and began to mend the roads and repair the walls; but the emperor dying soon afterwards the works were stopped.

On our way home we visited the 'Bell Temple,' where there is a bronze bell, eleven feet in diameter, fourteen and a half feet high, and about four inches thick. It is said to be the largest bell in the world that is hung. From rough measurements I calculated the amount of bronze in it at 300 cubic feet; this would make its weight about 160,000 lbs.

The bell is covered, inside and out, with Chinese characters, all of which are close together, and none more than half an inch long. It is said that the characters were cast on the bell, but this seems almost impossible. The whole inscription is a prayer for rain; and during a drought the princes and chief ministers come to this temple and pray for rain, remaining on their knees until the prayers are answered, a duty which they perform much as the emperor does his at the Temple of Heaven.

It is said that the tones of the bell are supernatural, and have the power of bringing rain. This superstition in all probability rests on a substratum of fact, for the vibrations of this mass of metal may cause the precipitation of rain from an overcharged cloud, just as the report of a cannon will sometimes bring on a threatening shower.

The largest bell in the world is that of Moscow, but this still rests on the ground, and has never been hung. It is nineteen feet high, with a circumference of sixty-three feet eleven inches at the rim, and its weight is computed at 443,772 lbs. Bells are usually cast with approximately the same proportions; a bell of the same shape and proportions as the Moscow bell, with the height of the Peking bell, would weigh 177,534 lbs. My rough calculation, as just stated, makes it 160,000 lbs., and it is not likely to be less.

What is Peking like? was a question that I knew I should often be asked on my return to England, and I determined that I would, if possible, be able to answer it; but the more I saw, the more hopeless seemed the task. I took a note-book out one day to try and write down what there was to be seen, but, as I began the task, I was nearly knocked down by a camel lumbering along with a load of brick tea.

I remarked to a friend, an old resident, that nothing but a series of coloured pictures or photographs could ever give an idea of Peking as it is: 'No,' he replied, 'and even then you would not get the stinks.'

There are still, as in old Marco's time, the streets so straight and wide, and the plots of ground, on which the houses are built, are still four-square; there are many open spaces inside the walls, large gardens, and trees; but its grandeur seems to be gone, and if the old Venetian were now to return the only part of his description that he would still adhere to would be that 'it is impossible to give a description that should do it justice.'

There are still extensive remains of drains, but their place has long been taken by open sewers in some of the streets. The smells that pervade the city at all seasons of the year are abominable, and the black dust that sweeps in clouds about the streets is probably the most filthy in the world, not excepting even that of London. In dry weather, this dust lies deep in all the streets, and in wet it is turned to a horrible black mire.

From the walls of the city, near the Observatory, a fine view is obtained over Peking, with the examination hall just below. The examinations, at which 13,000 candidates had presented themselves, were just over.

The public examinations are one of the most remarkable institutions in the country. In every city hundreds and thousands of candidates present themselves yearly to pass for their degree. Each is shut up in a cell, about five feet square, which he cannot leave for two days. He is then liberated for a day, and again shut up with a fresh paper of questions.

Very little bribery and personification takes place at these examinations, but amid the universal corruption that prevails throughout the Chinese administration, it would be quite impossible entirely to avoid unjust dealings.

A story is told that on a certain occasion the examiners of some provincial capital were dining with a high military official; and during the dinner a letter was handed to one of the former. By Chinese etiquette a person receiving a letter in company must hand it to the host if he asks to see it. On this occasion the military man requested permission to look over the document; this was at first refused, but the demand, repeated in a peremptory manner, was eventually complied with.

The letter was from the father of the examiner,

saying that he had received a large sum of money from a certain person who intended to be a candidate at the next examination.

The military official read the letter, and called out to his servant, 'Bring the chaff-cutter.' The instrument was produced; and the officer put the examiner to death with his own hands, cutting him across the belly, this being the legal punishment for an examiner convicted of malpractices.

The officer immediately wrote to Peking, demanding a legal punishment for his crime, but he received for answer that his conduct had been exemplary.

November 8.—Winter was approaching, and speculation was now rife as to when the Tien-Tsin river would close, for during the winter it is fast frozen up, and the cold is so bitter that the ships lying at Tien-Tsin are roofed over, to protect the sailors from the severity of the climate, which seems, except for a few short weeks, to know no moderation.

It was time for me to leave, unless I wished to run the risk of being kept through the winter, with the choice of making a long tedious, cold, and miserable land-journey from Peking southwards.

Chin-Tai was told to hire carts, and make preparations for going to Tien-Tsin; but owing to the fact that the examinations were just over, it was very difficult to find them. Chin-Tai came in mournfully, and said there were plenty of carts that would take me to Shan-Hai-Kuan, and he spoke in such a reproachful way that I really felt as if I ought to suggest going there instead. Perseverance was, however, at length rewarded, and I began my return journey on November 9, and halted for breakfast in the familiar inn where the cockroach had come out of his crack in the table to look at me, six weeks before.

November 9.—My poor grey was none the better for the last four weeks' travelling; the hard work had made him a little shaky about the legs, and in the dark, as we were trotting to Ho-Se-Wu over some rough ground, he came down, and before I could clear myself he dragged me across the road. The stirrup leather fortunately came out, but now we had a hunt for it, and for my hat, which, though a white one, could not be seen in the pitchy darkness. An old donkey-man coming along assisted in the search, and, apparently being possessed of cat's eyes, he found both the lost articles. I told him I would treat him to a supper at Ho-Se-Wu, but finding he was not bound for that place, I gave him a few cash instead, and sent him on his way rejoicing.

Riding on, Chin-Tai began to sing and shout in the most exuberant spirits, and then asked me what I should like for dinner. I answered that as the carts were miles behind, and there were no cooking utensils, I was not prepared to be very particular.

He was silent for a moment, evidently thinking what he could do without an enormous saucepan, big enough to broil a turkey in, that he had insisted on buying at Tien-Tsin. He soon solved the problem, and bursting into another refrain, promised me soup for the first course, and asked what I should like next. I allowed him to revolve in his mind the extraordinary culinary feats that he was proposing for himself, when with a pæan of triumph he declared he could cook a beefsteak, and that he had a bottle of claret in the breast of his coat. I joined in his hilarity, and with light hearts we rode into the inn at Ho-Se-Wu.

Here I had a not altogether satisfactory experience of the Chinese institution of the kang.

When properly made, properly lighted, and properly

attended to, there can be nothing better for warming purposes than a properly constructed kang, but when the door of the fireplace is inside, instead of outside, this imperfection very soon discovers itself, for the smoke, as perverse as smoke usually is, persists in coming into the room in dense volumes.

In compliance with my demand a man came in with some shavings and a bundle of millet stalks. The first were lighted and the ends of some of the latter were held in the flames. As they burnt away the man kept pushing them in until they were consumed. He then took fresh ones, and so on, until the whole bundle had disappeared, a consummation that arrived in about half an hour without perceptibly affecting the warmth of the apartment. But the man was too lazy to do anything more, and putting his hand on the kang, smiled blandly, as much as to say 'How nice and warm it is now.' As a matter of fact, the most delicate thermometer that ever left the establishment of Mr. Casella would have failed to show a rise of temperature, but it was not worth while trying to make the people light the fire properly, as the operation would have taken a long time, and they had evidently made up their minds not to be bothered with it, so I professed satisfaction, and let the man go.

When a kang is thoroughly in use the fire never goes out altogether; the glowing embers remain, and the air inside, once the large mass is thoroughly heated, does not cool for many hours.

The kang itself, too, a mass of clay or brickwork, retains its warmth for a long time, but after having been out of use it is not possible to light it at once and warm up a room immediately.

November 10.—It was still dark when we started the next morning, and there was now a sharp frost, so after clearing the town I dismounted, until a smart double of about two miles, together with the rays of the rising sun, sent the blood tingling through my veins.

We halted for a couple of hours at Yang-Tsun, and leaving this at 12.30, rode the fifteen miles to the bridge of boats outside Tien-Tsin in exactly two hours. From this point we threaded our way leisurely through the crowded streets to the hotel in the European settlement.

A steamer was not going to start for Shanghai for some days, and I passed the interval very pleasantly amongst the hospitable Europeans, breakfasting on the French gunboat, or lunching with the officers of the English one. Tien-Tsin races occupied one day, and another was spent in coursing. Newspapers and letters filled up any odd corners in the afternoons, and the evenings were never very long.

Tien-Tsin in the winter is a glorious place for the man-of-war sailor, whether he be English or French. There can be no other harbour in the world where English sailors live so well and so cheaply. The officers used to complain that the men became so fat that they could do nothing. A sailor was seen one day to buy forty-six teal for a dollar (about 1*d*. a piece); geese, duck, and quail are all sold at proportionate prices.

By November 16 I had made the few necessary preparations for departure, and disposed of my ponies; and in the evening of that day, after dinner, embarked with the rest of my property on board the steamer 'Pao-Ting.'

November 17.—An American gentleman and his wife were the only other passengers. We left Tien-Tsin very early, and when I came up on deck we were already some distance down the river. The morning was still frosty, but there was a brilliant sun, and a walk up and down the deck was thoroughly enjoyable.

The 'Pao-Ting' did not stick on the mud quite so often as the good ship 'Chih-Li' had done on my upward voyage, but there was quite enough grounding and bumping to try the temper of the good-natured captain.

On this occasion Taku bar presented no difficulties, and at two o'clock we steamed towards Chi-Fu across the gulf of Pe-Chi-Li.

November 18.—When I ventured my head outside the companion, the ship was rolling heavily, and I found that snow was falling, so I retreated again to the cabin until we reached Chi-Fu, where we anchored at nine o'clock. The swell setting into the harbour made us roll so much that even here we were obliged to keep the fiddles on the dining table, and later in the day it was necessary to move across to a more quiet berth to finish coaling. This operation was rather a lengthy one, and it was not till nine o'clock in the evening that we were again under way.

November 19 *and* 20.—The ship rolled about all day in a heavy cross sea; but during the next night a change of course, or of wind, brought her head to the waves, and she was a good deal steadier. As we steamed southward we gradually left the stormy seas and entered a more peaceful region, where the sun came out and a gentle south-west breeze warmed the air; and on November 21, at 8.30 in the morning, we moored off the wharf at Shanghai.

CHAPTER IV.

A CYCLE OF CATHAY.

Cradle of the Chinese Nation—Their Settlement in Shan-Si—Characteristics of Chinese History—Manifold Invasions of China, but Chinese Individuality always predominates—Imagination essential to Advancement, but the Chinese have it not—Inventions ascribed to Them—Stoppage of Development—China for the Chinese—Foreign Help Inevitable—The Woo-Sung Railway—Defects of Chinese Character—Shanghai in Winter—Pigeon-English—Perverse Employment of it—The word 'Pagoda'—Preparations again for Travel—Detail of Packages—The Shanghai Theatre—Chinese Treatment of Chinese—The Chinese Performers—Sport at Shanghai—A Shooting Excursion on the Yang-Tzŭ—Names of the Great River—The so-called 'Grand Canal,' and its Humours—Shooting Scenes—The Mixt Court at Shanghai—The Yang-Tzŭ Steamers—Additions to our Following.

THE birthplace of the Chinese nation is veiled in mystery. Mr. Douglas, in an exceedingly interesting article in the 'Encyclopædia Britannica,' observes: 'Some believe that their point of departure was in the region to the south-east of the Caspian Sea, and that, having crossed the head waters of the Oxus, they made their way eastward along the southern slopes of the Teen Shan. But, however this may be, it is plain that as they journeyed they struck on the northern course of the Yellow River, and that they followed its stream on the eastern bank, as it trended south, as far as Tung-Kwan, and that then, turning with it due eastward, they established small colonies on the fertile plains of the modern province of Shan-se.'

Mr. Douglas also states that the nucleus of the

nation 'was a little horde of wanderers roving amongst the forests of Shan-se without homes, without clothing, without fire to dress their victuals, and subsisting on the spoils of the chase eked out with roots and insects.'

There were aborigines already here; but of them little is known; their remnants are said to exist at the present day amongst the Miau-Tzŭ of Kwei-Chou.

But the Chinese were the better race; they were also apparently already agriculturists, and as such in a higher state of civilisation. One result could but follow; the inexorable law of nature had its way; the inferior and less civilised race were pushed out by degrees, just as all the barbarous tribes still remaining are surely disappearing before the steady advance of the Chinese; as the New Zealand Maories and American Red Indians are dying away before the Anglo-Saxon race.[4] There is no record that the Chinese were ever a pastoral people, excepting that which lingers in some of the ancient characters of the language, and, as some say, in the wavy outlines of their roofs. However that may have been, they appear to have settled down as agriculturists in Lower Shan-Si.

Northern China had not yet been denuded of her forests; but though the climate may have been more favourable for agricultural pursuits than in the present day, the province of Shan-Si can never have been one that yielded a profusion of wealth without the steady application of labour.

Baron Richthofen remarks that 'the altitude of

---

[4] It will not do to argue from this analogy that so will the barbarians of Central Asia disappear before the European. The Anglo-Saxon cannot colonise there; if the Russians can, they have indeed a grand future before them.

its arable ground renders nearly the whole of it unfit for raising two crops a year.'

Neither is the climate so severe that labour in the fields cannot be carried on at all seasons.

The Chinese race, therefore, in its infancy found itself in a country where steady labour and thrift were necessary for life ; and here were perhaps the germs of the industry and exceeding carefulness so remarkable in the character of the Chinese of the present day.

Further, this was the order of things most suited for the production of a sentiment of equality amongst the people, for food was not too easily procured, and a sharp division between rich and poor would not immediately ensue. It is, therefore, not surprising that a strong democratic feeling should be another feature of the Chinese as they are.[5]

The dim history of those days throws but a feeble ray of light, but it shows us that civilisation advanced, and the existence of trade is proved by the establishment of fairs.[6]

The people now spread eastward, and in 2300 B.C. we find their capital in the neighbouring province of Shan-Tung, and their kingdom extending to the north and east of the present Peking, and as far south as latitude 23° N.[7]

But the southern climate seemed to soften the hardy northmen, and the varied conditions of life to destroy their cohesiveness. . . . We read of a

[5] In China all judicial affairs are conducted more or less in public. Even in the presence of the highest officials anyone can turn in from the street to see what is going on, no one trying to hinder him. A beggar will sit down and smoke his pipe in the presence of a magistrate, and sometimes join in the conversation unasked. The literary examinations are open to all; no matter how lowly a man may be if he can pass his examination he may become the highest magistrate in the land.

[6] *Encyclopædia Britannica.* [7] *Ibid.*

ruler in 1818 B.C. in whom were combined the worst vices of kings;[8] but the vitality of the people was still sufficient to make them rise against him and sweep away all traces of him and his dynasty.[9]

During the next eight hundred years we hear of little but internecine wars and consequent weakening of the kingdom.

Nigh two thousand years had elapsed since first the black-haired race had come from the north-west; three sovereign dynasties had reigned, of which the last was sinking amid the rivalry of feudal states, and China seemed rapidly disintegrating, when the Princes of Thsin, a state founded five centuries before with their capital at Chang-Gan in Shen-Si, conquering in succession the six or seven other states, restored (B.C. 251) a strong central power.

With the accession of new blood, China was reinvigorated, and this was one of the most flourishing epochs in the varied history of this marvellous empire: roads were made, canals were dug; and before long the powerful desert horde of the Hiung-Nu, who had long harassed the Chinese, were completely routed and driven into Mongolia; and in the year 214 B.C. the Great Wall was commenced as a protection against the inroads of these barbarians. The veneration of antiquity preached by Confucius now seems first to take root, for at this time 'schoolmen and pedants were for ever holding up to the admiration of the people the heroes of the feudal times.'[1]

This reverence for antiquity throughout the ages that follow, amidst scenes of strife and disorder, as well as during the intervals of prosperity, sank deeper and deeper into the nature of the Chinese, and in it

[8] *Encyclopædia Britannica.*   [9] *Ibid.*   [1] *Ibid.*

is to be found one of the causes of the present decadence of the nation.

History now repeats itself again and again with almost wearisome monotony; tumults and disorders, and the consequent weakness of the people invite assaults from the north, but time after time the vanquished Chinese seem only reinvigorated by their invaders, and we find that each fresh incursion is followed by a period of glory.

In 121 B.C. the Hiung-Nu were driven to the northeast of the Caspian; then succeeded the troublous time of the 'Three Kingdoms'; and in the fourth and fifth centuries of our era the Wei, a race of Siberian nomads, conquered and ruled in Northern China; but in the seventh century arose the Thang, the most glorious of all the native dynasties; under them Chinese rule extended to Turfan, Khoten, Kashgar, and even to the Jaxartes, whilst Chinese fame was so great that ambassadors came from the Caliphate, and even from Imperial Byzantium.

Thus the marvellous vitality of the Chinese disposed of successive races of invaders, either driving them far from their borders, or absorbing them and assimilating them when they could not be expelled.

But yet another army of barbarians appeared in the Khitans. These, however, never extended their rule very far south, although in 997 A.D. tribute was paid to them. Later, the Chinese invited a fourth horde, the Kin or Niu-Chih, to expel the Khitans. The Kin succeeded in this only too well, and in 1150 A.D. established themselves in the whole country north of the Yang-Tzŭ.

A new race, the Mongols, now came on the scene; they wrested province after province from the Kin, and the place of these knew them no more. This was in the thirteenth century, and in the brilliant light that radiated from these the most successful, the most glorious of all the conquerors of China, the feeble glimmer of Kin and Khitan was extinguished alike. This was the most celebrated era in the whole history of the Chinese Empire; but it was the Mongols, and not the Chinese who made it so.

The latter were known to Marco Polo as the people of Manzi, who, if they 'had but the spirit of soldiers, would conquer the world; but they are (quoth he) no soldiers at all, only accomplished traders and skilful craftsmen';[2] whilst Friar Odoric says: 'All the people of this country are traders and artificers.'[3]

True, both Polo and Odoric speak in glowing terms of the rich and noble cities of Manzi, of their wealth, magnificence, and luxury, but these were as nothing before the glories of the Great Kaan, whose subjects they were, and who was a Mongol. But the Mongol power waned, and by a turn in the wheel of fate the son of a Chinese labourer drove out the successor of Kublai. In more recent days, to quell rebellions in the south, the Chinese invited the aid of the Manchu Tartars, who now are seated on the Imperial throne.

Thus, through long ages of varied fortunes, the Chinese character has been formed; and it would be surprising indeed, if a nation that had survived so many and such great vicissitudes, had been conquered many times, and had each time risen superior to defeat, had absorbed one race of victors and driven out another, did not possess some characteristic that would mark it as a peculiar people—and this characteristic is the individuality of the race. It is, indeed, a matter

---

[2] Marco Polo, book ii., vol. ii., p. 166.
[3] Cathay, vol. i., p. 105.

for wonder that a people so numerous and covering so vast an area should everywhere appear the same; who, whether they are found in the north, the south, the east, or the west of their own huge empire, who, whether they are observed as coolies in America or Australia, or met as ambassadors in London or St. Petersburg, should universally possess the same thoughts and the same feelings, wear the same clothes, and eat the same food, should be imbued with the same habits of intense industry and thrift, and should act precisely in the same manner as they did many hundreds of years ago.

Where else in the history of the world can we read of three hundred millions of people thus amazingly unchangeable? and who can doubt that they must yet remain for many centuries an important factor in the Asian problem?

Of all qualities that conduce to the advancement of a people, imagination is perhaps the most important; without it a nation must remain stagnant, with it the limits of its forward march can never be reached.

No matter what branch of industry or science is examined imagination lies at the root of its advance.

Surely it was in one of the most mighty flights of imagination that the keen gaze of Newton, sweeping across the wild chaotic waves of theory that each in turn must have leapt up towards his searching intellect, singled out the exquisitely beautiful and simple one of gravitation to account for the most complex motions of the vast masses that roll through space.

What but the richest imagination could have enabled Darwin to conceive the descent of man? or how could Professor Owen without imagination have built up from some paltry fragment the form of a gigantic mammal? Who without imagination could from mere scratches on a rock have enunciated the theory of a glacial epoch? or how, without imagination, could the present marvels of electricity have been evolved from the twitching of the muscles of a frog? Of art it is hardly necessary to speak; no one can ever have attributed a want of imagination to either painters or poets worthy of the name.

Imagination and originality are more or less inseparable; an individual devoid of one will certainly be deficient of the other, and what is true of an individual will equally hold good of a nation.

In the Chinese character originality and imagination are conspicuous by their absence. The Chinaman is eminently a matter-of-fact person; sights that would be disgusting to a European have nothing unpleasant in his eyes, for everything is looked at from a utilitarian point of view. The beauties of nature have no charms for him, and in the most lovely scenery the houses are so placed that no enjoyment can be derived from it. If the unhewn log of a tree will serve as a beam in the wall, he does not think it worth while to spend money or labour in squaring it. A Chinaman may express the highest admiration for a pair of European candles, but if they cost a trifle more than his filthy oil lamp, he will rarely exchange the glimmer of his time-honoured institution for the brilliant light of a composite. A Chinaman will feel the texture of a European coat, and admit its superiority, but his first question will be, how much did it cost? In their pictures there is no imagination; they draw birds and insects as they see them, and really well. Animals also they attempt, but

their ignorance of anatomy renders their efforts in this
direction ridiculous ; but abstract ideas, such as have
made the memory of old European painters glorious, any
attempt to portray, Faith, Hope, or Charity, any effort
to rise above the level of every-day life, are things unknown in Chinese art. So in their sculpture, they
represent men, women, and children as they see them,
but that is all ; they can imitate admirably, but they
can imagine nothing. Their want of imagination
precludes almost all idea of badinage. On one occasion, when the door of an inn was blocked up by
inquisitive people, it was agreed that as long as they
kept outside, the door should remain open. At length
a boy ventured to put his feet over the door-sill.

'I suppose you think those are very fine boots of
yours,' was the foreigner's sarcastic remark.

'Yes,' replied the youth, 'they cost half a tael.'

The idea of being chaffed never entered into his
matter-of-fact mind.

Thus at almost every turn the want of imagination, and with it the absence of originality are
evident.

But the Chinese are credited with having invented
almost everything : how can this be reconciled with a
want of originality ?

In the first place there are a good many things
that the Chinese have never invented or discovered.
The principle of the pump, the circulation of the
blood, and the science of grafting are still unknown
to the Chinese. It has frequently been asserted that
they invented gunpowder ; but the late Mr. Mayers,
Chinese Secretary of Legation at Peking, has effectually demolished their claim to this invention.[4]

[4] Morrison gives 1275 as the time of the invention of powder and

The word 'P'ao' which now means 'cannon' was, it was asserted, found in old Chinese books of a date anterior to that in which gunpowder was first known to Europeans; hence the deduction was drawn that the Chinese were acquainted with gunpowder before it was used in the West. But close examination shows that in all old books the radical of the character 'P'ao' means 'stone,' but that in modern books the radical of the character 'P'ao' means 'fire'; that the character with the radical 'fire' only appears in books well known to have been written since the introduction of gunpowder into the West; and that the old character 'P'ao' in reality means 'Balista.'

So the word 'Chiang' means 'spear,' but the radical of the written character means 'wood'; the same word 'Chiang' means 'musket,' but the radical of the latter means 'metal.' [5]

Parallel cases are not wanting in other languages.

'Banduk' is the Hindustani word for 'musket'; yet we read in Marco Polo of Bendocquedar, the 'Soldan of Babylon,' a name which, as Colonel Yule points out, is *Bandukdar*, the Arblasteer.

Long before the invention of gunpowder, 'musket' was the old English word for a hawk used in the chase; when firearms were adopted for the same purpose, the name was handed on.

The mariner's compass, it is said, was known to the Chinese at a very early date; and it must be admitted that the early use of bank-notes, and the knowledge of printing, give the Chinese some claim to originality in ancient days.

It would be a deeply interesting study, and one

guns, and was aware that what they called 'P'ao' were machines for throwing stones.

[5] Mr. Baber was the first to notice this last fact.

well worthy of the labour, for anyone with sufficient acquaintance with the written language of China to investigate the ancient books, and from their internal evidence, and not from the prejudiced and superficial views of foreigners, to ascertain the history of the formation of Chinese character. It would appear, however, that originality, if they ever possessed it, has been stamped out, partly by the insane teachings of Confucius that everything ancient is sacred, and the still more insane idea that anything new, no matter what, is dangerous. Another cause for the disappearance of originality may be found in the preposterous system of examinations. Magisterial and official posts are awarded only to those who can pass the literary examinations; and until the examiners have been satisfied, no man, no matter what his rank or position may be, can hold any official position whatever; the 'literati,' or those who have passed high examinations, are the class most highly esteemed in China, and the desire to be numbered amongst them is almost universal. And what are these examinations? Examination only in the ancient classics, the obscure passages in which must only be explained in the orthodox manner.[6]

It is not difficult thus to realise that the Chinese character may have changed during the last few centuries, and that the originality and power of conception they may have possessed may have been crushed out by the worship of antiquity and the system of examination.

If this be so, the extraordinary stoppage of the early development of the people may be accounted for; for without originality, and devoid of imagina-

---

[6] See later under *May* 6.

tion, they must necessarily have stagnated, and have been arrested in the onward march towards a more perfect civilisation.[7]

Another feature in the Chinese character that may have assisted in some degree to retard their development is the intense desire of every man to do everything for himself. It is undoubtedly prompted by a sturdy feeling of independence, but carried to the excess in which it is seen in the Chinese it must be hurtful.

A Chinaman, if he can, will grow his own grain, grind it, or husk it, and cook it on his own premises. If possible, he will cultivate his little bit of cotton, and weave the cloth without assistance from beyond his household; all his clothes are perhaps made by his wife or family; and thus he is almost independent of any extraneous aid. We in Europe know that this is not an economical way of doing things; but the Chinese have done so for generations,—and

---

[7] Another reason for the stagnation of the Chinese people may be possibly found in the fact that all the talent of the country is absorbed in the service of the State. This is partly because of the contempt in which the non-official class is held, and partly because there is no entrance to official life of any kind except by competitive examination. Now, even in progressive countries, a system which would divert from private enterprise all those who help to make the country great, would have lamentable results. How much more must this be the case in one where enterprise of any kind is almost unknown, and which has, as it were, been asleep for centuries. In Western States, honour, fame, and dignities attend those who succeed, no matter in what walk of life; but in China none but the officials can hope for any of these.

If we look back at the history of our civilisation we find that all the great strides in science, and nearly all the greatest works of literature and art, have been due to private individuals. The discovery of America, the establishment of the Overland Route to India by Waghorn, the extraordinary development of newspaper correspondence, are but a few of the instances that will occur to anyone but slightly acquainted with history; and in our own country does not Government always look with distrustful eyes on any measure laid before it which would appear likely to interfere with, or to retard individual effort?

what was good enough for their fathers is good enough for them. Of course under these circumstances it is almost hopeless to expect any improvement in agriculture or agricultural tools, or any advance towards a use of machinery. Thus with the nation, at the present moment, it is the extraordinary idea and wish amongst some of the most advanced thinkers to begin their mining operations, smelt their iron with their own coal, and make their own rails for their railways, before they do anything else; they want to have China for the Chinese; they desire to do everything for themselves, and if possible to exclude foreigners. But how far they are from this, they little know.

True, the palmy days of the British merchants are over; the Chinese have at last learnt how to buy and sell without their aid, and they are fast ousting the foreigner from mercantile pursuits. We cannot of course but be sorry that the fine race of men, open-handed and generous, full of courage and enterprise, a type of all that is manly and thoroughly English, should die out and disappear, and mournful tales are told of the destruction in consequence of English trade. This is, however, but a superficial way of regarding the irresistible march of events. If the British merchant is ousted, it is because the Chinese can do things cheaper than the English; the result must be that we in England will get our tea and silk cheaper than heretofore, and that the people of China (if they buy it at all) will buy our cotton cheaper, and in consequence buy more. How then is trade injured: is it not rather on a better footing?

But although commercial pursuits may not be so profitable as they were, there must yet be a future for Europeans in China. Great as the opposition is at present, railways and telegraphs must certainly be laid down, and will for many years to come give employment to large numbers of Europeans, for owing to the want of originality in the Chinese they cannot hope to undertake the sole management of railways and telegraphs.

The Chinese may be taught almost anything— they are wonderfully quick at learning and imitating— and they would doubtless soon acquire the power of managing engines and telegraphs as long as all went smoothly. But in the moment of difficulty, if any fresh combination of circumstances should necessitate some original action, or even the smallest amount of reasoning, a Chinaman would be found unequal to the emergency. The Chinese Government have for a long time owned steamers, but the engineers are still European, and it will be the same with the railways and telegraphs. There are at present no railways in China. Some of the merchants of Shanghai instituted a short line between Shanghai and Woo-Sung, but it came to an untimely end, not so much on account of the absolute dislike of the Chinese to railways, as from some unfortunate circumstances connected with its origin. Rightly, or wrongly, the measure adopted irritated the Chinese Government, who declined to have the Woo-Sung railway forced upon them, and, when it came into their hands, contemptuously tore it up. During its construction, and in the early days of its existence, there was considerable opposition amongst the people of the adjacent villages, excited probably by the literati of Shanghai. There were even some attempts at suicide, the perpetrators being probably bribed to commit these acts. There was considerable method shown in the way that the attacks on the railway were carried out, and it may not be uninteresting to notice one in detail as an illustration.

There was a Chinaman living at Woo-Sung of a character so bad that, amongst the inhabitants of the place, he was known as 'The Pirate,' and of a reputation so evil that he dared not show his face in Shanghai. This man had a nephew who was a 'ganger' on the railway. Possibly bribed by the officials, or for some motives that never came to light, this man and his nephew incited the people of Woo-Sung and of another village to evil deeds. They proceeded to dig the ballast from between the rails, and pile it up on the line, in the hope of upsetting the train, but as great crowds of people collected on and around the line at this point, when the train arrived at the obstacles the engine-driver saw that something was wrong, and stopped.

The train was then attacked, but the engine-driver and guard repulsed the mob, captured the nephew of 'The Pirate,' locked him up in a carriage with another prisoner they had caught, and went back towards Shanghai. On the way thither more mobs collected, and one man attempted to commit suicide by throwing himself down in front of the engine; but the engine-driver was again able to pull up in time, and the would-be suicide was made prisoner, and, with the other two, conveyed safely to Shanghai.

The question must have presented itself to many people whether the Chinese are likely to succeed in their resistance to the Russians in Kuldja.

A careful consideration of the circumstances would lead to the conclusion that such a conflict would be disastrous to the Chinese. This is not due to any want of courage in the Chinese soldier, but simply to want of officers and want of organisation. With European officers, as under Colonel Gordon, we know how well the Chinese have fought, whilst, unlike most Orientals, they have not been utterly demoralised by a check; properly led they would make magnificent troops, for by nature the Chinese are singularly obedient to authority, and would not question the commands of those who had once established an influence over them. In this they are like other Easterns, but more than others their national characteristic renders them particularly incapable of military combinations. A Chinaman can learn anything, but he can conceive nothing; he may readily be taught any number of the most complicated military manœuvres, but place him in a position slightly different from that in which he has learnt, and he will be found utterly incapable of conceiving any modification to suit the altered circumstances. This national characteristic is the growth of centuries of a narrow education, its roots are deeply seated, and lie in the insane reverence for antiquity, which is almost the beginning and end of a Chinaman's belief. Prompt action, readiness of resource, ability to seize on the smallest advantage, or to neutralise a misfortune, and the power to evolve rapidly fresh combinations,—these are the qualities that make a soldier, and these are the very qualities that cannot co-exist with the Chinese want of originality. This is no unimportant matter, for it proves that, as they are, the Chinese cannot be feared as a military nation, but that with a large number of European officers, their almost unlimited numbers, their obedience to authority, and personal bravery, when properly led, would make them almost irresistible.

Further, there is in the Chinese mind a great dread of Europeans. Supernatural powers are popularly attributed to foreigners, and though they profess to hold the barbarians in contempt, in reality the feeling of fear predominates in their mind, although perhaps they would not own it even to themselves. But with good and skilled European officers they would, as they have done before, make magnificent soldiers.

Shanghai in the winter is a very pleasant place for Europeans ; the houses are comfortable, and good coal is burnt in the grates. The Bund, as the road along the river side is called, is fine and broad and kept in good order. There are always some of her Majesty's ships in harbour, and the officers enliven the place. There is a very good club, the members of which are most hospitable to wandering strangers, and the comfortable library full of books is a rare treat after a month in the saddle.

In the European Concession the roads of course are macadamised, and in the evening all the rank and fashion, youth and beauty of Shanghai turn out, on horseback or in carriages, on 'the Bubbling Well Road,' the Rotten Row of the place.

Shanghai boasts a racecourse, and a boat club, a drag-hunt, and a society for paper-chases on horseback, a volunteer rifle corps, and a volunteer fire brigade.

Pigeon-English is much used at Shanghai as a means of interchanging ideas between English and Chinese.

In every English merchant's house in China there is an abominable person called a 'Compradore,' who, in reality, does most of the work, and is the medium between the English merchant and his Chinese client.

At the first appearance of the English in the

country, the Chinese, who are naturally an imitative people, began to pick up a few English words, and soon constructed a language, which was an unnatural combination of deformed English words with Chinese deas and forms. The result was a jargon as hideous as it was illogical; but the English traders of the early days, finding they understood somewhat of this comic medley, instead of inducing the Chinese to make use of correct words rather than the misshapen syllables they had adopted, encouraged them, by approbation and example, to establish Pigeon-English —a grotesque gibberish which would be laughable if it were not almost melancholy. The English of the present day cannot do much to help themselves, but they might do more; for although it is to a certain extent true that Pigeon-English is understood, while the grammatical language is not, yet it is not possible to believe that when a glass of beer is poured out, even a Chinaman can more readily understand the idiotic expression 'can do' than the good English of 'that will do'; or that a Chinese boy would not in two days learn that 'upstairs' was the same thing as 'top side.'

But far from thinking it any shame to deface our beautiful language, the English seem to glory in its distortion, and will often ask one another to come to 'chow-chow' instead of dinner; and send their 'chin-chins,' even in letters, rather than their compliments; most of them ignorant of the fact that 'chow-chow' is no more Chinese than it is Hebrew; and that 'chin-chin,' though an expression used by the Chinese, does not in its true meaning come near to the 'good-bye, old fellow,' for which it is often used, or the 'compliments' for which it is frequently substituted.

Each of two polite Chinamen entering a room

together will urge the other to go first, and will then sometimes say 'Chin-Chin,' meaning thereby something very different to what an Englishman means when, in a letter, he sends his chin-chins to a common friend.

Pigeon-English has now become a fact that must be accepted, but it would be less deplorable if, instead of being admired, it were reprobated. There are, however, one or two words whose use it is almost impossible to avoid. One of these is *Ma-Fu,* a word that can no more be translated into English than the Hindustani *Ghora-walla,* of which it is an exact and literal translation, and which is used in exactly the same way. The word cannot be rendered into English, for a man who never grooms a horse can hardly be called a groom, and the literal translation 'horseman' means, in ordinary parlance, a man on horseback.

Pagoda is another word, the use of which is sanctioned by long custom.

This word is applied by Europeans to a peculiar form of tower, always called 'Ta' by the Chinese. These are high towers, generally erected in or near large towns, and are supposed to bring good luck to the places which they dominate. They are not used as watch-towers,—sometimes there is no means of ascending them; and a look-out or watch-tower is called 'Lou.' The derivation hitherto usually accepted is from the word *Dagoba,* though various others have been suggested. Littré, in his magnificent dictionary, derives it from the Persian *But,* idol, and *Kedeh,* temple. Stormonth, in his etymological dictionary, gives *pagão* ('pagan' in Portuguese) as the origin of 'pagoda'; and many attempts have been made to derive it from the Chinese language.

Colonel Yule has favoured me with a note on

the subject which can hardly fail to carry conviction:—
'It is a difficult word, but I do not think the origin can be Chinese. The word occurs early in the Portuguese books about India, too early to admit of a Chinese origin. And you will find that Chinese origins for those Anglo-Indian words are very rare. *Mandarin, Joss, Chop*, are none of them Chinese in origin.[8] Wedgwood gives the derivation from *pagão* 'pagan,' but this is inverting things. The Portuguese probably confounded the word in their own minds with *pagão* more or less, but that could not be the origin of a word they used only in the East. *Dagoba* is a real word, not Burmese but Pali, i.e., of the sacred Indian language used by the Buddhists in Ceylon and Burmah, which is a modification of Sanscrit, much as Italian is of Latin. *Dagoba* is, in Sanscrit, *Dhâtu-garbha*, 'relic-receptacle,' and the word is used in Ceylon; but I don't believe it is the origin of 'Pagoda.' *But-Kădăh* (or *Kedeh*) is also a real Persian word, and I was formerly inclined to think 'Pagoda' might be from this, shaped by the suggestions of *pagão*. The word is used by the old Portuguese writers in the sense of *idol*, as well as *idol temple*; you find 'Pagod' thus used also in old English travellers. It is likewise applied to the gold coin which was long the standard currency of South India. This, in its native shape, had figures of idols on it.

'I believe now that the real origin of the name is the word *bhagawat* or *bhagawatî*, 'deity' or 'divine,' which is current all over India with various special applications, and which appears in Marco Polo in the shape of *Pacauta*.[9] As regards the attempts to

[8] *Mandarin* is merely a Portuguese corruption of Sansk. *Mantri*, a minister of State.—(Y.)
[9] See 2nd edit. ii. 322, 330.

derive the word from Chinese, I may note the occurrence of the word in Barbosa (1516), whilst the Portuguese were not familiar with China till many years after.'

I was now making preparations for a long journey into the interior of China, and found plenty of occupation in getting stores of all kinds ready.

Mr. Baber, of the Consular service, who was a member of the Grosvenor expedition to Yun-nan, had invited me to accompany him to Ch'ung-Ch'ing. I eagerly availed myself of his invitation, but as yet formed no definite plans as to my future movements, only making up my mind that I would be ready for anything that might turn up.

I therefore prepared stores of all kinds, and arranged my provision-boxes in pairs, each pair to contain a complete supply for two months. Chin-Tai used to carry out my orders with amazement. First I had some large tin boxes, for soldering down, made to order, with strong wooden dovetailed coverings. Then I bought some small tin boxes to put inside these, but finding they did not suit I abolished them, and had others made. On trial these were found too large, and had to be reduced. Each time that Chin-Tai brought the things in, and saw me try them, first one way, then another, and finally carefully weigh every box, he would get more puzzled, till at last he shrugged his shoulders, and came to the conclusion that I was mad.[1]

---

[1] I had 6 boxes packed each with 30 candles (English candles, six to the lb.)
    1 tin box for tea, $5'' \times 5\frac{1}{2}'' \times 8\frac{1}{2}''$.
    4 boxes of matches.
    6 2-oz. pots of Liebig's Extract.
    2 packets of Marseilles compressed vegetables.
    1 bottle Worcester sauce.

The Chinese theatre at Shanghai, though a mongrel establishment—half Chinese and half foreign—is, nevertheless, well worth a visit, for the acting is *bonâ fide* Chinese acting, and the house is filled with Chinese, who come here to enjoy themselves in their own characteristic fashion. But the size, shape, and arrangement of the house are essentially European, as are also the lighting with gas, and the system of payment at the doors.

There is no such thing in China, properly speaking, as a theatre at which people pay. Theatrical performances are usually given by rich people to their friends; or sometimes the inhabitants of a street will combine and engage a set of actors; in this case the theatre is set up in the street, occupying the greater

 1 tin box for cigars, 10″ × 8½″ × 2½″.
 1 box of toothpicks.
 1 tin box of tooth powder, 3½″ × 2½″ × 1½″.
 1 small bottle cayenne pepper.
 Six other of the large boxes were packed each with:—
 30 candles.
 1 tin of salt, 5″ × 4″ × 3½″.
 1 tin of mustard, 2½″ × 4″ × 3½″.
 6 2-oz. pots of Liebig's Extract.
 1 tin for cigars, 8½″ × 6″ × 2½″.
 1 packet Marseilles preserved vegetables.
 4 boxes of matches.
 4 cakes of toilet soap.
 1 cake of yellow soap.
 1 cake of carbolic acid soap.
 2 little boxes of Brand's meat lozenges.

Each of these boxes, when finally packed and soldered down, weighed a little over 30 lbs.; quite enough for the mountainous countries.

The quantity of tea that I took was unnecessary, but I only had my northern experience to guide me; and in the province of Chi-Li, and beyond the Great Wall, tea can never be bought. In Southern, Central, and Western China, tea is always to be procured. The lids of the boxes were all screwed down, so that they could be opened and shut as often as necessary; and as I could not manage to get sufficient candles into the boxes without unduly increasing the weight, I took besides an extra supply.

part of it, and leaving scant room for a passing sedan-chair. But at Shanghai the Chinese have learnt European manners so far as to have a public theatre, to which anyone is admitted on payment.

It is a lofty, oblong building, that would be considered large even in London. A gallery, supported on plain wooden pillars, runs round three sides of it, and is divided into boxes, in which, when we entered the theatre, we could see family parties, smoking, drinking tea, and cracking water-melon seeds.

There were about half a dozen rows of people in the end gallery, those in the front having tables for their tea, sweets, &c. In the body of the theatre, the people sitting in the best places were provided with tables, but those behind were all packed close together.

The stall-keeper led us to the front, pushed aside the people that were standing about, without the least ceremony or politeness, walked up to one of the best tables, turned out the family party without asking their leave, and most obsequiously invited us to take our places.

The Europeanised Chinaman seems to acquire a supercilious contempt for his more conservative countrymen. This may be in a great measure owing to the fact that attendants amongst Europeans are so often taken from a low class.

On one occasion, a Chinese gentleman was visiting me, and I ordered one of my servants to get some tea; he told me that he would get tea for me if I wanted it, but that he was not going to wait on a Chinaman.

Of course this sentiment was more strongly expressed towards the unofficial classes, and for a magistrate of high rank my servants had a certain amount of respect.

VOL. I. M

A man with an enormous basket over his arm, full of water-melon seeds, walked about the theatre, continually filling the little dishes in front of the spectators with this incomprehensible delicacy. He was obliged to refill his basket (about the size of an English baker's) many times during the evening.

Tea, of course in Chinese fashion, is consumed by all. The ordinary Chinese fashion of making tea (except in the West, where the tea of Pu-erh is taken) is to put about a teaspoonful of tea into the cup, and pour boiling water on it. The Chinese drink it nearly scalding, and the cups are continually refilled with boiling water, fresh tea rarely being put into the cups. The object of putting a cover over the cup, instead of a saucer underneath, is to prevent the tea-leaves getting into the mouth. A Chinaman, before putting the cup to his mouth, always sweeps the surface of the tea with the cover, to push the floating leaves away from the side. He is very skilful in drinking, always holding cup and cover with one hand, and leaving just sufficient aperture for the infusion to pass without letting the leaves through.

The Chinese have a theory that if the water is properly boiling the leaves will not float on the tea, but if the tea has been made with water that does not boil the leaves will at first come to the surface.

The stage of the theatre was raised, as in Europe, and the orchestra, consisting of one wooden and two brass drums, was on the stage. There was a little painting on the background, but not much scenery.

The actors were dressed very gorgeously in silk and embroidery; and their faces were covered with paint, red, black, or white; the paint being apparently mixed with red lead or putty, laid on all over their faces with a scalpel, and highly polished.

Female actors are very rare in China; the women's parts are generally taken by men, who speak in a high falsetto.

The piece performed on the occasion of our visit was in a great measure conducted in dumb show; and unless one of the characters was actually speaking, the orchestra beat their drums with all their might, the noise made by them becoming almost unbearable.

All the performers seemed more or less acrobats, and no one remained on the stage more than a few minutes at a time; an actor would say a few words, posture a good deal, throw his legs about in a manner not far short of the celebrated Vokes, and then go out, some one else coming on immediately.

The performance on this occasion ended in a tremendous battle that lasted half an hour, during which four regular acrobats, naked to the waist, came on the stage and performed some feats that would not be considered very remarkable in Europe.

One of the company was evidently a sort of Mr. Toole; for directly he appeared, and before he said or did anything, the audience at once began to laugh in anticipation.

The theatre was very cool and well ventilated, but the awful drumming and noise was so nerve-shattering and so continuous that we none of us could endure it very long; and passing through a vestibule, where the nose was assailed with odours that seem necessary to a Chinaman's existence, we stepped into the fresh air of the street, where a number of chairs and jinnyrickshaws, the [2] carriages and cabs of the

---

[2] See p. 10. Giles states this word to be taken from the Japanese pronunciation of three characters signifying ' Man's—Strength—Cart.'

Shanghai Chinese, were waiting for their owners, or for the chance of being hired.

The English, of course, carry their sporting proclivities with them to Shanghai ; not only is there a society for paper-chasing on ponyback, but one of the merchants at the time of my visit had started a pack of draghounds, which gave capital gallops and plenty of jumping to an enthusiastic band of followers.

But, undoubtedly, the sport *par excellence* at Shanghai is the wild pheasant shooting, which with its concomitants of cheery companions, complete freedom, and life in a house-boat, is perhaps only to be equalled by woodcock-shooting in the neighbourhood of Corfu.

Most of the leading merchants have a house-boat; this is merely one of the ordinary shapeless, flat-bottomed, shallow boats of China, with two or three rooms built on it ; these are always very comfortably fitted up with beds, tables, lockers, &c., and have besides accommodation for servants, cookery, and dogs. It is usual to make up a party in a couple of boats, and go away for a week or two to the Grand Canal, and shoot over the plain between it and Shanghai.

This country was the theatre of war during the Tai-Ping rebellion ; for years afterwards it remained a desert, with nothing but ruined villages, and scarcely a single inhabitant. During this time the pheasants and deer increased and multiplied ; and not many years ago, it was possible for a good shot to bag forty brace of pheasants to his own gun in a single day ; now the country is being repeopled, villages are springing up, cultivation is increasing, and of course the game is diminishing.

I joined some friends in a trip, and leaving

Shanghai one night, we found ourselves the next morning steaming up the Yang-Tzŭ-Chiang.

Like the rivers of most Eastern countries, those of China do not bear the same name at every part of their course.

Near its sources this mighty river is known under various names. The Mongol name of *Murui-ussu* is given by both Huc and Prejevalsky; the latter gives *Di-chu* as a name in use by the Tangutans (as he calls the tribes of N.-E. Tibet); Burei-chu, or Bri-chu, corrupted by the Chinese to *Polei-chu*, is another Tibetan name. The Tibetans again at Bath'ang, and a little lower, call it the N'jeh-chŭ ('chŭ' is the Tibetan for 'river').

From Bath'ang to Fu-chou it has the appellation of Chin-Sha-Chiang, or Golden Sand River, from the quantity of gold dust amongst the sand in its bed. No other name is applied to so long a stretch as this; and the Chin-Sha is the name best known of all.

Near its mouth, where it opens out to a width of some miles, the Chinese call it the Yang-Tzŭ-Chiang, or Ocean River. Friar Odoric, writing about A.D. 1320–1330 of the Great River, calls it the River Talay (Dalai), which is just a Mongol version of the Chinese name, and would seem, therefore, to have been applied to it by the Mongols then ruling in China. The use of the word 'Dalai' in this way is, therefore, quite parallel to that of 'Bahr,' as applied by the Arabs to the Nile. So also the Tibetans apply the term 'Samandrang' (samudra, 'the Ocean') to the Indus and Sutlej.³

I have seen it stated that the name Ta-Ho is applied also.

This is to a certain extent true; for there is

³ Cathay, vol. i., p. 121.

scarcely a river in China that at some place is not called Ta-Ho, or Great River. Where an affluent enters a river, it is of most frequent occurrence to find the main river called Ta-Ho, and the affluent Hsiao-Ho, or Little River.

The French have invented a name expressly for themselves, and call it 'Le Fleuve Bleu'; and Prejevalsky has unfortunately adopted it.

A day and night's steaming brought us to Chin-Kiang; and thence we started in a native boat for the Grand Canal.

The Grand Canal of China is a work that has attracted much attention amongst Europeans, who have generally formed a vague idea of a magnificent highway, where great fleets of fine ships come and go, and where there is yet room for an unlimited increase of traffic. As a matter of fact, it is in many parts little more than a stinking ditch; it is already overcrowded to a degree almost incredible; and the water in it is often so low that a junk of very moderate dimensions may stick and entirely stop the traffic.

We entered the canal from the Yang-Tzŭ by a creek, ten yards wide, so full of craft that, to an inexperienced eye, it would have seemed impossible to get through, as the vessels completely blocked the waterway, none of the crews having apparently the least desire to make progress.

Our boatman, however, coolly charged them, paying equal heed to the oaths and howls of the people on the boats smaller and weaker than his own, and to the indifferent and supercilious glances of the occupants of the big and unwieldy junks.

Great and small they were somehow pushed on one side, two of the lightest being nearly shipwrecked in the process.

We joined our house-boat close to the city of Tan-Yang, where we were jammed for twenty-four hours, surrounded by an immovable block of boats; and during the next day it was only by the dint of strong language on the part of everyone in the boat and out of it, that we succeeded in advancing about a mile.

The sight was, however, extraordinary, and well worth coming to see.

Standing on the top of the embankment, the eye roamed over many miles of perfectly flat country, where numerous villages were hidden in clumps of bamboo, and where the canal, with its forest of masts, stretched away into the dim horizon.

About the whole scene there was a wonderful amount of life and animation : gay streamers from the thousands of masts, people shouting and pushing, and trackers on the banks calling to their companions in their boats.

Each boat contained a whole family of many generations, and often some half-dozen coffins besides ; for the Chinese are very particular about their coffins,—building them of rare and choice woods long before they die, and sometimes giving as much as a thousand taels for a good one ;—they carry them about wherever they go, and make them their constant companions in life, until in death they become their homes.

There was a huge war-junk in front of us, stuck in the mud, completely blocking the way ; and not until after a great amount of vociferation from the men in our boat could the people on the unwieldy vessel be induced to make any effort to get off ; at last she moved, we followed, and as she went aground again almost immediately our bow bumped her stern, whereat a hag, so ancient and so hideous as to pass all conception, put her head out of a window and made use of language more hideous than herself.

Presently it was necessary for two men to go behind, and pushing handspikes under the keel, to hoist the vessel by main force out of the mud. In the course of a few hours our boat succeeded in passing her, but only by dint of much violent abuse, rough dealing, and the liberal use of the almighty dollar.

'And this,' I thought, 'is the Grand Canal of China!'

For the tired merchant, or the hard-worked consular official, the novelty of the life on a trip of this kind, and the air and exercise, form a pleasant change at Christmas time after many months of busy Shanghai; and though the country is flat, there is still enough diversity and incident in the day's proceedings to make them amusing.

Walking over the wide plain not far from the canal, whose high banks conceal all but the tallest masts, we find nearly all the ground cultivated, and the young green crops coming up. In other parts we have to plod over a heavy fallow. Here is an old graveyard, covered with long dry grass and a few thorns, and it will be surprising if we do not turn a pheasant out of it. Now we come to the ruins of a village, with thick thorn bushes growing amongst the remains of the mud walls, from which a couple of cocks get up in a terrible bustle, cluck-clucking as they top the bamboo growing just beyond. The report of the gun brings some people out of a house that has been built amongst the ruins; they follow us some way, but soon get tired of walking after us.

Here is a creek with a great deal of soft mud in the bottom, and we must make a long detour to find a bridge. After crossing it, we come to a bamboo copse, some twenty or thirty yards square, on the site of an old cottage whose walls can still be traced amongst the undergrowth of brambles and thorns.

A peasant standing by says he saw a couple of deer go into it; but his information turns out worthless, though the thicket contains half a dozen pheasants.

By the side of another wide creek a bird gets up, and drops on the other side.

One of us calls out in polite Chinese to a peasant digging in his field, and asks him if he will fetch it.

'I do not understand any language but my own,' replies the man, shading his eyes with his hand, and looking across the water at our figures.

'But,' replies our speaker, 'I am talking your language.'

'Why, you don't say so; I never should have thought it; I thought no barbarian could ever learn our beautiful tongue.'

So we return to our boat, where a cold pheasant cut up awaits us, and seems so large that a hot dispute arises whether it can possibly be only one bird.

Besides the pheasants, a kind of small deer, commonly miscalled a hog-deer, is very abundant, but it offers very poor sport.

In the neighbourhood of Chin-kiang there are great numbers of enormous wild boar; some were brought in to Shanghai when I was there, weighing 360 English pounds. These boars have very small tusks, and some people, on this account, hold that they are the descendants of domestic pigs, that have at some time or another escaped from civilisation, and adopted the wild life of the forests.

We did not stay away long enough to find our trip monotonous, and returned to Shanghai satisfied, but not wearied with our sport.

The Mixt Court in Shanghai is very interesting to a stranger.

Offences are tried here before two judges, one Chinese and one foreign. One of the English judges took me with him one day, and I sat on the bench next to the Chinese official, who had the rank of Chih-Fu.

The room was fairly large, and the judges' table raised on a low platform. The space in front was divided into three portions by railings; the policemen, witnesses, &c., were on the right, and the prisoner was brought in to the centre division, led by his plait. He was obliged to remain on his knees during the trial.

This man had pretended that he was a broker, and had gone to the different European firms, from each of which he had obtained a sample of sugar, which he afterwards sold retail. He was convicted and sentenced to two months' imprisonment.

The Chinese official at this stage of the proceedings offered me a cigar, and tea was brought in; after which refection another prisoner was arraigned for driving a jinnyrickshaw without a license; and for which he received twenty blows with a stick.

The next had stolen a watch; and the last in a crowded thoroughfare had refused to 'move on.' It was a very amusing sight, and strangely like 'orderly-room' in an English barrack.

During my stay in Shanghai an event of some significance occurred.

Up to this period Messrs. Russell & Co., an American firm, had owned a very fine fleet of steamers, plying up the Yang-Tzŭ, and to Tien-Tsin.

Besides Messrs. Russell & Co. there was an English company that owned steamers not inferior to these; and there was also a Chinese company, whose vessels, though large and well built, were never favourites with either Europeans or Chinese. The last was strongly supported by the Government, especially by Li-Hung-Chang, and in the hopes of increasing their business and gradually getting rid of foreigners on the river, they bought the whole of Messrs. Russell & Co.'s vessels at a very high price.

From the fact that after the agreement of sale was made public, the shares of the Russell Company went up from sixty to ninety, it may be gathered that the Americans did not make a very bad bargain.

The Chinese traders were not at all pleased to see the boats pass into the hands of the Chinese company; for they feared that their Government would put pressure on them, and compel them to send their merchandise in the Chinese ships, where they knew that they would be compelled to pay more, and be less well served.

Before leaving Shanghai, Chin-Tai was instructed in the art of bread-making, so that, during the two months on the river, we were never reduced to chupatties.

I also obtained possession of a dog whose numerous good qualities, as appraised by his owner, would have made him cheap at any price. Baber and I laid in a considerable stock of provisions and delicacies for the voyage, amongst which two barrels of flour took a prominent position.

Before starting, I engaged another servant, also a Tien-Tsin man, and friend of Chin-Tai. His name was Chung-Erh, and, according to his own statements, he threw up a marvellously lucrative engagement, out of pure love and friendship for Chin-Tai.

CHAPTER V.

THE OCEAN RIVER.

Start up the Yang-Tzŭ—The Dog Tib, and his ethnological perspicacity—On Board the 'Hankow'—Transhipment and Arrival at Hankow—Manufacture of Brick-Tea—Tea in Chinese Inns—H.M.S. Kestrel—Boat engaged for Upper Yang-Tzŭ—The Lady Skipper and her Craft—Our Departure—Chinese Fuel—Our Eccentricities in Chinese Eyes—The New Year Festival—Chinese View of the Wind-points—Hsin-Ti—Entrance of the Tung-Ting Lake—Aspects of the River—Camel Reach—Vicissitudes of Tracking—Chinese Duck-Shooters—Great Bend—Wild Geese and Porpoises—Rice as Food—Ho-Hsueh and Sha-Shih—River Embankments—Tung-Shih—Meeting with H.M.S. 'Kestrel'—The Hills Entered—Walks ashore—Population dense only on the River—Passage in a Cotton Boat—The Telescope Puzzles—Arrival at I-Chang—The Chinese Gunboat.

At length the time came for our departure, and the cordial good wishes that I received from so many, whose acquaintance I had hardly formed, made me feel that I was leaving many good friends behind; it was not therefore without some regrets that, finally turning my back on Shanghai, I stepped on board the steamer 'Hankow,' on the night of January 23, 1877.

The steamers that ply on the Yang-Tzŭ-Chiang, between Shanghai and Hankow, are built in the style of the American river-boats; they draw scarcely any water, are very light, and are perhaps the most luxurious steamers in the world.

Baber and I were the only passengers, and so there was plenty of room for us and our luggage, of which there was by no means an inconsiderable quantity.

Before turning into the luxurious cabin I went to see the dog, whose name was 'Tib,' but he barked at me as an intruder, and the endearing epithets and biscuits that I lavished upon him producing not the slightest acknowledgment of good-will on his part, I left him to renew his acquaintance at a later date.

This dog had been almost entirely amongst Chinese, and either the appearance or the smell of a European was distasteful to him. The Chinese, who to a European nose always emit a peculiar odour, declare that they can perfectly well distinguish the smell of a European. There can be no doubt that 'Tib' could detect, even at a distance, a European by his smell, for he invariably barked at the French missionaries directly they entered the courtyard of my house at Ch'êng-Tu, although they were always dressed in Chinese clothes.

Anyone who has been long in India will recognise the smell of a Hindoo; and although it is not flattering to our vanity to admit it, it certainly seems as if we, as well as all other people, had an odour peculiar to ourselves.

January 24.—There was a Chinese steamer following us up the river; but our vessel was a little the faster of the two, and there was a merry twinkle in the captain's eye, as he stopped at each station and picked up all the Chinese passengers, leaving none for the vessel following. He had done the same on his last trip, and had so much annoyed the Chinese that they had invented a tale for the occasion, and had officially reported that our captain had sent press-gangs ashore and taken the passengers on board by force.

January 25.—It was a cold snowy morning, and the hills as we passed them were white; chill, heavy clouds were overhead; the wind whistled through the ship, and the dreary cry of the leadsman, which could be plainly heard in the saloon, made us appreciate the comforts to be found inside.

On the voyage up this river it is necessary to sound without ceasing; thus, as regularly as the hand of the clock touched the minute, the voice of the Malay quarter-master was heard in a kind of slow sing-song, 'No bottom,' 'By the mark five,' and so on.

January 26.—We anchored at three o'clock on the morning of January 26 for want of water, as the river was very low. The time of year for shipping tea occurs just when the river is at its highest, and then there is water enough for the great ocean-going steamers to run up to Hankow, where they take their loads of tea, and steam off direct for the London Docks.

At about nine o'clock the small steamer 'Tun-Sin' came alongside, to take some of the cargo about fourteen miles up the river, where it remained in lighters until another vessel came for it from Hankow; but as she could not take everything at one trip we remained on board the 'Hankow' all night.

January 27.—The decks were covered with snow when we looked out in the morning, and a heavy northerly gale howled mournfully. The work of shifting cargo advanced but slowly, and was not completed till five in the evening. We were able to appreciate the light build of these vessels, for as the bales of merchandise were moved about the deck below the saloon, the glasses on the table jumped from their places with the tremendous vibration.

When all was finished we said good-bye to the captain of the 'Hankow,' and embarked on the 'Tun-Sin.'

January 28.—Another cheerless snowy morning broke over the muddy river, and a damp mist almost hid the banks. Few boats passed up and down, and except one or two sea-gulls, circling round a melancholy-looking beacon, there seemed no life in the place.

We had anchored during the night, and had not been long under way in the morning when we ran on a mud-bank in the deepest part of the channel. The cargo was at once discharged into lighters kept for the purpose; but we did not get off the shoal till about midnight.

January 29.—We took the cargo on board again during the morning, and weighed anchor at 1 p.m. The sun came out for the first time during the voyage, and lit up the scenery as we ran along under the slopes and cliffs which here ran down to the river from the hills on each side.

January 30.—We arrived at Hankow [4] at about ten in the morning in a dismal pour of rain, and we thoroughly enjoyed the blazing fires of the hospitable consulate in which we were lodged.

Soon afterwards the rain turned to snow, which fell steadily during the rest of our stay.

Some idea of the magnificence of the Yang-Tzŭ may be formed from the fact that at Hankow, 680 miles from the sea, the river is still about 1,100 yards broad.

It is embanked with a magnificent bund, which is

[4] According to Sir T. Wade's system of orthography this should be Han-K'ou (the mouth of the Han), but the other spelling is now too widely accepted to admit of change.

the principal feature of this town. At the time of my visit the water was unusually low, being about thirty-five feet below the top of the bund. In the summer it rises sometimes even over this work, flooding the country and the town.

Under these circumstances, supposing the average velocity of the current to be six miles an hour (and it certainly is not less), upwards of a million cubic feet of water per second must pass Hankow.

Hiring boats for the journey to Ch'ung-Ch'ing was not altogether a simple matter. It was necessary to let our servants make all the arrangements before disclosing ourselves, for boatmen sometimes object to taking foreigners, and always try to overcharge them.

It was easier to settle our money matters. A firm at Hankow gave us a letter of credit on their Chinese agents at Ch'ang-Ch'ing, so we were not obliged to carry more silver than was necessary for the voyage.

During our stay in Hankow we visited the Russian factory, where brick-tea is prepared for the Mongolian market.

Bricks are made here of both green and black tea, but always from the commonest and cheapest; in fact, for the black tea, the dust and sweepings of the establishment are used.

The tea dust is first collected, and if it is not in a sufficiently fine powder, it is beaten with wooden sticks on a hot iron plate. It is then sifted through several sieves to separate the fine, medium, and coarse grains. The tea is next steamed over boiling water, after which it is immediately put into the moulds, the fine dust in the centre, and the coarse grains round the edges.

These moulds are like those used for making ordinary clay bricks, but very much stronger, and of less depth, so that the cakes of tea when they come out are more like large tiles than bricks.

The people who drink this tea like it black; wherefore about a teaspoonful of soot is put into each mould, to give it the depth of colouring and gloss that attracts the Mongolian purchasers !

The moulds are now put under a powerful press, and the covers wedged tightly down, so that when removed from the press the pressure on the cake is still maintained.

After two or three days the wedges are driven out, the bricks are removed from the moulds, and each brick is wrapped up separately in a piece of common white paper. Baskets, which when full weigh 130 lbs., are carefully packed with the bricks, and are sent to Tien-Tsin, whence they find their way all over Mongolia and up to the borders of Russia.

I was told that this tea could be sold retail in St. Petersburg, with a fair profit, at the rate of twenty copecks the pound.

The green tea is not made of such fine stuff, but of stalks and leaves.

The Mongolians make their infusion by boiling. In this manner they extract all the strength, and as there is no delicate flavour to lose, they do not injure the taste.

The manufacturer here set up a small steam-engine for the press, but found coolie labour cheaper.

He told me that the tea the Russians usually drink in their own country is taken direct to Odessa from Hankow by the Suez Canal; and in answer to an inquiry that I made, he assured me that even before the canal was opened it never passed through London.

A better price is given by the Russians in Hankow than the English care to pay. This is the real reason why the tea in Russia is superior to any found in London; for caravan tea is a delicacy even amongst the nobles of St. Petersburg.

Anything but very ordinary tea is rare in Chinese inns or houses; occasionally, however, a cup of tea has been given me with a delicacy of flavour and a bouquet that I have never met with elsewhere.

A very delicate tea is grown in Pu-Erh in Yunnan; it is pressed into annular cakes, and can almost always be purchased in the large towns of Western China, even in Ssŭ-Ch'uan; cakes of the Pu-Erh tea were often given to me as a present. But these are exceptions to the general rule, as the tea in inns and private houses is indifferent.

The brick-tea made for the Tibetan market is prepared entirely by Chinese at Ya-Chou. It also is made from dust and rubbish, and the manufacture is very similar to the process at Hankow.

H.M.S. 'Kestrel' was at Hankow, and a day or two before our departure she left for I-Ch'ang, now a treaty port under one of the clauses of the Chi-Fu Convention, carrying thither Mr. King, the newly-appointed Consul to that place.

The European officers of the Chinese Customs Service were also going up, so that Baber and I anticipated a merry meeting on our arrival. The 'Kestrel' left Hankow on February 5; the captain expecting to be back again in about three weeks. But the river was so low that it was eventually a very much longer time before the ship returned.

When the mysterious process of hiring the boats

had been accomplished by our servants, we went on board to look round, and to be introduced to the owner and skipper, who was a lady.

She declared herself capable of navigating the ship, taking the helm, working the *ulo*, and keeping the trackers up to the mark. Our subsequent experience showed that the last of these accomplishments was her strong point, for she had a tongue that nothing could withstand. The *ulo* is a kind of gigantic scull, that is worked by two or more people, sometimes from the stern, and sometimes at the side of the vessel.

The old lady introduced her little boy to us, who made a polite Chinese bow; and thus all the ceremonies were complete.

February 7.—Our large boat lay at the mouth of the Han river; but the small one came down in the afternoon to take us off, and we went on board with our servants, our dog, and our few remaining effects, including 70 lbs. of corned beef that Chin-Tai had bought for six dollars.

At 3.15 the last rope was let go; the Consul on the bund waved his hand; we pushed off into the stream, and started on our long journey. With a light westerly breeze we made our way over the current, and reached the large junk that was lying off Han-Yang.

Here we had to wait some time; for the old lady had suddenly discovered that a sail would not be altogether a useless article, and had sent to buy one. So we lay with the nose of our boat just ashore.

A steep mud slope about thirty feet high rose above us. This seemed to be a deposit for every conceivable kind of filth, and grubbing in the mire there were pigs, dogs, and miserable human beings who scraped a living by turning over the dirt with little rakes, and picking up scraps.

Crooked piles driven into the mud supported wretched hovels that overhung the river, and at the foot of the mud slope there were hundreds of all sorts of boats.

The men of the ferries in which the people crossed the Han river, or the Yang-Tzŭ, were inviting passengers to make use of their boats ; each man praising the excellences of his own craft, and trying to shout his neighbour down. Some of them would not start until they were loaded almost to the water's edge.

Sometimes a couple of richer men engaged a boat for themselves. Here one little boy was navigating a crowded vessel, and there a couple of big men were rowing one almost empty; but all were talking in the loudest tones, abusing one another, pushing each other about, and making a desperate noise.

Our old lady was slow about her purchases ; and after watching this noisy busy scene for some time, I looked round the boat to see what manner of craft was to be our home for so many weeks. She was about eighty feet long and eleven feet broad, and the main deck, if such a term is applicable, was about two feet out of the water.

The bows, for a space of twenty feet, were uncovered; aft of this a house about twenty feet long was built right across the deck, leaving no room to pass round the sides ; there was a small open space aft of the house; and right over the stern another high building, where our skipper lived, was piled up to a great height. The house was about seven feet high, and was divided into four compartments, giving us a living room and two bedrooms for ourselves, and a room for the servants.

There was a hold about three feet deep where we stowed away our heavy boxes.

We had a little American stove in the sitting-room. It used sometimes to get red hot; at others the chimney would get twisted, and the wind blowing down would send great tongues of flame darting across the room; of course it smoked occasionally; but these little vagaries made us appreciate it all the more when it burnt properly.

Our party now consisted of Baber and myself, a photographer whom Baber took up with him, Baber's chief servant Hwu-Fu, who had travelled some time with Baron von Richthofen; Baber's second servant, Wang-Erh, a giant of six feet two inches, who had been a soldier drilled by European officers, but who had never before been in the service of a European; my two servants, Chin-Tai and Chung-Erh, both over six feet high, and 'Tib,' a brown retriever. There was, in addition, an official sent by the Tao-Tai of Hankow to accompany Baber.

February 8.—We were still at Han-Yang when we awoke in the morning, and our skipper now said that the sail did not fit well, and must be altered before she could start.

So we tried to shake ourselves down; we made bookshelves of the doors of our sleeping cabins, and pasted paper over the cracks in the wall through which an icy wind was blowing.

The sacrificial cock was expended during the day, and his blood sprinkled on the bow of the boat; for without this ceremony, and the subsequent more serious one of eating the flesh of the bird, it would have been nothing less than sheer madness to make a start, at least so thought our skipper and her crew.

Towards the afternoon a fresh easterly breeze

sprang up; the old lady suddenly declared that the sail was ready, and we started at 2.15 p.m., but only made seven miles before anchoring for the night, or rather mooring to the bank, at a little village called Chuan-K'ou.

February 9.—The snow-storm was so heavy all the morning that the sailors would not leave their moorings, and we passed the time looking out of the window to see if there was a change of weather, and in trying to stop up the cracks about the door through which the snow was driving.

The river was full of boats and huge rafts of timber; some of the latter a hundred feet long, fifty feet broad, and ten or twelve feet deep, on which there were often half a dozen huts for the people in charge.

In the afternoon the snow cleared off a little, and with a strong wind we sailed to Chin-K'ou, where we ran into a creek for the night.

Our fireplace had not as yet proved by any means a success, and on thinking that the fuel was in fault we experimented on the Chinese mixture of coal and clay. The Chinese are too economical to burn coal alone, and mix coal-dust with a certain proportion of clay, making up round balls about as large as eggs. This burns well enough, and gives out a fair amount of heat, but it is, even in a house, a very unpleasant fuel on account of the dirt; and in our cabin, a gust of wind coming down the chimney, or a draught in an unexpected corner, used to blow this fine dust all over the room in clouds, and we came to the conclusion that we had not yet discovered a perfect fuel.

February 10.—We were always moored at night in a crowd of vessels; and of a morning, when all was quiet the whole place seemed to wake up suddenly. At six o'clock there was not a sound; but a few minutes later the crews of all the junks in the neighbourhood would arouse themselves with one accord.

Then commenced the shouting and jabbering of all the people getting under way. Presently another junk would come against us with a violent bump, and threaten to carry away the chimney of our stove.

This rouses the ire of our skipper and her crew, who all at once vociferate in the choicest terms that they can cull from their flowery language, the crew of the other junk returning the abuse, and amid the babel the shrill voice of the old lady is easily distinguished. Then it is our turn to run into something else; and so on, scraping and bumping, with all the timbers of the deck-house groaning and creaking, until we are clear of the crowd.

During the morning we ran to Teng-Chia-K'ou before a fair and strong wind.

Here the river turns round with a sweep, and in company with a number of other junks we were obliged to anchor in a creek; for these vessels can only sail with the wind on the quarter or astern, and in deep snow the trackers can make no way against the combination of a strong wind and swift current.

From the position in which we were moored, the wind now was dead ahead, and blew more snow in through the cracks of the door. We spent the morning in the manufacture of curtains, and sat all day with everything closed, wrapped up in ulsters, and reading or writing by the aid of candles.

The Chinese used to think us very funny people; we never could sit in a room without a fire; although they never used a fire at all except for cooking, and were quite content to remain with windows and doors open, almost in the open air, trusting to their wadded garments and thick-soled shoes to keep them warm. They saw no harm in a gale of wind blowing in their faces, but we were always draught-hunting, stuffing in some cotton wool here, pasting paper there, hanging curtains up, and taking an immense amount of trouble to keep out a little snow or a current of cold air; and as for our fire, we were perpetually fussing about it; if we found one kind of fuel did not burn, we were always worrying the servants to try something else, instead of doing without, like sensible people.

Then they never cleaned their places, why should we? but if we found an inch or two of harmless dust anywhere, or a pile of dirt in a quiet corner, nothing would satisfy us but having it removed. And notwithstanding all this, we, who felt the cold so much, were always taking off our clothes, and would in the morning sit for no conceivable object in a tub of cold water, instead of following their plan of keeping on the winter garments night and day, until the weather should begin to get warm. Then our clothes were preposterous—stupid, thin, tight-fitting affairs—as useless as they were hideous; no wonder we felt cold.

We certainly did feel cold; but notwithstanding the severity of the weather we adhered to our national customs, and at length, by dint of perseverance, we made our room tolerably tight, and managed to keep up a moderate degree of warmth.

During the day the wind moderated, and we started with the rest of the fleet.

At the best of times, the scenery can scarcely be said to make this part of the river inviting; and we did not find much to regret in the necessity for keeping the windows covered. There was nothing to see but sloping mud banks, and a dead level beyond, all white with snow; whilst a collection of miserable huts that there might be here and there, with a stunted and leafless willow, a few reeds or bits of long grass, just appearing out of the white covering, only served to lend additional dreariness to the scene.

In the evening there is the usual shouting and hallooing as we come to our moorings; bump succeeds bump as we crash amongst the mass of boats, until we ultimately make fast, and the men have their supper.

Then the loud cry of the hawkers, who go about amongst the craft selling bean-curd cakes, or little drops of spirit, makes itself heard above the shouts of the sailors; this gradually ceases, and nothing but the hum of conversation is left; by-and-by, as one by one the men wrap themselves up and go to sleep, this dies away, and all is still, save for an occasional word from some one less sleepy than the rest. The last voice presently is hushed, and as we sit reading or writing there is not a sound but the lapping of the water against the sides of the boat.

February 11.—The sun at last shone out again; the morning was clear and frosty, and after having been shut up in the dark for so long it was pleasant to stretch our legs ashore whilst the coolies were tracking the junk up, walking with bare feet and legs in the snow.

February 12.—This was the time of the Chinese new year festival, which lasts about ten days; there was much feasting, popping of crackers, and beating of drums all the morning, and the people were so well amused they did not want to leave. We sent out at seven o'clock to know when they were going; they replied 'immediately.' At eight o'clock we wanted to know how long it would be before they started; they answered 'no time at all.' At nine o'clock we said they really must get under way; they declared they were going to. At ten o'clock we threatened that they should have no new year's present unless they moved at once; they sent back to say that we were just off. At eleven o'clock Baber ordered Hwu-Fu to go to our accompanying official; but they said he would be left behind if we let him leave the boat. At twelve o'clock we began to make a real disturbance, when they let go the mooring rope, and we went on to a place called Hua-K'ou.

February 13.—This was the Chinese new year's day; and at about nine o'clock all the servants entered in their most gorgeous clothes to wish us a happy new year, and to receive the wonted tribute to the inevitable custom.

The captain then came in, with two of her men and her little child, for the same purpose.

A present of 5,000 cash was then distributed amongst the crew, that sum representing something less than 1*l*.

February 14.—We asked our captain this morning which way the wind was. She replied that the north wind was strong, but the east wind not so strong; by which she meant that it was about NNE., and made us almost think she had learnt mathematics, and understood the resolution of forces!

We were off at six o'clock with a strong wind, and bowled along merrily; but the inconvenience of heeling over was considerable, for our tables and chairs were ordinary land furniture, and not made fast in any way.

The wind held true all the morning, and we accomplished thirty-nine miles.

February 15.—In the middle of the day we reached Hsin-Ti, a somewhat anomalous place, for although it has not the rank of a town, it is of so much importance and trade that it is presided over by a Tao-Tai. There are in China only four other places, not towns, that are in a similar way the seats of a Tao-Tai.

Hsin-Ti (Sing-Ti of Blakiston's map) is a large straggling place about two miles long, but with no depth back from the river. Here there were a number of the picturesque Chinese gun-boats. Their form is exceedingly graceful, the upper works having a good sheer at the bow, and even more at the stern, on which a small house is built for the commanding officer. Over the body of these vessels, when at anchor, there is a little tent of blue and white striped cotton; and with their gay red banners and streamers they look very bright and cheerful.

We stopped here some time, as our old lady wanted to buy something; so we sent ashore for some white, blue, and red cotton to fasten round the cabin, and hide the rugs which we had hung up as draught-excluders.

February 16.—The view from our ship had been for some time limited to a bank of mud, thirty feet high, but about noon we arrived at a rocky bluff, where a little hill rose up out of the plain. From the top there was a fine view of the noble river here, three quarters of a mile wide, winding through a great plain, where broad lagoons lay stretched out amongst fields that were protected from the summer floods by extensive dykes and embankments. A village nestled among some few trees at the foot of the hill, and away in the distance to the south there was a fine mountain covered with snow

I walked on from here, picking up a few teal, although surrounded by a crowd of little boys, who exhibited the imitative power that is so prominent a feature in the Chinese character by the quick way in which they caught the words I used in calling the dog, and as I left I could hear them shouting to one another in excellent English to ' come here.'

February 17.—We passed the entrance to the Tung-Ting lake. The Chinese consider the river that flows through this lake the main branch, and the stream that comes from Ch'ung-Ch'ing a tributary. This is because they measure the magnitude of a river by the amount of traffic on it.

We both noticed that beyond this point the number of boats we passed was very much less than lower down.

During the summer, the river overflows its banks and floods the surrounding country. There are extensive lines of embankment from one to two miles inshore, and all the villages are behind the inner line. This gives a dreary appearance to the landscape; and the traveller, walking for hours without seeing a village or meeting a human being, might easily be misled into the belief that he was in an uninhabited country.

February 18–22.—These days afford nothing to record. The 20th was a fortunate day, for Camel Reach, twenty miles long, lay before us, running nearly due north and south, and, a southerly wind favouring us, we ran the whole distance before it to Shang-Chê-Wan.

There were a great many villages in the neighbourhood, which seemed in a very flourishing and well-to-do condition. The country was closely cultivated, the fields were protected by splendid embankments, and as the snow had now all melted the young crops coming up looked fresh and green.

February 23.—At Chien-Li-Hsien-Ma-Tou. We did not get under way in the morning, and on inquiring the reason, the sailors pointed out the masts of a junk about a hundred yards away, and asked us if we thought it could be safe to go on when there had been a squall violent enough to sink a junk.

February 24.—The next morning our crew still refused to move; they declared that the sight of the wrecked junk frightened them; and although there was a light and favourable breeze they said that they could not tell what the weather might be by-and-by, because it was cloudy, and they could not see the sky. They added that the other junks had not started, and it was quite contrary to custom to get under way before other people. The threat of an appeal to our official at length overcame their reluctance, and once we were off the rest of the fleet followed.

February 25.—There had been more snow during the night, but the morning was beautifully fine; every one was early astir, and the whole fleet was under way before seven o'clock.

It is a busy scene when a large number of junks are tracking together. Now an ambitious captain thinks he can shoot his vessel in front of another inshore, and tries to pass his tracking rope over the mast-head of his rival. This excites the jealousy of the crew, and if the tracking ropes foul, or the junks bump together, it rouses their anger.

The two captains then mount to the highest parts of the deckhouses, swear at one another, stamp their feet and shake their fists, both crews in the meantime shouting directions to the coolies on shore; but as they all talk at once, down to the smallest children, they are not generally very successful in making themselves understood. Then the confusion is tremendous; a track rope is unexpectedly tightened, and one or other of the vessels heels over so much that she is in danger of foundering. At last the junks shake themselves clear, but by pure good fortune, management having played a most insignificant part in the manœuvres. After they have been out of hearing of one another for some time the captains leave off swearing; but should accident again bring them together, the skippers at once mount to their elevated positions, and the commination service begins afresh.

Although the trackers are often a quarter of a mile away from the boat, and at that distance the people on board naturally find it very difficult to make themselves heard, and though there is often such a crowd of junks whose crews are all shouting together, that it would seem impossible for any coolie to distinguish the orders meant for himself, yet they never attempt to introduce a code of signals. It is customary, however, in the rapids, higher up the river, to use a drum, the coolies pulling as long as the drum beats, and stopping when it ceases.

February 26.—I was ashore in the afternoon about a mile below Parson's Point, and seeing some teal in a creek I went toward them. Just at this moment I unexpectedly heard a shot, and found four men with a gun twelve feet long, the bore of which was about one and a half inches in diameter.

They had a framework, covered in front with rushes, behind which they hid themselves until some unwary birds came near enough to shoot, when they fired away about a pound of rusty iron shot of all shapes and sizes. At the last discharge, though there had been some thirty or forty birds to fire at, they had only succeeded in bagging three, and besides these they had no more than five to represent the day's work.

I offered to buy them, and walked on with the men to the shore, where there were half a dozen regular gunning punts, each with a long gun. They had only a few geese and ducks amongst them all, and it seems difficult to understand how they can get a livelihood out of their sport, for their powder and shot, though of the most miserable description, costs what is to them a good sum of money, and they cannot sell geese to any but the very poorest, for no Chinaman cares to eat a wild goose, except to keep himself from starving.

February 27.—We made very slow progress in the morning, for even close in-shore the current at this part of the river is very swift, and the shoals run out half-way across. We grounded on the mud several times, and the trackers were so far off they could do but little good in the strong tide that swept across the shallows.

The country here is closely cultivated, protected by fine embankments, and the villages are very numerous. But the banks of the river are being swept away, and for a long distance we could see that a breadth of about six feet must have fallen in quite recently. There is a great bend between Last Bottle Reach and No Beer Channel, where two points on the river are separated by a neck of land not more than three quarters of a mile broad, the distance between them by the river being about fifteen miles. This neck must become more narrow each year, for the river sweeps down on to it at both sides. Blakiston represents it as one and a half miles across in his time.

There can be little doubt that in a few years it will be cut through, and become the channel.

The day was very pleasant; and the crops coming up, and the birds just appearing on the willow trees, together with a delicious feeling in the air, made us hail with pleasure the advent of gentle spring.

On a wide sandbank, across the river, I could watch the movements of thousands of geese. There were no people or boats on that side, but the birds appeared very uneasy. In their movements they put me very much in mind of swallows flocking at the approach of winter, and I wondered if they were preparing to leave the country before the hot weather.

Every now and then they would get up with a great clamour, fly across the river, wheel round and round, and then return to the sandbank; they generally began calling just as they rose from the ground, but on one occasion they did not commence their hoarse croak until they were well in the air; and at the distance of about half a mile the simultaneous flapping of some thousands of big wings sounded like the report of a heavy gun very far away.

There were a great many porpoises in the river; these creatures ascend the Yang-Tzŭ nearly up to I-Ch'ang, 900 miles from the sea.

February 28.—There was a fair wind in the morning, and as the men sat on the forecastle, eating, drinking, and talking incessantly, they 'whistled for the wind' as Europeans do; it is rather curious to find this practice so universal.

On an occasion like this, when the wind relieved them of their work, they used thoroughly to enjoy the unaccustomed treat of eating their meals in a leisurely manner.

They used generally to get up at a quarter past

five, and roll up their blankets, and take down the framework and matting with which the front deck was always covered in at night. The start was usually effected immediately after this, and by seven o'clock the cook and cook's mate had prepared a gigantic bucket of rice and a few vegetables.

The ship was then anchored for ten minutes, during which time the coolies would manage to eat each two or three basins of rice.

In the middle of the day, a quarter of an hour was allowed for a similar meal ; but at night, when work was over, they could spend as long a time as they liked over their supper.

Nine-tenths of the food of these coolies was rice boiled perfectly plain ; they would eat some chopped vegetables with it, cooked in a great deal of grease ; and when by chance we shot a gull, a crane, or other strange bird, it afforded them the rare luxury of meat ; but the proportion of rice to all their other food was so large that the amount of grease they ate was not very considerable, though all their little luxuries, such as a bit of ancient fish, or a lump of fat pork, were cooked in large quantities of grease.

Rice is a food that is not well adapted for men doing hard physical work, except where it is so cheap that large quantities can be eaten at a less cost than a smaller proportion of more nourishing food, and in travelling it is very striking to note that the very day on which the rice-growing country is quitted, some other grain at once becomes the food of the people ; rice is so bulky that even one day's carriage makes it too costly for any but the well-to-do.

The grease eaten by the coolies, far from being an unaccountable taste, is an absolute necessity ; no man can live without grease in some form or another, least of all those doing hard physical work on rice for their staple food.

About half our coolies were opium smokers, but whether it was owing to the active life in the fresh air, or to the weakness of the drug they used, it did not seem to do them any harm.

We stopped to-day at Huo-Hsüeh (Ho-Hia of Blakiston), a straggling town along the river side. Many of the houses are built on bandylegged-looking poles resting on the steep bank, and when the river is high, they must be unpleasantly near the water.

There are remains of a masonry embankment, but now it is little more than a heap of loose stones.

March 1.—We reached Sha-Shih in the evening.

March 2.—In the neighbourhood of Sha-Shih, the country is very carefully embanked against the depredations of the river, the embankments in many places being faced with stone; and although it is the general custom in China never to repair anything, there were large gangs of men employed in restoring these.

March 3.—The country now began to change in appearance; the level of the ground was well out of reach of the river, even in its highest floods, and hills appeared in the distance on both sides.

I went ashore and walked through the large town of Tung-Shih, where I was followed by a somewhat excited crowd of people, one or two mischievous boys throwing stones. After I had passed it, as I saw the 'Kestrel' in the distance coming down the river, I hired a sampan. All the inhabitants of the town now wanted to come with me, and as many as could jumped into the sampan, and settled themselves down for a pleasure trip at my expense. The boatman, however, did not like this any better than I did myself, and he soon turned them out. The captain of the

'Kestrel' could not stop, but threw us a rope ; but owing to the stupidity of my boatman, who had never seen a steamer before, one of her boats broke our mast off short, let the sail into the water, and threatened to capsize us. The boatman was quite scared, and looked on helpless, until with the help of Chang-Erh I pulled the sail into the boat for him, and even then he scarcely recovered his senses.

I only had time to accept an invitation to dinner and return to our junk.

When I called for the bill of the broken mast, I found it amounted to only an equivalent of about half-a-crown of our money, and the boatman was overjoyed with the magnificent payment of a thousand cash.

Baber and I then turned out our portmanteaus, and having made ourselves presentable by the addition of unwonted collars, we hired another sampan and set off down the river after the 'Kestrel,' ordering our captain in the meanwhile to go on as far as she could, for we knew that we should always be able to overtake her. We soon boarded the gunboat and heard all the news. They had found the river much lower than was anticipated ; indeed it happened that this year the water was unusually low, so much so that the captain of the 'Kestrel' had given up all attempts to reach I-Ch'ang.

We both of us thoroughly enjoyed the good companionship and hospitality of our naval friend, and were sorry when, at half-past ten, we were obliged to say good-bye and push off in the sampan.

The night was cold, and as the stars were shining brilliantly, we took a brisk walk for half an hour, to warm ourselves before settling ourselves into the boat. After we each had smoked another cigar, Baber suggested going to bed. The operation was a simple one, as we had nothing more to do than to exchange our sitting for a recumbent posture on the bottom boards of the boat.

Sleeping on a plank in a hat with a stiff brim is not altogether a very luxurious method of taking rest, but, nevertheless, like the 'sea boy on the giddy mast,' we found our 'eyes sealed up' by 'nature's soft nurse,' and our senses were soon 'steeped in forgetfulness.'

At about three o'clock in the morning the boatmen said they would go and look for our junk; but as we all knew she was up the river, and they started down, it was evident that they had some ulterior motive, in all probability a village at no great distance.

March 4.—At four o'clock, being very cold, Baber and I decided to look for our vessel ourselves, and disembarking we walked along the bank in the fine clear moonlight. Nearly everything was asleep, except the dogs, who barked furiously as we walked through a small town, and one or two early boatmen just lighting their fires. We reached our junk at half-past five, found a warm room and a cup of hot chocolate, and turned into our comfortable beds.

We had now left the vast and monotonous alluvial plain of the lower Yang-Tzŭ, and were fairly in the hills. The ground was well cultivated, and the crops, which seemed to be growing by magic, were very green. Temples and pagodas here are perched on the highest points. Comfortable-looking farmhouses nestle in the hollows, surrounded by small bamboo copses. Children in the dirt, with pigs and dogs, play about the doors, where the women sit sewing and talking. On the hill-sides there are little clumps of cedars and firs, or patches of long grass; dog violets are in blossom at the sides of the path, and the flowers of great fields of rape shine as brilliant streaks of yellow in the distance. The grand river, still half a mile wide, now clear and almost green, rolls below the cliffs of red sandstone, and numerous junks going up and down lend life and animation to the scene.

The old lady took a horse in the morning, and rode off to I-Ch'ang, and the boatmen, anxious to get there also as soon as possible, made a long day's journey to I-Tu, where we moored amongst a number of other boats.

Sitting after dinner, with open windows, a man in a junk alongside said something I did not understand, when, to my astonishment, Baber took a header out of the window, and 'went for that heathen Chinee.' The man, however, escaped, and when Baber returned through the door, he explained that the object of his wrath had called us devils.

Another man presently came, and resting his arms on the window stood calmly gazing at us. At last Baber politely asked him what he was looking at. Not in the least abashed, he quietly replied, 'I am looking at you sitting down,' an eminently matter-of-fact reply, very characteristic of the Chinese character.

March 5.—I took a walk in-shore to-day over the hills, about 600 feet high. Directly the river is left, even by half a mile, the thinness of the population becomes apparent. Here the cultivation was only in the valleys, all the slopes and the tops of the hills being covered with beautiful long grass and low scrub. During a walk of more than two hours I scarcely saw a house, and did not meet half a dozen people.

After a time I returned to the river, and through a telescope saw the junk sailing away before a fresh breeze.

I did not particularly wish to walk to I-Ch'ang, because I had heard from the officers of the 'Kestrel' that there had been some sort of disturbances there, and I had no wish to get into an unpleasant hooting crowd, if I could help it. So I told Chung-Erh to try and engage a boat. There was some difficulty about this, as all the boats belonged to fishermen, who did not care to do anything out of their accustomed ways; but I presently fell in with a small junk, full of traders, carrying cotton up to I-Ch'ang; they were very civil people, and took me on board.

They looked at my gun and cartridges, for which they did not care much; but my telescope was a source of great merriment. They knew well enough what it was, though one and all completely failed to manipulate it. First one man took it, and the others eagerly asked him what he saw. After having pointed the glass steadily at the sky for some time, he answered in a doubtful sort of way that he could not see much; at which his friends jeered him, and made him give up the glass to the next man, who took it with a most superior air, as much as to say, 'Ah, just let me show you how to do it!' But after putting it out of focus, and looking straight into the bottom of the boat, he tried to see the inside of the telescope, and passed it on with a shrug of his shoulders, distinctly under the impression that it was stuffed up. The third man, after I had again focussed it, chiefly poked it into the eyes of everybody else, and knocked their hats off, at which he was voted a nuisance.

Then the evening closed in, and under the shelter

of the straw covering we had tea, and smoked until we arrived at I-Ch'ang.

When I entered the cabin of our junk I was warmly congratulated on my safe arrival by a voice from a vast collection of opened newspapers. Careful search revealed Baber hidden in the product of three mails, and in answer to my question, he explained that his hearty reception was caused by my escape from the mob of I-Ch'ang, who at this time were very turbulent, so much so that the newly appointed consul had deemed it prudent to send out a strong escort to look for me.

I then learnt all the news. There was now a considerable European community at I-Ch'ang. The English consul and his vice; the chief of the Chinese customs with two assistants; the captain of a river steamer, who was up here to prospect, and three missionaries. The chief commissioner of customs had been the first to arrive, and after him the consul had come in the 'Kestrel,' to choose a site for the English settlement. At first they found the people civil and obliging; they were never annoyed in any way, and used to walk about anywhere and everywhere. The consul selected a piece of ground, made the necessary agreements, ordered the boundary stones, thought that everything was comfortably settled, and was going to mark out the concession, when the aspect of affairs changed completely.

There was amongst the richer classes, and especially amongst the literati, a strong anti-European feeling. A report was spread that land was to be taken without payment, and other slanderous tales were invented by which the minds of the easily excited Chinese population were inflamed. One day, without previous warning, the consul was unexpectedly mobbed and insulted, and after that no European was able to walk on shore without an escort.

Such was Baber's news, and I heartily congratulated myself on the fortunate rencontre with the traders' junk.

March 6.—We were obliged to stop at I-Ch'ang for a couple of days. The vessels wanted recaulking, some fresh rigging was required, and above all a new crew; for the navigation of the Yang-Tzŭ above I-Ch'ang is very different to the simple tracking below, and the shoals, rocks, and rapids, some of which are very dangerous, require a very skilful and practised crew.

We were escorted from I-Ch'ang by a petty official named Sun, and by one of the picturesque gunboats, the commander of which came to pay his respects to Baber, and knocked his head against the ground in humble manner. He afterwards moored his vessel by the side of ours, a proceeding which though it increased our dignity had its disadvantages, for the crew were most regular in their watches, and always relieved guard to the sound of a very powerful drum. No sooner had we closed our eyes in bed, than they would begin with a few gentle taps, which gradually swelled into a grand roll, ending in an extraordinary flourish. The noise would cease for a moment, and turning in bed, fond hopes would arise that it was over, when, with a sudden clang that penetrated into the inmost recess of the nerves, the drumming would recommence with redoubled violence. Even if habit be second nature, it would take a man a lifetime to get accustomed to the noises made on board a Chinese gunboat; but Baber metamorphosing a French proverb, remarked, '*Il faut souffrir pour être grand.*'

CHAPTER VI.

THE GORGES OF THE GREAT RIVER.

The I-Ch'ang Mob—Chinese Unreason—Departure from I-Ch'ang, and Adieu to English Faces—The Gorges Entered—Vicissitudes of Ascent—Scenery of the Gorges—Preparation for the Rapids—Ascent of Rapids—Difficulties of Nomenclature—Puzzles of Orthography—Yang-Tzŭ Life-boats—Dangers of the Rapids—View at Ch'ing-Tan—Ch'ing-Tan Rapid—Waiting our Turn—Pilots of the Rapid—Successful Ascent—Kuei-Chou—Its Means of Support Obscure—Mode of Tracking—Coal Districts Entered—The She-Skipper's History—Cave of the Cow's Mouth—Expedition in the Hills—Position of Pa-Tung—Long Gorge of Wu-Shan—Weird and Gloomy Aspect—Oracular Rock Disks—Cave Habitations—Rough Scrambling—Opium hostile to Bees—Town of Wu-Shan—Perplexing Eddies—Kuei-Chou-Fu—A Visitor, and his Monstrous Torrent of Speech—Method of Local Salt Manufacture—The Kuei-Chou Aqueduct—Credulity and Untruth—Civility of Ssŭ-Ch'uan Folk—Troubles in the Rapids—River Wall of Miao-Chi—Great Haul of Fish—City of Yun-Yang—Extraordinary Richness of Soil—Importance of Rape Cultivation—Chinese Skill in Agriculture overrated—Their True Excellence is in Industry—Ample Farmsteads—Rock of Shih-Pao-Chai—Story of Miraculous Rice-Supply—Luxuriant Vegetation—'Boat Ahoy'—The Ch'ung-Shu Flowers—St. George's Island—Cataplasm for a Stove-in Junk—The River of Golden Sand at Last—Fu-Chou—A Long Chase—Effort to Reach Ch'ung-Ching—The Use of the Great Sculls—Arrival at Ch'ung-Ching.

March 7.—The governor-general of the province had come up here to arrange matters with our consul; but he went away two days after our arrival, either because he would not take the trouble to arrange matters, or because he was afraid of the responsibility of failure. No doubt he thought that things were going wrong, and in plain English his departure would have been called running away.

When he left, of course all the people in the neighbourhood who had spare gunpowder let off guns.

At about ten o'clock the consul went ashore again with the Tao-Tai, attended by the other chief Chinese officials, and escorted by a regiment of braves.

They were at once surrounded by a yelling mob; and as the officials and braves were quite unable to quell the disturbance, they retired to a temple.

On the way they succeeded in making prisoners of two men who appeared to be ringleaders, and these they carried off.

When they were inside the walls of the buildings, one of the officials walked up and down, stamping and calling the people of I-Ch'ang by all the vile epithets he could think of.

The clamour outside now induced the officials to give up their prisoners. It had much the same effect as a pot of Liebig amongst a pack of wolves.

After awhile a retreat was determined on; and the whole party returned to the landing-place, amidst a shower of dirt, stones, brickbats, and tiles.

On the way the Tao-Tai lost his temper, and stamping with rage said to the mob, 'Here I am, why don't you kill me at once, and be done with it?'

The mob either had no reason in particular, or did not care to give one, and the party advanced without a reply to the question.

The Tao-Tai still showed a bold front, until he was suddenly met by a hideous old woman with a ladle of filth; this was too much, for the awful nature of a ladle of filth in China can hardly be conceived.

When they again reached the shore, we could see the performance from our boat. About a hundred little boys led the procession hooting and shouting

'Foreign devil!' Next came a dozen braves in red clothes armed with gingalls.

The vice-consul followed under the protection of a gigantic brave from the north of China, with whose enormous strides he vainly attempted to keep step.

After them the consul was walking with the Tao-Tai, who seemed rather glad to discard his red official canopy.

Behind all was the howling mob; and the remainder of the braves were scattered about amongst the crowd.

The party regained the boats without a very serious butcher's bill; the vice-consul lost a button from his coat; and one of the braves was cut by a stone. He smeared the blood all over his face, and with this ghastly aspect rushed to the Tao-Tai and demanded an indemnity of ten taels.

There are no people more easily led than the Chinese, by those who have fairly established an influence over them; ordinarily, too, they are exceedingly respectful and obedient to authority.

If instances were wanting, the way in which Gordon could do what he liked with his Chinese army, shows how powerful in the minds of a Chinaman is the instinct to follow those who can lay claims to his fidelity.

But it sometimes happens that in large towns some rich family may get more influence than the officials, especially if the latter are very corrupt or extortionate. This was the case at I-Ch'ang, where a family named Fu were believed to be the chief leaders of the people, and the instigators of the disturbance.

The Chinese are, moreover, eminently an unreasoning people; the movements of a mob everywhere are dictated rather by caprice than reason. It is very easy too to raise the devil of popular wrath, but it is generally a more difficult matter to allay it ; and later, although it was supposed that the Fu family were desirous of doing so, they were quite unable to quiet the populace.

The extraordinary ideas that penetrate a Chinese mob of course help to make their conduct inexplicable.

Here they had a notion that our consul was the brother-in-law of our Queen, and agreed that, for that reason, it would not be proper to injure him. Although it is difficult to trace the logic in this reasoning, it shows the respect of the Chinese generally for high authority, even under circumstances where it would be least anticipated.

The Chinese are very superstitious, and will readily believe anything that is told them.

There is a rock a little below I-Ch'ang said to resemble the tooth of a tiger, and therefore called the Tiger's Tooth. The Chinese are firmly persuaded that it will be quite impossible for foreigners to come to I-Ch'ang, because they say the tiger will eat the sheep; the word ' *Yang* ' meaning indifferently ' foreigner ' or ' sheep ' (there is a slight difference in the written character, but absolutely none in the pronunciation). This may appear ludicrous to a Western mind ; but an educated Chinaman can in all seriousness believe this, and many things much more marvellous.

March 8.—The officials at length managed to effect some sort of compromise between the rioters and the Europeans, and the boundary-stones were successfully put up ; after which the consul left for Hankow.

We saw the blood of the cock duly sprinkled on the bows of the boat, and our skipper and her crew were very busy making preparations : taking on board great quantities of ropes of all sorts and sizes, some of bamboo and some of hemp; strong stanchions had been put up on the gunwale on both sides, to act as thole-pins for large strong oars.

Then the forward rudder was arranged. This is a very strong oar, some forty feet long, which projects thirty feet beyond the bow. At the inboard end, ropes are fastened, so that some half-dozen men can assist in the steering; and thus a very powerful steering apparatus is formed.

I-Ch'ang seemed to be a cheap place for cabbages, for the crew brought on board an enormous cargo. There is a peculiarity in the market of I-Ch'ang that I never heard of elsewhere; for the price of things never varies, but when they are dear or cheap, there are more or fewer ounces to the pound!

Before leaving we found a carpenter who was able to fix glass into the windows of our cabin, and as we succeeded in buying a couple of panes we very much increased our comfort.

March 9.—This morning all English faces were left behind, and from this time, until my arrival at Bhamo in November, I saw no European save my travelling companions and the French missionaries.

With many a hearty shake of the hand we said good-bye to the customs officers. At 7.30 the mooring lines were let go, and as the entrance to the gorges loomed before us, we seemed to have cast loose the last rope that bound us to civilisation.

After having been so long slowly winding up the tortuous reaches of the river, gliding through the alluvial plain, where there is scarcely anything to relieve the monotony of the landscape, the sudden change in the scenery that appears beyond I-Ch'ang is very striking.

The river soon narrows to a width of from 400 to 500 yards. Steep spurs from mountains 3,000 feet high run right down to the water's edge; their sides, wherever they are not absolutely perpendicular, covered with long orange-brown grass, that seems to grow almost without any soil. On the more gentle slopes terrace cultivation is carried on, little patches of the most brilliant green, sometimes a thousand feet above the river, and looking almost overhead, showing the presence of some industrious farmer, who will not leave a square yard uncultivated if he can help it.

Sometimes the hills are broken into precipices, rising 300 feet sheer up from the water, beneath which the river runs with a glassy surface; at others there are loose piles of *débris* or gigantic masses of rock strewn about the bed, where the water dashes in wild confusion.

Now and then a cleft in the hillside discloses a tiny stream leaping from rock to rock amongst ferns, long overhanging shrubs, and brambles.

Once the steep slopes running up a thousand feet were crowned at the top by a grim wall of white cliffs 300 feet high and about a couple of miles long, and looking up a valley, pine forests could be seen on the northern slopes of the snow-capped mountains.

Nor is it the change in scenery alone that causes a feeling of strangeness, but the mode of travelling itself combines to give a sense almost of bewilderment.

Now there is no foothold for a goat at either side. The trackers come on board, and we have to row, five oars on each side pulled by ten lusty coolies, shouting to encourage themselves and mark the time. With each stroke of the sweeps the boat creaks and shakes, and from cliff to cliff, before and behind us, are echoed the regular cries of many boatmen, all urging their vessels against the rapid stream. Suddenly the cadence ceases, a confused babel of tongues swells in loud disorder, and looking out, we find the trackers are being put ashore, the crew of every boat struggling to get before that of another.

Every man with a different idea about the way something ought to be done, and proclaiming it as loud as he can, tries to shout down all the rest. The noise increases, and seems to peal from one end of the long reach to the other—when suddenly all on board is still; we glide smoothly along, not a plank or a beam giving out a note of straining, but away ashore, quite softened down by distance, we still can hear the regular cry of the coolies, as keeping step they draw us quietly along.

Now the towpath comes to an end, and the coolies must again come on board, but this time in a sampan, as here the vessel cannot run ashore. Now we cross the river to a path that runs up till the trackers look right over the mast-head. But one thing is never wanting at a critical moment, nor when the wild chorus of shouts is at its loudest; for above the din, whatever it may be, the shrill tones of the old woman at the stern rise in hideous discord.

In the afternoon we made fast to one of the big rocks lying about, near some level ground in the bed of the river, where people living in a few small temporary huts were doing a little trade by selling odds and ends to the boatmen who stop here to rest. Stepping ashore we find little choice in the walks. There is but one path, and that soon leads to a zigzag track up the mountain side. We follow up, but every now and then lose it, and have to clamber about with hands and feet from one rock to another, till we unexpectedly come upon a hut, perched on a tiny artificial plateau, surrounded by a few bamboos, orange trees, and a fir or two. Our sudden appearance startles a couple of fowls, who rush off cackling to a safe refuge by the fire inside. The never-absent dog comes out to see what is the matter, and does not cease barking until our retreating forms disappear behind some gigantic rock. Up we clamber, our protecting minions from the gunboat puffing and panting as they wonder why the mad foreigners want to be always going up hill. At length we reach a projecting point, where a bit of flat rock gives us a comfortable seat, and almost underneath us, a thousand feet below, the river, dwarfed by the distance, looks no more than fifty yards wide. To the south and west, the hills rise in masses one behind another; mountains backed up by mountains, higher and yet higher, one giant leaning lovingly on the shoulder of the next, till as we gaze towards the setting sun, with the eye of fancy we can see them range beyond range, stretching far over the borders of the Chinese empire, and at length culminating in the mighty peaks of the Himalayas.

March 10.—It was raining the next morning, and as the trackers could not move over the slippery rocks we remained at anchor, the fragrant perfume of bean flowers being wafted in at the window from the fields surrounding a little hamlet opposite.

The crops seemed almost to be growing visibly: but a few days previously everything was covered with snow, and now the trees were budding and the green wheat was a foot high.

In the afternoon the rain cleared off, and we reached the little village of Huang-Ling-Miao, where the valley has a wild and savage aspect, and where the track along the river bank is strewn with gigantic boulders brought down, when the river is in flood, from the far distant and unknown recesses of the mountain ranges.

March 11.—It was a dull morning when the usual stir amongst the boatmen awakened me. Soon after I was up, as we were rounding a projecting point, the current twisted the boat's head, and notwithstanding the tracking line, and forward steering apparatus, we spun round like a top until our bows pointed down the river. Matters, however, were soon righted, but we immediately afterwards anchored, as a storm of sleet and snow came down, wrapped everything in mist, and hid the view. Fierce squalls of wind from every point of the compass blew down our chimney, sent tongues of flames from the stove darting across the room, and made matters generally unpleasant.

It cleared up later, and in the afternoon we reached the lower end of the Ta-Tung rapid, where we were obliged to anchor again to await our turn.

In the meantime the usual bamboo tracking-line was cast off, and a strong hempen one substituted, and our old skipper, after much talking, concluded a bargain with the extra coolies required to help us up the rapid.

At the foot of all the serious rapids there are a number of temporary shanties erected—temporary, for the ground on which they stand is under water during the floods of summer.

Coolies who come up here for the winter and spring live in these, and make a livelihood by assisting the ascending junks to pass the rapids, for a large junk may require an extra hundred of coolies to haul her up.

Amongst these hired coolies there is always one who, owing to his skill, is a person of such importance that he is often saluted with an explosion of crackers when he first comes on board.

At length our turn arrives. We have now only five men left on the forward deck; four of these, picked for their nerve and experience, stand to the forward steering apparatus, and the fifth squats down with the drum between his knees; all give one anxious glance round to see that everything is right; the signal is given, the drum is beaten with a regular cadence, the coolies ashore shout as the rope tightens to their pull, and in a moment we are in the rapid. The water boils and foams about us, and leaps now and then up at the bow, as if it would engulph us; but we steadily ascend; inch by inch we make our way; the coolies ashore attending carefully to the signals given by changing the cadence of the drum.

Now it is interesting to watch the movements of the agile coolie, who was received with so much respect; he seems to combine the activities of a goat and a fish.

The bed of the river is strewn with granite boulders, some as large as a small house; the tracking line catches in an uncompromising corner of one of them, in an instant the naked coolie—for he has disembarrassed himself of every shred of clothing—is at the top, and the line is clear. Now, behind a ledge of rocks there is a backwater, and he has to swim across it to disentangle the rope from the mast of a fishing-boat anchored in the rushing torrent; and again, active as he is, he is on shore only just in time to save the rope from another rock.

Little by little, though it seems slow work, the end is approached. At last, after three quarters of an hour, we pass the two hundred yards, and glide round a rock, into a pool of still, calm water, where our coolies receive the congratulations of their friends, and we anchor for the night.

March 12.—The commandant of our gunboat was very fond of firing salutes in our honour; the starting gun awakened me early, and I heard the voice of his pet lark trilling merrily to the daybreak.

We were looking straight up the Niu-Kan gorge. In the distance a glorious mountain towered above us, seeming double its real height from the clouds that hung around its sides, and left only its summit clear against the sky. Cliffs two hundred to three hundred feet high bounded the river on either hand, the hillsides glowed in the rich colouring of browns and deep orange reds, and the huge boulders lying about gave a savage grandeur to the scene.

The people here call the river the Ta-Cha-Ho, which means the river of lees or dregs, a most appropriate name, for the whole bed is strewn with *débris* brought down from the far distant mountains.

For anyone who does not know the written language, it is most difficult to obtain the proper name of any place in China.

The country people have often a vile pronunciation; and even when a name is pronounced correctly, it is very difficult for a European unacquainted with Chinese to catch the sound. When I afterwards travelled by myself, I adopted and carried out a suggestion made to me by Baber, that I should have the names of the places written in Chinese either by the innkeeper or by some other local *savant*.

Baber himself translated for me with infinite care the names of all places between Ch'ung-Ch'ing, Ch'êng-Tu, and Sung-P'an-T'ing; and since my return to England, Mr. Johnson, of Her Majesty's Consular Service, has been most kind, and spared no pains in the translation of the names on the road from Ch'êng-Tu to Bhamo.

Thus I have been able to produce the names of all the principal places, written in Roman characters, on the system invented by Sir Thomas Wade.

A case occurred here that shows the difficulty of obtaining a name correctly. After having the name of a place repeated to me several times, I wrote it down as Tung-Ling; the old woman who told me drawing her hand round her throat, by which she meant that Ling, which might have a hundred different significations, was the Ling that would be translated into English by the word 'collar.' When I interrogated Baber, I found that, notwithstanding the very great care he had always taken, and his knowledge of the Chinese language both written and spoken, on his last visit he had called it Lio-Lin, or willow-grove.

On questioning the old woman, Baber found out that she called it T'ou-Ling, or head and collar, on account of a rock in the river, surrounded by a circle of foam, that looked, so she said, like a head and collar. Baber, who never spared trouble to get anything properly done, now sent ashore, and found that the real name was Kung-Ling, or amphitheatre of hills.

Thus, for this place no less than three very different names had been given, and those not by strangers, but by people belonging to the neighbourhood.

Orthography of names is always a thorn in the side of, and often a terrible stumbling-block to, travellers in Eastern countries. More especially is this the case in China, where the language is full of delicate sounds almost undistinguishable by an unaccustomed European ear. Until quite recently the confusion was made worse by the English pronunciation of vowels; and it will not be until Hunter's system is fairly established and understood that Englishmen will have any certainty of the correct orthoepy of the names even of places in our Indian Possessions. The foreign pronunciation of vowels is now happily recognised; and for Chinese sounds Sir Thomas Wade has invented a system of orthography which has the advantage of being the only system that is, or ever has been, in existence; on its other merits I am not competent to form any opinion.

It is to be devoutly hoped that all future writers will, as far as possible, avail themselves of Sir Thomas Wade's system, so that the identification of places referred to by more than one writer may be certain: a feat of literary gymnastics sometimes almost impossible, as, for instance, when one writes 'Show,' and another 'Hsiao' for the same word!

On arriving at the foot of the next rapid, a very ominous sight presented itself to us. Stranded on a rock, with the water boiling and foaming around it, was half of a junk, which coming down the river four days before had driven her stern on to the pitiless ledge. In a very short time the furious stream had broken off the fore part of the vessel, and left the remainder, an object of terror to the superstitious sailors. No lives were lost, and the greater part of the cargo was saved; but the grim and shattered relic, with a coil of rope and a bundle of cabbages still lying on the after-house, formed a warning to rash navigators in the dangerous rapid.

To make the scene more thrilling there were a couple of life-boats paddling about close in amongst the rocks.

These are not life-boats in our sense of the word, as to floatation, but they are as to saving life. Strongly built, they are manned by a picked crew of six soldiers,[5]

[5] The Chinese make little distinction between sailors and soldiers.

and stationed at the dangerous places, to rescue any unfortunates from a wrecked junk that may be struggling in the water. The boats are painted red, and have some characters written on them. The men wear the usual blue trousers, blue tunic, and the blue Ssŭ-Ch'uan turban. Over the blue tunic there is a yellowish drab coat without sleeves; and on the front and back of this is a white circle inscribed with characters in red, indicating the company or camp to which the men belong.

They seemed to manage their boat in a quiet sailor-like fashion, and paddled steadily beside us as we went up. When once the junk was absolutely in a rapid our crew also worked very quietly; there was then always one guiding spirit, and until we had safely passed, everything was left to his judgment. But the moment the danger was over the shouting and noise began again, everyone trying to make up by louder vociferation than usual for the few minutes of enforced silence.

Ascending a rapid in a big boat is in fact an operation that requires the very nicest skill and judgment, and the most prompt and ready obedience to the smallest signal given by the commander. The very slightest error, or the smallest delay in executing an order, would often be fatal, and bring about a serious accident. The old lady never attempted to take charge under these circumstances, but generally the chief of the coolies she had hired at I-Ch'ang was in command, though, on some occasions, a pilot came on board with the extra coolies at the rapids.

Often the vessel will be driven violently ashore, or on to a rock, by an eddy, and to deaden the shock a simple kind of buffer is used. This is a very powerful spar on the starboard side, loosely lashed to a stanchion on the bulwark. When in use the forward end is pushed a long way in front of and below the bow, and the united strength of three or four coolies, at the inboard end of the spar, takes the first shock and lessens the concussion of the boat, though often, notwithstanding this, the blow is very violent.

In this part of the river, the fishermen anchor their boats to a rock where the current is strongest, and where the water boils and swirls all round them.

There is only one man in a boat; he uses a long-handled scoop-net, which he pushes as deep into the water as he can, sweeping it down the river, and again to the surface, making about one stroke a minute.

A kind of carp is caught here, which is eatable, but not good, and which rarely exceeds eighteen inches in length.

We anchored at Ch'ing-Tan; and Baber and I went for a walk, or rather scramble, for the side of the hill sloped at an angle of 36°. Up this we clambered, with our hands, knees, and feet, to a height of 800 feet, where we found a zigzag that took us to the brow of a spur 1,300 feet above the river.

The view was magnificent; the setting sun just lit up the snow lying on the tops of the higher mountains; down below the shadows were of a deep and transparent blue, and the river seemed almost at our feet. The side of the mountain was cut into terraces; and little patches of wheat were springing up, but so small that they would seem insufficient to feed even the few inhabitants that are found here, living in miserable little shanties on a few yards of artificially levelled ground.

But it was already late, and we were compelled to tear our eyes away from the lovely scene, and use them for the more practical purpose of noting every spot on which we put our feet as we descended the steep track.

We picked up a precocious juvenile as we were scrambling down, and Baber being hot gave him his hat to carry; the boy promptly put it on his head with a grin.

Baber said to him, 'I hope your head is clean.'

'Yes,' said he, 'I have no lice, nor fleas either; you can look if you like,' and taking off the hat, he held down his head for inspection.

About half-way to the bottom there was a ruined temple, at which the boy pointed and said, 'Ah! I think that would be the sort of place for a foreign house.' Foreigners were daily expected at almost every town of importance; and they would everywhere have been welcomed by the people, when not incited by the slanderous inventions of the literati.

There were a great number of orange-trees here, and another fruit-tree, called by the Chinese Pi-Pa (*Eriobotrya Japonica*).[6]

We were now anchored at the foot of the Ch'ing-Tan rapid, the worst of all the rapids on the Yang-Tzŭ, and in the still night air we could hear it roaring and rushing only fifty yards above us.

It is said that this rapid was caused by an enormous mass of the mountain falling into the river, and the date is somewhere given as very ancient; our official, however, told us that it occurred in the time of the Ming dynasty (from A.D. 1368 to 1644).

March 13.—When we looked out in the morning, the steep slope of the water was so apparent that it seemed as if it were impossible any boat could ascend it. Rocks cropped up in most unpleasant places, a

---

[6] Well known in India as the Loquot, from its Canton name, and in Italy as the Japanese *nespolo*, or medlar.—Y.

broad sheet of white foam extended right across, and the very fish were jumping and leaping in their efforts to ascend.

Our accompanying official, Sun, sent to say that he had no intention of risking his valuable life in any boat up that awful torrent; and that we had better follow his example, and not only walk up ourselves, but send our valuables also by land.

We, however, came to the conclusion that all our goods were equally valuable, and that unless we regularly unloaded the ship, we could do little good; and as for ourselves, we determined that the excitement of going up was worth any risk there might be. We thought, too, that if we remained on board the people might be more careful than if we went ashore.

There was a long time to wait before our turn came, and we watched a small junk make several attempts to ascend before finally succeeding; whilst a crowd of people gradually collected who had come to see the unwonted sight of two foreigners going up the rapid.

The shore was strewn with gigantic boulders, amongst which knots of Chinamen in their blue cotton clothes sat and stood in every conceivable attitude; some were perched on the tops of the rocks, others at the edge of the water were catching fish about the size of sprats, and little ragged and dirty boys had arranged themselves in artistic groups that Murillo alone could have painted.

A steep bank rose up thirty feet, on which the town was built, but the level ground was so scarce that the houses were obliged to seek extraneous aid, and support themselves on crooked and rickety-looking piles.

Beyond towered the giant mountains above an

almost perpendicular wall of rock that rose many hundreds of feet straight up from the river.

The ship was now lightened as much as possible by the removal of some of the heavy cargo; and all the morning was occupied in laying out warps. One, 400 yards long, led straight up the rapid; and two other safety ropes were made fast ashore, so that if the first and most important should have parted, we should have merely glided back whence we came, always provided that we did not strike one of the vicious-looking rocks whose wicked heads rose above the foam.

Just at this time a little sampan with two rowers and a helmsman came down, and it was really a fine sight. As they entered the broken water the boat disappeared altogether from view, and the fearless yet anxious look of the steerer was quite a study. A couple of seconds, and they were through, and floating in the smooth water below.

Presently a most important functionary came on board, a serious-looking man, with a yellow flag, on which was written, 'Powers of the water!! a happy star for the whole journey.'

This individual must stand in the bows and wave his flag in regular time; and if he is not careful to perform this duty properly, the powers of the water are sure to be avenged somehow. Another method of softening the stony hearts of these ferocious deities is to sprinkle rice on the stream all through the rapid; this is a rite that should never be omitted.

At this rapid it is necessary to take a pilot, and at three o'clock the chief pilot and his mate came on board. They were gentlemanly-looking men, dressed in light grey coats, and they gave their orders in a very quiet but decided manner. The pilot's mate was certainly the most quiet and phlegmatic Chinaman I ever met; but these men have to keep their heads uncommonly cool. Directly they came on board our crew became very silent, with the exception of one hungry-looking coolie with a pair of breeches so baggy that he looked as if he could carry about all his worldly goods in them; but the severe looks thrown at him by the rest soon silenced him, and he seemed to subside into his capacious nether garments.

Just as all was ready a most ill-mannered junk put its head into my bedroom window, smashed it in, and threatened to do the same to the whole side of the deck-house. She was, however, staved clear, and eventually all damage was rectified with some paper and the never-failing pot of paste.

At half-past four our bows entered the foam. Everything creaked, groaned, and strained; the water boiled around us as we passed within a couple of feet of a black and pointed rock. The old ship took one dive into a wave, and water came on board at a rate that very soon would have swamped her; the drum was beaten and the flag waved; ashore the coolies (nearly one hundred of them) strained the rope, and their shouts could be heard above the roar of the foaming torrent; one line parted, and gave the vessel a jerk that made her shiver from stem to stern; but in ten minutes we were through, and anchored safely in smooth water.

Our small junk followed without much difficulty; the boat of our protector Sun received no more damage than the loss of her rudder; and our gunboat, a handy affair, making very light of it, we all at last found ourselves together above the dreaded spot.

Ropes were then to be coiled down, and our junk

made shipshape, before starting afresh and sailing through the Mi-Tsang gorge.

This is one of the most striking of all the gorges in the Yang-Tzŭ. Huge walls of rock rise up perpendicularly many hundreds of feet on either hand; the banks are strewn with *débris*; and where a gully or ravine opens up nothing is seen but savage cliffs, where not a tree, and scarcely a blade of grass, can grow, and where the stream, which is rather heard than seen, seems to be fretting in vain efforts to escape from its dark and gloomy prison. A fair breeze took us through the gorge, and we anchored for the night at the upper end.

March 14.—The next morning the commander of our gunboat fired his starting gun at six o'clock; and passing another insignificant rapid we arrived opposite the walled town of Kuei-Chou, whose officials came to visit Baber, bringing presents of fowls, ducks, and mutton, as a leg of goat, with a large bone and not much meat, was facetiously called.

The town of Kuei-Chou is a small place, but enclosed by a wall that runs up the side of a steep hill, and contains a considerable extent of open ground at the back, where, as is often the case in China, the people seem to like the idea of freedom, and build the greater part of their houses outside the wall. How the citizens of this flourishing place find a living it is difficult to say. There was very little cultivation around it, there were no junks stopping at it when we were there, and all the traffic passed by contemptuously on the other side. There were no fishing-boats belonging to it, and when there is neither commerce, agriculture, nor fishing, there are not many sources of wealth remaining. Baber said that it must live like a gentleman,—on its own private fortune!

In a report made by the delegates of the Shanghai Chamber of Commerce, they observe that many of the towns on the banks of the river thrive without agriculture, commerce, or other apparent means of livelihood. These delegates say that the people subsist upon piracy, and that when a junk is wrecked (as happens almost every day, when the river is in a flood), the coolies run away, and the inhabitants come down and appropriate the cargo.

This statement, however, should be received with caution, for it is not likely that piracy to such an extent would be permitted on the great highway of China. On this journey we ourselves passed several wrecks; in every case the cargo was safe ashore under the care of the coolies, who, in one instance, had built themselves temporary sheds of matting on the bank, and were living in a little encampment of their own, while the junk was being repaired.

I took a walk opposite Kuei-Chou, where the tracking path was cut out of the rock, and where in the steepest parts regular steps had been made. In many places the tracking line had cut deep grooves in the faces of the cliffs, and at one point, where a nasty projecting rock runs out into the river, rollers had been fastened for the ropes to work on. I came back just in time to see Sun and his boat swept down the stream. The tracking line had parted, an eddy spun the junk round, and the current carried her a mile, before the men on board could shoot her into the calm water close inshore.

During the day we went through two rapids, but in the low state of the water they were trifling, and the extra number of coolies employed was small.

The coolies fasten themselves to the tracking line in a very ingenious manner. They wear a sort of cross-belt of cotton over one shoulder, the two ends are brought together behind the back, and joined to a line about two yards long. At the end of this line there is a sort of button or toggle, with which one half-hitch is taken round the tracking rope. As long as the strain is kept up, it holds ; but if the coolie attempts to shirk his work, and slackens his line, the toggle comes unhitched, and his laziness becomes apparent to his comrades, and to the overseer or ganger who superintends the work.

This ganger is armed with a stick, and it is his duty, by shouting or gesticulating, to excite and encourage the men. He rushes about from one to another ; sometimes he raises his stick high in the air over one of them, as if he were going to give him a sound thrashing, but bringing it down he gently taps his shoulders as a sign rather of approbation than of wrath.

When all the coolies are harnessed, they walk forward swaying their bodies and arms from side to side, and shouting a monotonous cry to keep the time. Sometimes the path where they can track is only twenty or thirty yards long, then as soon as a coolie arrives at the end he casts himself off, runs back to the other end, fastens himself on again, and begins pulling afresh.

During the day we entered the coal districts. The people here do little more than scratch the surface, and the coal they obtain is not of a very first-rate quality.

Whilst we sat at breakfast Baber's headman Hwu-Fu, who was waiting on us, was in an exceedingly merry frame of mind, so much so that even in the august presence of his master he was unable to contain his mirth. Baber wishing that we should enjoy a share of the laughter, asked him what was the matter, and although he found his story so amusing that it was with difficulty he could tell it, yet Baber managed to extract from him some interesting episodes in the life of the old lady, the owner of our vessel.

She had been married some years previously, and was apparently able to exist in the society of her husband, until the river gods decided to wreck their vessel in a rapid.

The appalling spectacle so terrified the unfortunate man, that although he received no corporal injury, he died of fright.

The old lady shed a parting tear, and would have wiped the corner of her eye with her pocket-handkerchief, if she had had one; but soon after, finding the care of a big ship and a little child too much for her unaided self, she, whilst vowing to the shades of her departed spouse that it was an act of paramount necessity, and that no disrespect to him was meant, decided to take another helpmeet. Not being altogether destitute of this world's goods she found no difficulty, and at I-Ch'ang recommenced a married life. But whilst yet in the honeymoon at Hankow a slight difference of opinion ended in the husband falling down a well, calumny going so far as to say that she pushed him down; but, be that how it may, the lady returned home quietly, and would have been quite prepared for a widow's lot, if some meddlesome folk passing by had not pulled the man out, and sent him back not much the worse. The pair, however, thought that after this accident married life would possibly not be unmixed bliss; so giving him her blessing, or her curses, and endowing him with a small sum of money, the woman sent him to his home.

Again surrounded by a flattering crowd of admirers, she selected a husband for the third time, and they went back together to I-Ch'ang ; but here evil-minded people told such wicked stories, that the husband ran away, and returned to Hankow by the first opportunity. Since that time she had been unable to get another spouse, and remained a widow.

After this interesting story Hwu-Fu went away, but presently returned with more information. He said he had heard of a place called Niu-K'ou, or the Cow's Mouth, where there was a cave, and where, he said, in ancient days a very remarkable cow used to drink, and although he did not know what it was, he was sure that there was something marvellous about the place, because he had been told so by a man who was upwards of eighty years of age.

Baber and I, not at all unwilling for a day in the country, decided that we would investigate the mystery, and we organised a picnic accordingly ; and although soon afterwards it appeared that the oldest inhabitant after all knew only of a village called Niu-K'ou, where somebody had once said that there was a cave, still, as we had made up our minds for a walk, we did not alter our plans.

After this, Sun came to see Baber, and was very anxious that we should not leave the boats. He was, or pretended to be, afraid that something would happen to us ; but whether he thought that we should meet with some accident, or be attacked by the inhabitants, was not very clear.

March 15.—Ordering the junk to go on up the river, we started on our expedition at seven o'clock, but Sun could not resist a last warning. He said we should be lost among the mountains ; Baber said we had a compass and could find our way anywhere.

'Ah!' he answered, 'but you may fall down a precipice.'

'Then,' said Baber, 'you had better come with us and walk in front.'

He did not seem to understand this logic, and inconsequently remarked that we might be robbed by thieves.

'Not,' answered Baber, 'if I take two or three braves with me from the gunboat.'

'But there are wild beasts,' he added.

'I think not,' said Baber, who then politely waited for his next argument.

But he had exhausted his stock, so with a mournful sigh he let us depart.

Chin-Tai, Hwu-Fu, and Wang-Erh came with us, besides one or two people from the gunboat, and coolies to carry our guns and provisions.

We ascended a rather steep path by the side of a valley, where small farmhouses stood singly, each surrounded with bamboos and apricot trees, now in full blossom; and after crossing a saddle 2,600 feet above the river, a beautiful landscape lay before us.

We were standing on the crest of a hill that bounded a horse-shoe valley; on the slopes well-to-do farmhouses were standing by themselves amongst pine woods, through which a gentle breeze was rustling pleasantly, and in the distance fine ranges of mountains and vast snowfields could be seen. And although there was no cave or other curiosity, the view quite repaid us for our walk.

We had been told that there was a 'Tung,' or cave; but Tung may mean a variety of things, and amongst others an amphitheatre of hills, such as we were now looking down on; and the stories of the grotto serve to show the difficulties to be contended with when any information is wanted in this country.

We went down to a cottage, solidly built of mud, with a high gabled roof, where a polite old gentleman, the proprietor, made us welcome, and showed us into his chief room, a large and lofty apartment, with a mud floor not cleaner or more dirty than a Chinese room usually is. One wardrobe against the wall, a chair, a table, and a couple of small benches composed the furniture, and mottoes painted on long slips of red paper adorned the walls.

The kang of Northern China is an unknown institution in these parts, but some charcoal was burning in a hole in the floor, to which air was supplied by a small tunnel. We sat down here and had our breakfast, the family coming in, one or two at a time, to smoke a pipe and have a look at us; but they were all very civil and well-behaved, and in reply to Baber's question they said there were plenty of pheasants, but no deer, in the neighbourhood. We left these quiet folk, who had scarcely ever heard of a foreigner, and ascending again reached a point about 2,800 feet above the river. From here we walked on the crest of a long ridge, and presently met some people sent by the chief magistrate of Pa-Tung to look for and to pay their respects to Baber.

A very long zigzag took us to the riverside. We were ferried across, and on arrival at our junk, Sun professed himself immeasurably relieved to see us back again safe and sound after our perilous enterprise.

Pa-Tung seems placed in a very unsuitable position for a large town. Steep mountains about 3,000 feet high rise straight up behind, and, except the narrow paths and zigzags, there are no roads to it from anywhere. It is a long straggling place, with little depth, built about forty or fifty feet above the water. There is no wall, and the houses that face the river have little patches of garden in front of them running down the steep bank, and, as is usual in these waterside towns where level ground is difficult to find, many of them are supported on piles.

March 16.—We now entered the Wu-Shan [7] gorge, the longest on the Yang-Tzŭ. In many places there were no tracking paths, and it was necessary to lay out a line with a sampan, make it fast to the point of some convenient rock, and then haul ourselves up.

The weather and water had here regularly honeycombed the faces of the limestone rocks, and in many places they looked like slag from some gigantic furnace.

We anchored for the night at Nan-Mu-Yuan (teak wood garden), a little village, the main street of which is a flight of irregular steps running some five hundred feet up one side of a very steep ravine; while on the opposite side of the rivulet that comes tumbling down, orange and apricot groves are interspersed amongst little patches of beans and wheat, planted wherever a level spot can be found or made.

I walked through this picturesque but uncomfortable hamlet, and, followed by a considerable number of people, continued the ascent of the very steep hill, passing a temple situated on a projecting point, amongst a clump of willows and fir trees, where most of the accompanying crowd found their muscles less strong than their curiosity; then through one or two hamlets, whose occupants took little heed of the strange

---

[7] Wu-Ngo-Nü was the enchantress of Mount Wu. The twelve peaks of this mountain were once twelve sisters; they raise clouds in the morning, and cause showers at evening, thus detaining travellers that they may remain overnight in the neighbourhood. This once happened to an emperor, who caused a tower to be erected there, called Yang-T'ai.— *Chinese Repository.*

visitor, till after a scramble amongst rocks and thorns, I gained the top of a ridge, 2,500 feet above the river, that commanded a fine view of the mountain ranges. Here I found myself alone, for the last of the boys who had followed me had found the ascent so severe, and the evening so warm, that after divesting themselves of nearly all their clothes, they had given it up in despair, and were waiting for me a few hundred feet below. They watched me keenly, and must have been much disappointed that I went through no more remarkable performance than an examination of my barometer, and then again came down the hill.

March 17.—We were still in the Wu-Shan gorge; here a wild chasm in the limestone rocks, where on the left bank the strata stand in an almost vertical position, and on the right are inclined at an angle of 45° below, turning over to a horizontal position up above.

On looking at these gigantic masses, which by some unknown force have been thus torn, it is easy to see that it is by some wonderful convulsion of nature, and not by the steady disintegration of a running stream, that these deep rents in the mountains have been formed.

The gloomy aspect of this gorge, shut in by high limestone mountains and precipices, where vegetation was scarce, and where a narrow streak of dull leaden sky was all that could be seen above, was enhanced by the solitude in which we now found ourselves, for we scarcely saw another vessel. There was something weird and mysterious in that long silent reach, where there seemed to be no room for life, and it was not difficult to understand how the superstitious fancies had arisen that had attached some mystical fable to almost every point.

On the left bank of the river in this gorge there is a sort of rough circle on the face of the cliff, where several layers of rock have, as it were, peeled off. The people say that with each succeeding dynasty a fresh layer comes off within the charmed circle, and discloses, on the face of the rock beneath, some characters which are warning or prophetic, and serve either as a guide for the conduct, or as an indication of the future, of the family seated on the imperial throne.

Within the circle it was plain enough to see that several layers had gradually flaked away; and as several dynasties had also succeeded one another, the coincidence was quite enough to connect the two in the superstitious minds of the people. There were also markings on the face of the rock, and although these might bear some fanciful resemblance to some of the characters in the Chinese written language, yet no one pretended to be able to decipher them. This, however, mattered little, and as they had been told the story, so they received it, and handed it down in implicit faith to their successors.

In this gorge, in the precipices, and on the sides of the hills, there are many caves; and where it has been found possible to cut a path in the face of the rock some of them are inhabited.

During the day we passed the boundary of Hou-Pei, and entered Ssŭ-Ch'uan, or the province of the four streams, a province that has been noted by all travellers as one of the most beautiful, perhaps the richest, and for foreigners certainly the most pleasant in the empire. In the evening we anchored at the little village of Ch'ing-She-K'ou (Coloured Snake Mouth).

March 18.—The next morning we started for a

picnic up the narrow glen, which from its twists and turns has given the name to the village.

The peaches and apricots were now in full bloom; there were little patches of opium, the first I had seen, where the poppies were now about a foot high, and here and there tiny plots of wheat were growing on slopes so steep that it seemed as if it would be impossible to reap without assistance from a rope.

As we ascended, the valley became more narrow and precipitous; the path was very rugged and difficult, sometimes only just wide enough to walk on, and cut out of the side of the hill, which was everywhere exceedingly steep. The tops of the mountains were broken into crags and pinnacles of the most fantastic shape. In the gullies and cracks in the cliffs, a wild and luxuriant vegetation of brambles, ferns, long grass, and all sorts of shrubs was clinging to the sides, wherever these were not absolutely vertical. Wild yellow jasmine was in bloom, maiden hair, and other ferns grew in profusion, and pink primulæ peeped out from crevices in the rocks. The inhabitants, who are very poor, live in caves; indeed, there is absolutely no level space on which a house could be built. In our day's walk we only passed two or three of these family residences, and a population can scarcely be said to exist. The narrow foot-path runs in front of the caves, beyond which there may be a sheer drop of some hundreds of feet, and to prevent the children or pigs from tumbling over, a light paling is sometimes put up on the outer side of the path. Passing one of these caves the sudden and unexpected appearance of a foreign dog frightened an ancient fowl, that fluttered over the fence, flapping its wings and cackling, till its voice was gradually lost in the depths and drowned by the murmur of the invisible stream below.

The scene was very wild; here the gorge divided into two branches, each enclosed between gigantic cliffs; on the top of these a little slip of sloping ground would be seized on for agriculture, and this again would be backed up by another cliff behind; zigzag paths, or steps cut in the rock, leading from one patch to another.

We had a most difficult scramble over sharp stones, and after two hours and a half arrived at a cave where an old woman was at home, and where we decided to breakfast. This old woman kept bees in hives of primitive construction, and told us that she sold both honey and bees-wax, but that the bees had not yet swarmed; she also said that wherever opium is grown in any quantity it drives the bees away, but that here the quantity was too small to have any injurious effect.

There were some furnaces not far off where saltpetre was purified, but the process was not going on just at the time of our visit, and we were unable to learn the details of the manufacture. After our breakfast we continued our walk, but soon found that the hill sides became too steep for a path of any description, and that the track we were on descended to the torrent and crossed over to the other side, so we retraced our steps and returned to our boat.

March 19.—During the morning we remained at anchor, as there was no tracking-path, and without it we could not make way against the current and strong wind that was blowing. It cleared off towards the middle of the day, but our progress was very slow, for it was again necessary to lay out ropes and haul ourselves tediously up. Here and there it was just possible to track for a couple of hundred yards, but even then the coolies were obliged to clamber about on their hands and feet from one gigantic rock to another, and in the evening, when we arrived at the little village of Tiao-Shih, our run was only seven miles.

The limestone rocks here have been worn in the most astounding manner. In many places there are in the face of the rock innumerable long vertical grooves; the surface of these is highly polished by the action of the wind and weather, and they look exactly as if they had been scooped out with a gigantic cheese-scoop. In other places the rocks are split up vertically into long needles and stalagmite-shaped masses.

March 20.—The Wu-Shan gorge that we had entered on the morning of the 16th seemed interminable, for we were still in it, with no immediate prospect of getting out. At eight o'clock we were opposite the town of Wu-Shan, where there was rather a severe rapid. While we were in it, the tracking line broke, and as we had no safety ropes we were swept away. The men steering, however, skilfully managed to shoot us into a back eddy, which carried us several yards up the stream inside the rapid; but before we could be made fast, we were again taken into the downward stream. Like a teetotum we spun round, and once more driven down the river we were a second time shot into the back eddy. Still we had no rope ashore, and back again we went, spinning about each time in a way that would have made some people sea-sick. Whilst we were thus being carried up and down, a man was trying to swim to the land with a line. When at last he gained the bank, the rope caught in a rock at the bottom, and as no efforts would avail to clear it, a second coolie fastened another rope to his body, and jumping overboard swam away But while struggling with the foaming torrent, the loop slipped over his body unperceived, and when he arrived on shore he was astonished to find that he had lost the end.

All this time we were at the mercy of the eddies, being whirled round and round, and carried up and down. We made no less than six voyages backwards and forwards, and it seemed likely that we should make as many more, when by a stroke of good luck, the first line cleared itself from the rock, and we hauled ourselves at length out of the rapid.

The day's journey was a succession of rapids, some of them long ones. It is by no means a simple or easy matter to get a big boat up one of these. After the line is out, and the necessary number of coolies are hauling on it, the boat is shot across the river into mid-stream, until often the line is almost at right angles to the direction of the keel. The vessel is then put about, and she makes a shoot inshore, and so on, tacking as it were backwards and forwards.

We anchored for the night close to a junk that had been wrecked, and was now ashore for repair; the sailors and coolies were all living under mats, and the caulkers were busy with cotton wool on the bottom of the vessel.

I took a turn for half an hour before dinner, and noticed a little rice growing. There was some barley also, and a good deal of wheat. The rocks here are ground down by the wind and weather, and all the shore is edged with long banks of fine clean sand; and on this apparently fruitless soil, the Chinese manage to raise crops of wheat.

March 21.—We arrived at Kuei-Chou-Fu, the second place of that name we had passed on the river, and here Sun and our gunboat left us, and we were handed over to the charge of others.

Here also the Chinese agent of the firm of Major and Smith, of Hankow, called on us. As he entered he knocked his head against the ground to Baber, who lifted him up and asked him to sit down, but just as he was recovering from this act of condescension I came in and frightened him afresh. He again went down on his knees, and touched the ground with his forehead, and it was not without a great deal of persuasion that Baber at length prevailed on him to take a chair. The sensation of being seated seemed to be like touching the knob of an electric bell, for he suddenly launched himself into an ocean of words from which it was quite impossible to withdraw him; he talked so fast that he could not articulate quickly enough. He persistently addressed himself to me, making the most pathetic gestures every time that Baber told him I did not understand Chinese. On these occasions he would only give one glance at Baber, and again devote the whole of his energy in talking to me.

After a long time this monstrous narrative of some private business trouble began to pall upon us both, and at length Baber hinted that his conversation was not quite so entertaining as it might have been. This seemed to act like additional battery power to the electric bell, for seating himself more firmly in his chair, 'he argued high, he argued low, and then he argued round about him.' Baber now called for another tin of the fragrant weed, and we both smoked steadily, the pile of cigarette ends growing higher and higher in the cover of the tobacco box as the worthy man still rolled out the stream of his volubility.

One by one the servants came to the door and listened in wonder, the old woman was seen in front to take a peep through a crack in the walls, and then to steal away awestruck at the torrent of speech. As the evening grew darker the smell of dinner was wafted from the kitchen, but still that fearful man sat glued to his chair.

Baber now made a polite attempt to stop him with fair words, but he might as well have tried to stop an avalanche with a sheep hurdle, and at last, like the bishop in the ballad, ' Oh, bosh! the worthy Baber said, And turned him out.'

Our parting official, Sun, brought his successor to pay his respects to Baber. He was a funny little man ; he thought that the fact of his having lost five dollars betting at Hankow races, ought to make a very strong and favourable impression on us, and boasted of having made a trip to Shanghai for pleasure. He took a cigarette with a card mouthpiece, but tried to light it at the wrong end, which gave Sun an opportunity of displaying his superior knowledge of Western ideas.

Kuei-Chou-Fu is surrounded by a very good wall, in much better condition than that of most towns. The town is well situated on the slope of a high hill, and there are a good many suburbs, some permanent, and built on the high ground. But a very large population live in temporary huts of matting set up on the shingly beach. These are removed in the summer as the river rises and covers all the ground on which they are built. Kuei-Chou-Fu is the seat of considerable trade, and at the time of our visit there were a great number of junks at anchor off the town.

We found provisions here more plentiful than at any place we had visited since leaving I-Ch'ang, and were able to buy excellent vegetables and very indifferent beef and mutton.

March 22.—A very large revenue is derived from the salt manufacture which is carried on at brine pits situated about half a mile below Kuei-Fu, on both sides of the river, where on low, sandy, shingly banks, close to the water's edge, holes are dug. The water finds its way into these through the soil, becoming in its passage impregnated with salt, but not strongly, for the taste of salt in it is scarcely perceptible.

Bricks are made from the salt earth in the neighbourhood, with which dome-shaped ovens are built.

These have a door in front, and there is a hole in the top in which a shallow iron pan, K, is placed.

FIG. 4.

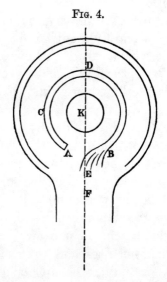

On the top of the oven, and concentric with the iron pan, a hollow in the brickwork makes a narrow trough A C D B. Above the back of the oven at E F the wall is covered with, and made up of cinders, slag, and earth.

The brine is first poured into the narrow trough at A, and running slowly round the top of the oven, discharges itself at E amongst the cinders, slag, and earth at the back. It permeates easily through these into the back wall of the oven itself, G H, fig. 5, and amongst the bricks of which it is built. Here the heat drives off the water, and leaves the salt deposited on and in the bricks.

After ten days or so the fire is let out, the back of the oven pulled down, the bricks from it carefully removed, and the oven built with fresh bricks.

The stuff that has now been taken out is broken with hammers and stones, and put into a large wooden bucket; more brine is thrown into this mass, which

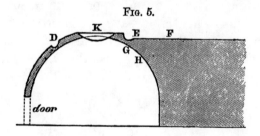

Fig. 5.

seems to be disintegrated by it, and now breaks up, forming with the water and the brine a black substance of about the consistency of freshly-made mortar. The water is poured from the bucket into the iron pan at the top of the oven, where it is evaporated, and very good salt produced.

We found the people at these pits extremely civil, very few troubled themselves about us, and our numerous and minute questions were patiently and politely answered. It is said that there are forty pits here, and that each pit produces one hundred catties (130 lbs.) of salt a day; this would make 890 tons of salt per annum.

The Government buys all the salt at a rate fixed

by itself, and then sends it over the country for sale, making an enormous profit. I subsequently learnt from a banker at Ch'ung-Ch'ing that the salt in the province of Ssŭ-Ch'uan brought to the Government a revenue of six millions of taels annually—roughly two millions sterling. The profit comes to about eighteen cash a pound, and at the rate of 1,600 cash to a tael, this would make the annual produce of salt in this province 237,946 tons, an amount that seems almost incredible.

In the afternoon we went to see an aqueduct by which water is brought into the town from the top of a hill behind, and which was reported to be something very curious. A narrow paved path, in many places made into regular flights of steps, leads up a steep hill, which is an enormous graveyard, the whole side of it, for a distance of more than a mile from the town, being closely covered with graves; some of these stand in the centre of a carefully levelled patch of ground, and are exceedingly solid and imposing structures of stone with doors closed by slabs on which long inscriptions are engraved; whilst others are the usual mounds of earth without a stone or writing of any kind.

The aqueduct we were following was simply a small stone gutter carefully laid along the side of the hill; a stream of clear water was running down in it. After a walk of about three quarters of an hour we reached a temple that had been indicated to us on the spot whence the water came. All the people said that 'Lily Pool,' the source of the rivulet, was here.

This temple was dedicated to a lily; but the pool was nothing more than a circular hollow thirty yards across, which, although it was no doubt full after heavy rain, was now as dry as Sahara.

We said to the guide, 'but where does the stream

come from?' for we were standing within a foot of the stone channel in which the water was running, and which had no connection with Lily Pool.

'From here,' he said, pointing to the dry hollow.

This answer was characteristic of the credulity of the average Chinaman. He had had enough walking, and simply said whatever first came into his head that he thought would prevent us going any further. He knew that if he were in a strange place he would accept, without for a moment questioning it, the statement of anyone belonging to the place; and it was almost impossible for him to conceive that we should not do likewise. For did he not live here? and was it not therefore obvious that he must know better than we, who were strangers from far away.

Of course, amongst a people where evidence is never weighed, and the wildest statements received without examination, it is only natural to find that lying is a common vice. Honesty in this case cannot be and is not the best policy; the people tell lies because lying pays; but when they shall be sufficiently far advanced in education to weigh evidence, then lying will begin to lose its value, and we may expect to find truthfulness amongst the people.

We who were accustomed to use the evidence of our senses, now followed the aqueduct for another quarter of an hour to a place where a jet of water spouted from a hole in the rock about 1,500 feet above the river. This aqueduct is nearly three miles long, and was built during the Ming dynasty (1600 A.D.), by some Fu of this town, ambitious of having a water supply all to himself. The rest of the inhabitants still have to content themselves with the river water, which fortunately for them is quite as good as the private tap of this arrogant magistrate.

At the outskirts of the town plenty of people came to see us, but no one said a rude word. The epithet 'foreign devil,' is never heard, because the people of Ssŭ-Ch'uan are very superstitious; they will not pronounce the word devil, for it is said to be unlucky; sailors do not like to make use of the word '*wind*,' for fear it should cause an adverse gale; nor will they talk about snakes, lest they should be bitten.

In the evening Sun sent us an invitation to a concert on board his boat, where after tea a little girl about eleven years old was introduced by her mother. She wore a scarlet coat trimmed with black; a blue embroidered cap, big ear-rings, and a large silver ring on one forefinger. She had a wooden guitar, on which she played an accompaniment, and a fan on which were written the names of the songs she knew; these were tolerably numerous, considering that she had learnt music only eighteen months. She was very self-possessed, and sang without affectation, but through her nose, and in a very high key. There was a distinct air running through her songs, although it was rather hard to follow, but there was not much melody in the whining droning kind of music that characterised the performance. Sun gave her a string of cash and a cup of tea, and then her mother took her away.

March 23.—It rained heavily during the night, and when we looked out we found that there was a fresh layer of snow on the mountain-tops, and that the water of the river had again turned to a muddy colour. Four coolies ran away in the morning, and we were delayed while the headman went to seek them. If it was the tongue of our old Jezebel that had frightened them, it was no matter for wonder. They were brought back repentant in the course of the afternoon, and we started again. We had now left the gorges, and the gentle slopes and open valleys were a pleasant change after having been so long shut up in the deep recesses where we could seldom see more than a narrow strip of sky.

March 24.—After an afternoon walk on shore, as the junk was out of the way, I took a voyage in our gunboat, and we came to a small rapid where a ledge of rocks made it difficult to track. The trackers let the line go, and the men remaining in the boat tried to row up ; but they were too few, and unlike the crews of most of these boats, though exceedingly energetic, they were remarkably unskilful, and had no idea of keeping time. The captain especially wielded his paddle with a vigour quite alarming in a sedate Chinese official, and made more splashing, and was more hopelessly out of time than any of his men. We then crossed the river, and a backwater carried us up a short distance ; after which we recrossed to find the trackers, who had taken advantage of the opportunity to hide themselves amongst the rocks to smoke or go to sleep. Much shouting and swearing ensued, particularly on the part of the well-meaning commander. We eventually arrived at our destination half an hour before our junk. When she was moored alongside the captain was most particular in the arrangement of the gangway, and when all was ready tried it himself before allowing me to put my feet on it ; but the dignity of the performance was altogether spoilt by my dog Tib, who smelling dinner rushed past the captain, and nearly upset him into the river.

March 25.—Near the village of Miao-Chi there is rather a difficult rapid with broken water right across the river. A stream that comes down from the mountain used to bring great boulders with it, and make the place very dangerous, so the government, aided by local subscriptions, built a solid wall 200 feet long, 45 feet wide, and in some places 12 feet high. This abutted at one end on to a reef of rocks, the prolongation of the upper bank of the stream, and turned the water so that it entered the river below the rapid. This was built only fifteen years ago, and was a creditable work of very large blocks of limestone, 12 or 14 feet long, and 1 to 2 feet wide and thick. The long stones were dovetailed into one another; but the mortar and cement, always a weak point in Chinese buildings, were bad. The edifice was adorned with what was no doubt a suitable extract from some of the classics, but the meaning of which was obscure. Though built so recently it was already falling to pieces, and no one seemed to care about repairing it. Many of the stones were loose and detached, and some had been swept away altogether. When we arrived here we found the whole male population of the place engaged in taking a shoal of small fish that had come up, and had been entangled amongst the rocks. There were upwards of twenty men in the river, few of these wearing anything but a turban, and a basket tied round their waists. They had only to put their hands into the water, and the fish were so thick that they took them out two, three, and sometimes half a dozen at a time. Those who were unprovided with baskets put their catch into their waistbands, if they had any, or into their mouths, if they had none; and when these receptacles would hold no more they walked ashore, deposited them in a place of safety, and returned to the water. Some of them merely flung them to friends, who gathered them up; and all of course were jabbering and shouting. The regular fishing apparatus of nets, and pots, like lobster pots, was left alone; and one old man, sitting in a corner of a rock with a scoop-net, seemed altogether out in the cold, as he caught nothing all the time we were looking on. By degrees the shoal visibly decreased, and when we left the number taken was very much less than at first.

We afterwards passed the city of Yün-Yang, enclosed by a fine wall, which does not seem to be very highly appreciated by the inhabitants, who nearly all live outside it, and leave the magistrate in his yamen almost alone in solitary grandeur.

March 26.—It rained heavily all day, but as there was a fresh and fair wind we made a good run. The country now changed considerably in appearance. The hills sloped more gently, and their sides were well wooded. Our view now was not so limited, and we could see little square enclosures, like towers, on the tops of the hills; these, they said, were erected in the time of the Tai-Ping rebellion, for the inhabitants to take refuge in; but as there could not by any possibility be any water supply, siege operations on an extended scale would hardly be necessary to bring a garrison to terms, and it would appear as if there must have been some other reason for their existence. It is very difficult in China to get any trustworthy information, scarcely any one knows anything; but no matter how ignorant, everyone has a complete answer ready for any question that may be put.

March 27.—Taking a walk by a narrow path paved with flagstones, which became a flight of steps wherever the sides of the hills were steep, and meditating on the melancholy fact that this was the almost invariable case in China, I thought that it must be one of the causes to which the want of observation in a Chinaman is to be attributed. For, brought up from childhood in a country where the paths are so dangerous that he can only use his eyes to pick steps for his feet, he is never able to look about him and observe what is going on.

The richness and verdure of this part of the country is almost inconceivable; the soil is bright red, and, where fallow, presents a delightful contrast to the fresh green of the young crops. The rape was now in flower, and field upon field of brilliant yellow rose one above the other. The terrace cultivation of rice occupied the bottoms of all the valleys, with patches here and there of wheat or beans. The houses looked comfortable and substantial, each enclosed in a clump of bamboo; handsome temples stood by themselves in groves of trees. Every here and there a species of banyan (without pendants) standing by itself, with perhaps a little niche underneath for burning incense in, was a graceful ornament to the landscape.

All these combined to present a scene of richness and fertility that I have seldom seen equalled, and which fully justified the praise that has been lavished by travellers on this beautiful province. And more striking than all is the fine open countenance of the people, who, though very independent, are undoubtedly the most pleasant and gentle of all the people of China. Baber and I went out for a stroll in the afternoon, and ascending a hill about 1,500 feet above the river we had a fine view of Wan and its surroundings. The rugged and wild mountains now were only in the distance, and around us the hills were of less height; all these were broken away at the top, giving their summits the appearance of solid towers with perpendicular walls. Descending by another road we found a great ledge of rock at the edge of the water. Here the industrious people, anxious not to lose an inch of ground, had made a bed of sand wherever there was the smallest hollow, and here little patches of wheat, sometimes only a couple of yards long by one broad, were absolutely growing out of the sand that had been strewn on the rock.

March 28.—The cultivation of rape in this neighbourhood is very extensive, and vast quantities of oil are prepared here from it; for the people have no butter, so that all grease required for cooking must be oil; neither have they any sheep, and therefore, in the absence of tallow, oil only can be used for lighting purposes. In this manner a large proportion is consumed in the province; but there is also a considerable export trade. On the right bank of the river, terrace cultivation of rice was maintained on every inch of ground, as the general shape of the gently sloping hills forms natural amphitheatres, where the water is easily led; but even where this is not the case this industrious people seem determined to grow rice, if there is any possibility of doing so. In the course of a ramble I came upon a hillock at the end of a spur, considerably higher than any ground immediately around it. To make use of this a stone aqueduct, raised seven or eight feet above the ground, and about 150 yards long, had been made, but the portion of land thus secured for use was so small that it seemed as if it would have been easier to level the hillock than to make this aqueduct.

Notwithstanding the industry of the Chinese and

their admirable system of irrigation and terrace cultivation, there can be very little doubt that the exceedingly high estimate in which their agriculture is held is very far from being deserved. This appears to have been derived from the French missionaries, for as early as 1804, Barrow speaks of the way in which it had been overrated; nearly all moderns who have been in China make the same observation, and yet there remains amongst Europeans out of China the conviction that the Chinese possess secrets unknown to, or unguessed at by Europeans.

But the real point in which the Chinese excel is in industry. It is industry that leads them to take such care never to waste the smallest trifle; and it is industry that makes it worth their while to gather up the last fragments. Industry again enables them to dispense with any other manure than the sewage of the towns; for a peasant will walk into the town, fetch his manure, and take it to his field himself. It is by industry that in the large plains the Chinese are enabled to keep their rice fields properly watered; for it is not possible to conduct the water by canals to every part and every level of a wide plain, it must therefore be lifted artificially, and all day long coolies are to be seen in the extensive plains raising water by the means of little treadmills.

But beyond their industry the Chinese can hardly lay claim to any superiority over other nations. They plough about as well as the natives of India, doing little more than scratch the ground. It is true that they can raise two crops on the same field, as, for instance, when they plant opium under rape, or yams beneath millet. But this is a system not altogether unknown to European farmers, and in the West Indies it is customary to grow yams underneath the sugar

canes. Some of Barrow's remarks appear to be worth quoting :
'They have no knowledge of the modes of improvement practised in the various breeds of cattle; no instruments for breaking up and preparing waste lands; no system for draining and reclaiming swamps and morasses.

. . . .

levelling the sides of mountains into a succession of terraces, [is] a mode of cultivation frequently taken notice of by the missionaries as unexampled in Europe, and peculiar to the Chinese, whereas it is common in many parts of Europe. . . . Of the modes practised in Europe of improving the quality of fruit they seem to have no just notion . . . . Apples, pears, plums, peaches, and apricots, are of indifferent quality. . . . They have no method of forcing vegetables by artificial heat, or by excluding the cold air and admitting at the same time the rays of the sun through glass. Their chief merit consists in preparing the soil, working it incessantly, and keeping it free from weeds.'

Thus wrote Barrow three quarters of a century ago; the Chinese are no further advanced than they were in his time; and it is hardly necessary to add anything to his remarks, except to observe that not only have the Chinese 'no just notion' of improving the quality of fruit, but that to this day they remain in complete ignorance of the science of grafting. To those accustomed to the appearance of European countries, the absence of hedges is at first sight strange, but in this country, as in many others, people recognise their own property by the divisions in the fields; and even where there are no marks, one man will rarely attempt to plough beyond his own land; boundary stones to properties are, however, usual. It is not to be supposed that disputes never arise, but when they do they are generally, or almost always, settled by the people of the place.

Some of the farmhouses here are very large and substantial; and plenty of opportunity is given for forming an opinion of them, owing to an eccentric custom of leading the high road through some part of the buildings, where half a dozen snarling dogs rush out yelping and yapping. The reason probably is that level ground is so scarce, that every yard must be utilised; and the ordinary avocation of threshing, or the women's occupations of grinding corn or sewing, can be carried on just as well on the footpath as elsewhere; and the number of passers-by is so small that they do not in the least interfere with the inmates. In one of my walks I passed a particularly fine farmhouse, and although in this case the road did not lead through it, I was able to examine it carefully. There was a large outer courtyard with sheds on each side. It was enclosed by a good wall on the side to the road; opposite there was a very large threshing floor and granary, well built and covered in; beyond this the house was arranged round three sides of another smaller open court. Three or four men were busy grinding flour at a large hand-mill; and I could not but be surprised that they did not make use of a good stream that was running past their house.

The Yang-Tzŭ would be admirably adapted for the stationary mill boats that are so prominent on the Danube, but in this rice country it is probable that the amount of grain to be ground is so small that it is hardly worth while to build water-mills. In the mountainous districts, where oats and Indian corn form the food of the people, water-mills are common enough.

The Chinese are never remiss in the care they take of the dead; and where the residences of the living are so good, it is to be expected that the graves will not be less sumptuous; and this is the case, some of them being magnificent structures built of large blocks of stone in an enclosure sometimes fifteen yards square, in which a few shrubs will be growing; the front is ornamented with carvings in relief representing birds, beasts, or men variously occupied. There are always long inscriptions, and the whole is generally sheltered by a fine banyan tree, or a few yews.

Whilst I was ashore, the junk ran aground, and knocked a big hole in her side; but Jezebel, looking at it with unconcern, remarked, between the whiffs of her pipe, 'cotton wool,' by which unusually laconic observation she meant that the hole was to be stuffed up with that material.

March 29.—We walked in the afternoon to Shih-Pao-Chai (Stone Jewel Fort), where there is a very curious rock, the top of which is about 150 feet above the ground on which it stands, and 300 feet above the river. All the faces of it are perpendicular, and on one side a pagoda has been built against the rock. We visited this, and went to the top, where we found a number of wooden figures carved and brilliantly painted and gilded. These figures represented the horrors of hell, and the punishments reserved for the wicked; the judges were life-size figures of men, with long black beards and mustachios, holding books in their hands. The executioners were horrible devils, with hideous faces, who were supposed to be thoroughly enjoying their odious task of torturing the victims. These tortures were of many kinds, and the most abominable that it is possible to imagine:

some of the sinners were being sawn in two ; some were being ground in a mill ; and others were arranged in layers between big flat stones, with their heads only projecting all round, while a demon stood above spearing their faces. There was no representation of Heaven. The eighteen saints with twenty-four attendants were in another room ; and at the entrance, the god of literature, a repulsive creature, was standing on the point of one toe like a ballet-dancer.

We met our official here, who took us to a room in the top story, and showed us a hole that had been the site of a miraculous draught of rice. He told us that in former days if a priest put a handful of rice into this hole, all the inmates of the Temple were able to sit down and take out as much rice as they could eat, no matter how many of the holy men partook of it. History does not relate whether it was ready-cooked. But in the time of Chia-Ching (the emperor who reigned from 1796 to 1821) a wicked and avaricious priest sold this heaven-sent food, on which the offended deities stopped the supplies, and all that now remains is the empty hole.

The only remark to be made on the story is that the deities must have been strangely ignorant of the Chinese character not to have foreseen the result.

A little higher up the river there is a place where for many years no efforts of the industrious Chinaman had availed to produce rice ; but on the cessation of the miracle in the pagoda, the gods compensated the country by fructifying this unfertile spot ; and since that time rice has always been produced here in quantity and quality equal to that in any other district.

It would be hardly worth while to repeat these

tales, but that they serve to show the credulity and superstition of the people; the official who was with us, and who was a well-educated man, believed implicitly in the story, and as the suggestion of a doubt would only have hurt his feelings, we listened to the narrative with becoming gravity.

March 30.—We went for a walk in the afternoon on the left bank of the river. The country was very hilly, and the valleys formed exceedingly picturesque amphitheatres, where the vegetation was almost tropical in its luxuriousness; the thick clumps of bamboo, and the rich green of the shrubs and trees, brought to mind the verdure of Ceylon or Singapore. We gained the top of a hill about 600 feet high; a beautiful panorama lay before us; and from our commanding position we could see our boats moored for the night at the other side of the river. When we had again reached the shore we saw a small boat not far off, and called out to the man in it to come and take us across; but the only occupant was asleep, and he was not aroused without much shouting. At length he awoke, and a long discussion ensued between him and our escort; the noise brought out some people from a village 200 feet straight over our heads, who joined in the argument, and a horrid old woman was heard to shout, 'Don't come, for when you get to the other side they won't pay you.' So, notwithstanding the offer of large reward, after about a quarter of an hour's talking, the man in the boat quietly paddled off and cut short the argument.

All this time the soldier who was with us from the gunboat was trying to make himself heard by his comrades on the other side, but meeting with little success, all began shouting together to try and attract attention, and the hills re-echoed the unwonted British hail of 'Boat ahoy!' It now became dark, and it was not until we had been amusing ourselves in this way for about three quarters of an hour, that at length some of our people were aroused to our situation, and we heard from the other side of the river a responsive cry But in order that the men coming to us might find us in the dark we continued shouting to guide them to our position. They discovered us at length, and after having waited on the shore for upwards of an hour, we embarked in a sampan and crossed safely to our dinner. But now that the people had been thoroughly roused, they became afraid that something dreadful would happen to us or them, especially the captain of the gunboat, who hearing us continue to shout after the first sampan had left, bustled across in a terrible hurry with his vessel and all his crew. No sooner had he started, than an envoy who had been sent by the Fu of the next town (Chung) to pay his respects to Baber, grew alarmed also, and collecting twenty men, nearly all the population of the immediate neighbourhood, went over to assist in the search, and it was a long time before the town returned to its normal quiet state.

March · 31.—Opposite Wu-Yang a considerable bend of the river commences. The neck is about five miles across, and as the distance by water is about nine we determined to walk over while our junk went round the bend. We landed about half a mile above Wu-Shan, and striking up a valley we made for a high hill in a south-west direction. A very fair paved path led at first gently up amongst small woods of fir trees. Beyond these comfortable farmhouses were enclosed in their clumps of bamboo, and all the valley was carefully laid out in terraces for rice cultivation, with here and there a small patch of opium, a few cabbages, or a field of wheat. We gradually ascended, until we found ourselves on a ridge 800 feet above the river. After enjoying the prospect for a few minutes we went down to a stream running over a rocky bed, where mosses and ferns were growing luxuriantly. A short ascent beyond brought us to a peasant's house, where we sat down outside, and talked with the proprietor about his affairs. He drew water for us, and gave us lights for our cigars, and although evidently exceedingly poor, for his cottage was very small, he at first refused the proffered string of cash.

There were two baskets here with their contents drying in the sun; these were the flowers picked off a tree in appearance very like an apricot; the blossom is a kind of long conical-shaped pod, on the surface of which there are a number of very small flowers full of yellow pollen. These flowers when dried are boiled, and the very poor people drink the broth instead of tea. The local name of the tree is Ch'ung-Shu.[8]

After this we continued our walk till we reached a point above the river from which we could see our convoy coming up, so we hurried down, and a sampan soon put us on board.

April 1.—In navigating this portion of the river, it is continually necessary to cross and recross from one bank to the other, partly to save distance by cutting off the angles of the numerous sharp bends, partly to get into the back eddies, and avoid the current, and partly because it is often impossible to track on one side, while there is a fair path on the other. It is this that constitutes one of the chief difficulties of the navigation, and leads to most of the accidents.

---

[8] An endeavour has been made, by consulting very high botanical authorities, to identify the flower here spoken of, but it has not been found possible to do so (Y).

The junks are always rowed across towards some place where a landing is practicable; as a rule there is scarcely any room to spare, and unless the exact point is gained, the swirls and eddies that often run violently amongst the reefs will drive the vessel against a rock. Amongst these tides it requires the greatest skill and nicety to shoot the junk exactly to the desired spot, and it is under these circumstances that vessels are often wrecked or damaged. During the day we met with two accidents in this way; on the second occasion a big hole was knocked in our ship, which as usual was repaired with cotton-wool and paper.

April 2.—Although in our walks we had frequently noticed little patches of opium, we had not seen it hitherto in any considerable quantity; but in this neighbourhood we saw it growing in large fields.

April 3.—The river was so low that when we arrived at the north end of St. George's Island we found that not only was there no passage inside it, but that a reef, rarely visible, was now plainly to be seen beyond. Our small boat, and that of our official, made no attempt to pass inside this reef; but our people, seized apparently with a fit of temporary insanity, thought that though our junk was much bigger, and drew more water than the others, they would try the inner passage. So, as we were sitting in our room discussing the probability of our arrival that night at the city of Fu-Chou, there was a sudden bump that nearly shook us out of our chairs, followed by a babel of tongues, in which as usual the shrill and jarring voice of the old woman was painfully audible. Rushing to the window we found the tracking line adrift, and the vessel spinning round like a top; then the scrunching and grating sound of the junk dragging herself over some sharp rocks was immediately followed by the sudden irruption of three or four coolies into our room, who, without any preliminary remarks, moved the furniture, lifted the floor boards, jumped into the hole, and taking all our boxes out, hastily passed them up to other coolies outside.

Looking down we now saw a large stream of water running in through a hole in the side of the boat, and we comprehended what had occurred. We had not, however, much time to alarm ourselves, for we were fortunately able to run on to a bank of mud before the vessel filled and sank, which she inevitably would have done in a very few minutes.

After having cleared the hold, the men set to work baling with buckets, and gradually succeeded in reducing the water, after which they repaired damages.

They first put on a kind of cataplasm of whitey-brown paper, mud and grains of rice,[9] over which they nailed a piece of wood, and stuffed the interstices with cotton-wool and bamboo shavings. As, of course, when the hole was made the planks were driven inwards, this patch was put on inside. The operation was a long one, and, extraordinary as the method may appear, it eventually proved tolerably effectual, although from the amount of baling that was always subsequently necessary, Baber suggested that our vessel should be called the Old Bailee.

We spent the afternoon walking round the island, and found some of the gold-washers, who above this are always seen in the sand and shingle beds washing for particles of gold. The quantity these men obtain is so small that it can repay none but a frugal China-

---

[9] The Chinese use a great deal of rice in this way. I have seen a kind of concrete for building purposes made of mud and rice grains.

man for the labour. Here the river is known as the Chin-Sha-Chiang, or River of Golden Sand, and this name is applied to it at least as high up as Bat'ang.

April 4.—Near the city of Fu-Chou, the principal river of the province of Kwei-Chou enters the Yang-Tzŭ. We landed for a walk on the left bank about four miles below this, and found ourselves in a little valley with a very pretty waterfall about thirty feet high, sparkling and leaping down amongst deliciously green verdure of moss, creepers, and ferns. We called it the Fountain of Egeria, and clambering up the hill on the west, reached what would be considered a good road in these parts, though it was little more than a long flight of steps of roughly squared stones about eighteen inches broad. After ascending 1,000 feet we gained a sort of fort, that had, so the inhabitants said, been built in the year 1804 as a refuge against robbers. The side towards the stream was guarded naturally by an inaccessible precipice, and a rough wall that was now falling into ruins had been built round the other three. Leaving this we struck off westerly, but the lay of the hills and valleys, and the direction of the paths, drove us up to the north away from the river. At last, from the top of a hill we saw a fine broad valley below us, with a paved road running along the bottom, where there were a great many people and coolies going in both directions. We knew that this would lead us to the water, so descending we followed it, and gained the banks between Fu-Chou and Li-Tu, opposite a joss-house marked in Blakiston's chart. With the aid of a telescope we could now just see the 'Old Bailee' disappearing round a point two miles higher up. We followed as fast as we could, by a path that wound about along the bank, now a couple of hundred feet up, now again close to the edge of the water, then with a sharp turn it would ascend a steep ravine for nearly a quarter of a mile, before crossing the little torrent at the bottom and returning to the river side.

A stern chase is proverbially a long chase, and we found it so in this instance. We walked as fast as we could, but each time the road gave us a chance of using our telescope, the ship, if visible at all, was further away from us than ever, and just as it was getting dark the wretched track came to an end, and left us face to face with a high cliff. Fortune, however, favoured us, for at this moment a small boat passed ; we hailed it, and prevailed upon the owner, by the promise of 200 cash, to take us up to our ship. She was much further away than we expected, but after we had rowed about an hour and a half we reached Li-Tu, and finding himself at home, the pious Wang-Erh, who used to accompany us in our walks, and who was always very glad to get back, was heard to exclaim, ' Oh! Mane, Buddha, I knock my head,' meaning thereby that he morally knocked his head on the ground in token of thankfulness, and when we sat down at half-past eight to our dinner we also knocked our heads.

The old woman now came to us, and said that the repairs to the junk had been so costly that she could not pay the crew, who were mutinous and hungry. When she was reminded that she had already received more money than she was entitled to, she appealed to our official, who, seeing a chance of getting some silver through his fingers, suggested that we should give him five taels, that he would pay what was necessary to the crew, and would be responsible for the whole. But the artifice was apparent even to our simple minds, so thanking him warmly for his disinterested offer we declined to give him so much trouble.

April 5.—Notwithstanding the complete consumption of the purse, all the people wanted to 'buy things,' as they invariably did at any town of importance, and we were late getting under way.

April 6.—We passed a remarkable temple cut out of rock on the left bank of the river, in which we could see two gigantic gods brilliantly painted and gilded ; and in the evening we anchored at a small village about four miles below Chung-Chou.

April 7.—The crew appeared to be up all night ; they were continually moving about, and were off by five. I heard the shouting and monotonous chaunt of the men in other junks even earlier, from which it was clear that everyone here hoped to reach Ch'ung-Ch'ing before dark. In many places it is impossible to track, and then the method of propulsion is by oars, or in some big junks by a very large scull, one on each side of the vessel. All the time that the coolies are at this work one of them chaunts a long story in time with the strokes, and at each stroke all the others join in a chorus of ' Hey-yea.' This will go on for ten minutes, when the story will end, and all will sing together, ' yoi hai ay-a.' The tone is continually varying, but the chaunting either of the story or the chorus never ceases. The method of employing the gigantic scull is quite unique. Every country uses it on a small scale, but I never heard of huge vessels being propelled in this way elsewhere. In any harbour in England dirty little boys may be seen sculling out of the stern of a boat. The Venetian gondolier also puts into practice much the same principle ; but here huge junks, of some hundred tons burden, may be seen with an enormous scull on each side, worked by as many as twelve or fifteen men. These sculls are supported at the fore part of the ship on a short outrigger, at the end of which there is a very short pin. This pin fits into a cup shaped hollow in the scull, and acting like a ball and socket joint just keeps the scull in its place. The men stand in a row, fore and aft, facing the water. At the end of the scull there is a strong leathern thong, which, fastened down to the side of the junk, keeps the end of the scull moving in a circle. This method, which is in fact an application of the principle of the screw, is no doubt the most economical way of applying the strength of the coolies; it is more frequently seen in use on junks coming down than on those going up the river, for in ascending there are such frequent changes to be made—sometimes tacking, sometimes laying out ropes, and only occasionally rowing—that these large sculls are not so convenient as oars; but in descending, when the middle of the stream is always kept, when rowing or sculling is the only method in use for driving the vessel, and when the whole crew is always on board, then the large scull is found the most suitable method of working the ship.

Notwithstanding the efforts of our crew, who worked really hard, we were unable to reach Ch'ung-Ch'ing this day, and anchored four miles below it, after a run of twenty miles.

One of our coolies, in stepping from one boat to another, fell down and broke his arm; poor fellow, he scarcely said a word, and if he felt much pain he bore it most quietly. A Chinese doctor was found to set it, although anatomy is a science the rudiments of which are as yet quite unknown to the Chinese, who are forbidden by their laws to dissect the carcases of any animal whatever.

CHAPTER VII.

CH'UNG-CH'ING TO CH'ÊNG-TU-FU.

Arrival at Ch'ung-Ch'ing—M. Provôt, Monsgr. Desflêches, and the French Missionaries; Their Cordiality—Official Visits—The Last of our Lady-Skipper—We are satirised in verse, and enabled to see ourselves as others see us—Mê, the Christian—Persecution of Christians—News of Tibet—Unfavourable Change of Feeling in Tibet as to Admission of Europeans—The Consulate at Ch'ung-Ch'ing Proclaimed—Consular Relations to Missionaries—Cheapness of Ch'ung-Ch'ing— General 'Fields-within-Fields'—Difficulties of Mr. Baber's Photographer—Exchange Troubles—Preparations to Start for Ch'êng-Tu—Elaborate Coolie Contract—Chinese Commercial Probity—Baggage Arrangements—Adieu to Baber—Joss-Houses and Paper Offerings—Characteristics of the Rice-Culture—Wayside Tea-shop—Pleasing Country—Poppy-fields—Yung-Ch'uan-Hsien—Everything Roofed in—T'sai-P'ing-Chên—Grand Thunderstorm—Paved Roads—Inns of Ssŭ-Ch'uan—Chinese Parsimony—Good Manners of Ssŭ-Ch'uan Folk—Jung-Ch'ang-Hsien—Characteristics of a Restaurant—Coolies at their Meal—Cooper's Dragon Bridge—Realistic Art—Want of Ideality in Chinese Character — Lung-Ch'ang-Hsien — Charming Inn —Wang-Chia-Ch'ang—Niu-Fu-Tu—Hsien-Tang—The Brine Town of Tzŭ-Liu-Ching—The Christian Landlord—The Brine-wells and Fire-wells—Mode of Boring and of Drawing the Brine—Further Details and Outturn of Salt—Politeness of the People—River Ch'ung-Ch'iang—Tzŭ-Chou—Gratuitous Diversion for a Village School—A Chinese Regiment—The School System—Sanitary Paradox of China—Red Basin of Ssŭ-Ch'uan—Transfers of Property—Yang-Hsien—Safflower Crop—Thickly Peopled Plain—Still Ascending Valley of the Ch'ung-Ch'iang—Thrift in Excess—Arrival at Ch'êng-Tu-Fu—Public Examinations in Progress.

April 8.—Early in the morning we reached the outskirts of the great city of Ch'ung-Ch'ing; and passing through a crowd of junks of all sizes, we hauled up to a position under the walls, where we very soon received a welcome batch of letters and papers from the agents of Messrs. Major and Smith. The Chinese merchants have an excellent postal system of their own: they arrange amongst themselves to send couriers or runners on foot at regular intervals, who travel very fast, and generally very securely. In this case the letters had been only fourteen days from Hankow, about six hundred miles by road. During the whole time I was in China I received every letter and newspaper sent me, except one letter, and that had been forwarded *viâ* Russia!

Soon afterwards Monsieur Provôt, one of the French missionaries, came to pay us a visit : a tall pleasant man, dressed in Chinese clothes, and with an artificial plait, for the missionaries in China invariably discard foreign clothes. He said that all sorts of conjectures had been rife about us amongst the Chinese. He asked Baber when he was going on to Yün-Nan ; and turning to me said he hoped that I should like living here. When he saw that we did not exactly understand the remark, he explained that it was the general opinion that Baber had been appointed a consul in Yün-Nan, and that I was to be consul at Ch'ung-Ch'ing. We hastened to undeceive him ; but even the missionaries could hardly believe in a gentleman travelling for his own amusement without any commission from the Government ; the Chinese certainly did not.

In the afternoon we received intimation that Monseigneur Desflêches, the Bishop, was coming to pay us a visit. He was a small vivacious man, and a true Frenchman ; he was most genial, and his expressions of delight and compliments to Baber knew no bounds. ' Ah, Monsieur Baber, it is you at last. How you are welcome ! ! Here is a grand thing that you have done ; ah! it is indeed a victory. Yes! yes! a victory indeed. See how at last we have this great river opened to foreigners, thanks to you and your Government.'

Nothing could have exceeded the sincere cordiality of his welcome.

Probably besides missionaries there were not more than twenty or thirty foreigners who had ever been here, and the arrival of a real consul, accredited by the English Government, was naturally a glad event to the missionaries. But for all that we could not but feel it a pleasure to be so warmly welcomed, and received with such true and hearty friendship. The Bishop talked for a long time; first he told us about his flock, his converts, and his trials, of which he made very light, dreadful though they had been; he praised the English and the English Government, and declared that our country was the only one in which there was any real religious liberty. He naturally expressed great pleasure that war had not broken out between China and England, 'for,' he said, 'if it had we should all have been massacred here.'

After the Bishop had left we received visits from several Chinese, some of them Christians, one of whom sent us a present of a jar of fermented liquor, made from a particular kind of rice; it tasted like beer, with a strong flavour of the better class of Chinese wine. The chief officials sent their cards to Baber, and promised a visit in a day or two.

April 9.—Baber went to pay official visits to the magistrates, and I subsequently joined him at the house of the French missionaries, where we were hospitably entertained with port wine and sponge-cakes. The missionaries were all very thin men, with drawn features and sunken eyes, and they certainly had not the appearance of getting much of this world's goods. They told us there were about one thousand Christians in the city, all of whom they clothed and fed, at one time when persecutions were going on.

We returned to our boat, as we had come, in state, with four bearers to each chair. The main street of the city is a very steep flight of steps, and is easy enough for the occupant of a chair going up, but coming down is very different. The chair is inclined at an angle of about 45°; there is no support for the feet, and the rider expects every moment to be ignominiously precipitated on to the back of one of the chair-coolies. This main street is so narrow that in places there is absolutely not standing room for a man between a chair and the shop fronts ; these are all poor and dirty ; and the place, though in reality rich and thriving, is by no means prepossessing in appearance.

April 10.—In the morning the Tao-Tai called on Baber.

He was a well-educated man, and had written a learned book on ancient characters. He had included in them copies of the inscriptions that are on the stone drums in the gateway of the Confucian Temple [1] at Peking ; before leaving he told us that he had issued a proclamation informing the people of Ch'ung-Ch'ing that we were decent peaceful folk. The Hsien also called, and the official who had accompanied us on our voyage.

April 11.—A very fat Christian, named Mê, came in to say that he had found a house that he thought would do for the consulate, so Baber went to look at it, and came back well satisfied.

April 13.—The old lady who commanded our vessel came in afterwards with her child, and kneeling

[1] *See* ante, pp. 127–128.

on the ground burst into a flood of tears, declaring that she was the most miserable and unfortunate woman in the world; that she was a lone widow with no one to take care of her; that every one conspired against her; that she was no match for the wicked people by whom she was surrounded; and although she felt she had gained a high distinction by being allowed to bring our honourable selves up here, still her misfortunes had been many, and she was out of pocket by the transaction; and in pathetic tones, she expressed her hopes that our noble and honourable excellencies would not allow her and her orphan child to die of starvation. As a histrionic performance it was certainly creditable, the old woman having extracted from us half as much again as any Chinaman would have paid her. With a tongue so fierce and foul that it inspired awe if not respect, I could imagine no one better able to look after number one.

We now said good-bye to our ship, in which we had lived nine weeks. Our goods were first of all moved, and after everything had gone we followed in chairs. The coolies carrying the chairs bustled along at a great pace up the steep and dirty steps; three soldiers were in front to clear the way; nevertheless a good number of little boys followed, trying to lift the blinds and peep in; but there was no hostile demonstration of any kind. Besides the officials, the people of this province are mostly either merchants or agriculturists, the literati—that generally highly-favoured class in China—being held in light esteem by the men of Ssŭ-Ch'uan; and to this is probably owing the fact that foreigners are always treated with great politeness, as wherever opposition to foreigners is carried to any great extent, it will generally be found to be owing to the influence of the literati class. There were of course some literati here, and so good an opportunity of showing their talents was not to be lost. So they wrote a poem in very bad rhyme, which Baber translated and headed, ' As others see us: '

'*AS OTHERS SEE US.*'

1. The Sea folk, once a tributary band,
   In growing numbers tramp o'er all the land.
2. English and French, with titulary sounds
   As of a nation, are the merest hounds!
3. Nothing they wot of gods, in earth or sky;
   Nothing of famous dignities gone by!
4. One of their virgins, clasped in my embrace,
   Told me last year the secrets of their race.
5. But all their deeds of darkness are as nought
   Compared with vileness by their fathers wrought!
6. I know their features, Goblins of the West!
   I know the elf locks on their devil's crest!
7. Cunning artificers, no doubt, but far
   Beneath our potency in peace, or war!
8. But now our opportunity is near;
   Learning and valour are assembled here;
9. Let all to the cathedral doors repair,
   Grapple the dogs, and never think to spare!
10. I read ye right! shall savages presume
    To harry China and escape the doom?
11. No! Let us all with emulous might combine
    To crush the priests, and save the Imperial line.
12. First slay the bishop, tear away his hide,
    Hack out his bones, and let his fat be fried!
13. And for the rest who have confessed the faith,
    Drag them along, and roast them all to death!
14. For when these weeds are rooted from the plain,
    What magic art can give them life again?

The author begins by inquiring why foreigners

should come to China; and though he shows an unusual amount of knowledge by stating that the French and English are different people, yet he denies nationality to either one or the other, who, he adds, are all mere dogs, and ignorant of the true religion. In line 6 he refers to the features of foreigners, which all Chinamen consider worse than hideous. Foreigners are usually also credited with red hair, which, in their eyes, is an abomination; hence the reference to elf locks. The author exhibits unexpected discrimination in crediting foreigners with being cunning artificers; Chinese generally think, or pretend to think, that we are ignorant of everything. In line 8, reference is made to the approaching examinations, when thousands of literati and students for degrees would be assembled at Ch'ung-Ch'ing. The last line refers to the popular belief that foreigners can after death return to life; and, once more showing more knowledge than might have been expected, combats this belief.

Mê, the Christian who visited us, owned a house in Ta-Li-Fu, and when Margary was there he visited Mê. At that time Margary was well treated; but after his sad death, his murder, though it took place at some distance, seemed to excite even the people of Ta-Li-Fu; and holding this extraordinary belief about the ability of foreigners to come to life, it somehow entered their heads that Margary was hidden with Mê, and that he had the sum of four thousand taels with him. In this frantic state of mind they stormed the house of the unfortunate Mê, and finding neither foreigner nor money pulled it down in revenge.

Baber's house was supplied with the amount of furniture usual in Chinese dwellings. A few stiff arm chairs and clumsy tables, and a couple of heavy bedsteads. The walls and ceilings were, of course, as dirty as can well be imagined.

The Chinese have no idea of convenience or inconvenience, and our chairs had hardly set us down when visitors came in. Amongst them a man named Hsuan, who came every day, and used to sit for two or three hours, talking incessantly the whole time; he had passed his examination, and hoped soon to be appointed a Hsien. Whenever he called he used to insist on my smoking one of his pipes; it would not have been exactly objectionable if there had been more of it, but, like eating water-melon seeds, the end seemed out of all proportion to the means.

April 14.—Monsieur Provôt came to congratulate Baber on his installation; and told us that the Bishop, Monseigneur Desflêches, had gone to Ch'êng-Tu to see the new Governor-General, who was expected in a few days, and to make complaint before him of the persecutions to which the Christians had been recently subjected in Chiang-Pei-Ling, a small city divided from Ch'ung-Ch'ing by the river Hsiao-Ho. Their houses had been surrounded, pillaged, and burnt, and the inmates driven away with circumstances of great cruelty; some of the Christians who ventured to return had been suddenly attacked and murdered, and the persecution had been continued for some weeks, during which time thirty persons had been killed, some being cut into pieces and thrown into the river. The officials had taken little notice of the outrage, not even holding so much as an inquest on the corpses, merely reporting to head-quarters that a slight disturbance had occurred, and the Tao-Tai, who, as was affirmed by some, secretly instigated them, never raised a finger to repress the outrages.

Monsieur Provôt gave us news of Tibet. He had received a letter from Monseigneur Chauveau, Bishop of Ta-Chien-Lu, who said that a report had spread all over Western China and Tibet of the expected arrival of British and Russian missions at Lassa; that this report had caused a most profound sensation; that the Lamas were urging the people to refuse admittance to foreigners, and that forces were assembling on the frontier.

There can be no doubt that a great change has come over the feelings of the Tibetans since the days when Bogle visited Lassa, and was so well received. There are two causes that may have combined to make the Tibetans afraid of Europeans. Firstly, our power in India has so enormously extended that the Tibetans say, with much justice, 'Wherever an Englishman comes he soon possesses the country; once we let an Englishman enter ours, we shall lose it.' The second adverse cause is the presence of the missionaries. In the time of Bogle there had been few attempts on the part of these to approach Tibet, and in those days the Lamas had no fear of foreigners upsetting their power and their religion. But since then there have been many missionaries on the borders; and these being the only foreigners the Tibetans know, they naturally fear for the supremacy of their faith.

In the days of Bogle and Manning, and even as late as the time of Huc, it appeared that among the Tibetans themselves neither Lamas nor people offered any objections to the approach of Europeans; but that all the opposition, great as it was, came entirely from the Chinese officials. Since that time, however, it would appear that the Lamas, who absolutely rule the people, have conceived a violent hatred to foreigners,

CH. VII. THE CONSULAR RELATION TO MISSIONARIES. 269

and have arrived at a determination to exclude them by every means in their power.

I had some musical boxes with me which I had bought in London, thinking that they would be useful as presents. I now found that out of six the three best would not work, and imagining that they would be rare curiosities, I showed them to our fat fried Mê, and asked him if he thought that there was anybody here who could repair them. 'Oh, yes!' he said, 'and you can buy them for three taels,' about the price I paid in London.

April 15.—A huge official placard was now posted on the door of Baber's house. It was to the effect that Baber had come here solely to look after trade; that he had no connection with the French missionaries; that people were to respect him; that any rioters would be severely punished, &c., &c. This was all very fine, but the details did not quite satisfy the exigencies of the uncompromising Baber, who visited the Fu in the course of the afternoon, and complained that the word England was not written big enough, and had not been put in a sufficiently exalted position on the paper! He also told the Fu that he would have done better to have left out the passage in which it was stated that he was not connected with the French missionaries. The Fu said he did not think that a matter of much consequence. 'Well,' said Baber, 'but suppose an English missionary comes, what will you do then? and I believe there is one now on his way.' The idea of two sorts of foreign religion was a complication that might well have exercised even a more educated mind than that of a Chinese Fu. So Baber promised to try and prevent the spectacle of two sets of missionaries preaching two different creeds.

The Fu, very anxious that the conversation should

not be overheard, conducted a great deal of it in writing, a process that must have made it somewhat tedious.

Whatever its demerits may have been, Ch'ung-Ch'ing was at least a housekeeper's paradise, as the following price list, supplied by Chin-Tai, will serve to show:

Yellow cow-beef (as the beef of an ordinary ox was called)	66 cash a catty.
Buffalo cow-beef	50 ,, ,,
Kid's flesh	120 ,, ,,
The best small chickens	100 ,, ,,
Old cocks (very tough)	90 ,, ,,

Eggs, 6 cash each. Cabbages, 2 or 3 cash.

The exchange made a penny about equal to 22 cash, and a catty is $1\frac{1}{3}$ lbs. avoirdupois.

The market, too, was well supplied with fish, liver, kidneys, carrots, turnips, peas, beans, and a vegetable that made a very good imitation of spinach. It was almost impossible to buy sheep, and the kid's flesh was about as tough and tasteless as it usually is in Eastern lands. But with a sack of potatoes, which the French missionaries presented to us, our feasts were by no means scanty.

April 16.—A great deal of noise ushered in the august presence of the Chen-T'ai, who came to pay a visit to Baber, bringing with him, according to Chinese custom, a miscellaneous crowd. In China, during an official visit, it is always necessary to admit almost anyone who wishes to come in, in order that the people may know everything that is going on, and that no conspiracies may be hatched. The name of our visitor, when translated into English, was *Fields-within-Fields*. He served with much credit during the Tai-Ping rebellion, and was present at the siege of Su-Chou, but he said he had never met Gordon.

He was evidently not a man of great intelligence,

for he asked Baber if in England we made glass, as they did in China, from rice. One of his attendants, with a grin, reminded the worthy Chen-T'ai that not even in the Great Central Nation had people learnt the art of making rice into glass. This did not disconcert him in the least, and saved Baber from the humiliation of being obliged to confess that we had not yet discovered this marvellous process.

My servant, Chin-Tai, had been suffering from rheumatism, and I had asked Mê, the fat Christian, if he could find a doctor. When General *Fields-within-Fields* had gone with his crowd, Chin-Tai came to me looking particularly happy. He told me that three doctors had been to see him, and that each had prescribed for him; I said that I hoped that there would not be the proverbial disagreement. 'No,' said Chin-Tai, 'with three doctors I am sure to get well very soon.'

April 17.—The town of Ch'ung-Ch'ing is built so crookedly, and with such tortuous streets, that the people are compelled to use the terms 'to the right,' 'to the left,' in giving directions about the way to any place. Ordinarily, in China, the towns are built with a certain amount of regularity, and the people say 'go north' or 'go south,' &c. They become so habituated to this that, even out in the open country, they use the same expressions, having, as a rule, not the most remote conception as to where the north point really is. This custom has had the effect of impressing on foreigners generally a most exaggerated belief in a Chinaman's knowledge of the points of the compass.

We went for a walk one morning on the other side of the river, and took the photographer with us, and left him to his own devices. When we returned home he told us that the people had thrown stones and bricks at the camera. He said that his attempts had not been very successful. The Chinese people believe that foreigners make a juice out of children's eyes for photographic purposes; they say 'A man, or a dog, or a horse cannot see without eyes, how then can that machine? If it has not got eyes of its own, it must have the eyes of somebody else.' Their logic is unanswerable, especially the brickbats and stones. The next time that Baber's photographer essayed his art, he went out under the guidance of the fat Christian Mê, who could talk to the people in their own dialect; the photographer, who was a Shanghai man, finding the language of Ssŭ-Ch'uan quite unintelligible.

I now began the preparations for my departure, and seeing that here, as everywhere else, money forms the sinews of war, I sent for the banker, on whom bills had been given me at Hankow. He professed himself quite ready to give me any amount of silver, and said he had correspondents at Ch'êng-Tu, so that I could get my money there instead of taking it with me. Of course this was a great convenience, so I arranged for him to give me what was necessary for the journey, and bills for the rest on the Ch'êng-Tu bankers.

Exchange is a matter that in China always gives a good deal of trouble. The tael is, properly speaking, a weight of about $1\frac{1}{3}$ oz. avoirdupois. The term 'tael' is a foreign one, the Chinese word being 'liang.' Almost every province, and often each important city in a province, has its own tael; thus a piece of silver that weighs a tael at Ch'ung-Ch'ing will weigh less than a tael at Ch'êng-Tu; and, as all payments are made by weight, it is necessary to have a balance for each place. Then the quality of the silver varies; and besides this, in making small payments, there is the further complication of the number of cash, or 'chen,' as the Chinese call them, to the tael; this is of course unavoidable. It costs less to carry a pound of silver one hundred miles than it does to carry the equivalent value of brass; and at places far removed from centres of civilisation, the tendency is, naturally, to bring more to an equality the value of the two metals, just as the values of all goods tend to equalise themselves relatively the greater distance they are carried. But, however unavoidable, the difficulty is none the less troublesome to a traveller, who thus has three things to look to: first, the quality of the silver; secondly, the weight of the tael; and, thirdly, the number of cash to the tael.

I now wished to engage another servant, and a man came to me, and went through the usual formality of knocking his head on the ground.

I asked him if he would be my servant, and travel with me.

Yes, he answered, he was willing to follow me anywhere.

'Very well then,' I said, 'I will pay you seven taels a month.'

Chin-Tai, who was acting as my interpreter, now conversed with him for about half an hour. The dialogue was reduced in English to the laconic statement, 'He says he wants fifteen taels.' This I declined, and Liu-Liu, for such was his name, went away; but he returned in a few hours, saying he had quite misunderstood my offer, that he had consulted his brother, and that if I would give him some advance for his wife, and would promise to pay his fare back to Ch'ung-Ch'ing when I had done with him, he would come with me. All of which meant that he had merely been trying, in accordance with Chinese custom, to get as much as he could. So the compact was sealed, and Liu-Liu (or the Willow) became my servant.

A day seldom passed without visitors. Mê returned triumphant from his photographic expedition; his part of the business had been satisfactorily accomplished; but the photographer's efforts can hardly be said to have been crowned with success. He could not show us much except some clouded glasses, and I never heard that any pictures were subsequently achieved. The banker came in again while Mê was with us, and Hsuan of the perpetual tongue.

There are no mules in this part of the province, and it was therefore necessary to look for coolies, but as I was able to send most of my goods by water, I did not require a great number of carriers. I had to buy a chair also for myself to ride in, because, in this province, a chair is the usual means of locomotion; and to have travelled otherwise would not only have been against the inexorable law of custom, but would have entailed a loss of dignity that might have been inconvenient. After I had once started, however, I rarely rode in the chair, except when entering or leaving a large town; in the country I invariably walked, or rode on a pony. The chair was, nevertheless, invaluable for carrying a few things in, for with four coolies, and no one riding in it, it could always travel very fast, and in the plains could even keep up with me when I was walking, so that when I arrived at my destination, the chair was seldom far behind, and I had not to wait an interminable time for all the odds and ends, writing and drawing materials, &c., &c., that I wanted immediately.

In a large city like Ch'ung-Ch'ing there was no difficulty in finding any number of coolies, and Chin-Tai soon found a coolie-master willing to provide for all my wants.

April 24.—An elaborate agreement that would have refreshed the heart of a lawyer in Chancery Lane was now drawn up between this coolie-master and myself: detailing specifically what I might and what I might not do; the places at which we were to halt; and how long we were to stop at them; and the extra amount to be paid, in case I wilfully delayed on the journey. The coolie-master on his side pledged himself to use all reasonable care and forethought for the safety of my goods, and to arrive at places specified within a certain time. But, unlike English documents, this charter once drawn up and verbally agreed to by the coolie-master and myself required neither witness nor signature; but being confided to the depths of my pocket, was as valid, according to Chinese usage, as the most formal document that ever issued from Lincoln's Inn.

This confidence that people in China have in one another is a feature in the character of the people that has been strangely unnoticed by foreign writers. Merchants in China rely implicitly on one another; indeed, if they did not, all business would come to an end at once. In my position I was over and over again compelled to trust the Chinese with large sums of money without receiving any receipt, and in other ways to rely on their probity to a far greater extent than I should have trusted Europeans, or Chinese if I could have avoided it. But I was never deceived in the smallest degree, nor did I lose anything during all the time I was travelling. Of course if I had set my wits against those of the Chinese I should have been taken in continually; and if I had tried to drive bargains, I should certainly never have succeeded. A Chinaman, if he is selling anything, will always ask as much as he thinks he can get, even if he knows it to be ten times the value of the article; but amongst the respectable Chinese there is a strong feeling of commercial morality. It probably arises not from any natural inborn virtues, but from the necessities of the case; for there is no reason to suppose that the Chinese race forms an exception to the general rule of humanity, the heart of which is declared to be by the highest Authority deceitful above all things, and desperately wicked. If a Chinese weaver adulterated his silk, it would be known at once, he would be a marked man, and his trade would cease. If the English manufacturer never sold his goods at a greater distance than one hundred miles from his doors, it is probable that he also would find the advantages of honesty in his policy. Necessity is not only the mother of invention, but the origin of all custom; and custom in time becomes not law but something even more binding.

When I called to say good-bye to the missionaries, I found that they were firmly persuaded that political missions from every quarter were being poured into Tibet, and that Baber and I were connected in some mysterious manner with the inscrutable purposes of these expeditions. When I assured them that I had nothing whatever to do with Government or Government missions, and that I was a private individual travelling for my own objects, they smiled incredulously, as if unwilling to be thought simple-minded enough to believe so foolish a story; and even with the proverbial politeness of Frenchmen, they could hardly help showing that they thought Albion was as *perfide* as ever ; and if reasonable Europeans could not believe it, how could it be expected that the Chinese would ? In fact, they never did ; from first to last I passed for an important official on some secret service, and was invariably treated as such.

April 26.—Everything was at length ready for a start, and I found coolies sitting about, waiting for their loads to be adjusted. A chair that I had bought was now fresh from the painter and decorator ; there were, besides, small chairs that were hired for the servants, and a pony about eleven hands high was ready to be saddled. Twenty coolies sufficed for my luggage ; besides these there were four coolies for my chair, four for the chairs of the servants, and one man, who glorified himself with the title of Ma-Fu, with the pony.

Baggage in this part of the world is carried in cages made of bamboo. Long bamboos run along the top of two sides of the cage, cross pieces connect the ends, and these rest on the shoulders of the coolies. This is a very convenient method of carrying baggage, as the loads can be packed and unpacked in a few minutes ; moreover, there is no jolting or knocking about, and the most slender box might be carried in this way for weeks or months without getting any harm ; on mules or in carts the baggage is terribly knocked about, and of these, carts are certainly by far the worst. I have seen new cartridge boxes, fresh from an English gunmaker, broken to pieces and rendered utterly useless in two days of cart travel. In fact, for mule or cart travelling, there is nothing like real solid English leather ; and a box or portmanteau well made of this material will survive even these terrible trials.

Dividing the luggage into portions of equal weight,

and arranging the loads properly, occupied some time; the coolies can measure the weight of a load very exactly by simply trying it. They are so accustomed to a certain weight that they can tell immediately whether it is too much. In doing this, they stand to the cross bars, take hold of them with their hands, which for this purpose they always cross, and just lift the load once or twice off the ground.

My fat friend Mê came to see me off; and with a hearty shake of the hand I said good-bye to Baber, whom I hoped to see in the course of a few weeks. Without this expectation I should not have parted with a light heart from one who had been a cheery companion for so many weeks. But 'Dieu dispose,' and I still have to look forward to the pleasure I hoped for so long ago.

I started in my chair, and as, with the exception of a short ride in the city, this was the first time I had tried this method of progression, I found that dignity and discomfort were in about equal proportions, for to one unaccustomed, the motion, especially in hilly countries, is very disagreeable. When I was well clear of the town I descended, and found walking preferable. The road ran for some distance by the side of the river, winding about amongst hills five hundred to a thousand feet high, sometimes shaded by hedges of pomegranates from the sun, which was now becoming powerful. The hill-sides were dotted with the white-walled Ssŭ-Ch'uan farmhouses in their clumps of bamboos, looking the very emblems of peace. Yew trees often sheltered fine large graves; and here and there, under a fine banyan, there would be one of the small religious shrines which the Europeans call *joss*-houses, from a corruption of the Portuguese 'Deos,' God. This term is applied to temples and shrines of all sizes : from the gorgeous buildings, the pride of some important city, to the roughly carved stone found by the wayside, which may be nothing more elaborate than a solid block, two or three feet high, cut into the form of a gable at the top, with a hollow chipped out in front for burning incense in.

At this time of the year, pious people bring paper money to the shrines and temples ; and in the neighbourhood of one of these the roads are strewn with such amazing quantities of this rubbish, that the traveller fancies himself again at school enjoying the sport of a paper chase.

Theoretically, real money is brought to these places, and put on the shrine as an offering. No doubt in the forgotten days of dim antiquity this was done ; but long ago the eminently utilitarian spirit of the Chinese conceived the idea of paper money, which is manufactured, in the vicinity of most temples, with a machine something like a gun-wad cutter, in imitation of copper cash—another proof, if proof were wanting, that the Chinese have now no religious belief whatever, and that their elaborate ceremonies are no more than customs hallowed only by their age.

The road was paved ; it was in excellent repair, and about six feet wide, very broad for a Chinese road. There was some opium ; but not one tenth of the land cultivated was planted with that crop. There was also a little wheat ; but nearly the whole ground was given up to rice. The water was now on the paddy-fields, and the country from the top of the hills looked as if it were flooded. I met very little traffic of any kind, and after a walk of twenty miles I found a fair inn at Pai-Shi-Yi-Ch'ang.

April 27.— The streets in this place were fairly clean, and not quite so narrow as is usually the case in China. They were all covered over with matting to keep out the sun ; and at the early hour of our start there were few people about.

Our course lay in a south-westerly direction, across a wide valley, amongst low undulating hills ; everything was very green, fresh, and, as nearly all the rice fields were under water, there was very little dust. The coolies in the fields were busy at work raising water from the lower to the upper terraces, sitting under the shelter of big umbrellas ; and from the top of a hill these looked like a number of gigantic mushrooms dotted about over the plain below.

The method in use is as follows :

A small trough is laid from the lower to the upper level. The trough is square, and open at both ends. An endless rope with floats on it passes over two wheels. The floats exactly fit the trough, and bring the water up. The wheels are turned by treadmills, on which the coolies work all day in the sun, nearly naked, sheltered by the big umbrellas, and fanning themselves the while. The use of fans amongst the Chinese is rather a novel sight at first. Everyone carries a fan ; the very chair coolies as they run along are fanning themselves.; the coolies resting by the roadside sit with fans ; travellers on horseback, and shopkeepers at their doors, none are without fans; and I very soon adopted the universal custom.

I halted for breakfast at a little wayside tea-shop, or restaurant : merely a roofed shed of mud, with no wall in the front, and the back partly open. A partition of mud, three feet high, divides the private residence of the family from the public part of the building ; and in one corner there is a fireplace for cooking. The main body of the room is occupied by little tables and narrow benches. In China benches are never more than three feet long and six inches wide.

Here the customers sit down, drink tea, and call for the dishes they desire; generally a little rice and chopped vegetables; or if particularly rich they may indulge in bean-curd cakes, or some of the innumerable sweatmeats always for sale, such as toffee flavoured with ginger, and hardbake made with walnut instead of almonds.

In the afternoon the road running in a westerly direction crossed the general line of drainage, and whenever it passed over one of the parallel ridges, which here ran about north and south, it ascended by a gorge to the summit, where a wall was usually carried for some distance along the crest. The road led through a strong gate, with places for a portcullis, and descended by another narrow gorge; the pass was thus made into a strong position that would be very difficult to force. In many places people were repairing these walls, though against what enemy it would be difficult to say.

My servant, Liu-Liu, who, like all new brooms, was very useful for the first day or two, managed to keep up with me, a success my other servants never achieved. He seemed, however, to share with the British tourist the delusion that the louder an unknown language is spoken the more intelligible it is. At times, when he thought I ought to understand him, the drums of my ears ran risks that long afterwards I trembled to think of. Directly we arrived at an inn, he used to send the coolie with the gigantic kettle of 'hot water,' for which it was customary to call, the tea always being understood. The Chinese always drink boiling tea if they can get it, and in a very short time I became so habituated to the hot drinks that

I preferred them to anything cold. In these paddy-field countries unboiled water is very dangerous, and it is therefore an enormous advantage to be able to get tea; for however poor the leaves may be, it is certain to be made with absolutely boiling water, and so all danger from drinking it is avoided.

April 28.—Our route kept rather to the south of west, and skirted another range of hills, running north and south, which ends near here, the road turning its flank. The landscape was exceedingly pretty; for the valley between the two ranges was not flat, but undulating, with charming clumps of trees and a great deal of wood on the hill-sides. The cultivation was still nearly all rice, but there was much more opium than I had noticed before. In the poppy fields the petals had now all fallen from the flowers, and the people were scraping a thick, black, viscous fluid from the outside of the seed-pods with knives.

There were great numbers of strawberries by the road-side. Like wood strawberries in appearance, they were very red, and looked delicious; but they tasted like grass, and were said to be poisonous. They have a yellow flower.

We halted for breakfast at Yung-Ch'uan-Hsien, a good-sized town, where the politeness of the inhabitants overcame their curiosity. The arrival of a European in European costume was an almost unheard-of event, and the process of writing without ink (for I always used a manifold writer) was one that invariably caused the deepest interest; yet in this town, at eleven o'clock in the morning, a very busy time of the day, I sat writing in an inn in the main street with my doors wide open, and not more than two or three people came to look at me, and these stood at a respectful distance.

After leaving this town we struck a river tumbling over some rocks in a little cascade, below which was a delicious-looking pool. We followed up a pretty stream for some distance to a town, where it was crossed by one of the roofed bridges, so common in China. The careful way in which everything is roofed here must strike the eye of any traveller: houses, gateways, bridges, triumphal arches, and, indeed, almost wherever it is practicable to put a roof, there one is sure to be; even the walls are often coped with glazed tiles, so that the timber-work, being built in the most solid manner, and carefully protected from the weather by an efficient covering, lasts an incredible time, even in a country where rains and snow are regular in their occurrence.

I was just preparing to enter the town in state, when I recognised one of our French clerical friends in a chair coming towards me; but the unexpected apparition of a foreigner, in a pair of knickerbockers, looking exceedingly hot, dusty, and untidy, so startled him, that for a moment he did not know me. He had been into the country to pay a visit to some Christians, and was now returning to Ch'ung-Ch'ing.

In the evening I stopped at a quaint inn at the village of T'ai-P'ing-Chên. The only room was a huge barn-like structure, with a loft over it; the ceiling had at one time been papered, but most of the paper was torn off, and hung down in filthy festoons. Under the floor there was a grinding machine and a store of grain. After I was safely established, somebody lit a fire in this place below me; and as the only escape for the smoke was through big holes in the floor of my apartment, I should soon have been stifled by the pungent fumes, had not the people politely put out their fire at my request.

In the night there was a magnificent thunderstorm. I discovered a window that would open, or rather a huge aperture closed by folding shutters; by standing on the bed I could just see out of it, and there was rather a pretty view. As a rule in China it is rare to find either windows or doors placed with any regard to scenery. The Chinese seem to take no pleasure whatever in a view, and their rooms always look out into filthy streets, or close courtyards, often roofed in. There is much to be said in a utilitarian point of view for this taste, because it keeps the rooms cool in summer and warm in winter; but it certainly grates against the feelings to find, in the midst of pretty and sometimes magnificent scenery, all the houses so constructed that nothing can ever be looked at except dirty walls.

There was a hillock with a solitary tree on it about four hundred yards distant, and as I stood watching the storm, I expected every moment that the tree would be struck, for the electric discharges were not more than one thousand feet distant. But the rain that descended like a deluge must have saved it, for it was still standing when I looked out the next morning. The forked lightning that came straight down all round was so continuous that I could almost read by the light; when I went to bed, my room was shaking with the continued crashes of thunder; I awoke in the middle of the night, and it was nearly as violent; and when I rose in the morning, it was still growling sullenly in the distance.

April 29.—The rain in most countries would have increased the difficulties of travel; but China is the land of contrasts; here the roads are all paved, and as in the plains of this fortunate province they are kept in excellent repair, the rain, instead of making mud, serves to cool the stones, which in the blazing sun become very hot for the feet of the coolies.

The rain had quite left off when we started; and the freshness of the morning was delicious. The watercourses were all brimming; streams ran over the road; the paddy fields were flooded; and the water was rushing in torrents from the upper of these into the lower. The road ran through an undulating, well-wooded country; but the landscape at this time of the year is always more or less spoilt by the paddy fields, which give the appearance of an inundation of muddy water, without any of the picturesque spots in a true flood. The perfect regularity of the fields and terraces is so monotonous that, however much they may be admired in a material sense, they can hardly be considered pleasing from an artistic point of view.

As the inns in this country are constructed for the accommodation of foot passengers and coolies, they are of a different character from those in the north, where nearly all the transport is by mules, and where, in consequence, the inns are arranged as much for the convenience of animals as for the comfort of men. Turning suddenly and unexpectedly out of the main thoroughfare of the town, if the traveller should be in a chair, a wide opened gateway through which he is carried is unobserved, and passing through what appears to be a short and narrow street, he is deposited at the door of the principal room of the inn. The narrow street is in reality the courtyard; and the low barn-like buildings on each side are the less important rooms of the hotel, where fat Chinamen may be seen through the open doors and windows, sleeping, eating, or smoking, and sometimes, though very rarely, reading or writing. The large state-room at the end is seldom occupied; for rather than expend the few extra cash for additional comfort, a Chinaman, unless he is tolerably rich, or an important official, prefers to live in the little pigsties at the side, and save his money. This is another feature in the Chinese character, and by no means a reprehensible one. Even to a man fairly off, the sum of one cash is a consideration, and to save it he would prefer an inferior article, or take a most astonishing amount of trouble. Amongst Chinese, three-quarters of the conversation is about cash; how Liu has saved one, or how Yang has been foolish enough to spend fifty on something that he could have bought nearly as good for forty-nine.

Every day I was more impressed with the gentleness of the people. After having been accustomed to find myself universally regarded as a fair and legitimate object of ridicule and wonder, it seemed quite strange to be able to come in of an afternoon and sit down to write for a couple of hours quietly before dinner. In many places in this part of the country I was left as much alone as I should have been in England; certainly much more so than a Chinaman, in his long coat, long plait, queer shoes, and huge spectacles, would be in any English market town.

At Jung-Ch'ang-Hsien I had the misfortune to rest at a hotel where the number of travellers was so great that the place appeared to be a small city.

The travelling Chinaman, when he has arrived at his destination, usually divides his time between eating, drinking, smoking, and sleeping; he seldom enjoys any excitement, not from lack of power so much as from want of opportunity. The afternoons and evenings of these people must be appalling in their monotony, and melancholy illustrations of the truth of Talleyrand's prophetic warning to the young man who did not play cards.

It would have been unreasonable to have expected the three or four dozen inhabitants of this inn to miss so rare an occasion of amusing themselves, and such an expectation would have been completely falsified.

My door was soon blocked up, and the little of the somewhat unsavoury air that had previously entered entirely excluded. The crowd remained gazing at me all the evening, and when I went to bed it had not altogether dispersed.

April 30.—Jung-Ch'ang-Hsien is on the left bank of a river ninety yards wide, which we crossed by a bridge just above a rapid. There were a good many boats, doubtless on their way to Ch'ung-Ch'ing. We followed down the right bank of the river for about a mile and a half, and then struck west, through an undulating country, towards a range of hills running north and south, the sides of which were more wooded than cultivated. A town beyond seemed to do a thriving trade in red terra-cotta goods. Tea-pots, snuff-bottles, pipes, and all sorts of odds and ends were exposed in almost every other shop. The streets were not crowded, and the people paid very little attention to me as I passed through.

A little further on I turned into a tea-shop for breakfast, and sat down at a table on which there were about half a dozen cups, with a pinch of tea at the bottom of each. The boiling water was poured into one of them immediately, and the refreshing draught was ready as soon as it was cool enough for my pampered throat. One of the waiters went round the room every five minutes or so, and filled up the cups with boiling water from a huge kettle. While my breakfast was being cooked I looked round. The tea-house, open along the whole length of its front, faced the road, but a wooden wall, coming down from the roof to within seven feet of the ground, kept out the heat and glare; and a thick straw matting, projecting from the top of the open part, cast a grateful shade across the road, tempting the voyagers to stop and have a dish of tea. As I sat facing the front, a short, benevolent-looking old man, with a grey beard and mustachios, stood behind a counter at my left. At his back there were a number of small square drawers, and above these some porcelain jars and bottles contained the various ingredients for preparing his savoury dishes. Some big wooden tubs for rice or grain were at his side, and a little child holding his hand joined in the gaze of wonder that some coolies, leaning against the front of the counter, bestowed on me, as over their trivial pipes they discussed my remarkable appearance. The cooking-place was on my right, with a smoke stack passing out through a hole in the roof. The centre of the room was occupied by small square tables, and there all my coolies were having their breakfast, and enjoying the unwonted treat of plenty of time to eat it in. That they found this a luxury I could guess from the way in which some of them dallied with their beans before commencing serious operations on the rice, instead of shovelling the latter down in the fashion of the boat coolies on our old junk. The people here seemed very fond of broad beans roasted. I watched several of the coolies commence their meal with a dish of these; one man in particular took them up one by one with his chopsticks, and chose them carefully from his little dish with the air of a gourmet, who feels that, having plenty of leisure, it will never do to

THE COOLIES AT THEIR MEAL.

throw away the opportunity of playing the epicure. Directly on my right, and near one corner of the room, a huge tub, kept warm by steam, contained the rice (boiled in some other place), and while I was looking on, a coolie came in with a fresh tub, taking away the other, which had just been finished. An attendant dips a large wooden ladle into the steaming tub, and takes out the rice; with an artistic turn of the wrist he puts it into a bowl, about as large as a small slop basin, and, giving it a dexterous pat, the clean white grains are piled up in a smooth and regular dome above the edge of the cup. This tub of rice gives plenty of work to the attendant. Another coolie demands a second portion. In an instant the waiter fills a bowl, walks quickly to the customer, and transfers the contents to the other cup without dropping a grain. The scene is full of life: the busy attendants with their bowls of rice, or pots of boiling water; others cooking, and more taking away the bowls and dishes that have been used. All the time coolies on their journey pass in front to and fro at the quick, half-walk half-run, sort of gait they adopt. Now a big chair, with red outside, and an official hat fastened behind, followed by a man with a red umbrella, proclaims an official of some importance; but the drawn blinds prevent my seeing what he is like. Now a very small and shabby two-coolie chair comes along, with a fat Chinaman half asleep and stupid with several hours of this unpleasant motion. Perhaps the coolies stop here for their food; but the sallow Chinaman sits stolidly without moving until they have finished. Most of the people at this time of day pass the tea-house, but some turn in for a little refreshment; and others, walking straight to where a tub of cold water is standing, rince out their mouths, and proceed on their journey. As a counter-foil to all this busy activity, across a field, I can see, about a quarter of a mile away, a clump of bamboos lazily waving their tops to the gentle breeze, and sheltering a house the roof of which just appears above a hedge of pomegranates and brambles. This is backed up by a fine clump of firs and willows, standing in bold relief against the liquid blue of a range of hills in the extreme distance.

About sixteen and a half miles from Jung-Ch'ang-Hsien we passed through another town; and at twenty miles, on gaining the summit of a low ridge, I could see the pagoda, which I felt sure dominated Lung-Ch'ang-Hsien, our destination. This is a seven-storied pagoda, with a good-sized tree growing on the top of it, which, at this distance, looked like the steering sails of an English windmill. Another three-quarters of a mile brought us to a town on a good stream twenty yards wide; this was crossed by a bridge of large stone piles, the tops of which were carved into the heads of gigantic dragons, gryphons, and other monsters. This must be the dragon bridge mentioned by Cooper.

Before reaching the city, we were made aware of its vicinity by several triumphal arches erected across the road. These triumphal arches are not only witnesses to the artistic feeling of the people of Ssŭ-Ch'uan, but they are, even in the excellence of their sculpture, characteristic of the realism so conspicuous a feature in the Chinese, and one that makes them so eminently a matter-of-fact people.

Neither in their buildings nor in their pictures is anything left to the imagination of the spectator, and the artists themselves seem devoid of this quality.

These triumphal arches that are so frequent

in Ssŭ-Ch'uan are generally of stone, and on the superstructure at the top are elaborate carvings in relief; these are most artistic in their execution, and represent officials administering justice, and various other scenes of domestic and public life, in which the expressions of the faces are caught with a wonderfully sympathetic spirit, and delineated with a masterly hand.

Yet in everything there is the Chinaman's want of ideality; his carvings represent nothing but what he has absolutely seen over and over again with his own eyes; he is quite incapable of forming an idea of anything beyond. His pictures are the same: insects of life size, magpies on willow trees, bridges, ponds, and hills, all realized, but with not enough imagination in the whole to produce even perspective. Even in the representation of Hell we saw the other day there was no imagination. The demons were people such as themselves, with painted faces; and the tortures such as might be inflicted by their own officials. Of heaven they have no idea, and that they never try to conceive. Everything they do is material and realistic, and imagination does not exist in their nature. From imagination springs the power of inception, or, in other words, originality; and, as might be expected, or rather as must follow by a natural sequence, the Chinese are remarkable for their want of originality. In the course of ages, as the necessity arises, as population increases, and life becomes more difficult, the law of the survival of the fittest may come into play, and the reign of intellect begin. But at present, with the want in the national character of the power of inception, they must be for a long time to come dependent on the aid of foreigners.

I came from Marseilles to Hong-Kong with two

Chinese who had been to Europe to learn European naval tactics, European ship-building, and European navigation. They were returning to their country no doubt highly instructed and much benefited; but one of them, by the permission of the captain, who wondered greatly, copied the log of the ship carefully every day. He was under the impression that if he should ever take a ship from Marseilles to Hong-Kong he would be able to do it by carefully sailing the same course.

I had a long ride through the main street of Lung-Ch'ang, and on arrival outside an inn found there was no room; but although I waited here some time, until the coolie, who was sent to explore, could return, the people of the town passed me without even troubling themselves to turn their heads. The next inn was but a few paces distant, and passing up the principal court I turned into a side yard that led to the room where I was lodged.

This was a remarkable contrast to my last night's lodging. In front there was a large courtyard, shut in by high walls. My room led into a smaller one, that opened out behind into another very narrow court, sheltered by a vine trailing over a bamboo trellis. My room was at least twenty feet high, and I appreciated the delights of quiet after the disagreeables of the previous evening. Ssŭ-Ch'uan has always been celebrated for the comfort, cleanliness, and size of its inns, which are generally far superior to those of other provinces, but this was, even in Ssŭ-Ch'uan, exceptional, and during all my travels I certainly did not find half a dozen others that could vie with it in any way.

The head man of the coolies came in, and we had a long discussion. He wanted me to go to the fire-

CH. VII.   ORANGE-GROVES AND POPPY-FIELDS.   293

wells at Tzŭ-Liu-Ching with only a few coolies, and let him go on with the rest by a different and shorter road. He said that we could effect a junction in four days, and that the road to the fire-wells was so bad that no coolies could travel fast on it; but I remarked, 'Twenty coolies can go as fast as one.' Then he said the inns were small, and there would be no room for so many.

He was, of course, only inventing excuses, and when I insisted on all going with me he made no further remark.

It is always customary to pay the coolies a portion of their wages every two or three days; and here it was necessary to go through one of the fearful cash operations. I gave Chin-Tai some lumps of silver, which, as usual, the money-changer found of a less weight than I made them myself.

May 1.—Leaving Lung-Ch'ang Hsien we crossed the river, and turning aside from the main road to Ch'êng-Tu struck into a bypath. But although not so wide, and in a few places out of repair, it was excellent for travelling on; winding about amongst low hills of sandstone, where there were many clumps of firs, and groves of orange-trees, that coming into blossom made the air fragrant with their perfume. There was not very much land under opium cultivation; wherever the fields had been devoted to that crop I noticed that the seed-pods had already been cut off the heads of the poppies. These seed-pods are threshed out with a wooden flail; the seed is winnowed, and the husks put in the sun to dry. I saw a little Indian corn also about a foot high.

We passed through one town after about eight and three-quarter miles, and went on to Wang-Chia-Ch'ang, a village fifteen miles from Lung-Ch'ang.

This was a wretched place, and my room would hardly have been called a good cattle-shed. The people here were in the middle of a fast, and for three days would eat neither eggs, fish, flesh, nor fowl; nor would they allow Chin-Tai to cook any of these in their cooking apparatus, or at their fires.

The sight, however, of others sinning only increased their self-glorification; and far from being scandalised, when I openly devoted myself to eggs and beefsteaks, they were highly gratified at being able to reflect on their own superior piety. In fact the pleasure was so great that no one in the place could deprive himself of it; as usual the door burst from the pressure, and all the people and children tumbled in on the top of one another. The scuffle roused my tired dog, who began to bark, at which the people, who had never before heard such a fearful noise, scattered and fled.

I marched in the afternoon to Niu-Fu-Tu.

The night was exceedingly sultry, and not a breath of air seemed able to penetrate even to the streets, much less to the recesses of the close and stuffy inn; but early in the night the distant growling of the thunder, and the frequent flashes of lightning, presaged another storm; and as I lay down under a sheet I exclaimed with the pious Wang-Erh, 'Oh, Mane Buddha, I knock my head!' and with good reason, for the tempest broke, the rains descended, and great was the fall of the thermometer.

May 2.—It was not with much regret that I left the gloomy and cavernous inn, where at three o'clock on a sunshiny afternoon I had been forced to light candles for my writing.

Crossing the river, here one hundred and fifty yards broad, we found the farmers transplanting the paddy into the rice-fields.

I breakfasted at a tea-house open at the back, where, in a sort of courtyard behind, there was a place roofed in, and raised like the stage of a play-house. This was very satisfactory, and I was able to eat, drink, read, or write with as much freedom as an actor in a theatre; it was pleasanter also for the spectators, who were able to see without treading on each other to any very great extent.

In the afternoon I marched to a place called Hsien-Tang, on the left bank of a stream one hundred and fifty yards wide. The boatmen here were very willing to impart their geographical information, but it was not worth much; all they knew for certain was that this stream was the same as the Niu-Fu-Tu river; but as, marching west, we had left that river flowing from north to south, and we were now on the left or eastern bank of another stream flowing similarly from north to south, I preferred my reasoning to their local knowledge. We here hired boats; and chairs, pony, coolies, and servants were safely put on board, just as a thunderstorm broke right over us; the thunder came in sharp cracks almost simultaneously with the lightning, and a drenching rain that fell made us thankful that there was a canopy over the boat that kept us all dry. The storm passed as suddenly as it had arisen. We ascended the river some little distance, disembarked, rode across a neck of land, and in fresh boats went up the same river to within an hour's ride of Tzŭ-Liu-Ching.

Approaching this town the number of tall scaffoldings around it at once attract the notice of a traveller: some right on top of the hills, others on the sides, and a few close down to the river. At a distance they look just like the tall chimneys of some manufacturing town in England. The town is prettily situated on the river, which is about one hundred yards wide, and is here bunded back; its banks are steep, and run straight up to little hills about two hundred or three hundred feet high, where, as the cultivation is not very close, there is a great deal of fresh green grass.

The inhabitants of this place have the reputation of being very rude, but I nowhere in China found more civil people.

The town is a wretched place, and its people bear all the indications of their miserable poverty. I had what seemed an interminable ride through narrow and more than usually dirty streets, all of them staircases of the steepest and worst description. The shops were very inferior, and the only novelty I remarked was a Chinaman sitting in an easy chair.

As a rule, a Chinaman sits in the usual high, stiff, straightbacked chair, so painfully familiar to any European who has penetrated into these regions. I never before saw anything like a lounge, but here there were low chairs with sloping backs, and a semicircular projection to fit into the neck, very like the cane chairs so much in use by Europeans in Singapore or Ceylon. Amongst the Chinese none but very old men use them, and a youth would be guilty of the most gross disrespect who should seat himself on an easy chair, or even loll about on an uneasy one.

My sedan-chair was put down for some minutes in the middle of the main street; a few woe-begone-looking people and children with pinched faces came to look, but seemed to take but little interest; and when we moved off and turned into a by-lane, not a dozen people thought it worth while to follow me to the inn.

This was really a fine building, with three courts separated from one another by strong gates. I had a capital room, opening on to a yard where there were a few flowers. The surrounding rooms were occupied by respectable well-to-do people; and the quiet of the place was most delightful after the noise and hubbub that there is usually in the courtyard of an inn, even when a crowd of men and boys are not fighting for a look at the foreigner. In most inns in this part of China the front court is more or less of a restaurant; people are continually coming and going, coolies shouting, customers quarrelling with the landlord about a cash, itinerant vendors of patties and cakes shouting out their wares, all at the top of their voices; while here there was nothing but the croak of the bull frog, and the distant bark of some unquiet dog, varied by the low hum of conversation in an adjoining room.

I found the dogs about here more savage than the ordinary Chinese cur, who usually beats a speedy retreat at the motion of picking up a stone. But there was a sense of independence and a democratic spirit about the dogs of this neighbourhood. They had no respect for anything, not even for good blood; and the life of poor Tib, whose valour was not equal to his breeding, was made very burdensome to him.

The landlord of this inn was a Christian, or, as Chin-Tai put it, 'he liked the French Joss.' He expressed great pleasure at seeing me, and after my dinner came to pay me a visit. Our conversation soon descended into the trivialities usual under similar circumstances. I asked him if he knew what was the annual produce of salt. 'Oh yes,' he said, 'a great deal.' 'But how many catties?' I continued. He thought that there would be a vast number. But did he not know what number? Yes, for there were a great many people always at work. 'But how many pits are there?' I said, trying another tack. He thought that there might be a thousand, but of these a large proportion were not working.

He then looked at all my things, asked what everything was for, and above all, he wanted to know the cost of each. Amongst my dressing apparatus there was a relic of European travel that could hardly be considered a *sine quâ non* in China, a railway key. He asked Chin-Tai what it was. Chin-Tai was quite equal to the occasion, and I was much interested at the readiness with which he evolved out of his own inner consciousness, a long and elaborate dissertation on the uses of an article of which, by no possibility, could he have known anything.

Eventually, when his curiosity was satisfied, I extracted from him, after much cross-examination, that salt went from here to I-Ch'ang, Ch'ung-Ch'ing, and Kwei-Yang-Fu, but not to Ch'êng-Tu-Fu. He told me that the people were wretchedly poor, and said that no foreigner had been here before except the French missionaries, who always dress, talk, and travel as Chinese. Before going away he informed me that he liked my cigars and my claret, and hinted that a small quantity of either one or the other would be a welcome gift.

May 3.—He came again in the morning to take me to see his salt wells, for he was part proprietor of a very extensive establishment.

We crossed the river by a good bridge, and after partaking of the inevitable cups of tea we proceeded to the works.

Here some of his people were engaged in boring one of the holes; this was already 2,170 feet deep, the average rate of boring being, if all went well, about two feet a day, but they said that they often broke their things, that accidents happened, and that it was thirteen years since this well had been commenced.

The jumper for boring is fastened to a bamboo-rope attached to one arm of a lever; the weight of three men who step on to the other arm raises the instrument, the men then leap nimbly off the lever on to some wooden bars fixed for the purpose, and the jumper falls.

Another workman stands at the mouth of the bore, and each time the jumper is lifted he gives a slight twist to the rope; the rope untwisting gives a rotatory motion to the jumper.

This operation is continued all day, the coolies employed showing the most extraordinary and untiring activity.

A few yards off was a finished fire-well, somewhat deeper than the one in progress; a bamboo-tube about three feet long had been put into the mouth of this boring, and some clay was plastered over the upper end to prevent the bamboo from burning. Up this well, and through the bamboo, the gas ascends from the bowels of the earth, and is lighted at the top; when the light was extinguished the odour of the gas was very powerful of sulphur, and very slight of naphtha; the latter smell was imperceptible when the gas was burning.

At no great distance was a brine-pit, which, I was informed, was two thousand and some hundreds of feet in depth, and about three inches, or perhaps a little more, in diameter at the top; immediately over the mouth was erected a scaffolding a little over a hundred feet high.

To draw the brine from this well, a bamboo-tube, a hundred feet long, open at the top and closed at the bottom by a valve, serves as a bucket. A rope, fastened to the upper end of this, passes over a pulley at the top of the scaffolding and round an enormous drum; this drum, turning on a vertical axis, was eight or nine feet high, and about twenty feet in diameter. Four buffaloes are yoked to this, and thus the rope is wound up. Near the end the rope is marked with bits of straw, like a lead-line on board ship, so that a man watching knows when it is near the end, and warns the drivers. The process of raising this bamboo once, occupied ten minutes. There is a driver to each buffalo. The bamboo being raised from the well, a coolie pushes the end over a receptacle, opens the valve with his fingers, and allows the brine to escape. When the water has been let out, the buffaloes are unyoked, and the bamboo and rope descend of themselves. This sends the drum round with a frightful velocity, which, in rotating, of course produces a violent wind. The 'break' for this is simplicity itself; a few strips of bamboo pass horizontally half round the drum, and both ends are made fast to the wall. These strips hang quite loose until a coolie, leaning against them, tautens them up, checks the pace of the drum, and stops it in a very few seconds. The brine thus raised is conducted to the evaporating-pans over the fire-wells I had already seen.

In this establishment, by no means the largest in the place, there are employed forty coolies and fifteen buffaloes, the latter in a stable kept beautifully clean (a most remarkable thing in China). They produce here 8,000 to 10,000 catties (10,000 to 13,000 lbs. avoirdupois) of salt per month; the proprietor pays no duty, but sells it for eighteen to twenty cash a catty ($\frac{1}{2}d.$ to $\frac{3}{4}d.$ per lb. avoirdupois); the purchaser then sends it away by coolies, paying duty at the

PARTICULARS OF THE BRINE-WELLS.

barriers, 300 cash (13½d.) per coolie-load, whatever that happens to be; it generally runs from about 160 to 200 catties (210 to 260 lbs. avoirdupois).

In some places they have the fire without the brine, and at a place about five miles up the river there is brine but no fire; the brine is therefore brought down from here in boats, of which I counted about one hundred lying by the bund constructed to keep a sufficiency of water in the river for these vessels.

At the top of the hill, close to the town, there is a fire-well without any brine; the principle of the pump being unknown, the method of raising the water is the clumsy and laborious one of a row of small buckets passing round two wheels, one at the bottom and the other at the top of a tower, of which there are a good many about in different directions. A blindfold mule going round and round at the top is the motive power; the water is thus raised twenty to thirty feet at a time, a trough leading from the top of one to the bottom of the next tower; in this case the brine was lifted seven stages before it finally reached the fire.

Some years ago some Chinese connected with a European firm attempted to introduce pumps; they only had their heads broken for their pains by the coolies, who declared that their labour was being taken away from them; since this no further innovations have been attempted. Baron Von Richthofen states that these wells are lined with tubes of cedarwood. I did not see any lying about, nor was I told of them, but my interpreter was nothing but a servant, and it was difficult to obtain technical information. Baron von Richthofen also states that when a portion of the rock is mashed, clear water is poured into the hole, and the turbid water raised by a bamboo tube.

The number of pits in this place must be greater than the thousand hazarded by the innkeeper. The produce of a thousand would be from fifty thousand to seventy thousand tons per annum; but as Tzŭ-Liu-Ching must supply from a third to a half of the salt manufactured in the province, and as, according to the statistics of the Ch'ung-Ch'ing banker, that amounts to 238,000 tons, the out-turn at these wells must be from 79,000 to 119,000 tons; from 1,200 to 2,300 pits would be necessary to furnish that quantity.

I found that the people of Tzŭ-Liu-Ching entirely belied their bad reputation.

I stood about the fire-wells for a couple of hours without being pressed upon in the least; and I never saw people anywhere with a more respectful demeanour.

May 4.—It rained heavily all night, and Chin-Tai, finding himself in very comfortable quarters, and treated as a person of much importance, wanted me to stop here; and held out as an inducement that the hotel-keeper would get me a bladder of gas, that I could then take it home to England, put a piece of cane in the mouth, and light it for the edification of the British public in general. Not even this was, however, sufficient to make me wait, especially as the hotel-keeper promised to send some bladders after me, a promise he fulfilled. I subsequently carried them about for a long time with vague ideas of analysis; but they grew small by degrees, and beautifully less, until they disappeared altogether; and the exact nature of the gas from the wells must for the present remain unknown.

We marched some distance up the river, which winds through green and grassy banks. Either there must be fewer people here than nearer Ch'ung-Ch'ing, for the land is not so closely cultivated, or else the inhabitants do not let their utilitarian spirit run riot to such a frightful extent, and allow something for the picturesque. This certainly makes the landscape more pleasant ; and as there is much less rice cultivation, it is really very charming. The roads, too, are less frequented ; instead of meeting coolies at almost every step, and tea-houses at every quarter of a mile, we only passed one tea-house all the morning, and it would not have been difficult to have counted the coolies.

I halted for breakfast at a tea-house in a small town, where the people were very respectful ; and although they all came to look on, none of them crowded round me. My food and method of eating caused much excitement. I think the Chinese have some reason on their side in the ridicule in which they hold our forks. They say, 'What barbarians ! to eat with a sharp prong, and run the risk of putting out their eyes, or digging a hole in their cheeks' ; and certainly it struck me this morning, when about fifty inquisitive pairs of eyes were watching my every movement, that a fork is not the most convenient implement wherewithal to eat a lightly-poached egg. Our civilisation, indeed, acknowledges this by always serving poached eggs on toast, without which the process would be almost impossible. But however much the Chinese may laugh at a fork, to our biassed minds at all events, chopsticks appear at least equally inconvenient ; and until the opinion of a Persian or a Turk can be obtained on the relative merits of the two weapons, it must remain a moot question, which is the most difficult for an uninitiated person to manipulate.

As it is with the chopsticks, so with everything else; it is almost impossible for a European to judge of the acts and thoughts of a Chinaman impartially, and without the bias of his own views and education; and on that account we should be the more careful as to the theories we propound ourselves, or the opinions we accept from others.

I stopped for the night at Wei-Yuan-Hsien, where the table of the inn was so filthy, that, habituated as I had become to dirt, I was obliged to get some clean paper and a pot of paste before I could venture to sit down to write.

May 5.—An excellent road took us over low sandstone hills, where the wheat harvest was being gathered from the fertile soil, that before autumn would yield another crop, and where the fruit already forming on the numerous Tung-oil trees heralded the approach of summer. After a pleasant march we arrived at the banks of the Tzŭ-Chou river, a clear stream, here one hundred and fifty yards wide, with very little current. On the opposite side the town of Tzŭ-Chou lay at the foot of a hill, the summit of which, some seventy or eighty feet above the stream, was crowned by a fine temple. Within the walls we could see some extensive yellow-roofed buildings.

We were on the right bank of the river, where there is an extensive suburb that stands on some little sandstone cliffs, about twenty or thirty feet high, where a number of fresh green trees droop gracefully to the water that rolls gently by. A ferry took us across to the town, which had a pleasant aspect, and was very quiet.

May 6.—There was some difficulty in finding a

place to breakfast in at the hamlet I selected for that agreeable entertainment. But at length the genius of Chin-Tai pitched upon the village school, and I was installed amongst the youth of the neighbourhood, who must have considered it an excellent diversion to see a foreigner, and a foreign dog, come and eat.

There were about a dozen little boys, from six years of age to twelve, all learning to write. They had a printed exemplar of the characters, which they placed underneath some thin paper, and traced through; their pens being, of course, reeds, and their ink what we always at home miscall Indian ink, but which is, in fact, the ink in ordinary use amongst Chinese of every class. There was a saucer on each of the three tables, and the boys rubbed it up when required. The old teacher came every now and then and patted them kindly on the head, or took hold of their pens and put them in a more correct position in their hands. There seemed no restraint; the children talked to one another, rose up and went outside, seemed to do much as they pleased, and looked very happy, as if their lessons were rather a pleasure than otherwise.

While I was waiting for my breakfast, a regiment of soldiers came by with some discordant horns and drums; they all had red jackets, with the big circle on the breast and between the shoulders, showing the district or regiment to which they belong. They were armed with spears having little square flags at the end, blue, red, and pink; their lance-poles were bamboo, painted with dark rings. They wore blue turbans, and were marching along in an irregular way, like a rabble. One or two of them had old muzzle-loading, single-barrelled muskets, and I also saw a thing that looked more like a very rusty and ancient duck-gun than anything else. They tramped along with a truculent air, pushing the coolies and people on the road out of their way. It must not be supposed that these were types of the regular Chinese army, for the men that were up in the north-west, fighting with Tunganis and Yacoob Beg, were armed with breechloaders of a very modern description, and had rifled cannon with them. However, I did not pay much attention to the soldiers, for schoolmasters, scholars and all, ran out to see the sight; and I, being entirely eclipsed by the more gorgeous show outside, was left alone; and an opportunity was thus given me to make a rough sketch of the interior of the building.

There was a large recess at the back, about twelve feet wide and two feet deep, in which there was a box with the remains of the burnt-out incense-sticks that pious people come and light here. Above this, and close to the roof, a quaint little god stood in a niche, and a few inscriptions in black on red paper aided to adorn this part of the room. There was no ceiling; a few dirty paper lanterns were suspended from the rafters, and in one corner some coffins in an unfinished condition were a cheerful addition to the ornamentation. On the other side of the schoolroom a wooden god and goddess, about half life-size, hideous creatures, painted red, yellow, and blue, were standing on a table; a few old cobwebby planks, gracefully leaning with an air of *abandon* against the wall, and a black stone tablet, with a long inscription on it, completed the decorations. A few little square tables, with the usual short, narrow benches, formed the accommodation for the scholars, which, in its simplicity, was quite in accordance with the teacher and teaching.

After reading and writing, the whole education of

the Chinese consists in the knowledge of the ancient classics, which in themselves contain many excellent doctrines, but are hardly sufficient to form the beginning, middle, and end of a man's education. Moreover, in these ancient classics, there are many exceedingly difficult and obscure passages; a certain fixed interpretation of these is prescribed by law; and woe betide the unfortunate candidate at an examination, who should venture to think for himself, suggest any new meaning, or cast additional light on that which has once been explained by the sages in a certain way, and of which in consequence any further illumination would be profane.

Can it be possible for any nation to devise a system which would more effectually crush out all germs of originality or thought from the mind of the people?

The show outside, however, passed, and the children returned to watch me once again. I completed my repast, lit a cigar, bid adieu, and gave my thanks to the kind old teacher, and proceeded on my way.

The road all day was on the top of a sandstone ridge running parallel to the river, of which a glimpse was now and then obtained. The sandstone beds are here in a horizontal position; and there are layers of yellow, green, and red; the red being very friable, and mixed with a sort of clay. This geological formation causes two peculiarities in the scenery: first, the tops of the hills are all scarped, giving the appearance of small low towers on the summits; this is caused by the sandstone falling away equally all round, as it does not do when the strata are inclined; and secondly, the terrace cultivation being in accordance with the lay of the strata, in every direction the eye can see perfectly horizontal lines of light or shade, dark strips of fallow ground alternating with bright yellow streaks of corn.

The road was excellent all day, not very much up and down, but I was surprised at the little traffic on it, and the fewness of the towns and villages. There were the usual triumphal arches at the approach to any place of importance, but none of them of so much finish as those I had seen before. We passed through only two large villages, and arrived at the inn at Nan-Ching-Yi, at five o'clock, after a march of twenty-eight and a half miles.

The room at the inn was very good and clean, with a clean straw ceiling, but unfortunately there was an open drain in the courtyard behind, with a most offensive smell. I rashly sent for a man to have the place cleared out, but he only succeeded in stirring up the filth; but with a saucer of carbolic acid under my nose I circumvented the stench. It would puzzle Mr. Edwin Chadwick to explain how it is that the Chinese can live and flourish in the fearful odours which surround them all their lives. Even if any of us have any faith left in the costly and magnificent systems of drainage that promise so much, it would certainly be shaken by a visit to the Chinese people.

May 7.—We were now fairly in what Baron von Richthofen calls the red basin of Ssŭ-Ch'uan, and a most appropriate title it is. The formation here is a layer of dark red clayey sandstone, and wherever the soil is bare the ground is of a rich dark red brown colour. The tops of the hills are nearly all on the same level, some three hundred or four hundred feet above the river; on their upper slopes there is a good deal of wood and coarse grass; and the bright green of a kind of low thorn contrasts pleasantly with the deep red of the clay. In the bottoms of the valleys, which are tolerably flat, all the ground is cultivated; but the formation does not seem well adapted for rice.

The villages and towns were very scarce, the country-houses less numerous, and the traffic on the road was not nearly so great as during the first few days after leaving Ch'ung-Ch'ing.

In this part of the country the property is generally marked by boundary-stones; I was told that the land changed hands very frequently; for, there being no law of primogeniture, when a man dies his property is divided, and some of the family soon become too poor to keep up their portion.

Whenever property is sold, the deed of sale must have the official seal, and for this, the usual charge is five per cent. But when a magistrate is leaving, he often puts up a notice to the effect that he will do the job cheap; then all the people who are thinking of buying or selling make up their minds at once, and come and get the seal, sometimes for two per cent.; and the departing official makes a nice little sum of money before giving up his office to his successor.

Disputes occur sometimes, which are settled by appeals to the public of the neighbourhood. The disputants fix on a certain market day, in the nearest market town, and invite all their friends, relatives, and anyone with local knowledge, to drink tea. Everyone being assembled, the question is discussed, and almost always settled amicably in this sort of congress.

Two miles from Nan-Ching-Yi we crossed a stream by a very handsome stone bridge, twenty feet wide, with three really elegant arches; and soon after we came again upon the river, which we followed for a little more than two miles, when we left it, striking it again four miles further on. Another mile and a quarter brought us to a gorgeous wooden triumphal arch, freshly painted in red, blue, and green; and this presaged the proximity of another town. The road now ran between low walls of mud and loose stones. The peasants behind them were gathering their crops of opium and wheat, and scarcely turned from their occupations to glance at our procession. After a march of eight and a half miles we found ourselves opposite the town of Yang-Hsien. The river was here about one hundred and fifty yards wide, with a current of two miles an hour, flowing between banks about twenty feet high; and after crossing at a ferry we passed through a quarter of a mile of suburb, and then the gate was reached; it was a nice, clean-looking place, with wider streets than usual, and with apparently not much trade; what there was seemed to be a general one in small articles, and I did not notice any speciality. About another nine miles brought us again to the river. We followed it for a couple of miles; and after a march of twenty-five miles reached the very small village of Yang-Chia-Kai.

May 8.—To my great astonishment Chung-Erh took it into his head to walk for an hour in the morning. I could not believe that it was the effect of my extraordinary example, for in China it is considered the height of eccentricity to walk when progression is possible in any other fashion. It is sometimes really piteous to see the coolies painfully ascending some steep hill, carrying a chair in which a fat Chinaman may be sitting, stolid and apathetic; and no matter how steep the hill, a Chinaman, however fat he may be, would never deign to get out for a moment to ease the unfortunate bearers, but would sit till they dropped dead; and even then I believe he would wait till a fresh lot came to take him on.[2]

Chung-Erh had asked me for an advance of wages the night before, and when I told Chin-Tai that I had given it to him, he reproached me for doing so. I asked him if he thought that Chung-Erh was likely to run away.

'No,' he answered, 'but he will lose his money.'

So it is possible that, having gambled away all his cash, in his last desperation he staked an hour's ride, and lost it.

Travelling over the same red undulating ground, the crops were much the same, with the addition of large fields of safflower, which here grows to a height of about three feet. Even at the early hour of our start the fields were full of men and women picking the heads off the flowers, which are used for making a dye; the safflower that comes from this valley being considered superior to any other in China.

The number of houses about here is not very great, and it was four and a half miles before we reached the first village. Another mile, and we debouched into a flat plain, about a mile and a half wide; and, almost immediately, a low pagoda, four or five stories high, that stands opposite Chien-Chou on the other side of the river, came into sight. This plain was thickly populated, the houses were very close together, and just where we entered it they were only separated from one another by about one hundred yards; many of them enclosed in mud walls, and sometimes two, three, or even four together shut in by a wall. Here the inhabitants used creaking wheel-

---

[2] Huc says that chair coolies have so much *amour propre* that they feel hurt if the rider gets out and walks. My experience by no means tends to corroborate the statement.

barrows, very similar to those of Shanghai, but with a much smaller wheel more to the front; in the undulating country through which we had been travelling they were not in use. The river we had been following flows through this plain, but we did not catch a glimpse of it until we had marched seven miles, when the valley narrowing, and the road ascending a little, a wider view was obtained. Soon after, mud walls on both sides surely indicated the outskirts of a large town; and, after nine miles of travelling, we arrived at a tall pagoda of eleven or twelve stories. Here there was a very large temple, and the first triumphal arch on the road. After marching through a considerable suburb, we arrived at the gates of the city of Chien-Chou, celebrated by Huc, who, in his most musical key, has sung the glories of its Kung-Kuan, or official rest-house. The walls of the city are in good repair, and it is a nice, clean-looking place, but at the time of my visit did not seem very busy. It stands on a small stream spanned by a fine roofed bridge, where a good many poor people are seated, with a few odds and ends for sale spread out on the ground before them; and where the inevitable beggar takes his stand, and prays 'your excellency' to bestow a few cash. The main street is a little more than half a mile long, and after leaving it we found very little suburb on the northern side. We now followed the river, a nice, clear stream one hundred and fifty yards wide, running between steep banks with no greater velocity than from one to two miles an hour, except at the rapids. There were a good many boats on it, trade being carried on with Ch'ung-Ch'ing.

From here we followed the banks of the river to Shih-Ch'iao, a dirty place, where the inn was so poverty-stricken that it boasted hardly any furniture

—it was the first house in China in which I had not found at least one chair—and the filthy table was not rendered more attractive by a foul oil-lamp. I converted the straw mat on the bed into a temporary table cloth, and sat down to wait for Chin-Tai, who had been left behind at Chien-Chou.

The walls of the room in which I was sitting were of lath and plaster between strong beams. The inquisitive little boys who were collected outside soon began to pick holes in this, and, had I remained here any length of time, there would hardly have been any wall left. Chin-Tai presently arrived with some capital fish and other stores. At Yang-Chia-Kai he had been able to buy nothing but eggs; but in Chien-Chou he appeared to have run riot in the provision market.

On the outskirts of the town there are some more brine wells, but not so deep or nearly so important as those at Tzŭ-Liu-Ching.

The road now left the river, and again wound about amongst low undulations. Three and a half miles from Shih-Ch'iao a range of tolerably high hills was seen to the north-west, and a mile and a half further we passed through a small village on a stream crossed by a very elegant three-arched bridge of red stone. Soon afterwards the road approached the hills, and, beginning to rise, took us to the summit of the ridge, about eight hundred feet above the level of the river.

Looking back to the south, there was a fine view over the wide expanse of undulating country we had been traversing. The gathering clouds cast varied shadows over the landscape, in which the prevailing red was modified by distance, and pleasantly contrasted with the deep green of the trees. The little village of Ch'a-Tien-Tzŭ, a few yards over the crest of the hill, was my resting-place for the night; here I found a really clean inn, with fresh whitewashed walls, an unusual absence of smells, and perfect quiet. High up above the plain it was nice and cool; and, with the pretty walks that there must be on the well-wooded hillsides, it would make a charming summer residence for anyone living at Ch'êng-Tu.

May 9.—The next morning we first crossed one or two more ridges of the range, which runs nearly north and south; and, finally, after a pull of three miles, arrived at the highest point, about seven hundred and fifty feet above Ch'a-Tien-Tzŭ. We had again struck some great highway—the road was literally covered with coolies coming and going, and the traffic seemed enormous.

The day was too hazy for a view, otherwise from the summit there would have been a very fine one over the Ch'êng-Tu plain. There was a considerable difference between the state of the crops on the higher and the lower ground; for up above a few flowers still remained on the poppies, and the people were scraping the black viscous matter off the pods. Just over the crest there was a small village, and from here to the bottom (a distance of three and a half miles, with a descent of about one thousand feet) was one succession of tea-houses. The traffic on a road may fairly be estimated by the number of tea-houses. Where there are many coolies passing, there are always numerous restaurants. Here the number was so great that it seemed almost impossible that they could all succeed.

Directly we were again in the plain, we met our old friends the squeaking wheelbarrows, and the little apparatus for raising water from one field to another was seen at almost every twenty yards. The number of these affairs was very great, owing to the size and flatness of the plain. Sometimes there were half a dozen of them close together; and at intervals, as far as could be seen, the effect of the big umbrellas dotted about the landscape was decidedly comical.

We presently came across a pole with a skull on the top, and underneath it was an inscription that informed all passers-by that the deceased had been executed for stealing silver.

At the village where we halted for breakfast, I ordered Chin-Tai to buy me some fresh paper to put over the table. But the thrifty spirit of the Chinese pervaded even my servants to such an extent that it became a positive nuisance. Thrift, like other virtues, is excellent in itself, but when carried to an excess becomes almost a vice. In this case, Chin-Tai expended every artifice that his intellect could devise, before he could consent to disgorge the few cash necessary. A feather will show the direction of the wind; and a trivial example of this kind serves to indicate the bent of the Chinese mind.

I sent Chin-Tai on to Ch'êng-Tu to look out for a comfortable place, and determined to spend an hour or two here, to give him time to make arrangements; but the house was even dirtier than usual; and, as well as the yard, was crowded with people, who all appeared to be quarrelling; whilst the noise and turmoil so far exceeded anything that I had been accustomed to, even in this land of talk, that I left it, and after half an hour's march found a grass field, about a quarter of a mile away from the road, with a delightful overhanging hedge, under which I sat for two hours, sheltered from the sun, and 'far from the madding crowd.' Here I smoked a peaceful cigar, only disturbed by the determination of the pony to roll on the ground with my saddle on his back, and the apparently equally obstinate resolution of the Ma-Fu to let him do it. I was quite at a loss to understand how a grass meadow could find a place in this rice plain. But here it was, and about three hundred yards away the foliage of a thick dark clump of trees was reflected from a paddy field that imagination might have converted into an ornamental sheet of water. To the right a wood of pines crowned the summit of a little knoll, and altogether the scene was very pleasant.

After this I went on to the capital deliberately, to give Chin-Tai plenty of time, and halted at every tea-house by the road-side, much to the delight of the chair-coolies, who dearly loved to stop every half-hour or so, and get a cup of tea, or a bowl of rice.

I was rather unfortunate in the time of my arrival at Ch'êng-Tu-Fu; for the examinations were now being held. These always bring thousands into the capital from every part of the province; and, in addition to this, the provincial governor-general was just leaving, and a new one being installed. Consequently the city was full of Fu-T'ais, Chen-T'ais, Sieh-T'ais, and T'ais of every description, not to mention the lesser lights of Fus, Chous, and Hsiens.

Every hotel was crowded, and after hunting up and down the town Chin-Tai had only been able to get a place in an exceedingly dirty inn outside the east gate.

Every official in Ch'êng-Tu appeared to have sent me a soldier; for there was a whole army waiting on me. I kept half a dozen, and sent the rest away.

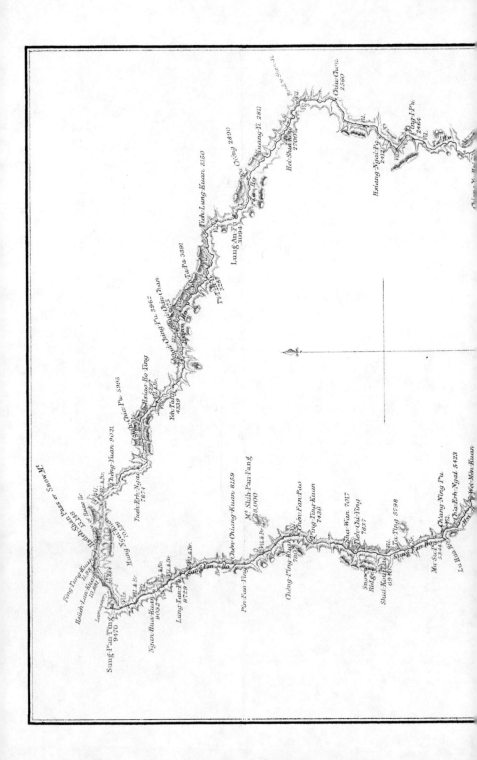

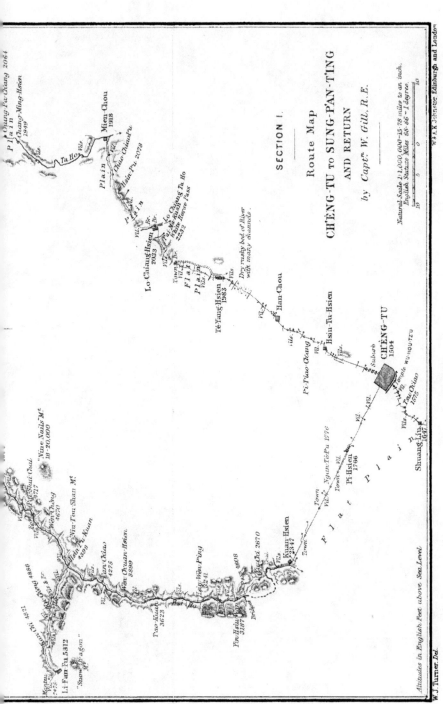

CHAPTER VIII.

A LOOP-CAST TOWARDS THE NORTHERN ALPS.

CH'ÊNG-TU TO SUNG-P'AN-T'ING.

Kindness of French Missionaries at Ch'êng-Tu—Arrangements with Mr. Mesny—Endeavour to take a House—Mystifications on the Subject—Pleasures of French Society—Proposed Excursion to the North—The Man-Tzŭ, or Barbarian Tribes—Preliminaries of Departure—Leave Ch'êng-Tu-Fu—Pi-Hsien—Engaging Official there—The Escort—Irrigated and Wooded Country—Halt at Kuan-Hsien—Scope of Excursion Extended—Frantic Curiosity of People, but no Incivility—Irrigation Works—Rope Suspension Bridge—Coal-beds—Yu-Chi Charming Inn—Yin-Hsiu-Wan, and Water-Mills—Hsin-Wên-P'ing—The 'Min River'—Macaroni-making—Wên-Ch'uan-Hsien—First Man-Tzŭ Village—Pan-Ch'iao—Traces of War—Relentless Advance of Chinese—Miraculous Sand Ridge—Hsin-Pu-Kuan—Rapid Spread of the Potato—Excursion to Li-Fan-Fu in the Man-Tzŭ Hills—Scenes that recall the Elburz—Carefully-made Hill Road—The 'Sanga' of the Himalayas—Angling—Village of Ku-Ch'êng—Peat Streams—Musk Deer—Arrival at Li-Fan-Fu—The Margary Proclamation—Tales of Local Wonders—The Traveller fain would see—The Lions of Li-Fan-Fu—Search for a Man-Tzŭ Village—Man-Tzŭ here a term of Reproach—The I-Ran Tribes and their Language—Ku-Ch'êng—Local Wonders again—Return to Hsin-Pu-Kuan—Resume Valley of Hsi-Ho (or 'Min River')—Wên-Ch'êng—The Himalayan Haul-Bridge in Use—Polite Curiosity at Ma-Chou—Grandeur of the 'Nine Nails' Mountain—Precipitous Gorges—Wei-Mên-Kuan—Difficulties of the Road—The Su-Mu, or White Barbarians—Alpine Scenery—Ta-Ting—Tieh-Chi-Ying—War with the Su-Mu—The Yak seen at last—Travellers' Disappointments—Glorious Mountain View (Mount Shih-Pan-Fang)—P'ing-Ting-Kuan—Expulsion of Man-Tzŭ—Maize Loaves—Wood Pigeons—Ngan-Hua-Kuan—Delicious Tea—Smoking in Ssŭ-Ch'uan—Country of the Si-Fan—Sung-P'an-T'ing.

May 10.—I sent Chin-Tai to the French missionaries with my card, to inquire at what time it would be convenient to them to receive me. But in the morning Monseigneur Desflêches paid me a visit. He made excuses for Monseigneur Pinchon, the bishop here, who, he said, was not very well. He welcomed me warmly to the provincial capital, and the charm of his manner and his cordial reception soon made me forget where I was, and I could almost fancy myself nearer the Arc de Triomphe than the gate of the city of Ch'êng-Tu. He promised to help me to find a better place to stop in than that I now occupied, which was simply disgusting. The walls were hung with cobwebs of the blackest description. There was a bedstead with some carving at the top, the interstices in which were nearly filled with dust and dirt; bits of string hanging from the beams had nearly lost their original character from the coating of filth that had accumulated on them, and every gust of wind brought down a shower of dirt from the roof on to my head. Under the bed I dared not look. This unwieldy piece of furniture had probably stood there for years, and according to Chinese custom, whenever the room had been swept during that time, the sweepings had been left underneath it. To clean the room would have taken at least a couple of days, and to have half cleaned it would by stirring up the accumulated abominations only have made matters worse.

I visited the missionaries in the afternoon, who received me most kindly, and treated me to a collation of wine, cakes, and sweetmeats. It was a great treat to join again in reasonable conversation, and hear the sound of a language I understood. At these entertainments the missionaries always showed themselves true Frenchmen; the ease of their manner and the sparkle of their conversation were in strange contrast with the associations of the place. The time passed quickly, and I was much astonished when I rose to take my leave to find that I had been here nearly an hour and a half.

May 11.—When at Shanghai, I had been in communication with Mr. Mesny, an officer in the service of the Chinese. He ultimately arranged to join me at Ch'êng-Tu, and subsequently travelled with me to Bhamo; and to his intimate knowledge of the language and ways of the people, I am mainly indebted for the friendly relations we always maintained with the Chinese officials. At present, he was still buried in the depths of the province of Kwei-Chou, although I was under the impression that he was well on his way to Ch'êng-Tu, and expected him every day.

Hearing nothing of him, however, I went away for a trip to Li-Fan-Fu, intending to return to the capital in ten days. But fresh circumstances arose, and eventually I extended my journey to Sung-P'an-T"ing, and Lung-An-Fu, and even then found myself back in Ch'êng-Tu before the arrival of Mr. Mesny.

For the present, I determined to take a house in Ch'êng-Tu for a month, if I could get one, for the rent of a house large enough for me was so small a sum, that it was quite worth while to take one for a few weeks, even if I had lived in it only for a couple of days.

During the day I received a visit from Monseigneur Pinchon, and afterwards sent Chin-Tai to see if he could find me a lodging inside the walls. He came back saying that there were twenty places. I was very cheerful at this unexpected plethora of accommodation, when Chin-Tai casually added, that only one of them would do, and that even that was not much of a place. I went off to see it, picking up on my way one of the missionaries' servants, who had told Chin-Tai that he knew of a house.

I looked at this one first, but it was in an hotel, and very small, so went on to the place that Chin-Tai had discovered. This was part of a private house, and would have suited me, but the missionaries' man said he knew of yet a third house, so I determined to see it before deciding. I could not go there straight, because things were not ready ; so I was taken to the shop of a Christian silk merchant, named Yeh, where I had a Chinese pipe and some tea ; Yeh exhibited all his silks, which were very much more expensive than similar ones at Peking ; he showed me the various colours that are made from the safflowers I had seen the people picking at Yang-Chia-Kai ; these varied from a light pink, through a rich orange, to a deep red. He showed me some black silk, saying that it would make a very elegant thing in coats, if I thought of adopting the costume of the country. I assured him I had no intentions of abandoning my nationality. He was evidently much distressed, and looked from the silk to my figure, and back again to his fabric, evidently comparing the two mentally, not much to the credit of Savile Row.

Presently Chin-Tai returned, and we went off to see the house, which was suitable in every way, and was offered me for fifteen taels a month, with two hundred taels premium, to be returned to me on my vacating the premises.

May 12.—I gave Chin-Tai one hundred taels in silver, and a bill for one hundred taels, and sent him off to settle the question of the house. The banker could not let him have the money for a day or two, but as the silk merchant was willing to buy the bill, this caused no difficulty.

Chin-Tai returned presently, with a sorrowful countenance, and said that the house belonged to two brothers ; that the elder had let the house without the consent of the younger ; and that the latter, who was not fond of foreigners, refused to ratify the bargain. Chin-Tai added with a sigh, that, if I had not gone to see the house myself, matters could have been arranged without the fact of my hated extraction coming to the knowledge of this inhospitable youth.

'*Tout vient a qui sait attendre*' ought to be translated in China, ' Nothing comes to people who cannot wait.'

Chin-Tai was again sent out to see what was to be done, with orders not to come back until matters were arranged. Quite late at night he returned with a favourable report; but there was still, he said, a good deal of talking to be accomplished, before any definite answer could be given.

The Tao-Tai of Ch'êng-Tu had heard of my difficulties, and presently sent me his card, with a polite message that he would get me a house if there was any further trouble; but whether this was merely an elegant formality, that he never supposed I should accept, and that meant nothing; or whether, under the impression that I was an important functionary, he thought that it was his duty to do all he could, I never exactly understood, for soon afterwards I learnt that the last house Chin-Tai had seen had been definitely taken, as I subsequently heard, by resorting to the mild subterfuge of informing the owner that I was an official from Peking.

May 13.—Even now I was not altogether without doubts, and as long as I was not fairly installed in my mansion, I could not quite believe that no unforeseen hitch would occur. My alarms were much allayed by the appearance of coolies early in the morning, who began taking away my things, and when I set off to breakfast with the missionaries, I began to think that I really might count on spending the next night under my own roof.

It was a delightful change from my own company

to that of some half-dozen lively Frenchmen. The mode of the meal, as they put it, was *moitié Chinoise, moitié Européenne*; one missionary was eating rice with chopsticks, and cracking jokes with a Chinese minister who also sat at table; another was washing down a Chinese dish with a glass of Tinto, which, contrary to usual custom, was taken in my honour. Excellent bread was on the table, for wherever a Frenchman is found there is sure to be good bread, and Chinese dishes succeeded others that might rather have come from the Boulevards than from a kitchen in Ch'êng-Tu. The meal passed very pleasantly, and afterwards I spent the greater part of the afternoon in the delights of hearing a familiar tongue.

When I took leave of the missionaries I went straight to my new house, which contained three rooms, and two dressing rooms, besides servants' quarters; it had a court in front, and was very clean for a Chinese house. It was supplied with three tables, half a dozen chairs, and two bedsteads, about the amount of fittings that make the difference in China between a furnished and an unfurnished house. I was now overwhelmed with a perfect inundation of what Chin-Tai was pleased to style policemen. Tinc-Chais they should be called, and were servants of the different officials in the city. All the magistrates in the place, the Fus and the Hsiens, seemed to send whole armies of their men, ostensibly as a compliment to me, and to take care of me; but in reality to keep a good eye on my movements, and still more to get some cash, whole mountains of which disappeared.

May 17.—I had made up my mind to visit a place called Li-Fan-Fu, which, from the accounts of the missionaries, was worth a visit. Amongst other things there was said to be an intermittent spring.

CH. VIII. PROPOSED EXCURSION TO THE NORTH. 323

I was told that this place was inhabited by the Man Tzu, or Barbarians, as the Chinese call them; and Monseigneur Pinchon told me that, amongst other pleasing theories, they were possessed of the belief that if they poisoned a rich man, his wealth would accrue to the poisoner; that, therefore, the hospitable custom prevailed amongst them of administering poison to rich or noble guests; that this poison took no effect for some time, but that in the course of two or three months it produced a disease akin to dysentery, ending in certain death.

Monseigneur Pinchon advised me to take my food from Ch'êng-Tu, and to avoid the temptations of feasting as a guest of this singular people. This superstition is almost an exact parallel to one related by Polo as in vogue amongst a tribe in Western Yün-Nan, *vide* Yule's 'Marco Polo,' 2nd ed. vol. ii. p. 64. It may be doubted, however, whether much more of the custom remains than the tradition.

There are altogether eighteen of the Barbarian tribes spreading over the country from Yün-Nan to the extreme north of Ssu-Ch'uan. Each tribe has its king—one of them a queen—and they live almost entirely by agriculture and cattle-keeping. The king usually derives a considerable revenue from his lands, and every family in his kingdom has to send one man for six months to work on his estate. In other cases he receives an annual amount of eggs, flour, or wheat from each household. He has absolute power over all his land, assigns certain portions of it to certain families, and, if they displease him, or he has any other reason for doing so, he displaces them at once, and puts others in their stead, all the houses and farm-buildings passing to the new comer.

One of these royalties, that of Mou-Pin, was at this time distracted by disturbances, a civil war, bandits, robbers, soldiers, and evils of every kind. The king had died not long previously, leaving a wife with three daughters, and a sister-in-law, who set herself up as the protector of an illegitimate infant son. There was at once a disputed succession, for by the law a female could not sit on the throne. The sister-in-law and the wife both wanted the ruling power. The sister-in-law succeeded in stealing the seal of State. She obtained some boy, who was permitted to go and pay his respects to the widow as sovereign, and who, while making his obeisance, managed to snatch the seal and escape to the sister-in-law.

A war then broke out, some people taking part with the queen widow, and others with the sister-in-law. As usual in such cases all the bad characters flocked to the place to feed on the booty; both the queen widow and the sister-in-law were obliged to take refuge in Ch'êng-Tu, and now the whole kingdom was given over to pillage, and the villanies always accompanying a civil war.

I sent Chin-Tai to the Tao-Tai, to inform him of my intended tour. He assured me that there was no such place as Li-Fan-Fu; but that there was a Chinese military station named Sung-P'an-T'ing, and that that must be the place that I wished to visit. He said he would send four Tinc-Chais with me, as it would not be proper for me to travel with less.

The banker came in in the evening, and brought me 150 taels of silver, chopped up by some neighbouring blacksmith into little pieces suitable for the small payments.

Before leaving Ch'êng-Tu, as it would be necessary for me to have intercourse with the officials on the way, Chin-Tai was ordered to buy a card-case for my Chinese visiting-cards, an affair of about the size and appearance of a small portfolio. Cards are sent, some hours before making a call in China, by a servant, to inform the people that they may expect a visit at such and such an hour, and to prepare them for the momentous event. People also, as in Europe, send their cards as a civility one to another without visiting.

The market prices at Ch'êng-Tu were much the same as at Ch'ung-Ch'ing. Beef sixty cash a catty; eggs six cash each; fowls ninety to one hundred cash a catty; wheaten flour forty, and fish fifty to eighty, for, as in Polo's time, still 'they catch a great quantity of fish.'

Twenty cash was equivalent to a little less than a penny, and a catty is equal to a pound and a third avoirdupois.

May 18.—As I thought I should only be away for a few days I took no more than eight baggage coolies, beside the chair coolies, and another pony. These were engaged only for the journey to Kuan-Hsien, for beyond that place the country is so mountainous that coolies accustomed to the plains will not or are not able to work in it.

On turning out in the morning I found a good many of my old coolies, and the same coolie-master. We sortied through the north-west gate of the city, near which there is a good deal of open space, and many gardens with nice trees, willow chiefly, and, of course, bamboos.

On leaving the gate a suburb extends about two thirds of a mile beyond, but it has no depth, and is little more than a line of houses on both sides of the road, behind which gardens and fields can be seen.

The yellow corn was waving in the breeze, the harvest was in full progress, and the rice-planting was still going on. There was less rice on this side of the city, more wheat and tobacco, and no opium whatever that I could see. The red clay of the soil had entirely disappeared, and in its place there was a kind of grey clayey sand, the city of Ch'êng-Tu having apparently been built on the extreme edge of the red clay.

There was not nearly so much water on the land as there was on the south of the city, although at each side of the road, which was not paved, there was a considerable stream; and I now saw no more of the pumping mills that had become so familiar.

There were more trees of all kinds; long rows were planted on the divisions between the fields and on the sides of the road, and the appearance of the country put me somewhat in mind of the neighbourhood of Peking.

I proceeded in great state with my four satellites, who shouted to everyone they met to get out of the way. Perhaps a poor man would come staggering along with an enormous load on a wheelbarrow, just where the track for these machines was very narrow, but where there was plenty of room for me at the side. Nothing, however, would satisfy my gentlemen, unless he cleared right out of the course; and once when one of these unfortunates was not quick enough, they upset the wheelbarrow into the brook at the side of the road. I remonstrated with them, but it had no effect whatever, as they had made up their minds to maintain their own dignity, however little I might care about mine.

Whenever I got on and off my pony, as much fuss was made about me as about a jockey mounting for the Derby : one man to each stirrup, another to the pony's head, a fourth to his tail, and the Ma-Fu to give me a lift, as if the animal was about eighteen instead of eleven hands high.

The road was very lively with many people carrying cocoons of silk, and many travellers riding, the latter with huge stirrups made of wood, into which they often thrust their heels instead of their toes.

I halted for the night at Pi-Hsien, where there was a large inn, which appeared to be a kind of barrack, for it was full of soldiers.

When first I came in I thought it a delightful place ; the room was open in front to a good-sized yard, beyond which was a covered square, with great gates opening into a further court, and I imagined that by shutting the gates I should be delightfully cool and quiet.

Four little rooms were entered directly from the large room, and in each of these there were about six soldiers, who behaved themselves very well.

It was not long, however, before it became necessary to open the gates, for the people of the town would very soon have burst them down. On they came, that curious crowd : first one barrier was passed, then another ; little by little the jabbering mob approached the door; soon they were in the room, and like a flood threatened to carry me away right through the opposite wall. At length, sorry as I was to disappoint their legitimate curiosity, it became necessary to turn the soldiers out of one of the small rooms, into which I retreated.

It was a filthy place, and none the cleaner for having been occupied by the braves, who seemed chiefly to have amused themselves by spitting about all over it ; and as I sat imprisoned, as if to mock me, a huge label that still stuck to my writing box, 'Grand Hôtel de Thoune a Thoune,' stared me in the face, and I could hardly help yearning, if not for the flesh-pots, at least for some of the comforts of civilisation.

The Hsien was at Ch'êng-Tu assisting to inaugurate the new governor-general, but his son called on me, and afterwards sent me a present of a couple of fowls, for which I had of course to give as much money to the man who brought them as would have paid for these muscular birds over and over again.

Soon afterwards I learnt that the Hsien himself had posted back from the capital on purpose to look after me; and he called on me after dinner, a man with a frank open face, very unlike most Chinese. He offered me everything he possessed, hoped I would stop a day, and he would take me about, and show me what there was to see, and said that if I came back this way he would put me up in his yamen. I thanked him, but said I could not stop just now, but hoped to see him again on my return.

Any official would of course go through the form of offering as much as this, but his face gave such a charm to all he said that I really think he meant it.

Before leaving he asked me all about my proposed expedition, and like the Tao-Tai of Ch'êng-Tu, assured me that in speaking of Li-Fan-Fu I must mean Sung-P'an-T'ing; so certain was he of this, that he took the trouble to write out an itinerary for me of the road to Sung-P'an-T'ing and Lung-An-Fu.

After he had gone, I turned into bed, and, notwithstanding the dirt that fell on my head from the matting above, where a healthy family of rats were steeple-chasing all night, I slept soundly.

May 19.—The Hsien gave me additional presents

of candied fruits, which were really excellent, and he insisted on sending me an escort of twenty soldiers. After some remonstrance I succeeded in reducing to ten the number of these useless but exceedingly picturesque braves.

Over the ordinary dress they wore a loose red tunic without sleeves; four of them were armed with spears terminating in an arrangement like Neptune's trident; and four others with weapons ending in short square swords. The heads of all the poles were adorned with large rosettes of blue and red with ends hanging down. The other two men bore flags, one in front and one behind.

The Hsien also sent his steward, a functionary of much importance. This man rode a pony, and gave me a good deal of assistance, praise that I can hardly lavish on the remainder of the procession, who were about as useful as the men in armour in a lord mayor's show.

When the soldiers left me they formed line to the left, and gave what I took to be a general salute; this they performed in divers and sundry manners, all laughing heartily.

This glorification of course was not achieved without the expenditure of a considerable sum of money; but as everybody seemed to enjoy the thing so thoroughly, it would, I felt, have been cruel to grudge them the pleasure.

We marched over the same beautiful rich fertile plain; and after about an hour the mountains appeared through the haze.

The whole country is a perfect network of canals and watercourses; and as the plain here begins rising rapidly (at least ten feet per mile), the streams are all very swift. The number of trees everywhere is enormous; the sides of the road are bordered with a small kind of beech, and also willows; there are often rows of trees between the fields, and clusters round the houses. Here is a line of fruit trees, oranges or apricots; there a temple enclosed by a wall with a number of fine yews; and in every direction the view is bounded by trees.

The beeches are used only for firewood, and for the manufacture of charcoal, which, as well as coke, is made in great quantities at Kuan-Hsien; and vast numbers of coolies are seen on the road carrying these in the usual way, or wheeling them in barrows.

There was no lack of tea-houses by the roadside, and I breakfasted in one close to the river, which, here sixty yards wide, and running swiftly over a pebbly bottom, looked a glorious place for throwing a fly.

A little higher up it was crossed by a neat trestle bridge in nine spans. The framework for the usual roof had just been put up over the roadway, and people were at work completing it.

At about twenty miles from Pi-Hsien the road passes through the heavy gates of a massive gateway, on which is built one of the three-storied buildings invariably erected over the gateways of the walled cities. There are no walls on either side, the gateway standing by itself in useless and solitary grandeur. I was unable to learn anything of the history of this building; but it certainly would seem as if in former times the walls of Kuan-Hsien extended to this point.

The road from here to Kuan-Hsien passes through a suburb, and underneath six very elegant triumphal arches, elaborately ornamented with carvings in relief, the work of the numerous stonemasons that are seen in this suburb engaged in chipping away at a soft grey sandstone.

The Hsien of Kuan-Hsien, as soon as he heard of my arrival, sent me more fowls, and sweetmeats in such quantities that it would have puzzled even a Russian to dispose of them; and to increase my dignity he lent me a number of red cushions from his yamen.

May 20.—I was obliged to wait a day at Kuan-Hsien as fresh coolies were to be hired, and reflecting on what the Hsien at Pi-Hsien had said, I began to think it would be worth while to extend my trip to Sung-P'an-T'ing. The trip was sure to be an interesting one; no European, not even the missionaries, had ever been to Sung-P'an-T'ing, and it was almost on the borders of the Koko-Nor district. I had expected nothing but opposition from the Chinese officials, but instead they were actually putting opportunities in my way.

I did not take long to make up my mind; and the only obstacle remaining was the want of money. I had left Ch'êng-Tu with a supply ample for the short journey to Li-Fan-Fu, but not enough to take me to Sung-P'an-T'ing. I therefore first sent to the Hsien, to ask him if he could and would buy one of my one hundred tael bills. He returned me a polite message to say that he had not that amount by him, and could not accommodate me.

But I had now made up my mind to go to Sung-P'an-T'ing; so I determined to send Chung-Erh back to Ch'êng-Tu to get the money; arranging for him to meet me again at Hsin-P'u-Kuan, where the road to Li-Fan-Fu turns off from the military road to Sung-P'an-T'ing.

I was making inquiries here about the intermittent

spring at Li-Fan-Fu, of which of course no one knew anything; but the hotel-master, anxious to gratify my taste for the marvellous, said that there was a very remarkable one at a temple just outside the city. When I arrived there, there was of course no spring; but I was nevertheless well repaid for the trouble of coming here, for the temple was at an exceedingly lovely spot.

It was a large place, surrounded by a wall, and its grounds were in perfect order, and well cared for; it hung over the water, which was dashing and foaming below, more a torrent than a river, but down which a few rafts managed to find a somewhat perilous passage.

To the north a fine valley, well cultivated and thickly wooded, ran up amongst hills also well-wooded, the buttresses of fine mountains behind, which plunged their tops into the clouds; while on the other side lay spread out, in all the richness of its verdure, the fertile plain of Ch'êng-Tu.

A well dressed priest, with a tall hat, showed me round, and gave me some tea.

The people of Kuan-Hsien do not enjoy a high reputation, and I found no reason to make my opinion of them an exception to the general rule. I was followed about by a gaping crowd, who exhibited more than the usual amount of the frantic curiosity of the Chinese people, who, notwithstanding their outrageous inquisitiveness, seem yet utterly devoid of the power of observation. I have looked at the faces of some thousands, and in scarcely one have I seen the smallest appearance of observing power. Where the eyelid ends, the forehead begins, leaving no room for the organs of this faculty. After I had returned from my excursion, my people managed to keep the courtyard clear; but in the door of it there was a little open latticework, and hour after hour it was blocked by heads, whose owners all that time can have seen nothing foreign save a bath-towel hung out in the sun to dry.

No one who has not gone through this process of being continually stared at, can thoroughly realise what it is; sometimes after arrival at an inn, when the fearful hubbub, which usually lasts about an hour, has somewhat subsided, and when at last the courtyard has been cleared, and the traveller fondly hopes the reign of peace is about to commence, he suddenly becomes aware of a whispering carried on somewhere near him,—a conversation carried on in a whisper is always disagreeable, but under these circumstances it is peculiarily irritating,—he lays down his pen, and listens, and the sound of a scraping noise outside the wall is heard; presently a finger is cautiously thrust through the paper that covers a little bit of window which he fancied far beyond the reach of escalade, and that well-known eye appears. He suddenly looks up, the eye disappears, a thud is heard on the ground outside, followed by the rumbling sound of some thirty or forty feet, as their owners scamper off, ashamed of having been found out.

Writing is recommenced, and the traveller is soon again absorbed in his work, when presently a scratching and scraping, accompanied by the same horrid whispering, discovers some one picking away the plaster of a lath and plaster partition. If one hole is covered up, another is made somewhere else, until at length even if people should appear underneath the floor it would not cause the least surprise.

May 21.—We still had our old coolie-master, who had hired the men necessary: baggage coolies, now reduced to six; eight chair-coolies and a pony with a Ma-Fu for myself, and two chairs and a pony for the servants.

Besides their reputation for turbulence, the people of Kuan-Hsien are said to be miserably poor; the latter they certainly are, for Chin-Tai was unable to change his silver. I did not find their turbulence exhibit itself in any other way than excessive curiosity, which was so great that not only were the foreigner, the foreign dog, and foreign clothes, objects of intense interest, but the wonder with which these were regarded was extended even to the servants, and a crowd of people, who apparently thought that a Chinaman who could perform the astounding feat of entering the service of a foreigner must bear in his body some outward and visible sign of the fact, followed Chin-Tai when he walked about the streets. Notwithstanding this insatiable inquisitiveness, I found them quiet enough, and no one said an uncivil word.

Leaving the west gate of the city the road ascended the left bank of the river, here about two hundred and fifty yards broad, a rushing torrent of beautiful clear water. This river debouches from the hills at Kuan-Hsien, where the valley is a mile wide; and immediately below this, the ingenious contrivances commence for dividing the river, and directing the numerous branches into the desired channels.

The works are most simple; large boulders, about the size of a man's head, are collected and put into long cylindrical baskets of very open bamboo network; these baskets are laid nearly horizontally, and thus the bund is formed. The streams into which the river is in this manner split up, irrigate the Ch'êng-Tu plain, and lower down again unite to form the Min River of geographers.

A little above Kuan-Hsien, there is a suspension-bridge across the river. Six ropes, one above the other, are stretched very tightly, and connected by vertical battens of wood laced in and out. Another similar set of ropes is at the other side of the roadway, which is laid across these, and follows the curve of the ropes. There are three or four spans with stone piers.

On account of my inquiries about the intermittent spring at Li-Fan-Fu, my people had now an idea that I wanted to see every drop of water that could be found trickling anywhere; and just outside the city, Chin-Tai said there was a remarkable stream at a temple. I visited the temple; but the water was merely a small brook that came down from the hills behind, and was conducted through the mouth of a serpent carved in stone. The temple was, however, well worth a visit; very large and beautifully kept, a very fine flight of stone steps in perfect preservation, where there was not a blade of grass between the flags, led through one or two low buildings, separated by courts paved with a very smooth concrete.

The gilding and paint on the decorations were quite fresh. There was everywhere a great deal of elaborate carving; carved figures on the roofs, and on every pinnacle; even the screens in front of the doors were ornamented at the top with some tracery; everything was in good order, and all the interstices kept clean and free from dust.

Altogether, in its cleanliness, it was very unlike anything Chinese.

In one place there was a tank with rockwork and beautiful ferns, where half a dozen tortoises were disporting themselves in their usual clumsy fashion.

A very civil priest offered me tea; but as I had a long journey before me I declined his invitation to remain.

The road following the river, at once plunged into the mountains, which rose about 1,200 or 1,500 feet. The first were of sandstone, and in this a couple of seams of coal, though only a few feet thick, gave plenty of occupation to a considerable population. The beds were here inclined 45°, and the strata ran up in a north-east direction, at right angles to the valley; these formations soon gave way to the inevitable limestone, here exceedingly rich; and large numbers of lime-kilns, and many coolies laden with lime, attested its value.

After following the river about eight miles, we turned to the right up a stream, where the vertical strata were well exhibited in some small cliffs, the strike being nearly north and south. The sides of the hills were almost too steep for cultivation, of which there was very little; but grass, flowers, shrubs, and trees were growing luxuriantly, and the richness of the verdure was charming.

About ten miles brought us to the village of Yu-Ch'i, where a covered bridge of wood on stone piers took us across the stream. The inn was remarkably clean, the people quiet and civil, and I sat writing with the doors open, no one attempting to intrude. At the back a window looked on to a fine steep hill; and some sweet-smelling blossoms in the courtyard made me ask myself whether I could really be in China.

We left this peaceful little place; and two miles and a half up the side of a ravine brought us to a small temple, 2,000 feet above Yu-Ch'i, perched on a saddle connecting two mountains which rose on either hand about 1,000 feet above it. Here I sat down, and pondered over the sad state of my nearly worn-out boots, while I waited for Chin-Tai, and watched the people passing by.

This road was much traversed. We met great numbers of coolies carrying timber on their backs; the logs were generally about eight feet long by ten inches square. Some were even larger, though these would weigh at least 200 lbs. There was evidently a great trade in timber, for at all the villages on the river there were large stacks.

Numbers of coolies were carrying roots of many descriptions, mostly medicines, and great numbers with baskets of the young shoots of the bamboo, which are cooked and eaten as a vegetable.

An easier road took us down another valley back to the river, and a descent of 1,450 feet brought us to the village of Yin-Hsiu-Wan, a very quiet little place with not enough inhabitants to crowd the inn; this could hardly be considered clean, but was very quiet. The window of my room, which for once in a way was in a position suitable for seeing out of, was right over the river; and, looking across a steep and wooded bank, I could see a fine mountain on the opposite side. The roar of the water made pleasant music in the evening air, and after the bustle and turmoil of the towns below, the peace and quiet of the mountain village was very enjoyable.

During the day I saw the first water-mill I had seen in China. I had begun to think that water-mills, like pumps, were unknown; but afterwards I found them at nearly every village in the mountains.

May 22.—All day we followed up the river by a very fair path, in which there was a good deal of up and down. The mountains here rise about three thousand feet; their sides are very steep, in places almost precipitous, and here and there there are cliffs, sometimes four hundred or five hundred feet high; but where they are not absolutely vertical, a luxuriant vegetation of grass, brambles, beautiful flowering creepers, jasmines, and ferns gets a hold in the crevices of the rocks. Small ashes, beeches, and other trees grow in profusion ; and the mountains are clothed in green to their very summits. Down at the bottom, if the valley opens out and leaves a little level ground, there is sometimes a patch of cultivation, and growing amongst the big rocks which lie tumbled about, there are quantities of a kind of barberry, just now in blossom, with a scent like wild thyme. Round every little village are fine clumps of trees, walnuts, peaches, apricots, and large numbers of Pi-Pa (*Eriobotrya Japonica* or *loquot*), the last now bearing fruit which, although the people here seemed very fond of it, appeared to me to have no taste whatever.

At every two or three miles ropes are stretched across the river ; the people make a sort of raft of two logs of wood, a line from this runs on the rope, and they cross on the raft ; rather an unpleasant operation in this foaming torrent, that falls one thousand feet before it reaches Kuan-Hsien, a distance that, taking all the windings into account, cannot be more than fifty miles.

I breakfasted at the little hamlet of Hsin-Wen-P'ing, built on exactly the same model as all the other mountain villages, with one inn, at which no one appeared to stop. It had only just been built, and the fresh clean wood panels of the wall and boards of the ceiling were quite a pleasure. The people treated me with the greatest civility, even taking the trouble as we passed the houses to keep their dogs from barking at mine. Some of them would come in and have a quiet talk now and then, or show me their curiosities in return for a similar exhibition on my part. Here they told me there were deer and wild boars in the mountains; that some of the latter were found weighing three hundred catties (four hundred pounds); and, as a proof, brought me a young one about a foot long which was striped longitudinally.

May 23.—Notwithstanding the fresh clean panels of the walls, which I had so enthusiastically admired on my arrival, I was horribly eaten of insects during the night, a process to which by this time I was tolerably hardened.

Directly we started we plunged into a wild gorge, the mountain sides running down very precipitously to the river which map-makers call the Min; the people here call it the Hsin [qu. Hsi-Ho?], when they have any name for it at all, which generally they have not.

The tops of the mountains were hidden in rain clouds, wreaths of mist hung about the lower slopes, and a steady rain did not tend to enliven the scene, or render the taking of notes more easy or more agreeable.

The road ran close to the edge of the water, the path being cut out of the rock, in many places propped up from underneath, or cut into steep and irregular steps which the rain made very slippery. The place was very desolate, and there was not a great deal of traffic, although every now and then we passed a good many coolies carrying loads of wood and roots; and at long intervals a small string of mules.

Early in the day my pony dropped a shoe, and I turned into a tea-house while the farrier was at work. Here a man was making a kind of macaroni, and I watched the process with much interest. He made a kind of very heavy dough of wheaten flour, with a little soda; the kneading process was most complicated; at times, as his unaided strength was not sufficient, he made use of a lever, like a long and very stout ruler. His table stood against the wall, in which there was a hole for one end of his lever, and he pressed on the other end with all his weight. The dough was rolled out over and over again, until at last he had a very thin, long-shaped sheet of stuff to his satisfaction ; this was then cut up into strips like tape, and the process was complete. These strips are boiled and eaten hot with some chopped chillies. Like most Chinese things it has very little flavour.

As Wên-Ch'uan-Hsien is approached the valley opens out, the sides of the hills are less steep, and there is some cultivation below. This town is a miserable place, and has a poverty-stricken air. The missionaries warned me to be very careful here ; they advised me to shut myself up in my chair and draw down all the blinds, for, as they put it, the inhabitants were very 'mauvais' ; but it seemed to me that, however vicious their inclinations might be, there were not enough people to put them into practice. I saw scarcely anyone about, andt he streets would have been absolutely deserted but for a few old women, who seemed ashamed of themselves for being there. The town is only about three hundred yards across, and we found a filthy inn in a wretched suburb on the northern side.

Here the Kuan-Hsien people left us, and the Hsien of Wên-Ch'uan sent his card, with a head-man and four Tinc-Chais. He also sent a present of fowls, ducks, and some tea pressed into an annular cake. The fowls and ducks were very welcome, for in this wretched town it was absolutely impossible to get anything, and without them I should have gone dinnerless.

I had been told that I could get yak beef here, as the mountaineers were said to keep yaks in a domestic state, and kill them for beef ; this, however, was a pure fable, invented to put me in a good humour. The tea was the celebrated tea of Pu-Erh; I was not at this time aware of its excellence, or I should have more fully appreciated the liberality of the donor.

Soon after starting we saw the first Man-Tzŭ village on the top of the mountains. I was walking ahead with two of the Tinc-Chais, and, pointing to the village, asked if it was not one of the Man-Tzŭ. 'No,' replied the man,' 'it's a village.' After which brilliant effort on his part the conversation dropped. The Man-Tzŭ build their villages in quite a different style to the Chinese; the houses are of stone, and the lower part is like a fort, with a few narrow windows like loopholes; there is a flat roof, and on part of this a kind of shed is erected also flat-roofed, and open to the front. There is a high tower in each village. These are usually square; but I once saw an octagonal one. I never succeeded in getting a very satisfactory explanation of these towers; some people told me that the possession of one was a privilege enjoyed by the head-man; but as I almost immediately afterwards saw three or four in the same village, this did not seem as if it were altogether to be relied on.

May 24.—The inn at Pan-Ch'iao, though small and dirty, was quiet, but the righteous soul of Chin-Tai was sorely vexed at the robbery of a coat by one of the lodgers. But it was not so much the loss of his coat that grieved him, as the injustice that permitted an inn to be kept by two women so wretched that he could not extract from them the value of the stolen article.

With much difficulty I tore him away from the scene of this disaster, and leaving Pan-Ch'iao we continued our journey. The river still wound about in a narrow gorge, and soon after starting the clouds lifted for a minute from the head of a fine snowy mountain. About two and a half miles from Pan-Ch'iao, the valley on our side opened out, and there was a little grassy plain, where a stream running down from the east joined the river. Here, hidden amongst the thick foliage of walnut trees, there was a little village, whose inhabitants cultivated the patch of level ground. It was a pretty place. There were a few apricot and peach trees by the roadside, and a couple of brilliant yellow birds were flying about amongst the branches.

Perched like an eagle's eyrie on the tops of the almost inaccessible hills, or like wild birds' nests, on the faces of perpendicular cliffs, there were many villages of the Man-Tzu; and down below, on the banks of the smiling river, there were the blackened ruins of many another once peaceful hamlet.

In one place, close to the ruins of some Man-Tzu buildings, that I could plainly see had been burnt not very long ago, there was a new and flourishing Chinese village, where the Chinese, having ousted the aborigines, had established themselves. A little further on there was a cluster of inhabited houses, built in the Man-Tzŭ style close down to the river, that had formerly been occupied by Man-Tzu, but had now been taken possession of by Chinese. I noticed that the Chinese, in one or two very new villages, were adopting in part the Man-Tzu style; but in these the high tower was always wanting, and the difference in the appearance of the new semi Man-Tzu villages and of the regular Man-Tzu buildings was most apparent.

The relentless advance of the Chinese was thus presented to the eye in a very striking manner; every village had its tale of battle, murder, or sudden attack by the barbarians on the peaceable Chinese. In imagination it was easy to fill the picture with living figures. I could in fancy hear the clash of arms, or see the flight of the Man-Tzŭ from their ruthless enemy, who left nothing but the smoking ruins of some once quiet hamlet to bear witness to the cruel tragedy.

The story as told me was always the same. How the Chinese came peaceably up the valleys, and were received by the inhabitants with every show of welcome; how unprovoked and unexpected attack was made on the new comers, who, at first fighting only for existence, ultimately secured the victory, and established themselves in the place of their treacherous foes. The Chinese, as at each successive village they narrated with never-varying details the events of every battle, dwelt with delight on the valour of their race, and the cowardly conduct of the barbarians, and never thought it possible that I should wonder what account these same barbarians would render, should they have the opportunity of telling their tale.

But the irrevocable law of nature must have its way; the better race must gradually supplant the inferior one; the Chinese will continue their advance, stopped only where the climate aids the soil in its refusal to produce even to these industrious agriculturists the fruits of the earth in due season.

But these mountains, whose heads are crowned with dazzling snow, into whose inmost recesses man has never penetrated, and whose rugged sides and mighty precipices must inspire awe in the most unpoetic soul, have not been without their influence on the minds of the inhabitants. Not only the shout of battle, but the miracle wrought by some Buddhist saint, the mystery attendant on some freak of nature, and even the gentle song of love, finds its place in the legends that cling to the sides of these romantic valleys.

Leaving behind us the melancholy records of a fast dying race, we crossed a little ridge, and my attention was called to a spot surrounded with all the halo of the miraculous.

On our left was a long ridge of loose sand, that fancy might conjure into the semblance of a gigantic snake; and hidden in its mysterious depths some marvellous creature even now resides. And with awe the tale was told me, how no effort of man has ever succeeded in clearing away that ridge of sand; for even if by dint of desperate labour during the day a portion is removed by nightfall, when the labourer returns to his work on the morrow, lo! all is as it was, and everything must be commenced afresh.

The fable has its origin in truth. No doubt there is a backbone of rock to this ridge of sand, and the wind coming out of the valley causes the drift, that even if cleared away would of course soon again collect.

There was yet something more wonderful about this place, and Chin-Tai told me an interminable story about a Fu, five dragons, and five swords; but it was very long, and he became so interested in it as to give it me more in Chinese than English, by which the moral, if there was one, is lost to posterity for ever. He, however, impressed upon me very strongly the fact that it was a miracle, and that as it was told him by some one who lived here it must be true.

Hsin-P'u-Kuan boasts a wall and gate, and is presided over by an official called a T'ing. He asked me to stop here all day, and placed his house at my disposal, an offer that he did not expect me to accept. He sent me the usual unromantic fowl, some potatoes which were very acceptable, and a piece of pork, which my servants gladly disposed of for me; for nothing short of absolute starvation would have induced me to touch the flesh of a Chinese pig, a peculiarity that afterwards obtained for me the title of a foreign Mahometan.

The potato is despised by the Chinese as food only fit for pigs and foreigners, but, introduced into the mountainous regions by the missionaries not much more than fifty years ago, the valuable properties of this useful root have already made themselves appreciated, and steadily, but surely, gaining ground, notwithstanding the contempt of the Chinese, it is destined at no distant day to take its place amongst the agricultural products of China. In all the mountain regions in Western China and Tibet potatoes are found, and as far as Ta-Li-Fu I was never without them during the whole of my journey.

Marching out of Hsin-P'u-Kuan the river was immediately crossed by one of the rope suspension bridges, that had by this time become familiar. The roadway on these follows the curve of the ropes, and at the two ends is rather steep. The bridges themselves sway about a good deal, especially if there happens to be any wind, and walking on them is something like walking on the deck of a rolling ship.

The volume of the river is here swelled by the tributary from Li-Fan-Fu; this is passed by a similar bridge, and leaving the main road to Sung-P'an-T'ing, the road to Li-Fan-Fu ascends the right bank of the tributary stream.

The scenery now changed entirely. At the bottom of the valley there was here and there a little flat ground, where fields of barley were divided by loose stone walls, the mountains rising up behind almost precipitously. With the exception of a few scanty blades of grass, these were perfectly bare, and standing like a long wall, almost unbroken even by a gully, presented a remarkable contrast to the magnificent verdure we had left behind.

In one or two places the Man-Tzŭ villages were now inhabited by Chinese; and up on the tops of the mountains, when the clouds lifted, the present dwelling places of these aborigines could be seen. Some of them put me much in mind of many a Persian hamlet lying hidden in the valleys of the great Elburz ; one in particular, close down by the stream, half hidden amongst trees, with a little patch of cultivation round it, and with the bare and rugged mountain rising like a wall behind, needed only a few tall, straight poplars to complete the likeness, and almost made me think I was nearer to the Atrek than to the Yang-Tzŭ.

In places the valley narrowed, and the hills running sheer down to the water, the road was supported from below, or rested upon horizontal stakes driven into the face of the rock.

The road was everywhere in an excellent state of repair ; great care was evidently bestowed upon it, and it must have cost much money and labour to keep it up. The Chinese are not as a rule in the habit of repairing roads ; but in a case of this kind a road left to itself would very soon cease to exist. In one place it had been found impossible to avoid a short tunnel, and when it had been necessary to cut steps in the rock, these were very regular, and carefully made.

There were a great many caves and caverns in the sides of the hill, some in very inaccessible positions, but I could get no information about them.

A little less than nine miles from Hsin-P'u-Kuan there is a bridge precisely similar in construction to the Sanga bridges of India, and to many others that subsequently became familiar to me in the mountainous regions between Tibet and Western China.

Seeing a very old man fishing with a rod and line, we purchased a couple of fish, which though not trout turned out delicious. I never found a specimen of the salmon tribe in Western China or Tibet; there are plenty of salmon in Japan, and in the Amur, but I have never heard of their being found in any river further south. I used always to examine the fish, but I never detected the smallest indication of the second back fin, which characterises the salmonidæ.

I now turned my attention to the rod and line, and found that the old gentleman used a reel as we do in Europe; but instead of a heavy brass affair he had a light octagonal wooden framework on which the line was wound; it had the advantage of a very large diameter, so that one turn wound up very much more than would be reeled up by one turn of our elaborate and costly machines.

The village of Ku-Ch'êng, where we stopped, was so small that it really hardly deserved a name, and the inn was a quaint and dirty place. One large room, whose walls were black with age and smoke, opened from the street. It was not provided with a window, and at the back there was a sort of cupboard, which did duty for a guest room or state apartment, and here, where in the brightest day the light of heaven never penetrated, I took up my abode. When the night closed in, the large room beyond was lighted by a fire of wood, and one or two oil lamps, whose fitful glimmering was just sufficient to cast a lurid glare on the faces of the strange figures seated round, smoking their long pipes and discussing the events of the day. One by one they put away their pipes, the conversation gradually dropped, and soon all were wrapped up in their blankets, and sound asleep.

May 25.—We continued our march up the desolate valley, where the cultivation on the little level patches of ground close to the water's edge only served as a foil to set off the bare and precipitous mountain sides. The streams that came down from these ran through deep and gloomy gorges, tumbling in little cascades between almost vertical walls of rock; many of them were of a brown colour, so like the peat streams of Scotland that I almost think there must be fields of peat in the unknown mass of mountains to the south. There was scarcely any traffic on the road, the villages were few and very small, and half hidden in walnut and willow trees, with a few apricots and firs.

We halted at a very little place for breakfast, where the people had a young bear they had brought down from the mountains; they told me that bears were not very numerous, and that they were found in the snowfields about twenty miles distant. I asked if there were any white bears, for I had been told that in some mountains to the west of Ch'êng-Tu there were white carnivorous bears; but the people said they had never heard of any. The musk deer is found in this neighbourhood, and I was offered here the first musk bag that I had seen.

Not far from here there was a place with an unusual aspect. It was like a large stockade, the walls of loose stones instead of wood, in which there were some long low buildings, where only a few people stood about. This was a barrack, and in the fighting time there had been a large garrison here; now it looked miserable and deserted, and was a post for twenty men.

About a mile before reaching Li-Fan-Fu the new pagoda that was built in 1876 is a prominent object. Soon afterwards a very short suburb is gained, and the wall of the city, which is in a somewhat ruinous condition, is passed by the eastern gate.

Some people joined our party here. They told Chin-Tai that the governor-general of Ssŭ-Ch'uan had sent them to warn all officials on the road that I was coming, that I was to be treated with proper respect, and generally taken care of. One of them put down a parcel in the room where the servants pack together, and the eyes of the inquisitive Chin-Tai alighting on it he brought it to me. It was the Margary proclamation.

The people of Ch'êng-Tu evidently thought that I was a consular officer, and that my mission was to see if this proclamation was posted. This was very natural, for the idea of any man travelling in discomfort, when he could stop at home in ease, is one that it is simply impossible for a Chinese mind to comprehend. Further, I came to Ch'ung-Ch'ing with Baber, a recognised consular officer; and finally it was one of the stipulations of the Chi-Fu Convention that these officers should travel about from place to place to see that the proclamation was posted.

The literati class is the one most esteemed in China, and partly to gain the credit of belonging to it, and partly to explain my otherwise incomprehensible habits of looking at everything, and asking what appeared to the Chinese innumerable insane questions, I always professed to be one of the literati class, went about openly with a note book in my hand, and declared that I was going to write a book. No protestations, however, on my part were of the least avail; even my servants, accustomed as they had been all their lives to the eccentricities of Europeans, never believed that I was simply a private individual, and probably never carried out my instructions to say so—not that it would have in the least availed, for had they sworn it most solemnly and continuously they would certainly not have been believed. Duplicity is in the nature of the Chinese, and being so *rusés* themselves, they naturally attribute sinister or at least hidden designs to others. I was thus surrounded with all the majesty and pomp of a high official, and however unpleasant might be the feeling that I was sailing under false colours, I could only console myself with the reflection that it materially added to my comfort, if not to my safety.

The Margary proclamation had been posted once already on the occasion of the visit of one of the French missionaries, who had been here not a very long time previously. The missionaries always dress in Chinese clothes, wear an artificial plait, conceal their European eyes with spectacles, and pass well enough for Chinamen without attracting much notice; but this one having a bright red beard, his non-nationality was very apparent; the fact of a foreigner being in the place was soon bruited about, and reached the ears of the Fu.

The Fu had not at that time posted the proclamation, and when he heard of the arrival of a foreigner his mind was filled with alarm, for he thought that a consular agent had arrived to see if the clause in the Chi-Fu Convention with regard to the proclamation had been faithfully carried out. In fear and trembling of being reported to Peking, he posted it, and by its side another of his own, exhorting the people to obedience and respect.

This country would run a very good race with the Holy Land (of the monks) for pre-eminence in local wonders. Chin-Tai promised me several for the next day. One from his description appeared to be a gigantic ammonite; another, fossil wood. Beautiful tales were in connection with these, which I never had the opportunity of thoroughly investigating. As for the intermittent spring, for which I had come all this distance, no one knew of its existence, although Monseigneur Pinchon had positively told me he had seen it here.

On the first day of the tenth month in every year the Sieh-T'ai fires a big gun at one of the mountains opposite; for if he did not do so the result would be serious. All sorts of bad luck would descend upon the town, many people would be killed, and tumults would arise, which would eventuate in plague, pestilence, and famine. This custom is probably a relic of the time when the Chinese were fighting and subduing the Man-Tzu. It is very likely that when this place was first occupied, they had a habit of firing a gun to remind the people of the presence of soldiers, and that the habit has since become a tradition clothed in the garment of fable.

The weather was very pleasant. The sun was shining all the afternoon, and there was no rain all day for the first time since leaving Kuan-Hsien. A strong wind came on after arrival; and the dust, dirt, and filthy rags of paper were blown from the window into my room, on to my table, and over everything. The view out of the window was very wild, and a huge wall of rock that reared itself up right in front was very much like the over-praised mountain of Murren.

I now had to make all my arrangements for sight-seeing; and, bidding Chin-Tai to find out the distances from Li-Fan-Fu of all the objects of interest, I told him that the two things I especially wanted to see were a Man-Tzu village and the gigantic ammonite of which he had spoken. The intermittent spring I had given up altogether; probably it was the only thing worth seeing in the place, and consequently the only one about which the people knew nothing. When Chin-Tai came back, he said that the ammonite was twenty miles away, but that the rest of the things were all quite close, and that I could thoroughly 'do' Li-Fan-Fu in a short morning. I therefore made up my mind to see what there was to be seen before breakfast, and return to Ku-Ch'êng in the afternoon.

May 26.—The next morning several official servants and an interpreter accompanied me in my expedition; and as the tracks in the neighbourhood were all impassable for ponies, mules, or chair coolies, we were obliged to walk on foot, much to the disgust of Chin-Tai.

My satellites went two or three before me, and the rest behind, ordering people out of the way and keeping them from pressing round; and as I marched along, the streets were lined on both sides by the people, all turned out to see the show.

First I was shown the rock at which they annually fire the gun, where, high up the side of the precipice, there was a place where some of the stone had broken away, leaving a hole with perfectly squared corners, about which there was something miraculous. I next saw a box, with some dead man's bones in it, but as I could not clearly understand the story in connection with it, my credulity was not very severely taxed. I was then taken up to the new pagoda, on a rock about two or three hundred feet above the river. The ascent was a little steep; and as Chin-Tai, since leaving Peking, had never taken a walk except on the two occasions when Baber and I made picnics from our junk, he was very soon out of breath.

The pagoda is a plain building of stone, but it is in a commanding position, from which a fine bird's-eye view of the city is obtained.

Li-Fan-Fu is situated on a little triangle of flat ground, at the mouth of a narrow gully. The river runs swiftly in front, and separates it from a wild and bare mountain, crowned with huge precipices that rise up some three thousand feet at the opposite side. It is enclosed by a wall, in many places broken down. This wall runs between the houses and the river, and then climbs a long way up the crests of two spurs which enclose the deep ravine running up at the back of the town; but as the houses are only built on the flat ground close to the river, the walls enclose a considerable vacant space. I counted the houses as well as I could, and at a rough calculation put them at about one hundred and twenty. The houses here, unlike those in other parts of China, are two-storied, generally built of stone below, with a wooden upper story and a balcony. Nearly all the roofs are flat, but a few of them are made of sloping battens of wood. There is a small suburb on the eastern side, but none elsewhere.

A rushing torrent comes down the ravine, flows through the town, and serves to turn numerous water mills, for as this is a corn and not a rice-growing country, there is a great deal of grinding to be done. The wheels are nearly always horizontal, and are enclosed in little low, round, flat-roofed houses, which look like small forts ; they have one little door, and are hardly high enough for a man to stand in.

In this place, and around it, under the command of the Sieh-T'ai, there are five hundred Chinese soldiers, and three thousand Man-Tzu ; the latter are scattered about amongst the Man-Tzu towns and villages.

There is another Chinese town called Cha-Chuo-T'ing, twenty miles up the river, which is the last Chinese station.

We now crossed the river by a rope suspension bridge, the roadway of which was merely a few hurdles laid down, and walking about half a mile up the river, we turned into a deep ravine and came upon a couple of houses, where, as an illustration of intermarriage, about which I had been making inquiries, I was shown a Chinaman who had a Man-Tzŭ wife.

The people who were with me evidently thought that now we should go back.

'But where is the Man-Tzu village?' I said.

'Oh!' they replied, 'that is too far, ten miles away up there.'

I looked, and saw a village on the side of the mountain, about four miles off, and two thousand feet above me. Fortunately the people had not time to make up a story, as they had simply trusted that the mere sight of the village perched up above me would be quite enough to damp my ardour, but to their astonishment I insisted on going there. Chin-Tai was the most disgusted of all, for the sun was shining, it was somewhat warm, and he appeared to be wrapped up in an infinite number of wadded coats. He very soon became a piteous object. The road was desperately steep, and very stony ; every step he took was a labour to him. He was compelled to sit down and rest every few yards ; sometimes he threw himself flat on his back with a groan. He several times declared that he was going to die, and I found it far greater trouble to look after him and make him come on, than to walk the distance myself a dozen times. My satellites kept their countenances fairly, but were evidently desperately amused, and I could easily see that amongst themselves they thought the whole proceeding eminently ludicrous. By dint of perseverance and much waiting, we at length succeeded in getting him to the top, and when there it was with true religious fervour that he 'knocked his head.' I found that the Man-Tzu people cultivated far more ground than I had thought. The upper slopes of the hills were all laid out in terraces, where barley and wheat were grown.

In this part of the country the term Man-Tzu is, amongst the aborigines, considered a term of reproach, they themselves preferring to be called I-Ran, or I-Jen: this is very unusual. The aborigines in the province of Kwei-Chou would rather consider I-Ran an insulting epithet; but here the term Man-Tzu is considered so bad that Chin-Tai would not let me go on, until he was perfectly sure I should make no use of the word 'Man-Tzu,' and in the conversations that afterwards passed in this village the term I-Ran was repeatedly used, and the word Man-Tzu not once.

This village was about two thousand two hundred feet above Li-Fan-Fu, but was not on the top of the mountain, which rose another thousand feet behind it. The houses were all of loose stones, with little windows like loopholes; the streets were not more than three feet wide, and everything was more filthy than usual. As we sat outside the entrance to the village waiting for Chin-Tai to drag himself up the last few yards, the clouds lifted from the head of a grand mountain to the south of Li-Fan-Fu, disclosing vast fields of snow. This is called Hsüeh-Lung-Shan, or Snow Dragon Mountain, and the people said there were fields of ice where it was too cold for anyone to live.

In the village we all had tolerably good appetites, and did justice to a huge loaf of Indian corn bread that one of the Tinc-Chais brought straight from the ashes in which it was baked.

I then asked to be taken to a house where I could sit down, and the village school was selected. Here I soon gathered a few people around me, who gave me what little information they themselves possessed. There are two kinds of I-Ran people, those living at Cha-Chuo and beyond having a different language to those who live here.

The I-Ran of this place are very like the Chinese in appearance; they wear the same dress, as well as the plait, but they have good teeth. The Chinese, as a rule, have vile teeth, ill-formed, irregular, very yellow, and covered with tartar. The I-Ran here all talk Chinese, as well as their own language; their writing is Chinese; and the school children were learning to write that language. The I-Ran of the west have quite a different writing, which appears to be alphabetic, but I completely failed in my attempts to get the alphabet; all they could say was that they had a great many characters. I, however, made one of them write me a couple of lines, and on comparing their writing, which is from left to right, with pure Tibetan, there is really no difference, and their statement that the number of characters was very great was probably some confusion. They gave me the numerals and a few words, in which the connection between the two languages is quite apparent, although they are very different.

I sat asking questions a long time, and the schoolmaster wanted me to stop and breakfast, but this I could not do for want of time, and I set off as soon as possible for the descent. As we left, a tremendous drumming and beating of gongs commenced in another village, about a quarter of a mile distant, on the opposite side of a ravine. There was evidently some excitement, and from the tones of those who were with me, I suspected that up here my room was considerably more desired than my company, although my satellites appeared anxious to conceal the fact. Chin-Tai went down with much greater ease than he had come up, but it was three o'clock before we regained the inn at Li-Fan-Fu.

By this time I had a tolerable appetite for my breakfast, which I very soon disposed of, and then I prepared to start afresh. The chair coolies now declared it was too late, and wanted to stay here for the night; the usual dispute ensued, which ended as disputes of this kind always did, and the coolies with a grumble shouldered the empty chair and moved off. As it was late, and I wanted to get on quickly, I hired an extra man for the chair in which Liu-Liu was riding. The coolie demanded five hundred cash for the job, and all the bystanders declared that he was absolutely sacrificing himself out of pure generosity. By dint of much argument he ultimately compounded for four hundred cash. When the bargain was struck the people all swore he was a lucky dog, and as he had no difficulty whatever in finding another to do the work for two hundred cash, I felt that they were right!

It rained lightly all the afternoon, but we arrived at Ku-Ch'eng about seven o'clock, and here I was pleased to find Chung-Erh with the money and a packet of newspapers.

May 27.—Marching down towards Hsin-P'u-Kuan, the clouds lifted from a magnificent snowy mountain in the east. It is called the White Cloud Mountain, and it must be fourteen or fifteen thousand feet high. For I subsequently found the snow line at this season at an altitude of thirteen thousand feet; and the summit as I looked on it was at least one thousand or two thousand feet above the line of snow. Moreover, the snow lies on this peak all the year round, and the limit of perpetual snow must here be at least fourteen or fifteen thousand feet above the sea.

At Hsin-P'u-Kuan I visited the T'ing, a funny little man who, hat, button, and all, did not come up to my chin; and as the fat little fellow sat on a chair, he could just reach the bar between the legs with his toes. He wanted me (or said he did) to stop a day in his house, but I said my coolies had all gone on, and I must follow. He had been here ten years; he said that there was not much snow in the winter, although the place is nearly five thousand feet above the sea; he showed me a marvellous fish five hundred years old, which every now and then goes up to heaven through the roof of the tank, but always comes back again. There is also another much larger fish here; but no one has ever seen it, because if they look on it they become blind at once.

Besides these wonders, which, I thought, were quite enough for one time, there were one thousand taels of silver in the tank. They all declared that they could see it, but I was assured that anyone trying to take it away died at once. The T'ing had not, however, yet exhausted his stock of stories, and said that if I liked to stop he could show me a cave, in which not more than twenty years ago a dragon lived that belched forth smoke at a regular hour every morning.

I assured him that ocular demonstration was quite superfluous, and I started again, as the clouds were gathering ominously about the hills.

We marched up the river, and now the beautifully wooded slopes and magnificent verdure had disappeared, and the rocks were very bare. The I-Ran villages were seen on the tops of the hills all the way to Wên-Chêng, which we reached just before a good downpour of rain commenced. This is a very small hamlet with a nice little quiet inn.

May 28.—It was a fine morning when we started; the sun dispelled the mists; and every now and then there was a glimpse of the grand snow-capped mountains, on which snow lies all the year round. Each has its legend, and in one of them, called, 'The Sacred Temple Slope Mountain,' a great dragon is supposed to reside. Chin-Tai had a great love of the marvellous; he used to collect all the stories about dragons, and other wonderful beasts, and was delighted to narrate them to me on the march.

In many places along the river ropes were stretched from bank to bank; the inhabitants manage to cross by these, and even carry goods. An opportunity was soon afforded to us of watching a man cross with a heavy sack; a sight that was all the more thoroughly enjoyed when he stuck in the middle immovable until another man came to his assistance.

There are always two ropes, one for going, the other for returning, so arranged that each has a considerable slope downwards. A small runner is first placed on the rope. This is a hollow half cylinder of wood, about eight inches in diameter, and ten inches long. The runner is placed on the rope (a very large twisted bamboo rope). The man takes a strong line, ties it round his body, and then, by passing it two or three times over the runner, makes a kind of seat for himself; it is then again passed round his body, and firmly secured. He is thus suspended close below the big rope; then, with both hands on the runner, he raises his feet from the ground, and shoots down the incline at a tremendous pace; of course, with a width of about one hundred yards, it is quite impossible to have one end so high that there shall be a regular slope all the way; so, notwithstanding the impetus he gets in descending, which shoots him some way up, the passenger always has to pull himself up the last few yards, which is done in the natural hand over hand fashion. This is a method of crossing a river that must require a considerable amount of nerve, when the torrent is roaring some two hundred feet below, dashing over most ugly and cruel-looking jagged rocks.[3]

The river still ran through limestone rocks, but in places on both banks there were extensive deposits of clay, *débris*, and sharp stones; the river has cut its way in one or two cases through this bed to a depth of nearly a hundred feet, and the perfectly horizontal layers are very plainly shown. Six and a half miles from Wên-Chêng there is a narrow gorge, where the road, scooped out of the side of the cliff, is closed by a gate with a roof and parapet. This gate was in course of repair, and the new and massive iron plated doors were not yet on their hinges. Immediately beyond this there is a square fort, through which the road is conducted.

Mao-Chou is very pleasantly situated, for just here the river valley opens out, and forms a little basin, about two miles wide, enclosed on all sides by high

[3] This is the *Chikā* of the Kashmir Himalya (see Drew's *Jummoo*, &c., p. 123).—*Y.*

mountains. Two pagodas on two hills dominate it, and bring it good luck. The Chou was away at Ch'êng-Tu paying his respects to the new governor-general; but the T'ing sent his card, with a duck, a fowl, and some sweets. He said he was very busy, and hoped I would excuse him coming to visit me.

After having been some little time in the inn, Chin-Tai said the people outside had never seen a foreigner before, and hoped that I would show myself for a few minutes. He said they were respectable shopkeepers, and not at all rude people; so I went outside, and asked them a few questions, after which they begged me not to disturb myself any more, but to return to my room.

After this a small military official sent to me to say that he was sure I should be much gratified at a visit from his little child; and although the most simple-minded must have seen through the transparent artifice, I admitted him and his child to a private view.

May 29.—The T'ing called on me in the afternoon; he was a Shan-Si man, and having been at Peking knew something about foreigners. I showed the pictures of Kuo and Liu, the Chinese ambassadors to England, for these happened to be in a newspaper I had with me; he was very much interested, and asked for some illustrated papers, a present I was very glad to make him.

The Ma-Fu went afterwards to his steward, and said that I was a great official, and, like all high functionaries, expected to live on the fat of the land without paying anything; and added that I had sent him to the steward to get food for the pony, or money to purchase it with. This appeared to the steward a very natural proceeding on my part, and he gave the Ma-Fu half a tael. Fortunately the affair came to my ears, and I was able to stop similar acts for the future.

May 30.—Before leaving Mao-Chou, Chin-Tai brought me a tale about two hundred thieves on the road. After discoursing me some time, he girded himself with a bloodthirsty-looking cutlass he had bought cheap in Shanghai, whilst I, much to his annoyance, contented myself with the armour of incredulity.

This was the first really fine day we had had since leaving Ch'êng-Tu. The morning was glorious; there was not a cloud in the sky, and a breeze from the snowy slopes kept the temperature pleasantly cool.

When we had marched two miles I looked back, and had a magnificent view of the Nine Nails Mountain, so called on account of its summit being broken into sharp peaks or points; but as for the number nine, it might just as well have been anything else. I had not before been able to appreciate the grandeur of this mountain, in which I could now see great fields of snow descending quite two thousand feet below its highest point. I stood and admired it for a long time, the people all wondering what I was looking at.

Very soon after leaving Mao-Chou, the mountains again close in on the river, which now runs through a series of narrow and precipitous gorges, great bare slopes and precipices running down to the water, and leaving scarcely a yard of level ground; except here and there, at the end of a projecting point, or up the bottom of a little valley, where a few flat acres are found and cultivated. The great mountain sides are ragged, and torn about in a marvellous manner, and huge masses broken from them lay strewn about. The road is cut out of the sides of the rock, and is often supported from below, or propped up for a few yards by horizontal stakes driven into the rock.

When a valley opening disclosed a view of the interior, the tops of the higher mountains to the west were seen to be well wooded; but there was no opportunity of seeing what lay to the east, as none of the valleys were sufficiently open.

We met a Man-Tzŭ Lama on the road, dressed in clothes of a coarse red stuff and a white hat, but as yet we had passed no regular Lamasseries.

We halted for breakfast at Wei-Mên-Kuan, a little village celebrated in the semi-fabulous history of ancient days. In the days of the Sung dynasty (extinguished in 1280 A.D.), one of the emperors had eight sons, the youngest of whom was sent as high military official to Wei-Mên-Kuan. The Mongols and Chinese were then at war, and some Mongols, commanded by a queen, came to this village, where a battle was fought, and the emperor's son taken prisoner. In accordance with the humane customs of the country, instead of leaving a captive to linger out a miserable existence in a dungeon, the queen was going to cut off the prince's head in a more or less gentle fashion; but her daughter, casting her eyes that way, saw that the man was of goodly proportions and noble face—in fact, altogether a godlike youth. She then and there fell in love with him, and her mother consenting, the wedding was celebrated with the pomps and glories necessary for such an occasion.

The inn at Wei-Mên-Kuan was the usual curious mixture of dirt and decoration; there were a couple of pictures of flowers, done in Indian ink with really an artist's touch; scrolls also on the walls, with proverbs on them, or extracts from some classic. In the centre of the ceiling there was a little rosette of red and yellow paper, with Chinese writing in a neat and small hand, but all covered with dirt and cobwebs.

In this part of the valley, besides the trees that had now become so familiar, there were a few of the soap trees (*Acacia Rugata*), in appearance something between an acacia and an ash; the fruit is very alkaline, and it is from this that the Chinese almost entirely make their soap, which they scent with a little camphor.

We halted for the night at Ch'a-Erh-^{Ng}ai, a village, if village it could be called, consisting of about two huts besides the inn. When I entered my room, I found five beds, on each of which was a filthy straw mattress. The people of the place were much hurt when I ordered the immediate removal of these hives of fleas; when I insisted on the dirt being swept away from under the bedsteads (which were merely two or three rough planks on trestles), it was too much for them, and they began to think me mad; and when I declared that I did not consider it a good plan to leave a heap of refuse in a corner, they gave me up altogether as a hopeless lunatic.

May 31.—From Ch'a-Erh-^{Ng}ai to Mu-Su-P'u the road is exceedingly bad, continually ascending and descending, and strewn with rocks and stones. Never leaving the river bank, it is sometimes three or four hundred feet above the water, and seeming for some distance to run along the face of the slope, the confiding traveller promises himself at least a few yards of level ground, when his hopes are rudely dispelled by a steep staircase of sharp and slippery rocks, where his whole attention is concentrated upon selecting a safe position for his feet. At this moment he becomes aware of a string of mules advancing from the opposite direction. All seems lost, for there is no room for animals either to turn in, or to pass one another; but fortunately round a projecting corner the road widens for a short distance, and the two caravans get clear, but not without a great deal of shouting and some danger. The difficulties of the treacherous road may sometimes be lessened by a gentle slope; this lasts but for a short time, and the temporary relief is more than compensated for by a scramble that almost requires hands as well as feet, and seems impossible for animals. The ponies, however, encouraged by the Ma-Fu, whose cries resound from rock to rock, bravely struggle on, sometimes slipping on the glassy stones, but seldom coming down. Again the road abruptly descends only again to rise, until at length, after many a narrow escape, the village is reached, and the tired animals relieved of their burdens enjoy a roll in the dirt and what scanty provender may be provided for them.

About six miles from Ch'a-Erh-^{Ng}ai the river receives a considerable affluent from the west, called the Lu-Hua-Ho. A six days' journey up this river is the home of another of the Man-Tzŭ tribes, the Su-Mu, or White Man-Tzŭ, as the people here call them, a tribe numbering some three and a half millions; and the Ju-Kan, or Black Man-Tzŭ, live in the interior, an indefinite number of days beyond.

The sovereign of the Su-Mu is always a queen. When the Tartars were conquering the land, this tribe happened at that time to have a queen for a sovereign, who gave the Tartars great assistance, and as an honorary distinction it was decreed by the conquerors that in the future the Su-Mu should always be governed by a queen.

The Su-Mu have been pillaged by the Ju-Kan, their houses burnt, and their villages destroyed. The Ju-Kan now wanted peace, and had offered an indemnity sufficient to rebuild the houses; but the

Su-Mu were eaten up with the desire of revenge, and their queen was now at Ch'êng-Tu praying that soldiers might be sent to punish the Ju-Kan. If she ever succeeded in her mission she probably will find herself in the position of the horse in Æsop's fable, who desired the help of man.

We marched all day through the same wild gorge, hemmed in with bare cliffs and ragged rocks, broken at the top into pinnacles and crags of fantastic shape. Now and then some of the valleys opening out to the right or left were less precipitous, and were well wooded on their slopes. Once or twice I caught a glimpse of a snowy peak, but nearly all day we were shut in by the steep hillsides, and could see little besides them and a narrow streak of heaven above. When the road descended to the river, there might be a few yards of level ground, where the barberry and other shrubs seemed to grow luxuriantly amongst the rocks; but the general aspect of the scene was barren and somewhat dreary.

June 1.—From Ta-Ting the road at once climbed up the precipitous side of the valley by a tolerably gentle slope; and we soon began to breathe the pure and invigorating mountain air.

As we ascended, instead of being shut in by steep hills and cliffs, the slopes became more gentle, though often high above us, at the very summits, there were again great precipices. Amongst the slopes and crags were many tiny plateaux, cultivated by a few people, who seemed to gather but scanty crops from the unfruitful soil. The sides of the valleys were either well wooded with pines, or covered with close and thick brambles, barberries, thorns, and all sorts of shrubs which were deliciously fre h and green. Many varieties of wild flowers grew luxuriantly, numbers of the purple iris in blossom, and acres of a kind of purple crocus; many sweet-smelling herbs shot up amongst the grass, and the whole scene is very fair to look upon.

As we ascended, we saw a great many cock pheasants strutting about, crowing loudly, quite innocent of fear, and unsuspicious of any harm from the hand of man. On this occasion their confidence was misplaced; and thenceforth my table was daily supplied with game, which varied the monotony of diminutive kids, shrivelled ducks, and emaciated fowls.

A steady pull of eleven hundred feet in four miles brought us to an inn close to the village of Shui-Kou-Tzu, where I found a room looking over the valley of the river. Here the air felt crisp and pure, and though the sun was shining brightly the thermometer at 9 A.M. was only 58° in the shade. Across the valley, a grand mountain ran down in precipices and steep and bare slopes, about three thousand feet, to the river; up a gorge to the left a deep green forest of firs crowned the summit; to the right, on a small plateau, a Man-Tzŭ village hung over the stream, with a little terrace cultivation at its side; in the background here and there a patch of snow was lying on the higher mountain-tops, and below in the bottom we could just hear the murmur of the invisible river as it tumbled over its rocky bed. The twinkling of the goat bells sounded pleasantly in the morning air, and after having been shut in for so many days in the close gorges, the place and all around it was very delightful.

After breakfast another steady pull of one thousand feet brought us to our highest point, and from here we had a fine view of the town of Tieh-Chi-Ying. This place is on a flat plateau, bounded on three sides by precipices, or exceedingly steep slopes, which fall down to the river fifteen hundred feet below. On the fourth side apparently inaccessible mountain crags rise abruptly behind it, the roads to and from it being cut out of the face of the mountain, making it a very strong military position. In the early days of the Chinese here this was a large and flourishing town ; the Chinese were at this time carrying on war with the Su-Mu, or White Man-Tzŭ ; but one fine day the latter, in vast numbers, managed to get over the summit of the mountain, and amongst jagged rocks and crags, where it would have been thought that hardly a goat could get a footing, they surrounded the town and cut off the water, which was led by a conduit from a mountain stream. The Chinese were either overwhelmed by numbers, or forced to surrender for want of water, and the place was burnt to the ground. The White Man-Tzŭ have now been thoroughly subjugated, and are tributary to the Chinese, but Tieh-Chi-Ying has not yet recovered, for inside the extensive walls there are now but a few houses. There is a garrison of five hundred soldiers.

This country is admirably adapted for a mountain tribe to defend themselves in, and it is only by the gradual pushing civilisation of the Chinese, and the inevitable consequences of superiority of race, that these hardy mountaineers have been steadily driven back.

We now descended to the river by a steep zigzag, where loose stones lay scattered about the narrow path, and then followed the river bank to Sha-Wan (The Sandy Hollow).

There was a freshness in the early morning air that now made us feel we were thoroughly in the mountains, and far above the oppressive heat of the steaming plains below. The road from Sha-Wan was very good, and the scenery most picturesque. On the right, crags and precipices rose into pinnacles generally crowned with clumps of pines; the northern faces of the hills were almost always well wooded with fresh green or yellow trees; and in the valley all sorts of shrubs grew luxuriantly. On the opposite bank of the river the hills sloped gently, and their sides were beautifully green with grass and shrubs. Presently on the road a string of yaks was encountered, to the intense gratification of my servants, whose lives had been a burden to them owing to my perpetual inquiries about this animal. I had, of course, seen them in the Zoological Gardens; but I had been looking forward with great interest to observing them in ordinary use. Now, I had almost commenced to look upon their presence as a myth, for I had been repeatedly told I should see them 'anon,' and I could not help recollecting certain fabulous salmon in the river Lar in Persia, where the inhabitants assured us that this fish abounded, and that great numbers were caught at the next village lower down. When we arrived at this next village, in answer to our eager questions we were told that there were none exactly here, but if we followed the river we should find some the next day. With exemplary confidence we continued our march, and were rewarded with the pleasurable information that our faith was fully justified, but that as all the nets were three or four farsakhs away, we must just go on a little further. Already in imagination we pictured ourselves feasting on some king of the waters; but still, like a will-o'-the-wisp, the lordly salmon was ever just before us, until we arrived at the sea, and then, as we could not again be requested to move on, the excuse was ready that we had come at the wrong season. So at Ch'êng-Tu the missionaries said I could buy the beef at Kuan-Hsien, and that at Li-Fan-Fu the yaks were in constant use. The people of Kuan-Hsien told me that if I had come last month, when they killed a Mao-Niu, or long-haired ox, as they called them here, I might have had some beef, but I should get any amount on the way to Li-Fan-Fu. At Li-Fan-Fu they said they could neither show me yaks nor give me beef, but that at Mao-Chou I could get plenty of the one, and see any number of the others. On arrival at Mao-Chou, I was informed that the people were too poor to afford meat, and that the animal was not used for domestic purposes; but at last we met them on the road, and I was able to bring them back from the mythical realms to which, with Chin-Tai's dragons, they had nearly been relegated. They have longer horns than the ordinary cattle; but otherwise, with the exception of the long hair, are in appearance and size very much the same.

We met numbers of coolies on the road laden with red deers' horns, some of them very fine twelve-tyne antlers. The deer are only hunted when in velvet, and from the horns in this state a medicine is made that is one of the most highly prized in the Chinese pharmacopœia; the antlers that are shed are collected and brought down to the plains for sale, where they are converted into knife handles, and used for various other purposes. These are sold at Kuan-Hsien at the rate of fourteen taels for one hundred catties (about sixty-five shillings for one hundred pounds).

There were plenty of pheasants about; these were in size and appearance like our English pheasant, and were without the white ring round the neck, characteristic of the bird generally found near Shanghai.

About eight miles from Sha-Wan, on turning a corner, a glorious view suddenly burst upon me; right in front was a perfect pyramid of virgin snow; the sun was shining brightly, and the brilliant white of the peak was all the more dazzling from the contrast presented by the deep shadow on the wooded flanks of the nearer mountains; this was Mount Shih-Pan-Fang (The Stone Slab House), and I stood almost spell-bound, lost in admiration. Long I gazed at this majestic peak, whilst my unsympathetic companion seized the opportunity to sit down and smoke a pipe, wondering the while what I could find to look at or admire.

During the morning we crossed the boundary of the Sung-P'an district, and a mile further arrived at P'ing-Ting-Kuan.

Every town and village now had a story to tell of some fight with the Man-Tzŭ, and numerous ruins, which from their appearance could not be very old, attested the truth of the statement that it was not more than eighteen years since there was fighting here. We arrived in the evening at Ch'êng-P'ing-Kuan, where on the bedstead there was a new piece of clean straw matting, to which the innkeeper, who had probably heard of my curious fancies, pointed with evident pride; but I knew too well the meretricious nature of this whited sepulchre. At my command the thin mat was lifted, and disclosed the ravening and wickedness below. There was a layer of straw and dirt underneath, that must have been accumulating since first the Man-Tzŭ left the place.

I received a visit from the Pa-Tsung, who commanded the garrison of eight soldiers. He came to pay his respects, and to apologize for not giving me a present. He asked for some medicine for his wife, who had a gathering behind her ear that had been open for two months. After leaving me with the most polite reverences, he managed to spare one-fourth of his army as a guard, who encamped outside my door, and made night hideous with the hourly gong.

The people here live entirely upon Indian corn; rice is of course far too expensive, as none is grown nearer than Kuan-Hsien. The maize is made into circular or oval loaves about an inch thick, and every coolie carries two or three of these on the top of his load.

June 3.—The morning air quite made my fingers tingle; and I thought with pity of Baber sweltering at Ch'ung-Ch'ing, with the thermometer at 100° in the shade. The road was now good and level, and kept close down to the waterside. The scenery was very beautiful; the tops of the mountains were crowned with dark forests of firs, and the valleys opening east and west disclosed a vast extent of pine-clad heights. The bed of the river was much wider, and bounded by slopes clothed with shrubs of many descriptions, amongst which wild flowers grew in profusion; in one spot there was a field of wild roses, one mass of blossom, and the air was literally laden with the delicious perfume. On the northern slopes were charming little woods of the freshest green; and the yellow flowers of the barberry, everywhere abundant, helped to give that warmth to the colouring that always seems to characterise a Chinese landscape. Five miles from Ch'êng-P'ing-Kuan a valley opening to the east gave a near view of the snow pyramid Shih-Pan-Fang, whose summit, as I now could plainly see, must be at least two thousand feet above the snow line. Pin-Fan-Ying is an important military station. It was quite new; all the houses were new, and the walls, which were exceedingly strong, had only just been built. Here there was a garrison of two hundred and fifty soldiers; and it is quite evident that, although, in these districts at all events, the Chinese have now conquered the wild barbarian tribes, still the reign of peace cannot be said to have fairly commenced.

At Lung-Tan-P'u, Chin-Tai came with the cheerful countenance indicative of misfortune, and declared that the dog had broken several fingers of one of my coolies. I was much relieved on ascertaining the nature of the accident, two or three slight cuts, caused by the chain being dragged through his hand. A piece of diachylon plaster, and an abundance of faith made everyone quite happy.

The simple confidence all the people about me reposed in my doctoring, and the intense pleasure they derived from it, would have been touching had it not been so sadly misplaced.

I shot some pigeons in the morning; they had a dark blue head, a purple green neck and throat. The back was light blue, with a broad white stripe across it; the tail above was dark blue, with a broad white stripe across the middle; the under side of the tail was the same as the upper; the breast was light blue; the wings, white underneath, had dark stripes on the upper side. They were much the same size as the English wood-pigeon, but did not fly so strongly. I have no doubt that regular search would have been repaid by many kinds of pheasants, but I saw only one species, and no other game-bird of any description.

June 4.—The landlord of the inn at ^{Ng}an-Hua-

Kuan had heard of my approach, and had cleaned up his house three days before in expectation of my arrival. Directly I came in he brought me a cup of the most delicious tea I ever tasted in China. It was of pale straw colour, like all tea taken by the Chinese, and the steam that rose from it diffused a delicate bouquet through the room. I was very glad to buy a packet of this from the friendly landlord; for as I had left Ch'êng-Tu with the intention of being absent for a few days only, I had brought with me a very small supply, which, like my candles and cigars, had now come to an end. I had been for a long time endeavouring unsuccessfully to make myself believe that I enjoyed the aristocratic water-pipe; but at last the conclusion was forced upon me that my tastes were vulgar, and venturing one day to borrow a pipe from a coolie, I never again resorted to elegant but trivial hubble-bubbles.

The coolie's pipe is the same all over China; but in Ssŭ-Ch'uan the method of smoking seems in accordance with the character of the people, who, being more independent in spirit, and less narrow-minded, are not so addicted to trivialities as the Chinese of other provinces. They, therefore, do not content themselves with the homœopathic doses of tobacco usually taken, but roughly roll a leaf of tobacco into a kind of cigar, and use the pipe as a mouthpiece.

We now entered rather a different country, the scenery everywhere indicating the proximity of the plateaux; the river valley opened out to nearly half a mile, and the bed itself became wide and shallow, the stream being broken up into several small channels. The mountains were now rounded, and separated by open level valleys, instead of the close narrow gorges which had hitherto been almost universal;

the main valley was all cultivated, whilst the hill-sides were cut into terraces, and crops grown all over them.

The Man-Tzŭ people had now been left behind, and we were approaching the country of the Si-Fan. These are a very wild-looking people. Some of them wear hats of felt, in shape like those of the Welshwomen, and high felt riding boots, and in their dress are much the same as the regular Tibetans. Now and then we met three or four, riding all together; and my truculent Tinc-Chais always made them dismount as I approached, in no way attempting to conceal their contempt for the conquered barbarians.

Sung-P'an-T'ing is on the right bank of the river, with an extensive walled suburb on the left; a hill runs down from the right bank, ending in a small cliff; and the wall of the town runs right up the side of the hill, taking in a great deal of open ground where barley and wheat are grown. The place seemed to have an enormous population for its size; and a crowd was collected in the streets; men and boys surrounded the hotel; the yard was full of people, many of whom were lounging about my room, even before I arrived. The hotel was unfinished, so it was at all events clean, and my room was large and light. There was no furniture in it; but a couple of chairs and a table were soon borrowed from somewhere; three planks put across two benches made a bedstead, and my upholstery was complete. For some time there was a crowd outside the windows, but a thunder squall came to my aid; and though a number of idle people and children remained gazing, everyone was very well behaved, and I had not the least cause for complaint.

CHAPTER IX.

A LOOP-CAST TOWARDS THE NORTHERN ALPS.

(*Continued.*)

SUNG-P'AN-T'ING BACK TO CH'ÊNG-TU.

Si-Fan Ponies—Reports of Game—Musk Deer—Si-Fan Lama—Language—Crops—Butter, Fish, Yak-Beef—Bitter Alpine Winds—Foreign Remedies Appreciated—The Traveller Quits the Valley ('Min River') and Turns Eastward—Si-Fan Lamassery—Herds of Cattle and Yaks—Desolate Hospice at Fêng-Tung-Kuan—Tibetan Dogs—Reported Terrors of the Snow-Passes—Summit of the Plateau of Hsüeh-Shan—Descent Begins—Forest Destruction—Verdure of Eastern Slopes—Splendid Azaleas—Slaughter by the Si-Fan—Luxuriant Gorges—Inhabited Country Regained—Chinese Dislike of Light—Shih-Chia-P'u—Miracle-Cave—Hsiao-Ho-Ying—Iron Suspension Bridges—Peculiarities of Chinese Building—Road to Shui-Ching-Chan—Gold-Washing—'Ladder Village'—*Mauvais Pas*—Lung-An-Fu—Approach to the Plains—Ku-Ch'êng—Fine Chain Bridge—Road to Shen-Si—Chiu-Chou—Chinese Fête Day—Moderate Drinking of Coolies—Preserved Ducks' Eggs, a Nasty Delicacy—P'ing-I-Pu—Boat Descent of River (Ta-Ho)—Great Irrigation Wheels and Bunds—Populous Country Again—Chang-Ming-Hsien—City of Mien-Chou—Road Resumed—Ill-smelling Fields—The Red Basin Again—Lo-Chiang-Hsien—Tomb of Pong-Tung—His History—Pai-Ma-Kuan, or Pass of the White Horse—Summer Clothing of the People—Method of Irrigation—Fine Vegetables—Roadside Pictures—Official Civilities at Tê-Yang-Hsien—Progress to Han-Chou—Marco Polo's Covered Bridge—Hsin-Tu-Hsien—Rich Country—Suburbs of Ch'êng-Tu—End of Excursion.

June 5.—Some Si-Fan ponies were brought down for me to see, and to purchase if I would. After the animals I had been riding they look enormous, and I estimated the height of the first I saw at thirteen and a half hands. He was a nice, strong-looking, grey pony, but without much breeding; and he was priced at thirty taels. The second was considerably larger, also a grey, but not so good-looking, although they asked fifty taels for him. I measured this one, and found he was only thirteen hands one and a half inches, which showed how my judgment had been misguided by the very small ponies I had been accustomed to.

I was informed that, at a place two days' journey to the west, there were great numbers of red deer; and I was promised excellent sport if I felt inclined to make an expedition. Wild sheep and goats were said to live amongst the crags and rocks in the neighbourhood of Hsüeh-Shan; and the people told me that on the road to Lung-An-Fu I should see plenty of hares, musk deer, and pheasants; a prophecy that was completely belied; for although there were a great many pheasants for the first few days, I never saw a hare, a musk deer, or any other game. There must, however, be a considerable number of the musk deer amongst the mountains; for the price of musk at Sung-P'an-T'ing was only three times its weight in silver. The musk deer are not shot, but trapped; for there is a belief that if one of them is wounded he tears out the musk bag, and so disappoints the hunter. It is possible that terror or pain, or both combined, may cause the animal to eject the musk, as the sepia under similar circumstances squirts out its ink, and as, on the authority of Æsop, the beaver is said to tear out a certain gland and cast it to the hunter.

A Si-Fan Lama, diffusing a powerful and unpleasant odour, came to see me. He was the second Lama in a Lamassery not far distant. He wrote me some lines of his language, and gave me the numerals and a few words, which were quite sufficient to show that it is a dialect of Tibetan. His name was Nawa; he was a powerfully-built, upright man, with a very firm mouth, and a haughty look about his eyes, and he had the appearance of one who knew how to command. His language sounded peculiar after having heard nothing but Chinese for so long. The letter R is rolled in a very pronounced manner, a striking contrast to the way in which this letter is slurred over by the Chinese, who in many cases cannot pronounce it, as, for instance, at the beginning of a word before A or I, when the R is changed into L. This confusion between R and L is rather curious to note; it seems inherent in the human palate. English children often substitute the L for the R; and according to Mr. Stanley, in Central Africa the same change seems to be of frequent occurrence. The Chinese cannot pronounce *rain* or *rice*: they always say *lain* and *lice*. Yet in other cases they are capable of producing the sound, as, for instance, in the word *I-Ran*.

The sound of the guttural Kh is also very frequent in the language of the Si-Fan, and there are other sounds that are quite impossible to catch. The word for the numeral *four* was repeated to me many times, and at last I could find no better way of writing it in Roman characters than *Hgherh*. *Fourteen* I put down as *Chu-ugurh*, and *forty*, *Hghtyetămbā*; but this orthography can convey but a feeble idea of the astounding noises the people make in their throats to produce these words.[4]

The crops here are nearly all wheat, oats, and barley, as it is too cold for Indian corn. There are also potatoes in the neighbourhood, and the market produced a vegetable like spinach. The principal food of the people is barley bread, and barley porridge,

---

[4] It seems probable that there was a mistake here between four and eight. Four, fourteen, forty in *Tibetan*, are *zhi*, *chu-zhi*, *zhi-chu*, or with the affix *thámpa* (full) *zhi-chu-thámpa*. Eight, eighteen, eighty, are *gyad* (or *gyä'*), *cho-gyad*, *gyad-chu* or *gyä'-chu*, or *gyä'-chu-thámpa.—Y.*

for which the barley is roasted before grinding it into meal. There is also some buckwheat, from which a heavy unleavened bread is made. The market produces the leaven and steam-baked bread, in the shape of dumplings, which seems to be universal wherever Chinamen are found.

Butter is made in the mountains by the Mongols, but it is not brought down here in any quantity, as this place is entirely populated by Chinese, who never make use of butter or milk in any form whatever. The landlord of the inn, however, had some, and made me a present of a circular cake about an inch thick, and six or eight inches in diameter, similar in shape, taste, and appearance to the cakes of butter found all over Eastern Tibet. The river produces a few little fish, very much like sprats in taste and appearance. Yak beef is plentiful, and costs forty cash a catty, and eggs cost seven cash each.

In the month of July there is an annual fair, when the Si-Fan, the Mongols of the Ko-Ko-Nor, and the Man-Tzŭ bring in their produce to sell. Skins of all kinds, musk, deer-horns, rhubarb, and medicines are the chief articles brought down, for which they take up in exchange crockery, cotton goods, and little trifles.

My landlord was a Mahometan, and his respect for me was much increased by my reputation for never eating pork or ham. He told me that he had been to the Ko-Ko-Nor, and that the journey occupied three months in going, and the same time in returning; the road, he said, passed over dreadful mountains, the very recollection of which made him shiver. In winter-time the cold is intense, and the wild winds that sweep across the frozen plateaux cut great gashes in the face or any part of the body exposed. He asked me to give him some medicine against the wind; and as Chin-Tai declared that the possession of a bit of diachylon plaster would render him exceedingly happy, I felt I could not deprive him of the pleasure, although I rather spoilt the effect by telling him I was afraid he would not find it a certain remedy.

The ignorant superstition of the Chinese attributes to the foreigner all kinds of supernatural powers, which are even extended in their minds to European goods. Amongst many Chinese the application of grease from a foreign candle is considered a specific for small-pox; and European sugar is almost a pharmacopœia in itself.

June 6.—We left the valley of the river which had been our constant companion for so many days, and, climbing up a gorge, we soon obtained a good view of the town. Ascending a little more, we crossed a ridge here only eight hundred feet above the river, to the valley of another stream, running nearly parallel to the main river. We now ascended this valley by a good and easy road, and kept up above the stream as far as the Lamassery, where Nawa was in readiness to receive us. The Lamassery was a low wooden building, very irregular in shape; about some of the chief rooms there was some coarse embroidery; round the largest of the chapels hung a number of rough pictures of saints, painted on a sort of cotton stuff; in one there was an image of Buddha, who here is known by the name of Khătye-Tăbā;[5] in front of him there were a number of lotus-flowers, and ten little brass bowls of water. They introduced me into the cell of the chief Lama, who acknowledged my presence by a slight inclination of the head; he was squatting

[5] Buddha is usually called in Tibet *Shakya Thubpa*, 'Mighty Sakya.' —*Y.*

before an immense pan of ashes, counting beads, and muttering prayers.

I did not stop here long. The Lamas, though exceedingly polite, were excessively dirty, and smelt horribly. This elevated plateau-land being ill-adapted for agriculture, but few Chinese are found, and we were now almost entirely amongst the Si-Fan. Their architecture is very much the same as that of the Chinese, but they do not turn up the ends of their ridges and gables; indeed, at a distance, the houses look very Swiss. On the hill-sides the roofs are made of planks, laid anyhow, with big stones on them to prevent their being blown off, just as in Switzerland.

The march was up the valley bounded on both sides by rounded hills and low mountains, all covered with grass and brushwood full of pheasants; but not a single tree or wild flower was to be seen. Here the Si-Fan keep immense herds of cattle and yaks that feed on the splendid pasture. We passed no village all day, a single house, surrounded by a little patch of cultivation, at about every mile and a half, being the only sign of a population.

A great many bees are kept, but at this season there was no honey. The valley of the Sung-P'an river, covered as it is with wild flowers, would seem almost to have been designed by nature for the production of honey.

For the last two miles and a half of the journey we did not pass a single habitation. We were obliged to stop at a solitary wayside hut, as there was not another roof for many miles; and as a heavy chilly rain came on, and wild gusts of wind swept down from the snowy heights, none of us were loth to take shelter in the hovel at Fêng-Tung-Kuan, or Wing Cave Pass, as it is most appropriately called. This place is not visited by any but a few of the poorest coolies, and the accommodation was suited to the requirements. It was a long low house of uncut, flat stones, between which the daylight was more apparent than the mortar; the single room, that constituted the public accommodation of this luxurious hotel, was sheltered by a gabled wooden roof, the ridge of which was left open in its whole length as an exit for the smoke of the fire, a most unlooked-for piece of thoughtfulness on the part of the architect. One end of the room, above which there was a loft under the gable, was divided off by a wooden partition; this portion formed the private residence of the hostess, and was on this occasion given up to me.

The people here keep very large savage dogs; in shape they are more like a colley than any other English breed, but much heavier about the head, neck, and fore part of the body. They have a very deep voice, and one of them would hardly let us enter the inn, if inn it could be called, for there was absolutely nothing to be purchased but a little buckwheat or barley bread.

Chin-Tai brought me awful tales of the terrors of the road that we were to traverse the following morning. He warned me that going up we must all be very quiet; anyone calling out or making a noise would be certain to bring on a terrific wind, a violent snowstorm, hailstones of gigantic dimensions, thunder, lightning, and every evil the elements could inflict. If a man on this mountain should express feelings of hunger, thirst, fatigue, heat, or cold, immediately the symptoms would be intensified to a very great degree. He told me that once a military official with an army of soldiers came to cross this mountain. He had with him his sedan-chair, to which about twenty men were yoked, before and behind, who could not get on without shouting. The troops also marching always made a great noise. This high functionary was warned that he should not attempt to cross the mountain, for if he did some fearful accident would befall him. He laughed however at the warnings, saying that he had the emperor's order, and must go on. So he went. A fearful storm of wind and snow came on; half his army perished; and he himself very nearly lost his life. Such were the tales about Hsüeh-Shan with which I went to bed; and if I did not shiver it was thanks to the quantity of clothing with which I covered myself.

The floor of the loft made a sort of ceiling to my apartment, but there was a large square hole in it, and through this, as I lay in bed, I could see the long opening in the roof, and the stars beyond when not obscured by clouds; and at intervals the rain came in for variety. Fêng-Tung-Kuan is eleven thousand eight hundred and eighty-four feet above the sea; it fully justified its name, for it blew a violent gale all night; but I put on a considerable number of garments, rolled myself up in three blankets, and neither the wind, nor the rain, nor Chin-Tai's weird stories disturbed my peaceful slumbers. The story of the general who cried 'Excelsior' was familiar to me; but whether it was told me before, *à propos* of this mountain, or whether I have read it somewhere, I am not sure. I have no doubt that it is a tale tacked on to many mountain passes.

June 7.—When the morning broke, low clouds were scudding across the sky, driven by the wind that howled amongst the crevices in the walls of the hut. A chilly rain that turned to sleet did not enliven the scene, and soon we plunged into the dank mists that swept over the summit of Hsüeh-Shan. One single partridge, startled from its bed, was the only living thing we saw as we made the dreaded ascent. The plateau, as the summit is approached, is bare and dreary; a climb of about one thousand feet brought us to the first sprinkling of snow, at an altitude of twelve thousand eight hundred feet above the sea; there was no snow on the path, but it was lying in little patches amongst the rocks, here all quite bare; a short distance more, and at an altitude of thirteen thousand one hundred and forty-eight feet above the sea we stood at length upon the summit of Hsüeh-Shan (Snow Mountain). At the very top there was a little hut without any inmates, but no one seemed anxious to remain here in the cold sleety rain; and quickly descending the steep path that leads to the west, we left the chill mists behind us, and soon reached a warmer climate.

Riding down another valley, that ran nearly east and west, on our northern side there was but little wood, all the slopes being covered with a rich green grass, but on the south, a serried ridge, whose summit was torn into wild crags and ragged pinnacles, bounded the valley, throwing out long spurs, where pine forests clothed the northern faces of the lower slopes, and masses of a shrub with white blossoms and a scent like our lilac grew amongst the trees in lavish profusion.

These forests are being cut down in a ruthless manner, and as of course no attempt is made to plant young trees, of the ultimate fate of these beautiful valleys there can be but little doubt. The trees gone, the rains will cease; and then these ranges will become dreary, bare, and useless masses, like the mountains of Northern Persia.

A march of ten and three-quarter miles, during which we passed only three small huts, brought us to the village of Hung-^Ngai-Kuan (Red Rock Pass), where the community lived in three houses. Here we halted for breakfast, and immediately afterwards heavy rain commenced, evidently no unusual occurrence, for the rich green of the dense woods that now surrounded us, and the wonderful verdure of the open slopes and valleys, were unmistakable signs that the climate of the eastern is much more moist than that of the western face of the great spur from the Himalayan plateau, which stretches to the south between the valleys of Sung-P'an-T'ing and Lung-An-Fu.

The ridges from each side every now and then threw out great masses of rock, ending in huge precipices over the valley; and between these green grassy slopes, with clumps of trees scattered about as in a park, ran up to the heights above. The bottom of the valley was wooded with low trees, and we marched all the afternoon through a thick copse, where there were not so many wild flowers as on the other side of the mountain. Here and there there was a house quite new, showing how recently the Chinese had reached this point.

The Si-Fan live only on the tops of the hills, and, as before, every opening has its tale of horrors. At one of them my attendants stopped, and said that here the Si-Fan had suddenly descended from their fastnesses, butchered five hundred soldiers in cold blood, and burnt all the houses without any provocation on the part of the Chinese.

'But what were five hundred soldiers doing here in the country of the Si-Fan,' I asked.

The question remained unanswered, and we

marched in silence to the village of Chêng-Yuan, three thousand feet below the summit of Hsüeh-Shan.

The inns on this road are only meant for the accommodation of coolies, and here I was obliged to displace the lady of the house from her den, as it was the only place that could be found; it had not much to boast of in the way of accommodation, but when a traveller gets four walls, and a dry roof overhead, there is not much cause for complaint.

The only provisions to be bought here were potatoes, which were sold at the moderate price of $1\frac{1}{2}d.$ per pound, and of which I laid in a large stock to take to Ch'êng-Tu.

June 8.—The river now ran for six miles through a narrow gorge, not more than one hundred yards wide, bounded everywhere by almost vertical cliffs, and clothed with the most dense foliage. In the valley there were azaleas fifteen and twenty feet high, covered with a mass of blossom as if prepared for a show at Kew. Wild peonies proudly planted their gorgeous flowers, and the delicate foliage of a small wild bamboo almost hid itself amongst the broad fronds of many a magnificent fern.

We passed three wretched huts, but except the little patch of garden round them there was absolutely no cultivation all the morning.

The affluent streams ran through exceedingly narrow precipitous gorges; but the foliage was so dense that it was impossible to see more than a few yards in any direction, except where the road rose a little above the river. Once I heard the roar of a waterfall, but though I was not above a few yards distant I was quite unable to get a view. The road was strewn with sharp stones, and although it never rose more than thirty or forty feet above the river, frequent ascents and descents of no more than this were sufficiently troublesome over the slippery and broken rocks.

We halted at Yueh-Erh-^{Ng}ai, a little hut by the wayside. It was but a mere shelter, the back wall being made of a few loose stones, on which the roof-beams rested. Here I breakfasted, and two of the Tinc-Chais seized the opportunity to lie down in a corner, where there was enough shelter from the draughts for the lamps of their cherished opium pipes.

Beyond this place the gorge became more narrow, and the sides of the hills more steep, until the river ran between two vertical walls of rock running up a clear five hundred feet, separated from one another by but a few yards, where with marvellous pertinacity the trees and shrubs, which still grew in rich luxuriance, continued to get a hold for their roots in every crevice in the cliffs. The road was dreadful, the descents were desperately steep, and the slippery rocky path was often blocked by great masses of pointed stone; where this occurred, as it often did, in places almost like an exceedingly steep staircase, not more than two feet wide, with a rough wall of rock at one side, and a precipice at the other, the travelling was not exactly pleasant for us who were going down; and it seemed as if the ascent must be an impossible task for either coolies or mules.

The river (for it had become one now) dashed in a succession of waterfalls over its uneven bed, now blocked by some gigantic rock, or almost stopped by the perpendicular cliffs that hem it in on either side so closely that it sometimes seems an easy jump across the top. It is quite impossible to give any idea of this extraordinary gorge; I could hardly have believed in the existence of a rift so narrow and so deep, and yet so wonderfully clothed with trees, ferns, and shrubs. On emerging from it, and looking back, there was nothing to be seen but a giant wall of rock; the chasm through which the torrent finds its way was nowhere visible, and it seemed almost impossible that there could be a road through that apparently impenetrable barrier.

This gorge comes to an end fourteen miles from Chêng-Yuan. The valley of the river then opens out, but it is still closed by high and rather steep hills; these are well cultivated, though there is a great deal of wood, and of the small wild bamboo largely used by the inhabitants instead of grass. In another three-quarters of a mile the village of Shih-Chia-P'u is the first collection of houses that can be, properly speaking, called a village since Sung-P'an-T'ing. The inn was of course small, but I enjoyed the luxury of a weathertight room. This apartment was the landlord's residence, prepared for me by two Tinc-Chais from Sung-P'an-T'ing, who had gone on before. The lighting arrangements were, as usual, defective; the walls and ceiling, nearly black with smoke and dirt, did not reflect a particle of light, and at four o'clock I was obliged to use candles, although outside it was quite light until about eight o'clock. The Chinese seem to have a positive objection to light, and it was always a mystery to me how they managed to pursue their vocations in the darkness in which they habitually live. The women may be seen working, and the men plying their trades of an evening by the aid of a filthy oil lamp that gives about as much light as a glow-worm.

I was invited to remain at Shih-Chia-P'u for a couple of days, and promised some venison if I should consent. For amongst the hills there are plenty of deer. The trappers from the villages set their nets, leave them seven days, and then return for their quarry; they had just gone off, and were expected back in a day or two. In the winter the deer are driven down here for food and water, but at this season they always remain up in the high mountains, where there are also pheasants and other game in abundance. Whilst I was at dinner Chin-Tai told me that during the morning Liu-Liu had gone to sleep on his pony, that the animal had stumbled, that pony and rider had rolled over together, and that the fall was only saved from being a nasty accident by the united efforts of a Ma-Fu, a coolie, and two Tinc-Chais, who fortunately caught the animal's tail just as it was disappearing over the brink of a yawning chasm. Divested of the halo of romance with which Chin-Tai loved to adorn his tales the affair would probably have assumed a less appalling if not a trifling aspect.

In one of the most gloomy recesses of the long gorge through which we had passed during the day, it is said that a long time ago a hermit took up his residence in a cave; but finding that, even for Chinese eyes, it was exceedingly dark, so dark that he could not even see to boil his rice, he fixed a mirror on the opposite side, which not only reflected the rays of the sun into the sombre dwelling, but (such was the holiness of the man) it had the additional useful property of reflecting the moon also, whether that luminary happened to be above the horizon or not. The hermit has long since been transported to a better sphere, but they say his looking-glass still remains, and the traveller who should have the misfortune to be benighted in this desolate gorge may still see the weird glimmer of the mirror on the darkest and thickest night.

June 9.—The country between Shih-Chia-P'u and Hsiao-Ho-Ying (Small River Camp) was not so picturesque, though perhaps more interesting than the wild gorges we had left behind ; for there are no woods or forests, but where the land is not tilled, it is covered with long grass and brushwood. The principal crop is Indian corn, which is cultivated on the slopes that are not more steep than 30°.

At a village three miles below Shih-Chia-P'u there was a patch of rice about twenty yards square. Beyond this, a little was grown near every village ; but the quantity was so small that rice can hardly be considered as one of the agricultural products.

The river valley opens out a little near Hsiao-Ho-Ying. This town is quite square, and is guarded from bad luck by a low pagoda, about two hundred yards outside the northern gate ; it is surrounded by a wall overgrown with creepers, more picturesque than useful, as the place is dominated by a steep hill, from which stones might almost be kicked into the street. The first properly built houses in the valley were seen here, all that we had passed previously partaking more or less of the character of sheds ; but every step we now took brought us nearer to Chinese civilisation.

At Hsiao-Ho-Ying there was a guard of twenty-five soldiers under a Pa-Tsung, who spared me one of his men to accompany me.

After leaving this the hills became more precipitous, there was less cultivation, and soon the river entered another short gorge, and descended rapidly by a series of leaps from rock to rock.

Five miles below Hsiao-Ho-Ying, at an altitude of about five thousand feet above the sea, there were the first cultivated bamboos; there were also the two kinds of wild strawberries: the bright red deceitful berry, so frequent in the Ch'êng-Tu plain, whose innocent and beautiful appearance is sadly belied by its poisonous properties; and another unattractive plant, whose pale pink fruit has all the delicate flavour of the wood strawberry of Europe.

There was a great deal of holly amongst the trees; this was not exactly the English holly, but in appearance very like the tree that grows in the neighbourhood of Chin-Kiang, on the lower Yang-Tzu.

Nine miles from Hsiao-Ho-Ying the river is crossed by one of the iron chain suspension bridges, so familiar to travellers in Western China, but of which up till now I had never seen a specimen. Seven iron chains extend from bank to bank; these are tightly stretched by powerful windlasses, bedded in a solid mass of masonry. The roadway is laid on these chains. There are piers at each end; and from the top of these (about eight or nine feet above the roadway) two other chains are stretched, one on each side. These two chains droop in the middle to the roadway, which is suspended from them at this point; but as these extra chains are intended only to prevent the structure from swaying about, and not as an additional support for the weight, the roadway is attached to them at this central point, and at no others. This method of applying the two side chains is rather unusual; for generally they are parallel to the roadway, about three feet above it, and are chiefly of use as hand-rails.

Yeh-T'ang was my night's resting-place, where I again found myself in an unfurnished inn. The walls were not yet completed, and the upper part was open to the winds of heaven; but the deep overhanging eaves prevented the rain coming in, and I was dry and free from insects. To the English architect it may seem curious that a house should have a roof and no walls ; but the Chinese always support their roof on a framework of wood, and when this is complete fill in between the timbers with wood, in a wood country, with stones up in the mountains, and often with lath and plaster, or with brick and mortar in the plains : the walls never supporting the roof.

June 10.—At Hsiao-Ho-Ying there was a ferry, and there was another at Shui-Ching-P'u (Crystal Village), where we halted for breakfast. These ferries are merely boats with one rope looped over another stretched across the river.

The valley from Yeh-T'ang to Shui-Ching-Chan is very open, and the road generally fair, though bad in places. Every now and then it would rise a couple of hundred feet above the river ; at other times it was scooped out of the side of the rock, or propped up from below in the usual way ; the hill-sides generally were not too steep for the cultivation of Indian corn ; and close down to the river the quantity of rice was rapidly increasing. This was now planted out in beds, from which a crop of opium had already been gathered. Round the villages there was a little wheat and tobacco. We had the same wild flowers by the roadside, but by no means in the rich profusion of the upper part of the valley. There were a few shrubs of barberry, some magnificent white lilies in blossom, and flowering pomegranates clustered round the houses.

About five miles below Shui-Ching-P'u there were a couple of men washing for gold in the river-bed ; but this can hardly be a profitable occupation, as in this valley I saw scarcely any other attempts to obtain the precious metal.

I was somewhat incredulous about the luxurious comfort that had been promised me in the hotel at Shui-Ching-Chan, and was not surprised when Chung-Erh, who had been sent on in advance, said that it was so bad that I could not possibly put up there. He had fortunately discovered a temple, where the priest had been willing to give up a nice, new, clean room, which was ready for me when I arrived. It was also the village school, and all the children in the place spent the afternoon and the greater part of the evening in a room next mine repeating their lessons aloud, in a sort of chorus, and before we (*June* 11) started in the morning they were already again hard at work. Our march led us through very much the same scenery, the river, exceedingly tortuous, everywhere running between peaks about two thousand feet high partly cultivated. In one place there is a very remarkable rocky promontory one hundred yards long, not more than ten yards broad at the base, and from fifty feet to one hundred and fifty feet high, projecting out from the hill-side, and almost enclosed by spurs from the mountain on the opposite side of the river. I stopped here a few minutes to make a sketch, and soon afterwards arrived at T'i-Tzu-Yi (the Ladder Village)—a well-chosen name for any village in these parts, where the roads are not much better than ladders. The people were still planting out rice, but it was nearly the last; and a little further down the valley this operation had been finished. Near here there were again one or two wretched creatures reduced to washing the sand for the scanty particles of gold to be found.

June 12.—Below T'i-Tzu-Yi the sides of the hills again became more steep; but still, wherever amongst precipices or steep slopes, a few roods of ground not steeper than 30° could be found, there was sure to be a patch of Indian corn. This is about the steepest slope up which a man can walk unaided by his hands. From the opposite side of the river the face of a slope of this kind has all the appearance of being nearly vertical, and the people hoeing on them look like flies on a wall. There are generally ten or twelve together, dressed in a line that would please the eye of a British drill-sergeant; and as they advance from the bottom upwards, seen from this point of view, it seems that they must slip down, and be precipitated into the river below.

The road as yet did not improve: rising up one side of a spur, and zigzagging down the other, by a desperately steep and slippery ladder of rock. Wherever there had been landslips, the track was strewn with gigantic rocks and sharp stones. All the projecting rocky points were exceedingly precipitous, and generally almost vertical on their western sides, the eastern faces of the spurs sloping more gradually, and clearly indicating the direction of the geological upheaval. The river twisted and turned in a most incomprehensible manner, and whenever it washed the foot of one of these cliffs, the road was scooped out of the face, or propped up in the usual fashion. In one place, instead of using poles, long stones were put horizontally into holes bored in the face of the rock; across these other stones were laid, and thus the road was formed. Here and there, there was only just room for the ponies' feet, and in one place, when I was looking at the scenery rather than at the pony, he stepped so close to the edge of a rotten bank as to elicit a shout of dismay from the usually phlegmatic Ma-Fu. This individual used to walk behind, and where the descent was a very steep one, over big stones or down a slippery staircase, he used to hold the animal's tail to prevent the glissade into space, that would inevitably have ensued on a false step.

Lung-An-Fu is situated at the foot of a spur thrown out by the mountains towards the river, the valley of which opens out considerably just before reaching the city. It is enclosed by a very long wall, running nearly a mile up both sides of the spur and across the top. This place rejoices in the majesty of a Fu, and the glory of a Tu-Ssu, who together are the representatives of the State, civil and military. They cannot together have much more to do than enjoy the emoluments of office; for though the streets were tolerably wide and clean, it seemed a very small place; and, as Chin-Tai remarked, there was a good deal of wall, but not much house. The officials were very civil; and whether it was owing to the vigilance of the swarm of Tinc-Chais sent me by the Hsien, or to the paucity of the population, the inn, such as it was, was kept perfectly quiet, except for two pigs domiciled behind my room, who passed the time between more or less successful mining operations in the wall, and protesting loudly against the cruel emptiness of their stomachs. The Chinese always starve their pigs for a day before killing them, and one of these suffered the last penalty even before I left.

An adventurous raft was here seen on the river, but considering the nature of the torrent, I was not surprised to find it alone in its ambition for the perils of shipwreck. The iron chain suspension bridges now became so frequent as to make it a matter for sincere congratulation that the Chinese had not discovered the irritating western system of the toll.

June 13.—Below Lung-An-Fu the river valley is

more open, and enclosed only by low hills, and by undulating closely cultivated spurs from the mountains behind. The hills retreat from the river at the bends, and leave flat plains nearly all planted with rice, where little hamlets and houses ensconced amongst clumps of trees give the landscape a peaceful, homely look.

The river here finds its way through a deep deposit of clay and rounded stones, from thirty to fifty feet thick; above which there are still the same limestone masses. This has clearly been the bed of a small ancient lake; gradually the river must have worn itself a deeper channel through the gorge below, or some slight earthquake shock may have rent one, by which the lake has been drained. Since then the water has cut its way through the bed of clay it had formerly deposited.

The appearance of the country indicates a great change of climate; the almost tropical wealth of verdure that characterises the upper part of the valley is not seen here, and everything shows that the rainfall must be much less.

The Indian corn was now four or five feet high, and the rice already from six inches to two feet. There were great numbers of the tung-oil trees, whose massive foliage gives a pleasant shade; and groups of coolies might be seen sitting under them, resting themselves, and enjoying the peaceful pipe. Apricots, too, were now in season, and, as well as great quantities of the Pi-Pa (or Loquot) fruit, were sold at stalls by the roadside.

The country produced plenty of vegetables; cucumbers could be bought for about a farthing, and tobacco, a great deal of which is grown in the neighbourhood, costs 280 cash a catty. It was a pleasant ride from Ku-Ch'êng. The road ran along the side of a steep and precipitous hill about two hundred feet above the river.

There were a good many trees to give shade; and at one projecting point, under a fine ash, an old woman had a table with a few of the little eatables and drinks that appeal irresistibly to the palate of a Chinaman.

About six miles from Ku-Ch'êng, a long point, from a spur, projects into the river, and nearly joining the other bank makes almost a horse-shoe bend. This is evidently the point where, in ages gone by, the two banks have been joined, forming, in what is now the valley, above a large lake. Here the limestone rock appears at the bottom, and can be seen underneath the deposit of clay and rounded stones which accumulated during the period of the lake.

As we descended the hills again became more steep, and in some places were broken into precipices, leaving less opportunity for cultivation. There were a good many mulberry trees nearly despoiled of their leaves. In the villages the silk cocoons were put out in great flat baskets to dry in the sun, and the women were seen sitting at the doors of their houses spinning silk.

At Kuang-Yi there was a beautiful new iron chain suspension bridge across a large stream that here joined the river. This was not yet quite finished, though we were able to cross it. The piers were very massive, and solidly built of stone; the dip of the bridge was hardly perceptible; elegant triumphal arches were being erected at each end, and when complete it will be a really graceful structure.

June 14.—Beyond Kuang-Yi the hills again close in, as if loth to let the foaming river escape; the population again is sparse, and for almost the last time the slopes are left partly uncultivated. There were some forlorn blackberries by the roadside, that seemed languishing for their companions of the forests; and throughout all nature there seemed to be a struggle for the mastery between the spirits of the mountains and the plain. The sun rose like a copper disc, threatening a hot and sultry day; but as the morning grew, the air was freshened by a gentle wind, that wandered amongst the trees, and lightly stirred the branches; and in the soft murmur of the foliage fancy might have heard a whispered farewell from the distant pine-clad heights.

On the road one of the Tinc-Chais pointed out what in England would hardly be called a footpath, leading straight up the side of a precipice, and with evident pride said that was the high road to Shen-Si.

We gained the entrance to Chiu-Chou just in time to escape a heavy pour of rain; and here the inn was of the dirtiest description. In one corner of the room half a dozen huge empty vats were piled; three bedsteads, where the usual pernicious mattresses sheltered untold myriads of fleas, lumbered the other angles. At one end a thing like a deep bookcase, with one shelf and no front, was sacred to the family deity; under this were the remains of all the earthen pots that had been broken in the household for generations; and there was an accumulation of filth among them that had collected during the same period of time. Another mouldering heap of corruption was under each bed, and what fostering rottenness might have been disclosed by the removal of the vats I dared not think of. There was, of course, the usual pigsty next door. I set the people to work to remove the filth, of which there was so much that in any other country there would hardly have been room for it in the village; but in China there is always room for dirt, if there is none for anything else.

This was another of the Chinese gala days, when a great festival is celebrated, when coolies like to idle, when servants expect gifts, when gongs are beaten and crackers exploded, and when all who can afford it eat something more or something better than usual. The coolies sent a message by Chin-Tai to the effect that they wished me many happy returns of the day, or in other words hoped I would give them presents. In accordance with custom, I sent them strings of cash that they did not deserve. Then came Tinc-Chais bringing gifts of walnuts and eggs preserved in lime. The coolies followed after, and like Oliver Twist asked for more. In due order I forwarded presents to the Tinc-Chais, who now returned, and severally knocked their heads against the ground. Then the gongs began to beat, the sound of crackers rent the air, and when the merriment was yet at its height, I retired to bed.

June 15.—We now entered the country of stone bridges, and a little below Hsiang-^{Ng}ai-Pa there was an exceedingly elegant, one-arched stone bridge. Ssŭ-Ch'uan is justly celebrated for its stone bridges, and we all began to realise the proximity of the plains. Nearly all the water was off the rice fields; the Indian corn was high; there were melons in the gardens; the climate was hotter; the grass by the wayside was rather burnt; and for the first time in this trip there was dust upon the road.

There are no vines in Ssu-Ch'uan, although the climate seems well adapted for them. The story goes that there used to be vines, but some wise ruler, thinking the people drank too much wine, ordered all the vines to be cut down; the order was ruthlessly carried out, and now, instead of good wine, the people drink spirits distilled from every kind of grain. Drunkenness is nevertheless almost unknown. During all my stay in China I scarcely ever saw a drunken man. I often used to see the coolies at breakfast taking their little 'chasse' of spirit; this they carry in stone bottles, which hold about as much as two sherry glasses, and they drink it out of cups not much larger than thimbles. Even this quantity is, however, a luxury they only indulge in now and then when they feel themselves very rich.

At breakfast I tasted one of the eggs the Tinc-Chais had given me, and thought it particularly nasty, although the Chinese consider eggs prepared in this way a great delicacy. Ducks' eggs are taken fresh, and steeped in a solution of lime and salt; the lime penetrates through the shell, turns all the egg quite black, and leaves a perceptible taste. After this the egg is encased in clay, and baked. In this way, with the clay outside, they will keep many months. The white becomes a jelly, and the yolk is in consistency like that of a rather hard boiled egg.[6]

At Hsiang-^{Ng}ai-Pa the river has at length escaped the trammels of the mountains, and though still a rushing stream, much encumbered by rapids, boats now navigate it, and can descend all the way to the Ocean Sea. The coolies had counted on an idle day, but Chin-Tai brought back the mournful news that the craft at this place were not large enough for us; so with sorrowful countenances they shouldered their loads and tramped to P'ing-I-P'u.

June 16.—This was a joyous day for the coolies.

[6] When Commissioner Yeh died a prisoner in Calcutta, several large chests of these eggs were found, which he had brought with him from China to solace the years of captivity.—*Y.*

We walked about a quarter of a mile to where a boat sufficiently large to accommodate us all was waiting for us. We were soon all packed and under way, and began the descent. There were rapids at about every half-mile, and the current was everywhere very strong. Many boats were tracking up, and the old familiar songs of the trackers resounded amongst the rocks. We seemed to fly past the shore, and several times in the shallows there was a scraping and bumping and a taking in of water over the bow that would have been alarming to weak nerves.

The first part of the journey was through narrow gorges, with precipices at each side; but at last the valley opened out for good, and we bid a final farewell to the mountains. The low hills which now bounded the river were of very red rock, well wooded; and though the slopes were easy, there was little cultivation on them. As in the basin of Ch'eng-Tu, there were a great number of trees in the plain, which was somewhat tropical in appearance. This was partly caused by some trees that had very straight and bare stems, and a bunch of foliage at the top, which in the distance looked like palms. There were great numbers of water-wheels for raising the water from the river; the appearance of twenty or thirty of these gigantic wheels, side by side, moving round in a most deliberate way, is very peculiar. These contrivances are exceedingly simple, but very effective; the diameter of the wheels is about twenty-four feet; bits of matting or light pieces of wood make the floats, and small bamboos about a foot long are the buckets. The current acting on the floats turns the wheel round, and the buckets, arranged round the circumference, lift the water to a trough. As the level of the river is of course liable to extensive variations, these wheels are not turned directly by the river, but by the water flowing through little canals, which at this season were considerably higher than the main stream.[7]

The bunding work all along the river is very extensive; the bunds being, as at Kuan-Hsien, of wicker baskets filled with boulders. A great deal of water of course constantly runs through, but this does no harm, and the dykes so formed keep back water enough to fill the irrigation canals. In some places the banks of the river are embanked, but in nearly every case the bund is made for the purpose of irrigation.

On we dashed, and a passing glance was all that could be obtained at the villages with their fields of tobacco; or the busy fishermen in their flat-bottomed punts, and a row of cormorants in the stern. We shot many a rapid in the morning's run, and anchored at Chiang-Yu-Hsien, where, as the Tinc-Chais were to be changed, I decided to go to an inn for breakfast. As I stepped ashore I saw a number of braves at target practice with bows and arrows, but I could not stop to watch them, as a crowd seemed to spring out of the ground, and I walked up to the town through a miscellaneous swarm of grown-up people and laughing children, who moved about amongst their elders' legs.

The distance was not far, but it was enough to show that we had now entered a different country. There were a great number of people in the fields and on the roads; crowds of boats were anchored off every considerable town, and the traffic on the river was very great. The towns were large, and full of

---

[7] There is a very careful and accurate detailed drawing (to scale) of one of these wheels in Staunton's account of Earl Macartney's embassy.

people; the fine open shop-fronts exposed rich goods for sale inside; and there was an appearance of wealth and prosperity, of life and activity, about the country, that contrasted remarkably with the miserable poverty of the mountains.

It was a sudden change from the quiet little mountain village that I left in the morning to the busy, noisy town where I breakfasted. I could, as I sat in the inn, hear all the going to and fro in the streets, itinerant vendors selling their wares and crying them out, and the constant chatter of the coolies in the restaurant hard by.

June 17.—Chin-Tai had already hired a boat for the next stage, but on arrival at the landing-place the captain wanted to take an extra cargo of passengers; there was a fierce battle over this; as usual all the people began shouting at one another; but matters did not take a favourable turn until the muscular Chung-Erh dropped one or two of them into the water, when the rest seemed no longer anxious to retain their seats, and the appearance of the four Tinc-Chais and superior officer from Chiang-Yu-Hsien, all in alarmingly gorgeous apparel, so overawed the captain that he actually held his tongue.

Huc declares that a pious Buddhist will not even kill the vermin with which his body swarms; my coolies went further, they wanted to sit in the bows of the boat, so that the breeze, instead of blowing the parasites into the water, would waft them gently about the vessel; some might even have visited me, but the coolies did not mind this, and objected to move until Chung-Erh again stretched himself. Then they went aft, except one, a confirmed opium-smoker, with a pimply red nose, who had wound his plait round his head, and arranged an enormous white lily in his hair, so as to bring the pure white of the flower into artistic contrast with the fiery hue of his rubicund organ. He pretended to be a boatman, until his complete ignorance of nautical manœuvres disclosed the imposture. He next endeavoured to conceal himself behind the legs of the captain, but the legs of the captain were thin, and the coolie was stout, and this attempt was not a complete success. He then tried to hide behind my chair but the broad brim of his hat bewrayed him, and he was ignominiously ejected and driven to the stern.

The river was now not so headstrong as it had been above Chung-Pa-Ch'ang, but, like a youth, had sobered down with increasing age, and the rapids, though numerous, were insignificant. After five miles we stopped at Chang-Ming-Hsien to get fresh Tinc-Chais; I did not go into the town, but landed to look at the crops, which were chiefly Indian corn, with beans and ground nuts.

The Hsien sent the usual anatomical fowls and ducks, an article that Chin-Tai facetiously called a ham, and some Tinc-Chais, whose beautiful clothes quite dimmed the lustre of the men from Chiang-Yu. As soon as the latter had knocked their heads, and the others had been installed, we started again.

I was amused at the artifices of some half-dozen of the coolies, who, taking up airy positions in the bows, again hoped they might escape my notice; the old villain with the lily held out till the last, and absolutely refused to get away until Chung-Erh beat him about the folds of his garments with a string of cash.

The river, now from seventy to a hundred yards wide, here ran through a broad flat valley, bounded by hills from fifty to one hundred and fifty feet high:

sometimes narrowing, a rocky point with a pagoda on its summit would project into the stream ; then the hills retreating, would leave an amphitheatre of cultivated plain. The slopes were of red rock, and were mostly covered with small woods, and notwithstanding the close cultivation of the plains, room was yet found in them for many groves of trees. In the evening our river journey came to an end at Mien-Chou, a large and important place, protected from the rivei floods by very extensive well-built river walls. The streets were nice and clean, and free from smells, and there was a really good inn. I attracted no attention, not so much as one little boy running after my chair.

Great quantities of beautiful vegetables were displayed in the market ; large cabbages as round as cannon balls, splendid turnips and brinjalls, and magnificent cucumbers in such quantities that it would be impossible to make use of them without the coolies, whose cast-iron digestions permit them to eat large uncooked cucumbers, skin and all. There was another vegetable unknown to me, in appearance, colour, and shape like a very large and perfectly smooth cucumber ; these are boiled, and the part between the rind and the centre is eaten. The principal food of the coolies was still Indian corn, and very few could as yet indulge in the luxury of rice.

The want of rain was beginning to be felt here, and the people were fasting, praying, burning incense, and beating gongs, whereby they hoped to bribe or terrify the deities.

June 18.—I used to put my clock on five minutes every day to counteract the tendency of my servants to get later each morning. They never discovered the artifice, and when we started from Mien-Chou it was barely light.

There was no one in the streets, and the roads were quite deserted, as we were before the earliest of the early coolies. This, however, soon changed; all became life and bustle, and everything showed that we had now struck a great highroad. Numbers of coolies going both ways, chairs and ponies and frequent tea-houses enlivened the scene.

The road first followed up a small stream in a flat valley, bounded by low undulations, rising to about one hundred feet. In the valley of the stream the crops were still Indian corn, beans, and ground nuts; of the latter the Chinese make oil, as they do of almost everything; they also eat them; and at all the stalls by the roadside there are little piles of some twenty of these, which are bought for a cash or two.

Rice was now extensively cultivated; the people were manuring their rice fields with liquid manure, and the smell of the country, far from being sweet, was abominable.

A great number of melons were grown in this neighbourhood, and in the gardens quantities of vegetables.

The road now left the stream, and entering the undulations followed a ridge about one hundred to one hundred and fifty feet high, from which little valleys sloped down on both sides, all laid out in terraces for cultivation, chiefly of Indian corn.

The want of rain here was terrible; the rice fields were quite dry, and the Indian corn in places looked rather burnt up.

The landscape was very pretty, the hills of a red clayey sandstone, and the houses, which were scattered about, all hidden in trees. The slopes were interspersed with little woods, the fields were particularly neat; and every now and then looking back the valley we had just left could be seen, the rice fields at this distance giving the appearance of an immense and rich flat pasture land.

A little inn received me at Hsin-P'u, where there was a sort of court with three walls and a roof; there was a screen in front, and though very small, being open, it was tolerably clean. I sat down here to breakfast, much to the astonishment of the Tinc-Chais, who thought it impossible for me to stop in such poor quarters.

About five miles from Hsin-P'u the road passes over another plain, which was very dry, and, crossing a low range of undulations, entered the valley of Lo-Chiang-Hsien. Here the country was at length well watered, and the crops very fine. The rice was getting large, and had arrived at the stage when the farmers, with their wives and children, go into the fields and stir up the mud about the roots with long poles and with their feet.

We left the main road here and followed the river, crossing its two branches by low bridges, impassable in the rainy season, when the river is crossed by a fine bridge lower down.

There was a suburb nearly three quarters of a mile long before the gate of Lo-Chiang-Hsien was passed, and then we entered a long, broad, well-paved, quiet street of private houses, with the garden walls on each side looking clean and nice. We went through nearly half a mile of this before entering the busy part of the town, and we had hardly emerged into the bustle and commotion when a turning led us to a very good and quiet inn.

The presence of a foreigner seemed almost unnoticed, perhaps because other Europeans had already traversed the road from Ch'êng-Tu to Mien-Chou.

The road all day was very good, level and well-paved, a great treat to every one after the difficult mountain tracks of the past month. The thermometer rose to 88°, but with a pleasant breeze the day was agreeable for travellers.

The Hsien sent the usual duck, fowl, and sweet cakes; and we changed Tinc-Chais again.

The grave of Pong-Tung, or Pong-Chou, is in the city of Lo-Chiang-Hsien. Pong-Tung was a celebrated man, who lived in the third century of our era, during the reign of the great Liu-Pi, a monarch who, from the countless stories associated with his name that are interwoven in the annals of this period, seems to have taken the place in Chinese history assigned to King Alfred in our own.

Pong-Tung at first attracted the attention of Chu-Ko-Liang, the prime minister of this sovereign, who recommended him to Liu-Pi; but the latter declined to put him into any high position, and only made him a Hsien. He was very discontented, and spent his time in making sonnets, eating, drinking, and smoking.

This came to the ears of Liu-Pi, who sent Chang-Fi to him with orders to cut off his head, unless he saw good reason for not doing so.

When Chang-Fi arrived, there were about one hundred people outside the yamen, waiting to have their causes tried, and Pong-Tung was amusing himself in an indolent way. So Chang-Fi said to him:

'What are all these evil reports that I hear of you, that you waste all your time in indolence, and writing sonnets?'

'What!' replied Pong-Tung; 'what am I to do? Why am I put in this inferior position of a Hsien, where I have nothing to occupy me? And why should I not write sonnets, if the king gives me nothing better to do?'

'But,' answered Chang-Fi, 'I see an immense crowd outside your yamen waiting to be judged. Nothing to do!' he repeated with some anger, 'how can you say that? there is work outside for weeks!'

Said the other, 'Work for weeks, forsooth! Perhaps so for those little men of inferior ability who frequent the court; but a man of talent does much in a short time.' Making a sign to an attendant, he summoned the first complainant, and, like a sharp barrister, seizing the points of the case, dismissed the parties satisfied in about two minutes.

One after another the people came in; he listened to their pleadings, attended to what their opponents had to say, and before very long the court of the yamen was clear, every one of the people having left perfectly satisfied with the justness of the decisions given.

Thereupon Chang-Fi rose up, and said, 'This is no place for you'; and, returning to Liu-Pi, he told the story. Pong-Tung was sent for, and eventually became the left prime minister of Liu-Pi, and before Chu-Ko-Liang himself. (In China the left hand takes precedence of the right.)

June 19.—The road for some miles to the west of Lo-Chiang-Hsien passed through an undulating country, where the want of water was sadly apparent; some of the rice fields being quite dry, and the soil cracked.

The highest point of these undulations is not much more than two hundred feet above the level of the plains, but they form the water parting between the basins of the river of Lung-An-Fu and the rivers that water the Ch'eng-Tu plain.

Pai-Ma-Kuan, or the Pass of the White Horse, is amongst these undulations. It is so called after an event in the life of Liu-Pi.

After the disastrous battle when Liu-Pi lost his wife, the king was mounted on a remarkable white horse; his enemies knew this, and were scouring the field in search of him, when his prime minister, Pong-Tung, or Pong-Chou, riding up, prevailed upon his master to change horses, on the plea that his was the faster of the two. The monarch, whose noble nature, if he had known that the white horse was the object of the chase, would never have consented to the exchange, escaped; Pong-Chou was killed, and buried in a temple at Lo-Chiang-Hsien, where his grave is still shown. A little further we came to where a great drumming, beating of gongs, shouting and chaunting was going on. Inside a number of candles and incense sticks were burning before several gilded images; there were about a dozen men and boys in the place, all more or less officiating; there was no priest, for the temple did not possess one, but an official servant belonging to an adjacent hamlet, who was well acquainted with the prayers and drill of the proceedings, was standing before the principal altar, reciting the formulæ and giving the signals for the others to say their 'Amen.' This was done by violent shouting, and beating drums and gongs. They seemed very well amused, and as I saw that clouds were gathering, I had no doubt their prayers would turn out efficacious.

All the people were now in their summer garments; the blue turban, which was so universal a month ago, had quite disappeared; most of the coolies were bare-headed, or wore enormously broad-brimmed straw hats. Their bodies were naked, the poles on which they carried their loads resting on their bare shoulders. Their loose blue cotton trousers reached no lower than the knee; a cotton cloth was twisted about their waist like a cummerbund; an enormous quantity of bandages were wound round the leg between the knee and ankle (I have counted as many as eighteen turns), and a pair of straw sandals completed the costume.

The better classes had straw hats of rather a finer material, white cotton coats, loose white cotton trousers, and cloth shoes. Fans and umbrellas were universal.

Seven miles from Lo-Chiang-Hsien the road enters the busy and fertile plain of another river; and after all the mountain travelling, where poverty was so painfully apparent, and there was so little life and activity, it was a pleasant change to be again in this country, where everything betokened comfort and prosperity.

The soil was a grey clay, and the crops much the same as before, but rice was gradually ousting the Indian corn, and as we advanced further there was little else.

Great numbers of trees were cultivated, and in many places along the roadside there were little groves that gave a delightful shelter. There was a remarkable absence of fruit trees; I had seen none since leaving the mountains; but peaches, apricots, and greengages were exposed for sale.

Again we heard the familiar sounds of the creaking wheelbarrows and treadmills, as coolies industriously pumped up the little remaining water from the lower levels, and from the reservoirs, of which there are a good number by the side of the road. These reservoirs were deep pits, about ten or fifteen yards in diameter, sometimes lined with concrete; now they were beginning to look wofully empty.

The gardens round the houses, and outside some of the large towns, were especially neat. I have never seen vegetable gardens so beautifully kept anywhere else. Luxurious cucumbers trailing over a kind of trellis-work of long rushes planted in perfectly straight lines in double rows, separated by about three feet at the bottom, and meeting near the top; little square patches of onions in rows, with never a weed between them; and all sorts of vegetables, each in a neat little oblong or square bit of ground. All the vegetables that are displayed in the streets come from these gardens; and at Te-Yang-Hsien the show was magnificent, indeed I never saw such a lavish profusion of beautiful fresh vegetables spread out for sale as there were here. Amongst others a kind of bean that produces a pod nearly a foot in length; this is boiled and eaten, pod and all. It is a little too hard for the European taste, but the Chinese like to hear the sound of their teeth in what they eat, whether it be a peach or a bean.

Great numbers of coolies were seen on this road carrying large baskets of vegetables northwards; and a coolie, with his brown body, blue trousers, and baskets of deep reddish-purple brinjalls, with a large bunch of fresh green chillies at the top, would make a study worthy of some of the old Dutchmen who were so fond of painting fruit and vegetables.

Every spot of ground is cultivated, and where there is a ditch, or small watercourse, running by the road, there will be a row of beans at each side, and sometimes a little rice in the water; this is quite safe, and no one thinks of damaging or interfering with it.

Passing a cottage by the roadside, screened by a clump of bamboos and a fine ash, the family may be seen inside seated at dinner. The father has just come in from his work in the garden or fields, he has nothing on his body, and there is a passing glimpse of little dishes and chopsticks as the simple meal is discussed.

Now there is a village, perhaps only two or three houses, and here there are sure to be some half-dozen naked little children playing about, or tumbling amongst the inevitable pigs that belong to every dwelling. Outside every house one or more little pigs are tied up, some by the head, some by the tail, some have a string round the leg, and others round the body; but a little pig tied up somehow there is to almost every door, not to mention the numbers that run squeaking about under the ponies' feet, while their elders lie grunting in the middle of the road.

Later in the day the women and girls bring their sewing and work to the doors, or sit in the shade outside their houses, some embroidering, others making the shoes—for in many houses all articles of clothing are made at home—and just now a great many were spinning silk.

Here is a quiet tea-house under some fine trees, and a couple of sedan-chairs are in the road in front. Their occupants are sitting at one of the tables drinking tea, or eating one of the dainty dishes that are always displayed so profusely, while the coolies enjoy a rest, smoke their long pipes, or perhaps spend a couple of cash on a cake of wheaten flour or whole Indian corn.

Passing through a big and straggling village, there

are as many as fifty coolies all together in some favourite tea-house that has a good reputation along the road; here the almost naked waiter has enough to do, the calls for rice (for the rice is now getting cheap enough even for coolies) and hot water being incessant; and the rattle of the chopsticks, and the gobble-gobble of the men as they shovel their favourite food into their mouths, can almost be heard.

Here, in a deep stream of water, crossed by a neat stone bridge, are a couple of buffaloes enjoying a bath, the string through their noses held by an urchin sitting on the parapet, who turns round with wonder at the sight of a dusty foreigner, with such very hot clothes, walking or riding on a pony when there is a comfortable chair behind.

Now we are again in the country of handsome and beautifully decorated triumphal arches (I counted as many as eight on the outskirts of Han-Chou), and a couple of beggars in ragged clothes lying asleep in the shade of one of these, with a few cucumbers at his head, would be a fit study for Murillo.

The towns get busier and more crowded as the capital is approached; they do not seem to have any speciality in the way of trade, half the shops in the main street being tea-houses or restaurants; the numbers of these, and the numbers of people that at all hours of the day are eating and drinking, is surprising.

In a few shops there are cotton goods for sale, and in others odds and ends, such as crockery, great numbers of wax candles, and tapers for burning at the altars.

Many look like chemists' shops, with little drawers and jars; others sell great quantities of fans; but, besides the eating and drinking and providing for the wants of this life, by far the greater number of the rest are given up to the accommodation of the dead; coffin-makers everywhere appearing to do a marvellous trade, and shop after shop is passed where the fearfully heavy and lumbering coffins of China are made in numbers that one would think would provide for the wants of half a dozen generations.

A good deal of furniture is made; and nicely carved bedsteads and cupboards, sometimes inlaid with ornamental wood, are seen, both complete and in process of manufacture.

In the busy thoroughfares there are sure to be half a dozen barbers' shops, and here you see the Truefitts of China, generally boys of about twelve to fifteen years old, shaving the heads, washing, oiling, and doing up the plaits of their customers, who, seated in a comfortable chair, smoke a long pipe and fan themselves while this luxurious operation is going on.

As we advance to the west the water in the country increases, until we again come to the streams running by the roadside, and here the crops look beautiful. The rice is getting on well, and the Indian corn in some places is nearly fit for harvesting; but the amount of the latter gradually diminishes, until very nearly the whole country is given up to rice.

Tê-Yang-Hsien is a fine large town, with broad, clean streets, where we found an excellent inn.

The natural politeness of the people here overcame their curiosity, and although I rode on the pony through the busy streets full of people, no one followed me, not even little boys. The Hsien sent me many presents, and satellites innumerable, with whom I marched out in state.

First came four Tinc-Chais, each with the summer

official hat, for there is a summer dress and a winter dress, and all over China this dress is changed on the same day by an edict in the Peking 'Gazette.' The summer hat is of light straw, conical in shape, the base being about one and a half times the height. At the top is a tassel of red silk, which a high official is always very careful to have arranged so that it spreads equally over every part of the hat.

One of the Tinc-Chais has a grey coat, tied in at the waist with a cummerbund, into which he has tucked the tails to prevent their dragging on his heels. The sleeves are always a foot too long, and the ends of these long loose sleeves are used indifferently as towels or dusters; on the march they may be tucked up to leave the hands free to use the fan.

Another wears no cummerbund, but has an umbrella strapped to his shoulder.

All four wear the ordinary Chinese dress of long loose coat, and trousers coming a little below the knee. Two of them wear bandages round the calves, and all have the usual straw sandals on their feet.

They walk in front of me, calling out to the people to get out of the way, and making the creaking wheelbarrows draw up at the side of the road as I pass; sometimes these will be loaded with a couple of huge fat pigs, lying on their backs, their legs straight up in the air, and lashed to the barrow with strings round their fat stomachs.

The steward's Ma-Fu rides behind me, and the steward himself, who is a person of much importance, rides next. The Ma-Fu has a huge straw hat, his coat is open in front, showing his breast, and he has nothing on his feet. The steward is well dressed, with a blue lining to the underside of the broad brim of his straw hat, and wears stockings and shoes.

Then comes Liu-Liu asleep on his pony, his mouth wide open, and his tongue sticking out.

My empty chair follows, and Chin-Tai sits asleep amongst his pillows behind.

I sent Chung-Erh on to get the house at Ch'eng-Tu ready for me, and he disappeared behind some trees, sitting all askew on a saddle with a broken tree, thumping his pony with a big stick, while the Ma-Fu ran behind.

Two branches of a river were crossed before entering Han-Chou; one by a bridge of stone piles, with stones laid lengthways for the roadway; the other by a covered wooden bridge, one hundred and twenty-eight yards long, where, on both sides, hawkers sat with little stalls, selling all sorts of things.

There are three inns in this town, but there were Chinese gentlemen staying in the best room at all three, and I had to put up with inferior accommodation, which a week before I should have considered the height of luxury. There was the usual abominable stench, but by the aid of carbolic acid I managed to counteract it.

I began to think that vile smells could not be so unhealthy as we civilised people imagine; for the Chinese, who spend their lives amongst sewers, and die in the odour of cesspits, seem to get on well enough.

They have certainly wonderful ideas with regard to dirt and the fitness of things; I met a man who was out collecting dung, which he picked up on the road with a sort of hoe-shaped instrument; he had the usual bamboo over his shoulder, with a basket at each end; in one of these he put what he picked up, and in the other he had sweet cakes for sale. No Chinaman would ever dream that there was any nasty idea about this, but would remark with Petrucio that 'dainties are all Cates.'

When the Chou heard that the inns were all occupied, he sent to offer me a house; but it was now six o'clock, so I thanked him much, and said I could manage for a few hours where I was.

The Chou sent more fowls, more sauce, cakes, and Chinese macaroni, of which I had by this time such quantities that I might have set up a restaurant with an excellent chance of success.

June 20.—We marched from Han-Chou over the rich and fertile plain, every step we advanced bringing us more and more into the well-watered country; and after about six miles, rivulets full to the brim ran along the sides of the road, which crossed a stream every ten minutes. Some few of these brooks were almost large enough to be called rivers, and there was plenty of water in all of them.

As I passed under the numerous covered wooden bridges, with the hucksters sitting at the sides selling their goods, I thought of old Marco Polo, who must have marched from Mien-Chou over the identical road I traversed.

At Hsin-Tu-Hsien I was met in the suburbs by an official, who brought me the card of the Hsien with his compliments, and conducted me into a magnificent house just inside the gates on the left-hand side.

This was the Kung-Kuan, or official rest-house for any officers of State who may travel this way. It was a very fine building, with courts and good trees; and there were three or four servants about the place eager to attend to my wants.

Huc has described the Kung-Kuan (or 'palais communal,' as he called it) at Chien-Chou, and if this was not quite up to his flowery description, the contrast to the mountain huts to which we had been accustomed was sufficiently strong to make his brilliant eulogium not very inappropriate.

The Hsien then sent embroidered cushions for the chairs, and a Chinese red table-cloth. These are not put over the table as we put ours, but are hung in front to give an imposing appearance as a visitor enters.

Soon after, bedding also arrived; and, as the place was so clean and quiet, I really regretted that I was not going to stay here for the night.

After receiving the inevitable duck and fowl, masses of sweetmeats and bundles of macaroni that would have puzzled a Neapolitan, the process of changing Tinc-Chais was gone through for the last time.

First comes Chin-Tai with a book, in which the names of all the individuals are written. Well knowing that there are plenty of curious eyes about, I examine this attentively, as if the Chinese characters were all familiar to me as English writing, and I signify to Chin-Tai that the departing Tinc-Chais may come in. They all enter, kneel on one knee, knock their foreheads on the ground, and then go away.

In a few minutes more, another book, with the names of the new Tinc-Chais, is brought. These are introduced in their turn; they also knock their heads, and then retire.

Then Chin-Tai goes away and gives pieces of silver to the headman, and strings of cash to the Tinc-Chais. Having received these, they again solicit the honour of being admitted into the presence of the great Excellency, that they may thank his honour for his munificence. This I permit, and when all is over ' I extend my hand to them thus,' and feel almost as great an impostor as Malvolio.

Leaving this, we marched the remaining distance over a beautifully-watered country, where the crops looked splendid.

Six miles from Hsin-Tu-Hsien, a low red spur comes down to the road, the country is again slightly undulating, and not quite so well watered. I noticed a great quantity of chillies with their long green leaves and white flowers, but otherwise there was no change in the crops until we arrived at the outskirts of Ch'ĕng-Tu.

On this side the suburbs are about one mile and one third long, and consist of a row of houses on each side of the road, with extensive-walled vegetable-gardens behind.

Here were the usual eating and drinking shops, and the number of cakes and pies made of wheaten flour, bean-flour, and flour from all kinds of grain, seemed greater than ever. Indeed, it is impossible to conceive how all the quantities of these, and of sticks of sugar-cane, and a kind of blanc-mange made from beans or rice, ever find mouths to eat them.

At length we turned into the familiar quiet street. A few paces more brought us to our door, where Chung-Erh was waiting to receive us. The trip was finished, and I was not sorry to find myself once again in the little house, and to have some rest, and time for quiet writing.

END OF THE FIRST VOLUME.

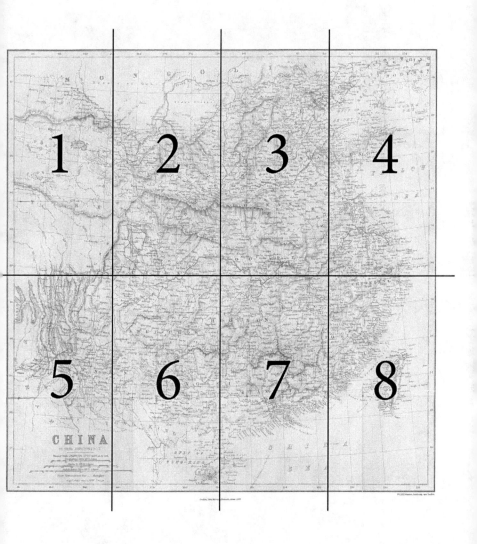

Key to following pages.

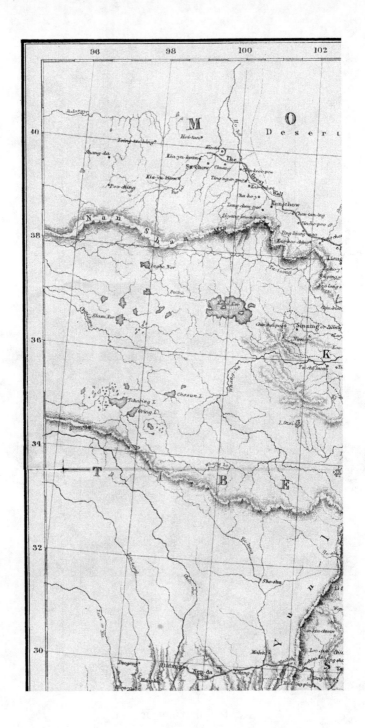

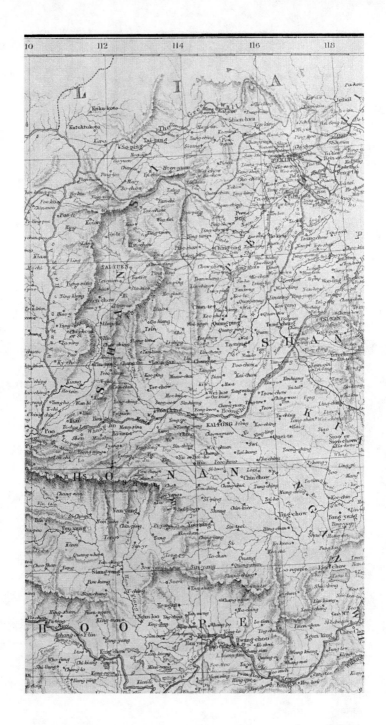

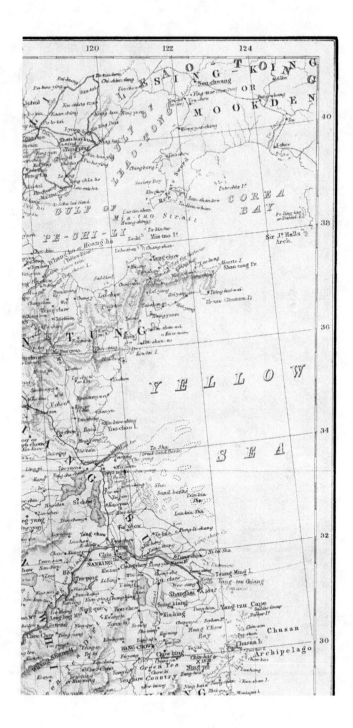

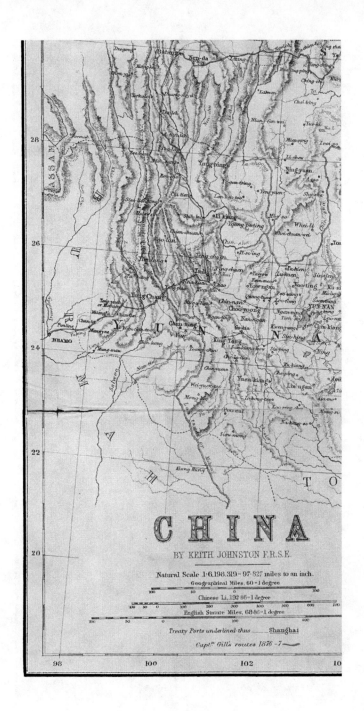

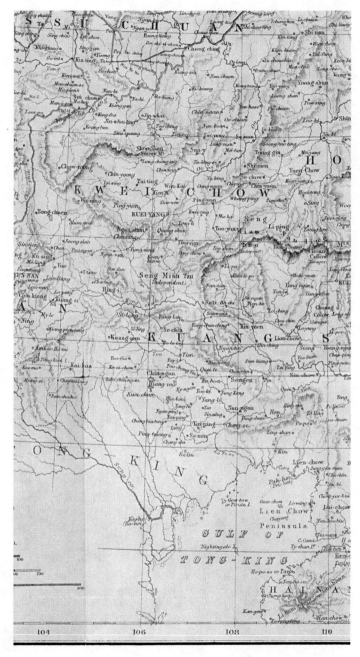

LaVergne, TN USA
19 November 2010
205506LV00001B/5/P

CAMBRIDGE LIBRARY COLLECTION

Books of enduring scholarly value

Travel and Exploration

The history of travel writing dates back to the Bible, Caesar, the Vikings and the Crusaders, and its many themes include war, trade, science and recreation. Explorers from Columbus to Cook charted lands not previously visited by Western travellers, and were followed by merchants, missionaries, and colonists, who wrote accounts of their experiences. The development of steam power in the nineteenth century provided opportunities for increasing numbers of 'ordinary' people to travel further, more economically, and more safely, and resulted in great enthusiasm for travel writing among the reading public. Works included in this series range from first-hand descriptions of previously unrecorded places, to literary accounts of the strange habits of foreigners, to examples of the burgeoning numbers of guidebooks produced to satisfy the needs of a new kind of traveller - the tourist.

The River of Golden Sand

William Gill (1843–1883) was an explorer and commissioned officer in the Royal Engineers. After inheriting a fortune from a distant relative in 1871, Gill decided to remain in Army and use his inheritance to finance explorations of remote countries, satisfying his love of travel and gathering intelligence for the British government. He was awarded a gold medal by the Royal Geographical Society in 1879 for his scientific observations on his expeditions. This two volume work, first published in 1880, is Gill's account of his expedition from Chengdu, China through Sichuan, along the eastern edge of Tibet via Litang, to Bhamo in Burma, a region little explored by westerners before him. Gill describes in vivid detail the cultures, societies and settlements of the region, and their political and economic systems. Volume 2 recounts his travels across the plateau to the upper reaches of the Irrawaddy, partly retracing Marco Polo's route.

Cambridge University Press has long been a pioneer in the reissuing of out-of-print titles from its own backlist, producing digital reprints of books that are still sought after by scholars and students but could not be reprinted economically using traditional technology. The Cambridge Library Collection extends this activity to a wider range of books which are still of importance to researchers and professionals, either for the source material they contain, or as landmarks in the history of their academic discipline.

Drawing from the world-renowned collections in the Cambridge University Library, and guided by the advice of experts in each subject area, Cambridge University Press is using state-of-the-art scanning machines in its own Printing House to capture the content of each book selected for inclusion. The files are processed to give a consistently clear, crisp image, and the books finished to the high quality standard for which the Press is recognised around the world. The latest print-on-demand technology ensures that the books will remain available indefinitely, and that orders for single or multiple copies can quickly be supplied.

The Cambridge Library Collection will bring back to life books of enduring scholarly value (including out-of-copyright works originally issued by other publishers) across a wide range of disciplines in the humanities and social sciences and in science and technology.

The River of Golden Sand

The Narrative of a Journey through China and Eastern Tibet to Burmah

VOLUME 2

WILLIAM JOHN GILL

CAMBRIDGE UNIVERSITY PRESS

Cambridge, New York, Melbourne, Madrid, Cape Town, Singapore,
São Paolo, Delhi, Dubai, Tokyo

Published in the United States of America by Cambridge University Press, New York

www.cambridge.org
Information on this title: www.cambridge.org/9781108019545

© in this compilation Cambridge University Press 2010

This edition first published 1880
This digitally printed version 2010

ISBN 978-1-108-01954-5 Paperback

This book reproduces the text of the original edition. The content and language reflect
the beliefs, practices and terminology of their time, and have not been updated.

Cambridge University Press wishes to make clear that the book, unless originally published
by Cambridge, is not being republished by, in association or collaboration with, or
with the endorsement or approval of, the original publisher or its successors in title.

THE
RIVER OF GOLDEN SAND

SECOND VOLUME

MAN-TZŬ HOUSES

THE
RIVER OF GOLDEN SAND

THE NARRATIVE OF A JOURNEY THROUGH CHINA
AND EASTERN TIBET TO BURMAH

WITH ILLUSTRATIONS
AND TEN MAPS FROM ORIGINAL SURVEYS

By Capt. WILLIAM GILL, Royal Engineers

With an Introductory Essay

By Col. HENRY YULE, C.B., R.E.

TREADMILL WATERMILL

IN TWO VOLUMES—VOL. II.

LONDON
JOHN MURRAY, ALBEMARLE STREET
1880

All rights reserved

CONTENTS

OF

THE SECOND VOLUME.

CHAPTER I.

'A RICH AND NOBLE CITY.'

PAGE

Account of Ch'êng-Tu, as given by Marco Polo—And by Padre Martin Martini—Description of the Modern City—The Rivers, and Probable Changes—Destruction of Documents—Arrival of Mr. Mesny—Political Aspects compel Change in Traveller's intended Route—Decision to travel Homeward *viâ* Bat'ang and Bhamo—The Bridges of Ch'êng-Tu—Chinese Temples and Notions of Worship—Ancient Inscribed Stone—The Heaven-Tooth—The City Walls—Visit to the Great Monastery of Wen-Shu-Yüan—Its Buildings and Curiosities—Chapel of Meditation —Antiquities of Ch'êng-Tu—Memorials of Early Events—The History of the Emperor Liu-Pi—Invitation to a Picnic—Notable Guests—The Dinner and its Peculiarities—Table Manners and Customs—Threatenings of Drought—False Notions about the Traveller—The Provision of Currency for the Journey, and Tedious Banking Business—Adieux at Ch'êng-Tu—'Organisation of Departure'—Concluding Notices of Ch'êng-Tu and the 'Province of Four Waters' (Ssŭ-Ch'uan) 1

CHAPTER II.

THE ANCIENT MARCHES OF TIBET.

To Shuang-Liu—Superstition as to Fires in Towns—Vast Fertility of Country—Personal Criticisms on Travellers—Chinese Idea of Foreigners in general—Hsin-Chin-Hsien—First Sight of the

CONTENTS OF

PAGE

Mountains —Ch'iung-Chou—Memorial of the System of 'Squeeze' —Fine Bridge on the Nan-Ho—Great Coolie-loads—Ta-T'ang-P'u —Tibetan Embassy— Pai-Chang-Yi — Indications of Opium-smoking—Paved Road in Decay—Ya-Chou-Fu—Its Commercial Importance—Cake-Tea—Coolies employed on the Traffic—Close and Beautiful Cultivation — Fei-Lung Pass—Wayside Anecdotes—Hsin-Tien-Chan—Sea of Straw Hats—Yung-Ching-Hsien—Traffic from Yün-Nan—Invitation to Breakfast—Ching-K'ou-Chan—Limestone Scenery—Cooper's Mistake about Tea-oil Trees: in reality Rhododendrons—Huang-Ni-P'u—Suspension Bridge—Mountains begin to close round—The Little Pass—Alleged Terrors of the Hills—The Great Pass, and the T'ai-Hsiang Pass—'Straw-Sandal Flat'—Ch'ing-Ch'i-Hsien—Divergence of Road to Yün-nan by Ning-Yuan—Circuitousness of our Route—Pan-Chiu-Ngai—The Insect Tree and the Wax Tree—Mode of Production of the Wax—The Mountain Crops—San-Chiao-Ch'êng — Hua-Ling-P'ing—River of Ta-Chien-Lu—Bridge Swept Away—Lêng-Chi—Hill People—Lu-Ting-Ch'iao and its Iron Suspension Bridge—Dominoes—Character of People of Ssŭ-Ch'uan—Toll upon Coolies—Hsiao-P'êng-Pa—Scantier Vegetation—Fragment of elaborate Ancient Road—Arrival at Ta-Chien-Lu 37

CHAPTER III.

'THE ARROW FURNACE FORGE.'

Ta-Chien-Lu—Legendary Etymology of the Name—Native King—Indian Rupees Current—The Place and People—'Om Mani Pemi Hom!'—A House found for us—The Local Government —Transport Arrangements—The Lamas and the Dalai Lama—The Prayer Cylinder and the Multiform Mani Inscriptions—The Lama Ambassadors—Menaces of our Fate if we entered Tibet—The Servants begin to Quail—Chin-Tai, his Greed and his Tempers—Heavy Provisioning for the Journey—Contrast of Tibetan and Chinese Habits—Of Tibetan Simplicity of Fare with Chinese Variety—Tariff at Ta-Chien-Lu—Preparations to meet Cold, and Needless Laying-in of Furs—Kindly Aid rendered by the late Bishop Chauveau—An Interpreter found through his Help—Curious History of Peh-ma, this Christian Interpreter—Civility of the Chinese Officers—Difficulties about Baggage Cattle—Negotiations for Carriage—The New Ma-Fu—Visit to a Lamassery—Currency for the Journey—Search for a Horse—Parting Gifts and Purchases—The Tibetan's Inseparable Wooden Cup—Tib left behind—Fresh Selection of Nags—Fatality of Small-pox in Tibet 76

THE SECOND VOLUME. [7]

CHAPTER IV.

THE GREAT PLATEAU.

I. TA-CHIEN-LU TO LIT'ANG.

PAGE

The Departure from Ta-Chien-Lu—The Loads Distributed—The Cavalcade described—A Supper of Tsanba—Village of Cheh-toh—A Wreck of the Crockery—Tibetan Salutation—Delicious Air — Half-bred Yaks — Splendid Alpine Pastures — Wild Fruits—The Sacred Cairn—Direct Road to Chiamdo—Ti-zu—Delightful *al fresco* Breakfast—A Land of Milk and Butter—Hot Spring—The Buttered-Tea Churn—Tsanba—Tibetan Houses—Hospitable Folk—An-niang—Wild Flowers—Roadside Groups from the Fair—A Gallop over Turf—Village of Ngoloh—Description of Quarters—An Imperial Courier—Halt at Ngoloh—Pass of Ka-ji-La—Magnificent Alpine Prospect—The 'King of Mountains'—Pine Forests—Wu-rum-shih—The Margary Proclamation—Illness of the Servant Huang-Fu—Descent towards the Ya-Lung River—Appearance of Green Parrots—The Octagon Tower—Tibetan Horse-shoeing—Ho-K'ou or Nia-chu-ka, on the Ya-Lung River—Tibetan Aversion to Fish, and its Origin—Lights of Pine-Splinter—Ferry over the Ya-Lung—Re-ascent towards the Plateau—Mah-geh-chung—Our Quarters there—A 'Medicine Mountain'—La-ni-ba—Crest of the Ra-ma Pass—Post of Mu-lung-gung—Female Decorations—Lit'ang-Ngoloh—Description of House and Appliances—Serious Illness of Huang-Fu—Domestic Sketches—Holly-leaved Oaks—Tang-Go Pass—Gold-washing—Cha-ma-ra-don Forest left below—The Great Dogs and their Change of Garb—Deh-re Pass—The Rhubarb Plant—Rarefied Atmosphere—Pass of Wang-gi—Patience of the Ma-Fu—Ho-chŭ-ka Village—Abundance of Supplies—Shie-gi Pass—The Surong Mountains—Treeless but Flowery Heights—Lit'ang in View—Arrival there—Huang-Fu's Illness 116

CHAPTER V.

THE GREAT PLATEAU—*continued*.

II. LIT'ANG TO BAT'ANG.

Departure—Quit the Lit'ang Plain—To-Tang—Dreary Aspects—Desolate but Healthy Region—Lassa Pilgrims—Characteristic Plant—High Pass of Nga-ra-la-Ka—Cold Granite Region—Alarms of Robbers—Dzong-dä—Pine-clad Valleys again—La-ma-ya—The Cairns and Mani Inscriptions—The Man-ga-La—Stupendous Alpine Scene—The vast Snowy Peak of Nen-

da — Tibetan Mode of Building — Surpassing Scenery — The
'Yün-nan Bridge' — Charming Halt at Rati, or San-Pa — Reminiscences of Huc and Gabet's Journey — The Guide's Head
turned with Wonder — Sublime Aspect of the Mountains —
Rhododendrons — Beautiful Valley of Pines — Unpeopled Tract
— Ta-shiu — Manifold Uses of Raw Hide — Striking Picture —
The J'Ra-ka-La — Savage Desolation — Our Guide — Our Quarters
at Pung-cha-mu — Descent towards the Chin-Sha River — Wild
Independent Tibetans — Ba-jung-shih — Welcome on Arrival at
Bat'ang — Chao the Magistrate, and Favourable Impressions —
Abbé Desgodins — Discharge of our Suite — The Name of
Bat'ang — The Plain — Possibilities of Navigation on the Rivers
— Earthquakes — The Great Lamassery — Number of such
Causes that fill them — The Lamas a Curse to the Country —
Eccentric Chinese System of Accounts — Property in Alpine
Pastures — Slavery and Cattle-holding — Gradual Depopulation
of Tibet — The Tibetan Chief of Bat'ang — Hostility of the
Lamas — Alleged Muster to bar our Advance — Quarters at
Bat'ang — Cordial Hospitalities of Chinese Officials — Lively
Banquet at Shou's — Noisy Game of Morra — The Wondrous
Bath — A Night Alarm 158

CHAPTER VI.

REGION OF THE RIVER OF GOLDEN SAND.

I. BAT'ANG TO SHA-LU.

Departure from Bat'ang — Adieux to MM. Desgodins and Biet —
Broad Features of the Geography, and Relation between Road
and Rivers — The 'Tea-Tree Hill' — Chinese Rendering of
Foreign Names — Valley of the Chin-Sha or Golden Sand — A
False Alarm — Journey along the Chin-Sha — Cross and Quit it
— Fine Scenery of the Kong-tze-ka Pass — Lama Oppression,
and Depopulation of the Country — Quarters at Jang-ba —
Growth of our Escort — Armed Opposition on the Lassa Road —
Kia-ne-tyin — March through Woods of Pine, Yew, and Juniper
— Clean Quarters at Dzung-ngyu — The River of Kiang-ka — Hospitable People — Desolated Country — Chŭ-sung-dho (Cooper's
'Jessundee') — Nieh-ma-sa — Continued Depopulation — Ma-ra
and Tsaleh — Conversations with Magistrate Chao — 'The Boy
the Father of the Man' — Watershed between the Chin-Sha
and the Lan-T'sang — Camping Ground of Lung-zung-nang —
The Bamboo Seen Again — The Gorge of Dong — Parallel Fissures from North to South — The Town of A-tun-tzŭ (Cooper's
Atenze) — The Oppression of the 'Lekin' — Evil Character of
A-tun-tzŭ — Prevalence of Goitre — Water Analysis — Farewell

THE SECOND VOLUME. [9]

PAGE

Visits from Chao and the Native Chief—Ceremonies of Adieu—Prevalence of Opium-smoking in Yün-nan—Patience Exercised—Long and Moist March—Passes of Mien-chu and Shwo—Difficulties of March in Darkness—Torchlight—Reach Deung-do-lin—The Word 'Lamassery'—Abbé Renou—T. T. Cooper—Carrier Difficulties—Taking it Easy—Change of Climate and Scenery—New Glimpse of the Chin-Sha—Sha-lu Village—Big Tibetan Dog 205

CHAPTER VII.

REGION OF THE RIVER OF GOLDEN SAND—*continued*.

II. SHA-LU TO TA-LI-FU.

Hazels, Pines, Currants, Rhododendrons—Pass of Jing-go-La—Ka-ri—Houses there—Verdant and Peopled Region—N'doh-sung—Domestic Decoration—Rapacity of the Fu of Wei-si—Chinese Agriculture Reappears—Hospitable Welcome of Ron-sha—Toilsome March in Rain—Reception at La-pu—Our Lodging there—Friendly Activity of our Conductor—Walnut Trees, and Marco Polo's 'Cloves,' *i.e.* Cassia Trees—Tibet passing into the Background—Village of Jie-bu-ti—The Banks of the Chin-Sha Regained, and Re-appearance of Orange-trees and Parrots—The Mu-su People—March down the Great River—Lamas still Seen—Mu-khun-do—The Pu-Erh Tea, and How to Make it—Ku-deu—Fine Position—The Mu-sus—European Aspect of some—Honesty of Carriers—Drunken Escort Officer—Ch'iao-T'ou—Tea-Houses Reappear—San-Hsien-Ku—Festive Reception—Shih-Ku, or Stone-Drum Town—Adieu to the River of Golden Sand—Mount Chin-Ku-P'u—Changed Scenery—Dense Population on Rice-plains—Inns begin again—Oppression of the 'Lekin'—City of Chien-Ch'uan-Chou—A Henpecked Warrior—Fair Words of the Chou—Road through Populous Rice-lands—Lake Basins—I-Yang-T'ang—Rudeness of Local Officer at Niu-Chieh—Wretchedness of the People—The Rude Officer Abashed—Position of Niu-Chieh in Lake-basin—Probable Changes in Water-levels—Dilapidated City of Lang-Ch'iung—Civil Landlord—Pu-Erh Tea—Sulphur Spring—Opium-smoking—Damp and Dreary Aspects—The Erh-Hai, or Lake of Ta-Li—Road along the Lake Shore—Arrival at Ta-Li-Fu 249

CHAPTER VIII.

IN THE FOOTSTEPS OF MARCO POLO AND OF AUGUSTUS MARGARY.

I. THE LAND OF THE GOLD-TEETH.

PAGE

Ta-Li-Fu Province, the Carajan of Marco Polo—The Lake and Environs—Père Leguilcher—The Plain of Ta-Li—The (so-called) Panthés, and the Name—The Mahometan Rebellion—The Mahometans of the Province—Presents of Local Delicacies—Occupations at Ta-Li—The Tao-T'ai—The Ti-T'ai, General Yang, a Remarkable Personage—The Old Troubles about Carriage—Illustration of the 'Squeeze'—Departure from Ta-Li—Adieu to P. Leguilcher—Yang and Yang—Marco Polo's Cakes of Salt—Paucity of Present Traffic on Road—Devastated Country—Diminution of Population—Yang-P'i River—Imaginary Bifurcation in Maps—Chain Suspension Bridge—Perversities of the Path—T'ai-P'ing-P'u—Lofty Hamlet of Tou-P'o-Shao—The Shun-P'i-Ho—A Treat of Bread at Huang-Lien-P'u—Mr. McCarthy's Servants—Dearth of Population—Traces of War—Chestnut and Oak Woods—Descent to Plain of Yung-P'ing-Hsien—The Town Destroyed—We Lodge at Ch'ü-Tung—Topographical Elucidation—Inn at Sha-Yang—Unendurable Smoke — View of the Mekong or Lan-T'sang River—Chain Bridge Across it—Desperate Ascent—Buckwheat Porridge—Ta-Li-Shao—Pan-Ch'iao—Rice Macaroni—Polo's Salt Loaves Again—His 'Vochan' and the 'Parlous Fight' there—Yung-Ch'ang-Fu—Difficulties about Transport—The 'Squeeze' again—A General on the March—A Quarrel Imminent, but the General is Drawn Off—Stones and Beads Brought for Sale—The Yung-Ch'ang Market—Difficulties of Marco Polo's Itinerary from Carajan to Mien—Mr. Baber's Solution—Recent Plague on the Road 299

CHAPTER IX.

IN THE FOOTSTEPS OF MARCO POLO AND OF AUGUSTUS MARGARY—*continued*.

II. THE MARCHES OF THE KINGDOM OF MIEN.

Departure from Yung-Ch'ang—Graves of Aborigines—Hun-Shui-Tang—Fine Ponies and Mules—Fang-Ma-Ch'ang—Pestiferous Valley of the Lu-Ch'iang or Salwen River—Passage by Chain-Bridge—Steep Ascent to Ho-Mu-Shu—Greedy Host—Needless Difficulties of Road—Old Custom of 'Wappenshaw' and Military Tests—The Lung-Chiang or Shwé-li River—Salutes by the

	PAGE
Way—A Celt for Sale—Ch'in-T'sai-T'ang—The City of Têng-Yüeh or Momein—Things better managed in Ssŭ-Ch'uan—The Chi-Fu Convention—Mule-chaffering—H'siao-Ho-Ti—Nan-Tien—Reception by a Shan Lady—Her Costume—First Burmese Priests—Change of Scenery—Kan-Ngai, or Muang-la—The Chief—Passage of the Ta-ping River—First Burmese Pagoda—Lovely Scene near Chan-Ta—Chan-Ta (Sanda) and the Chief there—Oppressions of Chinese—Festival at Chan-Ta—Shan Pictures by the Way—Shan and Kakyen Figures—Road-side Scenes in Ta-ping Valley—Bamboos and Birds by the Way—Lying Litigants—T'ai-P'ing-Chieh or Kara-hokah—Reach Chinese Frontier Town of Man-Yün (Manwyne)—Visit from Notorious Li-Sieh-Tai—Treatment at Man-Yün—The Pa-I People—English Goods in Bazar—Letter of Welcome from Mr. Cooper—Scene of the Murder of Augustus Margary—The Kakyen Country—A Shot at the Party; only Tentative—Kakyen Huts—Meddling with the Spirits' Corner—Fire got by Air-Compression—Buffalo Beef—Grand Forest Scenery—Bamboos and Potatoes—An Imprudent Halt to Cook—A Venture in the Forest—Perils from Ants and from Bullies—Would-be Leviers of Blackmail—Benighted—A Welcome Rencontre—Cooper's Messengers and Stores—A Burmese *Po-é* or Ballet—Embark on Bhamo River—The Irawadi Disappointing—Kindly Welcome from English Agent at Bhamo—Alas, poor Cooper!—Bhamo to Dover—The Journey Ended	347
APPENDIX A	413
" B	426
" C	427
INDEX	433

ILLUSTRATIONS and MAPS to VOL. II.

MAN-TZU HOUSES	*Frontispiece*
TREADMILL-WATERMILL	*Title-page*
ROUTE MAP. SECTION II. FROM CH'ÊNG-TU TO TA-CHIEN-LU	*To face page* 37
HORIZONTAL SECTIONS BETWEEN TA-CHIEN-LU AND BAT'ANG, AND BETWEEN TA-LI-FU AND BHAMO .	„ 77
ROUTE MAP. SECTION III. FROM TA-CHIEN-LU TO BAT'ANG	„ 117
ROUTE MAP. SECTION IV. FROM BAT'ANG TO A-TUN-TZŬ	„ 205
ROUTE MAP. SECTION V. FROM A-TUN-TZŬ TO TZ'U-KUA	„ 235
ROUTE MAP. SECTION VI. FROM TZ'U-KUA TO TA-LI-FU	„ 273
„ „ VII. FROM TA-LI-FU TO BHAMO	„ 311

Errata in Vol. II.

Page 52 line 5 from foot, *for* 'Chin-Sha-Ching' *read* 'Chin-Sha-Chiang.'
„ 53 „ 7 from foot, *for* 'Sun-P'an-T'ing' *read* 'Sung-P'an-T'ing.'
„ 166 „ 5 from foot, *for* 'Le-ka-ndo' *read* 'La-Ka-Ndo.'
„ 81 „ 4 ⎫
„ 185 „ 10 ⎬ *for* 'Sieh-T'ai' *read* 'Hsieh-T'ai.'
„ 196 „ 27 ⎭
„ 229 „ 4 from foot, *for* 'above the Dong' *read* 'above Dong.'
„ 230 „ 4, *for* 'nestled' *read* 'which nestled.'
„ 253 „ 5 ⎫ *for* 'Chin-Chŭ' *read* 'Chiu-Chŭ.'
„ 258 „ 23 ⎭
„ 259 „ 11 ⎫
„ 259 „ 15 ⎬ *for* 'Kung-Chŭ' *read* 'Shio-gung-Chu.'
„ 262 „ 4 ⎪
„ 264 „ 2 ⎭
„ 269 „ 13 insert full-stop after diameter; the word 'rising' commencing a new paragraph.
„ 269 „ 16 alter semicolon to comma after the word 'woods.'
„ 278 „ 4, *for* 'have' *read* 'had.'
„ 280 „ 19, *for* 'Chin-Ch'uan' *read* 'Chien-Ch'uan.'
„ 308 „ 1–3, *for* 'thereby' *read* 'although;' *and for* 'Tao-Tai. According' *read* 'Tao-Tai, according' &c.
„ 312 „ 7. The date '*October* 5' should be inserted here.
„ 316*n*. The German map is one compiled by Berghaus.
„ 327 „ 9 from foot, *for* 'T'ieh' *read* 'T'ien.'
„ 327 „ 4 „ „ *for* 'Yang-Pi' *read* 'Yung-P'ing.'
„ 330 „ 10, *for* 'Yung-Chang' *read* 'Yung-Ch'ang.'
„ 360 „ 7 from foot, *for* 'veiled' *read* 'vailed.'
„ 368 „ 25, *for* 'Chin-Ch'êng' *read* 'Kan-Ngai-Chiu-Ch'êng.'

THE
RIVER OF GOLDEN SAND.

CHAPTER I.

'A RICH AND NOBLE CITY.'

Account of Ch'êng-Tu, as given by Marco Polo—And by Padre Martin Martini—Description of the Modern City—The Rivers, and Probable Changes—Destruction of Documents—Arrival of Mr. Mesny—Political Aspects compel Change in Traveller's intended Route—Decision to Travel Homeward *viâ* Bat'ang and Bhamo—The Bridges of Ch'êng-Tu—Chinese Temples and Notions of Worship—Ancient Inscribed Stone—The Heaven-Tooth—The City Walls—Visit to the Great Monastery of Wen-Shu-Yüan—Its Buildings and Curiosities—Chapel of Meditation—Antiquities of Ch'êng-Tu—Memorials of Early Events—The History of the Emperor Liu-Pi—Invitation to a Picnic—Notable Guests—The Dinner and its Peculiarities—Table Manners and Customs—Threatenings of Drought—False Notions about the Traveller—The Provision of Currency for the Journey, and Tedious Banking Business—Adieux at Ch'êng-Tu—' Organisation of Departure '—Concluding Notices of Ch'eng-Tu and the ' Province of Four Waters ' (Ssu-Ch'uan).

MARCO POLO thus describes the plain and city of Ch'êng-Tu-Fu:

'When you have travelled those twenty days westward through the mountains, as I have told you, then you arrive at a plain belonging to a province called Sindafu, which still is on the confines of Manzi, and the capital city of which is also called Sindafu. This city was in former days a rich and noble one, and the kings who reigned there were very great and wealthy.

'It is a good twenty miles in compass; but it is divided in the way that I shall tell you.

'You see, the king of this province, in the days of old, when he found himself drawing near to death, leaving three sons behind him, commanded that the city should be divided into three parts, and that each of his sons should have one; so each of these parts is separately walled about, though all three are surrounded by the common wall of the city. Each of the three sons was king, having his own part of the city and his own share of the kingdom, and each of them in fact was a great and wealthy king. But the Great Kaan conquered the kingdom of these three kings, and stripped them of their inheritance.

'Through the midst of this city runs a large river, in which they catch a great quantity of fish. It is a good half-mile wide, and very deep withal, and so long that it reaches all the way to the Ocean Sea—a very long way, equal to eighty or one hundred days' journey; and the name of the river is Kian-Suy. The multitude of vessels that navigate this river is so vast that no one who should read or hear the tale would believe it. The quantities of merchandise also which merchants carry up and down this river are past all belief. In fact it is so big that it seems to be a sea rather than a river.

'Let us now speak of a great bridge which crosses this river within the city. This bridge is of stone; it is seven paces in width, and half a mile in length (the river being that much in width as I told you), and along its length, on either side, there are columns of marble to bear the roof—for the bridge is roofed over from end to end with timber, and that all richly painted; and on this bridge there are houses, in which a great deal of trade and industry is carried on. But these houses are all of wood merely, and they are put up in the morning and taken down in the evening. Also there stands upon the bridge the Great Kaan's Comerque, that is to say, his custom-house, where his toll and tax are levied; and I can tell you that the dues taken on this bridge bring to the lord a thousand pieces of fine gold every day, and more. The people are all idolaters.'

Ritter thus writes :

'Father Martin Martini, who gives us his account of China from the time when the Ming were still reigning, previous to the conquest by the Manchus, and might well have good information regarding Tsching-tu-Fu, since the Jesuits had a mission in that city, which was only abandoned by the fathers in consequence of the advance of the Manchu army, says :

'It is a much frequented commercial city; the palace of the king was magnificent; it was four miles in circuit, having four gates, and was placed in the centre of the town. From the southern gate extended a broad street, containing many arcades artistically built of stone.

'Throughout the city are navigable canals, revetted on each side with square and cut stones, and crossed by many stone bridges.

. . . .

'One of the rivers, the To (or Tu-Kiang?) says Father Martini, is a branch of the Min-Kiang, excavated and led out of its course by the order of the Emperor Yvo [presumably Yau, the famous semi-mythical emperor, *circa* B.C. 2300], as a remedy against the outbreaks and inundations of the Kiang.

. . . .

'Thus many of the broad pieces of water and lakes, which in the neighbourhood serve as moats and trenches to the city, have been artificially excavated.'[1]

The city of Ch'êng-Tu is still a *rich and noble* one, somewhat irregular in shape, and surrounded by a strong wall, in a perfect state of repair.

In this there are eight bastions, four being pierced by gates. It is now three and a half miles long by about two and a half miles broad, the longest side lying about east-south-east, and west-north-west, so that its compass in the present day is about twelve miles. A stream, about thirty feet wide, runs through the city from west to east; parts of this are embanked with perpendicular revetments on either side.

At one point it is spanned by three bridges close together, each of stone with a single arch. The one in the centre has at one time evidently been larger and of more importance, for on the other side of the road that lies between the water and the houses, almost buried in the buildings, there is a stone lion with his back to the brook. This has clearly been the former end of the bridge, so that the houses must have advanced some yards since this was built. This bridge, which is near the southern gate of the imperial city, probably led in former days to the broad street spoken of by Martini.

The city is well laid out, the streets, straight and at right angles to one another, well and carefully paved.

One of them is very pretty, and runs by the side of the stream that flows through the city. Looking in at the doors of the fine shops on the right, respectable old gentlemen can be dimly discerned in the

[1] See Ritter, iv. 415–416.

semi-obscurity smoking their long pipes. Overhead a bamboo matting, or a bit of trellis-work covered with creepers, shelters the street from the glare of the sun ; while on the left hand is a strip of garden a yard wide, enclosed on either side by trellis-work, covered with scarlet-runners, whose small red flowers form a pleasing contrast to the fresh green foliage, and through the leaves the brook is seen sparkling in the sun. The shops in Ch'êng-Tu are very good, with handsome fronts ; every description of goods is sold in them ; there is especially a very large trade in silk, and Ritter quotes Martini as saying :

'In the river Kin, which flows on the southern side of the city, they wash the silk, which thereby attains an extraordinary brilliancy.'

The main river still runs at the south side : it is about a hundred yards wide, and crossed by many bridges ; one of them, ninety yards long, has a roof, and, as is the case on nearly all covered bridges, hucksters sit down under the shelter on both sides, as in the days of the old Venetian traveller, and sell whatever they can to passers by.

There are still large numbers of junks on this river, which come up from Ch'ung-Ch'ing, and possibly some from the 'Ocean Sea.'

It is difficult to account for the great difference between the state of the city as it was in the time of the early writers, and the present condition of Ch'êng-Tu.

The hills, however, that enclose the plain of Ch'êng-Tu are of sandstone, and are of course easily worn away by water.

The drainage of the basin is by a river of considerable size, which must in the course of five centuries have deepened its bed at its point of exit from the plain where it is closed in on both sides by the sandstone hills. At the same time it would seem probable that the *débris* brought down by numerous streams from the surrounding mountains would rather have tended to raise than to lower the general level of the plain itself. Anyhow, when we consider how very flat the plain now is, we should, without the aid of the historian, be almost driven to the conclusion that it was in former ages the bottom of a lake.

Martini, in the passage above quoted, tells us that some of the ponds, lakes, rivers, or canals were artificial; and the river full half a mile in width spoken of by Polo may, in reality, have been a shallow fleet crossed by a causeway, or even by a long bridge such as he describes.

In the course of the last five centuries, as the bed of the river at its exit has been deepened, the plain has gradually been drained: and thus will nature have performed her part of the change.[2]

It is an historical fact well-known at Ch'êng-Tu that the city formerly covered a very much larger area; for in olden days, the temple of Wu-Ho-Tzu, now a mile or two outside the city to the south-west, was within the walls.

Since the days when Marco Polo travelled this way, the times have been turbulent indeed: the city has been pillaged, lawless bands have roamed with fire and sword across the fertile plain. In the early part of the Ming dynasty (commenced A.D. 1368), the whole province was overrun by a brigand named Chang-Shien-Chung; he went about ravaging and

---

[2] The fact that an actual bifurcation of waters seems to take place near Ch'êng-Tu (see Richthofen's *China*, p. 327)—one branch flowing south, as the Ta-Kiang, Min-Kiang or what not, to Siu-Chou-Fu, and the other south-east, as the To-Kiang or Chung-kiang of maps, to Lu-Chou—renders change in the distribution of the streams about the city highly probable.—*Y.*

destroying everything, and is pictured as a devil incarnate; amongst other things he destroyed all the books, so that the ancient written history of the place is lost; there is therefore nothing improbable in the total disappearance of the fine works spoken of by Polo. Thus may the hand of man have combined with nature to change completely the appearance of the city of Ch'êng-Tu.

June 21.—On the day after my return to the provincial capital, I called upon the French missionaries in the afternoon, and when I went home I found that Mesny had at length arrived from Kwei-Yang-Fu, where he had been living for many years.

Now the very serious question presented itself, whether I could carry out my intention of travelling through Kansu to Kashgar.

My whole difficulty lay in European politics. Supposing that I had found myself unable to proceed any further towards Kashgar than Urumchi, I could have passed through Russia, if there had been no danger of England being entangled in a war with that country.

But with England and Russia at war, this of course would have been impossible; and if unable to enter Kashgaria, I should have had no choice but the dreary journey in mid-winter back to Peking; and even should the road to Kashgaria have been clear, the mountain passes would not have been open, and I must have waited north of the Himalayas until the spring.

This would not have deterred me for one moment, but for the critical state of affairs between our country and Russia; in the event of war it was equally my duty and desire to be somewhere within hail, and I could not feel myself justified in running the risk of being buried for so many months in Central Asia.

This was the more disappointing, as I had everything prepared for this journey, provisions, clothes, and about three thousand taels in silver. I was very loth to give it up; but after anxiously reading every word in the scanty items of European news that were available, and after thinking over the matter night and day, sorely against my will, and with a heavy sigh, I at last determined to come home with as much speed as possible, but at the same time to travel by some new road.

The only route left was that by Bat'ang and A-tun-tzŭ; for the objections that applied to the Kashgar route applied equally to the only alternative, a journey *viâ* Lassa, which might or might not have been practicable.

The die was cast at length. I made up my mind that I would travel with the utmost speed *viâ* Bat'ang. My desire to get on was ably seconded by Mesny; and considering the nature of the country, and the difficulties always to be encountered, the journey actually was a very fast one, and we had the satisfaction of thinking that during the whole sixteen weeks we never lost a single hour.

June 24.—The mosquitoes had already often sounded their warning notes, and although they had not yet given me any trouble, Mesny had been so devoured that I thought it advisable to see about mosquito curtains. The Chinese have a capital arrangement for travelling curtains. The top is made with a little triangular pocket at each corner. The ends of four light bamboos are joined together by two brass tubes, and the other ends of the bamboos inserted in the small pockets stretch the top of the curtain. One nail in the wall or the ceiling is all that is required, and the curtain can be put up or taken down in a few minutes. The bamboos being of no great length are easily carried. I bought one of these, and found some regular Indian mosquito-gauze in a shop in the city

Fig. 1.

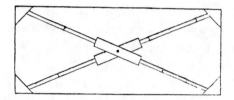

with which some curtains were made that served me in good stead. I used them almost every night throughout my journey, and they effectually kept out not only mosquitoes, but insects of many other kinds.

As Mr. Wylie[3] in recent days had said that Polo's covered bridge was still in its place, we went one day on an expedition in search of it. Polo, however, speaks of a bridge full half a mile long, whilst the longest now is but ninety yards. On our way we passed over a fine nine-arched stone bridge, called the Chin-Yen-Ch'iao.

Near the covered bridge there is a very pretty view down the river.

On the left there is a brick wall, by which the river runs, and here all the houses built close to the edge have wooden projections overhanging the water. On the right the bank is shelving, and there is a pretty flat landscape, with crops and plenty of trees, and of course a temple adjacent.

These temples do not correspond in any way to European churches, for in China people do not go to church in Western fashion.

[3] Colonel Yule's *Marco Polo*, vol. ii., p. 31.

But if a Chinaman wants anything in particular, such as wealth, children, success in business, or the like, he goes to a temple and makes a bargain with the deities, promising to give money, build a bridge, or a triumphal arch, or do some other good deed, if he gets what he wants ; not altogether an illogical proceeding. The men of Kiang-Si, like the Pharisees, make long prayers ; they take their incense-sticks or candles, light one before one god, and while this is burning, pray with many words that this particular deity will cast blessings on their kitchen. They then move on to another god, and with another bit of incense exhort him to look favourably upon some money transaction in which they have just embarked ; a third is supplicated for blessings on the family; and so on till they have exhausted their candles and list of desires.

On the other hand the Shen-Si man says little, but sits down and waits until the other has finished, when he, with one candle, invokes all the gods at once, and says, 'Oh! all ye gods of whom my friend of Kiang-Si has just been asking so much, I pray you to give me all the blessings he has begged for.'

Near this there is a stone, on which there is an ancient inscription, which I was told contained references to the Christian religion.

Thinking that I had alighted on another stone like the celebrated one at Si-^{Ng}an-Fu in Shen-Si, I obtained a copy of the inscription. It was exceedingly difficult to translate, but Mr. Douglas, of the British Museum, most kindly undertook the task. It contained no reference whatever to the Christian religion.

Translation.

'The Lord bestowed his liberality upon the world, the moving heavenly bodies, and the animals (lit. the seven kinds of animals). Afterwards they fell into Hell, and there was no one to bring them back, or to become incarnate to help them. Then the Lord of Devas himself came down on earth, and having spread universal peace (those he came to save) went with speed, and in fear and trembling, to the Heavenly Abode of Buddha. Thus the chief and least (lit. heads and feet) of the sufferers were all equally pure with those who had been joyfully and readily obedient to the Lord of Devas. To meet this difficulty the Lord of Devas separated those who had been saved from purgatory from those who had not entered it. But while timidly meditating on this, he thought how he was to act towards the animals. Presently he reflected that in the eyes of heaven pigs, dogs, serpents of the desert, and all animals which move secretly, appear unclean, and the thought of this caused him grief as though a spear pierced his heart, and he considered who there was to save those who might compliantly return. He reflected that there were those perfect ones who had obediently followed the examples (set them), and He perceived that such were restored animals. So from time to time he constantly liberated (others).

'Then he took every sort of incense and flower, and all kinds of food, and went to the abode of the world-honoured one, and exhibited them. Having worshipped the seven encircling streams, and having reverently waited on and nourished (them?), he entered and sat for awhile in the abode of the world-honoured one, where all were pure and good. With regard to the past business connected with the animals he readily looked to the world-honoured one to compassionately save them.

'Having finished this consultation he left the matter to the care of the world-honoured one, and ascended from the (mountain) top. As he did so a brilliant brightness illumined the world of the "ten quarters"; the wants of the people were supplied, and all nature smiled. The Ancestor then addressed Buddha, and said, "The Lord of Devas ought to know clearly [here follows a sentence of eleven characters, the meaning of which is not clear] that Tathagata has now received baptism, and has acquired complete prophetic power, having been purged from all vain passion, and placed beyond its reach. That the animals delight in the abode set apart for them, where they are able to contemplate a rich destiny. That those (men) who recite a book (of prayers), and establish the 'fragrant' precepts by displaying them (in practice), shall obtain everlasting life, and hell shall be removed far from them, while those who by contemplation cultivate virtue shall be born as rulers of the prisons (of hell). That the inhabitants of the world shall leave it empty, and having been liberated shall return to the gate of the heavenly region, where they will regard the 'Portico of Past Life.'" The Lord of Devas then returned to the pure abode, and declared that he had redeemed the laws and precepts of the world.

'The world-honoured one then received from the lips of the Lord of Devas this dharani.'

On our way home we went to see a stone called the 'Tooth of Heaven;' it was merely a bit of sandstone in the shape of a tooth; there was a little house built over the entrance to it, but the roof did not cover the stone itself, for they say that if the stone were covered, the God of thunder would commit some fearful devastation on the town.

There were a great number of the common white butterflies about the vegetable gardens. The Chinese say that when these appear in great numbers there will be much sickness.

The existing city walls were built only in the time of the second or fourth emperor of the present dynasty (1662–1722 and 1736–1795), the place having been entirely destroyed about two hundred years ago. Ch'êng-Tu, as it now is, is divided into two parts, the Chinese and the Tatar cities, both enclosed by the main wall. Not quite in the centre of the Chinese city, but rather towards the west, is the imperial palace, a rectangular open space enclosed by massive walls about twenty feet thick. This was built towards the end of the fourteenth century by the first or second emperor of the Ming dynasty, the Ming emperor employing one of his family as governor or king of the provinces in this part of China. The buildings inside this are now used as the examination hall.

The city of Ch'êng-Tu still bears on its face all the evidences of wealth and prosperity; the people are well dressed, and some of the temples in the city are richly endowed.

We paid a visit one day to the Wên-Shu-Yüan (Literary Book Hall), a very fine temple near the north gate.

This monastery was built some time during the Sung dynasty (from A.D. 960 to 1279). It was then called the Chin-King-Sze; it fell into decay during the Mongol occupation, and was rebuilt by the second emperor of the present dynasty, the famous Kang-Shi (better known in the form of Khang-Hi), who reigned 1662–1722. This emperor richly endowed it with lands; but notwithstanding its wealth, it seems to have been predestined to misfortune, for it was again neglected, until the time of Kia-Ching (Kia-King), the fifth emperor of the present dynasty (1795–1820) when it was rebuilt by public subscription with stone instead of wooden pillars. Since that time it has gone on increasing in wealth and magnificence, and is now one of the richest in the country. To have the right of living at this monastery it is necessary to be a priest of a particular sect; but besides the priests, there are resident here a number of students qualifying themselves for holy orders; altogether there are about one hundred and fifty inmates.

A remarkable air of refinement and cleanliness pervaded the place. The courtyard was laid with smooth-cut flagstones, not one out of its place, and not a weed or blade of grass permitted to grow in the interstices; all the buildings were in perfect repair, and a man was walking about the court with a cross-bow. His employment was to shoot stones at the sparrows that infested the roofs, and which, if left to their own devices, would do serious damage. Immediately on the right of the entrance was a very clean reception room, and whilst preparations were being made to escort us over the establishment, we were refreshed with the usual cups of tea. We were not kept waiting above a couple of minutes, and then we were invited to proceed.

The refectory, a long wooden building on the right-hand side, opened into the court; here were twenty-five tables, each prepared for six people. For each person was laid one pair of red wooden chopsticks and three porcelain bowls, one for rice, one for vegetables, and one for tea, no meat of any description ever being permitted here; everything, the tables, bowls, and chopsticks were beautifully clean, a most surprising thing in this country, where usually dirt reigns supreme. Passing this, we entered a chapel, where, at the end, the repulsive countenances of a number of huge and hideous images were partially obscured by a kind of throne for the prior, whence he discourses on the religious classics to the students.

On either side of the chapel was a reception-room. The general arrangement of these rooms is almost always the same, and whether a private house, a ya-mên, or a temple, the description of one stands as a representation of all the others: no furniture in the middle of the room; along two sides are arranged, in symmetrical though inartistic order, the usual heavy, stiff, uncompromising, and utterly uncomfortable arm-chairs of China; between each two is a little high and square table, all corners and angularities, like the Chinese character. At the end of the room is the kang or raised dais, ten feet long, four feet broad, and two feet high, where in the centre is placed a small table, six or eight inches high, between two cushions of the most brilliant scarlet—these are the seats of honour; and footstools of wood for those seated thereon complete the furniture.

For ornament, a few bronzes, or the roots of trees carved into representations of impossible dragons, are arranged behind the kang, while from the ceiling hang paper lamps, some of them really artistically painted, and arranged just low enough to knock off the hat of a foreigner. In China, etiquette rules that in polite society the hat is kept on the head, and at a dinner party it is amusing, when all the guests are intimate and of the same social standing, to see the alacrity with which permission is always asked and given to exchange the official hat for the little skull-cup, which each person's servant has somewhere secreted about the capacious folds of his garment.

A collation of tea and cakes, sweet but nasty, was looked at rather than partaken of, while the monks gave us what history of the building I have been able to relate, sitting, as etiquette ordains, with their backs quite stiff, on the extreme edges of their chairs, and with their bodies slightly turned round to their guests.

From this we ascended to the upper story, where the principal room was a magnificent chapel filled with gifts and curiosities, a very fine and richly-decorated altar, rubbings from ancient tablets, a great deal of blue and white china, pictures painted on glass from Canton, and, amongst other things, a present from a young lady of a piece of embroidery, entirely worked with her own hair. This represented the goddess of mercy sitting under a bamboo, the leaves of which were really most admirably represented.

In this chapel also the contributors to the building, maintenance, or decoration of the temple are immortalised, their names being written in gold on black tablets and put under a glass case. Here also is the library, where huge cupboards are filled with books of the religious classics, which form the unique and dreary study of the inhabitants.

We passed on to another chapel set apart for meditations. Here the priests and students, in yellow robes and with shaven heads, come at least once a day, and lighting an incense-stick before one of the images, sit down at the side of the room and meditate, trying to work themselves into a state of religious ecstasy, in which they shall be entirely withdrawn from impressions from the outside.

A few of them appeared to be really in this state of semi-unconsciousness; but the majority, though trying to look as if they did not see us, could not resist a sidelong glance every now and then. They remain in this state about half an hour at a time. The impression formed upon my mind by the appearance of those who had succeeded in their extraordinary task was rather a painful one.

Passing through another chapel, where a number of beautiful red and yellow lotus-plants were growing in pots, where a tailor was at work in a corner, and in which were the portraits of all the deceased priors, we again came to the gate, where a number of huge and hideous figures—the guardians of the place— were grinning horribly, and where the monks with exquisite politeness bade adieu to their unwonted guests.

We went from this, along a road between walls that enclosed magnificent vegetable gardens, to the grave of a concubine of Shu-Wang.

Shu-Wang ('King of Shu') was the aboriginal king of this country before its conquest by the Chinese, and he lived in the time of the Chinese emperor T'sin-Shih-Hwang-Ti, the builder of the Great Wall of China, in the third century B.C.

This grave is an artificial mound of yellow clay, about one hundred yards long, running north-west and south-east, and about twenty yards broad; its two ends being raised about ten feet above the other parts.

At the south-east extremity, half buried in the clay that has fallen on it, is a huge limestone disc. Neither its diameter nor its full thickness are exposed, but, judging from the segment, its diameter must be about sixteen feet, and there is a thickness of three feet visible; how much more there may be I cannot say. Near the circumference of the stone, there is a circular hollow about six inches across, but it is very irregular, and I should say was accidental. The stone has evidently fallen from its place, so that any examination as to its position was useless. But it must have been a great labour to bring this enormous slab from beyond Kuan-Hsien, the nearest place where the limestone is found.

We had by this time collected around us a considerable crowd of dirty little boys, who made somewhat of a clamour as we walked home; but an elderly person in the streets rebuked them for their rudeness, and the noise diminished considerably.

Near the Wan-Li-Ch'iao is the temple of Wu-Hou-Tz'ŭ (Military Marquis Memorial Chapel), erected to the memory of the prime minister of Liu-Pi, Chu-Ko-Liang, or, as he was familiarly called, Kung-Ming-Sien-Sen.

At the latter end of the Han dynasty (third century of our era), a little boy being at that time emperor, the country was thrown into disorder by ambitious ministers who plotted for the throne.

At this time a certain Liu-Pi, though a member of the imperial family, was in very straitened circumstances, and was at one time driven to making a livelihood by selling straw sandals.

He fell in with two men also of poor position, one Chang-Fi, a pork butcher, the other Kuan-Yu, a seller of bean-curd cakes. These two counselled Liu-Pi to seize the throne; and they then formed themselves into a confederacy, calling themselves the Three Brothers.

They had little to start with, but by great bravery and force of character they eventually succeeded in establishing Liu-Pi as emperor.

Liu-Pi, however, did not actually reign over the whole of China; for there were two other kings, one in the south, and one in the north: all three regarded one another as usurpers. Hence this period is known as that of 'The Three Kingdoms.'

But Chu-Ko-Liang, hoping to consolidate the power of Liu-Pi, arranged a marriage between Liu-Pi and the sister of Cen-Chien, the king of the southern state,[4] who had his capital at Chin-Kiang, near Nan-King.

Cen-Chien was strongly averse to this proposal; but Chu-Ko-Liang obtained the co-operation of the queen mother, and Liu-Pi went to Chin-Kiang, and married the girl, notwithstanding the opposition of Cen-Chien. An iron pagoda still in existence at Chin-Kiang is said to be the place where the ceremony took place.

But Liu-Pi was by no means firmly established, and Chao-Tsiao,[5] the northern king, went to war with him. Cen-Chien pretended to assist Liu-Pi, and sent an army to him in the hopes that he would leave his capital unguarded, when Cen-Chien would have marched in himself with another army he had in readiness. But Liu-Pi managed to hold his own, and his bravery, fine spirit, and great talents have endeared his memory to the Chinese.

A characteristic story showing how he valued those who assisted him is told of him. In one of his battles, at which his wife and infant son were present, he was worsted, and was obliged to flee, leaving his queen and child. But Tsao-Yun (Tze-Lung was his

---

[4] This southern state was called the State of Wu. (Cen-Chien is the Sun-Kiouen of De Mailla's *Hist. de la Chine*, vol. iv., in which the history is related at tedious length.—*Y*.)

[5] Called in the French histories T'sao-T'sao, the founder of the dynasty of Wei or Goei.—*Y*.

popular name), one of the generals of Liu-Pi, found the mother and child sitting by a well. The queen handed her infant to Tsao-Yun, and jumping down the well put an end to her life. Tsao hid the child in the breast of his coat, and gallopped off to rejoin Liu-Pi, whom he succeeded in reaching after many dangers. When the monarch heard what had happened he flung the child on the ground, saying ' You worthless bit of flesh, is it you who have dared to risk the life of one of my best generals.'

Liu-Pi afterwards removed the capital of his kingdom to Ch'êng-Tu.

His son succeeded him, but was as feeble as his father was powerful; he was sustained, however, by the talents and energy of Chu-Ko-Liang, but was the last of his race; the province being subsequently overrun by numerous barbarous tribes, who established many petty independent kingdoms.

The emperor and prime minister now lie side by side; for the great Liu-Pi was buried in the temple of Wu-Hou-Tz'ŭ, erected to the memory of Chu-Ko-Liang. Round the temple there are some very pretty gardens, and it is the fashion to make up a party and bring out a good dinner from a restaurant and eat it here. It is a charming spot for a picnic; and somehow or other the small square tables and benches in a kind of balcony overhanging a pond of lotus-flowers, full of tame fish and tortoises, looked so European, that when I visited the place I expected every moment to see a waiter with a table napkin over his arm, asking monsieur what he would have for *déjeûner*, and suggesting *un bon filet aux pommes* and a pint of Medoc.

From this we went to a place celebrated for something or another, but what it was was never made very clear. There are five stones one on top of another, but though these are conspicuously shown on the Chinese map of Ch'êng-Tu, we failed to discover anything very remarkable about them.

There is also near this a big mound, said to be a grave of someone, but no one could tell us of whom.

We were invited to a picnic at a temple not far from the Wu-Hou-Tz'ŭ, at which place it was agreed that our party should meet.

Though the sun was powerful there was a little air moving outside the city, and the heat was by no means oppressive.

We were about an hour reaching the Wu-Hou-Tz'ŭ. Here our friends were waiting for us, and we all went on together to a temple, built during the seventh century, by the great poet Tu-Fu, as a country residence.

There are, as in the temple of Wu-Hou-Tz'ŭ, a number of rooms, covered passages, corridors, and pavilions, furnished with little tables and chairs, where the people of Ch'êng-Tu come to picnic.

The grounds are large, containing fine trees and great numbers of large bamboos, that everywhere cast a pleasant and grateful shade. There are ponds with tortoises and fish in great numbers, and a couple of dwarfs with enormous heads earn a livelihood by selling bread and cakes for the people to feed the fish with.

We first went into a nice large cool room, where all the woodwork was painted black; but as the upper half of both the long sides was entirely window, there was no sombre impression. All the windows were open, and the eyes rested on the fresh green foliage, which almost completely excluded the midday glare, whilst the breeze gently rustling the bamboo leaves, and the occasional caw of a rook or a magpie, produced a pleasant feeling of repose.

We found the company assembled. There was a very fat, heavy-looking man, a civilian with the rank of Fan-Tai, by name Wei, whose manners were polished to the highest degree, and who would have been profoundly shocked at the smallest breach of the intricate etiquette of the Chinese. In remarkable contrast to him, a tall thin man, with the rank of Chen-Tai, was walking about. His face differed much from the usual Chinese type; he looked as if he was more of a man than the Chinese generally appear; and although his face and manners betokened a love of ease, there was none of the listless, apathetic appearance about him so often seen in this people. His name was Yang, and he had a nephew who had been apprenticed to my friend the Christian silk merchant, Yeh.

In the time of the Tai-Ping rebellion this lad ran away and joined the army. He was active and intelligent, and soon brought himself to the notice of Li-Hung-Chang, who recommended him to Ward, an American in command of a body of the Chinese troops. Ward took a fancy to him, and promoted him, and when Gordon left China, this boy, now grown to be a man, succeeded him in his command. He eventually died of wounds.

The silk merchant, who always had a few stray locks of hair escaping from his now scanty plait, and looked in consequence exceedingly hot, was also of the party; his name was Yeh, though amongst his friends he was known by the familiar sobriquet Chin-Tsai; and Ngien, the secretary to the Bishop, whose French it was possible to understand, completed the number of the hosts.

After our hot ride (in chairs) we sat down, and the grateful beverage was soon introduced. Mesny and I were pressed to take seats on the kang; but among so many we left it unoccupied, and sat down on the chairs at the side of the room.

A basin of hot water and a piece of rag were brought in; an attendant, whose hands must have been made of cast-iron, dipped the rag into the almost boiling water, and wrung it out several times. He brought it to me, and I wiped my face and hands in correct Chinese style. The rag, or as Huc calls it, a linen table napkin,[6] was dipped afresh and wrung out for each person present.

Mesny then opened the conversation by asking everyone he did not know, 'What is your honourable name?' 'What is your honourable age?' 'Where do you come from?' and in return answered similar questions with true celestial politeness, and although I did not know a dozen words of Chinese I could see what was going on.

The secretary then proposed to take me round the temple, and we walked about looking at the tortoises, the ponds, the dwarfs, and the idols.

He showed me an isolated building in one place, with four very large images of Buddha in the centre, and upwards of a thousand pictures of the head of Buddha on the walls. We then came back, and after a time signs of dinner appeared in the form of a *zakouska*, for before seating ourselves at the round table a bowl of soup and four little puddings, with minced meat and onions inside, were handed to each person.

I did not know how to manage these things; but I watched the others take up a pudding, put it into the soup, partially break it, and so eat it. I did the

[6] Huc, *L'Empire Chinois*, vol. i. p. 184.

same, but there was too much garlic for my taste. This appeared to me quite a meal in itself; but my Chinese friends finished their four puddings, and looked upon this exactly as the Russians do upon the little bit of salt fish or caviare they take to whet their appetites. The pudding to put into the soup also is quite a Russian custom.

Soon afterwards, at about half-past four, we sat down to a very extensive dinner. To every man was a pair of chopsticks, one little piece of paper, one little saucer of soy, one china spoon, one saucer of watermelon seeds and kernels of peach stones, and one cup about as big as the bottom of an egg cup (without a handle).

At a given signal everyone at once dipped their chopsticks into the centre dish and commenced operations. The silk merchant was very polite to me, and always assisted me if he saw I was not sufficiently skilful with my chopsticks.

The guests thus went through about twelve dishes that were on the table, some sweet, some sour, some raw, and some cooked. They were much the same dishes that I had seen at Shanghai or Peking; shrimps raw, duck or ham cut into little bits, sugar-candy, lotus root, walnuts cooked in soy, giblets, with preserved eggs, shrimps, and other things, all equally flavourless. A servant then came in, and removing two or three of the nearly empty bowls, brought in others; and so on, dish succeeded dish in somewhat weary monotony; duck appeared in two other forms, fowl came on twice, tripe was dressed in two ways, and a dish of peaches stewed in arrowroot was given in the middle of dinner. There was one dish of really excellent mutton, and of course at least half a dozen of pork in different forms. The greatest delicacy was minced pork, dressed with something sweet, and wrapped up in a huge lotus leaf. To our Western ideas the mess the table and floor get into on an occasion of this kind is horrid; there are no plates; when the dishes are brought in, if they are solids they are piled up as high as possible, and if they are soups the bowls are filled to running over. In helping himself with a chopstick the most skilful will now and then drop something, and to eat the gravy the spoon is dipped into the central bowl, and then put down wet and greasy on the table.

The *débris* also collects on the table more or less, though a person accustomed to these things does not leave much, for he spits or throws it on to the floor. Bread is not offered until the end of the meal, and when I asked for some, earlier during the entertainment, a whole baker's shop of loaves was brought in for me. The drink was a very palatable fermented liquor made from rice, and was taken hot.

Directly two guests have taken wine with one another the cups are filled by the attendants. The silk merchant was very anxious on my account, and asked me to drink with him after each course, and seeing that the mutton was the thing I really liked, he had it specially left for my edification.

The waiters were all naked to the waist, and the guests would have been the same if Mesny and I had not been present; but out of deference to us they kept on a thin garment over their bodies.

The last dish of all was a bowl of what Europeans call 'conjee'—rice boiled almost to a pulp, and served up with the thick rice-water. In ordinary society a bowl of plain rice takes the place of this; but at these grand entertainments it is customary to have conjee instead.

After this the guests laid their chopsticks across the empty bowl, rose up and saluted one another, and then again putting the chopsticks on to the table the dinner was over.

I gave each of the gentlemen a Manilla cigar, produced a penknife, showed them how to cut off the ends, and offered them a light from a box of wax vestas, at which they were much delighted. The general and my French-speaking friend lighted their cigars; but the Fan-Tai and the silk merchant put away theirs for some other opportunity.

While the servants were clearing up the mess we strolled about the grounds. The general, pacing up and down smoking the cigar, had far more the air of an Englishman than a Chinaman; but the secretary, although he seemed to like the smoke, did not quite manage it *a l'Européenne*. We loitered about some time, and many amusing stories were told.

July 7.—At my request, the banker now came to pay me a visit to talk over money matters, for I had much more silver here than I required for my journey. He told me that he had no correspondents in Western Ssŭ-Ch'uan or Tibet, and I was therefore obliged to ask him to give me the three thousand taels in bulk. This weighed two hundred and six pounds avoirdupois, rather a serious consideration.

The banker said, on looking at my bills, that these had not been drawn on him, but on some other bank; he added that, if I had given notice on my first arrival at Ch'êng-Tu I should have received the usual interest of 1 per cent. *per month.*

The drought was now becoming very serious, even in this province; at the time of my visit there was as yet no scarcity of food, but in the neighbouring provinces the famine eventuated in the awful calamities that have filled the readers of our daily papers with horror; and even in Ssŭ-Ch'uan the drought, though not so disastrous as elsewhere, was in 1878 very dreadful.

In the fertile plain of Ch'êng-Tu itself, the rice crop never fails, even in the driest season; for the brimming brooks that course by the roadside and sparkle in the sun derive their supplies from the streams which, descending from the snow-clad heights, are never-failing, and unite to form the considerable river of Kuan-Hsien. There the impetuosity of the turbulent torrent, which dashes and foams over its rocky bed, is curbed by the irrigation works that divide the river into numerous streams, and those, meandering through the beautiful plain, and subdivided into canals and yet smaller ducts, and finally pumped up by the simple treadmills, leave not an acre of land without its perennial supply of water. Thus, even at a time when all the horrors of famine and pestilence were desolating the lands that lay just beyond the surrounding hills, this favoured spot was still enabled to present a scene of comfort and tranquillity.

All the officials, dressed in their robes of state, formed a procession and went to a temple to pray for rain; and, in order to appease the offended deities, a general fast was ordered. It was forbidden to kill meat of any kind, and the people were compelled to live on vegetables, not even eggs being permitted. Fortunately, however, the officials were unable to prevent the fowls laying, so eggs could surreptitiously be purchased. Chin-Tai, also, notwithstanding the prohibition, managed to smuggle a very young chicken into the house, and my own farmyard still contained three or four leprous-looking ducks, the relics of the official gifts I had accumulated.

July 8.—Mesny, who had paid a visit to Pi-Hsien, told me that the people there at first thought that it was I who had returned; but finally arrived at the conclusion that it was another foreigner; for they said I looked nearly sixty years old, and Mesny did not look thirty. To European eyes, our ages did not appear very unequal, but the Chinese invariably thought me very much older than I actually was. This was because I wore a short beard; and to a Chinaman's ideas a man who wears a beard looks old, most Chinese being quite smooth-faced.

There was also a report at Pi-Hsien that I had died at Sung-P'an-T'ing. This was, I suppose, because I did not come back when I was expected; and there may have been a lurking wish somewhere father to the thought.

We had great difficulty in hiring coolies and ponies; the latter were almost impossible to get. The Nepalese embassy was expected with tribute for Peking. There was an embassy bringing tribute from Tibet, and another going up from here into Tibet, all on the same road that we were going to follow to Bat'ang.

These officials and embassies show little mercy, and press every man and beast they come across into their service. The owners of ponies were afraid that if they came with us to Ta-Chien-Lu, they would, as soon as we had left, be forced into the service of some of these people, with very little pay or none at all, and there was naturally a strong objection on their part to come up.

The banker came in the afternoon, with a load of silver done up in thirty packets of one hundred taels each. Each of these packets had to be opened, each bit of silver examined to see if it were good, and each packet weighed. The other banker, who had supplied me with the first amount of silver, came as my friend, 'to watch the case for me,' as a lawyer would say; for the manipulation of this vast sum of money was sure to create one or several keen discussions.

The grand question of weight first came on, and a hot dispute arose as to what scale should be adopted. I naturally suggested a big one, for I had a lively recollection of paying the Ch'ung-Ch'ing banker about sixty or seventy taels, on account of difference in weight. Something about the scale was written on one of the bills; but this, instead of settling matters, only appeared to cast oil upon the already flickering flame of angry passions.

I sat passive, until I was asked whether the Ch'ung-Ch'ing banker had not given me a weight, to show what sort of a tael he meant; then I bethought me of a string of cash, carefully and neatly tied up in paper, which had been given to me with my bills by the banker in Ch'ung-Ch'ing. I had undone this packet to see what it was, but up to this moment its use had remained a mystery; it now flashed across my mind that it might be the weight intended. It was produced, and there seemed some chance of coming to a final settlement; but when it was put into the scale against one of the packets, the banker threw up his hands to heaven and swore he was undone. 'No,' cried one of the clerks, 'but the string of cash *has* been.' I was then asked who had untied it. I said that no one had meddled with it except myself. They did not like to say that I had been adding extra weight to this, but they looked it!

The actual and physical thermometer was now at 93° in the shade outside, and somewhat higher in the room, while the moral temperature of the disputants was rapidly rising to boiling point, as they wiped the perspiration from their brows.

The second clerk took off all his clothes, as far as his waist, and said that nothing was written about weight on one of the bills.

'No,' replied one of my friends, 'but there is the weight sent with the bills; they all come from the same place, they are all for you, and you will have to give the weight, the full weight, and nothing but the weight.'

The banker then suggested I should take what he had prepared for me, and write to Ch'ung-Ch'ing for the rest; and my friend the merchant seemed to think this a good way out of the difficulty.

I should have laughed them to scorn, but it was too hot for violent physical exertion, and I merely shook my head solemnly.

At last some one suggested something; I did not quite know what; but they hurriedly took the paper off the string of cash, and some writing was disclosed inside, over which they all bent as eagerly as a betting man over the first Derby telegram.

A rush was made for the scales, the wrapping up paper weighed, and the banker gazed in triumph round the admiring circle.

I discovered that usually the weight is taken paper and all, but in this case it was written inside that the paper was not to be included. This paper weighed a little over half a tael, which, taken thirty times over (once for each hundred), made a difference in the banker's favour of nearly twenty taels.

This put out all the fires at once, and we took a drink all round of tea.

The work was then proceeded with; the banker signed his name on each large piece of silver, and with a bland smile told me that, if any of them should turn out badly, he would be only too glad to exchange them. This, he said, well knowing that he would see my face no more; but for form's sake I allowed him to continue the useless and tedious operation.

At length the banker and my friend the merchant were satisfied, and agreed that the banker's weight was short by three taels, and that there were about fifteen taels weight of inferior silver, which must be changed, so I handed over the three bills and bid adieu to my honourable friend.

The merchant stayed behind, and confidentially told me that the watches produced in my honourable country were very good.

I answered that he was very kind to say so.

He then said that if my great excellency would sell him a watch, he should never know how to thank me; so I told him that I would give him one in return for the kind way he had assisted me.

Chung-Erh now asked me what the price was in England of the watches that I was giving away; I told him five or six taels, and he also wanted to buy some. I said I could not spare them. He went away very sorrowful, not because he had great riches, but because he hoped to have made some by selling the watches at a high price.

July 9.—The last day in Ch'êng-Tu was a busy one; we paid a final visit to our kind friends the missionaries, and then all our acquaintances came to say good-bye to us. A young man to whom I had given some European candles brought me an ancient book and a stone for rubbing up Indian ink. The Chinese are very particular about the stones used for this purpose; they take a great deal of trouble to get the finest possible, so that the ink may not be gritty; and an ink stone of really high quality is looked upon as a very distinguished gift.

All the baggage was weighed and divided into forty portions, for forty coolies were required to carry it; and a bargain was eventually struck with the coolie-master to supply us with sixty coolies to take us, our chairs, our baggage, and our servants to Ta-Chien-Lu, at the rate of 3·2 taels per coolie.

In all the principal towns in China there are firms who make it their business to let out coolies; a traveller always applies to one of these firms, and he is then tolerably sure to get respectable men, and to be fairly treated, the firm being, in a certain measure, responsible for the coolies supplied by them. A written contract is always made, and one copy given to the hirer, stipulating for not more than a certain number of stoppages, and undertaking to make the journey in a certain number of days.

Of course the coolie-firms make a very large profit, paying the coolies a miserable proportion of what they receive themselves; but, nevertheless, the traveller would scarcely gain by hiring coolies for himself, even if it were possible, which in most cases it would not be. For he would have no hold on the coolies, and he would certainly get a very worthless lot, who would in all probability rob him.

Small as their pay is, the coolies dearly love to dawdle, even though they are paid for the journey and not by the day; they like to stop at every wayside tea-house, and a long day's march is an abomination in their eyes. Throughout my journey it was always one continued fight with the coolies, or muleteers, to get them over a fair amount of ground.

There was an excuse invariably ready for halting short of the proper distance: the road was bad, the mountains were steep, there was no inn, it was going to rain, or the sun was too hot.

The attractions of a big town were irresistible, and sometimes it seemed almost impossible to get them away.

These, after all, were but very minor evils, and I was altogether exceedingly well treated; my goods were very fairly taken care of; boxes and portmanteaus were never thrown about in the wanton manner of European porters, and during my whole stay in China, I was never robbed of the smallest thing.

The head coolie Fu-Tu, who had come with me from Ch'ung-Ch'ing, again appeared, bringing with him several of his former coolies, notably the old opium-smoker with a pimply nose. Chin-Tai had formed a strong friendship with Fu-Tu, who was the only person I ever heard say a good word for my factotum.

July 10.—It is not customary in China for a servant to ride in a chair with more than two coolies, and an official is forbidden to permit his servant to do so; but I did not feel myself bound by Chinese customs, for my boys, both over six feet, would always have been miles behind me if I had not allowed them three coolies each.

The organisation of departure, as Huc is pleased to term the disorder of a start in China, was now complete; the baggage was all packed, the coolies' bickerings gradually were settled as they moved off one by one, and at length, on the morning of July 10, we left the provincial capital for the confines of the Province of the Four Waters, a sufficiently interesting country, to judge from the account in Ritter:

'Tsching-tu-Fu,[7] the ancient capital of Szütschuan,[7] lies near the eastern foot of the sublime masses of perpetual snow and ice of the Yün-Ling. It had already been visited by Marco Polo, and named by him Sindin-fu, as he took his journey south-westwards towards the then depopulated Thibet. He had succeeded in getting thus far from the north, i.e. from Singan-Fu, the capital of Schensi, crossing the parallel chain of the Tapa-Ling by a made road, of which we will speak presently. This city (Sindin-fu) and its environs might well be considered as belonging to the most remarkable alpine regions of Asia, which have been famed for their high state of culture from the earliest times, such as Kashmir, Katmandu, and Asam. Being, moreover, itself the seat of royalty, and possessing an indigenous civilisation, it cannot have failed to exercise an influence on the progress of history, though we are hardly in a position to prove it. Towards the end of the thirteenth century the narrative of Marco Polo leads us to this city, which had formerly been the capital of an independent kingdom of the same name, but had been fearfully plundered and laid waste by the Mongols. According to Chinese history, it was carried by assault in 1236, at which time one million four hundred thousand people were said to have perished in the capital, and an equal number in the provinces. Notwithstanding this former devastation, the Venetian mentions it as a large and magnificent city, once the residence of rich and powerful monarchs.

'The Chinese geographer, Lou-Houa-Tchu, had already, in extolling its great antiquity, quoted the Yuking, a celebrated chapter of the Schuking, or the ancient description of China (2300 B.C.). It is spoken

[7] The German orthography.

of as lying under the constellations, the Twins, and Cancer; its highlands spread over the west like the tiled roof of a high house (forming the protecting province of Inner China, as Khorasan was called the Shield of Iran).

'The oldest name of the city, about the commencement of the Christian era, was Ytscheou, and that of the country Chou.[8] In the tenth century the kingdom was styled Chou, and the city Szütschuan, which name has since been borne by the province. It first became a province of the Chinese empire during the dynasty of the Sung, about 1000 A.D., and towards the middle of the thirteenth century fell under the dominion of the Mongol emperors.' Ritter then goes on to quote the description by Marco Polo, with which this volume commences, and that of Martini, partially cited at p. 3. The latter 'also mentions among the notable things of the vicinity that yonder lofty mountains produce rhubarb, musk, and yaks with fine tails. . . The father goes on to observe the mountains of the province are rich in iron, tin, lead, loadstone, and salt' (and describes correctly the process of boring the brine wells).

. . . . . .

'Martini represents some of the mountains adjoining the city as reaching to the clouds; one of them, the Cinching, is said to cover above a thousand stadia, and to be the fifth in rank among the most famous mountains of China, and on which the *Shin-sien* or 'Immortals' congregated. Another mountain, named Mount Pin, was said to be sixty stadia (36,000 feet?!) in height; it contained the source of the Kiang; and from Mount Tafung descended an immense waterfall.

[8] Rather *Shu*, v. p. 17 *supra*.—*Y*.

On Mount Cung-King were apes in size and figure resembling a man. On Mount Lungan were to be seen the ruins of a palace which the king of Chou frequented for coolness in the summer.'⁹ . . . And so forth.

⁹ Ritter, iv. 417.

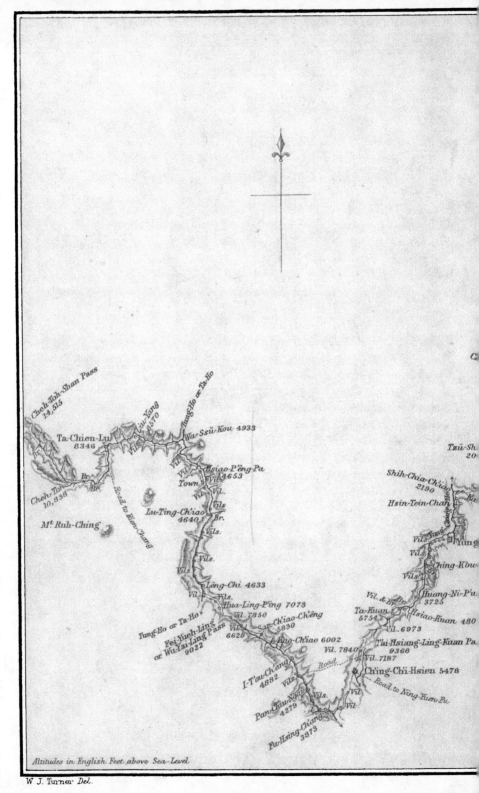

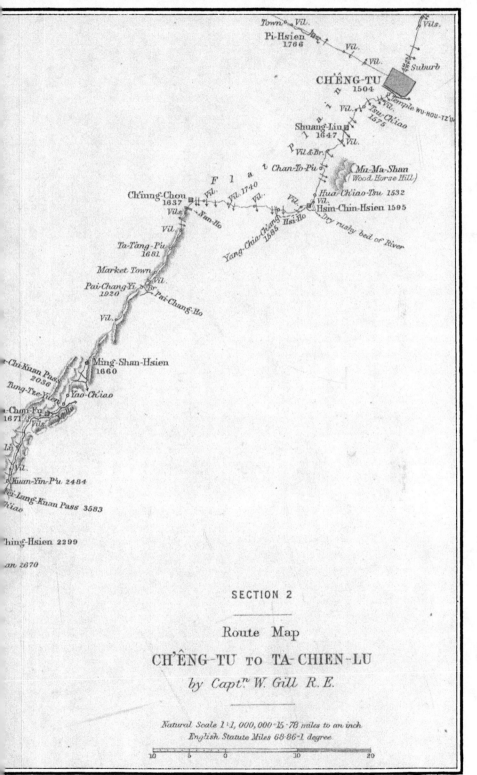

CHAPTER II.

THE ANCIENT MARCHES OF TIBET.

To Shuang-Liu—Superstition as to Fires in Towns—Vast Fertility of Country—Personal Criticisms on Travellers—Chinese Idea of Foreigners in general—Hsin-Chin-Hsien—First Sight of the Mountains—Ch'iung-Chou—Memorial of the System of 'Squeeze'—Fine Bridge on the Nan-Ho—Great Coolie-loads—Ta-T'ang-P'u—Tibetan Embassy—Pai-Chang-Yi—Indications of Opium-smoking—Paved Road in Decay—Ya-Chou-Fu—Its Commercial Importance—Cake-Tea—Coolies employed on the Traffic—Close and Beautiful Cultivation—Fei-Lung Pass—Wayside Anecdotes—Hsin-Tien-Chan—Sea of Straw Hats—Yung-Ching-Hsien—Traffic from Yün-Nan—Invitation to Breakfast—Ching-Kou-Chan—Limestone Scenery—Cooper's Mistake about Tea-oil Trees: in reality Rhododendrons—Huang-Ni-P'u—Suspension Bridge—Mountains begin to close round—The Little Pass—Alleged Terrors of the Hills—The Great Pass, and the T'ai-Hsiang Pass—'Straw-Sandal Flat'—Ching-Chi-Hsien—Divergence of Road to Yün-Nan by Ning-Yuen—Circuitousness of our Route—Pan-Chiu-Ngai—The Insect Tree and the Wax Tree—Mode of Production of the Wax—The Mountain Crops—San-Chiao-Ch'êng—Hua-Ling-P'ing—River of Ta-Chien-Lu—Bridge Swept Away—Lêng-Chi—Hill People—Lu-Ting-Chiao and its Iron Suspension Bridge—Dominoes—Character of People of Ssŭ-Ch'uan—Toll upon Coolies—Hsiao-P'eng-Pa—Scantier Vegetation—Fragment of elaborate Ancient Road—Arrival at Ta-Chien-Lu.

July 10.—The march to Shuang-Liu was over the busy, fertile plain, entirely given up to rice-cultivation. In the gardens there were melons, cucumbers, all sorts of vegetables, and patches of Indian corn. The country was beautifully watered; little rills brimming with water coursing by the road-side, or among the fields; and, as elsewhere on this plain, there were numerous detached farmhouses embowered in trees and bamboos.

The road was crowded with coolies and passengers proceeding in both directions, and the number of tea-houses was in proportion.

At the end of our journey Chin-Tai told me with a melancholy smile that he could buy neither fish, flesh, fowl, nor eggs, and wanted to know what I should like for dinner.

The fast was now being kept very religiously, and the south gates of all the cities were shut.

In China, wood forms a very large proportion of the material used in the construction of houses. After a continuance of hot and dry weather, the timber-work having been baked day after day is like tinder, and would blaze up on the slightest provocation. A fire would soon spread over a town, and with the inadequate means at disposal for extinguishing it, the damage would be very great. Fires are therefore intensely dreaded during a drought, and as it is supposed that the fire-god can only enter by the south gates of the cities, these are kept closely shut.

Perhaps the superstition has its origin in some old sun-worship; and the sun being always in the south, the fire-god is expected from the same quarter.

July 11.—But as we left Shuang-Liu, the Hsien, treating us as persons of great distinction, opened the south gate to let us through; this saved a long round, and was almost the highest honour he could have paid us. If fire had subsequently broken out, it would certainly have been attributed to his rash and foolish act.

West of Shuang-Liu the road still led us over the flat and level plain, where the amazing fertility of the soil was apparent in the magnificent crops of rice that now, from two to three feet high, presented to the eye a vast expanse of the richest green.

Riding through a town where, as it was market-day, all the streets were crowded, I was much edified by the remarks passed by the crowd upon my person.

I wore a helmet, and one man said, 'Does not he think himself a swell with a hat like a ram's horn?' 'Yes,' replied another, 'but look at his nose, he might be an official with that nose.'

The Chinese are great physiognomists, and always admire a good-sized nose; generally their own noses are perfectly flat, without any bridge, and by saying I might be an official, the man meant that my nose was good, and that therefore I ought to possess some talent that would fit me for an official position.

Another man said that I had tremendously long legs. The Chinese always wear such loose baggy raiment, that, in appearance, the length of their legs is very much diminished.

The observations that are made are not as a rule very flattering, and forcibly illustrate the old proverb about listeners. I once heard of an English gentleman of whom an educated Chinaman remarked with the intention of being highly complimentary: *Why, he is not so dirty as a Mongol.* A Mongol never takes his clothes off all the winter, eats fat and grease by the pound, wipes his fingers on, and drops messes all over his leather coat, and is about as greasy and dirty a personage as can well be imagined. On another occasion, an Englishman was told that he did not smell so bad as a Man-Tzŭ. However little it may flatter our Western vanity to admit it, there can be no doubt that every nation has its peculiar odour; but on this point I have already remarked.

It may seem impossible for us to understand how

such remarks can be made seriously, and without the smallest offensive intention ; but this is only another proof of the difficulty of understanding the Chinese. To judge of a Chinaman's character, we must look with the eyes of a Chinaman, and put ourselves outside every conviction that we have formed, even about ourselves.

A Chinaman from his earliest infancy is brought up to believe that besides the Chinese nation there are in the world only some few insignificant barbarians. The chief knowledge of foreigners was originally derived from intercourse, peaceful or warlike, with the Man-Tzŭ, Tibetans, or Mongols ; and even now the number of Chinese who have been in contact with Europeans is very small. When, therefore, a Chinaman had to form his idea of foreigners, there was nothing very wonderful in his comparing them with the Man-Tzŭ, the only type of foreigners known to him ; nor was this idea after all very much more erroneous than that prevalent not so very long ago amongst many English people, that frogs formed the principal part of a Frenchman's diet, or the opinion that is even now indulged in by many of our home-staying countrymen, that all foreign cookery is greasy.

At Hsin-Chin-Hsien, where we halted, the coolies all fell out, because some declared that the loads had not been properly distributed; and it was not until everything had been re-weighed, and all the baggage re-arranged, that they were satisfied. A fair load for two coolies on level ground is one hundred and sixty pounds, but in mountainous countries goods cannot be carried in the cages between two coolies, and the loads are then much less. There is, however, another class of coolie in the mountains, who carry enormous loads on their backs. These, however, travel very slowly, and are not usually employed by travellers.

The dislike of the coolies to carrying extra weight was natural at any time, and in the intense heat that now prevailed, I could not help feeling for the poor fellows as they ran along, the perspiration streaming from their bodies. There was but little shelter from the glare of the sun; the temperature rose as high as $99\frac{1}{2}°$ in the shade, and at ten o'clock at night, in the yard of the close and stuffy inn at Hsin-Chin-Hsien, where not a breath could stir the stifling air, the thermometer was still at 90°. This heat was not rendered more endurable to us inside by the bugs that swarmed about the inn.

July 12.—Everyone was glad to get up and away before sunrise, and we started at half-past five, the thermometer even at that hour being no lower than 85°.

It was with intense pleasure that, at half-past two in the afternoon, when the sun was darting its most fiery rays upon us, we caught the first sight of the mountains on the horizon, and our minds dwelt with pleasure on the snow-fields, and awful glaciers, so vividly depicted by former travellers in the regions we were now approaching. Soon afterwards, as if to cheer us, an easterly breeze sprung up, and the thermometer falling to 93°, the weather felt quite pleasant.

The streets of Ch'iung-Chou, though very wide, seemed poor and dirty, and could in no way compare with those of some of the fine Ssŭ-Ch'uan cities. The inn, however, was very comfortable; we had fine large rooms, with a yard behind, and there was a spacious open court in front, with a vine trailing over part of it.

July 13.—We saw only half the city, for as the south gate was not open when we marched out in the morning, we were obliged to retrace our steps to the east gate, by which we had entered. The walls of the city were in good repair, and solidly built of large blocks of red sandstone uncut; but in places where repairs had been recently executed, fine blocks of cut and squared stone had been let in.

Just outside Ch'iung-Chou there was the commencement of a triumphal arch, with all the carved stones lying about by the side of the road. Some inhabitant of the city intended to have erected this, but when half done, the magistrate forbade it, on the plea that it would bring bad luck to the city, but in reality, because he hoped to extract a large bribe from the people who were building it, for they were very rich. The latter, however, declined to come to terms, and the basement of the triumphal arch remains a monument to the system of ' The Squeeze.'

Not far from here was an ancient pagoda, which had fallen into ruins; some ambitious people determined to rebuild it in a magnificent manner, four hundred and fifty feet high, with a spiral staircase up the centre; but after attaining the moderate height of about forty feet, all the money was spent, or stolen, and the project came to a conclusion most lame and impotent.

A short distance from Ch'iung-Chou we came to the river called Nan-Ho (Southern River); the bed where we struck it was about one hundred yards wide, but following down stream about two hundred yards, we crossed by a remarkably fine fifteen-arched bridge of red stone, two hundred and forty yards long, and nine and a half yards wide, with a somewhat boastful inscription on a tablet, proclaiming it the finest in Ssŭ-Ch'uan.

After passing the river, we entered an undulating country, the hills of a reddish-yellow clay, and well wooded, principally with pines in small clumps; the road, running in many places between hedges, would have put me much in mind of some of the Hampshire scenery, if it had not been for the rice in terraces. The cultivation was only on the flat ground, the slopes being everywhere given up to trees. The road was exceedingly tortuous, winding about and twisting in a most perplexing manner, following the summit of a ridge from one hundred to two hundred feet above the valleys. We met coolies carrying logs of wood, sometimes as much as two hundred pounds in weight. These enormous loads are carried about ten miles a day, the wood being principally for coffins, which, when made from a particular and much-prized species of tree, cost sometimes as much as 300*l.* or 400*l.* Here nearly all the women had feet of the natural size, and many of those whose feet were cramped, had not squeezed them in nearly to the usual extent; but those seen about were mostly of the poorer class, for the richer folk do not permit their women to walk about much in public.

Coolie-hire was very cheap, as nearly all our chair coolies engaged other men to relieve them by turns.

At Ta-T'ang-P'u we breakfasted at a tea-house, where, amongst the tempting dishes exposed for sale, there were some fish cooked in oil. A box of sardines which, after half has been consumed, has been forgotten in a cupboard for a month, and has been subsequently exposed to a hot sun and plenty of dust for a week or two, would give a very good idea of the greasy mess displayed to entice the hungry coolie to a meal. One came up whose sole remaining fortune was one cash, and signified his desire to become a purchaser. The old woman who kept the place offered him one fish.

But even if a Chinaman has one cash only, he will not lavish it without enjoying the pleasure of a bargain.

My coolie was no exception, 'No,' he said, 'that is not sufficient, those fish are very poor things. I cannot afford a whole cash for such a trivial animal.'

The old woman at first refused to let him have a greater quantity, but discovering a head that had in the lapse of ages become separated from its body, she offered it to him. He examined it with the eye of a critic, turned it over from side to side, and at length agreed to take it. When he rejoined his companions, he said to them, 'See what a fine bargain I have made for one cash only, I have bought more than a fish.'

At Pai-Chang-Yi, we found one of the embassies that were just now on the Tibetan road established in one of the inns. The ambassador, if he can be dignified with such a title, was an official of very inferior position.

The Chinese always send petty officials as ambassadors, in order to show their immense superiority to foreigners; it is as much as saying 'Oh! anything is good enough for a foreigner'; and it must have gone sorely against the grain to despatch two men of high rank to England.

The people here are not such early risers as in the north of China; as we marched through Pai-Chang-Yi[1] in the morning none of the shops were open, and there were very few people about in the streets.

A Chinese town with its shops all shut up is even

---

[1] *Pai* means a hundred. *Chang* is a measure of ten Chinese feet.

more dreary in appearance than Regent Street on Sunday. The shop fronts—when open there are no fronts—are made of dirty wood, from which the paint has long worn off; and everything looks shabby to the last degree.

These habits of late rising are in consequence of the opium smoking that is very prevalent in Ssŭ-Ch'uan.

Whatever other effects the habit may have, it is certainly very troublesome to travellers who want to make an early start, or to accomplish a good day's march.

The town of Pai-Chang-Yi seemed principally to thrive upon the petty trade carried on with the thousands of coolies that pass through. The most noteworthy articles exposed for sale were copperas, borax, and alum, all of which come from some place near; and in the streets we saw a man carrying a load of pig-iron.

July 14.—The road to Ming-Shan-Hsien ran along the top of a ridge about two or three hundred feet above the valley. Every now and then, between the hills, there was a good view of some fertile little plain laid out for rice cultivation and beautifully green. The slopes were well wooded, or cultivated with Indian corn, and the tops of the ridges, where sufficiently flat, were covered with rice-fields. There did not seem to have been a very great scarcity of water, although some of the fields were still very dry.

The road was most unpleasant for walking on. It was paved, and the original intentions of the constructor had evidently been excellent, for most of the stones had been cut quite flat, and as the ground was tolerably level, if the execution had been as meritorious as the conception, the road would have left nothing to desire. But the contractor, in order to save money, had made use of the rough stuff he found lying about, and had put in at every foot or so an uncut boulder that thrust itself above the general level in a most obtrusive and unpleasant manner. No doubt by bribing the road surveyor, he had obtained a good report of his performance.

Between Ming-Shan-Hsien and Ya-Chou, the country was more broken, the smaller ridges giving way to detached hills of red clayey sandstone, all still well cultivated and wooded.

We followed a little stream through a miniature gorge, and ascending a branch, gained a saddle four hundred feet above the plain, whence we had a fine view of the Ya-Chou valley. This is about two miles wide, is quite flat, and bounded on each side by mountains from eight hundred to fifteen hundred feet high. The river bed is about two hundred yards wide, though at this season the water is not more than forty or fifty yards across. The stream ran at the rate of about four miles an hour, and was now very shallow. A little lower down, the valley closes in on the left bank, and steep, red hills, clothed with deep green foliage, hang over the water, forming little cliffs, and making a very pretty picture. We ascended the river about three miles, and crossed it in a ferry, just below the city of Ya-Chou-Fu, where we found a particularly nice inn with an open court in front.

In front of some of the houses, before reaching Ya-Chou, we saw a few vines, trailing over a trellis-work above the road. There was also some tea put out to dry, of which a little grows here. At Ming-Shan-Hsien some very celebrated tea is grown, but only in small quantities.

Ya-Chou is a place of great importance, as it is the starting-point of all the commerce to Tibet, to which place tea and cotton are the chief exports.

The most remarkable trade of this place is its commerce in tea, vast quantities of which are sent from here through Tibet, and up to the very gates of our own tea-gardens in India. The tea for the Tibetans is merely the sweepings that would elsewhere be thrown away, the poor Chinese in Ya-Chou paying seven or eight times the cost of this for what they drink themselves. It is pressed into cakes about 4 feet long × 1 foot × 4 inches, each of which is wrapped in straw, is called a *pau*, and weighs 24 lbs. The average load for a coolie is about ten or eleven of these packets. I have seen some carrying eighteen— that is 432 lbs. Little boys are constantly seen with five or six *pau*—120 lbs. These men wear a sort of framework on their backs, which, if the load is bulky, often comes right over the head and forms in rainy weather a protection from the wet. Each of them carries a thing like the handle of a spud, with an iron shoe and point at the end, and when they rest themselves the handle is put under the load, the point into the ground, and thus they relieve their backs from the weight. A coolie gets 1·8 taels to carry six *pau* (144 lbs.) from Ya-Chou to Ta-Chien-Lu, 150 miles over an exceedingly mountainous country; a distance usually accomplished in twenty days. The pay would seem barely enough to keep life in them under their tremendous loads. They eat scarcely anything but Indian-corn bread, made up into round cakes nearly an inch thick, and from six to ten inches in diameter.

July 15.—Beyond Ya-Chou we left the main river at once, and crossing a little ridge gained the valley of a tributary. At first this was nearly two miles wide, and quite flat, covered with rice-fields, and bounded by ridges of sandstone about five hundred feet high, behind which a chain of mountains rose fifteen hundred feet above the valley. The sides of the hills and mountains were cultivated with Indian corn right up to the tops, except on the slopes that were too steep, where there were patches of wood, firs, acacia, ash, a tree that is called here water-oak, and many others whose names I do not know. There were also a few orange, walnut, and tung-oil trees, and one or two banyans. I did not see many fruit-trees, but there were plenty of small apples, peaches, plums, and greengages exposed for sale, all horribly unripe. The Chinese here never let their fruit ripen. They are afraid that the rain will injure it, or that it will fall to the ground or spoil; and they have thus become so accustomed to unripe fruit, that they really prefer a peach that eats like an apple to a soft juicy one.

The country here had not suffered so much from want of water, and everything was beautifully green; the road was well shaded with clumps of trees and bamboos, and we had a pleasant walk to breakfast.

July 16.—The road from Kuan-Yin-P'u ascended a stream, running along the side of a spur that was closely cultivated to the very top. I never before saw ground so closely cultivated; as far as could be seen in any direction, every inch of ground had been brought under the plough, except here and there, where the hill-side might be broken into a small cliff. In some places the Indian corn was growing where it would seem an utter impossibility to carry on any agricultural operation. The sides of the hills generally were not very steep, 15° to 30°, but here and there there were steep bits and broken cliff. There were plenty of trees about, some standing singly, others two and three together, and in small and large clumps; they lined the road-side, and surrounded all the farm-houses, but there were no woods properly speaking.

Along this busy road there were plenty of large tea-houses, full of coolies, for this is the main road to Yün-Nan, and the traffic on it is very great. Tea and cotton seemed the chief exports from Ssŭ-Ch'uan, and in return the coolies were bringing in wood and indigo.

The road ascended by a zigzag to the Fei-Lung-Kuan, or Flying Dragon Pass, 3,583 feet above the sea. As the air had now become fresh and cool, I walked to the summit, and one of the coolies was heard to remark:

'He walks up hill because he is merciful to his men and beasts, but what can he walk down hill for?'

The Tinc-Chais were very officious, keeping in front, and calling out to everyone to get out of the way of their great excellencies who were coming along the road. On one occasion, when a coolie staggering along under a weight of two hundred pounds, with his head bent to the ground, did not move quickly enough, one of the Tinc-Chais flipped him with a little whip.

'What are you hitting me for?' called out the man.

'Hitting you! I was brushing a mosquito off your ear, and you are ungrateful enough not to thank me,' replied the Tinc-Chai.

One of my coolies bustling along failed to give the usual shout of warning, and with his load nearly upset a man standing in a narrow place.

'Do you come from the dumb man's city?' he asked with some warmth

On observing us pick flowers, put them away, and ask questions about all sorts of things, a man by the wayside remarked :

'What wonderful people these are! No one like them has been here since the time of the Three Kingdoms.'

This was highly complimentary, and it was very remarkable to find common coolies by the wayside conversant with the facts of an historical era so remote, for the Three Kingdoms flourished in the fourth century. The great Liu-Pi reigned over one of them, and his name, and those of his favourite ministers and generals are still looked up to as those of men whose talents or valour were almost fabulous.

At Hsin-Tien-Chan (New Inn Stage) it was market day, but as fasting was still being kept up, there was no meat of any kind for sale. The chief articles exposed in the market were cotton-stuffs, iron tools, other small iron goods, and vegetables ; and the usual amount of bread and cakes were to be seen at the shops of the innumerable bakers, who seemed perpetually at work baking or steaming. The Indian-corn cakes are baked on a gridiron over a little bit of charcoal fire ; the other bread is baked in an iron basin, the baker moving it, turning it round, and over and over all the time, to prevent its being burnt.

The weather was hot ; the people were now wearing broad-brimmed straw hats about two feet and a half in diameter, and as the streets were now all densely crowded with men, women, and children, jostling and pushing each other about, all these hats more or less overlapped. From the back of the pony upon which I was mounted no other part of the body or of the dress of the people was visible, and I seemed to be looking down upon a surging sea of straw hats, through which the Tinc-Chais cleared the way with shouts and gesticulations.

The towns here are not particularly neat: the streets are dirty and ill-paved, and present a great contrast to some of those in the neighbourhood of Ch'êng-Tu. The gardens are not so well kept, or so tidy, but they grow much the same vegetables. The cucumbers especially are very abundant, and I saw some sun-flowers, of which the people eat the seeds.

At Yung-Ching-Hsien there were some grand Lamas on their way to Lassa, but they could not get coolies or ponies, and with an enormous quantity of baggage were waiting here. The unfortunate Hsien was obliged, in the meantime, to entertain them and their retinue in the official house, or Kung-Kuan, and pay for everything they ate and drank.

The people in the bazaar now used to try and frighten our servants, telling them that we were all certain to be killed somewhere.

A man appeared who said that he had been recommended by Bishop Chauveau to the late Mr. T. T. Cooper, and had travelled with him until he was imprisoned; he did not tell us in what capacity he served him, but he said that on one occasion he was beaten by the official servants, and that he left Cooper in consequence. He asked us after the welfare of two of his companions, who had stayed with Cooper through his difficulties, and though we knew nothing about them, Mesny told him that they were both now in high official positions, and that if he had not deserted his master in the time of difficulties, he would also have been in a post of honour!

Yung-Ching-Hsien is situated on a plain between two branches of the river, one of which we crossed by a ferry, which, like all ferries in China, is free, and the expenses of which are paid for by the produce of a certain portion of land set apart for the purpose.

July 17.—As we marched through the town in the early morning, the shops were just opening, and I was not very favourably impressed with the place; but I had perhaps been spoiled by the very fine cities in the Ch'êng-Tu plain, for certainly even these towns of Western Ssŭ-Ch'uan would compare favourably with those in the north of China. The streets are wide and fairly paved, though there are a great many round stones used, which are equally disagreeable for man and beast. The unfavourable impressions are also partly owing to the abominably shabby state of the houses, which never seem to be painted, whitewashed, or repaired. All the woodwork is black with dirt, the paint is rubbed off, and everything looks dreadfully dilapidated. This is most apparent before the shutters are removed, for the fronts are quite open, and in the day all the inhabitants of the town collect in the main street, and in a measure conceal the imperfections; but in the morning, when the greasy shutters close the fronts, and there are only a few sleepy coolies about, or an early pieman selling his hot cakes, all the dirt is seen in its full glory.

We met a long train of mules bringing opium from Yün-Nan, and others carrying brass. There was a man with a cargo of parrots, which, he said, came from the mountains in the interior; but he did not know much about it, as he did not get them himself. We afterwards found the home of these birds in the neighbourhood of the Chin-Sha-Ching, south of Bat'ang.

A man standing in the market-place said, 'Do you see those men?' 'Yes. Autumn and Spring,' replied another, alluding to our apparent ages, for

Chinamen always thought that I was an old man. Another said we were Siamese. He probably came from Yün-Nan, and had heard of Siamese there.

The coolies found a cheap breakfast very early, and we halted at a tea-house kept by an old lady, who had a voice so shrill and sharp that it might be heard all the way to Ch'êng-Tu.

She called out to all the passers-by to come in.

'Come in here, come in here. I have beautiful rice, very clean and quite hot; anything else you may like to eat, and much cheaper than you can get it further on. Come in and breakfast. See that nice place opposite, where you can put down your loads, or if you don't like that you can put them in here. Come in, come in, bread just baked, and everything of the best.'

We breakfasted at Ching-K'ou-Chan (Dark Gorge Stage), so called because a little above it two hills much more steep than any of the others, precipitous at the top and covered with very deep green trees, come down to the stream, one on each side, and form a short gorge. Here the inn was in the regular mountain style, the best room over the stable, with a window overlooking the valley.

About six miles from Yung-Ching we left the main river, and ascended a tributary up a very pretty valley bounded by sandstone hills. The red sandstone formation presented a remarkable contrast to the limestone of the valley of Sun-P'an-T'ing. The limestone is always broken into sharp crags and pinnacles, leaving tremendous precipices. The streams find their way through long and gloomy gorges, sometimes winding for miles between perpendicular walls of rock, scarcely broken by a chasm. But the sandstone hills are round or flat-topped; they are rarely steep, are much cut up by streams and valleys in every direction, are broken into detached hills, and are sometimes almost undulating.

The stream we ascended was clear as crystal, and was bounded by hills that were cultivated with Indian corn. There were the same trees as before, and in addition there was a tree which grows to a height of forty feet, and bears a flower very like a gardenia in appearance, size, and smell. This tree flowers twice a year, but produces no fruit. There is another species of the same tree, which flowers only once a year, the flower is inferior to this, but the tree bears fruit. The Chinese name for it in this part of Ssŭ-Ch'uan is Tzh-Tzh-Hua.

I was looking out for the tea-oil tree, of which Cooper speaks so much in his 'Pioneer of Commerce'; but it is quite certain that he mistook the rhododendron for the tea-oil tree. The tea-oil tree was as familiar to Mesny as an oak to an English farmer, and during all our journey we never saw one. The tea-oil tree is called by the Chinese Ch'a-Yo; it has a large flower like a rose; some trees bear red, and some white flowers, but the petals are thick like those of a gardenia, the leaf not so large as a laurel, and quite smooth.

At Huang-Ni-P'u there was some tea of native growth which had not been fired; it was exceedingly fragrant, and though the odour was somewhat of fresh, new-mown hay, there was, besides, the true flavour and aroma of tea.

At Huang-Ni-P'u there were many signs of the proximity of the mountains; the coolies, although they grumbled much at the lowness of their pay, hired other men to do their work; these latter carry their loads on their backs, the universal custom in the mountain districts. The chair coolies, also at their own expense, hired helpers with ropes to drag the chairs in the steepest parts, and on one occasion the Ma-Fu hired an assistant; but this must simply have been a piece of swagger on his part, as he had no more severe task to perform than to hold the tail of the pony in difficult places.

There are copper and iron mines in the neighbourhood of Huang-Ni-P'u, but they are not worked, because they say that no one has the capital necessary to open them. Copper, however, seemed cheap, for in the villages, instead of the usual little wooden basins which, each with a bit of rag in it, are arranged in a row of half a dozen before the eating-houses, there were copper basins instead. These basins are for the people to wash their faces in, when they arrive hot and dusty.

July 18.—As we ascended the river we had followed thus far (Yung-Ch'ing-Ho), the mountains began to close around us, and the amount of cultivation seemed to decrease with every step in advance. The river was bounded by hills which were much higher than those we had as yet passed through, and now, too steep for agriculture, were densely wooded with trees and undergrowth. We crossed a stream by a suspension bridge, the roadway of which was laid on five iron chains, the links ten inches long, and made of round iron three quarters of an inch in diameter. There were as additional supports two other chains, one on each side, which also acted as hand-rails and guards. The droop of this bridge was very slight. The river was crossed immediately afterwards by a similar bridge, forty-five feet long, with seven instead of five chains for the roadway.

The limit of the Indian corn and bamboo was

near here 4,132 feet above the sea; for between this point and the summit of the T'ai-Hsiang-Ling-Kuan, or the Pass of the Great Minister's Range, there was, with the exception of one tiny patch, absolutely no cultivation, the hill-sides being clothed with a rich and brilliant green foliage of trees and undergrowth, which completely obscured the red colour of the granite, of which the whole of the mountains here are formed.

Quitting the river, we ascended a tributary for a short distance, and then crossing a little spur, a descent of fifty feet brought us back to the main stream. The road over this spur is closed by a gate called Hsiao-Kuan (Little Pass), and the village at the foot of it bears the same name.

Whilst sitting at breakfast Chin-Tai came in with a sad look about his face, and after a cough that was the invariable prelude to a miraculous narrative, he began:

'The road goes over a very big Shan.' (He never succeeded in learning the English word 'mountain.')

'Yes,' I said, 'so I understand.'

After a moment's pause he continued, 'We must not make much talk on the big Shan.'

'Indeed,' I said. 'I suppose a great wind would come if we did.'

'Yes,' he answered; 'there was once a big military official who——'

'Ah,' I interrupted, 'he was advised not to go up with his army?'

'Yes; all the people tell him that——'

'That is enough,' I said, 'I have heard the story before;[2] and as these mountains seem to have been so fatal to military officials, you had better go and knock your head that you are but a humble civilian.'

[2] See vol. i., page 382, at Fêng-Tung-Kuan.

Beyond the Little Pass, the road followed the valley, and was one of the worst I ever travelled on. Now zigzagging up the side of a mountain, the path was cut in steep steps over sharp pointed rocks, and now winding along the side of a gully, some stream was crossed by a ford or a bridge. Everywhere the wooded hills rose above us some one thousand or two thousand feet, very steep but never precipitous. Sometimes we were down at the level of the stream, at others far above it, but the steady ascent always continued. After a time again leaving the main valley we ascended a steep spur by a long zigzag, and reached its crest at Ta-Kuan (Great Pass), 5,754 feet above the sea. Near here we passed an unfortunate pony that had fallen down under its load, and was left to die by the roadside. I wanted to shoot it, and put it out of its misery, but was told that its owner would be sure to come up, and accuse me of having killed a fine and healthy animal. After this the road again rejoined the river without descending appreciably, and another long pull of four thousand feet brought us to the summit of the T'ai-Hsiang-Ling-Kuan, 9,366 feet above the sea.

Directly we crossed this, the landscape changed entirely, the mountain sides being all green grassy slopes, very little cut up by valleys, and not so steep as those on the other side. There was no wood, no cultivation, and little undergrowth, but the ground was covered with beautiful rich grass and many wild flowers.

The rain had been falling on the eastern slope, which, from the luxuriance of the foliage, appears to possess a much damper climate than the western face; but as the ridge arrested the clouds, we were now in tolerably fine weather, and from the little tea-house close to the summit we could see the city of Ch'ing-Ch'i at the foot of a steep spur 3,888 feet below. This tea-house rejoiced in the name of Ts'ao-Hsieh-P'ing (Straw Sandal Flat), and was doubtless so called on account of the numerous straw sandals expended in the passage of this terrible mountain.

Another wearisome zigzag led us down a very steep spur; on the way, the first cultivation was a patch of tobacco, and a little lower, the familiar fields of Indian corn and beans again covered the hill-sides.

Ritter quoting a Chinese itinerary thus speaks of the ascent, and of the city of Ch'ing-Ch'i: 'From Yung-Ching-Hsien the road soon becomes very difficult. The way is blocked by the great mountain, Hsiao-Kuan, in the profound gorges of which heavy torrents of rain continually fall, and the flanks of which are always clothed in mist and fog. Wild mountain streams fall headlong down from it. But now the road passes over a second and far more formidable mountain, the Hsiang-Ling, where the snow is so deep in winter and spring as to render it impassable.

'The road down to the town of Ch'ing-Ch'i is very steep. Frightful storms reign here; whirlwinds suddenly arise by day or night with great violence, destroy everything, and make the houses shake; but the inhabitants are now accustomed to these phenomena.'

From near the summit of Ts'ao-Hsieh-P'ing, a cross road, without inns or accommodation, makes a short cut across the mountains. Most of our baggage coolies went that way and rejoined us at I-T'ou-Ch'ang; but I did not succeed in finding out exactly where the two roads came together.

July 19.—At Ch'ing-Chi-Hsien we again found

the principal inns occupied by Tibetan embassies, and there was no room for our coolies, who were obliged to put up at an inn outside the gate. The market produced some excellent potatoes; but there was no meat, as the people were still fasting, although there was rather an excess than a want of rain, so much so that in the morning the coolies declared they could not start, as the road would be impassable. It was fortunate that we did not listen to them; the road certainly was very bad, and even dangerous in places, but we managed to traverse it, which we certainly should not have been able to do had we delayed a few hours.

Ch'ing-Ch'i seemed a wretchedly poor place; when we started there was no one in the streets, the shops were all shut, and the city generally bore a miserable aspect. I have a very vague idea of the day's march, for everything was shrouded in mist and fog. This is amply compensated for by the remarkably vivid impressions retained of the road; in many places there was no road at all—we had to cross ravines, where the torrents, swollen by the rain, had altogether carried away the goat-track that did duty for a path, and sometimes in these narrow gullies it was almost impossible to get the chairs round the sharp corners. The soil was a soft sticky clay, so slippery that in many places I was absolutely unable to walk in European boots; the chairs were continually bumped about by projecting rocks; the coolies stumbled, the rain fell, and altogether it was anything but a lively performance, as may be gathered from the fact that we were six hours covering the nine miles to Fu-Hsing-Ch'ang.

On leaving Ch'ing-Chi we descended to a stream, crossed it, and ascended the hills bounding it on the other side, until we gained the crest of a ridge that separated it from another valley. From this point the main road to the province of Yün-Nan leads to the south-west. We left it on our left, and crossing the ridge followed up the stream to Fu-Hsing-Ch'ang.

Although the spiral of Archimedes possesses many beautiful properties, as a method of reaching one point from another it is not a curve to be recommended. Yet such had hitherto been the direction of our road, for after nine days' marching we were in a straight line, almost as far from Ta-Chien-Lu as we were at Ch'êng-Tu; but even the spiral approximates to its centre by degrees, and from this point our road began to bend round towards its destination.

In China the roads are kept in order (or supposed to be) by the magistrate of the district; but although he has the power to order men to repair it, if he does so it helps to make him unpopular, so as long as he thinks that no high official is likely to come his way he leaves the roads alone. This is the great high road from China to Tibet; embassies are continually passing; and the commerce and traffic is very great, and yet there were one or two places that must certainly have been impassable after the heavy rain that fell steadily all day and night, for when we left Fu-Hsing-Ch'ang it was still raining, though it cleared up afterwards.

July 20.—We ascended the valley, the mountains on our left sloping steeply down to the stream in remarkably smooth slopes, those on our right being broken into hills from three hundred to five hundred feet high. There were little patches of rice in the bed of the river; but on both sides the cultivation of Indian corn was carried on as usual right up to the top, and except by the roadside and round the houses there were no trees. Above Pan-Chiu-^{Ng}ai the valley opens out a little, and is very picturesque; on the east, high mountains throw out low, gently sloping spurs, and on the west, others steeper but not so high close in on the stream, but have not the same smooth unbroken slopes as lower down, the sides being more cut up by small ravines. Here the trees commence again, and there are fine detached clumps on the hills and round the villages. In this part of the valley there were great numbers of the celebrated insect trees. These are in appearance very like orange trees, with a similar leaf, but they have a very small white flower that grows in large sprays, and now that the trees were covered with great masses of blossom, their strong smell, which was not very sweet, pervaded the air.

It is on this tree that the insect is bred that produces the celebrated Pai-La, or white wax of Ssŭ-Ch'uan. These trees are chiefly grown in the neighbourhood of Ning-Yuan-Fu, and the eggs are thence transported towards the end of April to Kia-Ting-Fu, where they are placed on the wax-tree, which is something like a willow. Here the insect emerges from his egg, and the branch of the tree on which he is placed is soon covered with a kind of white wax, secreted. It is this white wax that is so celebrated, and is one of the most valuable products of Ssŭ-Ch'uan. These eggs cannot be exposed to the heat of the sun, and whilst being carried from the breeding to the producing district the coolies travel only in the night, when the road is said to present a very remarkable appearance, as the coolies all carry lanterns. Ordinarily in China no travelling is done at night, and as the gates of all towns and cities are closed at dusk, and are never opened for anybody, no matter who he may be, travelling at night is rendered impossible. But during the time for bringing the eggs to Kia-Ting-Fu all the city gates are left open night and day,—probably the only exception in China to the rule of shutting the gates at dusk.

In the 'Journal of the Agricultural and Horticultural Society of India,' vol. vii., there is an account of the wax insect, derived principally from two Chinese writers. The insect is said to feed on a shrub *Ligustrum lucidum*; the trees are cut down every four or five years, and as a general rule are not stocked until the second year after this operation. The nests of the insect are about the size of a fowl's head, and are removed by cutting off a portion of the branch to which they are attached. The sticks with the adhering nests are soaked in un-husked-rice-water for a quarter of an hour, when they may be separated.

The nests are then tied to a tree. In a few days they swell; insects the size of *nits* emerge. From *nits* they attain the size of lice; and having compared it to this, the most familiar to them of all insects, our Chinese authors deem further description superfluous.

The description of the rest of the process agrees well with what I learnt myself, and with the writings of Baron von Richthofen.

But, until the researches of the German traveller furnished us with the details of the process, even Chinese writers seem to have been unaware that the insect was reared on one kind of tree and the wax secreted on another, although the Chinese authors quoted above knew that the insects were removed from one tree to another.

Baron von Richthofen, speaking of the insect-tree, says:

'It is so valuable that it constitutes a separate article of property distinct from the soil on which it grows'—like the olive-trees in Cyprus.

The 'Comptes Rendus' for 1840, tome x., p. 618, are quoted by the editor of the 'Agricultural and Horticultural Journal of India' as stating that the wax insects are raised from three species of plants: Niu-Tching (*Rhus succedanea*), Tong-Tsing (*Ligustrum glabrum*), Shwui-Kin, supposed to be a species of Hibiscus.

The knowledge that more than one species of tree was employed seems therefore to have already existed, as also the knowledge that the nest was removed from the tree on which it was reared to the tree on which the wax was secreted; but until the researches of Baron von Richthofen threw a flood of light on Western China, no one appears to have combined the two facts, or to have been aware that one species of tree was favourable to the growth of the insect, and another to the production of wax.

The Chinese authors above alluded to also state that:

'When the insects emerge from their nests, they with one accord descend towards the ground, where, if they find any grass, they take up their quarters.

'To prevent this, the ground beneath it is kept bare, care being also taken that their implacable enemies, the ants, have no access to the tree. Finding no congenial resting-place below, they reascend.'

Baron von Richthofen, however, states: 'The insect has no enemy, and is not even touched by ants.'

The attempt of the insects to descend from the trees, as described by the Chinese writers, would seem to confirm Baron von Richthofen's theory that the insect secretes the wax when in an unhealthy condition.

The editor of the 'Agricultural and Horticultural Journal' quotes Mr. R. C. Brodie as stating that, 'although in appearance the substance resembles stearine or spermaceti more than bees'-wax, it comes nearest to purified cerin.'

Monseigneur Desflêches at Ch'êng-Tu gave me a specimen, which is now in the British Museum.

On this road we continually passed long trains of coolies, carrying tea on their backs, climbing mournfully and with measured tread the desperate and staircase-like tracks. There was something very sad in the aspect of these men; they seemed more like beasts of burden than human beings; they never smiled, and scarcely ever said a word; and as our lively Ssŭ-Ch'uan coolies, ever ready with some banter, passed them, they would stand on one side, with rigid countenances that scarcely relaxed into an expression of wonder as the two strange foreigners came by. These coolies, who do the chief part of the mountain transport, are quite a different class to the comparatively well-paid coolies of the plains; they carry the tea as far as Ta-Chien-Lu, beyond which point that extraordinary and hardy animal the yak is almost solely employed.

July 21.—From Pan-Chiu-Ngai we continued our ascent of the valley, the hills on each side being cultivated with Indian corn to the very top; those on the left running up steeply to a height of about one thousand feet above the stream, while on our right a huge mountain threw down gently sloping spurs to the valley. As we ascended, a little buckwheat and oats appeared, but beans and Indian corn, as usual, formed nine-tenths of the crops; the amount of cultivation gradually decreased, and the hill-tops, becoming steeper, were covered with dense green foliage of small brushwood and low jungle.

Beyond San-Chiao-Ch'êng the road is carried above some cliffs that form a narrow gorge, but in a short distance it again descends to the river, now little more than a mountain torrent. The steep sides of the narrow valley, through which the stream dashes, are clothed with a thick jungle of brushwood and wild roses, and rise from two hundred to four hundred feet above the road.

The people had told us that there were no more regular mountain passes before reaching Ta-Chien-Lu, but mile after mile we ascended, in continual showers of heavy rain. The road was broken into rocky steps, sometimes so steep that it seemed as if neither ponies nor coolies could possibly mount, and sometimes so slippery that I was quite unable to walk in European boots, nothing but the straw sandals that the coolies wear giving any hold on these steep paths. At last, after a long clamber up many a weary zigzag through a dank mist that shrouded everything from view, we gained the summit of the pass called sometimes Wu-Yai-Ling (which means the 'Range without a Fork') and sometimes Fei-Yueh-Ling ('Fly beyond Range')[3] nine thousand and twenty-two feet above the sea. From here, as the clouds lifted for a few minutes, there was a fine view in both directions. The valley on the northern side was rather more open, and the hills less steep, and we descended about a couple of thousand feet to the town of Hua-Ling-P'ing, perched among many walnut and other trees on a little plateau about five hundred

---

[3] Correctly, I believe, *Ling* is the pass or *col*, not the *range*.—*Y.*

feet above the stream, where there was a small but very comfortable inn.

Ritter again quotes the Chinese itinerary from Ch'ing-Ch'i : ' Soon one must cross a torrent, which, from the force of its rushing stream, has been called the double-edged sword of the philosopher,' &c., &c., &c., and thus speaks of the Fei-Yueh-Ling : ' The Fei-Yueh-Ling, a colossal mountain, of which the wild fantastic rocks heave themselves up nearly perpendicularly over its foot, where, in the time of the T'ang dynasty, lay the town of Fei-Yueh-Hsien, from which the name is taken. Its flanks are clothed with clouds, its lofty ridges are buried in snow all the year round.[4] The way across, over huge rocks and through wild chasms, is one of the worst in China.'

July 22.—The last shower fell as we left the town, the day then cleared up, and in the valleys the damp heat was again almost oppressive after the chilly air of the mountain-tops. We descended the stream until some people met us with the pleasing intelligence that the bridge by which we ought to cross had been washed away in the morning, and that we should probably have to wait until it should please the river to subside. We went to look for the remains of the bridge, but there were absolutely none to see, and the muddy torrent roaring and foaming over huge rocks and stones was evidently quite impassable. Our guides and the inhabitants of the place, with one consent, now tried to frighten us, and assured us that there was no road ; but not heeding them, we found a track through a field of Indian corn, which, leading above a little cliff that bounded the stream, led us down to a village, whence the road was very fair to

[4] The line of perpetual snow must be seven or eight thousand feet above its summit.

the junction of this stream with the river of Ta-Chien-Lu.

The valley we descended is not so picturesque as that at the other side of the pass ; the mountain sides rise up to a height of about one thousand or fifteen hundred feet above the stream, at an angle of about 30°, and are well cultivated nearly to their tops. There is no wood on them, but in places a good deal of low jungle, brushwood, thorns, and briars. The villages on the side of the stream are almost hidden in thick clumps of trees, mostly walnuts and the white-wood tree used for fuel ; there are also a good many apricot and plum trees. Rice-cultivation commenced about four thousand feet below the summit of the Fei-Yueh-Ling Pass, and bamboos appeared a little lower down.

There was a market town not very far from the *embouchure* of the stream, where the people all came out, and after wishing their great excellencies a respectful welcome, assured us that it would not be safe to leave their town, as the roads were bad, and the journey so long and tedious, that we could not hope to reach the next village before night ; but, not deceived by the bold effrontery with which the whole population of the town joined in telling this enormous lie, we anticipated an easy ride to Lêng-Chi. The custom of two such distinguished people as ourselves, with a retinue of sixty followers, had excited the cupidity of the inhabitants, who naturally wished to enjoy the unwonted luxury of guests who paid.

We found Lêng-Chi at no great distance ; it stands on the banks of the river Lu,[5] which here, forty yards broad, runs down to Kia-Ting-Fu. The village is very pretty to look at from the outside, but

[5] Called in Map and Itinerary Tung-Ho, or Ta-Ho.—*Y.*

its beauty is decidedly skin-deep. The inn was filthy, close, and stuffy, and filled with ravenous mosquitoes. The rooms were so small there was hardly space to sit down in, and with the smells that here were not less than elsewhere, it was not altogether a luxurious retreat. The houses in this neighbourhood are built of rough stones, picked up in the beds of the streams; many of the roofs are of loose boards of wood, with stones on them to prevent the wind blowing them away; but some of the better houses are roofed with the regular Chinese tile.

There is a native chief at Lêng-Chi, and there are a few aborigines in the adjacent mountains. We were informed that these were the most southern of the outlying barbarians; but it is not probable that this statement was strictly true.

July 23.—When we left Lêng-Chi the sun had scarcely risen, and the mountain tops were shrouded in mists; but for a minute the clouds lifted from a wooded peak; at that instant the first rays of the rising sun flashed on it, tinged the summit, and seemed to fill the very air with its glory; it was a vision of beauty, but lasted only for a moment, the clouds again gathering over the mountain and hiding it from our view.

It was a delightful change at last to find a level road. Here the path was very good, and ran along the side of the hill about five hundred feet above the muddy river. On each side, steep mountains, with grassy slopes, or waving fields of Indian corn, where there was but little wood, rose some couple of thousand feet above the river, and once or twice an open valley giving a view into the interior disclosed the distant pine-clad heights.

The villages were very pretty from the outside,

surrounded by and almost hidden in the thick foliage of walnut and other trees, but, as at Lêng-Chi, on entering them more than the usual amount of squalor and dirt became apparent.

At Lu-Ting-Ch'iao the river is crossed by an iron chain suspension bridge, of one hundred yards span. The roadway is laid on nine iron chains, and there are two other chains at each side for hand rails; the links are of seven-eighth-inch round iron, and are about ten inches long, but those underneath are much eaten away by rust. The roadway consists of planks laid across, which were originally lashed to the chains; but all the lashings were now adrift, and the planks quite loose, with wide gaps between them. There is a deep pit at each end of the bridge, into which the chains are brought, and where, if they get slack, they can be tautened up with powerful windlasses. I crossed with a good many people, and there was very little vibration; but Mesny, during the afternoon, walked over by himself, and found it swayed about a good deal.

On arrival at the bridge I was directed to cross it; I dismounted, and walked across to examine the structure, and pace its length, and I did not take much notice of what my people were doing with the ponies. These animals were rather frightened at the loose planks, but the men, instead of letting them go slowly and put their heads down to see what they were doing, dragged at the bridles, and attempted to pull them over by main force. The poor brutes, in consequence, could not see where to put their feet, one false step was made, both the animals started, and in a moment all their eight legs were in the openings between the planks. By the aid of a number of coolies, however, they were lifted up bodily from their perilous position, and reached the other side more frightened than hurt.

There was an archway at the end of the bridge which seemed to be the principal seat of trade as well as of amusement; for here there was a large party of coolies playing dominoes, with pieces not very different from those used in Europe.

After having crossed and waited about some time searching for the inn, we found that our men had taken up our quarters at the other side; so leaving the ponies here we recrossed to the left bank of the stream, where we found a delightful inn, large and comfortable, with two good bed-rooms, besides the sitting-room; but of course with the invariable bad smells from piggeries and other foulness. There was an upper story also, and for the first time in China I heard people walking about overhead.

The people here took very little notice of us; as they are accustomed to the constant presence of Tibetans, and as all foreigners (including these) are classed together, we did not attract much attention. But the dog was still an object of much curiosity, and a good many people came about on his account. Chin-Tai, finding access to our room incommoded by the people, told a man rather sharply to get out of the way, and not come staring at us; to which the man replied, not exactly 'a cat may look at a king,' but something very like it, for he said that he might look at the Emperor of China, and he supposed that an Englishman was not better than that. Whereupon high words ensued, and something like a fight, and Chung-Erh came in breathless with rage and excitement with no other purpose whatever than to tell me that he had made up his mind to kill one of the men of Ssŭ-Ch'uan before he left their country.

Both my boys had taken a violent dislike to the people of this province, chiefly because on one occasion when Chin-Tai had bargained for a fowl for one hundred and twenty cash, and had been obliged to go and fetch the money, he had been told on his return that the price was one hundred and thirty. This rankled in his bosom ever after, and now neither he nor Chung-Erh missed any opportunity of abusing these people. We foreigners found them peculiarly agreeable, and there is no other province in China, except perhaps Kwei-Chou, where Europeans would meet with such invariable civility; but the people are a little touchy, they have their own way of doing things, and if they are treated brusquely they take offence. Hence Chin-Tai and Chung-Erh, with the blustering and unpolished manners of the north, did not succeed in cultivating pleasant relations with them.

Airing ourselves at the inn door, we entered into conversation with a man, who told us that the bridge was three hundred Chinese feet long, and had thirteen chains. On inquiring the reason of this gratuitous information, we were told that our reputation for asking questions had preceded us, and that the bridge had been measured for the first time within the memory of man expressly for our gratification.

We then strolled off to see a young bear, which a man was particularly anxious to sell to us, but he did not succeed in inducing us to set up a travelling menagerie.

July 24.—The official at the bridge sent to ask us to bring all our coolies together, so that he might know which were ours; for a charge of ten cash is levied on each loaded coolie—from which tax we were exempt. This is one of the perquisites of the official of the place, and he is supposed to keep the bridge in repair with it, but of course, unless some other much higher officer is expected, nothing is ever done to it. The fortune, however, that the official accumulates in this way cannot be very large; for unless the river is in flood, every one crosses by a ferry—a much more tedious operation, but one whereby a few cash are saved. In a strong wind, moreover, the bridge sways about a great deal, and the heavily laden coolies prefer the ferry to the bridge.

The road from Lu-Ting-Ch'iao ran along the side of a mountain on the right bank of the river, keeping generally about five hundred feet above it, but descending once or twice to reach a village or cross a torrent. The river valley now closed in, the hill-sides became more steep, and the cultivation almost entirely disappeared; but in the bottoms of the valleys there were still some tiny plots of rice, the last we saw for many a long and weary day. The little agriculture carried on on the slopes produced as usual chiefly Indian corn and beans, with small quantities of pearl barley.

Two miles before we reached Hsiao-P'êng-Pa, we saw the first aboriginal village on the opposite side of the river. The style of architecture was very similar to that of the Man-Tzŭ villages in the Sung-P'an-T'ing valley; but here the roofs were gabled instead of being flat.

Beyond Hsiao-P'êng-Pa the river ran between precipitous mountains, with here and there wild bare slopes running down sharply to the stream; the road was not very good, in some places ascending long and steep inclines or steps, and at others rounding a bluff at an angle rather too sharp to be easy for a chair. Seven miles beyond Hsiao-P'êng-Pa the road crossed a torrent by a covered wooden bridge, and an icy breath that suddenly saluted me, made me look up the narrow gorge, and between the clouds that rolled up the mountain sides some snow was visible lying on a peak at no great distance.

Wa-Ssŭ-Kou, where we slept, is situated at the junction of the stream that comes from Ta-Chien-Lu with the main river. Both streams here flow through narrow gorges, and at their junction there hardly looks as if there was room to pitch a tent; but the Chinese do not mind being crowded, and have managed to find place for the few houses that make the village.

July 25.—The valley, for the first ten miles beyond Wa-Ssŭ-Kou, is closed in by steep hills, whose rugged sides have been rent by the rigours of the climate, and torn into cliffs and precipices, that overhang the roaring stream. As Ta-Chien-Lu is approached the valley is more open, but the ground and river-bed are everywhere strewn with great boulders, and the water leaps down in a succession of falls over huge masses of rock. At this time the rains had filled it, and it thundered down a mass of foam, falling nearly three thousand feet in the twenty miles from Ta-Chien-Lu. Here and there, a rift in the rocks on either hand disclosed some torrent hurling itself headlong from the heights above, and with a last mad leap flinging itself down some hundred feet into the boiling river. There was little grown up this valley besides Indian corn, potatoes, and cabbages. The potatoes were particularly good, and were not much inferior to any found in Europe.

As the stream is ascended, the grass and low green brushwood, that have here supplanted the cultivation on the slopes, become more scanty, and the bare rocks thrust themselves grimly through the half-starved vegetation ; the climate changes rapidly, and the wild clouds that sweep after one another across the mountain flanks are driven by boisterous and chilly blasts, that rudely buffet the traveller, softened by the gentle breezes that breathe in the quiet vales below. The walnut trees at last give up their struggle with the climate, and leave some of their more hardy brethren to fringe the stream and road, and stand between the houses and the fierce winds of winter. The road is in harmony with its savage surroundings, but in three or four places there are remains of what appears to have been a fine, ancient road, fifteen feet wide, evenly paved, and on which all the gradients were easy. I suddenly came upon this, after my wretched pony, nearly worn out with fatigue and starvation, had almost fallen down over an unusually cruel staircase—(the Ma-Fu had gambled away all his money, and had no means of buying provender for his wretched brutes); but the poor animal's sigh of relief soon evaporated, as, after about fifty yards, this good road came to an end, and the rough and heart-breaking path recommenced. In all probability, in the palmy days of old, when the great Liu-Pi despatched Kung-Nung to the Arrow Furnace Forge,[6] there was a good level paved road all the way from Ta-Chien-Lu, but like everything else in China it has fallen into decay.

Notwithstanding its turbulence, there are some fine fish in this stream. It is difficult to understand how they can ascend it, unless they possess the powers of the fish spoken of by the boastful Yankee, who guessed that 'our fish, sir, can swim up Niagara.' At Liu-Yang we bought a fish, between two and three feet long, in very fine condition, with brown flesh, and though not connected with the salmon tribe it was excellent eating.

[6] See beginning of Chapter III.

Ta-Chien-Lu was the end of the first stage in our journey; and as we knew we should be compelled to stay there a few days, we sent a man on in advance to find us a good inn; but the miserable place into which we were shown on arrival was worse than usual. It had not even the appearance of an inn; there was a long narrow restaurant that opened direct from the street, and at the end of this, partitioned off by only a screen, there was a dark and dreary cell. The greater part of the end wall of this was occupied by an altar, where there were idols and the ashes of innumerable incense sticks; and at each side there was a place like a cupboard. One of these cupboards was a trifle larger than the other, and boasted a hole, in which there still remained about half a window frame. A great mass of rock rose behind, blocking out the light and air, and it was truly a miserable place. But the only good inn was already taken up by the Lamas, and we had no choice but to stop here.

We received a warm letter of welcome from the late Bishop Chauveau, who said he would have found us a house if we had written to him beforehand. But as we had intended to remain only long enough to get provisions and transport, we had not thought it worth while, imagining that we should be sure to find a good inn in the border town between Tibet and China.

We heard soon after of some rooms, which I went to look at; they were an improvement on those we occupied, but as we thought we might find something better still, we determined to remain where we were for the night. Before dinner we called on the missionaries. Bishop Chauveau received us with every expression of cordiality and friendship, and we spent a delightful half-hour before returning to our gloomy lodging.

CHAPTER III.

'THE ARROW FURNACE FORGE.'

Ta-Chien-Lu—Legendary Etymology of the Name—Native King—Indian Rupees Current—The Place and People—'Om Mani Pemi Hom!'—A House found for us—The Local Government—Transport Arrangements—The Lamas and the Dalai Lama—The Prayer Cylinder and the Multiform Mani Inscriptions—The Lama Ambassadors—Menaces of our Fate if we entered Tibet—The Servants begin to Quail—Chin-Tai, his Greed and his Tempers—Heavy Provisioning for the Journey—Contrast of Tibetan and Chinese Habits—Of Tibetan Simplicity of Fare with Chinese Variety—Tariff at Ta-Chien-Lu—Preparations to meet Cold, and Needless Laying-in of Furs—Kindly Aid rendered by the late Bishop Chauveau—An Interpreter found through his Help—Curious History of Peh-ma, this Christian Interpreter—Civility of the Chinese Officers—Difficulties about Baggage Cattle—Negotiations for Carriage—The New Ma-Fu—Visit to a Lamassery—Currency for the Journey—Search for a Horse—Parting Gifts and Purchases—The Tibetan's Inseparable Wooden Cup—Tib left behind—Fresh Selection of Nags —Fatality of Small-pox in Tibet.

TA-CHIEN-LU means 'Arrow Furnace Forge,' and was so called in the time of the great Liu-Pi. During the third century the barbarians from Tibet invaded China, and advanced as far as Chung-Chou. Liu-Pi drove them back; but they made fresh inroads, regaining the country as far as Ta-Chien-Lu. Then Liu-Pi sent against them his redoubtable warrior, Kung-Nung, who, coming here, forged an arrow-head; he shot this at a rock, and called the place ' The Arrow Furnace Forge.'

After that the barbarians retreated to Bat'ang, and never since have advanced beyond Ta-Chien-Lu, which

HORIZONTAL SECTIONS

along Captⁿ Gill's route
between

TA-CHIEN-LU & BAT'ANG
and
TA-LI-FU & BHAMO

by Cap^{tn} W. Gill R.E.

Horizontal Scale, 1:1,000,000 = 15·78 miles to an inch
English Statute Miles, 68·30 = 1 degree

Vertical Scale, 1:120,000 = 8,353 feet to an inch
English Feet

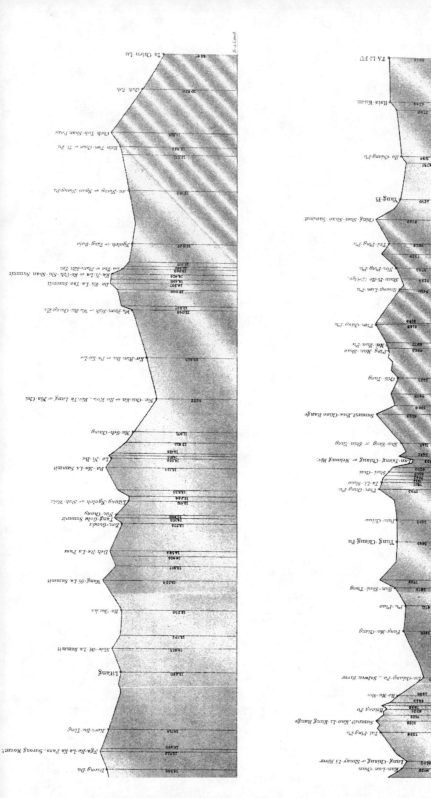

may now be considered as the boundary of China, for up to this point the people are directly governed by Chinese; but beyond this there are native chiefs who, subject to China, rule over the people.[1]

There is a native king here whose territory extends to Ho-K'ou, a few days' journey to the west. Although he enjoys the rank of king, he is obliged to pay an official visit twice a month to the Chinese chief magistrate (a Kiun-Liang-Fu). The king always refused to see Europeans, because he was afraid that the formalities of an inferior to a superior that are exacted from him by the Chinese officials would be demanded also by foreign visitors.

It seemed very strange to us to find the Indian rupee in use here. The Tibetans and mountaineers of these countries find themselves so cheated by the Chinese in their money dealings, that they have abandoned the cumbersome method of making payments by weight, which lends itself so easily to every kind of trickery, and have adopted the rupee, which has now become the current coin of the country. There is no coin less than a rupee, and for small payments it is cut up into little bits, which are of course weighed by the careful Chinese at Ta-Chien-Lu; but the Tibetans do not seem to use the scale, and roughly judge of the value of a piece of silver. Tea, moreover, and beads of turquoise are largely used as a means of payment, instead of metal. These rupees come in thousands all through Tibet, Lassa, and on to the frontiers of China, where the merchants, who eagerly buy them

---

[1] There can be little doubt that, at a date many centuries later than Liu-Pi, the 'ancient Marches of Tibet' came within fifty miles of Ya-Chou-Fu. See notes in *Marco Polo*, 2nd ed. ii. 37. One may add a suspicion of the Chinese legendary etymology of Ta-Chien-Lu, which is in all probability only a representation in Chinese character-syllables of the old Tibetan name sometimes written *Tachindo*, and by H. della Penna, *Tarchenton.—Y.*

up, are, by melting them down, able to gain a slight percentage. Only those who have gone through the weary process of cutting up and weighing out lumps of silver, disputing over the scale, and asserting the quality of the metal, can appreciate our feelings of satisfaction at again being able to make purchases in coin; and it was very pleasing, and somewhat flattering to our national vanity, to see the portrait of our Sovereign Lady Queen Victoria on the money we used. The rupee is the current coin as far as Lu-Ting-Ch'iao. Below that place the rupee may be met with, but does not pass current. The value of a coinage is thus practically demonstrated to the Chinese; but it is probably not so much their conservative instincts that prevent them establishing a coinage for themselves, as the knowledge that a Government mint would only open another door for the cheating, bribery, and corruption that infest the land.

At the time of our visit, we found it difficult to obtain a large number of rupees; for the embassy that had just arrived from Peking, and was on its way to Lassa, had bought them all up; but Monseigneur Chauveau contrived to find one thousand for us amongst his friends and acquaintances.

Ta-Chien-Lu is situated in a small open valley at the foot of mountains enclosing it on all sides except to the east, and is surrounded by a wall in a poor state of repair.

The brawling stream that divides the city into two parts is crossed by a wooden bridge, and a good many trees grow about the banks. The streets of the place are narrow and dirty, the shops inferior, and in them are all sorts of strange wild figures,— some dressed in a coarse kind of serge or cotton stuff, and wearing high leathern boots, with matted hair or long locks falling over their shoulders ; others in greasy skin coats, and the Lamas in red, their heads closely shaved, twisting their prayer-cylinders, and muttering at the same time the universal prayer, ' Om Mani Pemi Hom.' [2]

Both the women and the men wear great quantities of gold and silver ornaments, heavy earrings and brooches, in which are great lumps of very rubbishy turquoise and coral. They wear round their necks charm-boxes ; some of gold, others with very delicate filigreework in silver. These are to contain prayers.

Some of the women are good-looking, and all are utterly unlike the Chinese in every way.

July 26.—In the morning the Bishop sent us a bottle of milk, with a message to the effect that he was sorry that the only house he knew of was occupied. This was unfortunate, for he had assured us that we should be lucky if we could make all our arrangements and start in ten days. This was a serious blow to my hopes of pushing on ; but unexpected delays are the invariable fate of travellers, and must be accepted with the best grace possible. We therefore moved into the house I had already looked at, which was not an inn properly speaking, for the rooms only were let, and no food or lights provided. The house was kept by a man who called himself Chinese, but did not look like one ; he was

---

[2] The pronunciation of the syllables of this ejaculation, wherever I heard it, was as follows:—The *o* in the first syllable very long, almost like *aum*, and the *m* very nasal. The *a* in the second syllable like the *a* in the English word *father*. The *i* in the third syllable like the *double e* in the English word *knee*. The *e* in the fourth syllable very broad, almost like the *ea* in the English word *pear*. The *i* in the fifth syllable like the *double e* in the English word *knee*. The last syllable like the first, with the addition of the initial aspirate.

probably a half-breed, the son of a Chinese father and native mother. The place was a great improvement on the other; it was light, was not so infested with fleas, and was considerably larger.

We had two fair rooms, and one very small one, all on the upper floor. The stairs were outside, and led to a veranda that ran along the front of the house; there was a small courtyard below enclosed by a high wall, in which there were several sheds and tumble-down buildings.

A pieman used to come every morning and spend his day in the yard, and the number of cakes he disposed of was astounding. The yard seemed to be the favourite resort of anyone in the place who wanted light refreshments, and all day long the rattle of the dice could be heard without ceasing, as the coolies staked their small amount of money against the pieman's wares. The gambling was conducted with dice thrown into a cup; I was never completely initiated into the mysteries of the game, but the balance of the odds was no doubt largely in favour of the pieman.

The whole of one wall of our principal reception room was occupied by an altar, which was the home of several idols. Every morning the pious people of the house used to come and fill six little brass cups with water; and every evening these were removed, and an oil lamp was lighted in their place. The regularity with which these rites were performed was worthy of a better cause, but the people evidently thought that some dreadful evil would befall if they were omitted.

Ta-Chien-Lu is under the jurisdiction of a Chinese Fu, who has the further title of Kiun-Liang-Fu (Military Provision Store Keeper), because he is in charge of the provisioning of the soldiers in the district. The name of the official who held this position at the time of our visit was Pao.

The chief military officer had the rank of Sieh-Tai, and was in command of three hundred soldiers quartered in the city. A land-tax is imposed on the occupants of the soil, who are obliged to bring a certain quantity of grain to the Kiun-Liang-Fu, by whom it is distributed to the soldiers in his district, which extends to Ho-K'ou, the country beyond being in the jurisdiction of Lit'ang. His income is derived principally from presents or bribes, according to the almost universal system in China; and the position of Kiun-Liang-Fu at Ta-Chien-Lu is a very remunerative one.

The Kiun-Liang-Fu collects his taxes, and imposes his *corvées* through the native king; the latter are chiefly the supply of food and transport for travelling officials.

The transport is often a very serious matter, and is usually called the Fu-Ma (men and horses). On this high road to Tibet, officials are continually passing, and embassies without number; all of them are provided with Fu-Ma; and as no payment is exacted, it becomes a very serious tax upon the people of this poverty-stricken land.

It was very difficult to make it understood that we intended to pay our Fu-Ma; hence there was, at first, great unwillingness on the part of the mule and pony owners, to let us have animals, and we were considerably delayed in consequence. We obtained them eventually through the steward of the native king, who collected them from the petty farmers or mule owners in the neighbourhood.

All the trade of the place is done on yaks, hence

there are no forwarding houses where, as in Peking, large numbers of mules are kept, or where, as in Lower Ssŭ-Ch'uan, coolies may be hired. Beyond Ta-Chien-Lu, instead of hiring our own animals without official assistance, we were obliged to apply to the magistrates for the regulation Fu-Ma. This was always an annoyance; the people of course thought that, like others, we should not pay them, and naturally disliked lending their animals, for long and arduous journeys to foreigners, for whom they cared nothing.

Everybody here, from the Kiun-Liang-Fu to the lowest coolie in the streets, believed that we were on our way to Lassa; and it was simply impossible to convince them of the truth of our assertions that we were going to Ta-Li-Fu. The Chinese officials professed their willingness to assist us, but at the same time asserted their inability to protect us against the avowed hostility of the Lamas of Tibet. Properly speaking, the Lamas are the priests of the Buddhist religion; but the Chinese, always very loose in their nomenclature, apply the term somewhat indiscriminately to laymen who profess the strict tenets of the Buddhist faith. The Lamas throughout Tibet wield a power that is as tyrannical as it is absolute; huge communities live together in the Lamasseries or monasteries, and it is said that they form one third of the whole population of Tibet.

The head of the Buddhist faith is the Dalai Lama, resident at Lassa; and he is supposed to be an incarnation of a divine being. When the Dalai Lama dies, the true believers in the Buddhist faith consider that his spirit has entered into the body of a young child. Search is then made over the whole empire for a child who is recognised by certain mysterious marks, the secrets of which are known to the Lamas. There seems to be very little doubt that this search is honestly carried out, the Dalai Lama often being chosen from the house of a peasant. The Dalai Lama who was living when Huc was at Lassa, was selected from a poor family resident at Ta-Chien-Lu. Those who have seen a Dalai Lama speak in raptures of the singular beauty of his countenance, and in all probability he is chosen in accordance with the laws of physiognomy, so that a mild and contemplative disposition is found in the head of the Buddhist faith. This is well for the Lamas, for if a man of energy, with ideas of reform, should ever succeed to this extraordinary position, their power would probably receive a blow from which it might never recover.

The Lamas shave their heads, they are filthy in their person, and their dress is poor. They wear a garment of a coarse red serge or sackcloth. This has no shape, but is simply an oblong piece of cloth, thrown over one shoulder, the other being generally bare; for the Lamas, not less hardy than their lay brethren, seem absolutely impermeable to cold. The Lamas wear another length of cloth wound two or three times round the waist, which forms a skirt, reaching to the ankle. Many of them are barefooted, others wear high boots of red cloth with the lower parts made of leather. A yellow scarf is sometimes worn round the waist, and, with a string of beads and a prayer-cylinder, completes their costume.

The prayer-cylinder, or prayer-wheel, as it is often most inappropriately called, is usually about three or four inches in diameter and in length; the mystical invocation, 'Om Ma-ni Pe-mi Hom,' is written on the outside, whilst a small weight at the end of a short string keeps the affair in rotation; and all day long, not only the Lamas, but the people may be seen muttering the universal prayer, and twisting their cylinders, invariably in the same direction with the hands of a clock. One or more great cylinders, inscribed with the sentence, stand at the entrance to every house in Tibet, and a member of the household, or a guest who passes, is always expected to give the cylinder a twist for the welfare of the establishment. At almost every rivulet the eye is arrested by a little building, that is at first mistaken for a water mill, but which on close inspection is found to contain a cylinder, turning by the force of the stream, and ceaselessly sending up pious ejaculations to Heaven, for every turn of a cylinder on which the prayer is written is supposed to convey an invocation to the deity. Sometimes enormous barns are filled with these cylinders gorgeously painted, and with the prayer repeated on them many times; and at every turn and every step in Tibet this sentence is forced upon the traveller's notice in some form or another.

A string, called a Mani string, is often stretched between the two sides of a tiny valley, and hundreds of little bits of rag are tied to it with the prayer written on all. At the top of every mountain there is a cairn made of stones cast there by the pious, thankful to have escaped the dangers of the mountain roads, and on each stone the prayer appears. Many sticks are planted in the cairn, with a piece of rag or cloth at the upper end, on which of course the prayer is written; and by the roadside are heaps of flat stones with the inscription roughly cut on them. These are especially frequent in the valleys; sometimes only a few hundred yards apart, they would appear to serve as a means for marking the road when covered by deep snow drifts, as well as for some pious purpose. Sometimes the road passes between walls of flat stones, on every one of which the sentence may be read by the passing traveller. A light pole, from which a piece of rag flutters, inscribed with the prayer, is placed at the top of every Tibetan house; and wherever a traveller may go he is constantly reminded that he is in the home of the Buddhist religion.

There must be some deep meaning attaching to a torn piece of cloth. The same idea is seen in Persia, where, at the summit of the mountain from which the pilgrim's eye first lights on the sacred shrine of the Imam Reza, the bushes are covered with hundreds and thousands of little pieces of cloth, which each devout pilgrim leaves as a memento of the blissful moment.

The Lamas in Tibet wield a power unequalled by a similar class of people in any other country, and every position of importance outside Lassa seems to be filled by a member of this strange community. There is a Lama at the head of every embassy; and the Chinese always insist on a very high Lama from Lassa residing at Peking; this is partly as a hostage for the safety of the Chinese officials resident at Lassa.

Those of the Lama ambassadors whom we saw seemed woefully poor; they always had something to sell, and were ready to dispose of any article of clothing, equipment, or adornment, except the prayer cylinders, which were very difficult to buy. It is said that in Peking the ambassadors part with nearly everything, and no doubt a rare collection of curios from Lassa might be made in that wonderful city.

One of these Lama ambassadors, who had completed his term of service at Peking, and was now on his return journey to Lassa, was stopping in the room adjoining ours. He was a Chinaman; but having been long resident in Tibet, and having become deeply versed in the knowledge of the Buddhist faith by long and intense study of the religious books, he was made at length a Lama by the Tibetans, rose to almost the highest rank amongst them, and was sent eventually as ambassador to Peking.

The Chinese officials resident in Tibet are not permitted to take wives with them, the ambassador resident at Lassa being no exception to the rule. The officials and soldiers, therefore, when in Tibet take to themselves Tibetan wives. The children thus become entirely Tibetan; and when the Chinese officials return to China they usually leave their family behind them. The Tibetans in this are wise in their own generation, for if they permitted the Chinese to bring their wives with them, and raise Chinese families, the country would soon become altogether Chinese.

The Lamas now made no secret of their intentions to oppose our entry into Tibet; they had already given orders that if we attempted it we were to be starved out; all the people were forbidden to supply us with food for ourselves or with forage for our horses, or to assist us in any way. We did not ourselves hear very much of this, but of course all sorts of idle tales were spread about the place, and our servants, always willing to gossip, lent a ready ear to every silly rumour. These were very rife, and if absolute threats were not thrown out, hints were not wanting that neither our own lives nor those of our servants would be safe in Tibet. Menaces of this sort would under no circumstances have met with much attention from us, and as we had no intention of crossing the Tibetan frontier they had no effect whatever on our peace of mind.

Mesny's boy was the first to quail before the prospects of fatigue, not by any means imaginary, and still more before the idea of terrible dangers, altogether visionary; he wanted to go home, but being a poor feeble thing was terrified at the idea of returning by himself, and persuaded my man Huang-Fu that he would never get back alive if he ventured beyond Ta-Chien-Lu. Huang-Fu then made up his mind to desert, and Chin-Tai also became faint-hearted, or said so, which came to the same thing; but, probably, although he was by no means a courageous person, his discontent was chiefly caused by the addition to our party of an interpreter, who would have the management of all money matters. This did not suit Chin-Tai, who every day became more greedy of gain, and his avaricious propensities were carried to such an extent that if I ever employed anyone else in the smallest money transaction, the loss of the squeeze, which he now seemed to consider as his sole and absolute right, so stirred his bile that he was in an ill temper for the rest of the day. It was very soon evident that no love would be lost between Peh-ma, our interpreter, and my servants; they had already begun to quarrel, and one day Chung-Erh, in a violent passion, went out and abused Peh-ma in the language of Shimei the son of Gera. 'Who pays you to curse me?' said Peh-ma, who was a heavy powerful man. What the result would have been, if our attention had not been attracted, it is impossible to say, for the Tibetans are a very independent people, and will not brook insults from anyone, high or low.

This greed for money was the root of all trouble and discord amongst our servants. When pay-day for coolies, Ma-Fus, or muleteers came round, the servants first of all disappeared, and when after some hours of absence they came back they were always out of temper;—Chin-Tai, because he had been obliged to divide his spoil, and the others, because they thought that their share had not been large enough. To arrange matters, there must always have been a fearful squabble, but they used wisely to get out of my hearing to fight their battles, which, to judge from the inordinate length of their absence, must have been right royal ones.

One day Chin-Tai came to me, and said sulkily that he, Chung-Erh, Huang-Fu, and Mesny's servants, had consulted together, and determined not to come with us any further; this was in great measure a fabrication on his part, because Chung-Erh soon afterwards said he had arrived at no conclusion in the matter. Little more was said at the time, but a day or two after Chin-Tai changed his mind, having apparently calculated that, once in the Chinese province of Yün-Nan, he could, by an extra squeeze, make up for temporary losses in Tibet. The kind old Bishop talked to Huang-Fu, who was a Christian; and, whatever may be their merits or demerits, the Chinese Christians have a very profound respect for their bishops. I employed at different times a great many Christians; they always served me faithfully and well; and on the only two occasions when I told Chin-Tai to find me servants, he chose them from among the Christians, although he professed to hold them in supreme contempt.

July 27.—We had some visitors during the day, who came partly from civility, partly from curiosity—an officer in the Chinese army, and a young official, who said that orders had been received from Ch'êng-Tu to supply all our wants.

We were told that the journey to Bat'ang would occupy eighteen days; that the intervening country was little better than a desert, the higher portions of which were covered with wide fields of snow, and that until our arrival at Bat'ang it would be quite impossible to buy food of any description. In accordance with these gloomy prospects, Chin-Tai was soon in his glory, laying in an amount of provisions that would have sufficed to stock a troopship. He at once bought one hundred pounds of beef, which he salted, and butter in quantities that would have puzzled a Laplander, or even a Tibetan. And the amount that a Tibetan will eat is startling! The chief food of a Tibetan is *tsanba*, or oatmeal porridge, generally mixed with a large proportion of butter; and buttered tea, that is, tea with enormous lumps of butter in it. In their food, as in all their ways and customs, and even in their buildings, the Chinese are in striking contrast to the pastoral people found on their frontiers. In the habits of these there always remains a trace, and often something more than a trace, of the nomad life; whilst in China Proper, and amongst the Chinese, everything betokens the ancient and high civilisation of a people that have taken root in the soil.

In every city and almost every village in China inns are found, an indication of a people accustomed to live in houses, and who, when obliged to travel, must have a roof to shelter them; the very coolies, poorly as they are paid, never sleeping in the open, but invariably expending some portion of their small earnings for night accommodation. Amongst the Tibetans, and the Man-Tzŭ, or barbarian population in the mountains, this is not the case; the people all originally leading a wandering life, the idea of inn accommodation has not penetrated into their habits. A Chinaman will, under no circumstances, sleep outside if he can help it; in Tibet the master of a good house will as often as not be found passing his night on the flat roof; whilst the hardy people in the winter time can sleep with their clothes half off, and with their bare shoulders in the snow. In China no house is complete without its table, chairs, and bedsteads, rough and clumsy though they often are; in Tibet these accessories of life in a fixed habitation are always wanting. Amongst the Chinese, mutton can rarely be obtained at all—they themselves think it very poor food; the love of a Mongol for a fat-tailed sheep is proverbial, and the natives of Tibet are not behind them in this taste. Although not exactly forbidden by their religion, the idea of killing an ox is very repugnant to the agriculturists of China, because, they say, it is ungrateful to take the life of the useful animal that draws the plough, and in the large towns the butchers are nearly always Tartars. The Chinese, as they never were a pastoral people, never kept flocks and herds; milk and butter are therefore practically unknown to them, while Tibet may safely be called a land flowing with milk and butter. As a rule, the Tibetan does not drink much milk: partly because it is all made into butter, and partly because, owing to the filthy state of the vessels, milk always turns bad in a few hours; but the traveller who makes his tastes known can always obtain an unlimited supply. Tea is often brought to him made altogether of milk without any water at all. The Tibetans also eat sour cream, curds, and cheese, and this brings a Tibetan bill of fare to an end, which, in its constituents and in its simplicity, bears the stamp of the nomad pastoral race.

The Chinaman, on the other hand, loves variety. In every tea-house by the wayside, owing its existence to no more opulent class than the coolies on the road, there are always several little dishes of some sort. Beans simple, beans pickled, bean-curds, chopped vegetables in little pies, macaroni of wheaten flour, macaroni made of rice, these—and in the large towns and cities, dozens of dishes made of ducks, pork, fish, and vegetables, rice-cakes like muffins, leavened bread of wheat-flour, sweetmeats, and sweet cakes—are to be seen at every turn; and of one or perhaps more of these every coolie will, when he can afford it, give himself a treat and vary his food, the main portion of which is rice, where it will grow, and in the high lands bread made from whatever grain the climate will produce.

At the time of our visit to Ta-Chien-Lu the exchange was 1,350 cash to a tael, and the general cost of provisions:

Beef	40	cash	a catty
Butter	114	,,	,,
Eggs	9	,,	each
Wood	2	,,	a catty
Charcoal	80	,,	,,

In the course of a day or two, when it was found that the threats of starvation produced no visible effect on our peace of mind, the people who surrounded us commenced to empty on our heads their vials of dismal tales about the rigour of the climate; these were repeated so continuously and so uniformly that I began to wish I had brought a copy of 'The Ride to Khiva' with me, that I might see again what measures Burnaby took to guard against the excessive cold.

I made inquiries about a fur coat, and Chung-Erh soon brought me one, which he said I could have for four taels. It was quite new, and not yet made up, and Chung-Erh told me in a mysterious whisper that the Lama next door was very hard up for money, and was willing to sell it. The necessity for a whisper was not very apparent as there was a wall between us, and even if there had not been, the Lama did not even talk Chinese, much less English, of which he had probably never heard a word. I soon decided to become a purchaser, and Chung-Erh was so pleased to have the handling of a little silver that he quite brightened up, and notwithstanding the dreariness of the miserable rainy day, he was more cheery than I ever saw him before, and when waiting on me at dinner, in his solicitude for my welfare, he encouraged me to a second helping, saying, 'Oh! do have some more, won't you?' as if he were my host instead of my servant. The skins that I had purchased from the indigent Lama were now to be turned into a suitable coat, and Chung-Erh was entrusted with the execution of the project. He brought it back made up with black cloth, like the coat of a coolie. This was a most improper tint, so, determined to be on the right side another time, I ordered the cloth to be changed for brilliant scarlet, the official colour. Another poverty-stricken Lama came with a pair of fur leggings, which I bought for a tael and a half; and after having finally invested in snow spectacles for myself and suite, I felt myself thoroughly prepared to defy the thermometer, although I could not help thinking, even then, that these preparations were more fitted for an Arctic expedition than a ride of three hundred miles to Bat'ang.

My forebodings were fully verified; for during the whole journey we found no snow except at the top of one granite mountain, where a few little patches were lying by the roadside. As for provisions, milk and butter were lavishly bestowed upon us by the owners of every house we stopped at. Scarcely a day passed when we could not have bought a sheep if we had wished it, but we received so many as presents that it was rarely necessary to make a purchase of any kind. Fowls and eggs are always dear and difficult to procure in Tibet, for grain of any kind is far too expensive to be wasted on fowls. The true reason of the doleful picture drawn for us lay probably in the fact that a true Chinaman, unless he can get rice, is under the impression that he is being starved; and as of course no rice can be found in Tibet, and as our information was principally drawn from the officials, we were gradually deluded into the belief that semi-starvation was before us.

I had already found it very inconvenient to carry about the enormous bulk of silver that I had with me, and as very pure gold comes from Lit'ang to Ta-Chien-Lu, Monseigneur Chauveau, who lost no opportunity of assisting me in all my troublesome transactions, found a trustworthy merchant, from whom I bought a considerable amount. It is cast into ingots about three inches long, and instead of the uncouth lumps of silver, it was quite a pleasure to handle these dainty morsels of pure and glittering gold.

I could not help reproaching myself for the trouble that Monseigneur Chauveau took to supply the wants of an utter stranger, for I felt that I should never have an opportunity of repaying any part of it; but I little thought that in less than a year this noble-hearted missionary would be no more.

An interpreter was necessary for us, as none of our party knew a word of Tibetan, and Monseigneur Chauveau found us an excellent one. He was a Christian, whose history tended to show that his conversion was not altogether due to conviction. He had an uncle, who took to the eccentric habit of living alone on the tops of mountains and other unpleasant places, and to such an extent did he carry this propensity that all the people said he was a very holy man, and as whatever 'everybody' says is always true, he really must have been fit to be numbered with the twelve apostles.

As in Tibet, people of all kinds bring gifts to holy people, the uncle of the worthy Peh-ma found that living about in stony places was not altogether an unlucrative profession; and in course of time he was in a position to build a Lamassery, which, under the guidance of this virtuous person, soon grew and flourished, and eventually ranked high enough to number amongst its inmates a living Buddha.

Peh-ma, at this time, knew nothing of his uncle; but one day wandering about with no particular object, he stumbled upon this Lamassery. Uncle and nephew recognised each other immediately, and Peh-ma was at once taken in, and turned into a Lama.

Being a man of natural ability, he soon made progress in the remarkable cultus of this profession; and as the uncle grew old, the living Buddha and his partisan, fearing that Peh-ma would take his uncle's place and deprive them of the powers they expected on that holy person's metempsychosis, decided that, as Peh-ma and his uncle had been joined in life neither should they be separated in transmigration. But Peh-ma hearing of it, and not appreciating the idea, ran away and became a Christian, in order that his enemies might no longer fear him; for as long as he remained a Lama he was dangerous to them, and his life was consequently in jeopardy.

The motive was, perhaps, not altogether a worthy one, but he was now a poor man, tilling a little bit of ground which he rented from the Bishop, thirty miles distant from Ta-Chien-Lu, and at the request of Monseigneur Chauveau he came down to enter our service. When he was introduced I was at once attracted by his fine, open, manly face. There was a determined look about his mouth, and a frank honesty in his glance that made me sure he would be a trusty servant. Physically, he was a remarkably well-built, powerful man, with very broad shoulders. He wore the plait like most of the civilised border Tibetans, and had a heavy dark moustache. His looks did not belie him, and, like all the servants recommended to us by Monseigneur Chauveau, he turned out thoroughly trustworthy.

Our Chinese friends also, in their way, were very civil, and used to send us elaborate repasts on huge trays. These used to arrive early in the morning, and were warmed up, in correct Chinese fashion, for our evening meal. One dish, in particular, was sent us several times. This was a *plat* fit for the most exquisite of Chinese gourmets; it had no name in particular, but it used to come in a huge circular and very deep iron or tin dish. The solid portions of it consisted of a huge lump of fat pork, a duck, and a fowl. Besides these there were little bits of ham, sea-slugs, shark's fins, bits of gelatine, mushrooms, and a dozen other ingredients swimming in the gravy. More than a day is required for the preparation of this *chef d'œuvre*, which is served up with rice, bread, and some sweet cakes. I have had considerable experience of this most *recherché* dish, and although it always appeared to me greasy and flavourless, like all Chinese food, this was amply compensated for, if I happened to be dining with Chinese friends, by the infinite pleasure beaming on the faces of my companions, when they set to work on it with a cup of soy and a pair of chopsticks, the savoury morsels disappearing one by one, and the rice following with a peculiar gurgling noise the Chinese always make when eating rice.

We found considerable difficulty in arranging about our baggage animals; we were exceedingly averse from making use of the assistance proffered by the Kiun-Liang-Fu, for knowing how much trickery and extortion is invariably practised by Chinese officials, we were well aware that whatever was paid to the magistrate but a scant proportion would reach the mule owners. On the other hand, the Bishop warned us that if we should succeed in finding animals for ourselves it was more than possible that, as the muleteers would be without official protection, the Lamas would prevent the inhabitants from furnishing them with supplies; and as, under the most favourable circumstances, there was no chance of engaging mules for a further distance than Lit'ang, the delays at that place would probably have been very great before fresh animals could have been hired. We ultimately decided to take the official Fu-Ma, and I fondly thought that when we had signified the same to our friend Pao, our troubles were for the present at an end. But I counted without the wily steward of the native king, whose duty it was to collect the animals from the neighbourhood. Presently an innocent message came from the Ya-mên that all arrangements had been made, and that the yaks would come any day we might like to name. This was not at all what we had expected, and we asked how many days yaks would be on the road.

The people gave us no more satisfactory reply than that they would go very fast. So we determined to make inquiries from an independent source.

We were now overwhelmed with offers from pony proprietors and coolies; this was as usual, for if anything is wanted in China, no matter of what nature it may be, the prices at first demanded are always outrageous; but directly one set of people finds that their neighbours are entering into some arrangements, they lower their rates, and try to underbid the others. As we had made a partial agreement with the officials, we could not accept these generous offers, which might, or might not, have been serious.

We then went to pay a visit to the Kiun-Liang-Fu. The Yamên was a poor place for the official residence of a magistrate; the entrance was across a filthy half-covered yard, with the staircase in one corner, underneath which there was the usual Chinese collection of stinking refuse, in comparison to which a European dust-bin would be purity itself. The reception-room was, however, large and tolerably clean. There were a few scrolls on the walls, and some small panes of glass in the windows.

The Kiun-Liang-Fu, or Chih-Fu, Pao Ta-Laoye,[3] after asking me my honourable name, expressed his pleasure at seeing me, and thanked me for my visit.

I then thanked him for a very handsome present he had made me of some really exceedingly valuable old blue-and-white china cups, and I explained to him, as well as I could, the blue china mania in England. When he learned that the manufactures of his

---

[3] 'Ta-Laoye' means *Great Excellency*, and 'Pao Ta-Laoye' would be properly rendered by *His Excellency Pao*. In conversation 'Ta-Laoye,' or Your Excellency, is often used by itself without the proper name.

country were so highly prized in ours, he was much gratified, and we shook hands (our own, not each other's).

He then asked me my honourable age, and found that we were both of the same year. This was a cause of great congratulation, and we shook our own hands more warmly than before.

He told me that he had heard that I was a great *savant*, continually taking notes. He wanted to know how many books I had written, and said he thought I had a remarkably literary cast of countenance.

Without stopping to inquire on which particular feature my hitherto undiscovered abilities were stamped, I hastened to acknowledge the compliment, which was in fact the greatest he could pay; for in China literary people are always held in the highest estimation, no matter what their rank may be.

I had assumed this character as it enabled me to take notes and write openly without creating undue suspicion; but if I had thought that the polished Pao meant anything deeper than the usual courtesy, I could not have helped a blush for the gross hypocrisy I was practising.

The possession of these rare mental qualities not being disavowed, Pao next praised my person, and said that I looked as if the climate agreed with me; he asked if I did not find the weather very cold and the place very dull; and when, not to be behindhand with fair words, I said that with such an excellent host as himself, no place could be dull, there was a fresh outburst of fervent handshaking.

He then offered to give me some bottles of Chinese wine for the journey; but I said I could not well carry it, and promised him in return some brandy.

Having sufficiently bandied about our sweet speeches, we turned to business and the absorbing topic of the Fu-Ma. Our conclusions on this subject were not quite so satisfactory, for Pao could only say that he did not think the journey to Bat'ang would occupy more than twenty-five days.

Mesny and I then sipped our tea with due formality, and were conducted to our chairs by our polite host, from whom we parted, amidst copious bows, with the warmest protestations of friendship.

Whilst at breakfast a young man, who had already paid us one visit, came in for a gossip, bringing with him a letter that Baber had sent me by special courier. This letter was in a white envelope without any red about it, and this being, to a Chinese, a sign of mourning, the young gentleman had scarcely dared give it to me, and the messenger who brought it was much afraid of some ill-luck befalling him. The young man sat and talked as we finished our meal, and asked for cigars, which I gave him, although I was quite sure that he would not enjoy smoking them.

Later on the kind Bishop came in, and when he heard we were to have oxen as baggage animals, he held up his hands and expressed his pity; for he assured us that a month, or more likely six weeks, would be consumed on the journey with these slow-moving beasts.

We therefore sent a polite message to Pao, firmly declining to have anything whatever to do with yaks, and begging him to supply us with a sufficient quantity of ponies and mules; to which he returned the characteristic Chinese reply that he would see about it to-morrow.

August 2.—A man came in to visit us one day,

and after many preliminary inanities, remarked that he had a son ; as he seemed unable to get any further in his narrative, we warmly congratulated him upon his fortunate possession. Thus encouraged, he observed that the youth would be invaluable to us in any capacity we might employ him, and at once introduced a boy of remarkable, though unprepossessing, appearance, dressed in a costume in no way peculiar, except for a pair of enormous English sea boots. We declined his services ; but as I went out for a stroll a short time afterwards, Boots followed me, and arriving at a temple he insisted on acting as cicerone. On my return I told Chung-Erh to give him a few cash, and asked where he had found those boots. It appeared that they were relics of poor Cooper. The boy was very proud of them, believing that when he had them on the spirit of an Englishman had entered into him, and that he was treated with distinction in consequence.

The boy was not easily rebuffed ; and Mesny being in possession of a pony that Pao had given him, Ting-Ko (for such was his name) constituted himself Ma-Fu.

The next morning I went out before breakfast to enjoy the soft balmy air and the unwonted sun, which was very pleasant after the cold rainy weather of the previous day. After I had unsuccessfully exercised my ingenuity in endeavours to kill with a stone the only snake I ever saw in China, I returned home, and on my way encountered Ting-Ko riding Mesny's pony, in high delight, and hugely proud of himself ; so much so that, although some distance away, he could not refrain from calling out and riding up to me, when he made some remarks which, if I had been able to understand them, would no doubt have been very much to the point. He introduced a man, a wild-looking Tibetan, who walked home with me, and with whom I kept up an interesting dialogue by always repeating the last few syllables of each of his sentences. This method of conversing seemed to give him the greatest satisfaction, and we parted the best of friends.

On my return I found that I had missed a visit from Pao, who had been entertained by Mesny, and who had made a minute inspection of everything lying about. He admired my alarum very much, and, greatly to his credit, was so much interested in the pictures and Chinese writing in Marco Polo that he wanted Mesny to get him a copy.

When he had gone, the native king's steward arrived; he had been looking after the horses and mules, and reported that all arrangements had been satisfactorily completed; that all our animals were to be horses or mules, and that we should do the regular stages. When we asked for the written agreement, he said that he would go and get it.

Much pleased at the prospects of satisfactory arrangements, I went to see the Bishop, who told me he had seen this written agreement, and that it contained a clause to the effect that the journey was to be made in twenty-five days, and that we were to have nine yaks. I did not need the Bishop's advice to refuse this absolutely, and hastened back. I found that in my absence, knowing that the interpreter was with me at the Bishop's palace, the steward had seized the opportunity of presenting the agreement (written in Tibetan) to Mesny, at the same time asking him for an advance of money.

Mesny had simply refused to look at the paper without the interpreter, and when I came back, bringing Peh-ma with me, we sent him to say that a proper paper must be written out, with the conditions; first, that all our animals should be horses or mules, and secondly, that the journey should be done in eighteen days, and we told him to add that, if the agreement was not brought soon we should return to Ch'êng-Tu.

This was a severe threat, for Pao had received orders from the governor-general to help us in every way, and we knew that he was afraid that we should go back to the provincial capital, and report our inability to proceed.

It was not until the next day that the steward, who seemed to have been touched by a present of some steel pens and a few old copies of the 'Illustrated London News,' came back with the agreement written out for horses and mules only, but even now it was not quite as we wanted it, for the time was fixed for twenty-two days. After a long fight we compromised for twenty days, which was reasonable. If we had said twenty days at first we should have been forced to compromise for twenty-two or twenty-three. No doubt the king's chief steward had a friend owning a drove of yaks, for which he had no immediate use; and the steward, anxious to do his friend a good turn, or possibly himself, had done his best to impose the oxen upon us.

Monseigneur Chauveau, never at a loss for some fresh method of obliging us, had been at infinite trouble to find two trustworthy Ma-Fus from amongst his flock; and in the course of the day two men came to be engaged. They presented a strong contrast to one another in appearance.

One of them, named Shuang-Pao (Double Gem), a silent and grave man, scarcely ever said a word.

The other, Chang-Shou-Pao (Long-lived Gem) was always laughing, whistling, or singing, and even in the most depressing circumstances of wind and rain, would trip along beside me in the most cheery manner. Shuang-Pao was a musk hunter, and Chang-Shou-Pao hunted the red deer for their horns in velvet.

Walking about the streets of Ta-Chien-Lu we attracted very little attention; even the Chinese boys did not follow us, and people scarcely turned their heads to look as we passed, though our costumes sometimes elicited a laugh. In this border town there are so many strange wild figures of different kinds that one more makes little difference. By the Chinese we were all classed together as barbarians, and a man who turned up one day, with a slight knowledge of the Bengali language, thought we were Nepalese,[4] and said our countrymen were the richest people in Lassa. He wanted Mesny to go there and establish himself as a watch and clock maker. This was very generous on his part, for he told us he had a monopoly of the business; he acknowledged that he could do no more than oil the clocks that were entrusted to him, and owned, with admirable candour, that he had never succeeded in making one go for more than a fortnight.

There are three large Lamasseries in the neighbourhood of Ta-Chien-Lu, and we went one day to visit one of them. For a mile or two we rode between stone walls almost entirely built of loose flat slabs, with the sacred inscription 'Om Ma-ni Pe-mi Hom' on each. On the way we met great droves of yaks, with enormous horns and heads like bisons, huge shaggy tails, and hair under their stomachs reaching to the ground. These were coming into the city in charge of some wild-looking, shaggy-haired fellows,

[4] The Chinese name for Nepal is Pi-Pon-Tzŭ.

with two or three of their large savage dogs. Yak is the Tibetan name for the bull, and the cow is called Jen-ma.[5] Europeans apply the word yak indiscriminately to both sexes, as do the Chinese their word Mao-Niu (Hairy Ox).

Ta-Chien-Lu, being situated at the very edge of the great Himalayan plateau, one day's march to the west brings the traveller to the glorious pastures of this magnificent table-land, and here the yak is naturally the almost universal means of transport. Very slow in his movements, and accomplishing but a few miles a day, this hardy animal is nevertheless the cheapest that can be employed; requiring no attendance, and no food that cannot be picked up on the mountain-side, or in the rich grass-lands of the upland plateau, the cost of keeping a yak is absolutely nothing. A caravan of yaks on the road will, when they arrive at a fine pasture, halt for a few days and let their animals feed; after which they will perhaps travel for three or four days more in the wild stony mountains, with scarcely any food until they reach the next grazing-ground.

We stood aside in the narrow path to let these lumbering beasts go past with their loads, and then proceeded up the valley: steep rugged hills running down on each side, and great rocks strewing the ground; it was a wild, desolate scene, closed at the back by snowy mountains, from which the clouds lifted now and then.

Crossing the arched bridge that spans the roaring torrent, we met a dozen Tibetan coolies carrying a huge log, keeping step to a kind of chaunt, by no means

[5] According to Jaeschke's dictionary the cow yak is *di-mo*, of which *jen-ma* is possibly a local variation. In Ladak it is pronounced also *bri-mo*. But Jaeschke gives also *zhou-ma* or *shön-ma* as 'a milch-cow,' which is more probably the word given to Capt. Gill.— Y.

unmelodious, and in which a sort of first and second could be distinctly recognised.

This Lamassery is finely situated on the slope of a hill, and is surrounded with many trees. Outside, the walls are whitewashed and well kept. There is a slight batter to them, and as they look very thick and massive, there would be something of the appearance of a fortification, if it were not that the windows are large, and outside many of them flowers were growing in pots. We entered a quadrangle, on the eastern side of which is the gate. This and two other sides are occupied by living-rooms in two stories, and the fourth —that opposite the entrance—is taken up with the principal chapel. This was not very gorgeous. There was a gigantic statue of Buddha at the end. The Lamas said it was all of brass, but it looked like clay coated with that metal. On each side of this was the tomb of a very sacred Lama, enclosed with iron-wire netting, on which a few scarves of felicity, called 'Khatas,' were hung. There were seven copper bowls of water before Buddha. We asked if any meaning attached to the number seven, and they replied that there were so many mysteries in it it was quite impossible of explanation. On each side of the chief chapel is a corridor leading into other rooms, into one of which they showed us. It was very dark, and, as far as we could gather, seemed to portray the horrors of hell. Outside it, hanging from the roof of the corridor, were skins of dogs, deer, bears, and other animals, roughly stuffed with straw. In many of these the sewing had burst and the straw protruded in a melancholy fashion, whilst the hair had fallen off in patches from all of them. Some of them were provided with glass eyes of awful dimensions, and they were fearful objects to look upon. To these also there was some mysterious meaning, but the Lamas would not tell us what it was. We were treated to a cup of tea each, and entertained by one of the chief Lamas, who, in his dress, did not differ from the others.

There were some fierce black dogs in the quadrangle, who, when we entered, gave tongue furiously, in a deep baying voice. These dogs had heads something like mastiffs, with an overhanging upper lip; they had shaggy tails, and some long hair about the head and neck. Here also a flock of enormous geese that were quite quiet before we arrived set up a loud cackling on our approach. In some parts of China geese are frequently kept as guards to a house, as they always cackle at the appearance of a stranger on their premises.

Early one morning, after a stroll outside the city, as we were sauntering homewards, we saw a flock of sheep. Mesny declared he had not eaten mutton for years; I had not tasted it for months, and our mouths watered at the sight of this unwonted food. From Ch'êng-Tu to this place we never had any other meat than chicken, and since our arrival at Ta-Chien-Lu our sole diet had been beef; for fowls were not to be bought, grain being so expensive that few people could afford to keep them. Wonderful for China even eggs were scarce; ordinarily, all over China, eggs can be bought in any quantity at a ridiculously low rate. Now, although by the aid of skilful cookery, we had thrown as much variety as possible into our meals, yet the *toujours bœuf* had given us a decided desire once more to taste the flesh of a sheep; so, calling the coolie, who was following us, we bade him address the gentle shepherd and demand the price of one of his flock, and in the meantime we sauntered home to breakfast. But the coolie, instead of doing as we told him, informed the Bishop that we wanted a sheep, and soon afterwards Monseigneur Chauveau sent us one of the fattest from his own flock.

We were told that most of our payments between Ta-Chien-Lu and Bat'ang would be made in tea and beads; so at Ta-Chien-Lu we bought a horse-load of the common inferior tea that we had seen carried by coolies all day long, and nearly every day, on the road from Ya-Chou; and we told Peh-ma to try and get some beads. We were somewhat astonished at the dirty-looking stones that he brought, and said were turquoises. They were of all sizes, some as small as No. 2 shot, others as large as No. 12 bullets. To me they looked the veriest rubbish, but as Peh-ma assured us that they would pass current as small coin, we bought three hundred and fifty for twenty-one taels.

The Tibetans, both men and women, are possessed of a taste almost amounting to frenzy for coral and turquoises; and the immense quantity of these that are used is surprising. The scabbards of their swords, the covers of their charm-boxes, their earrings or bracelets, all are ornamented with coral and turquoises. Quantity, however, is more regarded than quality, and in the whole of Tibet it would be difficult to find any pieces that would have any value whatever in the European market.

A sack of rice for our servants, another of wheaten flour, and a few dozen *khatas*, or scarves of felicity, completed the purchases that Peh-ma deemed it advisable to make.

The 'khata' is a great institution in Tibet. It is a little scarf, of some common material, that may be any colour except red, but is generally white gauze. Etiquette ordains that every present should be accompanied by a khata, and pious people visiting a Lamassery generally tie one to the rails in front of the image of Buddha.[6]

We now had to buy ponies, for up to this point we had ridden hired ones; and on making our wants known, everyone in the place who had an unsound animal to dispose of placed it at our disposition. Two horses were brought that seemed tolerable, and I took them out for a turn. The first was a good strong animal, $12\frac{3}{4}$ hands, quite a giant after what I had been riding, and now having something in front of me, I could hardly divest myself of the idea that I was sitting on the animal's tail. I then tried the other, but though pleasant enough to ride, he was very timid, and as his sides were like two planks nailed together, I decided to look for another if I could find one.

The Tibetans breed great numbers of ponies, and the pastures, that in their marvellous richness must equal anything to be found elsewhere, afford magnificent grazing grounds to vast herds of these animals, but though, like the Mongols, the Tibetans seem to pay no attention to careful breeding, the ponies found here are, generally speaking, better looking than those of Mongolia, and, unlike them, are very docile; I never saw one that showed the least sign of temper or vice. They seem to be as hardy as the people themselves, requiring no clothing, and scarcely ever being groomed.

The next animal brought for inspection had a sore back, another was lame—in fact, the people trotted out all the halt and the maimed, and I was quite prepared to be asked to pass judgment on a blind one very soon.

[6] On the *Khata*, and the manifold occasions of its use, see Huc, *Souvenirs*, &c., 1850, i. 86.—*Y.*

The offer of twelve taels that I had made in the morning for the bay, had been, I thought, accepted; but I found that the owner had changed his mind, and had gone off to the mountains without waiting for me to raise my price.

One of the Bishop's people came in one afternoon, and brought me a horse he wanted me to accept as a present, but as I could not have taken it without giving him something in return, at least as valuable, I declined it; for he valued it at thirty taels, the ordinary price of horses here being from twelve to fifteen taels.

Saddles were another difficulty. It was quite unusual to let out riding horses, and we found throughout our whole journey to Ta-Li-Fu that it was almost impossible to hire saddles for the servants. Ponies or mules we always obtained somehow, but our people were often obliged to ride on a mattress, or a roll of bedding on the top of a pack-saddle. It would have saved an infinity of trouble if we had bought saddles, but we never anticipated that, in a land where everybody rode, and where horses could be counted by hundreds, the want of saddles would be a source of trouble.

We were invited to breakfast one day with Monseigneur Chauveau, who said that he had not entertained European guests since the late Mr. T. T. Cooper was at Ta-Chien-Lu, just nine years before. The feast terminated with buttered tea, made with the Bishop's own butter. The butter to be bought at the houses in the country and in the towns invariably has a somewhat rancid taste, owing to the filthy vessels in which the milk is kept and the butter manufactured; but that made in Bishop Chauveau's establishment would have rivalled the produce of Devonshire and Alderney, as indeed it should, considering the wonderful pastures on which, during the short summer, the animals can graze.

In a cold climate, buttered tea, made with good tea and fresh butter, is not such a repulsive drink as would be supposed, and is admirably adapted for a people living at the great altitude of the Tibetan plateau.

In the summer time, when the climate is pleasant, much heat-giving food is not required, and the people can take their tsanba and tea with the least amount of butter; but when the howling winds of winter sweep across those dreary wastes of snow, they can only maintain their vital heat by large quantities of carbonaceous food, and butter is the most suitable of all that can be obtained. For animal food is most plentiful in the season when it is the least required; in the winter, the cattle and sheep can scarcely find anything to eat, and become miserably lean, out of condition, and totally unfit to provide the fatty food necessary for the people; while the butter, made in large quantities during the summer when the animals are at the height of their condition, is easily stored up for winter use.

This shows also why so little milk is drunk by the people. The winter is the season of trial, and it is for that time that all provisions are made; in the summer large quantities of milk or butter are unnecessary, and every available drop of milk is made into butter for the winter. In the long winter, again, milk must be exceedingly scarce, and thus drinking milk has never become one of the habits of the people.

The afternoon in the society of Monseigneur Chauveau was a most pleasant one, for though he had lived thirty-two years in China, time had not dimmed his interest in European affairs, nor his affection for his country. His courtly manners, those of a nobleman of the old French *régime*, were in striking contrast to the wildness of his surroundings, and would have made me forget that I was on the borders of an almost barbarous country, if his enthusiasm for the propagation of the faith had not kept it constantly in view.

He used to speak with great affection and admiration of the English, and of the religious toleration experienced under their rule ; and he looked forward with the keen eye of faith to the day when, the English being established at Lassa, the missionaries would be able to follow, and sweeping at last across those wild wastes of superstition carry the Christian faith to the very home of the Dalai Lama, shake the throne of that arch impostor, and strike with mighty strokes at the very root of the Upas-tree of Buddhism.

'Ah,' he said, 'my proper title is Vicaire Apostolique of Lassa, but I call myself by the less pretentious one of Vicaire of Tibet, for I feel that my eye can never look over the border into the promised land. But,' he added with flashing eye, 'I feel sure that my successor will reach the goal denied to me.'

Listening to him, as the colour mantled to his cheeks, I could not help sharing his earnest enthusiasm, and wishing that it might be he who should be the first to enter the haven so long desired. But a few short months elapsed, and he went to his last rest, bitterly mourned by his faithful little flock in those far-away regions, and deeply regretted by all who knew the nobility and grandeur of his nature. To me he was almost more than a friend. I owed him a debt of gratitude that nothing could have repaid, and never shall I forget his venerable figure as, standing at the door of his palace, he bade me a final adieu, and quoting the passage in which Goldsmith, who was his favourite historian, narrates the last speech of our unfortunate monarch Charles, said the one English word, '*Remember!*'

On our return to our house, several visitors came in to see us, bringing with them little presents; another inkstone was given me, and a bronze jar of what the Chinese call living water. This little jar was exceedingly ancient, and made of a certain bronze that has the peculiarity of causing the water put into it to become 'living,' as the Chinese call it. To produce this effect the jar must be filled with water, and from time to time, as evaporation goes on, a little more added, so that it is always full. At length a certain kind of moss forms on the edge, and the water is then said to be living; and a flower placed in it will retain its freshness for a very long time. That is the Chinese theory, but I cannot vouch for its accuracy.

Pao came in to wish me good-bye, and rather puzzled me by asking me what I admired most in China. Then all sorts of ornaments came pouring in for sale, for we had been inquiring about them—enormous finger-rings, barbaric earrings, brooches and buckles, some of silver and some of gold, and all set with huge lumps of coral and turquoise.

The greatest curiosities, and the best worth buying, were the charm-boxes, made of gold or silver. On the top of these there is generally some filigree, quite equal to any of European manufacture, and it is surprising how the handicraftsmen of Lassa, where these things are made, can with their rough clumsy tools produce work of such extreme delicacy. These boxes, which are invariably adorned with a lump of coral, are to contain a slip of paper, on which is written the usual formula, 'Om Mani-Pe-mi Hom.' No Tibetan is ever without one of these, no matter how poor or dirty he may be. A miserable yak driver, with perhaps no home, and no worldly possessions but a bit of serge for a coat, will invariably have a charm-box, which may be worth some twenty or thirty taels. It is very curious to see the women, always dirty, often ragged, and sometimes almost too poor to afford themselves clothes, wearing massive ornaments of silver or gold, and immense plates of silver in their hair. I bought a very heavy bracelet of solid gold, embossed with some emblematical design, and a good many finger-rings, ear-rings, and charm-boxes.

There is another article that almost forms a part of every Tibetan. This is the wooden cup, or Pu-ku, in which he eats his tsanba or drinks his tea. It is always kept in the bosom of his capacious garment, a space that serves not as a pocket merely, but rather as a portmanteau, in which he can carry about the whole of his not very extensive possessions.

These cups are made of different woods, and polished; no Tibetan is ever without one; he seems to be born, not with a silver spoon in his mouth, but with a wooden cup in his bosom. The cup never leaves him, night or day, as long as he lives, and would no doubt go down to the grave with him, if burial were the custom of the country. Some are supposed to have the valuable property of annulling the effects of poison, and others are lined with silver; but none of these priceless articles were to be bought to add to our collection of Tibetan curiosities, and we contented ourselves with plain, simple, but useful cups, like those of our coolies.

A prayer-cylinder also was brought, which I began to twist the wrong way, much to the consternation of the people; they were really seriously alarmed, for they seized my hand and stopped me immediately.

In the afternoon I made an agreement with eight chair-coolies, for whom I was obliged to provide four ponies, partly to carry their food, and partly to enable them to take an occasional ride to relieve themselves amongst the mountain roads. I took my chair to Ta-Li-Fu, chiefly because it was so very useful for carrying the small odds and ends I always wanted on the march, or immediately on arrival at the halting place. It would also have been invaluable in case of sickness. The other chairs were taken no further than Ta-Chien-Lu.

Poor Tib was also left behind; he had supported the difficulties of the road very badly, and much to the discontent of the coolies had been carried more than half-way from Ch'êng-Tu in my chair. Had I taken him further he would probably have broken down altogether, if he had not been killed by the savage dogs of the Tibetans; and as Monseigneur Chauveau offered him a comfortable home, I accepted it for the poor beast, though in China the presence of a dog is a great safeguard against thieves.

Later in the day a number of ponies were waiting for me, and after trying a few of them, I was just going to make a bid for one or two, when a man came in with a grey and a chestnut. The grey, though small, at once attracted my fancy, as he had more breeding about him than any of the others, and the shape of his head, and the way in which his tail was set on, were quite of the Arab type. I had him saddled, and taking him over the worst bit of road I could find in the neighbourhood, the way he came down hill on the stony path at once determined me to buy him. My Ma-Fu had followed on the chestnut barebacked; the saddle was changed, and I found that I liked the second almost as well as the first. On returning I asked the price of the two; the horse-dealer demanded forty taels, and I promptly offered twenty. After a time I went up to twentynine, and the dealer coming down to thirty a bargain was struck. The dealers here have a curious way of telling the price to one another by putting their hands together under their sleeves, and by signs well understood communicating the figure without the bystanders knowing anything about it.[7]

None of our coolies or Ma-Fus cared to engage themselves to come beyond Bat'ang; for the Tibetans have the greatest dread of entering China, on account of small-pox—a disease almost unknown in Tibet. A Tibetan once attacked by small-pox never recovers. The Chinese look upon this disease much as an Englishman does on a cold; they are generally ill for a few days only, and get over it, though of course there are a large number of severe and fatal cases. But when a Tibetan is attacked his family take him outside the village, out of the way, and put him under a tree, or in a cave, with some tsanba and cold water, and leave the poor wretch to die.[8]

[7] The Burmese have a similar method of bargaining, which is not, I believe, used exclusively for horse-dealing. (It is in occasional use almost all over Asia. See *Marco Polo*, 2nd ed. ii. 486.—*Y.*

[8] Colonel Prejevalski notices that the inhabitants of the Lob-Nor district have very similar customs.

CHAPTER IV.

THE GREAT PLATEAU.

I. TA-CHIEN-LU TO LIT'ANG.

The Departure from Ta-Chien-Lu—The Loads Distributed—The Cavalcade described—A Supper of Tsanba—Village of Cheh-toh—A Wreck of the Crockery—Tibetan Salutation—Delicious Air—Half-bred Yaks—Splendid Alpine Pastures—Wild Fruits—The Sacred Cairn—Direct Road to Chiamdo—Ti-zu—Delightful *al fresco* Breakfast—A Land of Milk and Butter—Hot Spring—The Buttered-Tea Churn—Tsanba—Tibetan Houses—Hospitable Folk—An-niang—Wild Flowers—Road-side Groups from the Fair—A Gallop over Turf—Village of Ngoloh—Description of Quarters—An Imperial Courier—Halt at Ngoloh—Pass of Ka-ji-La—Magnificent Alpine Prospect—The 'King of Mountains'—Pine Forests—Wu-rum-shih—The Margary Proclamation—Illness of the Servant Huang-Fu—Descent towards the Ya-Lung River—Appearance of Green Parrots—The Octagon Tower—Tibetan Horse-shoeing—Ho-K'ou or Nia-chu-ka, on the Ya-Lung River—Tibetan Aversion to Fish, and its Origin—Lights of Pine-Splinter—Ferry over the Ya-Lung—Re-ascent towards the Plateau—Mah-geh-chung—Our Quarters there—A 'Medicine Mountain'—La-ni-ba—Crest of the Ra-ma Pass—Post of Mu-lung-gung—Female Decorations—Lit'ang-Ngoloh—Description of House and Appliances—Serious Illness of Huang-Fu—Domestic Sketches—Holly-leaved Oaks—Tang-Go Pass—Gold-washing—Cha-ma-ra-don Forest left below—The Great Dogs and their Change of Garb—Deh-re Pass—The Rhubarb Plant—Rarefied Atmosphere—Pass of Wang-gi—Patience of the Ma-Fu—Ho-chŭ-ka Village—Abundance of Supplies—Shie-gi Pass—The Surong Mountains—Treeless but Flowery Heights—Lit'ang in View—Arrival there—Huang-Fu's Illness.

August 7.—At length the day of our departure arrived, and ponies and pack-saddles appeared in the yard.

First it was necessary to collect all the baggage together, so that the people could decide how many baggage animals would be required. I had estimated the number at twenty-five, but as the time sped by,

SECTION 3

Route Map

TA-CHIEN-LU TO BAT'ANG

by Captⁿ. W. Gill R.E.

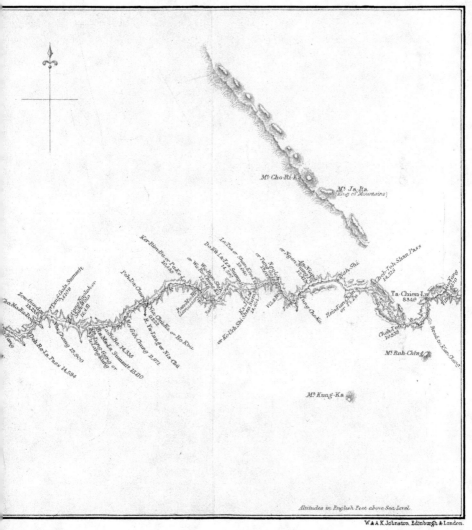

and box after box emerged from the recesses of the mysterious kitchen, it became evident that we should require many more. What Chin-Tai had in all those boxes I never was able to discover. We arrived at the conclusion that he was going into trade, for if they all contained provisions, as he declared, we must have been victualled for a year at least. I began to think that Chin-Tai was never going to finish his packing, but at length it was solemnly proclaimed that everything was ready, and the pony-drivers began to arrange the loads ; but upon this subject they had constant differences of opinion. First one man would collect three or four articles together, and exhibit his idea of a fair load ; then another would take them somewhere else, apparently for the pleasure of moving them ; and a third would finally disarrange the work of the other two. This did not tend to expedite matters ; but the three chief pony-drivers at length came to a satisfactory conclusion, and the overseer informed us that we wanted twenty-nine ponies for the baggage, four for the servants, one for the interpreter, and four for the eight chair coolies, thirty-eight in all.

The pack-saddles here are not like those used in the north of China, where a framework is put on the saddle. Here the load is lashed directly to the saddle with leathern thongs. The men do not seem to be very particular about lashing the packages tightly, which is a great convenience to the traveller, for with the other plan there is always so much lashing and hauling that it is quite an undertaking to get anything undone.

It was just noon as we rode away, and turned our backs on the Arrow Furnace Forge. We were a goodly party, and somewhat quaint to look on : Peh-ma riding first on a light-coloured chestnut, with a white tail, enveloped in all sorts of blankets and wraps; Peh-ma himself in a garment gathered in at the waist, where in a gigantic fold he seemed to be carrying all his worldly goods, and whence, later on, when I began to be unpleasantly sensible of the fact that I had not breakfasted, half a loaf was produced. Mesny dressed in a kind of patrol jacket, European trousers, Chinese high boots, and a Kuei-Chou pancake hat, and mounted on a very gorgeous Chinese saddle, with short stirrups, that drew his knees somewhere up towards the vicinity of his chin, would at least have attracted notice wherever he might have been. Huang-Fu had, by a quantity of bedding, clothing, and saddle-bags, raised the altitude of his small pony almost to that of an elephant; and he looked the picture of contentment, as he sat perched up with a long pipe and a red umbrella. Chung-Erh wore a large straw hat over a long black coat; and Chin-Tai, and Mesny's coolie (now raised to the rank of 'Boy,' *vice* Hsi-Sen resigned), with one of the pony-drivers, formed our mounted party. But Ting-Ko (or 'Boots,' as we called him), in the hopes of gathering up some fragments, came with us for one day's journey, as he said, and made himself generally useful. My laughing Ma-Fu, dressed all in red like a Lama, the other Ma-Fu, and our two spare ponies completed our caravan; while behind us the chair, with its eight coolies and four ponies, and the twenty-nine baggage animals, followed in beautiful disorder.

Our course lay up a valley, nearly due south, past the Lamassery we had visited a day or two before. On each side hills covered with low green brushwood sloped down to the river; in the valley the fields of oats and barley, nearly ready for the sickle, were divided by stone walls, and a good many fine large trees lined the edge of the water. To the south, right in front, a fine snowfield on Mount Ru-ching glittered in the bright sun; and to the south-east another mountain every now and then showed in patches of snow, as the clouds came and went from its lofty summit.

At a little less than four miles from Ta-Chien-Lu, the road leaves the main river, and strikes to the west, up the valley of a smaller stream. There is but little cultivation, and scarcely any wood, the hill-sides being all covered with a dense green undergrowth. The road is good, ascending steadily without any of those desperately steep zigzags that we now looked upon as almost a necessary part of a day's march.

On the way we passed the house of one of the Ma-Fus. It was a poor shanty, standing by itself in the middle of a little cultivation, but it was his home; and his wife, children, and dogs, all ran out to welcome him. The Ma-Fu found four eggs, which he brought as a respectful present, and was very pleased when we told him to stay for the night with his family. We rode on to the little village of Cheh-toh (the Jeddo of Cooper), where we halted for the night, and as there was no immediate prospect of the arrival of our baggage, we both had a bowl of tsanba, which we ate after the manner of the country. A large basin of oatmeal, two good-sized cups, a kettle of buttered tea, and two pairs of chopsticks were brought to us—these last, a Chinese innovation, never used by a Tibetan, who finds his fingers sufficient. A Tibetan first helps himself to what oatmeal he requires, the buttered tea is poured over it, he stirs it up with his fingers, adding oatmeal and tea to suit his taste, and then eats it. It is very like porridge, and, as in Scotland, the oats are grilled before grinding them into meal.

The village of Cheh-toh consisted of about four houses, one of which was the inn. From the road we entered a large room, which served as a general room, stables, piggery, and most things else; the horses were tied up here for a time, but were afterwards removed; and the place was not particularly dirty. This room was lighted only by the door, and as the floor, roof, and walls were all more or less black, and as the overhanging eaves projected far in front of the entrance, it was tolerably dark. From this a door led to the state apartment, in which there was no window, and where light was only borrowed from the darkness of the general room. It was by no means so bad a place as I had expected, though it seemed rather a sin to be sitting with lighted candles at four o'clock, with the glorious bright and unaccustomed sun shining outside. There was a good deal of rain hanging about, and we caught the tail of one or two showers, but they were trifling, and the sun shone so brightly that I felt my heavy clothes rather hot, and I almost wanted the sun-hat that I had put away as a useless weight.

The mules did not arrive till six o'clock; and then Chin-Tai told me that all the plates, dishes, cups, glass, and crockery had been smashed. I consoled myself with the reflection that as the plates and dishes were of iron, there was a certain amount of exaggeration, and went out to see the wreck. The muleteers had stopped behind us at Ta-Chien-Lu, to have a final drink and glorification before starting; and in attempting to make up for lost time, had only succeeded in producing the alarming catastrophe reported. They were now in a state of wholesome fear that they would be made to pay for the damage done, and when introduced into the august presence, to explain matters, they went down on their knees, thrust out their tongues, and repeated at intervals the word 'La-so.' Protruding the tongue as far as possible is a respectful salutation in Tibet, and 'La-so' is a term of respect used by inferiors; it means also 'be merciful.'

Our dinner did not appear until very late; at first there were no candles to be found, and Chung-Erh said everything was 'east and west,' a Chinese expression for a general state of disorder.

August 8.—The cooking things were again all 'east and west' in the morning, and the time consumed in getting ready was frightful. Chin-Tai had an extraordinary genius for putting two forks into three boxes, and dividing his cooking things amongst the greatest number of packages possible, so that every time he wanted to prepare a meal, he had to pack and unpack boxes enough to load a good-sized caravan. But in the course of time, and by the exercise of patience, everything was at last ready, and we made a start.

The road followed up the side of a stream, between grass-covered hills, sloping about 20°, capped with bare crags of limestone; the lower rocks were of granite, and great blocks lay scattered about in wild confusion. The morning was beautifully fine: there was a delicious feeling in the air, and looking back down the valley, there was a glorious view of a snowy mountain, whose edges were just lit up by the rising sun, and whose glittering pinnacles of ice and snow shone like points of brilliant light.

The road was broad and good; there was not much traffic, but we met great droves of yaks, and half-bred oxen, a cross between the yak and the ox. We ascended steadily by an easy gradient to the head of the stream; here the valley opened out, and formed a little basin, enclosed by bare and rugged hills, with a strip of green grass beside the tiny rill that trickled at the bottom. A few yards more, and reaching the summit of Cheh-toh-Shan, we at length looked upon the great Himalaya plateau. The pass of Cheh-toh-Shan (the Jeddo of Cooper) is 14,515 feet above the sea; and from this point, with the exception of a dip into the valley of the Ya-Lung-Chiang, the road is always at an altitude of 12,000 feet above the sea, until the descent into the valley of the Chin-Sha-Chiang is commenced at Pun-jang-mu. There was no snow here, but a few small patches were lying two hundred or three hundred feet above.

I had outwalked everybody, and I sat down to wait for the rest of the party. My merry Ma-Fu alone was with me, and he whistled and sang as he let Manzi browse on the delicious herbage.

Stretched at our feet lay a beautiful valley, closed on both sides by gently sloping, round-topped hills; a carpet of luxuriant grass covered the whole surface of the hills and dales; the richness of the pasture was astonishing, and thousands of yaks and sheep were feeding on the magnificent vegetation. The ground was yellow with buttercups, and the air laden with the perfume of wild flowers of every description. Wild currants and gooseberries, barberries, a sort of yew, and many other shrubs grew in profusion, but there were no large trees. There were a few gooseberries on one or two of the bushes, but they were quite unfit to eat. The Tibetans said that the gooseberries never ripened, but that the currants were sometimes eaten.

At the summit of the pass there was a huge pile of stones; and bits of rag inscribed with the sacred sentence fluttered from the heads of long poles set up in the heap.

At this altitude there is great difficulty in breathing. The Tibetans ascribe this to subtle exhalations, which they say rise from the ground; they call all high mountains 'Medicine Mountains,' and so universal is this custom, that the comparative heights may be roughly guessed at by the amount of 'medicine' attributed to them by the people.

In the winter many travellers are said to die here; the passage is greatly dreaded, and those who arrive safely at the top add a stone and a rag to the trophy, as a thank-offering for dangers escaped.

When the rest of our caravan arrived, one of the horse-owners, a sort of petty chief, with a sword-scabbard set with great pieces of turquoise and coral, tied a rag to one of the poles, and cast a stone on to the pile.

To my surprise Ting-Ko turned up again; he said he had come another day's march for the fun of the thing, and begged to be allowed to come with us to Bat'ang; when at last I consented, he grinned with delight, and seemed to think more highly than ever of his remarkable boots. From here, there is what is said to be a good and not very mountainous road to Chiamdo, which strikes off to the north-west, and passing through a populous country, and by a not very difficult route, reaches that place in fourteen days. There are no Chinese officials on this route, and probably no hotels, but it is much frequented by traders.

A descent of about three and a half miles brought us to a solitary hut, glorying in the name of Hsin-Tien-Chan, or New Inn Stage; the Tibetan name is

Ti-zu, which has the same meaning. Here we halted for breakfast, but it was such a miserable place that we at once passed a unanimous vote for a picnic.

A couple of boxes were placed beside a low table on the delicious fresh grass; a gentle breeze, and now and then a passing cloud, moderated the sun, which was almost strong enough to make itself felt, and helped me to give myself up for a while to day-dreams and the charms of scenery and climate.

This was one of those days, which come sometimes to a traveller, when he feels so thoroughly happy, that the pleasures of civilisation are forgotten, and he dreams of perpetually seeking fresh fields and pastures new, and of spending his life amongst the mountains. It must be confessed that these days are rare; but sitting outside the little shanty, the scene was so peaceful, and there was such an exhilaration in the air, that I thought I could contentedly spend the rest of my life in this lovely valley.

Tibet is truly a land flowing with milk and butter, for milk appears in liberal quantities in every form and shape. We had not sat down a minute before a cup of buttered tea was brought for each; this was followed by a bowl of whey; the people offered us tsanba (made with butter), and said they could get us cheese.

After breakfast we went to a hot spring about three hundred yards away. The temperature was 111° Fahr., and the stones in the stream running down were covered with a saline incrustation, but whether of soda or potash I cannot say, as the Chinese name is the same for both. Gas of some sort was bubbling up from the hot spring, which was quite black from sulphur; people come here to cure skin diseases, and they say it is very efficacious. We stayed long enough to immerse a thermometer, and then rode on after our caravan.

We continued our march down the valley, and presently came upon a little tent pitched by the hillside; here some Tibetans were lying about, their fierce dogs tied up to pegs in the ground, and innumerable herds of cattle and sheep grazing round them. At this season of the year the sheep are taken in great flocks from Lit'ang to Ta-Chien-Lu and Ch'êng-Tu, for sale. At other times, though there are always considerable numbers, such immense flocks would not be met with on the road. Outside the tent a quaint and wild group of Tibetans was gathered round the buttered-tea churn, making their mid-day meal. A bag of oatmeal lay on the ground by the churn, and one of the men was filling the wooden cup with a brass ladle. Like the wooden cups, the churn is almost a part of every Tibetan community. On entering a house at any hour, someone is certain to be seen making buttered tea in the churn; a mule, with a sack of oatmeal, and a churn, for every three or four men, forms part of every caravan; at a halt the churn is immediately produced; and, in fact, wherever there are half a dozen Tibetans gathered together, there the churn will be found.

This churn is a cylinder of wood about two feet long, and six inches in diameter. The butter is churned up in the boiling tea, and there is some art in doing this in such a manner as to make the ingredients mix properly.

The tsanba is prepared in various ways according to fancy; the meal is sometimes kneaded with the fingers into a stiff paste, and eaten like a cake, and at others it is mixed with sufficient tea to be almost thin enough to drink.

Further down we came to the first Tibetan house, at a distance looking like a strong castle, and up a little valley behind it were two or three others, together forming a small community or village, and from here to An-niang, houses, separated from one another by about a quarter or half a mile, stand singly on the right bank of the stream. These houses are great piles of loose stone with scarcely any mortar, sometimes three or four stories high; the roof is always flat, and a gable is never seen. With their little slits of windows, they are gloomy in the extreme, and looking as if they were half in ruins, give an idea of great misery. They are, nevertheless, very picturesque, and the view down the valley as the sun was setting would have made a lovely picture.

The dryness of the air we breathed, and the salt in the beef we ate, and in the buttered tea we drank, combined to make us exceedingly thirsty; and we sent a Ma-Fu to one of these houses to beg a cup of tea. The house was at the other side of the stream; but the lady herself, dirty beyond conception, and covered with jewelry, came out with a servant to regale us, and brought a huge jug of boiling tea made with milk without any water. The draught seemed like nectar to our parched throats; we paid for it with a few turquoise beads, and went on our way refreshed.

All this valley is covered with wild flowers, from one of which a paper like parchment is made; another has the valuable property of killing lice; caraway grows wild, and is also cultivated. Barley and oats grow well in the valley, but the people do very little but keep cattle, sheep, and ponies, of which there are great numbers, some exceedingly good-looking, with quite an Arab head.

Some of our followers declared they saw a herd

of deer; we could not, however, make them out with our glasses; but although we saw no game of any description, deer, wild sheep, wild goats, and hares are said to abound.

We halted at An-niang, a Tibetan name that means nothing in particular. The Chinese, in trying to approximate to the sound, call it Ngan- (quiet) Niang- (woman, mother) Pa (place).

Here the people took evident pleasure in showing their hospitality; they led us to the best house in the place; they immediately set the churn of buttered tea and bowls of boiling milk before us, and prepared the feast of Jael the wife of Heber the Kenite, but were kind enough to spare us the sequel to that entertainment.

August 9 —Marching down the beautiful valley by an excellent road, amongst the fresh green grass, buttercups, and wild flowers, we met large parties of Tibetans returning from a fair that had been held at a Lamassery a mile or two away.

Women as well as men were riding ponies *à la califourchon*, and as we approached they turned a few yards aside and dismounted.

As they stood about, or sat on the grass, they formed most picturesque groups. The men were wild-looking fellows, with long shaggy hair, whose garments, always with a bit of red about them, seemed to have no shape in particular, and were gathered in with a cloth tied round the waist, leaving a fold in which they carried an immense amount of property, not only in front, but at the sides and behind their backs.

The women, too, all wore something bright-coloured, and fastened up their hair with a circular disc of silver engraved with Tibetan characters, and set with coral beads. No matter how poor and dirty, they all wore this expensive ornament. Both men and women had necklaces of turquoise, coral, or coloured glass, from which they hung charm-boxes of gold or silver.

At a fair of this kind the people eat and drink, the men and women dance together, and make a few trifling purchases which they bring home in the capacious folds of their shapeless clothes.

One of the Ma-Fus had a friend in an adjacent house, and when we arrived we found a feast spread for us on a low table amongst the buttercups by the roadside.

The lady of the house again came out, and brought us a huge caldron of tea made with milk, bowls of tsanba, and some cream cheese without any taste, and of the consistency of indiarubber.

We thanked our hostess, who seemed delighted to have been able to entertain us, and leaving a few turquoise beads as a present, rode on.

The turf was so tempting, and the air so delicious, that Chin-Tai and Chung-Erh could not resist the excitement of a gallop, and were soon racing over the level plain. Mesny's boy, ambitious to emulate their example, essayed to follow them; but never before in his life having ridden a horse, he had a tremendous fall. He was carrying Mesny's somewhat crazy gun, and though he did not hurt himself, he lost one of the locks; this was not noticed at the time, but by the aid of the almighty rupee, a search that was subsequently instituted proved successful, and this ill-used weapon was put again in order.

After riding down the stream another couple of miles, the road turned sharply round to the northwest, and ascended another similar valley. Both these valleys are well populated, the gloomy Tibetan houses, with their flat roofs and little windows, standing, at intervals of two hundred or three hundred yards, close by the side of the stream. This method of living shows how different in character are the Tibetans and Chinese. The Chinese love crowding together, the tighter the better; and for the sake of living in a town or village, will walk a mile or two every day to their fields, rather than live in a house by themselves; for though it is true that in China the plains are dotted with detached houses, yet there are always villages at every few miles. But here there is scarcely ever such a thing as a village, properly so called, and the few that there are, are generally occupied by the Chinese, as was the case at Ngoloh, or Tung-Golo, as the Chinese call it (the Tung-olo of Cooper), where we halted, and where the Margary proclamation was posted in the portico of the Chinese hotel, in which we found chairs and a table. It was not much of a place, but better than anything we had anticipated. The people were as hospitable as ever, for no sooner had we arrived than they brought presents of cheese, butter, tsanba, mushrooms, radishes, and eggs.

Our room was rather dark, and the smoke descended from a hole in the roof, and penetrated through numerous cracks in the wall; but otherwise there was nothing to complain of, except the monotonous drone of a holy man seated in the adjoining kitchen, who burned incense, and gabbled prayers incessantly, at intervals striking together two discs of brass, which rang with a clear bell-like sound. There was a good yard, enclosed by high stone walls, with room enough even for our caravan, which was of a tremendous size, as all the people belonging to it had taken the opportunity of bringing a few pony-loads of tea to sell at Bat'ang. The animals were picketed in the yard in the usual Tibetan style, *à la* Royal Engineer, with a rope stretched along the ground, and a hobble to one forefoot.

They keep several big dogs to guard their mules and their property; and this is done so effectually, that one day when I sent Chung-Erh for a box, he came back and said the dogs would not let him have it, and I must wait till the dogs' master returned.

The clatter of an imperial despatch from Peking awoke the echoes of the slumbering village at 3 o'clock in the morning: a few dogs barked, a cock crowed, but in less than a minute the rattle of the hoofs was lost in the distance, and the place lapsed into its normal silence. These imperial despatches are carried by horsemen who travel night and day, and everybody and everything has to get out of the way. The messengers that awoke us for a moment reached Wu-rum-shih, a place about fifteen miles distant over a stiff mountain, in two hours. Imperial despatches are carried from Ch'êng-Tu to Peking in eight days, but not by one man.

August 10.—The muleteers came to us in the morning, and kneeling down and thrusting out their tongues, prayed for a day's rest, and said that the grass here was lovely, that the next marches were long and stony, and through a desolate country, where they should find no fodder for their animals, and that if we would consent to rest here they would forego the halt that had been promised them at Ho-K'ou; when we agreed they said 'La-so,' and went away joyfully to turn their animals out on to the hillsides.

The quiet of this little place was very pleasant; there was scarcely a sound in the early morning, and sitting in our room, I could hear the peasants in the distance singing a plaintive melody, as they went off to gather sticks, or drive the cattle.

A green hill with a fir wood on it rose behind the house, and thither I strolled with my gun in search of some kind of bird that I heard calling. I was told that this was not a pheasant, though it was as large as one, and had a long tail; the people said it was very shy, and on this occasion it did not belie its reputation, my purse and my gun being equally unsuccessful in procuring me a specimen.

The view from the hill was very lovely; the valley below, where smiling crops of barley, oats, and a little wheat were nearly ready for the sickle, was quite flat, and about four hundred yards broad; the fields were divided by hedges of the wild gooseberry, or by strong fences of well-made wattle. A post and rails ran down the side of the hill; a stream meandered through the bottom, and a broad road between hedges brought to mind many an English country lane. Beautiful green slopes running smoothly down shut in the valley on both sides, and a mile and a half away, by the edge of the stream, a gloomy Tibetan house, looking, with its tall and solitary tower, like some old mediæval castle, seemed to keep watch and ward over the scene.

Ngoloh nestled at our feet, a hamlet of about a dozen houses, built, in the usual Tibetan style, with flat roofs, adorned with Mani poles, and little patches of garden round the homesteads, in which the industrious Chinese had raised a few of their favourite vegetables.

The officer in charge of the place called on us in the afternoon. He told us that there were seventy or eighty families in the district, and that he had a hundred soldiers under him.

He said that during recent years cattle-plague had been very frequent; that during his youth he only recollected one occurrence of this terrible scourge; but that in the last ten years there had been eight visitations, and that in 1876 sixty per cent. of the cattle had died. He had no theory as to how or whence it came, but looked on it as sent by Providence. The symptoms are a watery discharge from the nostrils, drooping ears, and the indications of violent dysentery. Animals attacked would die in a very short time; but at the first signs of the plague, they are killed and buried.

Before leaving, the official warned us of robbers who infested the road beyond this place. He said he had received orders from Pao at Ta-Chien-Lu to furnish us with an escort, and that we should have ten soldiers to accompany us. We placed little faith in the existence of the robbers, and still less in the value of our escort.

August 11.—As we ascended from Ngoloh, the valley gradually narrowed, and the hills were more wooded; the road ran between hedges of wild gooseberry, small willows grew by the edge of the stream, and the slopes on the right were covered with a dwarf holly. We stopped for breakfast at a solitary hut, standing at the foot of a steep zigzag that leads to the summit of Ka-ji-La. The hut is called La-tza, or in Chinese Shan-Kên-Tzŭ, both names meaning 'the root of the mountain.' 'La' is the Tibetan for a pass over a mountain.

We sat outside, and enjoyed the scent of the pines and wild flowers borne to us from the hills above, and as our escort collected round us, we were presently astonished to observe on one of their coats the buttons of the 47th regiment. After this we noticed that half the men we met buttoned their coats with British regimental buttons. These find their way from the old clothing in India, through Tibet to the very frontiers of China.

From La-tza, we still followed the side of the stream amongst the pine woods, whose cones were of a deep indigo purple; but we gradually ascended above its bed, and could only hear the chatter chatter of the brook, two hundred feet below, as it leaped from rock to rock. The ascent was not very terrible, nor was the road anywhere very steep, and we at length emerged from the woods into an undulating plateau of beautiful grass covered with buttercups and wild flowers, amongst which the arnica plant was conspicuous. Here there were no trees, but thousands of yaks, mules, and ponies were feeding; and here and there there was a shepherd's tent, with a group of wild-looking fellows standing about, guarded by even more wild-looking and savage dogs.

The first crest we reached is called in Tibetan Ka-ji-La, or in Chinese Ko-Erh-Shi-Shan, 14,454 feet above the sea. The view when we reached the summit was superb. Looking back in the direction from which we had come, range after range of mountains lay at our feet, culminating at last in the most magnificent snowy heights, one of which raised its head about four thousand or five thousand feet above its neighbours. It was a magnificent peak, and at this distance looked almost perpendicular. Its name in Tibetan is Ja-ra (King of Mountains), and I never saw one that better deserved the name. Never before had I seen such a magnificent range of snowy mountains as here lay stretched before me, and it was with difficulty I could tear myself away from the sight.

Our road now lay for a couple of miles over an

undulating grassy plain; at the centre of this our escort left us, and marched back across the ridge of Ka-ji-La.

The official in charge of the ferry at Ho-K'ou passed us here. Pao had sent for him to give him instructions about us, and he was now on his return journey.

From the western side of the plateau, there was a still finer view, and as the day was fortunately very clear, our guide could show us where Ta-Chien-Lu was lying, at the foot of a grand snowy range.

We now descended a narrow valley between steep hills, well wooded with firs, some of very large dimensions. There was a shrub with leaves like a laurel, and white flowers like a wild rose; there were also great quantities of holly, and in the bottom, willows, wild gooseberries, and currants.

From La-tza to Wu-rum-shih, there was not a single habitation, and all the afternoon we marched through a forest of pines. Here there is a kind of moss, growing chiefly on the pines, but also on the hollies and other shrubs; it hangs in pendants two or three feet long, and sometimes at a distance gives the trees the appearance of weeping willows.

At the inn in Wu-rum-shih, we found a capital room, much larger than we had enjoyed for a long time. The Margary proclamation was posted in the inmost recesses of our apartment, but as it appeared nowhere else in the village, it can hardly have been the means of enlightening any large proportion of the population of Tibet.

The innkeeper, a Chinaman, brought us a present of eggs and vegetables, and my Ma-Fu, the Long-lived Gem, whose home was in the neighbourhood, brought more.

My servant Huang-Fu here reported himself sick,

with pains and aches all over his body; poor fellow, he was beginning to look very ill, but there was always something comical in his appearance, as he sat on the top of a pile of blankets on his pony, perpetually smoking a long pipe.

From Wu-rum-shih, the road commences its descent to the Ya-Lung-Chiang, or Nia-Chŭ, as the Tibetans call it, the only river of importance between Ta-Chien-Lu and Bat'ang.

We marched through mile after mile of pine forest. There were many fine walnut trees by the side of the stream we were following, as well as wild peaches, plums, and apricots. As we descended, the gooseberries and currants disappeared, but the hardy barberry was represented by two species. A few wild cherries, bearing fruit sour beyond description, many other berry-bearing shrubs, and a kind of birch, were interspersed amongst the pines. There was also a tree with a leaf something like that of a plane, but more delicate, and under our feet nature still exhibited the rich treasures of her floral wealth.

August 12.—There were numbers of green parrots flying about from tree to tree ; the proper *habitat* of these birds is, doubtless, in the warm climate of Southern Yün-Nan ; but during the summer, short though delicious, they fly up the two rivers, the Chin-Sha and Ya-Lung, and scatter about amongst the entrances to the valleys of the tributary streams ; they may be seen during one or two marches on both sides of both rivers, but no further; and as soon as autumn tinges the leaves, in all probability their green plumage disappears.[1] The people said that the mountains

---

[1] Lieut. Garnier notices the warm temperature in the valley of the Chin-Sha at its confluence with the Ya-Lung. The characteristic plants of Kiang-Hung on the Mekong, 4° further south, were found here. *Voyage d'Explor.* i. 502.—*Y.*

were the haunts of all kinds of game and wild beasts ; and we were told that a panther had recently killed a pig in the village of Wu-rum-shih.

Half-way to Ker-rim-bu, we again overtook the Ho-K'ou official, who had dismounted for a few minutes, and, with his party, was sitting by the side of the road, enjoying a jar of beer. As we approached he rose and advanced to meet us ; we joined him for awhile, and drank a glass of his liquor ; it was made from barley, and tasted like an extremely small home-brewed ale. Afterwards we rode on together to Ker-rim-bu (the Octagon Tower), or in Chinese Pa-K'ou-Lou, so called from a tower just outside the village, of which the lower part is cruciform, and the upper part octagonal.

Fig. 2.

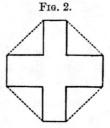

It is the fashion in Tibet to shoe the horses only on the hind feet ; but the Tibetans are not accustomed to ride their horses every day. Possessing vast herds, a horse is used for two or three days, and then turned out loose on the pasture land, while others are ridden in its place ; thus the hoofs have constant opportunities of growing, and, if much worn away during the few days' hard work, they are soon restored when wandering freely on the soft turf.

My grey pony began to feel the regular marching over roads, which in many places were very stony ; and when I mounted him at Ker-rim-bu, I found him lame from a bruised hoof. As horse-keeping was very cheap, I told Peh-ma to look out for another, and received the stereotyped reply, that if I had only mentioned it yesterday, he could have brought me any number of beautiful animals.

Nia-chu-ka (Ho-K'ou, 'River Mouth' in Chinese) is situated, at an altitude of 9,222 feet, at the junction of two streams with the Ya-Lung-Chiang,[2] and is surrounded on all sides by bare and precipitous mountains, that run sheer down to the water, leaving no flat ground for grass or cultivation, and very little for building.

Opposite the town, and dividing the river from the stream we had followed, a bare rock, seven hundred feet high, rises almost precipitously from the surging water; a pile of stones marks its summit, and the flutter of the pious rags that wave from the usual poles can just be discerned from the houses below.

Though at a greater elevation, the climate is warmer than that of Ta-Chien-Lu, which place is particularly cold, owing to the masses of snowy mountains that surround it on every side.

The town of Ho-K'ou consists of twenty houses only, and is a wretched place, garrisoned by a few soldiers under the command of a petty official. From Ho-K'ou there is a road much used by traders, which, generally following the valley of the Ya-Lung, reaches the frontier of[3] Yün-Nan, by a twenty days' march.

For a couple of marches on both sides of the Ya-Lung-Chiang, the main road between Bat'ang and Ta-Chien-Lu leaves the plateau, and descends to

---

[2] Ya-Lung is the Chinese syllabling of the Tibetan *Jarlung* ('White River,' according to Ritter's authority), the valley of which is regarded as the cradle of Tibetan monarchy.—*Y*.

[3] Probably near Yung-Ning-Fu.

the warm valley of the river, where there are no longer any pasture lands; baggage animals consequently fare but badly, for food is never carried for these hardy beasts, which, if well fed whilst on the plateaus, are accustomed to a few days of semi-starvation in the lower country, picking up any scanty blades of grass or young shoots they may have the good fortune to encounter.

At Ho-K'ou we found a very good inn, where we luxuriated in two rooms; it was kept by a half-breed, who brought us vegetables, eggs, and milk, presents which were supplemented by a fowl from the military commandant, and a remarkably good fish from the river.

Except where there are Chinese, it is quite impossible to get fish in Tibet. The Tibetans in this, as in everything else, present a remarkable contrast to the Chinese; for the Chinese are particularly fond of fish, breeding them in great quantities. The Tibetan aversion to fish arises from the method in vogue amongst them of disposing of the dead, the bodies being generally cast into the rivers and streams.

In these elevated regions, where little agriculture is carried on, no oil-producing plant will grow; oil is therefore very expensive, and pine splinters are the only form of lamp in use. They are not very satisfactory at any time, dropping sparks about most lavishly. One evening, when my india-rubber tub had been inflated in readiness for the morning, it very nearly came to an untimely end. This apparatus naturally excited the wildest speculation, partly because, as washing was a process in which the people could see neither pleasure nor profit, the use of the machine failed to penetrate into their ideas, and partly because an india-rubber tub, blown out by a pair of bellows, was in itself so extraordinary, that it could only be regarded as fit for a museum of curiosities.

The tub having been arranged in a corner, attracted the attention of some visitors, who, with pine splinters to light them, made a minute examination of this inexplicable engine, and holding the flaring torch above their heads, bent over the tub in eager curiosity. The sparks flying about would soon have burnt holes in my bath, but the danger was fortunately observed, and my morning ablutions were saved from an abrupt termination.

August 13.—There is a ferry at Ho-K'ou across the Ya-Lung-Chiang, but it is for the use of officials only; Pao had already given orders to the officer here to take us across, so we met with no difficulty.

The ordinary way of crossing is in a coracle, the shape of a walnut, made of raw hides stretched over wicker-work; and as the current is rapid, and the water broken, it does not look a very pleasant operation. Animals of all kinds have to swim; even the soldiers carrying the imperial despatches are obliged to leave their horses behind.

Our baggage animals swam across in the evening; we followed the next morning with our servants and horses, and waited in one of the few houses on the right bank of the Ya-Lung, opposite Ho-K'ou, whilst our caravan was reassembled.

The owner of the house in which we had taken up our quarters, begged me to give him a pigeon that I shot. He wanted to make a poultice for his servant, by pounding the flesh, and mixing it with some herbs or flowers gathered on the mountains. I quite failed to ascertain the nature of the disease for which pigeon's flesh is such a potent remedy.

Our animals were at length all collected, and we

started with two fresh soldiers ; those that had come with us from Ta-Chien-Lu leaving us here, and returning to their own quarters.

The road from Ho-K'ou again ascends to the plateau from the warm valley of the Ya-Lung. Climbing up the bed of a tributary stream, the same order of vegetation and trees is seen as on the eastern side of the river. At first, there is neither cultivation nor pasture land, but the road at once enters another dense forest of magnificent pines. By the stream there were wild peaches, wild cherries, and barberries : the peaches and cherries quite uneatable. As we ascended we found a few currants, and great numbers of a tree very like a walnut in appearance. Near Ho-K'ou, the parrots were numerous ; and I saw a grey striped squirrel, like the squirrels so common in India.

After about seven and a half miles of this forest-travelling in a narrow valley, where the foliage was so thick that our view was limited to little more than a few yards beyond us, the gorge opened a little, the hills on our right sloped gently, and left a few tolerably level little patches, where an acre or two of barley shut in by an oaken fence was nearly ripe.

Though there were no houses, the place had a name—Shin-ka. About a mile further there was one solitary hut, and a number of the religious cairns ; and another mile brought us to Mah-geh-chung, the Ma-kian-dzung of maps, spoken of by Huc as an important place.

There are about half a dozen wretched houses here in a little open ground, where there are a few patches of wheat and barley. We were lodged in the house of the official ; but, although the best, it was a wretched place.

There was a sort of shed that served as general residence for the family and their domestics, in which the Margary proclamation was posted; and beyond this, there was the only thing that by any possibility could be called a room. A wall of loose stones at the end, tumble-down wooden partitions at each side, a boarded floor and a roof, were all that entitled it to this dignity. The side partitions were full of holes and gaps, and failed to reach the end wall, from which the roof was also separated by several inches. Everything was out of the perpendicular, and black with the smoke that pervades every recess in all Tibetan houses.

I ascended at once to the flat roof by one of the usual staircases, which invariably consist of a rough length of a pine tree with notches cut in it. On the roof I found a shed, with a Lama reciting prayers from sheets of paper he held on his knee; the holy man, far from resenting my intrusion, smiled benignly on me, and politely moved to another smaller shed on the same roof. Here he continued muttering his pious ejaculations hour after hour, except when curiosity prompted him to come and look at what I was doing; and long after the evening had closed in, and dinner was over, as I sat in the room below, I could still hear the monotonous droning of the Lama's voice.

August 14.—As the 'Long-lived Gem' ran merrily at our side, he confided to us that we might think ourselves very lucky in our muleteers, who allowed our servants to ride the hired animals, even in the worst places; for he said that even when Lamas, or other people of distinction hired their horses, they were always compelled to dismount in the bad parts of the road. He naïvely added that the breakage of all my crockery was a very good thing, for the fear of having to pay for it had made the muleteers thus unusually obliging.

Above Ma-geh-chung, there is what the Tibetans and Chinese called a medicine mountain, and our road was over this. We marched up a beautiful valley, through a forest of noble pines. I measured the largest I saw, and its girth, at a height of four feet from the ground, was thirteen feet six inches. There were oaks also, poor scrubby things; indeed it was only here by the discovery of acorns on them that I was able to satisfy my mind as to their identity, for anything less like our oaks it is hard to imagine. The parrots were seen no more, but amongst the trees there was some large jungle-fowl, called pheasants by our people; pheasants, however, they certainly were not, they looked more like jays, though much larger, and made the same kind of chattering noise.

Gradually as we ascended, an open valley here and there showed us grassy hill-tops, where not even those hardy oaks and pines would grow. The forest became thinner, and the trees smaller, and at about four hundred or five hundred feet below the summit, the last of the pines and oaks were left behind.

The summit of Ra-ma-La is 14,915 feet above the sea, but none of our party seemed to experience the evil effects of the great medicine mountain. Of course those who walked found breathing difficult, but neither Mesny nor myself noticed anything unpleasant, unless we stooped, when we felt giddy; this appeared to me the more remarkable, because in other countries, at much less altitudes, I have found difficulty, in breathing even when sitting down.

The crest of Ra-ma-La may be said to divide the plateau from the valley of the Ya-Lung-Chiang, and once across it we again find the green pastures, the wild flowers, the great herds of yaks and sheep, and the profusion of milk and butter so characteristic of the upland country. The plateau extended for many miles, and the rich grass was covered with buttercups and other yellow flowers, that grew together in great masses; there were patches of red, purple, or blue, and the variety of colour was wonderful.

The droves of yaks, ponies, and sheep were tended by Tibetans armed with swords and guns; for the Tibetans, unlike the Chinese, always carry a long matchlock, and a sword, studded with turquoise and coral.

La-ni-ba (a Hollow between Two Mountains) was the name of a miserable hut at which we halted; a shower of rain put a stop to a picnic on the beautiful grass, and drove us in the room, where the floor was as mountainous as the surrounding country. The door and the holes in the roof let in a little light, and a great deal of smoke, the usual accompaniment to a residence in Tibet; and for ornament the Margary proclamation was fastened to the walls.

We were told that the houses and inns used to be much better; but that twenty years previously some tribes from the north, called Nia-Rung,[4] came down, invaded the country, and carried fire and sword throughout it. The houses were destroyed, and since then they have not been built up again in a good style.

From the next crest of Ra-ma-La, 15,110 feet above the sea, which we reached soon after breakfast, we should have had a very fine view, had it not been for the heavy clouds on all the mountain-tops.

[4] Or perhaps Nia-Jung—the letters R and J are very interchangeable. Nia is probably the same word as the Tibetan name Nia-Chu for the Ya-Lung-Chiang.

It was, however, tolerably clear to the west. They showed us where Lit'ang lay, about seventy miles distant, and an occasional flash of sunlight struck some snowy pinnacle on the high mountains surrounding that place.

From here our road lay along a grassy ridge, with deep wooded valleys running down; and seven and a half miles from La-ni-ba, we came to Mu-lung-gung, a military post, composed of a couple of poor shanties standing by themselves in a bleak spot, called by the Chinese Pu-Lang-Kung. Here our escort of two soldiers changed horses; but the appearance of the place was not sufficiently seductive to induce us to make a halt, notwithstanding its eminent piety, as evidenced by more than the usual number of scarves and bits of rag on the poles that bent before the strong wind that was now sweeping over the plain. After marching a short distance further, we caught sight of our resting place, lying almost at our feet, in a beautiful and sheltered valley, bounded by gently sloping wooded hills; a good-sized stream was winding and twisting at the bottom, meandering amongst corn-fields and flat meadows, where oxen, sheep, and ponies could be seen feeding on the beautiful grass. It was really a lovely view; but how different it would have been in the hands of the industrious agriculturists of Ssŭ-Ch'uan, who would raise crops from every inch of the hill-side.

We now descended a steep spur into the valley, and here there was rather a bad and difficult zigzag over loose, sharp stones: it was disagreeable enough to go down, and would have been much worse to come up, but there was no real difficulty in riding anywhere. The descent was not very long, and once down a level bit brought us to the village of Lit'ang-

Ngoloh. The largest house in it belonged to the father of one of our muleteers, a wealthy man in these parts, as he was supposed to possess property to the value of a thousand taels (about 300*l.*).

Our muleteers had ridden on ahead, and all the family, including the ladies, came out to welcome us; the damsels were dressed in their best, which included a considerable amount of dirt, and were covered with beads and jewelry. On each side of the head they wore a disc of chased silver about the size of a saucer, these meeting above formed, to a front view, an inverted v, thus ʌ. Another smaller disc was worn behind; and all were loaded with coral, and sham or real turquoise. A lock of hair, about an inch broad, was brought vertically down over the centre of the forehead, and cut off at a level with the lower part of the nose. They had necklaces of beads, and great silver ornaments, and charm-boxes were hung from chains of beads that seemed to be wound about all over their bodies.

We were led into the house with great pomp and ceremony, and as soon as we had been installed in the best room, the women quickly brought us buttered tea, milk, and sour cream.

The house was really a well-built solid structure: quite a palace after our recent accommodation, and betokened the comfortable position of its owner. The whole of the lower area formed a covered and extensive stable, divided by immense pillars of wood, that supported the ceiling and the house above. Instead of the usual notched log, a sumptuous ladder led through a spacious trap-door to the upper story. This consisted of a quadrangle, the floor of which was planked, the living and sleeping rooms being arranged round the four sides. The roof of these was flat, and projected far enough to shelter a large portion of the quadrangle and the inmates of the house from the rays of the sun in the summer, and from the snow in the winter. The roof was gained by another ladder, and was surrounded by a parapet; and a covered shed was erected on part of it, where piles of hay were stacked. The room we occupied was lofty and commodious, the back wall of solid mud, the others of wood. The floor was planked, and the windows looked upon the quadrangle. We had some difficulty in manufacturing a table, as the simple people themselves having nothing that they can want to put upon a table, are unaccustomed to the use of this article of furniture. There is, however, an object of household equipment of which we Westerns are still in ignorance.

Shaped like a table, eight feet long by two feet broad, and nine inches high, there is a large circular hole in the centre, in which a pan of charcoal or wood is put. This is the fireplace of the country, and two persons can sit on it, one at each end. Finding one of these in the room, we raised it on stools and packets of tea, and improvised for ourselves an excellent table.

I was always astonished at the miserable appliances for warming rooms that are used in Tibet: a wretched fire lit on the floor, emitting far more smoke than warmth, or a pan of charcoal, such as we found in this room, would seem to be but a poor protection against the frightful severities of the climate during the winter in these elevated regions; yet nothing better is ever seen, and it must be chiefly by clothing and food that the Tibetans keep themselves from perishing of cold.

August 15.—We halted here a day, to let the

baggage animals feed after the period of scarcity through which they had passed; and as I had sat up writing very late, I was still in bed when Mesny went out. When I finally got up, I had some difficulty in getting rid of the people while I bathed. The old man of the house took the most fatherly interest in all my actions, and remained in the room after every one else had gone. I pointed to my tub, but he evidently did not understand me; at last, however, he comprehended what I was going to do, and then his expression was most ludicrous, as he beat a retreat at a pace that I should have thought impossible for so old a man. After I had finished my toilet he brought me a great basin of hot milk in his own hands, and at breakfast he laid before each of us about a pound of beautiful fresh butter, quite free from the peculiar flavour that this butter generally had.

Huang-Fu was now seriously ill; it appeared that he had been unwell at Ta-Chien-Lu, and just before leaving he obtained from the Bishop a quantity of medicine, of the nature of which he was in profound ignorance. He did not know the amount to be taken, nor could he tell me for what class of ailment it was intended. For the last few days we had been doctoring him with chlorodyne and quinine; he was getting a little better, and this gave him such a profound belief in medicine generally, that he took a large dose of the Bishop's physic on top of it. The result was hardly satisfactory, and he was now so weak he could scarcely sit on a horse. Through all his trials he never deserted his pipe, and was now lying down in a corner smoking at intervals. We wanted to leave him here, where we thought he would be in good hands, until he should be well enough to return to Ta-Chien-Lu; but the people were afraid of his dying in the house, and would not consent. There was, therefore, nothing for it but to take him on, as best we could.

I found a very good pony here, a strong-looking bay, with black points, and after the farce of bargaining, always necessary, I bought it for sixteen and a half taels.

August 16.—I went out on the roof early in the morning to put out my thermometer, and found the old master of the house sleeping placidly under a shed, wrapped up in a heap of ragged skins. Presently one of the girls came up with a jug of hot buttered tea and a cup; she poured out a cupful, which the old man consumed, and then leaving the jug beside him, she retreated below. There were two girls here, one the wife, and the other the sister of one of our muleteers. The wife was always gorgeously arrayed with strings of beads, from which great gold and silver ornaments were suspended; she seemed to sleep with her jewelry on, for no matter how early, or how late, if we ever caught a glimpse of her she was still covered with these uncomfortable-looking accoutrements. We wanted to buy the complete set, but she would not part with them, because she said it would be like dying before her time, and very unlucky. She did not show herself very much, and always hid if she thought either of us were looking at her. The other girl was dressed quite plainly, and seemed rather to like being looked at.

There seemed to be a certain amount of polyandry, not to say promiscuousness about their arrangements, and I never thoroughly understood the degrees of relationship, which would have puzzled even so able a genealogist as Sir Bernard Burke.

Rain clouds were hanging about the hill-tops as

we left the hospitable village of Lit'ang-Ngoloh, and marched up the pretty little valley between the fields, divided by stone walls or hedges of wild gooseberry bushes.

We then ascended the hills at the other side, and entered an undulating country, where pines of the most beauteous form were disposed by nature in such lovely groupings, that they would have brought feelings of despair and envy to the owner of the noblest European park. Besides the pines there was a great quantity of a kind of holly-leaved oak; on the lower branches there were prickly leaves, almost precisely like our holly, whilst up above, there were small, smooth, rounded leaves without any serrations. My doubts as to the identity of the tree were dispelled by the presence of acorns, although otherwise I should hardly have recognised it as any relative of the English oak.

I was riding ahead with the 'Long-lived Gem' through the silent woods, when we started a musk deer from its lair; it bounded down the side of the hill, and disappeared in a thicket beyond. It was the only game I saw in all the journey.

A march of four and a half miles brought us to an open country called Niu-chang, which means the 'Cattle Feeding-ground'; and here vast herds of oxen, yaks, sheep, and ponies grazed upon the rich pasture. The summit of Mount Tang-go-La was a mile beyond, 14,109 feet above the sea, and from it a descent of six hundred feet brought us to the bed of a stream where some men were washing for gold. I failed to acquire any idea of the average amount obtained by a man during the day; sometimes they said they might get only $\frac{2}{100}$ or $\frac{3}{100}$ of a tael, while at others as much as $\frac{2}{10}$ or $\frac{3}{10}$, and now and then a large nugget falls to the lot of some lucky individual. I did not see any, but I was told that the gold was not in the fine particles in which it is found in the Chin-Sha, but in nuggets. This place was called Zu-gunda, and here there was a Tang (the Chinese name for a military post). These posts are arranged at short distances apart the whole way to Lassa, and a few soldiers live in each; a Tang consists only of a miserable hut, built of stones, with a flat roof. Horses are kept in all of these for the use of the couriers, who change at every Tang. A courier met us at Ho-K'ou on his way to Ta-Chien-Lu; he had been to that place, and he passed us here on his return journey to Bat'ang. Our soldiers, for we always had two with us, also changed their horses at each post.

A little farther down there were some of the black felt tents of the cattle-keepers, who bring great herds, and wander about from place to place during the three or four months of summer, when the ground is free from snow.

The muleteers told us that the regular stage ended at some huts a little farther on, called Cha-ma-ra-don; they grumbled when we insisted on making a longer march, and assured us that it would be quite impossible to reach another building before dark; but having had such wide experience of their assurances, we recognised intuitively by their manner the rare occasions on which they spoke the truth. This was not one of them; and without wasting time in argument, we ascended another valley, where the pines still clung to the sides of the green hills; but gradually the trees became thinner, a little more than a mile brought us above the level of the last pine, and now nothing lay before us but grassy rounded slopes, where there were numerous encampments of Tibetans.

These encampments are called in their language 'Ba'; their tents are always guarded by large and savage dogs, which, like the cattle, change their coats during the summer. Many of these had not yet lost their winter fur, and were really ridiculous objects to look upon, with perhaps a nice clean summer coat on the fore-quarters, while behind, great tangled masses of fur and dirt still clung to their bodies, or dragged along the ground.

We were told of robbers here who have a habit, that must be especially unpleasant in cold weather, of robbing a man of everything, down to the last rag on his back, and leaving him absolutely naked. There were four families living at the bottom of the mountain, whose duty it was to look out for thieves, and guard the road.

From the summit of Deh-re-La (about 14,584 feet above the sea) we had a fine view over one of the valleys, so characteristic of this part of Tibet. A small stream meanders through a little plain, enclosed on both sides by hills, rising sometimes as much as a thousand or fifteen hundred feet above the valley below. Nowhere is there any cultivation, nor is there a tree to be seen, nothing but gentle slopes and rounded tops all covered with grass, the richness of which is marvellous, and is only seen in places where snow lies for three quarters of the year. In these valleys there are quantities of the rhubarb so valued as a medicine; it is a fine-looking plant, of which there is a very good picture in Prejevalsky's book. Another herb that grows in profusion is something like a gigantic dock; its leaves are sometimes as much as two and a half feet long, and it throws up a straight thick stem, at the top of which is a large bunch of small yellow flowers; it is not used for anything, nor in itself is it particularly ornamental, but the masses of big leaves by the side of a stream look fine and handsome amongst the delicate grass and wild flowers.

From the summit of Deh-re-La, we looked across this plain to the mountains on the other side, and to the pass Wang-gi-La, a mountain of which Peh-ma now observed that, though it was not very high, there was plenty of medicine in it.

The summit of this mountain is 15,558 feet above the sea, and the excessive rarefaction of the air renders breathing difficult; but as the ascent is commenced from only a thousand feet below the top, and the road leads up by an easy gradient, the Tibetans do not realise its great altitude, and being quite unable to comprehend the sensations they experience, attribute them to noxious vapours, or other causes, and call the mountain a medicine mountain.

The sun was shining brightly, and there was a gentle breeze; as we marched across the little plain, the road lay along turf, which was simply perfect for a horse's foot; all nature smiled upon us, and I began to think that the happy valley of Rasselas must have been in Tibet!

About five miles up the valley, there was another Tang on the slope of the hill. Here two of our head muleteers met us, and going down on their knees, begged us to stop for the night, as they said the mules could never reach the next station. But notwithstanding the piteous way in which they cried 'La-so,' we pushed up towards the dreaded pass of Wang-gi-La. And now ominous clouds were gathering, rain soon began to fall, and a change came o'er the spirit of the scene. When we reached the summit a pitiless sleet was driving before a cutting wind, and there was a dreary view down a narrow valley, where the tops of the hills were shrouded in mists. It was very cold, and another degree or two would certainly have changed the sleet into snow. What a difference a couple of hours made in my estimate of things in general! Two hours before I had been living in a sort of heaven, and now the happy valleys had lost their charm, and a coal fire in an English house seemed infinitely more desirable.

I was in front with the 'Long-lived Gem,' who was as cheery as ever; his patience used to be sorely tried, leading two ponies amongst the rich herbage, for these animals appreciated only too highly the unwonted luxury of the food, and, stopping every other minute to pick a delicate mouthful, they dragged at the poor fellow's arms till I expected to find his shoulder dislocated; but Job himself could hardly have been more enduring, and he certainly was not a person of such a cheery disposition. Lower down, when the dismal sleet turned into rain, the Ma-Fu became alarmed for his boots, so taking them off, he put them where a Tibetan puts everything, in the capacious fold above his waist-belt, and trudged on in the cold and wet, as happy and smiling as ever.

The 'Double Gem' was not quite of such a bright temperament, though in every way a first-rate fellow. His only fault was that all day long he would mutter prayers to the goddess of mercy, in a droning kind of intonation, which became rather tedious to listen to.

There was nothing to vary the monotony of the march down the mountain; seven miles in the driving wind and rain brought us to the Chinese village of Ho-chŭ-ka, consisting of two or three miserable tenements, of which the house we stopped in was the best. This was a Governmental building, or Kung-Kuan, for the use of travelling officials, and as such, was of course out of repair, though the Margary proclamation had been posted for our edification. It was a wretched place, with but one small room; one end of this was occupied by the dais usual in a Chinese house; there was a hollow for a fire in the mud floor, but there was no window, and no hole for the escape of smoke; the roof leaked horribly, and the drops coming through brought large quantities of mud with them. Some wet sticks were brought, with which a little fire was made, that filled the room with pungent smoke.

But if our accommodation was not luxurious, we soon found that the tales, with which we had been frightened, of the impossibility of obtaining food, were, as usual, utterly untrue. First, some people brought dried fish; and immediately afterwards we were offered, for a rupee, twenty good fresh fish just out of the water, averaging about half a pound; another man brought us a dozen hen's eggs, and fifteen pigeon's eggs. Some mutton and a fowl of a certain age arrived as presents; the village produced one turnip, and two cabbages; Chin-Tai discovered some flour, and eventually we had a sumptuous repast.

Just before turning in, a cow strayed into the little courtyard, and apparently deeming my chair would provide an unhoped-for shelter from the mournful rain, endeavoured to make use of the rare opportunity, but was fortunately discovered before doing any other damage than the destruction of a paper lantern.

August 17.—The stream from the summit of Wang-gi-La falls at Ho-chŭ-ka into a river thirty yards broad; this river flows to the south, and in all probability falls into the Chin-Sha-Chiang, after being joined by the Lit'ang stream (or Li-chŭ). There were a number of small blue sea-gulls circling about, and a family of young divers with a parent bird.

We followed up the right bank of the river, for about seven miles, by a very good and level road, amongst rounded, smooth, grassy hills, without a single tree, until, at the only hut between Ho-chŭ-ka and Lit'ang, we left the river and ascended a small stream. A man came out of the hut to look at the unwonted sight, and a huge dog barked savagely, and tugged at his chain in a way that threatened to break it or tear it from its fastenings.

The road up the valley was good, and nowhere steep; we ascended through the same undulating, grassy country; great herds of cattle, and sheep, and good-looking ponies were browsing on the slopes, and the silence was occasionally broken by the whistle or cry of the herdsmen, or the deep bay of a dog belonging to one of the numerous encampments dotted over this magnificent plateau. As we proceeded, the rain clouds cleared off, and the sun now and then shone out in fitful gleams. The people say that here, as well as in the Lit'ang plain, it rains every afternoon in the summer, but that the mornings are generally fine. The pass Mount Shie-gi-La (14,425 feet above the sea) is only 1,170 feet above Ho-chŭ-ka, and the ascent to it is gradual and very easy. From here gentle slopes lead down about seven hundred feet to the plain. This is from eight to ten miles wide, and stretches out for many miles east and west. Opposite, a range of hills bounds the plain; behind it rises the magnificent range of the Surong Mountains, stretching as far as the eye can see to the east and west, snowy peak rising behind snowy peak, where even at that great distance vast fields of snow almost dazzle the eye as the sun shines on them.

A river winds through the centre of the valley, numerous streams run down from the mountains on each side ; at this season of the year, when covered with luxuriant grass and wild flowers, one can hardly regret that the excessive cold prevents anything else from growing; even the pines are wanting here, and since leaving the huts at Cha-ma-ra-don we had scarcely seen a tree. In fact, so unfruitful is the plain, that the people actually are obliged to use butter for lighting their rooms, as it is cheaper than either oil or pine splinters.

From the pass of Shie-gi-La, the house of the native chief is visible, but the town of Lit'ang lies back on the slope of a spur, and cannot be seen until close to it. We soon descended to the plain, where particles of gold were glittering in the sand in the watercourses, and by the road-side ; and after a march of seventeen miles Lit'ang came into view.

It is a cheerless place, and one of the highest cities in the world, situated at an altitude of 13,280 feet above the sea;[5] no cereals of any kind, nor potatoes, can be raised. Just round the houses a few half-starved cabbages and miserable turnips appear to be the only things that can be produced.

Although there are only a thousand families in the city, there is a Lamassery within the walls containing three thousand Lamas, and not five miles away, another of nearly equal size. Notwithstanding the miserable poverty of the people, the Lamassery in Lit'ang is adorned with a gilded roof, that cost an immense sum of money. The roofs of all the Lamasseries that I have seen are gabled like the Chinese roofs, and those at Lit'ang are no exception.

Huc finds that Lit'ang means the 'Plain of Copper,'

[5] Potosi is 13,330 feet above the sea.

but I could hear of no such interpretation. There are three hundred Tibetan and ninety-eight Chinese soldiers in the neighbourhood, under the command of a Shou-Pei.

August 18.—The Chinese magistrate, a Liang-Tai, called upon us in the morning: an uninteresting person, who said little; and after him the Shou-Pei came, who told us that he had met Baron von Richthofen near Ch'ing-Ch'i. We received presents from these, and from the first hereditary chief and the second native chief, called by the Chinese Chang-Tu-Sze and Fu-Tu-Sze.

My servant, Huang-Fu, was getting weaker every day; his illness had now developed into decided dysentery, but it was little use giving him medicine, as he took no care of himself, and was hopelessly indulgent in his diet; here, because liquor was cheap, he treated himself to large quantities of the fiery native spirit, and was naturally very much the worse. It was impossible to leave the poor fellow behind, for there was no one to look after him, although he was now so weak that it was necessary to lift him on to his horse, but once mounted, he would sit all day quite contentedly, and as long as he could have his beloved pipe, he never complained or murmured at his hard lot.

CHAPTER V.

THE GREAT PLATEAU.

II. LIT'ANG TO BAT'ANG.

Departure—Quit the Lit'ang Plain—To-Tang—Dreary Aspects—Desolate but Healthy Region—Lassa Pilgrims—Characteristic Plant—High pass of Nga-ra-la-Ka—Cold Granite Region—Alarms of Robbers—Dzong-da—Pine-clad Valleys again—Lama-ya—The Cairns and Mani Inscriptions—The Man-ga-La—Stupendous Alpine Scene—The vast snowy Peak of Nen-da—Tibetan mode of building—Surpassing Scenery—The 'Yün-nan Bridge'—Charming halt at Rati, or San-Pa—Reminiscences of Huc and Gabet's Journey—The Guide's head turned with Wonder—Sublime aspect of the Mountains—Rhododendrons—Beautiful Valley of Pines—Unpeopled Tract—Ta-shiu—Manifold Uses of Raw Hide—Striking Picture—The J'ra-ka-La—Savage Desolation—Our Guide—Our Quarters at Pung-cha-mu—Descent towards the Chin-Sha River—Wild Independent Tibetans—Ba-jung-shih—Welcome on Arrival at Bat'ang—Chao the Magistrate, and Favourable Impressions—Abbé Desgodins—Discharge of our Suite—The Name of Bat'ang—The Plain—Possibilities of Navigation on the Rivers—Earthquakes—The Great Lamassery—Number of such—Causes that fill them—The Lamas a Curse to the Country—Eccentric Chinese System of Accounts—Property in Alpine Pastures—Slavery and Cattle-holding—Gradual Depopulation of Tibet—The Tibetan Chief of Bat'ang—Hostility of the Lamas—Alleged muster to bar our Advance—Quarters at Bat'ang—Cordial Hospitalities of Chinese Officials—Lively Banquet at Shou's—Noisy Game of Morra—The Wondrous Bath—A Night Alarm.

August 19.—The road from Lit'ang was said to be infested with robbers, and we were furnished with an escort of twelve Tibetan soldiers—the men who had come with us from Ho-K'ou leaving us. In accordance with what the people said was the custom of the place, it rained all the morning, as we marched over the low ridges thrown out into the plain from the mountains on the northern side. This plain is a favourite summer resort of the Tibetans, and numerous encampments of the black tents were dotted about. Immense herds of cattle and sheep were browsing around them, and the quiet was broken by the deep bay of the watch-dogs.

Four and a half miles from Lit'ang a spur runs out to the road, and on its summit, about a mile distant, there is a large Lamassery. Close to the road, on the crest of the same low spur, there is a building containing hot sulphur baths. The water from these runs in a natural, underground channel, whence gas bubbles up through the crevices of the sandstone; and at the end of the spur steam may be seen issuing from the ground.

The river Li-chŭ, which joins the Chin-Sha somewhere in the neighbourhood of Li-Kiang-Fu, is about forty yards wide, and is crossed by a bridge called Chi-zom-ka; this is in four spans, the piles being of loose stones encased in timber. It is not a structure of any great strength, and no one ventures to ride over it. At the other side is a Chinese To-Tang, and here our escort changed horses.

We now left the Lit'ang plain, and, striking up a small stream, ascended a valley, where we discovered that the mountains on this side (the southern side) are all of granite. Very bare, dreary, and desolate they looked; the ground strewn with great boulders, which made travelling anything but pleasant. As we ascended, the rain, which had been falling on and off all the morning, turned into sleet and hail, which did not tend to enliven the proceedings. We followed up a stream with low undulations, not more than a hundred feet above us, all covered with loose stones; and after winding about a little, over stony ridges, and equally stony valleys, the road steadily rising all the way, we arrived at a To-Tang, where there were a few Tibetan soldiers.

The day was miserably damp and chilly, the steady rain only being varied by heavy squalls of hail; and although the place was half in ruins, and there were no coverings to the windows or doorways, we were glad enough to shelter ourselves from the inclement weather in the two rooms set apart for travelling officials. The place is called Jiom-but'ang, or the Flat Plain, and is one thousand four hundred and thirty eight feet above Lit'ang—five hundred feet higher than the summit of Shie-gi-La.

August 20.—I got up soon after four, and found it required some resolution to plunge into my bath. There was a damp mist hanging over everything outside, and the place looked dismal enough; but by and by the fog lifted, and I discovered that there was a plain a hundred yards wide and a mile long; and as it was a flat place where no other flat places existed for some miles, its name was not altogether unsuitable. It was an unfertile spot, however, and the animals found but little to pick at amongst the big stones that were scattered about. But even here, wherever there was grass, wild flowers grew profusely amongst it. The flat place ran about north-east and south-west, and was bounded on the north-west by a steep and broken ridge, fifty to a hundred feet high, quite bare, and covered with loose granite boulders and stones. On the other side a similar ridge ran up, on which, however, there was a little grass. A sprinkling of snow had fallen in the night on the tops of some of the mountains, and the white covering now looked very close. It was not a cheerful scene, as the dank vapours hung about and threatened to descend in the form of hail, rain, or something disagreeable.

The yard of the inn was in such a filthy state that it was almost impossible to move without sinking knee-deep in mud; and women as dirty as the yard, with clothes no cleaner, but covered with strings of beads and huge silver ornaments, were paddling with bare feet and legs in the black slime, engaged in various menial offices.

We were warned that, as there were many robbers about, we must all keep close together, with our mules, chair, horses, and horsemen. We had a large escort—some in front, some behind, and others in the middle. One man was carrying a red banner, on which prayers and invocations were so closely written that if only one half of them had been listened to by Buddha, it would have been perfectly safe to walk into a den of thieves.

Our sick people were all better to-day. A chair-coolie, who at Cha-ma-ra-don had lain flat down on the ground, swearing he could not move, was now quite sprightly; and Huang-Fu, with his pipe, was able to mount his horse without assistance.

On the summit of a ridge we met a party of pilgrims on their way to Lassa. There were men and women, each carrying a spear, and trudging mournfully and slowly over the rocky path.

As the sun got up the mists rose, and there was every promise of a fine day, or at least an absence of rain. The road, though somewhat stony, was not very bad; nor was the ascent very steep; but the scenery well merited the epithet 'desolate' given it by Cooper:—rough undulating ground, in every direction covered with loose stones and huge masses of granite, of some twelve or fourteen feet cube; low hills, backed by jagged peaks, their tops covered with a sprinkling of snow, but not sufficient to hide the barrenness and nakedness of the rock beneath. A plant was growing in the hollows, eminently characteristic of the scenery in all Tibetan plains. It runs up a straight stem, some two feet high. The leaves on it grow downwards and fold back over one another; and a number of these growing quite straight out of the ground looked more than anything else like the cleaners for the glass chimneys of oil lamps used in England. The people tear the outside skin off and suck the centre of the stick. It tastes something like sorrel.

Four and a half miles of march brought us to a plain called Nga-ra-la-ka, where, at the foot of a couple of ragged peaks, there is a little pond about a hundred yards across. A robbery had been committed two or three days previously near this place. The robbers stripped the unfortunate victim, and, being in a hurry to get the ornaments off his plait, cut it off to save time.

Another mile brought us to the dreadful summit of Nga-ra-la-ka, 15,753 feet above the sea. The mules were a few hundred yards ahead of us, and we heard the muleteers set up a shout of joy as they gained the highest point. They say that in foggy weather people often swoon here. Ting-Ko, the boy we picked up at Ta-Chien-Lu, seemed to feel the rarefaction of the air very much, and could hardly drag himself along. Here and there, just at the top, there were a few patches of snow lying in the road, but they were very small.

After passing the crests we descended over the same dreary wastes of huge granite blocks. All this mountain mass is of a very hard, whitish-grey granite, and is much colder than the sandstone ridge at the other side of the Lit'ang plain. A good many skulls of oxen were lying about here, and it can be no matter for surprise that great numbers perish in the winter months, when the whole place is deep in snow. There are no poles and no cairns to indicate the path. It must then be a matter of the greatest difficulty to find or keep the road; and a wretched animal, stumbling between two boulders, each as big as a small room, would have little chance of escape.

A little below the summit we found a pond, or small lake, about half a mile long, dignified with the name of Cho-Din, or The Sea.[1] We were told that some robbers had slept here the night before, and gone off before daybreak. They said they were robbers because they were not known; they had no visible occupation; and they went away very early—three certain indications of bad characters.

The native chief of Lit'ang had sent off parties of soldiers to scour the hills in all directions, directly he had heard of our approach. This was owing to the attentions of Pao at Ta-Chien-Lu, who was an excellent magistrate, and had sent most stringent orders regarding us. He had been at one time Liang-Tai at Lit'ang, and when he first came he took such active measures, and made such severe examples of the first robbers he caught, that during the rest of his term no more brigandage was heard of in his district; and though we were no longer within his jurisdiction, his name was still so highly respected that we were well taken care of.

We halted for breakfast at a place called Dzong-Dā, which means 'dry sea.' It is eight hundred and fifty-seven feet below the summit, and is a sort of marsh in a valley two hundred yards wide, and one

---

[1] *Cho*, or rather *Thso*, is 'a lake,' and is attached to the names of all the great lakes of Western Tibet, e.g. Thso-Pangong, Thso-Langak, Thso-Mapham,—the last two the Rákas Tál and Manasaráwar of the Hindus.—*Y*.

mile long, running up to the east, between low granite ridges, covered with loose stones. Here we were astonished by a swarm of mosquitoes, most unexpected assailants, for we had not seen a mosquito for weeks.

Dzong-Dā consisted of no more than one hut, a Tang, and we left it to march over the same granite waste for four miles, when, after crossing a stream, we suddenly struck the sandstone, and the scene changed as if by magic. We again entered the rounded grassy hills, and a little lower, descending a stream, the pine-clad valleys appeared, and the lovely landscape that had charmed our eyes in the sandstone at the other side of Lit'ang was again spread out before us. We descended rapidly, and every now and then caught a glimpse of a grand snowy range in the distance; the sun came out, the air was delicious, and the afternoon ride was most enjoyable.

Thirteen and a half miles from Dzong-Dā, the stream we were following was joined by another equal in size, the two together forming a fair-sized river; and here the welcome sight of barley met our eyes, the first cultivation we had seen for a long time; and another mile and a half brought us to La-ma-ya.

The house we stopped in was built by Pao, when he was Liang-Tai at Lit'ang. There had been an upper story, the lower rooms being for servants only, but when Pao left it fell into disrepair, and the upper story gradually disappeared. The present Liang-Tai, although he had money for the purpose, did not choose to spend it, and so the house was now in a half-ruined state.

August 21.—We left the river we had been descending, and striking up a narrow gorge, through which a little stream was trickling, again turned our backs on cultivation and trees. The road was very fair, though a little stony in places, and the ascent to the summit, through sloping hills covered with beautiful grass, was neither long nor difficult. We were still in the sandstone, and the hills were smooth and rounded. The pass is called Yi-la-ka, and is 14,246 feet above the sea.

Looking over the valley to the mountains on the other side, we caught a glimpse every now and then of a magnificent snowfield, as the clouds came and went across it. Down below us the road was marked out by the familiar religious cairns, immense piles, with 'Om Mani Pemi Hom' roughly engraved on every stone; and from a hill on which we were standing, these heaps appeared like some gigantic serpent twisting through the valley.

From here a zigzag took us down to another fine grassy plain, and at the bottom we again came upon the granite. The river Dzeh-dzang-chŭ winds through the valley, its water of that peculiar bluish-white, by which snow water can almost always be recognised. The river is crossed by a bridge near some hot springs called Cha-chŭ ka, and a road from here leads by a seven days' journey to the Chung-Tien district. The granite lies only in the bottom of this valley, for at the other side the hills are again of sandstone, with a great deal of quartz and friable slaty shale. I could see no difference in the Flora of the two systems, and, as for the Fauna, with the exception of the ever-present magpies and crows, there appeared to be absolutely none.

We had descended 884 feet, and now went up 250 feet to a pass called Man-ga-La; down again a little way, and once more up we found ourselves on another of these magnificent grassy plateaus, with a splendid panorama of snowy mountains.

Our escort halted here a little while, and we spent the time in getting the names of the different peaks; but the natives are so ignorant of these, that a mountain seen from one point can hardly be identified from another. Nen-da, and Gombo-kung-ka appeared to be the names of the two highest; of the former we were afterwards thoroughly satisfied, but the other remained doubtful. Here enormous fields of snow seemed close to us, huge icy pinnacles frowned above us, and we could not wonder at the superstitions engendered in the ignorant minds of those who live amongst these scenes. In the icy breaths wafted from that pure expanse of dazzling white, imagination could hardly fail to feel the presence of the spirit of the frost and snow, or in the fitful gusts that murmured through the gullies to hear the rustle of that spirit's wings. The ice-blue water of the stream below, as it dashed over its rocky bed, seemed to leap for joy at its escape from the frosty trammels that had bound it, and the spray that broke from the rocks in sparkling gems seemed in the very wantonness of mirth to cast defiance at the hoary giants above. It was a scene never to be forgotten. and we both gazed long, with mingled feelings of wonder and admiration. But time is inexorable—our journey was not yet finished, and mounting our ponies we continued our march up the Nen-chŭ river.

A little farther on the village of Le-ka-ndo nestled by the side of the stream. It consisted only of two or three houses, one of which appeared a remarkably good one, well built with strong walls of stone. Here again barley was growing round the village, and close down to the stream, but there was none farther up, though as we ascended the valley there was plenty of ground well fitted for the purpose. In the valley the usual wild currants, gooseberries, barberries, willows, and other shrubs were growing. Pine forests clothed the hills, and, as everywhere in the sandstone formations, the same beautiful grass and wild flowers. A couple of ruined villages were passed on the way, signs of the troubles in times not so very long ago, and after a march of little more than fourteen miles, we found another house built by Pao, precisely similar to that at La-ma-ya. These places are kept up for officials when they travel, and correspond to travellers' bungalows in India.

From here we had a glorious view of Mount Nen-da, and as the setting sun cast its last ray on the summit, I could well appreciate the solemn beauty of the scene. No words can describe the majestic grandeur of that mighty peak, whose giant mass of eternal snow and ice raises its glorious head seven thousand feet above the wondering traveller, who yet stands within five miles of its summit. He can but gaze with admiration, and appreciate the feelings of the Tibetans, that have led them to call it Nen-Da, or The Sacred Mountain.

I had intended to have made a sketch of this unrivalled peak, but I left it till the next morning, when I knew that I should have some spare time, and descended from the housetop to the room below. The smoke inside was almost insupportable, and it required all my resolution to settle down to write. It is a marvel that people who burn fires in their houses during ten months out of the twelve have not discovered the use of a chimney; but the most luxurious do no more than make a hole in the roof, and often there is no means of egress for the smoke but by the window or door. The smoke of the green wood, which is the only fuel, is most pungent, and after sitting in it for any length of time, it becomes absolutely painful to the eyes. The Tibetans, also, are hopelessly ignorant of the art of stone-dressing. The walls of the houses are made of stone, mostly flat pieces of the slaty shale,—so abundant in the sandstone,—stuck together with mud; but all the framework and party walls are of wood, and that even at Lit'ang, where there is plenty of splendid granite quite close, and where timber must be very expensive.

August 22.—The mules had been turned out to graze on the delicious pastures, and it was some time before they were captured and ready for a start. So I strolled out in the early morning and enjoyed the delicious air and lovely view. Clouds had gathered round the noble brow of Mount Nen-da, but vast snowfields still were visible, and enhanced the beauty of the little hamlet that shares its name.

It is on a little triangular plateau of grass; bounded on the south by the stream that separates Nen-da from its giant neighbours, it extends a mile and a half to the north, to the beautiful undulating grassy hills, and on the third side is the river Nen-chŭ.

Our escort came with us no farther than this, for they said we were now beyond the haunts of the bandits, whose existence was sufficiently real to prevent them returning by daylight.

The snowy air here must have been very healthful: Huang-Fu had now almost recovered; the mules seemed fresh and ready for another long march after their luxuriant repast; and all were in good spirits as we left the little village amidst the respectful salutes of our escort.

Marching up this valley, scene after scene of loveliness meets the unwearied eye. Forests of noble pines, grassy slopes and level plains, covered with sweet-scented flowers, where Nature, in one of her most lavish moods, seems to have compensated by the wonderful beauty of the scenery for the short duration of the summer.

The road ran along the side of the river, generally fifty feet to a hundred and fifty feet above it; but now and then descending to the water's edge; in one or two places passing over gentle grassy banks. The road was of the finest gravel, and the slopes dotted with pines and yews of the most delicate forms, grouped by Nature singly and in clumps in a way the hand of man has never rivalled.

Every now and then a valley opening on our right disclosed the vast snowfields of Mount Nen-Da, and with my glasses I could discern the blue glint of the ice. On the western side, which we saw at the end of the day, the snowfields seemed unlimited. With its spurs it covered the length of our day's march, and its summit is 20,500 feet above the sea.

Seven miles from Nen-Da there is a bridge across the river called the Yün-Nan Chiao (Yün-Nan Bridge), so named because it was built by Wu-San-Kwei, a general known as 'Pacificator of the West.' This hero, in the reign of the last emperor of the last dynasty, was a general in the army in the province of Chi-Li. Quartered somewhere in the neighbourhood of Shan-Hai-Kuan, it was he who invited the Manchu Tatars into the empire, upset the last dynasty, and established the present one. When the first emperor found himself firmly seated on the throne he sent Wu-San-Kwei, with the title of King Pacificator of the West, into Yün-Nan and the neighbouring provinces, where there were great disturbances. Coming here from Yün-Nan, to chastise the rebels, he called this the Yün-Nan Bridge. He was justly a celebrated man, and the blame he received from his friends for not taking the imperial power himself will not be awarded him by the unbiassed historian.

A little higher up the valley there is the print of his horse's feet in the rock. This is now carefully covered over with stones to protect it from the weather.

For eleven miles we marched through a sandstone formation, after which the rocks were of a rotten, friable granite; and here again the scenery began to change, though not so suddenly as before. After about fourteen miles we emerged into a wide, undulating plateau, with great loose boulders lying about. The grass was still luxuriant, but the trees had disappeared, and the scenery was totally different from that of the more temperate sandstone.

We halted at Ra-ti, the Tsanba of Cooper. The Chinese call this San-Pa, or The Three Plains. It is situated at the end of a charming little green plain about two miles long, with a width in the widest part of a little less. The village consists of eight families, living in two or three houses; but in the plain there is a large nomad population in black tents, who feed their cattle in the level valley, and on the sides of the gently sloping ridges that enclose it. These wandering shepherds remain here during the short summer, but at the approach of winter move lower down to some less rigorous climate. It is believed that more than half the population of Tibet live in these black tents; but giving this due consideration the population must be exceedingly sparse.

In our marches we rarely passed a habitation of any kind between the villages. The generality of these contained no more than ten or a dozen families, and the largest eighty, or a hundred families at most.

There were some Chinese at Ra-ti, and we stopped in a new Chinese house, built of wood, very roughly put together.

Here we met an old man who told us that he had formerly been a soldier, and that he had been one of an escort that had conducted a foreigner from Ch'êng-Tu to Si-Ngan-Fu in the year 1848. He said that there was only one foreigner, that he was a big man with a large nose, and that he had come from Tibet under an escort. We at first thought that the foreigner with the big nose must have been Gabet, for our old friend might well have taken Huc for a Chinaman; and he might easily have made a mistake about the date, as of course he spoke entirely from memory. But Huc and Gabet were taken to Macao, and not to Si-Ngan-Fu; and as this was a discrepancy we could not account for, we could only come to the conclusion that the foreigner had not come from Tibet, but that as those were the days of great persecutions some other missionary had been taken to Si-Ngan-Fu. We were loth to give up the idea that we had stumbled on traces of Huc and Gabet, especially at San-pa, the scene of the death of Ly-Kouo-Ngan, the Pacificator of Kingdoms.[2]

The old Chinaman told us, almost as Huc narrates it,[3] the story of the king of Tibet, who poisoned the three Dalai Lamas, and who was afterwards overthrown by the Chinese ambassador. He added, however, what Huc does not mention, that the ambas-

[2] *Souvenirs d'un Voyage*, vol. ii. chap. x. p. 501.
[3] *Ibid.*, vol. ii. chap. vi.

sador collected a vast amount of loot at the sacking of the king's palace; and it must have been some of this that the great Ki-Chan (Keshen of our English negotiations at Canton) confided to the adventurous Abbé.[4]

August 23.—Mesny bought a good strong grey pony here for fifty rupees, and left his other behind, as the poor brute was so lame that it could hardly move. This was the beginning of our troubles—for after this pony after pony succumbed to the long and continued marches.

It was a miserable morning and raining heavily as we left the village of the Three Plains, and though it cleared up a little soon afterwards we were enveloped in a Scotch mist nearly all day. After marching for an hour and three quarters over an excessively dreary plateau, where granite blocks strewed the ground in every direction, and where a few small shrubs, gooseberries, and currants, and a dwarf yew, grew amongst the stones, we reached the beginning of the final zigzag, and after ascending for another quarter of an hour by a path that was not very steep, we gained the summit of Rung-Se-La, or San-Pa-Shan, 15,769 feet above the sea. From here we had a magnificent view of fog in every direction; and the distance we could see was nearly a hundred yards all round!

We had a guide with us who had never seen a foreigner before. He was a half-breed, and rode a pony that had been mauled in the flank by a wolf. He was so much interested in us that he rode the whole way with his head turned round, and left the pony to find the road, which it did admirably. I at

---

[4] *Souvenirs d'un Voyage*, vol. ii. chap. viii.

SUBLIME ASPECT OF THE MOUNTAINS. 173

last began to think the guide was riding backwards. I was wrong, however: he was sitting the proper way, and none of my anticipations as to his head falling off came true. They say there are great numbers of enormous wolves here; and in the forests, on the western side, there are (so they say) every kind of wild beast—tigers, panthers, bears, wolves, and monkeys. The descent of the mountain was much more difficult than the ascent; but after three miles we escaped from the mist, and our toils were forgotten in the wild scene that lay before us. Bare crags towered above a sea of pines, and a weird forest of naked and blackened trunks seemed like the relics of some huge strife of the elements; indeed, it was not difficult to fancy the fierce conflict still being waged; and it only wanted the crash of thunder to complete the illusion. To the left a vast forest of pines rolled up the mountain-side, as though to storm its summit; but far above the highest, and laughing to scorn their efforts, the grim and savage rocks rose high towards the heavens. Thousands of dead stems in the van of the attack looked like the victims of this furious combat, whilst down below myriads of mighty pines seemed marshalling their hosts for a renewed assault. On the other side green grassy slopes looked calmly on at the desperate battle, whilst right in front a gigantic wall of rock, towering up nearly perpendicularly, as though ready to hurl itself into the fray, reared its stupendous head into the clouds that sometimes swept across its summit. Not ten yards from us, the blackened stems of two colossal pines twisted their withered branches into all sorts of fantastic shapes, standing like spectre sentinels over the struggle.

We gazed some time on the magnificent scene, and then, descending a steep and rocky path, plunged into the dense forest.

Here there were vast numbers of rhododendrons, called by the Tibetans 'Ta-ma.' After much inquiry I elicited a Chinese name, but as in all probability it was invented expressly for me, I did not put much faith in the title Yang-Ko-Chai.[5] There were also great quantities of the holly-leaved oak.

This is the forest of which Huc speaks in such raptures as the most beautiful he had seen in the mountains of Tibet [6]; but in enumerating the trees he miscalls the holly-leaved oak a holly,—a mistake very easy to fall into when acorns are not to be found.

When we arrived at the end of the valley where the stream joins the river, there was a beautiful open glade, where a sort of monument had been built in honour of the mountain. Here our coolies and people lit a fire under a tree and cooked their tsanba. It was a most picturesque group, though the pleasure was somewhat marred by the rain, which again came down heavily.

From here we marched up what was, perhaps, the most beautiful of all the valleys we had been in. It differed, too, from the others; for on one side there was a range of bare and ragged rocks behind the grassy hills, and on the other, tremendous and steep mountains frowned upon the glen. A dense pine forest covered their lower slopes, whilst up above crags and pinnacles of the most unwonted shapes crowned great precipices of bare rock. Here and there the pines came down and formed a belt across the green valley.

---

[5] The Chinese for Rhododendron is 'Chia-Chu-T'ao.'
[6] *Souvenirs d'un Voyage*, vol. ii. p. 499.

We marched through it very rapidly, the road running in almost a straight line due north to the village of Ta-shiu, which the Chinese call Ta-So, the first houses we had seen all day—for since Ra-ti we had not passed the smallest hut, and not even the vestige of a ruin. There were no black tents of the nomads, and between the two villages, a distance of twenty-four miles, there is absolutely no population. Here we took up our quarters at the foot of the Ta-So, or Ta-shiu range, the last of the mountains before Bat'ang. And here I received a letter from Monsieur Desgodins, with the welcome intelligence that the chief magistrate of Bat'ang had already arranged for mules and horses, and that we should not be unnecessarily delayed.

The house we stopped in—if house it could be called—was like a disused battery on a rainy day. The place did not possess a window, and looking out at the door a narrow path, about a foot deep in sludge and black mud, led between two rows of tumble-down huts built of roughly-squared logs, their roofs covered with timber and earth. Their size, their construction, the nature of the aperture doing duty for a door, and their general appearance, were just like field powder-magazines. These were the outbuildings of the official residence and best house in the village of Ta-shiu—an important place, for there were as many as fifteen or sixteen families in it.

The grazing ground was so good here, and there was so little at Pun-jang-mu, our next halting-place, that the muleteers begged permission to let their mules feed till late in the day. They said 'La-so' very humbly, and, as if the request had not been granted they would have started late just the same, we made a virtue of necessity, and graciously gave the required consent.

August 24.—The animals, however, were ready earlier than we had expected, and the men came in with their raw hide thongs to pack the loads. In this country raw hide takes the place of the bamboo of Southern China. Everything is done with raw hide and thongs of raw hide. When the doors tumble to pieces—as they always do—they are looped up with raw hide thongs. If the legs of a table are all loose—and I never found them otherwise—they are secured in the same way; and it was our impression that if there were any surgeons in Tibet, they would tie on the arms of a man, if he had lost them, with raw hide thongs. They become wonderfully skilful with these things. A muleteer takes a box, he passes his thong round it once or twice, and with a turn of his wrist, it is lashed up in a manner that would puzzle anyone but a Tibetan muleteer or a Davenport Brother to undo. Leather is unknown, because tanning is an expensive process, requiring more capital than a Tibetan can lay out.

We left the beautiful valley, and plunged at once into a wild gorge, the tops of the hills broken into jagged points of the most fantastic shapes. We found a capital road, the ascent was not difficult, and we soon left the trees behind us, the last to defy the severities of these tremendous altitudes being a few bold and hardy yews.

The valley and hill-sides were strewn with stones and rocks, amongst which the usual quantities of wild flowers of every colour were growing, and after six and three quarter miles we found ourselves in a little circular basin, about a hundred yards in diameter, surrounded on all sides, except that by which we had come, by steep and ragged precipices, three hundred feet high. At the bottom there was a little pond of clear water; no opening was anywhere visible in the savage walls of rock, but up one side a desperately steep and rough zigzag led to the top.

As this basin was nicely sheltered, and only about five hundred feet from the summit, I determined to use the hypsometer here, and in the wondering gaze of guides and soldiers I set to work. The scene would have made a splendid picture:—the wild surroundings of bare rocks, and the still more wild-looking fellows grouped about, with their tall felt hats, their sword scabbards set with coral and turquoise, and long matchlocks, with prongs at the end of the barrel; Mesny with a long scarlet cloak reaching almost to the ground, the ponies with their queer saddles covered with felts and sheepskins, and the transparent water of the little pond reflecting the proceedings. We were 16,129 feet above the sea, and the summit of the pass was 540 feet above us.

The rocks here were full of iron, and affected the compass, how much I could not tell. I took a bearing as an experiment close to a rock, and, moving only a few yards, found a difference of a degree and a half. The tops of the crags were all yellow with iron, which sometimes produced remarkable effects.

The Tibetan name for this mountain is J'Ra-ka-La, but strange to say this is almost forgotten, and it is usually spoken of as the Ta-So-Shan; even Pehma and the Ma-Fu, whom we had questioned on the subject, had forgotten the native name, and it was not until after some conversation with the muleteers that they recollected it.

We reached the summit without much difficulty;

there was no snow anywhere visible, nor was there any view, as the mountains were all shrouded in heavy clouds. Here, on the razor-like edge of the ridge, there was a pile of stones; the pious of our party added to the heap, and knocked their heads in thankfulness to Buddha for the dangers happily passed.

Just over the crest of the pass there is a great basin two miles in diameter, and such a wild and savage scene I never before looked on—a very abomination of desolation. Great masses of bare rock rising all round; their tops perpendicular, torn and rent into every conceivable shape by the rigour of the climate. Long slopes of *débris* that had fallen from these were at the bottom, and great blocks of rock, scattered over the flat of the basin, lay tumbled about in most awful confusion amongst the masses that cropped out from below the surface; three or four small ponds formed in the hollows were the sources of the stream that, descending from the basin, plunged into another valley, and falling rapidly, soon became a roaring torrent, dashing through mile after mile of dense pine forest. The stillness of this place was very remarkable. The air was so rarefied that I could hardly hear the horses' feet only a few yards off, and when quite out of hearing of these, as I walked on alone, the silence was most impressive.

The road began badly, over the rugged stones of this desolate spot. It went on worse, as, descending sharply, it plunged into the enormous pine forest of which we now only saw the commencement, and it ended worst of all in a sea of black mud spread over the same unpleasant masses of rock.

The guide, who rode ahead, was a very remarkable figure. He had no head-dress whatever, and his hair fell in tangled locks over his shoulders. He had a very long nose, like many of the people here—a great contrast to the small features of the Chinamen. He had no hair on his face, and he was dressed in one garment of coarse sacking. A long matchlock with prongs and a huge cooking cauldron were slung at his back; the end of a coral-mounted sword projected from his clothes. His little pony was covered with felts and sheepskins, and at each side of the saddle were two great sacks. All the way down the 'Long-lived Gem' walked beside him, and narrated some wonderful stories, probably about ourselves, and his exclamations of surprise were continued. 'Arī-ī-ī,' he would say, dwelling on the final ī, and drawing it out for nearly a minute. Then, as evidently the Ma-Fu made some more than usual astounding statement, the guide would turn his head, and look at us with wide open eyes, and exclaim 'Eh-h-h-h, ī-ī-ī-ī.'

At two o'clock we came to some fields of barley, surrounding the village of Pung-cha-mu, or Pun-Jang-Mu. The Chinese are worse than the English, and as bad as the French, at getting hold of foreign names. Mu-lung-gung they pronounce Pou-Lang-Kung; this place they write and say Pun-Jang-Mu, and Ta-shiu in their mouths becomes Ta-So.

If the house we stayed in at Ta-shiu was like a battery before a siege, the Kung-Kuan at Pung-cha-mu was like the same work after a severe bombardment. This place was very Chinese, and a fearfully heavy roof had been placed on walls without foundations; everything had given way, and nothing was perpendicular, except what ought to have been sloping. Logs of wood, and planks, that once formed part of the building, encumbered the filthy entrance, which was feet deep in black slime. The buildings themselves, half toppling over, seemed only saved from a complete downfall by their excessive lowness; great beams, that served no architectural purpose whatever, thrust themselves out obtrusively in all directions, and helped to complete the likeness to the interior of a siege work. We were, however, well accustomed to curious places, and settled down philosophically, contented with a roof overhead, and much gratified at the unusual absence of smoke.

Huang-Fu was now getting quite strong again, but Peh-ma was sick to-day, with the invariable symptoms of pains all over him. I never knew one of my followers ill, who was not afflicted with this extraordinary ubiquity of dolours. I was not fond of practising unlearned experiments, even on the 'corpus vile' of a Tibetan, but Mesny courageously dosed our interpreter with a handful of pills, which, with a large proportion of faith, seemed to relieve his mind, if not his body.

August 25.—To my mind the forest we marched through from Pung-cha-mu was more deserving of Huc's panegyrics than that to which he applied them. The holly-leaved oaks were here fine trees, and the rhododendrons reached a considerable size; gooseberries and currants at first abounded, though, as we approached the Chin-Sha, these shrubs, which belong to the plateau rather than the valleys, soon disappeared. We were now leaving behind us the grassy upland, and descending steadily towards the valley of the River of Golden Sand. By degrees everything betokened the approach of a warmer climate; the parrots again darted from bough to bough, the rich grass and lavish profusion of wild flowers was no more seen, the houses were surrounded by patches of cultivation, and the more delicate trees were interspersed among the hardy pines and oaks.

This forest is said to be full of wild animals and monkeys. Soon after our arrival at Pung-cha-mu, a bear had ventured near to the village, killed one of our baggage animals, eaten half of it, and buried the remainder. Two Tibetans lay in wait all night over the buried portion, but the bear did not return to finish his meal.

The rocks here are of an exceedingly rich limestone; huge masses lie about looking like marble, and quantities of quartz.

Not far from Pung-cha-mu there is a model of a boat, fashioned by some magic hand in the rock, possessing the useful property of keeping travellers in the right path. The guide who was with us failed to find this misplaced vessel. Nevertheless, we did not lose our way, and had it not been for the solemn assurances of our conductor, I should have been inclined to attribute our good fortune more to the fact that it was quite impossible in the dense forest to wander from the path without being entangled in the thick undergrowth, than to any mysterious virtues of the ship of stone.

We were told before we started that a drove of two hundred oxen was expected from Bat'ang, and as there are parts of the road where one animal can by no possibility pass another, it was necessary to send a man on in advance to keep the narrow places clear. We met the caravan soon after leaving, but in a spot where the slopes above and below the narrow path were not too steep to prevent the lumbering beasts clambering out of our way.

They were driven by men from a part of Tibet beyond Bat'ang, who wore hats of felt, shaped very much like the ordinary high European hat. Our guide said that they were wild, independent fellows, with but little respect for authority of any kind, and ever ready with their knives to resent an injury, fancied or real. They always refuse to give their horses or oxen, even to the highest officials, without regular payment, and have been known to take summary vengeance on one who attempted to evade it. It was only for such exalted persons as ourselves that they would allow their animals to be driven off the path. The men of our escort were not slow to take advantage of the opportunity of a little brief authority; they showed but scant mercy, and the frightened beasts crashed amongst the thick undergrowth, some of them dropped their loads, and others only just saved themselves from rolling down the almost precipitous bank into the torrent we could indistinctly hear below.

The road was very fair; it followed the side of the hill, and descending much less rapidly than the brawling stream was soon six or seven hundred feet above it.

Six miles from Pung-cha-mu, looking down the valley in the north-west direction, in the far distance, there was a hill with a very remarkable knob at the top. The people said it was at a place three days' journey beyond Bat'ang, called Tang-ye, where they said there was a wild, independent tribe, who owned allegiance to neither Tibetans nor Chinese.

After a march of nine miles we reached the village of Ba-jung-shih, called by the Chinese Hsiao-Pa-Chung. By this time we had left the dense forest of pines and oak, and the hill-sides were rather precipitous, bare in places, or covered with a low scrub; while in the valley there were yellow-plum-trees, gooseberries, currants, and wild mint, the latter reaching a height of four or five feet. There is a hot

WELCOME ON ARRIVAL AT BAT'ANG.

spring close to this village, and three miles further the river enters a narrow gorge with precipices about one thousand feet high on both sides. The bed of the river is exceedingly steep, and the water falls and tumbles in cascades, much after the manner of the stream at Ta-Chien-Lu. The road now was execrable; cut out of the side of the rock, or propped up by a pole in the old familiar style, and sometimes running along the side of a very steep slope of loose *débris* brought down from the mountains, it helped to remind us that we had left the rolling plains.

As we descended, the road became worse, and staircase-like tracks, such as we had not seen since leaving Ta-Chien-Lu, sometimes compelled us to dismount. The sun now pouring down into the narrow valley, the exercise made the thick clothes we wore rather oppressive; everyone, therefore, commenced to discard his extra coverings; one garment after another was thrown off, till we could no longer recognise each other's figures.

At the outskirts of Bat'ang a few soldiers and official servants were drawn up, who kneeling on one knee bade us a respectful welcome, and presented the card of the chief magistrate, 'Chao' Ta-Laoye,[7] the Liang-Tai. Scarcely had this important ceremony been concluded than another party, not less numerous, saluted us in the name of Shou, the chief military official, a Tu-Ssu, and when these were passed a retinue of the native chief of Bat'ang announced the wishes of their liege lord that our stay in his territory might be propitious.

We were conducted straight to the house prepared for us, and we immediately received an exceedingly kind note from Monsieur Desgodins, accompanied by

[7] 'His Excellency Chao.' See *supra*, p. 97.

a loaf of excellent bread, a bottle of wine, and some peaches.

We had barely washed our hands when the Liang-Tai 'Chao' was announced, and the chief military officer with him. After complimenting me on my literary ability, he said we could hardly get away in two days, but that he would have everything ready if we remained three days. This announced rapidity of action was as unexpected as it was gratifying, and I could scarcely believe that the performance would equal the promise.

Chao was a man with a very agreeable countenance, bearing on it the signs of his active mind; unlike most Chinamen, both he and the military official here, cut their nails short. I remarked on this to Monsieur Desgodins, who answered, 'Ah! but they are real workers.' Chao always appeared to be at work, writing, or reading despatches, and his physical was almost as great as his mental activity. He was a remarkably small eater; I scarcely ever saw a man who ate less, and I had many opportunities of forming an opinion.

Monsieur Desgodins always spoke of him as a model magistrate, who endeavoured to deal fairly with all classes: altogether he was a remarkable man, and a bright contrast to the generality of Chinese officials.

After the usual complimentary questions regarding our honourable names and honourable ages, Chao asked if we had any rifled artillery in our portmanteaus. This question was not put as a joke, but meant in all seriousness. On our replying in the negative, he said, 'Ah! if you only could have given me one or two I soon would have made these Tibetans say "La-so"; now I often have to say "La-so" to them.' This unpremeditated question and remark did more to show the true nature of the relations between the Chinese and Tibetans than anything else I had seen or heard.

When he had left, the native chief came in. He was dressed in Chinese costume, and wore the red ball; he is a heavy-looking person, and does not look like a Tibetan in the least. His name is Loh-chung-wan-tun, and he has the Chinese rank of Sieh-Tai. After he had gone, we called on Monsieur Desgodins, who was living alone. He is a most interesting and intelligent man, and has made many valuable observations on the geography of a totally unknown country, which he has had rare opportunities of studying.

August 26.—The first business to be undertaken in the morning was the payment of the muleteers, chair coolies, Ma-Fus, and interpreter, and the presentation of gifts to the many soldiers who had accompanied us.

We had bargained with the muleteers to do the journey in twenty days, and as they had accomplished it in nineteen, they were well entitled to the extra payments that made their eyes glisten with delight, and they thrust out their tongues further than I could have deemed possible. They afterwards returned, bringing a present of a jar of spirit made from what is here called black wheat. In the process of manufacture, sticks of juniper, a shrub that grows in profusion on the mountains, are thrown in, and the taste of the liquor is very much like that of weak gin. When the gift had been accepted, they begged that if any other of our countrymen should pass this way, we would recommend them, and they would serve them as well as they had treated us.

The chair coolies were so pleased with their rewards, that they determined at once to take me further; this was a great convenience, as they had now fallen into my ways. The Ma-Fus, however, wished to return to their families, for which I was sorry, as they had both been exceedingly good servants; Peh-ma also returned to his little bit of land, though they all three seemed doubtful of reaching their houses in safety, on account of robbers. I subsequently learned that they would have been willing to come with us to Ta-Li-Fu, had it not been for Chin-Tai, whose overbearing manner they were unable to endue.

Now, every person who could possibly frame an excuse brought presents; first they came singly, then collectively, till I began to suspect they were like an army on the stage, and refused to pay for any more of the so-called gifts.

The Abbé Huc declares that the name Bat'ang means 'The Plain of Cows,' though in what language he does not say.[8]

The Tibetan name of the place is Ba, a word that has no meaning; the Chinese have added their favourite termination T'ang, which may mean either a place or a post station. The name 'The Plain of Cows' would certainly be inappropriate, for the plain, such as it is, is nearly entirely given up to cultivation.

There is a fable connected with the origin of the name, which is probably without any foundation of truth, but the existence of which clearly shows that

---

[8] Without presuming to have an opinion, let it be said, in justice to Huc, that Jaeschke's Tibetan Dictionary gives *Ba* or *Bha* = 'cow,' and *thang* = 'plain.'—Y.

the word 'Ba' can have no meaning. The story goes that once upon a time an old man and his wife wandering over the mountains with one sheep, their sole possession, in search of a habitation, came upon this place, and enchanted with its position, warmth, and fertility, decided to settle here, but could not think of a suitable name. Whilst they were discussing the question, the sheep began to bleat, and they agreed at once that the animal had decided the matter, and called it 'Ba.'

Bat'ang is well situated on a stream; it is about half a mile from the left bank of the river, and in a somewhat commanding position overlooks the plain.

The plain of Bat'ang, described by Huc as 'La magnifique, la ravissante plaine de Bat'ang,' lies at an altitude of 8,540 feet above the sea, but notwithstanding this, its sheltered position, and distance from the mountains of perpetual snow, render the climate temperate and agreeable. In January 1877 the lowest reading at six a.m. was + 10° F., but that was exceptional, the mean for the month at six o'clock being + 22° F. The mean for July 1877 at two p.m. was + 83° F., although on one occasion + 93° F. was registered at that hour by Monsieur Desgodins. At the time of our visit it was very warm, and the number of house-flies that swarmed in the houses, the streets, and even in the fields, was very remarkable, and equally disagreeable. The plain is not more than two to three miles long, and one to two miles wide, and is enclosed by almost bare and precipitous mountains. There is scarcely a tree to be seen in its length and breadth, and it is nearly altogether given up to the cultivation of wheat, black wheat, buckwheat, barley, and Indian corn. Before the earthquake in 1871 the vines of Bat'ang were celebrated, and the native chief used to make great quantities of wine; but since that year, the vines have either not been replanted, or they have not had time to grow, for now there are very few, and wine is made only in very small quantities. Mulberry-trees grow well in the vicinity, but silk is not manufactured, as killing the cocoon of the silkworm is a mortal sin.

The Bat'ang river, a rapid stream twenty-five yards wide, winds through the plain, and joins the Chin-Sha-Chiang five miles below the town. Neither river is navigated, though there are a few boats on the Chin-Sha, a little lower down, at a place called Niu-K'ou. These, however, are only used for local purposes, and do not venture more than a few miles. I subsequently had many opportunities of forming an opinion, and certainly the river Chin-Sha is, generally speaking, not navigable above Shi-Ku, though there are many long, broad, quiet reaches where boats could be used. The boats in use at Niu-K'ou are mostly made of about eight raw bullock hides stretched over a wooden framework; one man only sits in them, and he steers from the bow with a paddle. They descend the stream in this way, and are carried back by land, the boat being sufficiently light to be easily carried on a man's back. There are also a few wooden boats at Niu-K'ou, that can be tracked a short distance against the stream, but these only ascend a few miles.

In 1871 Bat'ang was visited by a frightful series of earthquakes, which, lasting over many weeks, devastated the whole neighbourhood. In the town itself not one house was left standing, and the loss of life was awful; there was not one family in which there was not one dead. The traces of this appalling calamity are still to be seen for many miles around this ill-fated town. The hill-sides are rent and torn, and huge slopes of *débris*, hurled from the mountains, have in many places buried and obliterated the ancient paths.

The town is now perfectly new, and every house is fresh; of these there are about two hundred, containing three hundred families, who are chiefly remarkable for their reputed immorality.

Close to the bank of the little river of Bat'ang, in the midst of the waving corn-fields, like the monks of old, the Lamas of Bat'ang have built their Lamassery, and sheltered by the golden roof that cost upwards of 1,000*l*. thirteen hundred Lamas live in idleness.

Lama is the Tibetan word for 'monk,' and means in their language 'superior person'; the French use the term 'Bonze'; and the Chinese in the west have another name, which, translated into English, means neither more nor less than 'criminals whose lives have been spared,' although the word in its application has lost somewhat of its signification. The story is that when the religion of Buddha was first established here, it was impossible to find inhabitants for the Lamasseries that were built, until they sent to these institutions all the criminals condemned to death. The Chinese now use this phrase for their priests, and see nothing incongruous in the epithet.

The number of Lamasseries throughout the country is astounding. At the small town of Ta-Chien-Lu there are three. At Lit'ang—a town of a thousand families—there are three thousand Lamas in the principal Lamassery; and outside the town is another building containing nearly as many. At Bat'ang, where there are only three hundred families, the Lamassery contains thirteen hundred Lamas. The traveller may march for days, passing by only a few straggling villages, containing at most ten or a dozen houses, and yet every now and then he is sure to hear of some huge Lamassery not far from his road.

Whatever may have been the difficulty in filling these institutions in the early days of the Buddhist religion, it is only too easy now.

Parents who have a son, or sons, with whom they can do nothing, or whom they cannot afford to maintain, send their useless offspring with a gift to the nearest Lamassery.

If a man gets into debt and cannot pay, he enters a Lamassery, where he is safe from any assault on the part of his creditor.

If anyone owes money to a Lamassery, as soon as the Lamas can get no more interest from him, they seize his land, he soon follows his possessions, and becomes a Lama.

The idle member of a family will turn Lama and enter a Lamassery ; he can then return to his relations for short periods of amusement or distraction, during which time he lives at their expense.

All those who, having committed crimes, wish to escape their deserved punishment, enter a Lamassery, and shelter themselves under the cloak of their assumed sanctity.

The Lamasseries are further peopled by the country-born children of the Chinese soldiers of the garrisons in Tibet. When these return to China, the foreign wives and children (see p. 86) are left behind, and the latter, in that case, generally enter a Lamassery.

Occasionally a Chinese soldier will take his wife

back with him ; but to do this requires an amount of moral courage not often found. For, instead of being admired for his constancy, he will meet with nothing but the gibes and sneers of his companions for his folly and ill taste in burdening himself with a barbarian woman and her children.

The Lamas and Lamasseries are enormously rich. They certainly possess the greater part of the cultivated land in the plain of Bat'ang, and now must own nearly half of the country. Their wealth is daily increased, partly by legacies—for a dying man generally leaves something to the neighbouring Lamassery—but still more by usury. Being the only people in the country who have any property, a man in want of money always applies to the Lamas, and then his fate is sealed, as surely as when some spendthrift in London commences dealings with the Jews.

The rate of interest they exact for loans, even when real property is mortgaged, is fatal to the borrower. Interest mounts up, and left unpaid, interest on interest, till at last, utterly crushed by the extortion of his creditors—his land gone, and with nothing left—the unfortunate debtor mortgages himself and his services for some temporary loan, and ultimately becomes a Lama.

The Lamas do not spend all their time in the Lamasseries entirely given up to devotion. On the contrary, theirs is a life of freedom. Whenever so inclined, they leave their Lamassery, return awhile to their families, or to almost any house they choose to enter, spend their days as they please, and take anything they fancy away with them.

The Lamas assist in no way in the maintenance of the State ; their lands are free from taxation ; they never lend their horses or animals for the public service, and do not pay one iota towards the Government expenses. They scarcely work in their fields themselves, as every Lamassery possesses hundreds of slaves. Thus the Lamas, by profession celibates, but in practice profligates, live in idleness and immorality —a curse to the country and the people.

The chiefdom of Bat'ang is governed by its native chief, under the immediate supervision of a Chinese official, who is paramount in the place. The taxes are collected by the native chiefs, who pay the imperial taxes to the Chinese official, who in his turn remits them to Peking. Not very long ago, the system of accounts adopted by the Chinese Government being imperfect, it was customary to send the whole of the imperial taxes in bullion, at least as far as Ya-Chou; the pay of the Chinese officials and soldiers being also sent in bullion back to Bat'ang. In recent years, however, knowledge of book-keeping has advanced sufficiently far amongst the Chinese to enable them to abolish this clumsy process, although the novel experiment has not been quite so successful as might have been expected, for the predecessor of Chao, in the office of Liang-Tai of Bat'ang, was a careless, indolent person. He spent the money himself, or allowed the native chiefs who collect the taxes to spend it, and neglected for some years to render an account to the governor-general of Ssŭ-Ch'uan, to whom he was responsible.

This governor-general must have been a careless person also, and probably the Liang-Tai trusted to this weakness in his character; but he suddenly demanded not only the year's taxes but all the arrears. The Liang-Tai failed to produce them; he was deposed, and Chao appointed in his stead.

Chao was at first afraid that he would have been called upon to pay the arrears, and knowing how ill the miserable people whom he was to rule could afford any extra taxation, he refused the appointment, until it was clearly and plainly agreed upon that he should not be expected to produce anything more than the taxes for the period of his service.

Besides the imperial tax that is sent to Peking, there is another tax for the native chief. The assessment is per village. The amount that each village has to pay was settled a hundred years ago, according to the number of families residing in it. Since the date of this assessment the lay population of the country has diminished fifty or sixty per cent., and as it continues to diminish the tax becomes yearly heavier, and is now almost unendurable: so much so, that from this part of Tibet the people emigrate in considerable numbers to avoid the pressure of taxation and the hated rule of the Lamas.

It would almost naturally be supposed that the enormous plateaus of grazing land did not belong to anyone in particular, but were, so to speak, common property. This, however, is not the case: all the pasturages, and many of the forests, being the property of some individual or individuals, for people often club together and buy up a pasturage or a mountain.

The cattle-owners in Tibet,—the richest of whom possess two or three thousand head,—rarely look after their cattle themselves, but employ cattle-keepers, who take the animals up to their pasturages, live in black tents, and look after them. If a family has not sufficient cattle to employ a cattle-keeper, it will club with another family or families. These then elect a head or chief, who decides which pasturage shall be taken first, and settles all matters relating to the common weal. The cattle-keepers are not paid a fixed salary, but for each cow in their charge they have to render to the owners a certain quantity of butter per annum; the remainder of the milk and one half of the calves by each cow being their payment. They let the animals loose during the day, and towards evening the herdsmen call them with a peculiar cry. Each animal knows the voice of his own herdsman, and eagerly comes for the lump of salt with which all are regularly regaled. The beasts are then caught and tethered for the night, the encampment being guarded by fierce and savage dogs.

The cross between the yak and the common ox is the most esteemed of the different kinds of cattle. It grows larger than the others, and can sustain the warmth of the lower valleys better than the true yak, who is most in his element among deep drifts of snow.

Slavery is a great institution in Tibet. There are rich families who own five or six hundred slaves. These are hereditary, and are often treated very cruelly. A family always counts its riches in slaves and cattle; but in Tibet proper, more by the number of slaves than by the head of cattle.

In this part of the country a man with three or four hundred head of cattle is rich, while one who has only twenty or thirty is considered poor. Even the agriculturists reckon their fortunes in mules and ponies, and not in land: for in a family there is rarely enough land to support the whole. One or another of their number then undertakes the trading, and has charge of the mules, yaks, or ponies, used as beasts of burden, which thus become the measure of the family fortune.

In Tibet, on receiving a visit, it is always the

custom, after the first few complimentary phrases, to ask how many cattle, or mules, the visitor possesses; but if the host knows that the number is very small, he will refrain from asking the question, lest he should confuse his visitor or make him ashamed.

Tibet is being gradually depopulated : partly by the oppression of the Lamas, who are detested by the people as much as they are feared ; and partly by emigration to Yün-Nan. Empty and deserted villages are constantly seen, and Monsieur Desgodins informed me that, even during the short time of his stay in the country, the decrease of the population in those parts well known to him had been enormous.

As the lay population diminishes by emigration, the land that the emigrants leave behind does not go to increase the fortunes of the remainder ; but, on the contrary, these are the more impoverished, for nearly the whole of this land passes to the Lamasseries, and being no longer available as a source of taxation, the burden on the remainder, who still have to pay the same amount, is increased.

What the end of the country is to be it is difficult to foretell, unless it falls into the hands of the English or the Russians. It can hardly be considered a desirable acquisition, although the people are very pleasant to live amongst.

Some of their customs can hardly fail to recall those chronicled of the people of Israel. In Tibet it is usual for the men to go on to the house-top to pray ; and, as I sate writing in a room below, I used constantly to hear the monotonous chant of the pious folk above. Bringing milk, and 'butter in a lordly dish,' to strangers, and the payment of the cattle-keepers as Jacob was paid, are points of strong resemblance in the habits of the two nations.

The first native chief of Bat'ang is of Chinese extraction, but as his family came from Yün-Nan ten generations since, he may fairly be considered as a native of the soil. His name is Loh-chung-wong-tun, and he was born in the year 1844. His elder brother is the second chief. Before the earthquake that destroyed Bat'ang, neither of these brothers had as yet succeeded to his inheritance. The elder married a daughter of the then second chief, and thinking that the prospects of succession were better in that family than in his own, he was adopted by his father-in-law, and became heir to the second chieftainship, resigning to his younger brother his father's title.

The younger brother, Loh-chung-wong-tun, also married two daughters of the second chief, but not being received into the family, remained his father's heir. In the earthquake both the chiefs were killed, as well as the two wives of Loh-chung-wong-tun, and the two brothers succeeded to the two chieftainships simultaneously.

Loh must have had good reason to be satisfied with the family of his late father-in-law, for two more sisters remaining, he took them as his wives, in place of the others. He is a heavy-looking man, pitted with small pox, and has a son, born in 1859. He is given the rank of Sieh-Tai by the Chinese, and wears the red ball and peacock's feathers; but he is, nevertheless, inferior to the lowest Chinese official, and is never permitted even to sit down in the presence of the Liang-Tai, or of the first military Chinese official; and if he has business with the Liang-Tai, he must stand at the door, unless invited into the Yamên.

He is overwhelmed with debt, principally to the

Lamas, though he borrows money from anyone who will lend it to him. He was reported to owe a hundred taels to a private Chinese soldier. He dislikes Europeans, though outwardly he is civil enough, and pretends to be friendly. He is one of the chief enemies to the missionaries, and is said to be the tool of the Lamas, to whom he owes hundreds of taels. His income is not very large, and he has to support two wives, a mother, sisters, and a numerous family. It might, however, suffice for a careful man in this land of poverty; but he is a regular spendthrift, his fortune is squandered, and his debts increase.

There can be no doubt that the Lamas were very strongly opposed to our entry into Tibet. Rumour, with her thousand tongues, had already been abroad, and the Lamas were expecting an attempted entry on the part of both Russians and English, which they were determined to resist. Whether the Chinese power in Tibet is as feeble as it is represented, it is difficult to say, but in all probability the Chinese have but a slender hold on the Tibetans, who, if it were not for the convenience of trade, might cease the payment of tribute. It was perfectly clear from the manners of both Chao and the military official, that they were exceedingly uneasy at the idea of any attempt to enter Tibet proper. Chao said that if we wished to go, he was bound by the letters he had received to do his best, but that it would be quite impossible to enter peacefully, and if we insisted on making the attempt, we should, in all probability, be obliged to fight our way. That he was really concerned for our safety was sufficiently proved by the fact that, even after he was quite satisfied that we really were going to A-tun-tzŭ, and not to Lassa, he not only came with us himself an eight days' march, but brought the native chief, and an immense escort with him. That this was mere espionage is quite impossible; he could have sent half a dozen spies, if he had been so minded, who would have reported our most minute actions as closely as he could have observed them himself; and certainly, without some grave reason, he would not have put himself to the trouble and discomfort of this journey.

Whilst we were at Bat'ang, it was reported that a powerful Tibetan chief, named Peun-kop-pa, living at Lassa, had been exiled to the province of Yün-Nan, whither he was supposed to be wending his way, and this report was carefully spread by the Tibetans and Lamas at Bat'ang. It was manifestly untrue; because the Tibetans have no power to banish any of their people to Yün-Nan, and an old Lama, whose words were reported to us, had said 'Ah! the Government of Lassa had heard of Englishmen and Russians coming to Tibet; if they do come, Peun-kop-pa with a few hundred brigands will be somewhere on the road.'

We were also told that the Lamas had ordered out six thousand men to guard the frontier, and our informant said that he had met the messenger sent by the Lamas to report our arrival.

Numbers are of course always enormously exaggerated, and the six thousand was not, in all probability, as many hundreds; but there can be little doubt that the Lamas, whose power is almost absolute, had made up their minds to resist any attempted advance on our part; and even if open hostility and violence had not been attempted, they would have simply starved us out. The whole country is in their hands, and if they had forbidden the inhabitants to give us food for ourselves and people, and forage for our animals, or to receive us into their houses, or supply us with transport, their orders would have been obeyed to the letter.

At this time I was much too anxious with regard to European politics to wish to spend the winter in Tibet, and even if this had not been so, it is a question whether I should have been justified, under any circumstances, in running the risk of provoking a conflict between the Chinese and the Tibetans.

It was a great relief, not only to Chao, but to all our retinue, when at length they became certain of our intentions. Our short stay in Bat'ang was very pleasant. We were lodged in an excellent house, where we had two good rooms, besides a room for the servants. The house was built in the form of a square, and like the whole of Bat'ang was constructed of mud. A large and massive door, on which our honourable names were posted on a red board, led into the stable which occupied the whole of the ground floor, except a portion of one side, where there was a very small room. A large stack of firewood was packed against one wall; our four horses, and one or two others, occupied another; two or three pigs, and some donkeys, were wandering loose about the place, which did not appear altogether as dirty as might have been expected. The house was one of the best, and instead of the usual notched trunk of a tree, there was a regular flight of steps, with a hand rail, leading to the upper story. This was in the form of a hollow square, the rooms arranged round the sides, with a passage in front of them, where there was a gaily painted balustrade, over which we could look down into the stable below. This quadrangle was without a roof, so that the centre of the stable was unprotected from rain or snow. One corner of the upper story was occupied by a fine large kitchen, where pots and pans, all kept exceedingly bright, were displayed on a dresser, almost as in an English house; from another corner a notched log of wood, resting in one angle of a square hole in the roof, formed a staircase by which the top of the house could be reached. This was flat, and on one side there was a shed with a pile of dry grass under it, where during the night half the inmates seemed to sleep.

The morning after our arrival at Bat'ang we received invitations to dine with Chao.

The Chinese, in the matter of issuing and accepting invitations, are as ceremonious as in all else. A card is usually received by the guests, inviting them for about four o'clock; but Chinese etiquette lays it down that they must wait for three notices before setting out. A second invitation is sent later, praying the guests earnestly to come at once; but until the third, pressing them to be quick, has been duly received, they are not expected to leave their houses. Sometimes the ceremony is rendered more intricate by the issue of four, instead of three notices; the dinner hour is quite unconnected with the hour named in the invitation, and that again has nothing to do with the hour at which the guests are expected.

When ladies are invited by one another, the first invitation arrives at daybreak; this is not much more than a polite message to come at once; but as the day wears on, the importunities to be quick grow stronger, until the last and final appeal is made late in the afternoon.

But on this occasion the hospitality of Chao was extended far beyond meaningless ceremonies, and he did all in his power to render the visit an agreeable one. Just before dinner he heard that I did not eat pork, and he had a sheep killed and cooked by Monsieur Desgodins' servant expressly for me; the fact that the other guests and the rest of the dinner were kept waiting an hour or two was of no consequence. He also borrowed knives and forks from the Abbé, and invited him to keep me company.

The next evening we dined with the military official, Shou, who, forewarned of my peculiar tastes, had prepared an elaborate repast for me of fowls, mutton, beef, and vegetables, without grease or garlic.

Directly we arrived, tea was handed round, followed by soup made of mutton to suit my tastes, with the usual dumplings *à la Russe*. After these dishes, which were served like the Russian *zakouska* before taking our seats at the table, there was a lull, during which Chao, who was a guest, disappeared. Shou was a confirmed opium-smoker, and was unable to eat his dinner without a few preliminary whiffs; but in order not to absent himself from his guests longer than he could help, he had asked Chao, who was evidently a most intimate friend, to go into his room to prepare his pipe for him. Chao also had been an opium-smoker, but during the last year and a half had entirely given it up, having been persuaded to do so by Monsieur Desgodins.

When Colonel Shou had smoked his opium, he returned radiant; and taking a pair of chopsticks in his hand, he came to me, made me a profound salutation, and placed the chopsticks on the large round table in the centre of the room. The same process was repeated with a little cup, and thus my place was indicated. This lengthy ceremony was religiously carried out with each of the guests, who in the meantime had obtained permission to remove their official hats, and had replaced them by little skull-caps brought by their servants. Before sitting down my host came to me, and with many elegant expressions thanked me for some trifle I had sent him, as if it had been a gift of inestimable value; and his politeness was carried so far as to demand my pardon for venturing to remain *en déshabille*, that is, without his official hat, whilst tendering his acknowledgments.

These laborious ceremonies having been duly observed, we sat down to table, and then the guests, under the influence of food and wine, began to unbend. I now had an opportunity of testing the truth of a fable narrated to me at Ta-Chien-Lu, to the effect that the fowls of Bat'ang, though excellent in themselves, when boiled were always tough, on account of some peculiarity in the water. My Chinese friends said that this was not the case: that the cocks, though few in number, were good, however cooked; but that the hens, of which there were plenty, were inferior and tough; but Monsieur Desgodins explained with a sly smile that these good Tibetans seldom parted with their cocks, and that they did not sell their hens until they were too old to lay eggs.

We now commenced the serious part of the entertainment; toasts were drunk in rapid succession, each guest challenging the others in bumpers of rice wine. Fortunately for all our heads the cups were not much larger than thimbles, for it was 'de rigueur' to drain them, when they were immediately filled up by watchful attendants, who always poured a liberal allowance amongst the *débris* of walnuts, ground nuts, and the inevitable water-melon seeds, with which the table was strewn. Towards the end of dinner Mesny and our host engaged in a most noisy game which I failed to understand, but the players held up fingers, counted numbers, and shouted words at one another, as rapidly as possible; one or the other always lost in about half a minute, and was expected to drink a cup of wine, and drain it dry.[9] Mesny, being a teetotaller and privileged guest, drank nothing stronger than tea. After a time Chao joined in the game. Then the old soldier began to talk of his campaigns. 'He fought his battles o'er again, and thrice he slew the slain.' After this he challenged Mesny to another turn, and as he was now perspiring under the effects of the exertion and the liquor, he made his subaltern officer drink his wine for him. This the young man was nothing loth to do, so the fun became fast and furious, each shouting louder and louder, and all the domestics crowding at the door, astonished at the novel and amusing sight. It seemed as if the dinner would never end, as dish after dish came in, and pipes were smoked in the interval. At length the rice appeared, the guests solemnly saluted the host with chopsticks, and the party broke up about nine o'clock.

Monsieur Desgodins engaged us two more Ma-Fus; there was some difficulty in finding these in the hurry, and when he sent them he remarked that he feared that they were not endowed with vast intelligence, but that he hoped they would prove faithful. His fears and his hopes were both fully realised.

The native chief sent a message one morning to ask permission to pay a visit to us, in order that he

---

[9] The Italian *morra*; *micare digitis* of old Rome. 'Morra is the pastime of the drinking-shop in China as in Italy, and may perhaps be reckoned among the items of culture which the Chinese have borrowed from the Western barbarians' (E. B. Tylor, in *Cont. Review*, May 1879).—*Y.*

might see my bath. When he arrived it was brought in folded up in its little bag. It was taken out, and as it expanded under the influence of the bellows, his surprise and delight were unbounded; and when it was solemnly filled with water, he evidently wished to see the last act of the drama, but did not venture to say so. After this he gave us presents of finger-rings, which he took off his own finger.

Before leaving Bat'ang, Chao also gave me some really beautiful presents,—some valuable china cups, and some pieces of jade.

The second night after our arrival, we were both suddenly awakened by a rattling and crash at our window; we both shouted, jumped up, and struck a match. Chung-Erh and Chin-Tai came in half-naked and with drawn swords; but all was now quiet. We found the shutter of one window burst open, and the lattice-work outside broken, but failed to discover the culprit. Both Chao and Shou seemed very anxious when we told them of it; a strong guard was put outside our house the next evening at sunset, and poor old Colonel Shou himself visited the sentries three times during the night.

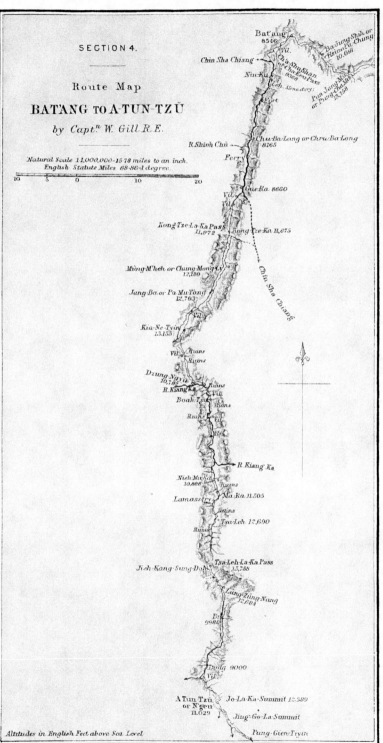

CHAPTER VI.

REGION OF THE RIVER OF GOLDEN SAND.

I. BAT'ANG TO SHA-LU.

Departure from Bat'ang—Adieux to MM. Desgodins and Biet—Broad Features of the Geography, and Relation between Road and Rivers—The 'Tea-Tree Hill'—Chinese Rendering of Foreign Names—Valley of the Chin-Sha or Golden Sand—A False Alarm—Journey along the Chin-Sha—Cross and Quit it—Fine Scenery of the Kong-tze-ka Pass—Lama Oppression, and Depopulation of the Country—Quarters at Jang-ba—Growth of our Escort—Armed Opposition on the Lassa Road—Kia-ne-tyin—March through Woods of Pine, Yew, and Juniper—Clean Quarters at Dzung-ngyu—The River of Kiang-ka—Hospitable People—Desolated Country—Chŭ-sung-dho (Cooper's 'Jessundee')—Nieh-ma-sa—Continued Depopulation—Ma-ra and Tsaleh—Conversations with Magistrate Chao—' The Boy the Father of the Man'—Watershed between the Chin-Sha and the Lan-T'sang—Camping Ground of Lung-zung-nang—The Bamboo Seen Again—The Gorge of Dong—Parallel Fissures from North to South—The Town of A-tun-tzŭ (Cooper's Atenze)—The Oppression of the 'Lekin'—Evil Character of A-tun-tzŭ—Prevalence of Goître—Water Analysis—Farewell Visits from Chao and the Native Chief—Ceremonies of Adieu—Prevalence of Opium-smoking in Yün-Nan—Patience Exercised—Long and Moist March—Passes of Mien-chu and Shwo—Difficulties of March in Darkness—Torchlight—Reach Deung-do-lin—The Word 'Lamassery'—Abbé Renou—T. T. Cooper—Carrier Difficulties—Taking it Easy—Change of Climate and Scenery—New Glimpse of the Chin-Sha—Sha-lu Village—Big Tibetan Dog.

August 29.—Before leaving Bat'ang the native chief sent soldiers out over the mountains to look for robbers or others who might wish to molest us, and when we left he came with us himself at the order of Chao, and brought a considerable escort with him, so that, with mules and muleteers, Chao and the chief, besides soldiers Chinese and soldiers Tibetan, as well as coolies and servants, we were more like an army than a private party.

About a quarter of a mile outside the town we found the good old Shou waiting to bid us adieu. We dismounted from our horses, and made the usual salutes; he then asked us into a house by the roadside, where we drank tea and stayed a few moments.

Soon after leaving him we met Messieurs Desgodins and Biet, who escorted us a little way; and then Peh-ma and the two Gems were by the roadside with a tray of wine, which we tasted.

Half a mile further we met the native chief, and all rode on together, till Messieurs Desgodins and Biet bade us adieu, and turned their horses towards Bat'ang. During my short stay Monsieur Desgodins had been a delightful companion; full of intelligence his conversation had been most interesting, and it was with great regret that I parted from him. Monsieur Biet has now taken the place of the late Monseigneur Chauveau, and is the Bishop at Ta-Chien-Lu.

The great plateau that extends over the whole of Central Asia throws down a huge arm between the Chin-Sha-Chiang (the River of Golden Sand) and the Lan-Ts'ang-Chiang, gradually diminishing in altitude as it extends to the south. The northern portion of this arm partakes more or less of the characteristics of the main table-land, but even in the latitude of Bat'ang the difference is apparent, and it becomes more striking as Ta-Li-Fu is approached. This arm is not more than thirty-five miles wide in the latitude of Bat'ang; and as the crest is generally about five or six thousand feet above the river, it is little more than a ridge of mountains running nearly due north and south between the two streams. A-tun-tzŭ lies on the western slope of this huge rib, the road from Bat'ang crossing the crest at the pass of Tsaleh-La-ka, 15,788 feet above the sea. This is the main road to Yün-Nan, and is so conducted, in all probability, partly for the sake of passing through the important town of A-tun-tzŭ.[1] It might be expected that, as the road to Yün-Nan again returns to the Chin-Sha valley, south of A-tun-tzŭ, there would be another and easier road, by following the valley of the great river, instead of leaving and returning to it. But in all probability there is no road down the valley of the Chin-Sha; the river appears to run through a succession of deep gorges, much as it does between Ch'ung-Ch'ing and I-Ch'ang, and as the Lan-Ts'ang does near A-tun-tzŭ, as well as further south, where the same river is crossed by the road from Ta-Li-Fu to Burmah.

Moreover, the road from Deung-do-lin to Tz'ŭ-kua keeps to the eastern face of the ridge, or, in other words, to the Chin-Sha basin; but near Deung-do-lin one glimpse is all that is gained of the river, a few miles distant, evidently tearing through an exceedingly steep gorge. The road then leaves the river to the east, and by two exceedingly difficult passes, crosses two very elevated spurs thrown out to the east from the main ridge, which still runs north and south. The crossing of each of these spurs is at least as difficult as the passage of the main ridge; for the valleys dividing them are four thousand feet deep, and their sides excessively steep. In crossing these spurs, the road passes no town whatever, and there is clearly no reason why it should not follow the river, if there was a practicable route; the probable conclusion is that the river, at all events between Deung-do-lin and La-pu, flows through

[1] The Atenze of T. T. Cooper.—*Y.*

narrow gorges, where there is neither a road nor a possibility of navigation ; and it would seem reasonable to believe that the case is the same between Bat'ang and Deung-do-lin, and that all the rivers running nearly due north and south of this region, maintain the same characteristics of rapid streams in deep and narrow rifts.

We descended the Bat'ang river, a little stream of clear water twenty-five yards wide, for five miles, then leaving it we crossed a low spur that divides it from the Chin-Sha, a muddy turbid river one hundred and seventy to two hundred yards in breadth. The name of this spur illustrates the difficulty of getting correct nomenclature, and the danger of trusting to the meaning of a Chinese name for an indication of the nature of the place. We asked a Chinaman,

'Pray, sir, what may the name of this hill be?'

'Great excellency,' he replied, 'it is called the Tea-Tree Hill.'

'The Tea-Tree Hill! and are there any tea-plants here?'

'Of course,' he said; 'why not, when it is called the Tea-Tree Hill?'

We looked about us, and seeing no indications of the useful shrub, we said so.

'Oh, well,' replied our informant, 'then there were some once; they were probably killed in the earthquake.'

Still our sceptical minds were not satisfied; and as soon as our Chinese friend was out of hearing, we inquired of a Tibetan, who said the hill was called Cha-keu, a Tibetan word meaning nothing in particular, and he was subsequently good enough to write the name for us in our book. The Chinese are hopelessly inaccurate in their conversion of foreign names into Chinese, and that Cha-keu was translated Ch'a-Shu, or Tea-Tree, was not surprising.²

The valley of the Chin-Sha is somewhat dreary. The steep and broken sandstone hills descend at a very steep angle sheer down to the water, leaving,— except at the *embouchure* of some small stream,—no place for cultivation. The path running along the edge of the river is strewn with stones. The hillsides are nearly bare, and every now and then there are long, steep slopes of loose rocks and *débris*, or small precipices by the road, or up above it, the result of the convulsions that destroyed Bat'ang, and tore the mountain-sides in all directions.

We breakfasted at Niu-ku, where there were a few leather boats, by which some of our coolies relieved themselves of their burdens. Here some despatches arrived for Chao from Lassa; he swallowed a mouthful of food in a hurry, and set to work to answer them before proceeding on the march.

Our cooking things were either in front of us or behind; and, much to our astonishment, the people of the house produced three little tin plates with an English alphabet round the edge, and a picture in the middle of Sir Henry Havelock, K.C.B., on a prancing horse. By what route they had come here we failed to find out.

In the afternoon, as I was riding in front, I suddenly heard a shot; and, looking up, I saw half a dozen wild-looking fellows with guns, behind a rock.

---

² The Chinese syllables representing foreign names must have *some* meaning, but the meaning is a mere accident, or at best forms only a kind of punning *memoria technica*. This Tea-Tree Hill is mentioned in a Chinese itinerary of the road to Lassa, and has led the excellent Ritter to cite this point as the extreme known western limit of the tea-plant (the tea-plant of Assam was not known to him then, i.e. in 1833). See the *Erdkunde*, iii. 237, and iv. 201.—*Y.*

For a moment the idea of an attack came into my head, but it was soon dispelled by a Chinese soldier dashing forward, for I knew that no Chinese soldier would voluntarily expose himself to danger. It is customary for large parties of a wild tribe of barbarians, sometimes two hundred or three hundred strong, to descend to this spot from the opposite side of the river, and attack the caravans of passing officials or traders. If these are Tibetans, or half-breeds, they do no more than rob them; but if they are Chinese they take them prisoners, and keep them until ransomed. There is a little square fort here for some guardians of the place; and it was these who had fired a salute in our honour, for which act of homage they of course expected a few rupees.

On arrival in the evening Chao came to take tea with us, and after him many people bringing gifts; every man, woman, and child in the place brought something. These were succeeded by head villagers, who bore a remarkable likeness to some who had previously appeared as ordinary villagers; and boatmen were followed by chief boatmen, whom I should certainly have mistaken for their coolies; but when a petty officer arrived who already had unmistakably headed a deputation of his men, and when I had enough provisions to feed an army, I at last cried out 'Hold, enough!' and refused to give any more rupees in return for a jar of impossible whisky, three very suspicious-looking eggs, and grass for a pony.

August 29.—We halted at Chu-ba-lang, a village on both sides of the river, each portion containing from ten to fifteen families. The place is celebrated for a fierce dragon that lives on a mountain on the other side of the river. It is said that if anyone were to rouse him, or enter the wood in which he dwells, the river would at once overflow. Chao said this must be true, for no one in the memory of man had ventured into the wood, and the river had never seriously overflowed.

August 30.—This is the place at which the passage of the Chin-Sha is usually effected by traders or travellers between China and Tibet; but there is a stream that runs into the river on the western side a little lower down, and as the bridge across this had been carried away, we marched four miles further, and crossed below the *embouchure* of the stream.

Here we found Chao and the native chief waiting, as the animals were not all across. It was a gay and busy scene. The rumour had spread all over the neighbourhood that a chair was coming, and as such a thing had never been heard of before, the natives had come many miles to see the rare sight. These now stood about in their strange costumes; and the red saddlecloths of the Tibetans, the native chief in a red coat, and the animals being driven into the boats, formed bright and picturesque groups in the brilliant sun that now came out. We sat down for some time on the sandy beach; and Chao, recognising one of the Ma-Fus, and not knowing in what capacity he served us, asked him what he was doing.

'I am the chief of their excellencies' stables,' replied the Ma-Fu, as if he had a dozen or two of assistant grooms.

'And what is your establishment?' said Chao, who thoroughly appreciated a joke.

'Great and honourable sir, I have one young man under me,' solemnly replied the Ma-Fu, not in the least abashed.

When we had crossed, we marched down to Guera, where there were five or six families. We then left the river, and, striking up a valley, clambered up about three thousand feet. As we ascended we gradually left the dreary, barren, and steep slopes behind us. The hill-sides were again wooded, and at the summit we came upon a charming little grassy plain, with pines, oaks, and poplars. From here there was a fine view, and we could see the Tea-Tree Hill, with a long stretch of the Chin-Sha-Chiang.

At a small village on the roadside the native soldiers turned out to salute us, and a man ran up with a basket of eggs; and directly we arrived at Kong-tze-ka, Chin-Tai reported with a grin that plenty of people had come with presents. Kong-tze-ka is 11,675 feet above the sea, and it is a village of twenty families. A small chief lives here, subject to the chief at Bat'ang, and his house, a very comfortable one, was placed at our disposal.

August 31.—The scenery round Kong-tze-ka would make the fortune of a Swiss hotel proprietor, if only it could be moved a few thousand miles in a westerly direction. I ascended to the housetop in the early morning, and thence my gaze roamed over a valley of wondrous beauty, enclosed by grassy hills, where masses of primrose-coloured flowers brought to mind many a bank in England. On the hill-side the pines and holly-leaved oaks contrasted their deep green with the brilliant yellow of the flowers. Here and there lovely slopes of the freshest grass were dotted with trees of the most graceful forms and delicate hues. In the distance rolling mountains filled the background; and the stream was heard rushing four hundred feet below. Herds of sheep and cattle luxuriated in the pasture, and round the village there were fields of wheat, barley, buckwheat, and peas. The usual wild gooseberry formed natural fences, and a road led along the hills right through lovely woods of pines, yews, and juniper. The road followed up the side of a stream. As we ascended the hills became less steep, and presently we entered a charming plain about half a mile to a mile wide. Here the French missionaries have built a house on some property they have bought, and for a summer residence it is hardly possible to imagine a more delightful spot. The house is situated at the foot of a gentle spur, thrown out from rugged mountains behind. A waving field of barley surrounds it; a meadow of grass, yellow with a carpet of flowers, lies beyond, where a stream meanders by a few large trees, and where great herds of cattle, sheep, and ponies stand up to their knees in the luxuriant herbage. Opposite the house the valley is closed by spurs of bright red sandstone, from a range of higher hills behind. Just above the little building a Lamassery stands on a grassy knoll; and two Lamas dressed in red, crouching under the hedge of the missionaries' enclosure, scowled at us as we passed. Some time back the missionaries' house was destroyed at the instigation of the Lamas, but Chao rebuilt it at his own expense.

In this valley the usual piles of stones are capped with flat slabs of white marble. These heaps are of a pyramidical form; at a distance the white tops have the appearance of a row of English bell tents; and looking through my glasses, I almost expected to see a red coat pacing up and down in front.

The Bat'ang chief told us that not very many years ago there were one thousand families living in this plain, but that now there were not more than three hundred; and the numbers of ruined houses we passed attested the truth of the statement.

The chief used to complain to us bitterly of the Chinese yoke; he said that the constant requisitions by the Chinese of horses and men, ground the people down to such an extent that they could not live, and emigrated to Yün-Nan. This no doubt is the case to a certain extent, but the oppression of the Lamas is far worse than that of the Chinese, and is the chief cause of the depopulation of the country.

The little village of Jang-ba, called by the Chinese Pa-Mu-T'ang, lies at the end of the plain. Here we breakfasted in the house of a native officer called a Ma-pen, whose rank is that of a Chinese Ch'ien-Tsung. The Bat'ang chief told us that this officer had gone away to Lassa without leave, and that if he did not return soon he intended to depose him. This he would have done before, but some ties of relationship induced him to give him grace.

The house had a particularly dirty exterior; and we were pleasantly surprised to find an unusually nice room, with couches covered with some sort of carpet. The morning had been very fine, but before we had finished breakfast heavy clouds were gathering ominously, and promised us a wet afternoon. Chao and the chief started off before us, and following them we rode over an undulating plateau, about thirteen thousand feet above the sea, whence we could see a high range of snowy mountains to the south-east, their tops hidden amongst heavy clouds. We had not long left the shelter of the village when the thunder began to roll in the distance, and a heavy downpour of rain descended.

I noticed that our escort, which had been gradually increasing, had by this time reached formidable proportions, and that now there were some two hundred men and officers with us. The officers wear felt hats, of the shape of our tall hats, but rather lower, with a broader flat brim. The men and officers are generally armed with a long matchlock, with prongs to rest on the ground, and two swords, one for cutting and the other for thrusting. All are mounted on Tibetan ponies, and carry great rolls of blanket and felt on their saddles.

I was riding on ahead of Mesny, and in the wind and driving sleety rain took but little heed of a man muffled in a big cloak crouching down on the opposite hill. A few yards further I suddenly found Chao and the native chief behind a knoll surrounded by about one hundred horsemen; and a quarter of a mile distant, on the opposite hill, some three hundred Tibetans were encamped, who had come out to oppose us, if we should attempt the road to Lassa. When first they saw us coming they had fired off warning guns, although Chao had sent to them to say that we were going to A-tun-tzŭ.

The boundary of Tibet Proper is five miles from Pa-Mu-T'ang, and the same distance from Kia-ne-tyin, where we slept. In Tibet Proper the country is subject to the temporal sovereignty of Lassa. In Bat'ang, though the spiritual supremacy of the Dalai Lama is acknowledged, the chief does not recognise the king of Tibet—or Nomekhan, as Huc calls him.

When we had safely passed the encamped Tibetans, but not till then, Chao followed us, and with two hundred soldiers we marched in the drenching rain along slopes cut up by deep ravines and valleys, where the red sandstone, breaking through the rich grass, contrasted with the dark hue of the pines and oaks on the hill-tops.

The village of Kia-ne-tyin stands in a tiny patch

of barley. Our escort bivouacked in an adjoining field, and wrapped up in felts and rugs paid little heed to the cold and rain.

The best house had been prepared for us, but it was a miserable place; the entrance was deeper than usual in mud and slime of the blackest, and the ascent to the upper story was by a notched log of wood, in which the notches were nearly all worn out. The floor of our room was made of hardened clay and rubbly stones; a hollow in it was the only fireplace, and here a few sticks were burning. A hole in the roof ostensibly served as an exit for the smoke, but in reality rather admitted copious showers of rain; and a window, two feet square, without any covering, let in violent gusts of wind. A pile of straw was arranged on one side, and two doors that had been taken down and laid across logs of wood served for beds. Such as it was, we were glad enough to get into the shelter, and crouch over the little bit of fire, where we were soon joined by Chao and the native chief, who was permitted, for the first time in his life, to sit down in the presence of the Chinese Liang-Tai. When Chao left, he was carried by a man pickaback through the sea of mud to his house. On the way the man fell down, precipitated Chao in the most undignified manner into the slush, and bruised his leg rather severely.

September 1.—It rained steadily all the night, and was still raining when morning broke. We were up at the usual hour, but after our early breakfast we were kept waiting a weary time with everything packed up; for the journey being short, all the people wanted to start late. The rain cleared off as the day wore on, and we marched through a red sandstone formation, where there was also a great deal of deep red clay, and where, after the rain, the road though good was slippery. The ground was very undulating, and much cut up by ravines in all directions. We wound about amongst woods of pines, yews, juniper, and a tree the Tibetans call Chien-a-ragi. Sometimes there was a little grassy valley enclosed by low hills, then the stream would run between slopes running sharply down to it, the water often hidden amongst the trees. We descended gradually, and as we left the higher altitudes the slopes became more bare, the red rocks and clay appeared more and more between the trees, and at the end of the march, the road ran down a stony valley between bare and craggy hills.

I was assured that the woods were full of monkeys; but my scepticism was apparent, and it was with an air of triumph that my people called my attention to what they said was a monkey. The truthful field-glass, however, discovered nothing more inhuman than a boy; but still, the repeated assertions of the people that the woods are full of monkeys are probably not without foundation.

Dzung-ngyu, or, as it is also called, Dzongun, is a large village of forty families, situated in a pleasant little plain one hundred and fifty to five hundred yards wide, enclosed by hills of sandstone. The climate is temperate, as it lies on the river of Kiang-ka, and although at an altitude of 10,792 feet above the sea, crops of wheat and barley ripen in the warm sun, and serve to maintain the three hundred families who live in the plain. The house in which we stayed, where two light and clean rooms with boarded floors were placed at our disposal, belonged to a petty chief, and was a remarkable contrast to our lodgings of the previous night; the want of tables and chairs was a drawback, but these are articles of furniture unknown in Tibet. The usual butter and tsanba were laid out for us on arrival, and the master of the house brought mushrooms, eggs, and peas for the horses.

September 2.—We changed all our transport animals here, but Chao and the secretary of the native chief saved us all trouble. The secretary was a most active fellow; he counted the packages, made lists, divided the loads, and arranged everything.

We left Dzung-ngyu a motley crew. Our baggage was carried by every description of animal, two and four footed—young men and maidens, old women and children, horses, mules, donkeys, and oxen were pressed into the service.

The native chief and Chao started first, and we followed as soon as Chin-Tai had succeeded in packing his abominable cooking things, an operation that always consumed an incomprehensible amount of time.

The river of Kiang-ka, or the Vermilion River of Cooper, takes its rise in about latitude 30° 20', and passing the town of Kiang-ka, waters the plain of Dzung-ngyu, where it is about fifteen yards wide. The basin of the upper portion of this river is composed nearly entirely of red sandstone and red clay, which gives the water the remarkable red brown colour observed by Cooper.

We followed the twists and turns of this tortuous stream through a rather dreary valley, where the view was limited to about half a mile of steep, high, and almost bare hills, and after a few miles we found Chao and the Bat'ang chief waiting for us in the house of another petty chief, where a few carpets were spread on the floor, and where butter, cheese, and tsanba were laid out on a low stool. It is a strong instinct of hospitality that prompts the master of a house thus to put food and drink ready for his guests; nor is this done with any niggard hand, a huge circular pat of butter about an inch thick, a cake of cheese of the same size, and great jars of oatmeal with gaily painted wooden covers, invite the travellers to partake freely of the best the household can produce; the tea-churn is not far distant, and taking his wooden bowl from the fold of his coat, a Tibetan soon makes an ample and luxuriant repast.

Chao had begged a little brandy of me to make a lotion for his bruised leg, and now he told me that, thanks to my gift, the wounded limb was decidedly stronger than the uninjured one.

From here we marched over an execrable road; and although Chao had taken care that the worst places were repaired, the track was so narrow and fearfully stony, that it was necessary to dismount once or twice where steep and slippery steps had been cut out on the face of the cliffs. Wherever the valley opened a little, there were a few houses close down by the river, all built, as at Bat'ang, of rammed earth, and embowered in clumps of walnuts, peaches, and weeping willows. The number of inhabitants was small; and the frequent ruins were sad proof of the diminution of the population, and the oppressive rule of the Lamas. We had been making inquiries for Jessundee, with the intention of saying a few words if possible to the old man, who had treated Cooper so well, but in Chŭ-sung-dho we failed at the time to recognise the name and so we lost the opportunity. Chŭ-sung-dho means the 'meeting of three waters'; here we left the muddy river, and ascending the bed of a beautiful clear stream, we plunged into a desperate gorge shut in by walls of bare rock eight hundred and nine hundred feet high. Here the native chief pointed out some wild oxen at the top of an almost inaccessible cliff, and a little further on, half-way up the mountain-side, there was a cave that our people told us was inhabited, although it seemed impossible to believe that any human being could clamber to it. About three miles from Chŭ-sung-dho the perpendicular cliffs gave way to slopes, where, though the hill-sides were still very steep, the road was somewhat better, and we could see a little more than a few square yards of heaven.

Nieh-ma-sa is prettily situated just where the oaks and pines again commence; and here we found a resting place for the night in a house in a different style of architecture to those in the neighbourhood of Bat'ang. The houses are here built on piles about six feet above the ground, some of which always sink and throw everything askew. Walls are made between the outer piles with loose, uncut stones, altogether devoid of any description of mortar. The space thus enclosed forms a stable, above which are the living rooms. The walls of these are made by laying the trunks of trees horizontally one above another; as these logs never fit, the houses are remarkably well ventilated, and must be most uncomfortable winter residences. In the house to which we were conducted there was a room on each side of the roof, with a box of China-asters in flower outside the window; these rooms were reserved for us, and had it not been for the wind, which made sad havoc of the candles, we should have found ourselves luxuriously lodged.

We soon all met on the roof, where a high fence of thin poles was erected, on which the good man of the house had hung his straw to dry. The chief was generally the first to visit us ; he used to come and write the names of places in our book in Tibetan. Then Chao would come in, and, graciously permitting the chief to sit down, we would all have a long chat over a friendly cup of tea.

September 3.—From Nieh-ma-sa we continued the ascent of the valley, passing ruins here and there, but few inhabited houses ; indeed, the whole population of the country from Ta-Chien-Lu, including the towns of Lit'ang and Bat'ang, would hardly suffice to fill one Chinese city. Here, again, there were forests of oak and pines, but although the hills were no longer bare there was not the glorious verdure to which we had been accustomed. There was a very fair road to Ma-ra, a village of only three families, situated in a little plain two hundred yards in length, and about the same in breadth. Here our horses were again changed, and the operation was successfully performed, with the usual amount of noise, while we were at breakfast. After leaving Ma-ra we marched through a wood of oaks, pines, and poplars, with many wild gooseberries, and currants, bearing long branches of black fruit. The hills and mountains were everywhere quite steep and precipitous, leaving very little space for cultivation, and the most charming spot in all the valley was occupied by a Lamassery, the only institution that seems to flourish in this country. This was high up the hill on the opposite side, and looked quite a large village. Tsaleh was our destination. This is usually written Tsali, but the native chief pronounced the word several times, and it certainly is Tsaleh. The native chief made us a present of a great quantity of the very yellow butter for which the place is celebrated. We salted this, put it into a tea-churn, which was previously carefully cleaned, and carried it for many days after we had left the land of butter. The pasture that furnishes such excellent butter is not less advantageous for feeding sheep. The mutton of the valley is renowned, and in order that we might judge of it fairly, Chao sent in some of the best.

We arrived in good time, and soon received visits from Chao and the native chief. Just after their arrival some one outside shouted out that the servants' dinner was ready.

'Run along,' said Chao, 'and don't forget that your master is tied up here, and you must come back and untie him.' By which he meant that he could not with dignity go away without servants. Chao was most considerate towards those under him, and it was not difficult to see that his servants were very fond of him. As we were sipping our tea, and discussing the events of the day, he told us a story of an official at Chêng-Tu, a very greedy person, who always kept his servants on very short rations. So one of them blackened his mouth, painted a false moustache, and thus disfigured, came into his master's presence.

'What do you mean by this?' said the magistrate, 'your mouth is in a pretty state.'

'Great excellency,' said the servant, 'I thought you cared nothing for the mouths of the little ones.'

We used to consume long hours in fruitless attempts to buy swords, guns, or ornaments at a reasonable price. We found that everything good came from a place called Turkai, or some such name, lying to the north of Lit'ang and Bat'ang, and eleven days' journey from both. The chief's best horses came from there, so did his saddles; all the jewellery, except the Lassa work, is said to be made in that town, and no swords or guns of any value are turned out from any other manufactory. Altogether it ought to be an interesting place, and well worthy of a visit.

September 4.—The village of Tsaleh is 12,690 feet above the sea, and is said to be a very rainy place; but although it had rained all night it was fortunately a fine morning. The muleteers had told us that, if wet, it would have been useless to start, as the mountain Tsaleh-La-ka was very difficult at all times, and quite impossible to pass in rain.

The morning felt very chilly, as, in order to prepare for an early start, we turned out into the keen air, and watched the people of the village wading about in the mud in their long leather coats, which, as I remarked to Mesny, probably lasted from the day of birth to the hour of death.

'How can that be?' said he, 'do people never grow in this country?'

In reply I pointed out a touching sight that was at that moment presented to our view.

An old man was performing the morning toilet of his son, a boy of about nine years of age. A huge coat, big enough for the father, was thrust upon the child, the sleeves were turned back till they were not more than a foot or so too long, the skirts were then drawn up, and a girdle being tied round the child's waist it was tightened up till we expected to see the boy drop into two pieces. This process providing the usual substitute for pockets, the father drew a parcel about the shape and size of his head from his own capacious fold, and thrust it into the child's bosom, with several articles, amongst which there was of course a Pu-ku, or wooden bowl. Then the boy was ready for anything. As he grows bigger his pockets will become smaller, but otherwise his coat will fit him well for the rest of his life.

From Tsaleh we continued our ascent of the stream, as ominous clouds were gathering amongst the mountain-tops. The valley was entirely without population, and we passed only one ruin before halting for breakfast.

At first the road was fair, and through woods of pines, oaks, and poplars.

There were long stretches of dead pines on one or two of the slopes, and the usual gooseberries, currants, and briars grew in the valley. We ascended gradually into the rain, but it was curious weather, at one moment it was raining, the next the sun was shining, and soon after we would have both at the same time. Once there was a magnificent rainbow down at our feet in the valley below, and the effect was very beautiful. After marching four miles we found ourselves amongst ragged peaks and slopes, broken into spires and pinnacles, where the road became very rocky, and we again entered the region of rhododendrons, and soon after we commenced the final zigzag that took us to the summit of Tsaleh-La-ka. There were a few small patches of snow at no great distance from us, but none on the road. The crest of this mountain, 15,788 feet above the sea, is the water-parting between the Lan-T'sang and Chin-Sha rivers. It marks the boundary between Yün-Nan and Bat'ang, and here the jurisdiction of Chao and the native chief comes to an end. But Chao was afraid that the Lamas of A-tun-tzŭ, who are directly under the king of Tibet, and are very hostile to foreigners, might try to annoy us, and being determined to see us safely through all difficulties he came with us to A-tun-tzŭ.

There cannot be much disputing about boundaries here, and no one runs much danger of being cursed for removing his neighbour's landmark. There can hardly exist a sharper line of demarcation, for the top of the mountain is like the edge of a knife.

We descended by an exceedingly bad zigzag for about half an hour, and then followed the stream to a little grassy opening, where we found our retainers near the remains of a hut, with a fire of sticks and a churn of buttered tea. The people here have a name for every opening in the forest, and this, being particularly small, rejoices in the remarkably long title of Jieh-kang-sung-doh. The rain held off for a little just as we arrived, so, seating ourselves on waterproofs on the grass, we breakfasted as well as circumstances would permit. The Tibetans do not seem to share in the superstitious dread of the Chinese for mountain passes—they sing and shout as they go up without any fear of evil consequences, and they regularly whistle tunes. The Chinese are unable to whistle, or, at all events, have never acquired the art.

From here the road took us to our camping-ground, through a pine-forest very like that on the western side of J'ra-la-ka. The spot was charming for a camp, or would have been but for the rain, which effectually deprives camp-life of its pleasures. A rivulet came down from the mountains through a dense forest of pines and oaks, and just at its junction with the main stream there were a few hundred yards of open space covered with grass and wild flowers, and though there were not even the remains of a hut, it was called Lung-zung-nang. Tents had been brought by Chao, and we found the native chief in a good-sized marquee, in which he and twenty men were going to pass the night. The tent that had been brought for us was not ready, so we sat down for a while with the Bat'ang chief, until our modest residence was prepared. This was a *tente d'abri* of one thickness of cotton, ten feet by eight, with many holes in the sides, and nothing to close the front; but as the rain did not come down very heavily, and there was no wind, we were fairly watertight all night. The servants and followers slept as best they could under trees, or elsewhere, and the place had probably never before seen so many horses and people encamped at once, for altogether, with Chao and the chief, and their retainers and baggage, we numbered fully one hundred animals. We turned in early, and were lulled to sleep by the pattering of rain on the top of the tent, the chattering of the brook close beside us, and the more cheerful sounds of the crackling of numerous fires outside, where many picturesque groups of men, smoking, drinking, or sleeping, could be seen as the pine logs blazed up in the dark night.

September 5.—It was a long time before we could prevail upon anyone to start in the morning, and our time was beguiled with fearful stories of the dangers of the road before us. At length, however, everything was ready, even Chin-Tai and his cooking things, and we continued our descent. My Ma-Fu was now a beautiful sight as he marched ahead of me in the rain. He was six feet high, and always out of breath; he wore a rough felt hat, with his plait twisted round it, a red serge coat, and trousers reaching to his knees. His legs and feet were bare, and he trudged along with his boots in his hand. A gun was slung at his back, and he was further armed with a pair of field-glasses and a couple of swords. He carried the remains of a Chinese umbrella over his head, but as there was little left besides the framework, it hardly seemed a useful article of equipment.

We presently found a bamboo, a poor miserable thing, but we had not seen one since leaving Ta-Chien-Lu, and we hailed it as the first sign of a return to a warmer climate.

Two and a quarter miles over a villainous road brought us to the entrance of the gorge of Dong, called by Cooper 'Duncanson Gorge.' The river here runs between walls of rock, rising up almost vertically from the stream, whose bed is but a few yards wide; the cliffs, however, are not altogether continuous, but are broken in places by exceedingly steep slopes clothed with dense foliage of pines and oaks, which seem to find sufficient nourishment in the crevices of the almost perpendicular cliffs. The road led us amongst trees, many of which had just been cut down to render the path practicable for us, but the branches of those remaining threatened every minute to knock us over, and made us stoop low over our horses' heads. We crossed and recrossed the torrent several times, and now and then the track was actually in the water. A huge sentinel rock marks the entrance to the gorge of Dong, which is two and a half miles long, and ends most suddenly in a little grassy opening, covered with trees, where the stream, as if weary of its headlong descent thus far, now ripples pleasantly and gently in a wide bed. After leaving the gorge the road is very fair, and rising above the river crosses a spur which divides it from another stream, and from this point the two rivers run for two and a quarter miles, nearly parallel to one another, only half a mile apart, and separated by a very steep and rocky ridge. A mile and a half beyond their junction the road is but a narrow track, eighteen or twenty-four inches wide, about two hundred feet above the stream, and it runs along the side of the hill, which is here at a slope of about 60°.

All the ponies with one accord used to insist on walking at the extreme edge of the paths. At this point Chung-Erh's pony, putting his foot over the edge, lost his footing; Chung-Erh was fortunately able to jump off, but the pony rolled down, and was lost to view among the bushes. A number of people clambered down to help it, but the poor brute was beyond all help, quite dead.

Soon after this the stream was joined by another running also parallel to it, and separated by another steep and narrow ridge. It is interesting to notice that all the great rivers, the Chin-Sha, the Lan-Ts'ang, and the Lu-Chiang, run nearly north and south, separated at comparatively short distances from one another by steep and high ranges of mountains, and that here their tributaries partake of the same character. It is as if some violent convulsion of nature in ages gone by had cracked and split up the surface of the country with huge rents all parallel to one another.

Up to this point we had been riding in perpetual rain, through continuous woods and forests, when most suddenly, thirteen miles from Lung-zung-nang, we seemed to walk out of the rain-clouds into sunshine. The firs and oaks seemed to disappear as if by magic. The hill-sides, still steep, were covered only with grass and shrubs, and the hollows of the slopes were dotted with houses, and laid out in terraces for the cultivation of barley and buckwheat. The air was still moist, but instead of the chilly feelings we had experienced all the morning, the climate was now like that of a hothouse. Soon the valley of Dong lay before us ; we descended to the level of the stream, and about a mile and a half from the village, at a level of about nine thousand feet above the sea, we came again upon a magnificent grove of very fine walnut trees. The valley of Dong is surrounded by mountains, and at the bottom there is a deposit of clay and stones, through which the river has cut a channel some twenty feet deep.

The little village stands amidst fields of buckwheat, barley, and sorgo, and there were a few stalks of Indian corn in a garden. We were taken to the house of an official, and found capital quarters, where a huge black dog with a mastiff head bayed fiercely in a deep voice, and tugged at its massive chain as it recognised the approach of strangers.

We were regaling ourselves with some most delicious Yün-Nan tea when the chief came in. He told us that his doctor, who is also a relation, had met with an accident similar to that of Chung-Erh, but he had rolled down as well as the horse, though fortunately he was not much hurt.

No sooner had we started from Dong than the rain again came down, and descended on us without intermission until we arrived at A-tun-tzŭ.

We again mounted one of those steep and dreadful roads, which were now becoming somewhat wearisome, and for three hours we toiled over the accustomed rocks and stones to the summit of mountain Jo-ka-La, 12,389 feet above the sea, 3,389 feet above the Dong. From this the road improved, the valley opened, leaving a little grassy space, where there were plenty of sheep and cattle, and further on there was a patch of cultivation, and a hut. Five-

and-thirty minutes of very steep descent down a slippery zigzag brought us at length to the end of our first stage on the journey homewards, at the Chinese town of A-tun-tzŭ (Cooper's Atenze), nestled in a little valley between high hills. We had made the journey very fairly—one hundred and seventy miles in eight days, a performance that reflected great credit on Chao, who had made all the arrangements for us.

Just outside the town the chief Lama came to meet us in a costume that would have put a beef-eater to shame; he had a wonderful red garment, the mysteries of which I had not time to penetrate, as I was fully employed in observing and admiring his hat. It can only be described by a sketch, and when it is added that it looked as if made of wood, and was gilt all over, a faint idea of the magnificence of the costume may be obtained. We visited him a day or two afterwards, and he wrote out the sacred ejaculation for us on a slip of paper. He told us that he was appointed by the spiritual authorities at Lassa, but was subject to the temporal rule of the second chief of Bat'ang.

On our arrival we were taken to a Chinese lodging-house, where there was only one very small room, and no place for our baggage. The landlord, who was a most civil fellow, brought us tea, sugar, and cheese in the Yün-Nan style, but as the accommodation was so insufficient, we were obliged to send out and look for better. There was a magnificent old Chinese vase about two feet high standing in one corner. I offered to buy it, and the old man said he would name a price the next day; he afterwards said he did not want to sell it, but eventually agreed to let me have it for six taels; he changed his mind again afterwards, and I came away without it. The lodging-house keeper directly we arrived went down on his knees, and knocking his head on the ground thanked us as Englishmen for having been the means of abolishing the Lekin taxes. These Lekin taxes are levied by magistrates on all goods passing through their towns, and are terribly oppressive. A trader passing through a dozen towns would pay the tax at each. The hotel-keeper had heard of the Chi-Fu Convention, and thought that the Lekin would be abolished forthwith throughout the whole of China. He said that the Tibetans would come and trade here, whereas before they had been prevented by these heavy dues.

A better lodging was soon found, in a house where there was a very large room, besides good accommodation for servants.

Chao paid us a visit almost immediately after our arrival, and sent us a dinner that he had ordered to be prepared, as he was afraid our cooking things would be late.

As time wore on, we began to have serious doubts as to the arrival of our beds; and we contemplated with not much satisfaction the prospect of a night with nothing but hard boards to lie on, in clothes more or less damp, and boots thoroughly wet. The night was very dark, but at every sound we rushed outside or to the window, and tried to penetrate the obscurity; time after time we were disappointed, and had quite given up all hopes, when at length the welcome news came of the arrival. Our mattresses were soon laid, and we turned into bed for a well-earned sleep. The baggage did not arrive till 9.45, and our people had had a most unpleasant time of it, crossing the mountains in the dark; but, strangely enough, the whole way from Dong they had had no rain.

A-tun-tzŭ is a Chinese town, and nearly all the people in it are Chinese; but, through long residence amongst Tibetans, they speak Tibetan better than their own language. They are not altogether Chinese in appearance, and the women were certainly better looking than any we had seen since leaving Ta-Chien-Lu. The immorality of the place is said to be very great, even worse than that of Bat'ang, the reputation of which town is about on a par with that of the worst in Eastern Europe.

The prevalence of goître in these districts is frightful; the Chinese attribute it to the salt, but whatever the cause, at least one third of the population are afflicted with this hideous disease; the swellings in the throat of some of the people being of appalling dimensions. It is said that the Chinese are not so liable to this malady as the Tibetans, possibly because they have not lived here so long, possibly because they never drink cold water.[3]

[3] At Deung-do-lin, where there was a beautiful stream of clear water, and where goître was as prevalent as at A-tun-tzŭ, I procured a bottle of water. I believed at the time that it was drawn from an unpolluted source, but I am afraid that I was deceived. This water reached London in safety, and was minutely analysed by Mr. Bernard Dyer, F.C.S., &c., &c., who showed it to contain:

	Grains per gal'on.
Total solid matter in solution	8·68
Loss on ignition (chiefly organic matter)	3·92
Chlorine	0·42
(Equal to chloride of sodium)	0·69
Nitric acid	0·059
Free (actual or saline) ammonia	0·077
Organic (albuminoid) ammonia	0·119
Oxygen, absorbed by oxydisable organic matter	0·406

After an elaborate, but by no means complimentary description of the water, and the effects that it would be likely to produce, Mr. Dyer concludes:—'In short, as far as I can judge, any peculiar properties this water may possess are to be attributed solely to the presence of a large quantity of organic filth.' An awful warning to future travellers to be careful whence they procure the water they destine for analysis.

The houses in A-tun-tzŭ are nearly all built in the form of a quadrangle, with the stables below the living rooms, and with flat roofs; but the evidences of Chinese civilisation are not wanting; some of the walls are whitewashed, and tables and chairs can be obtained.

At Bat'ang I had determined to shoe all the horses, but when I gave the order I was told that the animals were not accustomed to shoes, and that some illness would be certain to follow so unusual an operation; much against my better judgment I listened to the advice. The results were disastrous, as half our horses were now lame from their hoofs being worn away. I determined, therefore, that the horses should now be properly shod; unfortunately, however, there was no one to do the work, or else the farrier had gone away somewhere else; as usual, I was consoled with the assurance, the value of which I fully appreciated, that there were plenty of shoeing smiths at the next village.

As we were now almost out of Tibet I was very anxious to buy a prayer cylinder, but the people had a superstitious objection to parting with them, and it was difficult to prevail on anyone to sell one. They had a curious superstition also about their wooden bowls; they said that if they sold the bowls from which they had eaten to a foreigner, their country would fall into the hands of the nation whose representative had bought them.

We paid our farewell visits of ceremony to Chao and the native chief, and I was very sorry to say good-bye to the excellent Chinese magistrate, who had taken such good care of us. Our visits were returned with all the rites attendant on so solemn an occasion. At these visits, Chinese officials are always in full dress, with their official hats on. There are usually cakes and fruits on the table, but they are seldom offered, being more for show than anything else; water-melon seeds of course there are, and these delicacies can seldom be resisted by a Chinaman, even under the most serious circumstances. Tea is always produced, but the visitor does not drink it until he takes his leave; then he rises from his seat, and holding the cup in both hands, raises it to his forehead, lowering his head at the same time. He then sits down again, while the host, who has performed a similar ceremony, calls for the horses or chairs of his guests. After this, the guest sips a little tea, rises, and walks to the door. When there, he clasps his hands, and stooping, brings them to his knees; he then straightens his legs, bows his head, and brings his clasped hands to his forehead, thus completing the complicated movements necessary for making a Chinese bow. The host follows his guest into the outer court, where similar salutations are exchanged, the horse is mounted, or the chair entered; but the ceremonies are not yet complete, for now again the clasped hands are brought to the bent head; after which, the rigours of Chinese etiquette having been complied with, the guest moves away.

On our return I made Chin-Tai turn out all the provision boxes, the number of which had been increasing during the last three weeks instead of diminishing. It nearly broke his heart to part with some ancient hams and joints of beef that had accumulated in quantities sufficient to stock Noah's ark, but I succeeded in reducing by six the number of useless boxes we had been carrying. Still the muleteers declared that I must have six more animals than I ever had employed before, and the talking that ensued at-

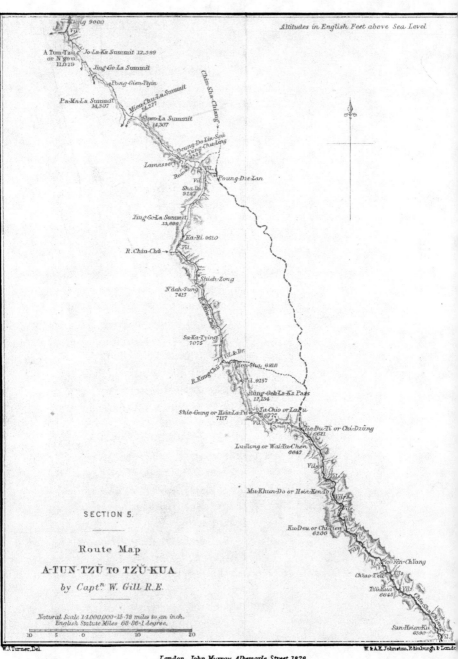

tracted the attention of most of the people in the town, who dropped in casually one by one, to see what was going on. The chaos that reigned it seemed impossible to regulate, but order if not harmony was at length attained, and the greater part of our baggage was sent off the day before we left ourselves, as there was said to be no halting place between A-tun-tzŭ and Deung-do-lin. Of the distance no one could give us any more exact information than that when people went there they started very early and arrived very late, and that at this season of the year, as the days were short, it could only be done by riding very fast.

September 9.—As we had such vague ideas of the distance before us, we were anxious to make an early start; but we were now in Yün-Nan, the province of China in which more opium is smoked than in any other, and in which it is proportionately difficult to move the people in the morning. There is a Chinese proverb to the effect that an opium pipe is found in every house in the province of Kwei-Chou, but one in every room in Yün-Nan, which means that man and woman smoke opium universally.

When sleepy-looking people at length appeared, they were all liberal in their beautiful promises. The son of our landlord had some mules for the few things left behind, and he vowed that wherever we went, no matter how far or how fast, thither his mules should follow. The chair coolies, with the empty chair, protested that the distance and difficulties of the road had no terrors for them, and that as long as they could drag one leg before the other they would struggle on, even in the dark.

Huang-Fu had quite recovered, but Mesny's servant was now unfortunately very ill; we had ordered a litter to be prepared, but no one had believed that we should start until two or three days after the appointed time, for that is the Yün-Nan 'take-it-easy' style of doing things, so the litter was not ready. Fortunately the chair coolies were willing fellows, and with their aid a makeshift was improvised. After we had been waiting upwards of an hour a man strolled leisurely into the yard, with the air of one who had all the day before him. He carried a saddle over his arm, and was leading a hungry-looking donkey. He tied the animal up to the wall, put the saddle on its back, and carefully buckled one strap. This severe labour having been satisfactorily accomplished, he produced a lump of tsanba made up into a cake from the omnium gatherum of his bosom, and broke a piece off and gave it to the donkey. He put the remainder back into his coat, and stood placidly watching his animal enjoy this sumptuous repast. When the donkey had quite finished, and signified the same by looking round for more, the man fastened another buckle, rewarded his enduring animal for its remarkable patience with some more cake, and sat quietly down to smoke a pipe. The tsanba and tobacco fortunately held out until the operations were completed; and then there was such an evident air of satisfaction on the man's face, that we could do no less than congratulate him on the rapidity with which he worked, and he clearly thought that his exertions deserved something more than mere congratulation.

At length it was proclaimed that all was finished, and we thought we were really about to start, when someone discovered that the men wanted their breakfast. Who the men were it was impossible to say. I had noticed Huang-Fu for the last hour, alternately behind his pipe and a bowl of food. My Ma-Fu had been so busy with tsanba that he had left all the horses out in the rain. An enormous pan of rice that had been in the kitchen early in the morning was now all finished, and still the people wanted to eat. 'Everything comes to those who know how to wait,' and we had by this time been sufficiently exercised in the virtue of patience to observe with some amount of philosophy the steady progress of the hands of our watches, although it was with some misgivings that we saw those uncompromising machines indicate the hour of nine, as we emerged from the doorway of the house into the chilly rain. As for the rain and fog, except for five minutes when the sun made believe he was going to 'please again to be himself by breaking through the foul and ugly mist of vapour that did seem to strangle him,' rain fell and fog enveloped us incessantly the whole day; and so let us have done with that subject, as the worthy Marco would say. As the opposite sides of the narrow valleys we travelled amongst were nearly always hidden by the fog, the beauties of the scenery hardly repaid us for the dreariness of the day. The necessity of pushing on rendered note-taking anything but a pleasure, and my note-book soon became in a state of sop. After a steady ascent of 1,271 feet, we reached the summit of Jing-go-La, 12,300 feet above the sea, where the usual pile of stones was sheltered by a cherry tree, and the prayers were gradually distilled by the rain from the holy rags that fluttered in the dank breeze. We did not immediately descend, but merely following a contour along the side of the mountain we presently struck up another valley, and reached a little open grassy space in the dense forest of pines and oaks, where there was an empty hut which had at one time been a guard-house. We then again entered the forest, and here the trees were all festooned with drooping moss, in threads some three or four feet long, and so thickly were the trees clothed with this garment that it was sometimes difficult to know one kind from another. A march of twelve and a quarter miles brought us to another hut, with a little patch of barley in front of it. It had also been a guard-hut, and was now occupied by a hunchback about four feet high, who appeared to live by himself in this desolate spot.

We had a guide with us, who wore a coat made (he said) of the skin of a wild sheep, but the hair was of a deep sienna colour, and much longer than any I have ever seen. He said we should pass no more habitations of any kind before reaching Deung-do-lin, and as the Ma-Fus said they had some bread for themselves, and there was a fire of sticks inside the shanty, we decided to eat our sandwiches and hard-boiled eggs here.

Up to this point, from the summit of Jing-go-La, our road had been very good, but after another half-mile it became rather stony. We ascended a steep path by the side of a torrent, through a forest of pines, oaks, and rhododendrons, to the top of Pa-ma-La (14,307 feet). Here, the snow on the mountain-tops was not more than five hundred feet above the road; it had fallen the night before, and the thickness of the sprinkling, combined with the icy state in which the rain was dropping, warned us that in another month travelling would be unpleasant in Tibet. We did not yet fairly commence our descent, and some people we met told us, in reply to our usual query about the distance, that we had still another mountain to cross.

The road soon rose over a long spur that divided two streams. This was also called a mountain—Mien-chu-La—and its height was about 14,227 feet. From this, we again descended a short distance, and at last a gentler ascent over a grassy plain led us to the summit of Shwo-La, 14,307 feet above the sea, the water parting between the basins of the Lan-Ts'ang and Chin-Sha rivers. All these mountains are called collectively in Tibetan N'geu-La-ka, or the mountain of A-tun-tzŭ, and by the Chinese Pai-Na-Shān. The character for *Pai* means 'white,' and that for *Na* 'to bring.' This is probably an attempted translation of some old Tibetan name. A very stony and somewhat steep descent amongst rather bare hills brought us to another pine-clad valley. Here we found some traders encamped, who, in answer to our questions, could give us only the stereotyped reply, that it was quite impossible to reach Deung-do-lin before dark. Still we descended, and still, fortunately, the road was good, till at five o'clock some people told us the distance was at most twelve miles. By-and-by an old man with a donkey said it was not far, and soon we came to some cultivation and a house. But we found a veritable pigsty, knee-deep in filth and sludge, and two dirty little children came out, and said that there were houses *just below.* So on we went till we came to a place where the road bifurcated, and it was very doubtful which track we ought to follow. Guides and muleteers had long been left behind, and we had no one with us but a mounted Ma-Fu, who knew no more of the road than we did ourselves. The daylight was fast fading, a mistake would have been fatal, but it was necessary to make a choice, and we chose the lower road, which soon took us to the right bank of the stream. It now became dark in good earnest; anything faster than a careful walk was impossible; and we began to think we had taken the wrong road, especially as we saw lights and heard the barking of a dog high upon the hill on the opposite side of the stream. Still we went forward ; we could see nothing, but we could hear the torrent roaring sullenly below. The track was narrow and desperately rough ; at times, as far as we could judge by the sound, we seemed to be right over the stream, and at last the Ma-Fu, who was in front, came to a halt, and declared that the path went no farther. Looking behind us, we now saw a light in the valley above, and as, about half an hour before, Mesny had fancied that he had seen houses there, we began to think of turning back. Just at this moment there was a faint glare of light in front, but as fireflies had already been mistaken two or three times for the lights of a house, we took a steady look to assure ourselves, and at last a flare showed definitely, and there was no longer any doubt about it. It was neither firefly nor glow-worm, but it disappeared again soon afterwards, and we bade the Ma-Fu shout ; but he had as much voice as a mouse, so setting our lungs to work we had the pleasure of hearing the hills echo on each side, and re-echo many a lusty British halloo. Soon afterwards a brilliant light appeared, then another and another, and it became evident that people were out looking for us. We remained where we were, the lights gradually approached, and we found that an official sent on from A-tun-tzŭ had already made one expedition in search of us, and had now brought nearly all the villagers with him, and quantities of pine-splinters. Our road was now more brilliantly illuminated than Piccadilly. For a mile in front we could see the torches flaring at short intervals, lighting up some gigantic rock or the trunk of a tree, or showing for a moment the black depths of the valley below. The smell of the turpentine and the burning wood was very pleasant, and the prospects of something better than the forest for a roof were more pleasant still.

At a quarter-past eight we joyfully found ourselves in a capital house surrounded by a number of people, who expended every effort in making us comfortable. A large iron pan was brought; soon the pine splinters were blazing, and we enjoyed the grateful warmth. Villagers brought presents of rice, pork, and food for the horses; bread, butter, cheese, and delicious tea appeared, and subsequently some rice was offered us. Cushions and rugs were laid on the beds, and after a cigar I was able to enjoy a delightful sleep.

September 10.—In the morning the Ma-Fu brought some eggs and cooked them. The people of the place supplied us with unlimited bread, butter, and tea, and we seemed to get on rather better without servants and baggage than with those luxuries. My Ma-Fu on foot, and driving three horses, arrived in the night, but no one else appeared until about ten o'clock, when Chin-Tai and Chung-Erh rode in. They said they had travelled all night, and at daybreak they had found a house where they had had some tea. Huang-Fu and his pipe came next. He reported that in the night he had tumbled over a cliff; that the mule had lain on top of him for an hour, and that he had been unable to extricate himself until it moved. He said the mule was none the better, and himself none the worse, and that he had not broken his pipe. My chair soon followed, and I was able to wash my hands with soap, and do some writing.

Deung-do-lin is a little village situated at one side of a ravine, where, on the opposite side, right on the top of a high hill, is the Lamassery of Deung-do-lin-

Tz'u. The word Lamassery is of European origin, and is formed much as Nunnery is derived from Nun; it can have no connection with the Arabic word 'Serai.'[4] The Chinese word for a Lamassery is 'Lama-Tz'u,' and 'Deung-do-lin-Tz'u' merely means the Lamassery of Deung-do-Lin.

Monsieur Renou, who was the first missionary in Tibet, came this way and stopped at the Lamassery, where he was well and hospitably received by the chief Lama, a man of very superior learning and intelligence, and the living Buddha to boot. Monsieur Renou possessed a large telescope, to which this Lama took a great fancy, and wanted to buy it, or take it, whether Monsieur Renou would or would not. But the latter would on no account part with it at any price; at last, however, yielding to importunity, he made a bargain that if the head Lama would teach him to speak Tibetan in ten months, he would make him a present of the telescope. This was at once agreed to, and setting to work in right earnest the extraordinary spectacle was witnessed of a Roman Catholic missionary, the pupil of a living Buddha. Since this time, the inmates of this Lamassery have always been friendly towards foreigners; and while we were at Deung-do-lin the chief Lama came down to see us, and brought us several little presents, tea, incense sticks, and medicines.

September 10.—In all probability if poor Cooper had taken this road instead of that by Wei-Si, he would have succeeded in his enterprise, but the Chinese authorities at Bat'ang took very good care to

---

[4] I should not speak quite so confidently. The word *Serai* was adopted by the mediæval Mongols for 'Palace' or 'Edifice' apparently, and became the name of more than one of their royal cities. But I do not know of any use of the word Lamaserie or Lamassery before Abbé Huc's.—*Y.*

keep him in ignorance of it, and to despatch him by the valley of the Lan-Ts'ang, where the physical difficulties were only equalled by those due to the hostility of the Lamas, and the open opposition of the officials.

It was late in the day before the baggage arrived. Instead of having been carried on animals, it had been brought by men, women, and children, utterly unaccustomed to carrying loads; the biggest and heaviest boxes had been carried by the smallest children, or weakest women, while some great strapping fellow had been burdened with no more than a hat-box.

We changed our carriers again here; those from this place could be taken no further than Sha-lu, only nine miles distant, and we sadly missed the excellent Chao, with his careful organisation. The petty officials we had with us were very helpless; our journey henceforth was a great contrast to our march to Bat'ang, and everything seemed to be in a perpetual state of 'east and west.' In these countries there are no regular coolies to be hired, and animals or porters can only be obtained from the officials or petty chiefs. The animals are often away in the mountains grazing, and the people, accustomed to have their services pressed without payment, are always unwilling to do anything, for it is next to impossible to make them realise the idea of distinguished people who pay.

The house we were in, though large, was in an exceedingly bad state of repair; the roof leaked in dozens of places, and the rain came through until there was scarcely a spot in the room clear of the drops, which brought a quantity of mud with them in their passage through the roof. Everything was out of the perpendicular; the roof of another part of the house had already fallen in, and if the floods have since descended very often on that house, it certainly must have fallen before now.

Everything arrived before night, except Mesny's boy ; we were very uneasy about him, and we sent out search parties over and over again ; but they never went further than the nearest corner, where they hid themselves for an hour or two, and returned with the favourite Chinese expression ' Li-Ta,' of which the accepted translation is ' coming immediately,' but which, as far as my experience goes, always means ' not coming for hours.'

September 11.—The muleteers had come to us in the evening, begging and praying for an early start ; this was something quite new, and we unsuspiciously got up at an uncomfortably early hour. But we soon found that this was a most unnecessary proceeding. The men had no intention of even writing out the agreement until they had talked over matters ; they could not be expected to discuss such a grave subject till after breakfast, and it was quite impossible for them to eat before smoking their opium.

When the paper was eventually prepared, they began to think about looking for animals, and they then discovered that there were none. By-and-by a man strolled casually into the place where our luggage was collected, and our officials, whom we had been roundly upbraiding for their dilatoriness, pointed proudly to the fact that a carrier had already arrived. After looking around him for a few minutes, this man took hold of a box to see how heavy it was, and with the perversity of an English railway porter, seized it by a feeble fastening, never intended to bear any weight; he then sat down and deliberately lighted his pipe. While he was thus agreeably employed, an old woman came, and, after regarding the box attentively some time, did precisely what the man had done before, and then went out. At length, however, five head people arrived, and succeeded, after much hot argument, in dividing the baggage into five portions, an achievement that was immediately celebrated by the consumption of much tobacco. At last, a very small child appeared, the biggest box was put on its back, and it trudged away; then a man went off with something light, and a woman followed with a heavy case; no man, woman, or child took more than one package, even though it weighed only a few ounces; and no box, however heavy, was carried by more than one person; but the heap of baggage grew smaller by degrees, and beautifully less, and at intervals on the narrow path along the hill-side, we could see our property slowly disappearing in the distance.

During the morning Mesny's boy came in declaring he was now quite well, only frightfully weak. The remedy for his illness had certainly been of a kill-or-cure nature. He had been dosed *ad extremis* with pills and chlorodyne, and he had finished up by passing a very rainy night in the open forest. Our people said his illness had been brought on by eating butter. He had been seen to buy half a rupee's worth—about a pound—and melt it down; and the stuff had disappeared immediately afterwards.

When about half our baggage had gone, we were tired of waiting, and started down the valley, the road winding in a serpentine track two to three hundred feet above the stream, generally following one of the contours. The scenery and climate changed very rapidly, for we seemed to get out of the rainy district into a dry and barren one. The wild gooseberries, currants, and oaks had all disappeared; the high and exceedingly steep hill-side being nearly bare, except at the tops, where pine forests appeared every now and then when the clouds lifted from the summits. The spurs ran down very steeply to the streams, but their tops were level, and well cultivated with barley and buckwheat. There were plenty of houses here and there, with walnuts, pear trees, and wild peaches; the fruit of the last not eatable even by the Chinese.

Soon we caught sight of the valley of the Chin-Sha-Chiang, where the river appeared to run in narrow gorges between high and bare mountains. The village of Poung-dze-lan was on a spur close to us; this place is sometimes called Pong-Sera, and in most maps is wrongly shown on the left bank of the river. We were not yet, however, to rejoin the River of Golden Sand; and two huge mountain passes were still to be crossed before we could follow the banks of the mighty stream. Our road turned abruptly away, and plunging into a narrow valley, we ascended some little distance to the house of the chief, who rules from Deung-do-lin over all the intervening country, and for some distance to the south.

During the day we passed a good many ruins by the road-side and amongst the houses. In all the ruined villages there were the remains of a high square tower, deserted and crumbling away, built exactly like those so invariable in the Sung-P'an-T'ing district. Some of the villages had been partially rebuilt, and inhabited houses and unroofed walls were sheltered by the same walnut or persimmon [5] tree, but

---

[5] *Persimmon* is the American name of *Diospyros Virginiana* of the United States, and has apparently been applied by Americans and Englishmen in China to the kindred *Diospyros Kaki*, producing the 'date-plum' (or 'keg-fig' of Japan).—Y.

in no case had the towers been repaired, and they stood fast decaying monuments of a people rapidly passing away.

Welcome evidences of approaching civilisation were again apparent, as we saw some sloping roofs, and passed over a bridge provided with the luxury of a hand rail.

Outside the house where we stayed, in the village of Sha-lu, a beautiful stream of clear water leaped merrily down towards the river, turning as it went a great prayer cylinder in a small building that stood under a grand old walnut tree; and a string stretched from bank to bank served to carry a number of Mani rags that fluttered in the breeze, and after passing through a fine portico, where there were benches for people to sit on, the guests on ascending the staircase were expected to give a turn to another cylinder that stood at the corner of the steps.

The name of the chief was Wang, and he held the Chinese rank of Chien-Tsung. He remembered the missionary Renou, who used often to come and pay him a visit. His house was very large, and we had a comfortable, though small, room on the ground floor, with a wide verandah, where we could put our table and sit. There was a square open court in front of the verandah, and above the buildings on the opposite side we could again see the parrots flying about amongst the pines and oaks.

The chief had a huge dog, kept in a cage on the top of the wall at the entrance. It was a very heavily built black-and-tan, the tan of a very good colour; his coat was rather long, but smooth; he had a bushy tail, smooth tan legs, and an enormous head that seemed out of proportion to the body, very much like that of a bloodhound in shape with overhanging lips. His bloodshot eyes were very deep-set, and his ears were flat and drooping. He had tan spots over the eyes, and a tan spot on the breast. He measured four feet from the point of the nose to the root of the tail, and two feet ten inches in height at the shoulder. He was three years old, and was of the true Tibetan breed.

CHAPTER VII.

REGION OF THE RIVER OF GOLDEN SAND.

II. SHA-LU TO TA-LI FU.

Hazels, Pines, Currants, Rhododendrons—Pass of Jing-go-La—Ka-ri—Houses there—Verdant and Peopled Region—N'doh-sung—Domestic Decoration—Rapacity of the Fu of Wei-si—Chinese Agriculture Reappears — Hospitable Welcome of Ron-sha—Toilsome March in Rain—Reception at La-pu—Our Lodging there—Friendly Activity of our Conductor—Walnut Trees, and Marco Polo's 'Cloves,' *i.e.*, Cassia Trees—Tibet passing into the Background—Village of Jie-bu-ti—The Banks of the Chin-Sha Regained, and Re-appearance of Orange-trees and Parrots—The Mu-su People—March down the Great River—Lamas still seen—Mu-khun-do—The Pu-Erh Tea, and How to Make it—Ku-deu—Fine Position—The Mu-sus—European Aspect of some—Honesty of Carriers—Drunken Escort Officer—Ch'iao-T'ou—Tea-houses Reappear—San-Hsien-Ku—Festive Reception—Shih-Ku, or Stone-Drum Town—Adieu to the River of Golden Sand—Mount Chin-Ku-P'u—Changed Scenery—Dense Population on Rice-plains—Inns begin again—Oppression of the 'Lekin'—City of Chien-Ch'uan-Chou—A Henpecked Warrior — Fair Words of the Chou—Road through Populous Rice-lands—Lake Basins—I-Yang-T'ang—Rudeness of Local Officer at Niu-Chieh—Wretchedness of the People—The Rude Officer Abashed—Position of Niu-Chieh in Lake-basin—Probable Changes in Water-levels—Dilapidated City of Lang-Ch'iung—Civil Landlord—Pu-Erh Tea—Sulphur Spring—Opium-smoking—Damp and Dreary Aspects—The Erh-Hai, or Lake of Ta-Li—Road along the Lake Shore—Arrival at Ta-Li-Fu.

September 12.—The mules were all away feeding in the mountains, and the muleteers only came in late in the afternoon to prepare for the next journey, which, they said, was a long and difficult one; but the day was so pleasant, the house so comfortable, and the people so civil, that we easily put up with the delay.

September 13.—Mesny's boy was now much better; he had been doctored with port wine during the last day or two, which, he had the bad taste to declare, he did not like half so well as pills. He was now set up on a mule, and performed the journey in safety.

We marched up a narrow valley, with very steep and wooded slopes on both sides, the stream at the bottom running through a thick jungle of briars, hazels, small poplars, and currants. I had often seen the hazel-nut tree before, but was never sure about it until now, when I found some nuts nearly ripe. There was a wild currant in this valley, with elliptical berries, as large as two ordinary currants. These were hardly ripe, and rather hard; they had the full flavour of the English red currant, but were very sour. Another small tree had exactly the leaf of the English oak, with long thorns on the branches. There were two kinds of rhododendrons, one just like that common in England, the other had broader and rounder leaves. A march of six and a half miles brought us into a dense forest of pines and magnificent rhododendrons, the latter growing twenty feet high. There were also plenty of wild strawberries.

The road, though steep, was very good, and for the first six miles was not stony. It then became rather rocky for a mile, and ended in a couple of miles of steep zigzag, which brought us to the summit of mount Jing-go-La, 13,699 feet above the sea, and 4,412 feet above Sha-lu. There was a very cold wind at the top, with rain-clouds hanging about, but we escaped with only a very slight shower. This mountain is a spur running east and west from the great range that divides the Lan-Ts'ang from the Chin-Sha-Chiang. I think the general name of this range must be properly Jing-go-La, for one of the mountains we crossed on the 9th had this name, although they gave a different one for all the mountains collectively. Arrived at the summit we had to descend an exceedingly bad, steep, and stony road, through a forest even thicker than on the northern side, and we could rarely see more than a few yards in any direction. We passed neither house, shed, nor hut, until sixteen and a half miles from Sha-lu, when we came across some woodcutters busily at work cutting down and chopping up the magnificent trees for firewood and splinters, which take the place of candles, and are the only means of lighting adopted here. There was one little hut in this place, but nothing else till we reached the village of Ka-ri, situated in a very small open space, where there were a few fields of buckwheat and cabbages.

The entire population turned out to meet us, and respectfully saluted us, in pure Tibetan style, by putting out their tongues, and afterwards each family (there were only four) brought a present of pork, rice, or wine. We were lodged in the house of the chief man, who at once brought milk and tea for us, and barley for the horses. He fortunately did not bring me any incense-sticks, for I had by this time received enough to set up a High Church establishment in London.

The architecture in this neighbourhood shows that the inhabitants have advanced a stage in civilisation beyond the pure Tibetans of the upper plateau.

When first an uncivilised man builds himself a shelter he is contented with a flat roof of mud; but as he advances he finds the inconvenience of the dirty drops that percolate through the covering of his house, and he casts about for some means to obviate the discomfort. His next idea is to place some sloping planks over the flat roof to shoot off the rain, and on these he puts a few stones to prevent the wind blowing them away.

The device is simple, but effectual, and this is the system on which we found many of the houses roofed that lie hidden amongst the spurs of the ridge on the western bank of the Chin-Sha.

The house at Ka-ri was not built altogether on this principle. The roof of the lower story was flat, and made of mud, but the upper story that covered half the lower was roofed with wooden battens, slightly sloping.

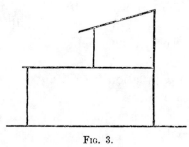

Fig. 3.

The room in which we lodged was the upper one; it had no wall in front, and was open to the flat roof of the room below.

September 14.—The early morning was delicious. I stepped outside, and whilst I was enjoying the lovely view, an old woman came on the roof to perform the devotions for the house. On the top of a wooden post there was a small clay fire-place, the chimney of which was an old black earthenware teapot. Some yew branches were thrust into this fire-place, and lighted, and some other ingredients were poured into the teapot, and the morning prayers were thus complete. Properly speaking, incense should have been burnt, but either it was too expensive, or else it was thought that the gods would not find out the difference.

From Ka-ri we followed the stream by an excellent and easy road to its junction with Chin-Chŭ, a beautiful clear river sixteen to twenty yards wide, coming from the west, and flowing rapidly through a pretty valley. This was very narrow, cliffs on both sides in many places shutting in the stream; but in the bottom open flats, two to five hundred yards long, and one to two hundred yards broad, were cultivated with buckwheat and a few cabbages. The hill-sides rose on each side very sharply, varying from slopes of thirty degrees to steep and precipitous cliffs, the mountains generally running up about a thousand feet above the stream, and covered everywhere with trees, mostly holly-leaved oak. The road ran by the river-side through a jungle of oaks, like the true English oak, holly-leaved oaks, poplars, thorns, barberries, peaches, wild plums, and magnificent walnut trees. The last only could be called trees, the others being little more than shrubs. Everything was very green, plenty of good grass growing in the open spaces between the trees. The country was more thickly populated. We passed a good many houses and small villages, though the amount of cultivation was still small.

After marching fourteen miles we came to a village of eight families, called Shieh-zong, on the opposite side of the river, where there was a bridge, and a quarter of a mile lower down we found a low table in a field, and two stools with cushions. A group of people were collected round a fire under a fine walnut tree, where a big pot was boiling. At our approach they all rose to salute us, headed by the son of the officer whose house was ready for us at the next village, and as soon as we were seated, tea and eggs were placed before us, and food, that had already been prepared, was given to our horses.

Before the after-breakfast cigar was finished, all the mules came up. Their loads were quickly off their backs, and they enjoyed a good roll, while the mule-drivers, lighting another fire, made their meal of tsanba. An easy march brought us to another bridge, and crossing it, between two rows of people, who respectfully saluted, we rode over the few intervening yards to the village of N'doh-sung (six families), past another line of people drawn up, who bowed and gave us a welcome to their humble home.

We found the son of our host at his door waiting to receive us, and he led us through a spacious court, surrounded by two-storied buildings, to a beautifully clean and new room, about forty feet square, on the upper story. The whole front was open to within two feet of the floor, but provided with shutters to close it. Two rows of wooden pillars, roughly carved at the top into a sort of capital, ran down the length of the room to support the roof. One wall was of plaster, and ornamental paintings had been commenced, but as the house was not yet finished the decorations were not complete. The artist employed had evidently not been a person of much skill, for although the intention was good, the execution was villainous.

There was a hexagonal pattern in black, with a white border, on a grey ground. There was a dull red cross in the centre of each hexagon, on which the imagination of Huc would have discovered proofs of a large Christian community. One row of hexagons was supposed to be equilateral, the next row with two long sides, but all the lines were crooked and the figures irregular, so that the beauty of the design was lost. A kind of frieze had been attempted along the

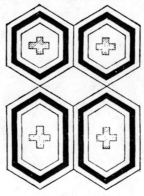
Fig. 4.

top of the wall, but it was so badly done that I utterly failed to evolve the pattern. The general arrangement of the colouring was really excellent. It was interesting to notice this first attempt at mural decoration among an uncivilised people, and the origin of the pattern afforded a wide field for speculation.

The house belonged to the native officer of the place, who held the Chinese rank of a Pa-Tsung. The Chinese magistrate at Wei-si, a Fu, was of a different stamp to Chao, and was notorious for his love of money, and the oppressive 'squeeze' that he extracted from all under him. He had heard that the Pa-Tsung of N'doh-sung was a wealthy man, and had demanded a large sum of money from him on some frivolous pretext; this officer had in consequence made over his house and property to the village, as he was afraid that the Fu of Wei-si would confiscate them; and he had repaired to the prefectoral city to try and come to some arrangement with the rapacious magistrate. At the time of our visit he had not returned, and it was his son who had so worthily done the honours of the paternal roof; for although the property had been vested in the village, in reality it remained in the possession of the Pa-Tsung.

September 15.—From N'doh-sung we followed the right bank of the river, and everything pointed to the fact that the highlands with their rich pastures and herds of cattle were rapidly being left behind. The first patch of Indian corn was just outside the village; the chirp of the grasshopper was heard in the fields; the river, instead of being a mountain torrent tumbling headlong downwards, was a smooth, though rapid stream; the sun was shining in a clear sky; and the soft and balmy air had lost the crispness of the mountains. Here men and women were at work in the flat fields by the river; and half a dozen naked urchins suddenly appeared scampering across a bridge to look at the foreigners. Passing through the villages, grown-up people and children ran out in a hurry, and gazed with a stony Chinese stare, though here and there some Tibetans still received us with the respectful salute.

Cairns by the road-side, or rags fluttering from the tops of poles, still reminded us that we were in the land of Buddha; but the ever-increasing amount of buckwheat and Indian corn growing in fields that had already yielded a crop of black barley, showed the presence of the agricultural Chinese.

Steep and high hills, sometimes broken with precipices and cliffs, rose up a thousand feet on both sides; the slopes were well wooded with pines and holly-leaved oaks; and at the bottom of the valley, which was very narrow, the spurs running out, all ended in flat points about a hundred yards wide, cultivated with Indian corn, and buckwheat, another crop appeared, called by the Chinese Hung-Pai, the grain of which is small and red, very like sago, and is used for making flour. Lower down, a crop for which we could get no Chinese name, was called by the Tibetans M'beh ; and at last the hearts of our Chinese followers leapt with joy as, down by the river-side, a field of true rice, in Tibetan M'jeh, was an earnest to them of the good things below.

Houses and villages were scattered about the valley, and an excellent road showed the existence of some traffic, though during the day we met no more than four donkeys ; this, however, was something, as for many marches, with the exception of our own long and straggling baggage-train, we had seen neither coolies nor animals.

The road now ran, like some English country lane, through a wood, where the murmuring river, sparkling in the sun, could be seen through a jungle of thorns, barberries, and briars, and where peaches and plums mingled their delicate leaves with the deep foliage of the firs and holly-leaved oaks. A few pears, pomegranates, and persimmon trees grew about the villages, and long groves of splendid walnuts just now were yielding a rich harvest to their proprietors.

Close to the village of Sa-ka-tying (three or four families), our host of N'doh-Sung appeared, and, saluting us as we dismounted, led us to a couple of arbours erected by the road-side, where, sheltered from the rays of the sun by fresh green branches of oaks, chestnuts, and firs, improvised tables and stools had been provided with cushions. Fresh milk and tea were laid before us, and the villagers of Sa-ka-tying brought gifts, and welcomed their unaccustomed guests.

Our ride to-day was a triumphal march, and as we approached Ron-sha (twelve families), the villagers all turned out, and most of them saluted as we passed. The road to our house, the passage to our room, and the apartment itself had all been decorated with freshly gathered pine branches; and walnut leaves were strewn over the ground, and on our beds. All the good things of the village were placed at our disposal, and our followers revelled in the luxuries of pork and rice; but not to let us forget that there were here some Chinese amongst the population some inquisitive eyes could be seen peering through the thick foliage that embowered our dwelling. Treated with such distinction, and attended to by such thoughtful hosts, it would have been thankless to have grumbled at the swarm of house-flies that pervaded the place—immolating themselves by dozens in our tea directly it was poured out, and half-concealing the food upon our plates—or to have noticed the ping of the mosquitoes, which was heard for the first time for many a long day.

September 16.—Leaving the river Chin-Chŭ, we ascended a stream, the valley of which for the first half-mile was flat, and cultivated with rice and buckwheat; but beyond this there was no longer any room for cultivation, the steep hill-sides running right down to the stream; and after another mile and a quarter, the road plunged into one of those dense woods of which we had seen so many. The trees here were most puzzling; everything looked like what it was not, first the oaks had been mistaken for hollies, and now the chestnuts looked like oaks.

Before we had gone very far the rain came down. The road was exceedingly bad and steep, and it seemed as if we would never reach the summit. As we ascended the rain became heavier and the mist thicker, the road grew worse, and still we mounted. Never before, during all our weary travels, had we clambered thus, for hour after hour; and it was not until we were 5,200 feet above Ron-sha, that at length we gained the summit of Rûng-geh-la-ka, 12,134 feet above the sea, where, enveloped in mist and pouring rain, we could see scarcely anything. We then descended five thousand feet, by another steep and slippery road, to the valley of the Kung-Chŭ, where there is the village of Shio-gung, called by the Chinese Hsiao-La-pu (or Lesser La-pu). The stream we followed from the summit must at times become a roaring torrent, for near its junction with the Kung-Chŭ, trees covered with mud for a height of five or six feet, and with stones in the lower branches, showed the level to which the water must occasionally rise.

At La-pu, called by the Chinese Ta-Chio, we were taken to the house of the chief man in the place, and were immediately waited on by the nephew of the Fu of Wei-si, in whose district La-pu is situated. He was a very young man, named Sun, and a tremendous talker; but though his tongue wagged incessantly, his conversation was more than usually sensible; he was a very active young fellow, and told us that he had already arranged for coolies to take us a two days' journey, because this district, which extended from Shio-gung, or Hsiao La-pu, to Jie-bu-ti, or Chi-Dzûng, and contained three hundred families, was an agricultural one, in which there were no animals to be procured.

Shio-gung contains twenty-five families, and is the seat of a native hereditary officer, whose title is that of Mu-kwa (or Moquoor of Cooper); at Ta-

Chio itself there were four of these native officers, called Mu-kwa, and thirty families.

The military officer, a Chien-Tsung, also came in to see us; and a dish of woodeny peaches, a plate of beautiful fresh walnuts, and quantities of sunflower seeds were put on the table, the latter in this part of China generally taking the place of the water melon seeds, and as a diet being equally futile and frivolous. Our two guests, however, thoroughly relished them, and no parrot could have been more skilful than these two men in the consumption of these delicacies. The seed was put in at one corner of the mouth, the kernel extracted, and the husk ejected at the other corner, with the rapidity of cartridges from a Martini rifle; saucer after saucer disappeared, and a small mountain of husks was formed at the feet of each of our friends. The people at this part of Yün-Nan seem to divide their time between opium-smoking and eating sunflower seeds.

The Chien-Tsung had a very evil face, but both he and Sun, who was a great dandy, with excessively long nails, were exquisitely polite, never made use of a wrong expression, and talked in the stilted language affected by the Chinese literati. They both visited us again in the evening, and remained a long time, trying to find out something about us, as the despatches they had received had been most contradictory. In reply to an observation that the Margary proclamation was nowhere posted in Yün-Nan, Sun said that a copy had been sent to the head man of every village, but that they had put them away in their boxes for fear of their being spoilt—like the Dutchman, who left his sheet anchor at home, for fear of losing it.

When the Chien-Tsung took his leave, he apologised for not having done anything more for us, and expressed his hope that on our return to our honourable country we would present his compliments to the Queen.

September 17.—We were lodged, *à la Chinoise*, on the ground floor of the house of the chief of the village, a large building, with a square open court, on one side of which there were stables; at the other side, in front of our room, the court was divided from a long shed by a low wooden partition, with a grating above. A vast number of people were collected here, and peering through the bars with expressionless faces, gazed at the coolies squabbling over the baggage, amongst which little Sun was running about, displaying an activity utterly foreign to the manners of the ordinary Chinaman : leading ponies himself, pushing the people about, and abusing them in a tone that considerably helped to swell the din prevailing in the place. The bearers and horses were so numerous that the yard, large as it was, seemed filled; and the crowd was further swelled by idlers and loafers, of whom, in China, there is always an unlimited number ready to spare a few hours on the smallest provocation.

Little Sun did his work thoroughly well. He found horses and saddles for our servants to ride, and coolies for the baggage; he started them in good time in the morning, and drove them along the road, so that they arrived early enough for all the baggage to be handed over before five o'clock, after a march of nineteen and a half miles. He rode with us to our destination, and altogether took very good care of us.

We were sorry to take leave of our friendly conductors, who had contrived to make the journey thus far very agreeable. They, too, seemed rather sorry that they had not consented to take us further; but they had doubtless been deterred by the fear of being squeezed by Sun, who, in all probability, kept a goodly proportion of the rupees paid him for the day's work.

We crossed the Kung-Chŭ [1] by a spar bridge, and the road followed the right bank of the river, which was, like a peat stream, of a clear brown colour, another indication of the gradual lowering of the country around, and a remarkable contrast to the Chin-Chŭ, the sources of whose bright blue water are probably in snow.

The river runs through a perfectly flat valley, varying from two hundred yards to half a mile in width, and is very similar to many a Tibetan valley in the plateau above: a fine stream winding through a flat plain, bounded by steep and high mountains sloping sharply down to it; but in one case, glorious pasturage cheers the heart of the horseman, as he looses the bridle, and lets his steed with unshod hoofs fly over the elastic turf; but in the other, close cultivation and paddy fields prohibit any attempt to traverse the valley, or deviate from the rocky or muddy path, at the foot of the hills. But this valley supporting three hundred families, would barely suffice in Tibet for ten or twenty; and the numerous villages and houses, all looking prosperous and well, were cheerful sights compared with the miserably sparse and poverty-stricken population of Tibet. The valley was chiefly cultivated with rice, in fields that had already produced a crop of wheat or barley. There was a good deal of Indian corn and buckwheat, and late in the day we saw a few patches of tobacco. Magnificent walnut

---

[1] In Captain Gill's map and minute itinerary, *Kung-Chŭ* is the name of a tributary of the Chiu-Chŭ, passed on the 15th near Ron-sha, whilst the river now mentioned, and followed to its junction with the Chin-Sha, remains anonymous.—*Y.*

trees surrounded the villages, and grew by the roadside in very great numbers. I did not see any of the 'loupes,' so well known in the Caucasus—excrescences that grow on walnut trees, and are so valuable for veneer. There were chestnut trees also, the fruit not yet ripe ; and the Kwei-Hua, a tree 'with leaves like the laurel, and with a small white flower, like the clove,' having a delicious, though rather a luscious smell. This was the Cassia, and I can find no words more suitable to describe it than those of Polo which I have just used.

The hills on each side were densely wooded with pines and holly-leaved oaks ; but the trees seemed to change in appearance with the climate, and approximate to one another in a very remarkable manner. There was now a rhododendron with leaves which, instead of being as usual hard and shiny, were quite soft, and unless I had found one solitary flower, I should have doubted the identity.

As we proceeded everything looked more and more Chinese : people in turbans standing about in great numbers, and turning out to stare ; pomegranate bushes near the doors of the houses ; great pumpkins trailing on a trellis in a garden, or over the road ; fine garden crops of beans and other vegetables, so dear to the heart of a Chinaman, all proclaimed the fact that Tibet was becoming more and more remote ; though still there were a good many Tibetan cairns by the road-side, and prayer-rags floated from poles in the villages, where some of the walls of the houses were of wood, and some of mud, and where the roofs were still of sloping wooden battens.

The road skirted the foot of the hills, and was generally good ; though in one or two places the mud was very deep, making our horses flounder about, and splash the dirt and water in every direction. We followed the Kung-Chŭ[2] to its junction with the Chin-Sha-Chiang, which we found the same muddy river we had left at Bat'ang. The clear stream of water shot itself into the brown fluid, and appeared to be making a vain attempt to retain its purity, but after a few feet gave it up as hopeless, and was lost in the dirty liquid.

The village of Jie-bu-ti, called Chi-Dzûng by the Chinese, and containing twenty-five families, was at this junction, and just as we arrived a salute of three matchlocks was fired by some out of a half-dozen ragged soldiers drawn up in a line, who welcomed us by grovelling on the ground at our approach. Here Sun had ordered an extensive repast to be prepared for us at the house of the chief man, and our host brought some delicious rice-cakes. These are made by steaming (not boiling) rice, until it is almost reduced to a pulp; it is then thoroughly pounded in a mortar, made into round cakes six inches in diameter, and about an inch thick, and allowed to dry slowly. When eaten the cakes are cut into slices, toasted, and served up hot; in this manner they are very like the British muffin, but without butter are much better than that homely article of food would be.

The house was a most comfortable one, with excellent upper rooms, and a nice little garden, where there was the first orange tree we had seen since leaving Ssŭ-Ch'uan, and where a good many green parrots were flying about.

The remainder of our journey, though short, was very bad; a wet stiff clay gave our horses a hard task, and a steep spur that projected into the river, and over which we passed at a height of about four

---

[2] See Note on last page but one.—Y.

hundred feet, by a steep and sticky zigzag, did not make it any easier. A heavy shower came on as we mounted this spur, dignified with the name of Mount Lu-jiong-la-ka, and my note-book fared but badly, as the rain melted the glue in the binding, and threatened to wash out the writing. The shower did not last long, and the sun, bursting out, lit up a double rainbow in a valley opposite, where dense black clouds were helping to increase the already swollen waters of the River of Golden Sand.

At Lu-jiong (fifty families), where we slept, we found a commodious though airy apartment quite open along the front. The roof was, however, water-tight, so, notwithstanding the heavy rain that fell all night, we were comfortable enough. The luggage all arrived in good time, and was counted and handed over by five o'clock. Then little Sun took his leave, and rode back to La-pu.

At Lu-jiong, we found some of the Mu-su people (or Moosoo), whose language our Tibetan Ma-Fus could not understand.

September 18.—The Chin-Sha-Chiang was here about eighty or a hundred yards broad; it was now so full that the bed was partly flooded, and trees were rising out of the water; it was a fine, swift, and muddy river, rushing through flats varying in width from a quarter of a mile to a hundred yards; these were cultivated with rice, where the soil was not too light; and where that favourite crop could not be raised there was Indian corn, small millet, or beans. Forests of pines and holly-leaved oaks clothed the sides of hills, which varied from five hundred to a thousand feet in height, and sloped sharply down, ending abruptly at the edge of the perfectly flat little plain between them and the river. Now and then they shot down to the water, altogether cutting off the plain; and here and there they ended in rocky and precipitous spurs, where the surging stream swept by at the foot of the cliffs. Numerous villages and houses were passed, and the scenes became more and more Chinese. Here a solitary Lama, in his red coat and yellow girdle, standing by a pile of stones, still reminded us of Tibet; but the numerous vegetable gardens, with their neat crops of turnips and beans, a few gigantic and gaudy sunflowers, a little patch of carefully tended Indian corn, pumpkins trailing over a trellis-work, and straight lines of tall poles covered with fresh green crops of climbing beans, showed us that we were now truly in the land of the careful and vegetable-loving Chinese.

Seven miles from Lu-jiong there was a covered bridge; presently we saw a buffalo standing in a field in the usual stolid manner of these uninteresting animals, and soon after we met a regular Chinese coolie carrying two loads on a pole over his shoulder.

The road was bad all the morning, through deep and stiff clay, which made cruel work for my unfortunate grey. He was far the best horse in our stud, and his performances fully carried out the promise his looks gave me when first I saw him. The bay and the chestnut were both sick and footsore from want of shoes, and were sorely out of condition.

We breakfasted at Mu-khun-do (*Chin.* Hsia-Ken-To) in a Chinese house, where an inquisitive Chinese crowd first peered in at us through every available crack, and then, gradually edging in by degrees, filled the little room and made it unpleasantly hot; for the thermometer was again getting amongst the eighties. There was an old man here who had been to Lassa, but it was long ago; he saw me smoking a cigar, and said that at Lassa people smoked 'hookahs,' the body made of brass, the long stem made of wood. He said that fire was put in at the top, and that it was handed round from one person to another. In fact, he described the water-pipe exactly, using the word ' hookah.'

Although we were now amongst the Chinese, we found that the people still used milk in one form or another; here it was like Devonshire cream.

In this part of China, nearly everyone drinks the celebrated Pu-Erh tea. This is pressed into annular cakes, and makes a strong black tea, which, according to the Chinese, does not affect the nerves. In other provinces it is handed round, after a heavy dinner, to assist in the digestion of the greasy messes that have been consumed, taking the place of the European 'chasse.' To make it properly, however, is a serious undertaking; the people of Yün-Nan first roast it very slightly, and then pour boiling water on it. But the true gourmet lets this stand for a minute or two only, when it is thrown away, and fresh boiling water poured over the leaves. This now stands for half an hour or more in a thick earthen pot over a few glowing sticks of charcoal, which just serve to keep the water from cooling. It was only on great occasions that this elaborate process was carried out by ourselves; but the true connoisseur of the Pu-Erh district never omits an item in the preparation. Cakes of Pu-Erh tea are often given to travellers as presents, and we had accumulated a considerable quantity. When made properly, or, indeed, if simply prepared with boiling water, the tea is delicious.

After we had emptied our last cup we bade adieu to the master of the house, and passing through the yard, where some of the familiar big flat baskets of

Indian corn were put out in the sun to dry, we mounted our horses and rode away through the village, where there were some houses of sun-dried bricks, the first we had seen, and beyond, under a clump of trees, some coolies eating bread instead of the favourite tsanba. This could not now be procured, much to the discontent of the Ma-Fus, who said they were already feeling ill from the change of food and climate, and could go no further than our next halting-place. They changed their minds eventually, and came on to Ta-Li-Fu.

We now marched over an excellent firm road, but men and animals were all tired, and we did not get on very fast. The journey was pleasant: sometimes we rode between hedges, with fine crops at each side; sometimes the road skirted a wood of pines and holly-leaved oaks, and now rising over a little spur that threw itself into the river below, we obtained a fine view of the fertile valley from the summit.

About eleven miles from Mu-khun-do, we saw the first *bonâ fide* Chinese village on the opposite side of the river, with whitewashed walls, gables, and roofs of tiles. Near this the river was a quarter of a mile broad, as it was much swollen, and covered a great deal of the flat low-lying ground bounding it; nevertheless, considering that it runs many thousands of miles before it reaches the sea, the magnificent volume of water is quite astonishing.

We arrived at Ku-deu after a march of twenty-eight and three quarter miles. This town was entirely destroyed during the Mahometan rebellion, and is now quite new, and very poor.

We were first taken to a temple, which was certainly swept and garnished, but there was no place to cook in, nor a room for the servants, nor were there any bedsteads, tables, or chairs. We were next shown into a room, open along the front, where piles of grain in one corner, baskets in another, and logs of wood in a third, did not leave much room in the middle for furniture, even if there had been any; so we sent cards to the officer of the place, a Wai-Wei, and asked him to find us a lodging. He came immediately, in his official hat, to wait on us; and soon found an upper room in a house close by, where we made ourselves tolerably comfortable.

The situation of Ku-deu, or Chi-Tien, is very fine. The mountains recede on all sides, leaving a circular basin, two to three miles in diameter, rising up steeply, broken with many spurs and ridges, with fertile and well-watered valleys running up between them, and covered with dense pine woods; the hills form a picturesque background. The plain is dotted with numerous villages in fine groves of walnut trees, a thriving population cultivates rice in all the flat ground, and at this distance from Ta-Li-Fu, the traces of the Mahometan rebellion are fast disappearing. The Chinese have not as yet advanced in sufficient numbers up the valley to cultivate the slopes of the hills, which are still covered with virgin forests, but a little lower down the river fresh land was being submitted to the plough.

The river was nearly a mile wide, and had the appearance of a fine lake; this was partly owing to its being in flood, but even in ordinary times the width must be considerable.

September 19.—At Ku-deu we found a good many Mu-su people, and it is possible that what I have called Ku-deu may be the Ku-tung mentioned by Mr. Baber. Referring to the Lamas who came in to write the Tibetan names in our book, I wrote in my diary, 'These Lamas had not their heads shaved; and one with moustaches looked more like a Frenchman than a Tibetan.' Baber, referring to the Ku-tung men, says, 'I felt in the presence of my own race.' Baber could obtain no information about them, except that Ku-tung was north of Ta-Li-Fu; it seems, therefore, at least possible that his Ku-tung men may have been of the Mu-su (or Mossoo) tribe.

We tried to find someone here to shoe our horses; but the state of the poor animals' feet was pitiable, and one or two of them were in such pain that they could not bear to have their feet touched. It was grievous to watch the poor beasts, and it was evident to us that some of them would not get much further.

September 20.—A fresh lot of coolies and mules struggled with our baggage for an hour or two in the inn-yard, and then went off with it by ones and twos, straggling over the road, and covering about a couple of miles of ground.

It has always remained a mystery to me why I was never robbed, or why some of these carriers did not leave their loads behind them. My money was tolerably safe, because it was always supposed to be in the carefully locked up English-made cartridge-boxes, which were small and heavy; and the two dirty old skin-covered trunks, that were thrown about in the dirt, and apparently never looked after, were not suspected to contain the quantities of silver that I had in them. From Ku-deu to Ta-Li-Fu, the carriers were changed almost every day; sometimes they did not arrive at their destination till late at night, and it was often quite impossible to count up the packages. The end of the journey was so near that a robbery, except of silver, would not have entailed a very severe loss, and I was myself the more careless in consequence, and yet not the smallest article was lost or stolen during the journey.

The Chinese carry their respect for the dead to an inordinate degree; an unfortunate rencontre with a funeral was mainly the cause of Baron von Richthofen's misadventure near Ch'ing-Ch'i; and we expected that, now we were advancing towards a thoroughly Chinese population, some of the carefully kept graves, so universal in China, would soon be seen. We were not disappointed; the first was not very far from Ku-deu, and then, by the side of one of the last Tibetan cairns, there was a Chinese tablet erected, on which a long inscription was written.

The river was here reduced to a breadth of one hundred and fifty yards, and enclosed by steep sandstone hills, throwing out flats into the water. A level and good road led us, through small woods of oaks, walnuts, and chestnuts, to Pai-Fên-Ch'iang ('white wall'), where we had arranged to breakfast. Ominous rumours had already reached us that Huang-Fu, his pipe, and his attendant soldier had all passed on, and on arrival it was found too true. This was a sad blow for the rice-eating people with us, who had expected to find a meal prepared for them. We did not waste much time, and soon after leaving, the officer with us turned into a friendly house in some village on the way, and got a glass of whisky. He found, apparently, several friends along the road, for by the time he joined us at Ch'iao-T'ou he was in a state of intense alcoholic stupidity.

A large stream empties itself into the river at Ch'iao-T'ou, across which there used to be a bridge; unfortunately for us it had been washed away, and with the tantalising sight of Ch'iao-T'ou not five hundred yards from us, we were obliged to march two miles up the stream to the next bridge, and then turn our horses back again.

We had arranged to stop at Ch'iao-T'ou, but on arrival we found that Chin-Tai and Huang-Fu had gone on five miles to the next village, where there was a good house. Our officer was not a useful person, he could only incoherently repeat that fifteen li were as good as thirty; and having many times given us the benefit of his opinion on the subject, he presently subsided into a state of melancholy. Notwithstanding his drunkenness, he turned out not very wrong, for it was a march of nearly ten miles to Tz'ŭ-Kua, where we found Huang-Fu, who, in a tone of voice that would have made the deaf adder stop his ears, poured out a torrent of words, all of which meant that at Pai-Fên-Ch'iang he had found a house with nothing in it but an old woman who could not speak; and with somewhat spirituous tears in his eyes he apologised for having found it impossible to prepare rice. Whisky must have been very cheap here, for this was the only occasion on which I saw any signs of drunkenness.

September 21.—Bad news was brought in the morning of our lame horses; one was lying down by the road-side, and absolutely refused to move, and the other could hardly be driven along; we never saw anything more of these unfortunate animals.

The servants, chair coolies, *et hoc genus omne*, came in early by twos and threes, and kneeling down, knocked their heads on the ground, ostensibly wishing peace and prosperity—for this was the third and last great feast in the Chinese year—but in reality expressing their expectation of large presents.

On the road we passed an eating-stall, and for the sake of luck, nearly all our party patronised it, re-

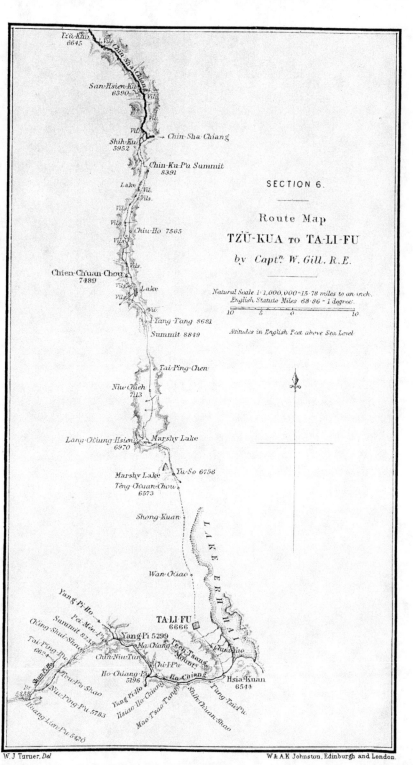

joicing in this first sign of the frequent wayside teahouse.

About a mile and a half outside San-Hsien-Ku, an official, whose rank entitled him to a brass ball on his hat, came out to meet us. I was walking on ahead of everybody; and at the sudden appearance of a hot and untidy foreigner, in a curious hat and strange garments, he was too frightened to do anything but open his eyes and mouth. I alarmed him still more by taking the red paper that he had in his hand and pretending to read it, and something serious must have occurred if one of my people had not opportunely arrived, and explained the situation.

On arrival outside the village, we found the people collected at a table, where incense was burning in our honour, and at the house of the chief man we found a great feast laid out, of cakes, Indian corn, bread, pears, pomegranates, walnuts, chestnuts, and the inevitable sunflower seeds, and to our great surprise Huang-Fu had arrived in time to make some excellent tea for us, and prepare rice for the servants.

At Shih-Ku there was a small river, which was crossed on the most simple rafts of the unsquared trunks of five pines, joined together at the ends by two pieces of wood passing through holes in the logs. There was no superstructure whatever. Outside the town the officer of the place met us, saying he had prepared his own house, and conducted us to a tumble-down and empty temple on the top of a hill a hundred feet high. This officer talked better Chinese than any we had lately met, and being more thoroughly Chinese probably oppressed the people more, so that they would do nothing for him. He admitted himself that he had no authority over them.

and he abused them for a set of savages. While we were in this place, damp as to our bodies, and discontented as to our spirits, on account of the wretched accommodation, the people brought presents of rice, and a fowl.

The name of this place, Stone Drum, is derived from a disc of stone at the entrance of the town. This disc is about four feet in diameter, by a little less than two in thickness. The people here pretend that it is upwards of two thousand years old, but all the characters on it are quite modern.

After dinner the servants reminded me again that they were blowing the trumpet in the full moon, and I did not depart from the traditions of the occasion.

We found some difficulties in getting coolies. The officer told us that this was owing to our having given the money for the payment of the last to the head men, who had kept the greater part for themselves; and he hinted that if we would give him the wages of the next he would see it fairly distributed. We felt that his large practical experience of the method of applying the squeeze enabled him to speak with some authority, but still we did not avail ourselves of his sticky fingers.

At the time of our visit to Shih-Ku the Chin-Sha-Chiang was so swollen by the constant rain that it was more like an immense sea than a river, and it was impossible to form a just idea of its normal breadth. We finally left it at this place, and crossed into the country drained by the Lan-Ts'ang or Mekong.

September 22.—From Shih-Ku we ascended a small stream running down a narrow valley between hills of red clay and sandstone, closely wooded with pines. There were no large trees; but as a good many gigantic old trunks, blackened by age and the weather, were lying about, there must at some time have been a wholesale destruction of the fine ones.

Not far from Shih-Ku there was an application of water-power that would be considered simple even in this land of simple appliances. It was a kind of pestle and mortar, probably for husking rice. A long log of wood was scooped out at one end into a trough, and the other end was made into a sort of hammer-head. This log was pivoted about the centre, and the trough was placed so that a stream of water filled it; the extra weight then overbalanced the hammer-head, and the trough end descended; in falling it upset the water, and thereby became less heavy than the hammer-head; the latter then fell on the grain beneath, and lifted the trough end into the stream, and so on.

The coolies here adopt a new arrangement for carrying their loads; they wear a wooden collar, to which a basket is fastened which rests on the back as they stoop forward under the weight; and no matter whether the load be a bed, a portmanteau, a bag, or a box, into the basket it must go. A strap, whose two ends are fastened to the basket, is led through two holes in the collar; the bight is very often, especially by the women, passed over the brow, so that nearly the whole weight is supported by the forehead; others will hold this bight in their hands, or pass it across the chest.

The road, at first fair, led through a dense jungle of briars and small trees, whose branches threatened every minute to scratch our eyes or tear our noses. The ascent was neither difficult nor long, and seven and a half miles brought us to the top of Mount Chin-Ku-P'u, 2,439 feet above Shih-Ku, and 8,391 feet above the level of the sea. This was the water-parting between the basins of the Chin-Sha and the Lan-Ts'ang, and was another point on the same ridge that we had already crossed, once at Tsaleh (15,788 feet above the sea), and a second time at Shwo-La (14,307 feet above the sea). Here the scenery changed altogether, and was very much like the country to the east of Ya-Chou that we passed through soon after leaving Ch'êng-Tu-Fu. We looked down upon undulating red hills which enclosed a long valley running down to the north; the slopes were chiefly covered with pine woods, but with here and there patches of cultivation. The lower parts were well cultivated with buckwheat and Indian corn, and contained a good many villages and houses. We descended to the head of this valley, where there was a small pond, and another valley running down to the south, at the head of which there was a charming little lake, a mile long. The lake was enclosed by low wooded spurs, the red soil showing through, and in pleasing contrast to the deep green of the trees. There were patches of buckwheat here and there, whose pink flowers amongst the pines and red soil were as usual in full bloom. Below the lake a fine plain commenced between the hills; this was entirely laid out in ricefields, and was, in places, more than two miles wide. Numerous villages and houses were clustered at the foot of the hills at each side of the plain; all these were quite new, and it is wonderful how soon this district seems to have recovered from the devastations of the Mahometan rebellion.

Nothing can be more striking to the eye of a traveller in Western China, and for aught I know in other parts of the empire, than the dense population of a rice plain, as contrasted with the absence of houses amongst the hills, and with the number of the inhabitants in plains not suitable to the growth of rice. Miles of country may be passed over amongst the mountains where there is scarcely a single habitation. Some spur is crossed, or a corner turned, when suddenly a rice plain is disclosed to view, where the villages are so large, and so numerous, that it seems as if they covered a greater area than was left for cultivation. For one brought up amongst European ideas, it is at first very difficult to realise how small a portion of land will support a man in this country, where two or three crops can be produced in a year. One acre of wheat will, in Europe, support two men; one acre in China will probably support twenty.

All the way down this valley the road was frightful; after a long dry season it would be good enough, but now, after the rain, it was, for the whole eleven miles, deep in the most sticky mud and clay.

Near a temple by the road-side, where a small stream came down from the hills, we noticed the first arched bridge we had seen since leaving Ta-Chien-Lu; and at Chiu-Ho we found the first regular inn we had stayed in since crossing the boundary of Ssŭ-Ch'uan.

September 23.—We had the usual trouble about coolies and horses here; the head men said that there were plenty of coolies, but no horses or mules, and they then sat down to look at us.

' All right,' we answered, with nonchalance equal to their own, 'then we will just ride off to Chien-Ch'uan-Chou, and ask the Chou there to send for our things,' and suiting the action to the word, gave orders at once to saddle our horses. This somewhat frightened them, for if we had carried out our threat, the Chou would certainly have found a means to get our things somehow, and that without the payment of a single cash, as they knew from long experience. To show that I was in earnest, I ordered the chair coolies off. They at first did not like going without the two ponies they have for riding. I told them they could do as they liked, but that if we saw anything more of them during the day, Mesny would get inside and remain there until he arrived at Chien-Ch'uan-Chou— now Mesny weighed about fourteen stone. Those chair coolies disappeared with most unusual alacrity.

We had succeeded in hiring some little ponies, as our own were nearly all knocked up, and one of the attendants, who came with the hired animals, was a lady, whose sex would have been hard to distinguish, but for the huge earrings she wore. Half our coolies were almost always women or children, and there seemed as little difference between the avocations as there was between the appearance of the gentlemen and ladies of these regions. The people told us that in these districts of Li-Kiang the oppression of the officials was terrible. The Lekin tax levied by these, to fill their own pockets, on all merchandise passing through the district they govern, and which has been abolished in the Chien-Ch'uan and Wei-Si districts, is here still enforced. The people declared that they had neither clothes to wear nor rice to eat, and the poverty was really dreadful. The native Mu-su chief, who should have come to see and assist us, thinking that, like the Chinese officials, we should take animals and coolies without payment, did not visit, or do anything to help us; and the poor fellows accompanying us declared that if it were once known that we always paid, we should never have the least difficulty in procuring any number of men or animals.

Our road for eight miles was as bad as the day before, and the weak, wretched little pony I was riding, even with my light weight, floundered about amongst the stones, plunged into the mud, and played every awkward trick it was capable of.

We followed down the same valley, laid out everywhere in rice terraces, and bounded by low spurs of red gravel, clay, and sandstone. The hills were covered with grass and small pines; and at a distance of three and a half miles the first pagoda, standing on an outlying spur, indicated the proximity of Chinese cities. At a village about two and a quarter miles short of Chien-Ch'uan-Chou, the house of the officer who, at the time of Margary's murder, was the Chen-Tai of T'êng-Yüeh, was pointed out to us. The governor-general of Yün-Nan, on account of the troubles consequent on the sad affair, frightened this man so much, that he was said to have poisoned himself.

After eight miles the road improved a little, though it was but indifferent up to the very gates of Chien-Ch'uan-Chou, the first walled city we had seen for months.

The walls of the city and the gates were in good repair, and if they suffered much, have been entirely restored since the Mahometan rebellion, but the streets through which we passed were poor and wretched, with miserable houses. Here the old familiar Chinese sights again appeared—fruit-stalls, eating-stalls, with the favourite bean-curd cake; stalls where hats, bits of ribbon, and other little articles, dear to the housewife, were displayed in as tempting a manner as possible, and the usual crowd of inquisitive Chinese that soon gathered around us.

We first went to an inn, the excellencies of which had formed the theme of conversation of a man with us for the greater part of the journey, but it was altogether a miserable place, and as there was a temple next door more or less prepared for us, we went to it. Here there was neither a cooking place nor any room for servants, so Chin-Tai was sent out to make inquiries.

Our messengers were not long in returning with the news that they had found quite a palace, with admirable culinary arrangements. So we moved into it, and although the palatial part must have been amongst the cooking pots, still we found a room with a strong family likeness to many a one we had already occupied. It was upstairs, and was provided with bedsteads. Chairs and a table were rapidly produced, and we soon made ourselves at home in the usual style.

The house we were in stood all by itself just beyond the east gate of the city, and was the sole remaining building outside the walls on any side.

The Chou of Chin-Ch'uan did not get on with his superior officer, the Fu of Li-Kiang.

He disapproved of the Lekin tax, and that no doubt was one cause of their disagreement. He had been here only one year out of the six for which he was appointed, but had found his position so unpleasant that he had resigned; perhaps partly also on account of the poverty of the people, in whom there was no juice left to squeeze. He said that the misery here was frightful, and that the people groaned beneath the weight of the excessive taxation.

There was a great deal of fighting here during the Mahometan rebellion, and the city was taken and retaken several times. We asked the landlord what he did when the rebels were here.

'Oh,' he said, 'I kept quiet, and did nothing.'

'Don't you believe it,' said his wife, who was

standing at the top of the stairs. 'He went over to the white flag; like a fool, he was always fighting, and got wounded all over his body for his pains.'

Of course there was no contradicting a lady, and the worthy fellow beat a retreat rather sheepishly, like many another brave man, more afraid of his wife's temper than of swords or bullets.

September 24.— Coolies were not wanting, as there was a whole army waiting in the courtyard. When everything had been packed some time, we asked why they did not take their loads and go. They seemed as much amused at the idea as it was possible for such miserable-looking people to be, and replied that they were waiting for the head men, without whom they said they could do nothing. The head men used to indulge in the abominable Yün-Nan habits of opium-smoking all night, and sleeping all the morning. When they eventually arrived, they looked at the luggage in a stupid sort of way, and then seemed to think they had done enough for that day; the coolies in the meanwhile sitting placidly in the mud, in the listless manner of people too oppressed to care for anything.

The Chou came in about half-past seven in the morning to pay us an official visit, and in his conversation piled up compliments on promises, in the elegant language of the Chinese literati. He told us he had sent a circular the night before to the officer at Niu-Chieh, informing him that their two distinguished foreign excellencies, the British Imperial Commissioners, would be at Niu-Chieh in the evening; ordered him to have the Lekin office cleaned out, prepared, and furnished with beds, chairs, and tables; and to have horses and coolies all ready for the next morning, so that there might be no delay; and telling him that, on the arrival of their excellencies, he was to attend to them, and see that all their wants were at once complied with. The Chou told us that the Lekin office was the best house in the place; that there were three rooms in it, besides accommodation for the servants; and before leaving he called two Tinc-Chais, and in our presence ordered one to go on with Huang-Fu, and see that everything was properly done, and told the other to look after our baggage, and remain with it all the way to Niu-Chieh. This was 'beautiful language,' as Joey Ladle would say, but we had too much experience of the meretricious nature of Chinese official speeches, to indulge in dreams of regal splendour in store for us at Niu-Chieh.

We set off through the city, which confirmed our previous impression of poverty and general misery. We saw potatoes in the market for sale, but nothing else that attracted any attention.

Mesny had hired a pony with four white stockings; curiously enough the Chinese have a rhyme about horses with white stockings something similar to the old English one, but although according to our theory one is harmless, two are doubtful, three suspicious, and four certainly bad, the Chinese say that with one or three, the horse is all right, but that if he has two or four white stockings, he is sure to be weak.

Our road to-day was a most irritating one; it was over a perfectly flat plain, but twisted and zigzagged about amongst the paddy fields, first one way, then another, and it was impossible to say where it was going to for ten yards ahead.

It had at one time been paved with blocks of stone, which, now all displaced, were lying about in a sea of mud and slush, in a state of frightful confusion. But if the road was irritating, the ponies were far more so; they floundered about, and put their feet into every possible hole; just when they were wanted to move a little faster, on a bit of comparatively good road, they would almost stop; whenever I took out my note-book, mine invariably began to trot, would jump, put its foot with a splash into a mud hole, rush into the hedge, if there was one, threaten to tear out my eyes with the thorns, and play any and every trick whereby it could spoil my writing, or bring my note-book to a greater state of decay than had already been caused by rough usage and the weather.

We marched for eight miles over the plain, which supports an enormous population, for we passed villages at almost every quarter of a mile, many of them very large. The crop was nearly altogether rice, but besides this there was a good deal of buck-wheat, some beans, and a grain called by the Chinese *Paidza*; it is something like rice, and, like it, grows in water.

At the eastern side of the plain there is an extensive lake, into which the river runs. The geographical notions of the people were somewhat vague: they said that one stream that had a name came into the lake, and that another without a name flowed out, and they would not for a moment admit that they were the same river.

September 24.—The plain of Chien-Ch'uan-Chou is similar in structure to the Ch'êng-Tu basin, and the plains of Ta-Li-Fu, and Lang-Ch'iung-Hsien. Surrounded on all sides by high hills, the central basin is fed by numerous streams, and drained by one river that rushes out through a narrow gorge. The city is now some distance from the shores of the lake, but as the geological formation is entirely a soft sandstone, it is evident that the outflowing river must continually deepen its channel, that the lake must formerly have stood at a much higher level, and that it will in course of time be altogether dry.

It is quite possible that the outflowing river makes a sweep to the west round the spur of I-Yang-T'ang, and eventually flows into the plain of Lang-Ch'iung-Hsien; otherwise it would be rather difficult to account for the large inflowing stream at the latter place. The difference of levels between Chien-Ch'uan-Chou and Lang-Ch'iung-Hsien, which amounts to 519 feet, favours this theory.

After eight miles we left the plain, and we were thankful to escape from its road of nubbly stones. We ascended a valley through a broken country, where the streams had cut deep channels through the soft clay and sandstone, which imparted their red colour to the roads, the houses, and the water; even the people wading through the mud, carrying huge baskets of pears to the market at Chien-Ch'uan-Chou, seemed in some degree to partake of the ruddy tint of the soil.

The ascent of a spur, though not a steep one, was frightful work for the ponies in the pouring rain: now in a soft place a fore-foot would sink beyond the knee in sticky clay; now the hind-legs of the unfortunate animals would slip from under them, and in the struggle to regain a footing the mud would be splashed in showers over our heads and faces, disfiguring the leaves of my note-book, and covering the flanks of the ponies with sufficient soil to grow a crop of cabbages.

A halt at I-Yang-T'ang was a welcome relief to the panting steeds, and bowls of rice and hot macaroni in the restaurant, which was almost the only building in the place, were appreciated by their riders after the cheerless journey.

From I-Yang-T'ang we again ascended slightly to the summit of a spur flung out from the Shwo-La range, and then descended by a road which, in the evil qualities of slipperiness and muddiness, transcended that by which we reached our breakfast place. I tried to walk, but my European boots had no hold whatever on the greasy clay, and I was forced again to mount my unfortunate pony, if the term *to mount* can be considered in any way appropriate to the act of gaining the back of an animal so small that when in the saddle the rider's feet are nearly in the puddles below.

We gained at length the plain of Lan-Ch'iung-Hsien, in its aspect and formation similar to the plain of Chien-Ch'uan-Chou. The road was, like the latitude and longitude of the amateur sailor, 'as before,' and remained so until we reached Niu-Chieh, where we found Huang-Fu smoking his pipe in the doorway of a deserted and tumble-down-looking place, which proved to be the Lekin office of which the Chou had spoken at Chien-Ch'uan. He said that the Tinc-Chai sent with him had disappeared at the first convenient house. Just then the officer of Niu-Chieh came up; he treated our arrival and appearance as an excellent joke, declared that no circular had been received by him from Chien-Ch'uan-Chou, hinted that we were base impostors, and getting a light for his pipe from Huang-Fu, sat down and declined to notice us except by a supercilious stare.

We did not ask his permission to take possession of the Lekin office, but mounted by a rickety staircase from the shed below to the upper floor, which our friend the Chou of Chien-Ch'uan had described as containing three sumptuous apartments.

One long room, where a couple of wooden pillars indicated the imaginary lines that divided it into three equal portions, was furnished with a crazy bedstead, and ornamented with some big stones that were lying casually about amongst the usual dirt and filth. Here we took up our quarters, as the inhabitants of the only eligible house declined to admit us.

It was no wonder, poor creatures; they were accustomed to the visits of hungry officials, who take up their quarters uninvited, eat their food, destroy their furniture, and enforce their labour without payment; and it was only natural for them to think that we should come and do likewise.

Some of our baggage arrived in good time, and as the head men had a favourite trick of driving away the unfortunate carriers directly they had deposited their loads, in order that they might the more easily retain the whole of the wages of these miserable people, we ordered two or three of the coolies to remain in our room with the things they had brought.

The people below us now formed numerous little camps, where they lighted fires on the ground, and our room was soon filled with the pungent smoke of the damp wood that came up in dense volumes through the yawning cracks between the floor boards.

The officer of the place suddenly appeared at about nine o'clock, saying that the circular had just arrived, he had read it and sent it on; and now his manner towards us was as servile as before it had been insolent.

Later in the evening, when I walked to the other end of the room, I discovered that the two or three coolies we had ordered to remain had now become about fifty; they were crowded together, lying in heaps one on top of the other, and when the time came to make a clearance, it was with amazement that I watched them disentangle themselves and file off one by one. Amongst others there was a woman with a baby on her back which she had been carrying all day besides the load allotted to her.

Descending into the place beneath was a matter of no small difficulty; people were all huddled together, even on the stairs, and for a moment I could not help thinking of a London ball—but what a piteous travesty! on the ground, men, women, children, and babies in arms, were so numerous that it was almost impossible to walk without treading on them. Some were sleeping; others smoking or trying to dry their soaking clothes over the wood fires. The occasional flare of some dry splinter in the reeking atmosphere served but to make darkness visible, for the walls and ceiling were black with dirt and the smoke of years. It was one of the saddest scenes I ever saw. The poverty and misery of the people, and the hopeless state of almost brutishness in which they live, were painfully visible in the listless, expressionless faces, which were now and then lit up by some fitful flash that burst for a moment through the heavy smoke. I returned again to the upper room, and the trifling discomforts to myself were forgotten in the recollection of the grievous scene below.

September 25.—The officer called in the morning and asked for a present, saying that if he had received the circular in proper time he would have prepared a house, and treated us in a more worthy manner.

To which we replied that our honourable countrymen always paid for everything, and made presents besides when they were deserved, and that although they could support indifference, they knew how to resent insolent behaviour.

Much abashed, he bowed to the ground, and went away.

Some of our luggage did not arrive till the morning; and from the window of the Lekin office we watched the lazy Yün-Nan people coming into market, for this was market day in Niu-Chieh.

The people bring all the materials necessary for erecting their booths with them—four pegs to drive into the ground, four upright bamboos, to which four others are attached round the top, a light bamboo mat for the roof, and small bamboos strung together for the table on which their wares are exposed. All this weighs scarcely a pound, and the shed is complete in a very few minutes. We were told that out of the ten thousand families living in the plain, ten thousand people came to the market here; and although, as is usual in dealing with Chinese estimates, a divisor is certainly necessary, yet the very great number of large villages we passed on the road, and saw on the plain, the people met at every step with baskets of pears, small red chilies, vegetables, and other things, showed that the population was enormous.

The unfortunate officer of the place either had not given up all hopes that we should relent and bestow *largesse,* or else he was afraid of being reported to the Chou of Chien-Ch'uan, for as we marched out a salute of three guns was fired, and he was waiting outside the gate in official clothes to pay his final respects.

The town of Niu-Chieh stands at the head or northern end of a magnificent rice plain, about three miles wide, and nine miles long, running north and south, and bounded by gentle rounded grassy hills of red clay and sandstone. At the south it is divided into two, by a spur from the mountains that close it at that end; and on the western side, a third plain runs back to the north-west, at the back of the western hills. This third part is almost entirely occupied by a large lake, or swamp, which extends across the road leading to Lang-Ch'iung-Hsien, a city lying at the foot of the western slopes. This road is a causeway through a swamp, which forms the southern end of the lake; here a good deal of cormorant fishing goes on, and we saw many people punting about, with cormorants perched on the side of their boats.

The lake is fed partly by the stream we had followed from the summit above I-Yang-T'ang, but mainly by a considerable river that flows in at its southern end, and which, as I have remarked (page 283) may quite possibly be the stream that is fed from the lakelet above Chiu-Ho, and which we had crossed just south of Chien-Ch'uan-Chou.

Near Lang-Ch'iung-Hsien this stream is embanked, and its level is some feet above the plain. The river escapes, at the extreme south-eastern corner of the plain, through a very narrow defile in the soft sandstone, where the water rushes down a foaming torrent, falling two hundred feet in two and three quarter miles.

From the nature of things it is obvious that a river of this kind must rapidly cut away the bottom of its bed; and in course of time this plain will be drained without any artificial aid; the lakes and swamps will disappear, the causeway will cease to exist, and some traveller coming here in 2379 will perhaps wonder how the historian of 1879 could have spoken of causeways in a plain that he finds quite dry, just as we have wondered at old Marco, the Venetian, speaking of a bridge full half a mile long in the city of Sindafu.

Five and a half miles from Niu-Chieh the stream was spanned by an elegant arched bridge ; and hence one road led to Têng-Ch'uan-Chou, and another to Lang-Ch'iung-Hsien, lying some miles off our road.

But as the distance to Têng-Ch'uan-Chou was variously estimated at from seventeen to twenty-five miles, as much of the baggage had not started when we had left Niu-Chieh, and as heavy rain, which had not troubled us as yet, could be seen all round, holding out prospects of another wet afternoon, it became doubtful whether we should be wise in attempting to reach Têng-Ch'uan-Chou.

'Where is Lang-Ch'iung-Hsien ? ' I asked ; 'and how far is it ? '

'Oh ! ' replied all the people with one voice, 'we have arrived there.'

'Show us, then, the city,' I said, and one man who professed great local knowledge, by way of pointing out its exact position, waved his hand with a circular sweep round his head.

There were some houses nearly hidden amongst trees, a couple of miles away ; I concluded that this must be the city, and as all the people said that we should find excellent accommodation, we then decided for Lang-Ch'iung-Hsien.

The houses among the trees turned out only a small village ; we passed through it, and mile after mile of road was left behind us, still there was no indication of the proximity of a city, and it was not until we had marched altogether thirteen miles that the gate was reached. Before we arrived a little light rain was falling, and no sooner had we a good roof over our heads than it came down heavily with a thunderstorm that lasted all the afternoon.

The wretched city was miserably poor, and had

suffered much from the attacks and ravages of both parties during the rebellion. It had been taken and retaken many times; its walls were level with the ground; and the population, which had formerly numbered six hundred, was now reduced to two or three hundred. The gate was gone; and the arched gateway alone was left to remind the visitor that it ranked as a Hsien. The road still passed underneath from mere force of habit, but it might just as well have gone anywhere else.

The best house in this dilapidated city was a Chinese inn, that did not at first hold out much hopes of comfort. The only room was on the ground floor, and was very low and small; it opened directly on to the main street, and one half of it formed a passage for men, animals, horses, dogs, cats, pigs, fowls, and ducks from the road to the back yard and stables. Tables or chairs there were none. The planks on trestles, which formed our bedsteads, were higher than an ordinary table, and turning-in was very much like going to bed in a bunk on board ship.

The Hsien was a man of high literary attainments, and was at Ta-Li-Fu, assisting at the examinations; but his representative sent his card, with presents of a lean duck and an old hen, and what was much more to the point, a table and two chairs. About a thousand candidates from this neighbourhood present themselves annually at the examinations at Ta-Li-Fu; and as each man on an average takes three animals, one to ride, and two for his servant and baggage, when we came into the district, nearly all the horses and mules in the country were already engaged, and we found great difficulties in hiring transport.

The landlord of the hotel was a most civil old fellow, and was much interested in all our things;

he looked at my candles, declared he had never seen anything like them before, and held up both thumbs as a sign of the very highest approbation. My mosquito curtains were pronounced extraordinary, but my tub was, as usual, the greatest marvel. He took my two barometers for a pair of spectacles, and gently hinted that an empty wine bottle would be a real prize.

We were not much incommoded by the people; a dozen or two crowded round the door, and one or two, more venturesome than the rest, came into the room, but heavy rain kept away all but the most inquisitive.

The coolies who had brought our things from Niu-Chieh were paid in the evening; and a mighty dispute arose, each man declaring that if the money was given to any other than himself, it would not be fairly divided. Eventually, after a noisy quarrel that lasted about half an hour, I gave the silver to one man whom the others appeared to distrust less than the rest; and the people departed, though hardly in peace.

Our old host must have been a rare connoisseur in tea; there was a little bit of charcoal burning on a small hearth, and here he kept a pot of the most delicate Pu-Erh tea. He was up all night, continually making a fresh brew; and from the moment of our arrival to that of our departure he carefully watched our cups, and always kept them filled with hot and delicious tea; and in the evening, when the last inquisitive idler had gone, when the last sounds of day had died away in the street, and nothing could be heard but the patter of the rain, when the doors were shut and barred, and the candles were lighted, and when the eyes of our good old host were beaming with the pleasure he felt in anticipating our wants, there was something comfortable and homely in our surroundings, and, notwithstanding the meanness of the apartment, I carried away a more favourable impression of this queer little place than I have of many a pretentious European caravanserai.

September 26.—There is a spring near Lang-Ch'iung-Hsien impregnated with what the inhabitants call natural sulphur. The water that evaporates leaves a yellow sediment, which is collected twice a year with the greatest care ; and it is held in such high estimation as a cure for many diseases that it sells for its weight in silver.

The hotel-master made us a present of a considerable quantity wrapped up in a piece of paper, on which was written a list of all the sicknesses for which the mineral was a specific. To judge from its length, no Dr. Dulcamara in Europe ever possessed such a panacea.

As we penetrated further into Yün-Nan, we did not find the lazy habits of the opium-smoking people improve, and the long and weary watching for coolies became a part of the day's proceedings. Then when the last odds and ends had been packed up in the last box, when even the cooking pots had been finally stowed away, and when the servants, all ready, were sitting listlessly cracking sunflower seeds or gazing vacantly into space, I used to find the hours very tedious ; but here I was much diverted by our host, who, standing on his doorstep, discoursed to an admiring audience on all the wonderful things the foreigners had and did. Amongst other things, he told the people that each of us had a pen that we could carry in our pockets, and that we also had knives by which we filled them with ink. This was his way of describing the simple operation of cutting a lead-pencil.

After paying our hotel bill, I made a present to the old man of silver equal to the value of the sulphur; and, finally, as a parting gift, I presented him with an empty wine bottle. I really think that of all I gave him he liked the empty bottle best; he looked at it as fondly as a blue-china-maniac would at an old bit of his crockery; he handed it round, took out the cork, examined the label, and even held it up to his baby for admiration; and the last that I saw of him, as I went out at the door, he was still toying with this precious gift.

The only two fine hours during the day came to an end as we marched out at the gate of the city, and the rain recommenced and did not cease the whole day.

In the vicinity of Lang-Ch'iung-Hsien the plain is little better than a marsh, across which the road, in an unaccountably good state of preservation, follows the river through a narrow defile, and debouches into another circular and similar basin, and, were it not for the high embankments on both sides, it would again be lost in the swamps, fens, and small lakes which cover the plain.

The rainfall this year had been unusually great, the country was frightfully wet, and the landscape as we splashed over the wet roads was dull and cheerless: villages standing in the swamps, or surrounded by water, with two or three ruined houses in the outskirts; people poling about in punts, or cormorant fishing; a huge pelican flapping its great wings or floating motionless on the water; the hills all shrouded in mists and rain clouds; the road by the river-side bordered with trees, and stretching out straight to the front across the marsh until lost to view in the distant haze; and the continual drop-drop of the rain from a leaden sky, all combined to make the scene a very dreary one.

In the upper parts of all these plains a good many trees of different kinds grow at the foot of the hills; but the plains themselves, and the villages, are nearly altogether bare. Here the only trees were growing by the edge of the river, and marked its course amongst the rice fields that covered the flat surface. Rice is the only crop, and this is grown wherever the water is not too deep.

We were now in the country of the prickly pear; for several days we had seen a little of it, but it was not until we reached the plain of Têng-Ch'uan-Chou that we noticed it in considerable quantities.

September 27.—The road, as well as the country, was nearly altogether under water, but the mud was less, and we could get on a little faster than usual; and, passing over the lower end of a spur, the lake of Ta-Li lay spread before us. In fine weather it may be very beautiful, but its beauties were not apparent through the mists that shrouded everything.

The lake of Ta-Li, or Erh-Hai, is about thirty miles long, and varies in width from about four to twelve miles; its eastern shores seemed to be bounded by mountains, which run straight down to the edge of the water. On the western side, down which we marched, a wide and very flat plain extends from the margin of the lake to the foot of the western mountains; this plain is almost entirely covered with rice, but, owing to the late continual rains, the crop was entirely lost, and I subsequently saw the young rice, on which the ear had hardly formed, being sold in the streets of Ta-Li as fodder for cattle. It was sad, indeed, in this frightfully poverty-stricken land, to think that so large a population would lose nearly all they had to depend upon until the next crop. The poverty was awful, the result of the terrible ravages during the Mahometan rebellion. At almost every step the ruins of some cottage were passed, where, in the place of a peaceful family happily living under a comfortable roof, wild thorns, briars, and huge rank weeds flourished between the remains of the walls, on the tops of which great prickly pears flung up their spiny foliage. What a contrast to smiling Ssŭ-Ch'uan, where, as Richthofen remarks, everything betokens peace!

At this northern end of the lake stands Shang-Kuan (The Upper Barrier), a small village, but being in a strong military position, fortified with a double wall.[3]

The direct road to Ta-Li-Fu runs along, or very near, the borders of the lake; but as this was altogether under water, we were obliged to follow an upper road. As I looked again upon the familiar junk, I could not help wishing for a comfortable steam launch, in which the journey to the other end could be done in little more than a couple of hours. The day will come, no doubt, not only for steamers, but also for railways; and judging from the crowds of coolies, mules, and horses travelling in both directions, there can be little question that either one or the other would be a paying concern.

Numbers of military students were flocking to the examinations at Ta-Li-Fu, and I laughed to myself

---

[3] This is the Hiang-Kouan of Lieut. Garnier's narrative; passing which, with notable boldness and adroitness, he escaped from the grasp of the Mahometan King of Ta-Li.—(See *Voyage d'Exploration, &c.*, i. 515-516.)—*Y.*

as I passed them by twos and threes, all carrying bows and arrows. The highest military officers have no more difficult subject than the stretching of a bow, or the lifting of a heavy weight, on which to satisfy the stern examiner. This in the days of breech-loaders, hundred-ton guns, and staff colleges!

Round some of the villages a good many of the people, men, women, and children, were engaged in stretching the cotton before weaving it. Two strong pegs are driven into the ground, about fifty feet apart; between these, a double row of thin sticks, two or two and a half feet long, are driven upright into the ground, about three feet apart, the rows being separated by about a foot. In each hand the operator carries a stick about two feet long; at the lower end of each of these is a reel of cotton. He or she walks up and down quickly, passing both reels inside two of the sticks, outside the two next, inside the next, and so on to the end, where the cotton on both reels is passed round the strong peg. In all this process the hands are never crossed; and, at a little distance, ten or twenty people, all walking backwards and forwards, separating and bringing together what look like little white balls, have a most comical appearance.

In one village the people were preparing indigo, but I saw none growing in the fields. I saw also here the first large bamboo; there was but one growing by itself, and it was only large in comparison with the little wild bamboo of the mountains, which is hardly larger than grass.

The great pagoda that stands on a projecting spur outside the city of Ta-Li-Fu is visible from a great distance, and long before the city is gained its height deceives the traveller into the belief that he has reached his journey's end. The longest lane, however, has a turning, and dreary as were the last few miles of march in the pouring rain, over the poverty-stricken and half-ruined country, we at length rode up to the north gate of Ta-Li-Fu.

It was closely barred, for the spirit of the waters is supposed to flee at the sight of the north gate shut against him.

We entered at the east gate, and the interior of the city presented a sadder scene of desolation than the country round. The streets were wide, but half in ruins, and bore the same aspect of poverty that was everywhere apparent.

The city was full of candidates for the examinations; it was difficult to find a suitable lodging, and we were obliged to content ourselves for the night in an exceedingly small room in a very dirty inn.

The Hsien sent to apologise for the poorness of the place, and begged us to excuse it on account of the examinations.

The people of this inn had at first refused to prepare a room for us; the Hsien had made every effort to find us a lodging elsewhere, but not succeeding had brought pressure to bear on the innkeeper, and had even gone so far as to imprison the unfortunate man, who then had consented to clear out a room for us. A seal had afterwards been set on the door, to prevent anyone else attempting to take possession.

The Hsien promised that he would try and find us better accommodation the next day. This was almost a necessity, as the continual rain penetrated into nearly all our boxes, and we wanted if possible to dry our things, an impossibility in the wretched apartment we were in.

CHAPTER VIII.

IN THE FOOTSTEPS OF MARCO POLO AND OF AUGUSTUS MARGARY.

I. 'THE LAND OF THE GOLD TEETH.'

Ta-Li-Fu Province, the Carajan of Marco Polo—The Lake and Environs—Père Leguilcher—The Plain of Ta-Li—The (so-called) Panthés, and the Name—The Mahometan Rebellion—The Mahometans of the Province—Presents of Local Delicacies—Occupations at Ta-Li—The Tao-Tai—The Ti-Tai, General Yang, a Remarkable Personage—The Old Troubles about Carriage—Illustration of the 'Squeeze'—Departure from Ta-Li—Adieu to P. Leguilcher—Yang and Yang—Marco Polo's Cakes of Salt—Paucity of Present Traffic on Road—Devastated Country—Diminution of Population—Yang-P'i River—Imaginary Bifurcation in Maps—Chain Suspension Bridge—Perversities of the Path—T'ai-P'ing-P'u—Lofty Hamlet of Tou-P'o-Shao—The Shun-P'i-Ho—A Treat of Bread at Huang-Lien-P'u—Mr. McCarthy's Servants—Dearth of Population—Traces of War—Chestnut and Oak Woods—Descent to Plain of Yung-P'ing-Hsien—The Town Destroyed—We Lodge at Ch'ü-Tung—Topographical Elucidation—Inn at Sha-Yung—Unendurable Smoke—View of the Mekong or Lan-T'sang River—Chain Bridge Across it—Desperate Ascent—Buckwheat Porridge—Ta-Li-Shao—Pan-Ch'iao—Rice Macaroni—Polo's Salt Loaves Again—His 'Vochan' and the 'Parlous Fight' there—Yung-Chang-Fu—Difficulties about Transport—The 'Squeeze' again—A General on the March—A Quarrel Imminent, but the General is Drawn Off—Stones and Beads Brought for Sale—The Yung-Ch'ang Market—Difficulties of Marco Polo's Itinerary from Carajan to Mien—Mr. Baber's Solution—Recent Plague on the Road.

TA-LI-FU is an ancient city, and was formerly a place of great importance, though now it is little better than a ruin. It is the Carajan of Marco Polo.

It stands at the southern end of a basin, about thirty miles long, entirely enclosed by high mountains. This basin is similar in structure to the plains of Lang-Ch'iung-Hsien and Chien-Ch'uan-Chou, and, like them, is nearly altogether occupied by an extensive lake.

Marco's description of the lake of Yün-Nan may be perfectly well applied to the lake of Ta-Li : 'There is a lake in this country of a good hundred miles in compass, in which are found great quantities of the best fish in the world.'

The fish were particularly commended to our notice, though we were told that there were no oysters in this lake, as there are said to be in that of Yün-Nan ; if the latter statement be true, it would illustrate Polo's account of another lake somewhere in these regions 'in which are found pearls (which are white but not round).'

Before the Mahometan rebellion the plain used to be well wooded, the villages were embowered amongst noble trees, and the landscape must have been as beautiful as any in China ; but now there is not a tree left standing in the length and breadth of the plain.

The lake of Ta-Li is called Erh-Hai, and the city stands about two miles from its edge ; in former days the level of the water must have been much higher ; and it seems possible that when the city was founded it stood on the shore of the lake.

One source of the river that feeds Erh-Hai is near I-Yang-Tang, whence we had followed it to Ta-Li-Fu, and it is possible that the stream that takes its rise in the little lake near Chiu-Ho (see p. 276) is also a feeder of Erh-Hai.

The gorge through which the river enters the northern end of the plain of Ta-Li is not a narrow one, but that by which it escapes is similar in all its characteristics to the gorge of Lang-Ch'iung-Hsien ;

and below Hsia-Kuan the river rushes and boils, with here and there a waterfall, over a rocky bed, falling twelve hundred feet in fifteen miles, in a way that would disabuse the most credulous mind of any belief in the ascent of boats from the Lan-Ts'ang-Chiang.

The appellation of 'Snowy Mountains' has popularly been given to the summits around Ta-Li-Fu, but snow lies on them for only six or seven months in the year, and the altitude of the highest peak is in all probability not more than twelve or thirteen thousand feet.

The line of perpetual snow in this latitude cannot be lower than between eighteen and nineteen thousand feet.

Ta-Li-Fu lies at an altitude of 6,666 feet above the sea, and the climate is always pleasant; but at the time of our visit, a most unusual amount of rain had fallen, so much that that irrepressible person the oldest inhabitant had never recollected so wet a season. In the city we constantly heard the sound of falling houses, and Monsieur Leguilcher, the Provicaire, living at Ta-Li-Fu,[1] told us that a fortnight previously, in the plain of Têng-Ch'uan-Chou, he had been going about in a boat over roads on which he had always previously travelled on horseback.

The people and officials were now all praying for fine weather; and one morning during our stay the Tao-Tai, in all the glory of official robes, headed a procession, which proceeded solemnly to the city walls, where they fired a gun at the sky, as a sign of anger and displeasure, by which they seriously believed

[1] It was Père Leguilcher who joined the late Lieut. Garnier on his daring journey to Ta-Li-Fu in 1868.—*Y.*

they would frighten the rain god into a more kindly frame of mind.

It was rather a remarkable fact that at the city of Yun-Nan-Fu, only twelve days distant, there was a severe drought; a fast was proclaimed, the south gates shut, and all the solemn rites, such as we had seen as we were leaving Ssŭ-Ch'uan, were being performed to obtain that rain which here had produced such disastrous effects.

There are some quarries in the neighbourhood of Ta-Li-Fu, where very beautiful marbles are found, so curiously marked, and stained by nature with such diverse colours that, when cut into flat slabs, a landscape of mountains and trees appears on the face. Monsieur Leguilcher made me a present of a very rare specimen, framed, and when hung up, it might at a little distance, easily be mistaken for a painting.

There are said to be very rich gold and silver mines within a few days' journey of the city. Marco Polo, speaking of this neighbourhood, observes: 'Gold dust is found in great quantities, that is to say, in the rivers and lakes; whilst in the mountains gold is also found in pieces of larger size.'

There are now about three hundred villages in the plain of Ta-Li-Fu, the largest of which does not contain more than two or three hundred families; while before the rebellion the population of the villages averaged seven or eight hundred families. In Ta-Li-Fu itself there are from two thousand five hundred to three thousand Chinese families, and one thousand five hundred to two thousand native families, for the Chinese are strangers here, though they outnumber the natives; the latter have a great dislike to foreigners, amongst whom they include the Chinese.

There are now no suburbs to the city, but outside the south gate the ruins of an old suburb extends for more than a mile by the roadside. Baber states that a suburb was said to have stretched as far as Hsia-Kuan; I never heard the report myself, but join with Baber in doubting its veracity.

Over all the neighbourhood the ruin of the country occasioned by the rebellion of the Hui-Hui, or, as Europeans call them, Mahometans, is grievously apparent. This rebellion lasted over many years, during which the most desperate fighting took place in almost every town within fifty miles of Ta-Li-Fu, the great centre of the movement, and the seat of Tu-Wên-Hsiu, the so-called Sultan Suliman.[2]

Towns and cities were taken and retaken by each side alternately, acts of frightful cruelty were perpetrated, and retaliations still more cruel followed.

During all these scenes of war and bloodshed M. Leguilcher remained in the province, and his life during this time would form a thrilling narrative of hardship and adventure. Once he took refuge in a wood, when he built himself a hut of small trees; after a time he discovered they were cinnamon trees, and he used to vary his diet by eating his house.

[2] 'The word Panthay has received such complete recognition as the national name of the Mahometan revolutionaries in Yün-Nan that I fear it will be almost useless to assert that the term is utterly unknown in the country which was temporarily under the domination of Sultan Suliman, otherwise Tu-Wên-Hsiu. The rebels were and are known to themselves and to the Imperialists by the name of the Hui-Hui, or Hui-Tzu (Mahometans), the latter expression being slightly derogatory.

'The name of "Sultan," utterly foreign to the ordinary Chinese, was never applied to their ruler, except perhaps by the two or three hajis among them.

'The name "Suliman" is equally unknown. The Mahometans of Yün-Nan are precisely the same race as their Confucian or Buddhist countrymen; and it is even doubtful if they were Mahometans, except so far as they professed an abhorrence for pork. They did not practise

At another time he had taken refuge in the mountains, with fifty or sixty Christian families. After a battle, a band of the defeated party came his way, and would have robbed or murdered them, but he bought the good-will of the chief with an old pistol and ten percussion caps.

The Chinese always maintained that there were a number of Europeans with the rebels, but M. Leguilcher told us that beyond a few people who came from Rangoon, and knew no words of any European language save *Padre* and *Capitan*, there were no foreigners whatever with them.

During the rebellion a horrible epidemic like the plague appeared, that first of all attacked the rats. These animals used to die about the houses for a few days, and then they would migrate in vast numbers from the towns into the fields. After this, the disease seized upon the miserable population, and carried off an enormous proportion of the people.

Another fact worth recording noticed by M. Le-

---

circumcision, though I am not sure if that rite is indispensable; they did not observe the Sabbath, were unacquainted with the language of Islam, did not turn to Mecca in prayer, and professed none of the fire and sword spirit of propagandism.

'That they were intelligent, courageous, honest, and liberal to strangers, is as certain as their ignorance of the law and the prophets. All honour to their good qualities, but let us cease to cite their short-lived rule as an instance of the "Great Mahometan Revival."'—*Mr. Baber's Report—China*, No. 3—1878.

The term *Panthé* is that recently applied by the *Burmese* to the Yün-Nan Mahometans. No one interested in the subject ever supposed it to be 'a national name' in use by the people themselves. Its origin is very uncertain; Sir A. Phayre thinks it has nothing to do with *Pa-thi*, the old Burmese word for Mahometans, which is probably a corruption of *Parsi*, Persian. The name *Suliman* was probably merely a formal style known only to the hajjis; but it is used in the Arabic proclamation which was circulated in neighbouring states, and is mentioned by Dr. Anderson, who appears to have heard it at Momien.—(*Report on Expedition to Western Yunan*, 1871, p. 150.)—*Y.*

guilcher was that, during the rebellion, when everyone was in a state of anxiety, never knowing at any moment whether he might not have to fly for his life, the births amongst the Christians were not more than four or five per annum amongst one hundred and twenty families, the normal number being fifty or sixty.

The Mahometan rebellion has been crushed, but large numbers of Mahometans, who may be known by the white turbans which they wear, but who are as ignorant of the Koran as they are of the Talmud, still remain in the province. They are not less discontented than they were before the rebellion ; all the elements of discord still exist, and a very small spark might rekindle a flame that would again cast its ghastly glare over all the horrors of a civil war.

We had scarcely established ourselves after our dreary march at the wretched inn in Ta-Li-Fu, when M. Leguilcher sent us a present of some beef. This was very acceptable, for the magistrates forbade beef to be killed, partly because the number of oxen in the district was so small that it barely sufficed for the agricultural necessities, and partly because there is almost universally amongst the Chinese a superstitious dislike to killing this animal. But in Ta-Li-Fu the pork-hating Mahometans found a way to provide themselves with meat, and M. Leguilcher was able now and then to obtain some portion of a slaughtered ox. Soon afterwards he came himself to welcome us to Ta-Li-Fu, and his friendliness and geniality were more like those of an old friend than the first words of a stranger.

The next day we moved into our new abode, a sumptuous apartment in an upper story, with a good reception-room on the ground floor, where we could receive official visits.

The Hsien sent his card soon after, with an army of soldiers and Tinc-Chais to wait on us; these were followed in rapid succession by ducks, fowls, sheep, geese, tea, pears, and pomegranates, presents from officials or from anyone who thought he could frame an excuse for sending one. The tea was invariably the tea of Pu-Erh, the pomegranates of Ta-Li-Fu are celebrated, and the geese deserve at least as high a reputation.

The hardened cream of the country was an invariable item of the gift, for in China it is not correct to send only one article; to do so is considered mean; three is the usual number, each object tied up with a piece of red paper, neatly arranged on a red tray, and brought in by an upper servant.

A day or two afterwards I sent a return present to the Tao-Tai, who fortunately had a weakness for cigars, the only things I now could give him. These were duly arranged with their red paper, and Chung-Erh was told to take them. He was, of course, much pleased at the idea of the large present he would receive from so exalted a functionary as the Tao-Tai, and went away to clean himself up for the occasion. But the simple fellow left the cigars in some room down below, and while he was adorning himself, one of the servants of the Hsien whipped them up, took them to the Tao-Tai, and received the silver to which Chung-Erh was looking forward.

When we were fairly settled down in our new hotel, we opened all our baggage, some of which had been shut for many days. The sight was awful; some of the boxes were absolutely rotting, and the things inside them wet and mouldy. We set pans of charcoal about, and soon gave our room the appearance of a laundry, with all our clothes hanging from strings stretched from the walls. The state of confusion became chaotic, and when Monsieur Leguilcher came in to visit us, he had to pick his way amongst the damp clothes, boxes, and masses of wet and mouldering paper that were scattered pell-mell over the floor, and on the tables, beds, and chairs.

During our stay in Ta-Li-Fu it rained incessantly night and day, and we scarcely left the house, except to pay the necessary official visits. We found that, notwithstanding the crowds brought into the city for the examinations, we excited but little curiosity, scarcely anyone following us in the streets. These, though wide, are very miserable in appearance, and the shops wretched; but the city is very interesting, for people of every type are seen, and the women certainly are better looking than the generality of Chinese and aboriginal women. Some of those walking about in the mud in Ta-Li-Fu were quite fair, a great contrast to the very dark mountaineers amongst whom we had been travelling.

We had omitted to study the Chinese almanac before starting on our round of calls, and found on arrival at the yamên of the Tao-Tai that it was one of those remarkable festivals on which the front gates are kept shut, and visitors are only received by the side door, and in unofficial costume; we therefore deferred our visits until a more auspicious occasion, when we were received by his excellency in all the dignity of full dress; we were regaled with cakes and sweets, wine that tasted like vinegar, and Havannah cigars.

This man was not here at the time of Margary's visit, but it was his predecessor who refused Margary the seat of honour. This was an insult, for thereby Margary had not equal rank with the Tao-Tai. According to Chinese etiquette, a visitor, unless a very inferior person, should be invited to sit on the kang. When the Tao-Tai visited us I gave him some champagne, over which he smacked his lips; but the glory of the thing was spoiled through Chung-Erh, who, afraid of frightening his excellency, opened the bottle outside; thus we lost the 'pop,' and the pleasure of witnessing the astonishment of the assembled satellites and miscellaneous crowd.

This was one of the few bottles of wine I had been carrying about as medicine, for which we fortunately never had any necessity.

General Yang, the Ti-T'ai, is perhaps one of the most remarkable men in China. He is almost a hunchback, but so active that the people call him 'The Monkey.' In the war, unlike most Chinese generals, who sit in their chairs in the rear, he was always on horseback under fire at the head of his men.

One day when he came to visit us he walked over from his yamên, a course of action that would shock the sensitive minds of most Chinese officials.

He has made himself so powerful and rich that he keeps two hundred soldiers at his own expense, and is more dreaded than loved by the Chinese Government, to whom, nevertheless, he is an excellent servant.

Baber credits him with the reputation of a Barabbas and a Bluebeard.

He is, undoubtedly, a man of a very violent temper, but his faults have probably been exaggerated, for those who knew him best used to say that they did not think he would be likely to chop off the head of a legitimate wife, if he could get one. At the time of our visit to Ta-Li-Fu he was very anxious to get a well-educated wife from a good family; he had a great mass of correspondence to conduct, and, afraid of treachery on the part of private secretaries, thought that a wife who could write his confidential despatches would be very useful. The good families, however, did not quite see it in the same light, and, notwithstanding the attractions of rank and fortune, Yang had not succeeded in forming a matrimonial alliance.

We sent all our servants home from Ta-Li-Fu, except the Peking boys. Ting-Ko, who had followed us unasked, was very sorry to leave, and begged to be taken on; but there would have been no possibility of sending him back to Ta-Chien-Lu from Bhamo, and so he went with Huang-Fu and his pipe to Ch'êng-Tu. The Ma-Fus returned to Bat'ang, and I had the satisfaction of hearing many months afterwards that all had arrived in safety.

We had of course the usual amount of difficulty in hiring animals, and it was not until our small stock of patience was nearly exhausted that we succeeded in making an engagement with a muleteer; and we were not even yet out of troubles, for shortly afterwards the Hsien sent to say that we had been imposed on; that the man who had agreed with us had neither money nor animals, and that if we advanced him any silver he would disappear, and we should see no more of him. From this we understood that the muleteer had refused to pay the Hsien a sufficient squeeze, and that the latter had therefore determined that we should employ another more amenable party.

We could, however, do nothing but submit to the dictum of the Hsien, who, by this means, brought his man to terms; and then was not in the least ashamed to assure us that he had now discovered the muleteer to be a model of virtue.

October 4.—During our stay at Ta-Li-Fu the rain had fallen without ceasing, and it was with much satisfaction that on the morning of our departure, when I looked out of window, I could see for the first time the lake of Ta-Li lying at the feet of the mountains, on which the first sprinkling of snow had fallen during the last few days. The sun shone in a clear sky, flecked here and there with fleecy clouds; the deep blue water of the lake sparkled as its surface was rippled by a gentle breeze; the morning was beautiful, and all nature seemed to rejoice in the pleasant change of weather. Out of all that remained of our stud, my grey was the only animal that was fit to take any farther; I at first rode a hired pony, and my new Ma-Fu walked on in front leading the grey. Much impressed with his own importance as keeper of the stables to their 'foreign excellencies,' he swelled with pride as he ordered everyone we met to move aside; if people were sitting harmlessly by the road, he made them stand up and salute; and he was not satisfied unless all riders dismounted from their horses, and paid proper respects. At last, as he was making us a perfect nuisance to all the passers-by, I was obliged to make him fall to the rear.

At the end of the suburb we halted. It was time for Monsieur Leguilcher, who had ridden thus far with us, to return to his solitary abode. Those who have never travelled in distant lands can little understand the feelings with which one stranger meets a fellow wanderer from home. 'A fellow feeling makes us wondrous kind,' and it was with no light heart that I bid adieu to our kind friend, for I could hardly venture to say '*Au revoir!*'

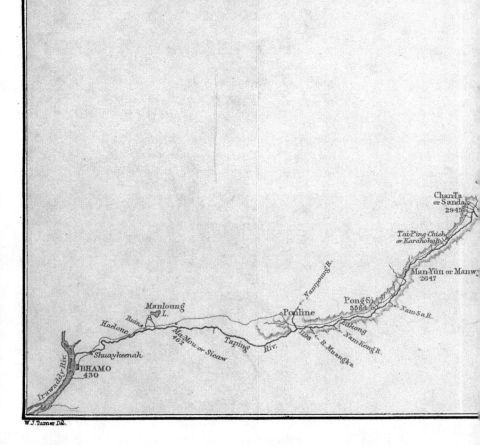

SECTION 7.

Route Map

TA-LI-FU TO BHAMO

By Captⁿ. W. Gill R.E.

Natural Scale 1:1,000,000 = 15·78 miles to an inch
English Statute Miles 68·86 = 1 degree

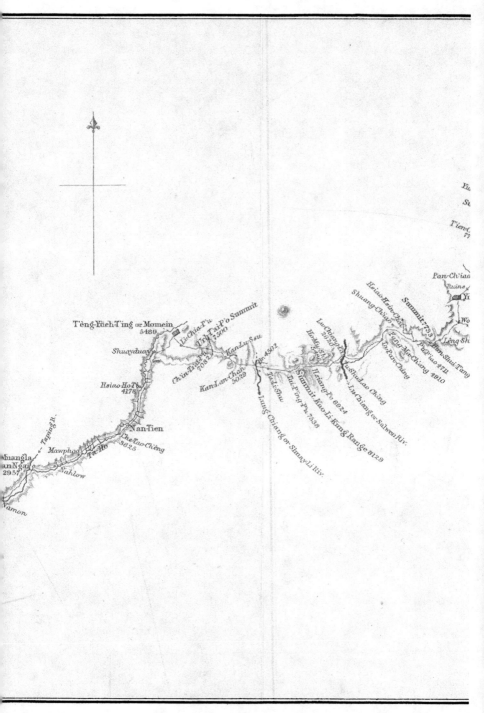

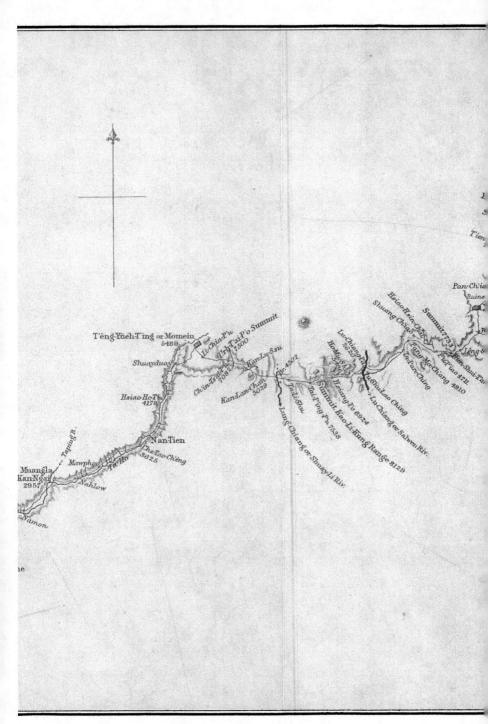

London: John Murray, Albemarle Street, 1879.

Altitudes in English Feet above Sea Level.

W. & A.K. Johnston, Edinburgh and London.

The lake at Ta-Li lies to the left, a couple of miles away; on the eastern side rounded hills sloped down to the water, leaving numerous little bays and inlets on the margin; and an inundated rice plain, in which many of the villages were quite cut off from the shore, stretched between us and the lake.

The first building on the road to Hsia-Kuan is a temple, the body of which was built by General Yang, and the former governor-general of Yün-Nan, who was deposed on account of the Margary outrage. Two wings have since been added by the people of the neighbourhood in honour of the founders. Its walls enclose a spring, to which some virtues attach, bridges have been thrown across the water, and though there is nothing imposing about the building, it is refreshing to find at last something that is not absolutely a ruin.

Hsia-Kuan is situated at the southern end of the lake, at the entrance to the gorge through which the river escapes, and through which the road from Burmah reaches Ta-Li-Fu. It is a poor place, half in ruins. The arch and brickwork of the southern gate had tumbled down with a good portion of the wall. These, however, formed a rather rough ramp, over which we rode to our inn, where we dined off some mutton given us by General Yang, which was so good that we both declared the General's name should be Mutton, and not Willow. The sound of the Chinese word Yang, which means Willow, is the same as the sound of another word Yang, meaning Sheep, though the written characters are quite different.

Before turning into bed we saw, as we believed, all the animals in the inn-yard, and comforted ourselves with the thoughts of an early start; but even yet we had not fathomed the depth of the cunning of these wily people, for when it became light we discovered that though all the baggage mules were safely in the place there was not a single riding animal, and we came to the conclusion that even if we should lock them up with us in our room they would somehow disappear before the morning.

The morning was beautifully fine, and as we stood at the window watching the sleepy people turn out and gradually open their shops, I remarked to Mesny that the salt, instead of being in the usual great flat cakes about two or two and a half feet in diameter, was made in cylinders eight inches in diameter and nine inches high.

'Yes,' he said, 'they make them here in a sort of loaves,' unconsciously using almost the words of old Polo, who said the salt in Yün-Nan was in pieces 'as big as a twopenny loaf.'[3]

We followed the left bank of the river which drains the lake of Ta-Li, and after little more than a mile it entered a defile like that from the plain of Lang-Ch'iung-Hsien. On the right bank a wall extended from the town of Hsia-Kuan to the entrance of the defile, where it ended in a blockhouse; but the interior of the work, as well as the greater part of the length of the wall, is so thoroughly exposed to enfilade and plunging fire from the road on the opposite side, that it would be of very little use against a force led by a commander possessed of the average amount of common sense.

It was market day in Hsia-Kuan, and we met great numbers of coolies and people coming in, nearly all laden with walnuts and sticks for firewood. The people met during the hour from ten to eleven, were counted, and out of a hundred and sixty-five foot

---

[3] Ramusio's version: see Yule's *Marco Polo*, 2nd ed., vol. ii. p. 48.

passengers, seventy-three were loaded with walnuts, forty-four carried sticks, and fourteen were bringing sacks, the contents of which were unknown, but which were probably walnuts. This hour was the most active, for afterwards we met but few people, and not more than fifty or sixty mules laden with opium and cotton. These last may be considered as representing the through traffic, and they came from Yung-Ch'ang.

Most of the trade comes from Ava. One of our muleteers, a black-moustached and whiskered Mahometan, had often traded thither, but had only once been to Bhamo. He said that there were forty marches from Ta-Li to Ava. Judging from what we saw, the through traffic on the road must be very small; but good government at Ta-Li, and the abolition of all Lekin and other oppressive taxation, would no doubt open up the trade.

The road generally was from a hundred to two hundred feet above the river, and very bad to boot. The river was a roaring, rushing torrent, falling one thousand four hundred feet in ten miles, with here and there a waterfall about ten or twelve feet high. The valley of the stream was very narrow, the hills generally running sheer down to the water; but the lower slopes were well cultivated with buckwheat, rice, and a crop noticed before, called paidza. The valley in its palmy days must have been well populated; but the towns and villages were now nearly all in ruins, and could contain but few inhabitants. A little below the very small village of Shih-Ch'uan-P'u, a very unpleasant descent began, ending in a bridge made by laying long slabs of stone from the banks to a rock in the middle; whilst, just below, the opening of a narrow glen gave a passing glimpse of a fine cascade, brimful after the recent rains. Here the walnut trees again appeared, but they now looked very autumnal; the leaves were very brown, and the nuts all plucked; there were a few persimmon trees with fruit nearly ripe. Much of the rice was nearly ready for cutting, and there were a few very fine large bamboos.

The road was very bad, in one place altogether washed away, and we were obliged to make a cross-country expedition over a field of buckwheat. Here, though the whole of the traffic was diverted through this field, scarcely any damage was done, all the animals following exactly in the same track. The Chinese, whether boys or men, never do wanton mischief, and in enlightened England a road suddenly taken through a field of corn would hardly leave the farmer so unscathed as here. At this point we overtook Chin-Tai, who had been sent on ahead, and who had been taken down the other side of the river to a bridge, now washed away, and we went on together to the little village of Ho-Chiang-P'u, where we found comfortable, though rather rough accommodation.

October 6.—We changed our baggage animals again at Ho-Chiang-P'u, but it gave us no trouble, for having made an engagement with our head muleteer to take us to Yung-Ch'ang, it was his affair, and for a wonder all the fresh mules were in the yard and ready for a start at a decently early hour. Our new bell mule was a beautiful sight, for the leader of a train of mules here not only has a bell round its neck, but is, in addition, adorned with an astonishing amount of finery. He has lovely waving feathers and all kinds of ornaments on his head, almost always inclusive of a bit of looking-glass.

We followed the right bank of the river, but gradually left it, cutting off the end of a spur that runs down between it and the Yang-Pi river, to which it is a tributary. For the first few miles the road was not very good; we then entered a fine plain, quite flat, and almost entirely cultivated with rice; but the evidences of wreck and ruin were painfully apparent in the large extent of ground laid out in terraces, but now uncultivated. These terraces had formerly been rice-fields, but the diminution of the population since the rebellion is so great that there are not enough hands to cultivate them. In Ssŭ-Ch'uan there are more people than that province can support, though there is not a square yard of cultivable ground untilled; the people are too poor to emigrate to this province, where there is room enough and to spare for the surplus population of Ssŭ-Ch'uan. An enlightened government would assist the people to come here, and it would well repay that of Yün-Nan to do so, but of course it would take time to recoup the outlay; officials are constantly changing, and no one cares for anything but the immediate present.

At Yang-Pi the civil official is a frightful tyrant, and puts the head men of the villages under his government into prison once a month, unless they pay him a good sum of money. Poor Yün-Nan! what with rebellions, opium-smoking, and bad government, the people have a hard time of it.

The road across the plain was very good, and ran between hedges, where several new kinds of plants appeared. There was a creeper, with a fruit like a melon, growing wild, which they said was good to eat when ripe. I picked also a wild lemon, the first I had seen in China; and many varieties of magnificent grass, some ten or twelve feet high, with stalks three quarters of an inch in diameter, and most graceful heads, were growing everywhere in great profusion. There is a peculiar kind of grass which invariably grows in the disused rice-fields; and, standing on the top of a mountain or a spur, miles and miles of now uncultivated terraces can be recognised by the white heads of this species of grass.

The little plain is bounded by well-wooded hills of red sandstone covered with beautiful green herbage; these are backed up by fine snowy-topped mountains, which, people say, are covered with snow all the year round; if this be true, the peaks must be at least twenty thousand feet high.

Baber says that 'the Yang-Pi river is represented in all maps as a bifurcation of the Mekong; but in so mountainous a country one is loth to believe that rivers can divide in this way.'

On the maps that I have examined, except Keith Johnston's,[4] I do not find this to be the case; but the source of a stream flowing north is shown on one or two maps so close to the source of the Yang-Pi river flowing south, that they can hardly be distinguished with a magnifying glass; some careless cartographer has probably joined them, and has thus given rise to Baber's remark. In any case, it may be taken as certain that there is no bifurcation of the Mekong (the Lan-Ts'ang).

Just outside Yang-Pi, an officer and some soldiers, who were drawn up waiting to receive us, fired three guns or crackers at our approach.

[4] An old German map, after D'Anville and Klaproth, 1843, is still the best map of Western China; both the Yang-Pi and the Shun-Pi rivers are clearly shown, with the range of mountains between them, although it must be admitted that it is very faulty beyond the boundaries of China Proper. The imaginary bifurcation is not shown on Garnier's map, nor on the map of Western China prepared by the Indian Topographical Department, nor on the German map above alluded to.

The mules came up after we had been here half an hour, and the muleteers declared that they could not finish the next stage, to which we replied by ordering our horses to be saddled at once. The military officer who had met us on arrival was waiting outside with three more crackers as we rode away, but the civil official was still in bed, and declined to be disturbed at that early hour.

We now crossed the river by an iron chain suspension bridge, of about forty yards' span, with nine chains, which was remarkably stiff and steady for one of these constructions, and, leaving the river, we at once commenced a very steep and rather difficult ascent of about two thousand feet; the road then improved. Another thousand feet brought us to the summit of Ch'ing-Shui-Shao, eight thousand two hundred and thirty-three feet above the sea, and we then descended one of the very worst bits of road we had encountered in our journey. It had once been paved with very large stones, now all misplaced, and the interstices filled with deep, stiff, sticky mud. Slippery banks at the sides and holes hidden by mud and slush made walking necessary for about a couple of miles from the top, after which a certain amount of improvement became apparent, and for the next few miles it was possible to ride. The morning had been very fine, with clear sky and bright sun, but very heavy clouds were gathering over the mountain tops as we ascended from Yang-Pi; these cleared off in the evening, and a brilliant planet shone out before we arrived, a star of hope that our mules might turn up some time during the night. Without their aid, however, a huge vessel of rice was cooked for soldiers and servants; walnuts and pears, poached eggs and wheaten cakes made us an excellent dinner, and, much to Chin-Tai's sorrow, when the mules arrived, at half-past nine, we declined the proffered dinner of several courses that he was anxious to prepare.

October 7.—There was no mountain on our way from T'ai-P'ing-P'u (The Peaceful Village), but the road, apparently out of very wantonness, went up about nine hundred feet.

The stream of T'ai-P'ing-Pu, which is bounded on its right bank by a spur from the mountain Ch'ing-Shui-Shao, runs into the Shun-Pi river at a distance of about eight miles from T'ai-P'ing-P'u, and the Shun-Pi river is crossed about a quarter of a mile above the junction of the streams.

Any ordinary person would imagine that the road would be taken somewhere near the edge of the water, but he would be quite wrong. This eccentric path rises steadily nine hundred feet to the crest of the spur, and then by a very nasty zigzag goes back to the stream.

At first we were unable to suggest any other reason for this monstrous behaviour than that the road-makers were afraid that, by leaving a few level miles, men and animals travelling would get out of training for the succession of mountain ranges that must be crossed. At the top, however, we discovered an unexpected village, Tou-P'o-Shao, and it is for the sake of the two or three huts that compose it that all travellers have to march up the hill and down again. The morning was beautifully bright, and the scenery charming : fine rolling mountains in every direction, whose sides, by no means steep, were well wooded with small pines ; there were many open spaces, some cultivated and some covered with rich fine grass, which, after the recent rains, was of the brightest green, and a huge range of mountains ahead promised us a hard day's work for the morrow. The sides of the valley were little cultivated, and we did not pass a single hut until we reached the very crest of the ridge where Tou-P'o-Shao is perched. Whence the half-dozen people inhabiting it draw their supply of water it is difficult to say; but there it is, situated in as lovely a spot as can well be conceived, and from a point a short distance beyond the Shun-Pi river can be seen flowing from the north.

The next village of three or four houses amongst the ruins of thirty or forty others is three miles further on, at the bottom of a heartbreaking zigzag, and here a couple of caravans were resting, one before undertaking the arduous ascent, and the other after having reached the level ground. The former was a train of fifty-six mules bringing cotton from Yung-Ch'ang, and the latter consisted of twenty-six mules taking salt thither from a place called Chao-Ho-Ching,[5] somewhere east of Ta-Li-Fu. Besides these we met no one on the road, except one travelling official and ten or a dozen other people, although this was one of the favourable seasons for travelling on the great high road from Bhamo to Ta-Li.

The country is scarcely inhabited. Besides the two miserable villages already mentioned, there are but two solitary huts between T'ai-P'ing-P'u and Huang-Lien-P'u, a distance of ten miles, and these two villages themselves contain but few inhabitants.

The Shun-Pi river is crossed by a suspension bridge of eight chains, six below and two at the

[5] Chao-Ho-Ching. I cannot find this name on Baber's itinerary from Yun-Nan-Fu; though he left scarcely a hut unnamed. It may be meant for Chao-Chou, a large and well-to-do town, fifty miles to the north of which, Mr. Davenport says, are the famous salt wells known as Pai-Yen-Ching (White Salt Wells).

sides. This is thirty yards long, and has been an excellent construction ; but the pier supporting one of the upper chains is broken down, and the bridge is consequently lop-sided, and something worse than rickety. The river runs, in a southerly direction, between beautifully green and uncultivated slopes that shoot down straight to the edge of the water, leaving no flat ground at the bottom ; the water, like that of all the other rivers seen lately, is of a very reddish-brown, from the red clay and sandstone of which the country they drain is formed.

The road winds along about a couple of hundred feet above it, and is very good from the bridge to Huang-Lien-P'u. It was, just at the time of our visit, being repaired on account of the examinations which were shortly to take place at Yung-Ch'ang, whither some exalted functionary from Ta-Li-Fu would shortly go.

Huang-Lien-P'u is situated about a quarter of a mile up a small stream tributary to the Shun-Pi river. Here an unexpected treat awaited us, for one of our men-servants, whose permanent employment was that of chief baker to Monsieur Leguilcher, said that he could buy some leaven, and that if we liked he could bake us some bread. We did like very much ; but even the thoughts of this luxury in store for us were not sufficient to reconcile us to the smoky atmosphere of the room, which was not rendered more pleasant by the fumes that came in at the window from a house next door, where the family were roasting their annual supply of chilies. Savages have been smoked out of caves with a few grains of red pepper on a fire, and our experience of the chilies led us to sympathize with the savages. Even the Chinese cannot stand this, and they can stand most things; in fact, it was so impossible, that our request to our next-door neighbours to desist was considered by no means an unreasonable one.

At about five o'clock in the evening we received a visit from a Christian who had been in the service of the missionary Mr. McCarthy, whom I had met at I-Ch'ang. This man and two others had accompanied Mr. McCarthy from Ch'ung-Ch'ing to Bhamo, and they were now returning to the former place. They had been twenty-four days on the journey from Bhamo; until the last three it had rained heavily every day, and now the two companions of our visitor were very ill, one of them having been twice already given up for dead; he was naturally very anxious about them, for if either should have died before they reached their homes, the foreigner, it would have been said, had killed them. We gave our guest some chlorodyne, and, as both his invalids seemed better the next morning, we advised him to wait a day or so in the healthy mountain air of Huang-Lien-P'u, and recruit his folk before proceeding on his journey.

October 8.—The road now led us across the range of mountains that divides the basin of the Shun-Pi river from that of the river of Yung-P'ing. Both these streams and the range of mountains between them run nearly north and south.

We ascended a very remarkable spur thrown out from this range some seven miles long, and scarcely a mile in breadth, with a deep gully on either side, in each of which a torrent was rushing down to the river of Shun-Pi. The formation was still the same red clay and sandstone, but after the dry weather the road was good enough, except just at the end of the ascent of Mount T'ien-Ching-P'u, where there was the usual stiff zigzag. It is worthy of remark that in the sandstone districts the roads generally follow the crests of the ridges, while in those countries where the geological formation is of the harder limestone or granite the roads invariably clamber up the bed of some torrent. The reason is obvious. In the sandstone the tops of the spurs are always more or less level, and offer an easy route, though the ascent to them is often very difficult. But amongst the limestone mountains the crests are torn into wild and ragged pinnacles; they are sometimes almost as sharp as the edge of a knife, and, as routes, are utterly impracticable.

The country was still almost uninhabited, and bore on its face sad traces of devastation. Long extents of slopes, laid out in terraces, once used for rice cultivation, but where now grasses and reeds were the only crops; and ruined villages, where rank weeds and prickly pears usurped the place of smiling vegetable gardens, bore pitiful witness to the havoc of the 'dogs of war.' At a distance of two miles from Huang-Lien-P'u a single hut with a patch of cultivation was the only sign of inhabitants, until a ruined village, Pai-T'u-P'u, was reached after another two and a half miles. Here, in the ruins that marked the site of a once flourishing village, where coarse grass and weeds grew amongst the few stones which indicated the positions of the houses burnt or sacked by one if not both parties during the rebellion, two or three huts had been rebuilt, and the busy Chinese occupants were hard at work reclaiming the soil from the weeds that overran it.

The hill-sides were mostly covered with long but rather coarse grass, and woods of pines, oaks, and chestnuts, where pheasants were heard calling. In these regions where the oaks and chestnuts grow close to and amongst one another, they seem to run into one another, and all sorts of varieties are seen that appear as if they were a cross between the two trees. First, there is the *bonâ fide* and unmistakable chestnut with the real chestnut leaf, and the nut encased in a thick husk covered with prickles; then we see trees with a leaf almost the same but slightly approaching that of the oak, and with some few leaves more like an oak than a chestnut, till we arrive at the real and true oak with an acorn and cup without any prickles. The fruit also varies from the chestnut to the acorn, some of the varieties being almost like the chestnut covered with prickles, and with only a little bit of the fruit appearing through the husk, while others bear fruit nearly like the acorn.

The next hut was two miles further on, by a temple where there had at one time been two presiding deities or dignitaries, one at each side of the entrance; one of these, however, had shared the fate of Dagon, and its place now knew it no more.

After a long but not difficult ascent of eight miles, we found ourselves at length on the summit of the T'ien-Ching range, eight thousand one hundred and forty feet above the sea, where a few wretched huts boast themselves a village, and glory in the name of T'ien-Ching-P'u.

Here a man joined our party, who told us that some time ago both his father and mother had died, and that, finding himself without money to bury them with, he had sold himself to a firm of traders at Ava, for to a Chinaman there could hardly happen a more fearful evil than to be unable to give father or mother a proper interment. He had been to Ava once, but as the firm had now given up business, or become bankrupt, he was free, and he offered himself to us as a travelling companion.

We passed a village of a few huts two miles further, but nothing else until we reached the fine plain of Yung-P'ing. The city of Yung-P'ing-Hsien was, we had understood, to have been our halting place, but now the muleteers said that it was a little off the road, so we did not go there.

We descended another spur from the western side of the same range, and soon the plain lay extended at our feet.

We asked a man with us if the city was on a river, or a little off it. His reply was eminently characteristic of a Chinaman : 'Oh,' he said, 'the city wall is destroyed, and now there are only houses.'

After a long conversation we prevailed upon him to say that the city was not on the river. Under these circumstances we were not surprised to see it built on both banks of the stream.

The road down-hill was very fair, but when we reached the plain it was awful ; in fact, there was no road at all, and in rainy weather it would hardly be possible to cross either river or plain. There is little cultivation but rice, and here we saw the first rice harvest, but again there were wide spaces of terraces which had not yet been recovered. The carcass of an old buffalo cow with a good many wounds in her body lay by the road-side, and near her were the remains of a calf ; and as Chin-Tai had seen a panther near the temple we had passed in the morning, there were probably a good many wild beasts about.

Ch'ü-Tung, a market town situated on the right

or western bank, was our destination, and we reached it after much floundering about in bogs and mud, where some of us had a fall or two, and where we all got wet in fording the river, which, now tolerably deep, would with very little more rain become impassable. The bed was now in several channels, but these evidently join together during heavy rains and form one large river. We met but little traffic during the day: a few people moving about from one house or village to another, a travelling official with his wife, and besides these nothing but a train of seventy-six unladen bullocks going to fetch salt from Chao-Ho-Ching. The fact that there are no loads for bullocks to carry shows what trade is worth at this season at all events. These were very fine animals, and looked fat, sleek, and clean. Some of them, instead of wearing their bells in the usual manner round their necks, had bells suspended from a kind of U-shaped bow, upside down, one end resting on each side of the pack-saddle.

The city of Yung-P'ing was entirely destroyed by the Mahometans during the rebellion, and not a single house was left standing. Now, although it still remains the prefectoral city, Ch'ü-Tung, which is on the high road, seems gradually to be ousting it from its position of commercial importance. There are already about two hundred families in Ch'ü-Tung, while Yung-P'ing itself can now boast of no more than three hundred. There are a great number of Mahometans at Ch'ü-Tung, as, indeed, there are all over the country; they are easily recognised by their white turbans. They certainly seem sufficiently numerous to render possible another outbreak of the deplorable rebellion that desolated this province. It would be a wise policy on this account for the

Chinese Government to assist emigration from Ssŭ-Ch'uan to Yün-Nan. It would not only relieve the already over-populated province, and supply labour for the now waste lands in Yün-Nan which cry out for hands to till them, but by gradually increasing the number of orthodox Chinese, the population of the so-called Mahometans would be lessened, and the fear of future outbreaks be by degrees reduced to a minimum.

We were lodged in a fine house at Ch'ü-Tung; it belonged to some general, and we enjoyed an unwonted immunity from smoke. The civil official of Yung-P'ing (a Hsien) came in to see us in the evening. He was from Chin-Kiang, and, of course, had seen foreigners before. He said that he had prepared a vast army of coolies and caravans of ponies for us; but he probably made this statement after having satisfied himself that we had already engaged transport as far as Yung-Ch'ang.

From Ch'ü-Tung we ascended two thousand seven hundred feet to the summit of another mountain called T'ien-Ching-P'u; and the fact that we went up three thousand feet and down again the other side, was becoming almost as monotonous to write about as the perpetual ascents and descents were wearisome to perform.

Baber remarks :—

'On the morrow, the inevitable climb awaited us. A winding track leads through a wooded glen to the foot of a steep ridge, which we only surmounted to find a most formidable range still barring our advance.

'Descending to T'ieh-Ch'ang, which means *Ironworks*, but contains neither works nor iron, being nothing but a squalid gathering of half a dozen huts, we found ourselves near the centre of a cultivated hollow; the stream which drains it seems to flow inexplicably into a bay of hills without any exit.'

He adds in a foot-note:—

'On referring to the route chart, it seems probable that the stream in question finds exit through a gap

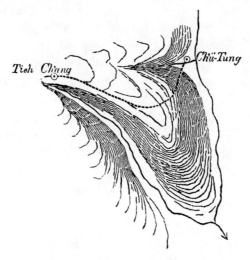

which was not visible from the road, and is the same brook that runs through Ch'ü-Tung. We failed, however, to detect any appearance of such a break from Hua-Ch'iao, or a little before it, where we rested for some minutes.'

The spur that is ascended from Ch'ü-Tung, is a spur from the western mountain of T'ieh-Ching-P'u, or the Hua-Ch'iao range; for it seems to be called by both names. The stream from Hsiao-Hua-Ch'iao (The Lesser Flowery Bridge) is not the brook that runs through Ch'ü-Tung, but it finds its way a little lower into the Yang-Pi river.

It is possible that the track we followed over the spur alluded to, is a little to the south of that followed by Baber, and so the features were more visible to us.

From Hua-Ch'iao, or Ta-Hua-Ch'iao (The Greater Flowery Bridge), a very steep descent led us to the plain of Sha-Yang. The road, beyond its steepness, was not altogether bad this dry weather, though it would become disagreeably slippery after rain; but, even as it was, it was almost impossible to walk down, and it was a case of either running or standing still. I found Chin-Tai at a poor inn, where he repeated his favourite phrase '*all have got nothing*,' by which he meant that the kitchen arrangements were defective. So while he went to find a better place, I sat down, and was able to note how the inquisitive Chinese were being gradually left behind. Here, in a large market town, although a good many people collected at the entrance to the inn, no one, not even a boy, passed the threshold, though I was sitting in a room some ten or fifteen yards back; and as I walked to the next place I seemed to excite but little curiosity. There are so many foreigners here, border tribes, wandering Burmese, &c., that as we all, including Englishmen, pass current under the one term, *barbarian*, little notice is taken of a fresh specimen of the genus.

We did not find the mansion that Chin-Tai discovered any great improvement on the one that we had left. The houses here are generally built with two wings to the main building, in the upper story of which is usually the best room. The gables of the wings are always left open on the side towards the main building, the latter being provided with windows, in a suitable position for the entrance of smoke from the kitchen, which is in one of the wings. As nothing but damp wood is ever burnt, and there is no exit but the open gables for the kitchen smoke, it pours into the upper room and there remains; for at the back there is neither window nor door, and on one or two occasions even our Chinese servants found the atmosphere unendurable.

The officials who visited us here were fairly driven away by it; but the magistrate came back in the evening after the fires were out. He was a really intelligent man, and instead of discussing the usual trivialities he asked questions about railways and steamboats, and seemed desirous of gaining information about foreign countries.

The people here called us foreign Mahometans, as we never touched pork. The presence of a large Mahometan population always rendered it comparatively easy to buy beef; and there were plenty of fat geese, so that we were never in any difficulty about food.

The traffic on the road from Ch'ü-Tung to Sha-Yang consisted of no more than twenty-five salt mules, bound for Yung-Ch'ang.

October 10.—As the muleteers had visited us in the evening with the respectful prayer that we would make an extraordinarily early start, it caused us no surprise to find that even after Chin-Tai had packed all his cooking things the muleteers had not finished smoking their opium. The first time we sent for them they were sleeping, the second time they were eating, and the third time they said that one of the animals had strayed and could not be found. The fourth missive, after some time, produced Chin-Tai's mule and our ponies; after which the men came and declared that they could not reach Pan-Ch'iao that night; which, considering the hour, was quite true. It had rained during the night and early morning; and although after we started the showers were very light, they made the road almost worse than a good downpour would have done, for that would have washed the round and slippery stones clean ; now they were as if they had been carefully greased for our benefit.

We were now in the basin of the Lan-Ts'ang-Chiang, known lower down as the Mekong river; but before reaching it we crossed a ridge about three hundred feet above Sha-Yang. The ascent was not very steep, but it was greasy enough to give our animals hard work. Here we met a train of forty-six mules carrying calico made in Yung-Chang. When the summit was gained, we at length saw the much-thought-of and long-talked-about Lan-Ts'ang-Chiang rolling at our feet; for the river seems to maintain the character Cooper gives of it higher up, and though there is not here another 'Hogg's Gorge,' yet the stream flows through desperately steep hills; and down the side of one of these a zigzag led to the river, 1,400 feet below the crest; it was a frightful bit of road, and had this been written at the bottom it would have been apostrophised in no measured terms; but what followed was so much worse that there is no bad language to spare for this descent.

The river, the bed of which is here 3,953 feet above the sea, is crossed by an excellent iron chain suspension bridge, in very good repair, and very steady. The bridge from the edge of one pier to that of the opposite one is about fifty yards long, supported on twelve chains below, and two above for handrails. The links are about one foot long of three-quarter inch iron, and the chains are fastened at the ends with shackles.

Near the end of the bridge a tablet was set up, on which there was some writing. One of our soldiers told us that he had escorted Mr. Margary to T'eng-

Yüeh, and he spoke warmly of the kindness he had received at his hands; in this he was like everyone who had met him; for Mr. Margary seems to have left a deeply favourable impression wherever he went.

The soldier told us that Mr. Margary had shot with a thing like a wine-bottle at the inscription on the tablet; as far as we were able to gather, he meant a pair of field-glasses.

Now commenced our day's work, and a hard one it was. The road at first led along the side of the hill; it had once been paved with great round stones, which now, half misplaced, lay about, leaving great muddy chasms. At the end of this was a village, and here the path left the river and went straight up a gorge, which, with a little poetic licence, might be said to be like the wall of a house. The muleteers had told us that we could never conceive the badness of the road, and they can hardly be accused of exaggeration. It was enough to break the heart of a millstone, not to speak of the unfortunate little ponies that carried our baggage or ourselves. We had to face it somehow, zigzag after zigzag, mile after mile of steps, sometimes a foot high, of round and slippery stones, and muddy bogs, into which the feet of the unfortunate animals would slip with a bang and splash the mire in all directions. But still, right overhead, the interminable track appeared; and when at length an ascent of two thousand three hundred feet brought us to the end of this desperate gorge, men and animals 'knocked their heads' each after his own fashion.

Here was the temple of Shui-Yin-Ssu, and a little tea-house, where all the men with us—two servants, two soldiers, a Ma-Fu, and a muleteer—had a good meal of buckwheat-porridge; each had two bowls, besides tea, rice, and cakes, and 100 cash (about sixpence) just covered the expense. This porridge is simply made by pouring boiling water over buckwheat flour, and mixing it up well with an enormous quantity of coarse brown sugar into a paste. The Chinese make a similar porridge of bean-flour; indeed it is hard to say what they do not make of beans; and how they would get on without this useful vegetable it is impossible to say. From this point the country seemed no longer to bear on its face the signs of the war which has worked such ruin in this province.

From here the road ascended easily in a valley well cultivated with rice, which at this altitude, 6,270 feet, was not yet ready for harvesting. In itself the track was still tolerably bad, but as the gradient was easy, and there were none of those abominable staircases, it seemed like Macadam compared with what we had passed, and after a march of about four hours Chin-Tai's mule at the door of a house was a pleasant sight to men and animals; and notwithstanding the porridge, soldiers, servants, and Ma-Fu did full justice to sundry bowls of rice all ready for them. After this everyone was in a good humour, and although our muleteers had made up their minds to stop here, as the people told us of a village five miles further on, we determined to take that bit off the morrow's journey.

The country now improved in appearance very much. There was much more cultivation on the slopes, chiefly Indian corn and buckwheat. The valleys between the hill-sides were covered where possible with rice; there were no traces of former cultivation fallen into disuse, there were not the same number of ruins about the country, and the villages were far more numerous not only in the valleys but on the mountains. The ranges of mountains that we had marched across had hitherto been almost unpopulated and uncultivated, and it was only in the valleys of the rivers that people, villages, and crops had been seen. But now it was a pleasant sight to see some snug houses nestled on the hill-side, or to watch a wreath of smoke curling up from the midst of some small wood high up above the road, showing that here at last everything was not given over to nature and wild beasts.

On our way to Ta-Li-Shao, we passed a train of forty-seven animals laden with salt for Yung-Ch'ang; the ascent was gradual, and the road very fair. We found Ta-Li-Shao, a group of about half a dozen cottages, to be 7,412 feet above the sea. A loft in one of them was free from smoke, and civil and obliging people did their best to make us comfortable, after one of the most severe marches on this road of difficult ascents.

October 11.—It was raining again in the morning, and the appearance of the clouds promised us a wet day. Before starting, a man from whom we were endeavouring to extract some scraps of information, told us that the road to Pan-Ch'iao was 'a good and level one down hill,' a remark that made us inclined to ask if he had any relations in Ireland.

We continued our ascent of the mountain, which was now very easy, only rising about four hundred feet in the couple of miles that took us to the final summit (7,795 feet above the sea), whence we overlooked the fine plain of Yung-Ch'ang. The road was amongst fine, rolling, wooded mountains, with open cultivated spaces, and a fair sprinkling of villages, and then commenced the descent of '*the level road down hill.*'

The first part was rather bad and steep over exceedingly slippery stones, but after about two miles from the top it became really very good, descending easily, and not being particularly sticky.

Pan-Ch'iao, where we halted for breakfast, is a large market town 5,692 feet above the sea, situated in the plain of Yung-Ch'ang, about a mile beyond the edge of the mountains. It seemed rather a wretched place; there was no hotel in it, nor did there appear to be a single house where we could have slept if we had been unfortunate enough to have wished it; there was a hovel that did well enough for breakfast, where rice was all ready for the people with us.

The inhabitants displayed a good deal of the Chinese curiosity, and during the hour and a half that we spent here, there was, notwithstanding the rain which fell steadily all the time, a good-sized crowd at the door, who stood staring in their usual vacant manner, but did not attempt to come into the room.

A new dish was set before us at this place, macaroni made of rice instead of wheaten flower; it was round, and looked very much like our European macaroni, but thinner, and instead of being tubular was solid. The salt here was in moulds about six inches high, for which there can be no better simile than old Polo's of twopenny loaves. The shape was something like the figure 8. Each was stamped, though in this case it was not the 'Prince's mark' that 'was printed,'[6] but a very ancient character, of which the signification is '*happiness*,' a way of wishing welfare to the purchaser. This salt comes from Min-Ching, in the magistracy of Yu-Lung-Chou.

[6] See *ante*, p. 311.

There is another shape that comes from Pai-Yen-Ching, a place mentioned before; for each locality appears to have its own shape and size. Here we found some large pears; we weighed one, it was a pound and a half (English weight), with a fragrance and slight flavour of a pine-apple; it was sweet and juicy, but a little hard.

This town is close to the left bank of the Pan-Ch'iao river, which waters the valley of Yung-Ch'ang, a perfectly flat plain, about five or six miles wide, entirely devoted to rice cultivation. Here again we came across the traces of the war: ruins around the villages and towns, remains of fortified towers, and on the lower slopes of the mountains some terraces fallen into disuse. This part of the country, however, seems to be recovering itself rapidly, for all the small valleys where the streams ran into the plain were well cultivated. The position of this river was contested for three years by the two parties. The Mahometan rebels on the right bank, and the Imperialists on the other being all this time separated only by the width of the stream—about twenty or twenty-five yards. The Mahometans built strong towers on their bank of the river, and with the aid of these prevented the Imperial troops from crossing. It is very interesting to find that this plain, the scene of that 'great medley' and 'dire and parlous fight,'[7] described by Polo, should in recent years again have been a position so hotly contested. But how the

[7] The old Venetian tells us that in this dire and parlous fight, the King of Mien, like a wise king as he was, caused all the castles that were on the elephants to be ordered for battle, and that the horses of the Tartars took such fright at the sight of the elephants that they could not be got to face the foe. Herodotus mentions that Cyrus in one of his battles used his camels to terrify the cavalry of the enemy, but with better fortune than waited on the wise King of Mien (*Herod.* i. 80).

valiant Nescradin ever managed to get two hundred elephants into China, unless there was some much better road than the one we had followed, must remain a mystery.

The soldier in whom we fancied we had discovered some Celtic blood was a wag in his way, for he volunteered the information that the next bit of road to Yung-Ch'ang was a '*twenty cash bit*'; for he said it was so bad that it wore out two pairs of straw sandals, each of which costs ten cash, and is supposed to see the wearer through the worst day's march. The same man told us that, in the year 1873, eight or nine foreigners had visited T'êng-Yüeh. He said that they bought all kinds of things, birds, insects, no matter what, and were in the habit of giving one rupee for a single specimen. The Chen-Tai of T'êng-Yüeh, hearing of this, imagined that they were simple folk being imposed upon by his wily countrymen, and he forbade his people to sell any more birds. No doubt the naturalists, whoever they may have been, would now be much amused if they could know why the supply suddenly stopped.

The city of Yung-Ch'ang is a sad spectacle of ruin and desolation. It appeared as if the greater part of the space within the walls had once been well covered with buildings, but now three quarters of it were vacant or under cultivation; for in many places crops of Indian corn were growing where there had formerly been houses. Notwithstanding this, the portion that had been rebuilt seemed very prosperous, and there was an amount of elegance, if such a word may be applied, about the shops that had not been seen since leaving Ssŭ-Ch'uan; the streets were very wide, and were full of well-dressed people, looking comfortable and well-to-do. Stalls at the side of the road were apparently driving a thriving business, and altogether there was an air of prosperity about it that was quite surprising. The restored portion was very small, but what there was in appearance far surpassed Ta-Li-Fu.

We were lodged in a real and very good hotel, where we had a comfortable upper room free from smoke. The landlord said that it cost him 3,000 taels to build; and the fact that a man could find it worth while to lay out so large a sum shows that the place must be reviving. Indeed we found traders here from nearly every province.

The Fu, on receiving our cards, told the servant that the examinations were going on, and regretted that he could not see us; for high officials who attend these examinations are strictly forbidden to receive or pay visits of any kind. He asked us to apply to the Hsien for anything we wanted. We had already sent cards to him; he promised to visit us the next morning, and in the meantime offered to supply all our wants.

October 12.—The mutton of the plain of Yung-Ch'ang enjoys a higher reputation than it deserves; and the Fu sent us specimens of it, and of the geese, which were better entitled to the praise bestowed on the sheep. The Hsien also sent the customary gifts, and three men to look after our luggage, for the examinations always cause a large influx of bad characters into a Chinese city.

When we asked the landlord if he could find us muleteers to take us to T'êng-Yüeh, he promised to get them within a very short time, but, as of course it was his interest to keep us in his house, we soon found it necessary to send Chin-Tai to look after mules for himself. This, however, was difficult, as it was only prudent to obtain a security for the honesty of the muleteers before entrusting our property to their mercy. When, as at Ch'ung-Ch'ing, Peking, or other great trading places, mules are engaged at a regular forwarding establishment, the house, which is well known, is responsible for them, and if anything is lost or stolen the firm makes it good. The first lot of muleteers that Chin-Tai discovered fell out with the landlord of the hotel, who probably demanded too high a squeeze; for a man in the position of the hotel-keeper has considerable power over these people. Merchants requiring transport naturally stop at the best inn, and the landlord can always make or mar the reputation of a set of muleteers. The next people that Chin-Tai found were reasonable in their terms; they appeared very respectable men, and as it was so difficult to arrange matters we agreed to dispense with the security, and, as it turned out, with no bad results. The abominable system of the squeeze upsets nearly all arrangements in China; but it is hopeless to try and alter such an ancient custom; even at the British Legation at Peking the porter has his squeeze on every article that passes the gate, unless it is brought in by some member of the Legation; for if the servants carry anything, the porter finds out the shop at which it was bought, and demands and gets his squeeze. No sooner were our affairs settled, than another set of muleteers offered their services, and were so anxious to come with us that they would not believe Chin-Tai when he told them that we had already made arrangements. They insisted on seeing us, and their crestfallen look, when they learned from our lips that they were too late, plainly showed that there was no lack of animals or muleteers, and that our difficulties were caused by the desire to extract just a little more out of the foreign mine of wealth so unexpectedly opened.

Some general on the march arrived in the town in the morning. One of his officers in advance came to the hotel we occupied, and, finding us in the best rooms, cursed the landlord in a tone of voice that reverberated through our apartment. Not daring to attempt any ejection of ourselves, he made great but unsuccessful efforts to take possession of the rooms occupied by our servants and baggage. The general had by this time arrived himself, and sat in the yard of the inn in his sedan chair.

The news of the turmoil soon reached the ears of the Hsien, who sent a polite message to the general, asking him to find a lodging for himself elsewhere; to which he gruffly made reply, that the Hsien had better find him a place if he expected him to leave the hotel where he was; and his minions thereupon commenced to turn out the occupants of all the minor apartments.

This was not very pleasant for us, for his soldiers, sharing the wrath of their commander, would in all probability have picked some quarrel with our servants, or have contrived to rob us of something. Our apprehensions were shared by the Hsien, who reminded the general that he would be responsible if anything of the sort occurred.

The general paid little heed to this warning, and ordered his goods to be unpacked, sitting, nevertheless, all the while in his sedan chair, as he no doubt anticipated that the officials of the place would arrange matters somehow without the loss of dignity which he would have suffered by consenting to move to another hotel.

In the meantime he sent for the landlord, who

came into our room, shut the door and barred it, and hid himself under the table in a great fright. He said that the general had been here the previous year, had broken his furniture, and nearly killed his cook; that some of his soldiers or people had stolen four pieces of silk from one of the people staying in the house; that he (the hotel-keeper) had been obliged to pay for this, and that his loss and annoyance from the great man had been almost insupportable.

The anticipations of the general were, fortunately, shortly realized, for the second military officer of Yung-Ch'ang, dressed in full uniform, rode into the yard, alighted from his horse, and, with many profound salutations, invited his great excellency to come and stay in his yamên.

The general was able to accept this invitation without any harm to his feelings; he graciously did so, and went off, much to the delight of the landlord, who declared that there could be no measure of the gratitude he entertained towards us for having saved him from this visit. He said that without our assistance he could have done nothing, and that besides the disagreeables consequent on the presence of the general, his staff, and riffraff following, he would have lost the custom of a civil official whom he expected, and for whom arrangements had already been made. Poor man! he was trembling with fright in our room, and it seemed likely that as one hotel-keeper had been put in prison at Ta-Li because he would not take us in, here another would be beaten because he had not refused to shut us out.

This sort of occurrence is by no means uncommon, for the officials, especially military ones, are most arrogant and oppressive.

October 13.—Many varieties of precious stones

are found in the mountains in the neighbourhood of Yung-Ch'ang, and besides this the sacking of Ta-Li-Fu had thrown great quantities of jewellery into the hands of all sorts of people, some of whom had not the faintest idea of their value; and continual visits were paid to us, and stones of every description offered for sale. A great deal of jade was brought in, some of it probably native; this stone is very highly prized among Chinese of all classes, and officials usually wear a great thumb-ring made of it. One man brought a pair of earrings made of malachite, for which he asked a price that would have bought a table in Russia, where that stone is plentiful.

Another brought some necklaces made of amber, something like the Roumanian black amber, but more opaque, and of a lighter colour; it looked something like brown agate, and we were offered one hundred and eight beads for 40*l*. We offered 13*l*., and if we had remained a few more days would doubtless have compounded for 20*l*., but in China no satisfactory bargain can be struck in a short time. This was a good necklace; all the beads were more or less similarly marked, and it would have been worth about 40*l*. or perhaps more in Peking, where officials give high prices for good necklaces. One hundred and eight is the regulation number, no one venturing to wear a necklace with one bead more or less.[8]

A man brought in a stone about the size of a small nut, perfectly clear, without a flaw, and of a faint amethyst tinge; this, no doubt, was crystal, or something even more valuable, and the man said that he had another much larger and better. We bade him fetch it, which he did; he returned with a stopper of an old scent-bottle, and the drop from a

---

[8] See *Marco Polo*, 2nd ed. ii., 330-331.

European chandelier, both of which were valued at comparatively high prices. Our ventures in stones were not very extensive, for as the Chinese, like all orientals, leave their gems uncut, it is impossible for anyone but an expert to judge of their value.

The market of Yung-Ch'ang was very well supplied; there was plenty of beef, and almost every kind of vegetable, including potatoes, and a root something like a Jerusalem artichoke; and the brilliant red of persimmons, the deep purple of brinjalls, and the brilliant green of chilies, contrasted picturesquely in the baskets of the sellers. The people said that the persimmons were sure to give fever to strangers, but that the pears were not only wholesome, but beneficial; one of the latter was found by Chin-Tai weighing three pounds. This was an enormous pear, but quite eclipsed in size by the pears of Kinsay, mentioned by Polo as weighing ten pounds, and those of Shan-Tung, quoted by Colonel Yule from Williams as reaching the same astonishing weight.[9]

October 14.—The itinerary of Marco Polo from Ta-Li-Fu is, as Colonel Yule states, full of difficulty:

'When you have left Carajan, and have travelled five days westward, you find a province called Zardandan.[1] The people are idolaters, and subject to the great Kaan. The capital city is Vochan.' ....
'The country is wild and hard of access, full of great woods and mountains, which 'tis impossible to pass; the air in summer is so impure and bad, and any foreigners attempting it would die for certain. ... After leaving the province of which I have been speaking you come to a great descent; in fact, you

---

[9] Yule's *Marco Polo*, 2nd ed. vol. ii. pp. 184–192.
[1] *I.e.* 'The Gold-Teeth.'

ride for two days and a half continually down hill. .... After you have ridden those two days and a half down hill you find yourself in a province towards the south which is pretty near to India. You travel therein for fifteen days through a very unfrequented country, and through great woods abounding in elephants and unicorns and numbers of other wild beasts. There are no dwellings and no people, so we need say no more of this wild country, for in sooth there is nothing to tell . . . and when you have travelled those fifteen days through such a difficult country as I have described, in which travellers have to carry provisions for the road because there are no inhabitants, then you arrive at the capital city of this province of Mien,[2] and it is also called Amien, and is a very great and noble city.'[3]

Colonel Yule assumes that the five days of Marco are from Ta-Li-Fu to Yung-Ch'ang-Fu, though the marches would be very long ones. Our journey was eight days, but it might easily have been done in seven, as the first march to Hsia-Kuan was not worthy of the name. The Grosvenor expedition made eleven marches with one day's halt—twelve days altogether—and Mr. Margary was nine or ten days on the journey. It is true that, by camping out every night, the marches might be longer; and, as Polo refers to the crackling of the bamboos in the fires, it is highly probable that he found no '*fine hostelries*' on this route. This is the way the traders still travel in Tibet; they march until they are tired, or until they find a nice grassy spot; they then off saddles, turn their animals loose, light a fire under some adjacent tree, and halt for the night; thus the

---

[2] *I.e.* Burmah.
[3] *Marco Polo*, 2nd ed. vol. ii. p. 91.

longest possible distance can be performed every day, and the five days from Ta-Li to Yung-Ch'ang would not be by any means an impossibility. After this five days came two and a half days down hill, and, as Colonel Yule points out, the itinerary is a continuous one. Dr. Anderson thinks that the descent of the river of T'êng-Yüeh is meant. The chief objection to this lies in the fact that there would be a very serious break of continuity, for Polo could hardly have overlooked the ascent from the Lu-Chiang or Salwen river, an ascent of five thousand feet above the bed of the stream, 'and one that is exceedingly wild and hard of access.'

From the Lung-Chiang or Shuay-Li also there is a steep ascent of three thousand feet, and the '*great descent*' could therefore not commence until the summit of Urh-T'ai-P'o was gained, on the third or fourth day from Yung-Ch'ang. From this place half a day's ride would have brought Polo to T'êng-Yüeh; but, unless his next two days ended at Man-Yün (Manwyne), the two and a half days down hill would have no meaning, for the river is followed to that place, after which the road takes to the hills, and is carried through woods and forests; it is therefore evidently at Man-Yün that the two and a half days ought to end. The two marches from T'êng-Yüeh would be desperately long ones, but perhaps not impossible.

Colonel Yule doubtfully suggests the valley of the Shwé-li, but this would not relieve us from the break of continuity, for between Vochan and this river there is the ridge of mountains above mentioned.

It seems reasonable to assume that the two and a half days should commence from Yung-Ch'ang; and Baber states that there is a road from Yung-Ch'ang,

CH. VIII.   MR. BABER'S SUGGESTED SOLUTION.   345

which is 5,600 feet above the sea, down the Yung-Ch'ang valley to the Lu-Chiang, 2,600 feet above the sea. From the nature of the ground the descent by this road must be gradual and continuous, and as the distance is forty-five miles it would seem to justify the expression of a '*great descent.*'

Baber also observes :

'The fifteen days' subsequent journey need not present much difficulty. The distance from the junction of the Nan-Tien[4] with the Salwen to the capital of Burmah (Pagan) would be something over three hundred miles. Fifteen days seems a fair estimate for the distance, seeing that a great part of the journey would doubtless be by boat.'

An objection may be raised that no such route as this is known to exist ; but it must be remembered that the Burmese capital changes its position every now and then, and it is obvious that the trade routes would be directed to the capital, and would change with it.

Altogether, with the knowledge at present available, this certainly seems the most satisfactory interpretation of the old traveller's story.

There are two roads from Yung-Ch'ang to Fang-Ma-Ch'ang. The main road, which does not pass over a mountain, is better than the other, but some miles longer. The main road, we were told, passes through the plain of Fu-Piau (P'u-P'iao of Baber?), which had been entirely depopulated by an extraordinary disease, of which the symptoms were like those of the plague, and which had, during the months of August and September, carried off upwards of a thousand people. Our informant added that now there was no one left except a few poverty-stricken

[4] The Yung-Ch'ang river.

wretches, who could not afford to move. A traveller who was stopping at the same inn with us at Yung-Ch'ang, and who left with us for T'êng-Yüeh, said that he had passed through the place in July; that at that time there were scarcely any inhabitants left, and that the dead bodies were lying about unburied. Now he said that the disease had ceased at that place, and had moved in a southerly direction to Niu-Wa, where it was raging. To a Chinaman, the idea of leaving a body unburied is very dreadful, and it would only be the most dire necessity that would permit such an atrocity. This disease is said to attack people passing through the country as well as the residents.

In describing the symptoms, the people said that a lump like a boil, about the size of half a small walnut, suddenly appeared on almost any part of the body; there was absolutely no attendant pain, and twenty-four hours was the outside that a person could live after the appearance of this lump.

Boccaccio thus describes some of the symptoms of the plague at Florence in 1348:

'Here there appeared certain tumours in the groin or under the arm-pits, some as big as a small apple, others as an egg; but they generally died the third day from the first appearance of the symptoms, without a fever or other bad circumstance attending.'

From Defoe also may be gathered that the plague of London was somewhat similar; but he was not himself an eye-witness of this terrible calamity, nor does he anywhere give a distinct account of the symptoms.

The city of Yung-Ch'ang itself, about 5,645 feet above the sea, is healthy enough, although there is at certain times a little fever.

CHAPTER IX.

IN THE FOOTSTEPS OF MARCO POLO AND OF
AUGUSTUS MARGARY.

II. THE MARCHES OF THE KINGDOM OF MIEN.

Departure from Yung-Ch'ang—Graves of Aborigines—Hun Shui-Tang—
Fine Ponies and Mules—Fang-Ma-Ch'ang—Pestiferous Valley of the
Lu-Ch'iang or Salwen River—Passage by Chain-Bridge—Steep
Ascent to Ho-Mu-Shu—Greedy Host—Needless Difficulties of Road
—Old Custom of 'Wappenshaw' and Military Tests—The Lung-
Chiang or Shwé-li River—Salutes by the Way—A Celt for Sale—
Ch'in-T'sai-T'ang—The City of Têng-Yüeh or Momein—Things better
managed in Ssŭ-Ch'uan—The Chi-Fu Convention—Mule-chaffering
—H'siao-Ho-Ti—Nan-Tien—Reception by a Shan Lady—Her Cos-
tume—First Burmese Priests—Change of Scenery—Kan-Ngai, or
Muang-la—The Chief—Passage of the Ta-ping River—First Bur-
mese Pagoda—Lovely Scene near Chan-Ta—Chan-Ta (Sanda) and
the Chief there—Oppressions of Chinese—Festival at Chan-Ta—Shan
Pictures by the Way—Shan and Kakyen Figures—Road-side
Scenes in Ta-ping Valley—Bamboos and Birds by the Way—Lying
Litigants—T'ai-P'ing-Ch'ieh or Kara-hokah—Reach Chinese Frontier
Town of Man-Yün ('Manwyne')—Visit from Notorious Li-Sieh-
Tai—Treatment at Man-Yün—The Pa-I People—English Goods in
Bazar—Letter of Welcome from Mr. Cooper—Scene of the Murder
of Augustus Margary—The Kakyen Country—A Shot at the Party;
only Tentative—Kakyen Huts—Meddling with the Spirits' Corner—
Fire got by Air-compression—Buffalo Beef—Grand Forest Scenery
—Bamboos and Potatoes—An Imprudent Halt to Cook—A Venture
in the Forest—Perils from Ants and from Bullies—Would-be Leviers
of Blackmail—Benighted—A Welcome Rencontre—Cooper's Mes-
sengers and Stores—A Burmese *Po-é* or Ballet—Embark on Bhamo
River—The Irawadi Disappointing—Kindly Welcome from English
Agent at Bhamo—Alas, poor Cooper!—Bhamo to Dover—The
Journey Ended.

BEFORE leaving Yung-Ch'ang all the civil officials sent us their cards, but the military men, evidently with *malice prepense* and aforethought, omitted the usual act of courtesy. This was the only occasion, during the whole time I spent in China, on which I was not treated with civility, if not distinction.

We followed the main road for a little more than a mile, and then plunged into a valley amongst the mountains, and followed up a stream between rounded hills covered with very fine grass. There was not a tree to be seen, and few shrubs. We met scarcely anyone, although it was market-day in Yung-Ch'ang. The stillness that reigned was not disturbed either by the flutter of a bird, or the hum of an insect, and walking on by myself in front of the other people I could not hear a sound of any kind.

The summit of the mountain was 7,733 feet above the sea, and a descent of nine hundred feet brought us to Hun Shui-T'ang (Troubled Water Station), a village of about six houses.

The practice of interment distinguishes the aborigines of this country from the Tibetan races in the north. On the road we passed a great many of their graves. These are circular towers, of sun-dried bricks, about six feet high and six feet in diameter, covered at the top with a mound of earth, on which there are usually some tufts of long grass. This shape of grave was now being largely adopted by the Chinese, but there were still a great many of the ordinary Chinese form.

A quaint little inn at Hun-Shui-T'ang produced an unlimited supply of well-boiled white rice, which made our people very happy; some eggs for ourselves, and some native pickled onions that would not have disgraced Crosse and Blackwell.

Our mules passed us before we left; they were a goodly array, for none were heavily laden, and there were about twenty spare animals. The examinations had brought hundreds of people into Yung-Ch'ang who could not afford to keep their animals there, and were glad enough to get anyone to take them home.

The mules and ponies were really first-rate animals. Though the mules were small, the ponies were larger than any we had seen since leaving Tibet; they were plentiful enough, and coming from Yün-Nan or Lower Ssŭ-Ch'uan, they might fairly be called either large or numerous, whichever reading may be adopted of Polo's words. As for the *riding long*, of which he speaks, all Chinese ride so short that it would be impossible to ride shorter. They put two or three thick felt things under the saddle, which itself is very high; on top of it they pile all their bedding, blankets, saddle-bags, &c., and as often as not their heels are in the stirrups instead of their toes; compared with a Chinaman, therefore, almost anyone would seem to *ride long*.

The muleteers were all tattooed about the legs, and most of them carried a small musical instrument like a miniature guitar.

The harness was admirable, made of plaited raw hide, very strong and durable, and in appearance smart and workmanlike; the breechings of the pack animals were exceedingly neat and well-made, and the breast straps were often covered with small bells.

From Hun-Shui-T'ang we continued our descent, by an excellent road, through an undulating country of rounded green hills, with a big range of cloud-capped mountains in the background. Crossing a stream about three thousand feet below the crest of the mountain we had passed, we ascended a short distance to the little market town of Fang-Ma-Ch'ang, surrounded by a broken-down wall, which was built during the rebellion, and with a new inn that may be commended to the notice of future travellers.

October 15.—The muleteers were anxious to cross the dreaded Lu-Chiang before the sun was hot; and everyone was, for once, ready at an early hour. We started amongst rounded, undulating hills, but soon entered a valley, which we descended by an easy gradient until we could see the mysterious river at our feet. A few low clouds hung over the valley, and as we stayed a few moments we could not but be impressed with a scene connected with so many weird associations.

Centuries had rolled by since Marco Polo spoke of the country 'impossible to pass, the air in summer is so impure and bad; and any foreigner attempting it would die for certain.' Already at Ta-Chien-Lu Monseigneur Chauveau, who had passed many years of his life in Yün-Nan, had warned us of this pestiferous place, and had told us that before the rebellion had destroyed every organisation in the province, it had been customary to keep a guard at certain places on the road to prevent anyone from attempting the passage during the unhealthy season. As we approached nearer and nearer, though the warnings were more frequent, the details of the story varied but little, and, incomprehensible though they appeared, we could not but give credence to the tales so oft repeated of the 'valley of the shadow of death.'

As it lay at our feet all nature seemed to smile, and invite the tired traveller to stay and rest. But it was the smile of the siren, for should a stranger venture there to pass the night, it would be with feverstricken limbs that, when the morning broke, he would attempt the escalade of the surrounding heights.

Even in autumn, the most healthy season, it is with bated breath that passengers hurry across at a favourable moment; and when the fiery rays of summer are darted on that low-lying valley, even the acclimatised inhabitants flee the 'infections that the sun sucks up,' and for months no living thing may venture there.

It is during an alternation of rain and sun that the poison is most rife, and then they say a lurid copper-coloured vapour gradually folds the valley in its deadly embrace.

But as we looked the sun rose higher, and gradually dispersed the clouds, and we were assured that the moment could not be more favourable for crossing.

The reasons for the extraordinary unhealthiness of the valley are not apparent; for although it is 1,300 feet lower than the Lan-Ts'ang, and nearly 2,000 feet lower than the Lung-Chiang or Shwé-li river, yet it is still 2,600 feet above the sea.

It was the finest-looking valley we had passed; instead of being perfectly flat, like so many others, the ground slopes gently on both sides from the foot of the hills.

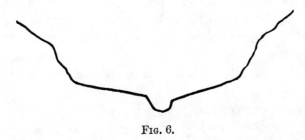

Fig. 6.

This formation is very favourable for the terrace cultivation, and here the rice harvest was well forward.

There are a few small undulating hills in the bottom of the valley, which is bounded by mountains well wooded or covered with long grass. There are plenty of villages, with a good many trees round them, and the landscape is more varied than any we had seen for some time.

From the rapidity of the river, and the undulating nature of the ground, it might have been supposed that this district would be healthy enough; but the secrets of the red miasma must remain hidden yet awhile in the recesses of the beautiful, but deadly vale.

The river is crossed by a chain suspension bridge of two spans, the second span in a line parallel to the continuation of the first, but about four yards from it on the same level. This system is probably adopted for the greater facility given for tightening up the chains; but it makes a mis-shapen affair of what would otherwise be a well-constructed bridge. The eastern span was about seventy-three and the western fifty-two yards long; each span is supported on twelve or fourteen chains underneath, and two above, the links being of three-quarter-inch iron, one foot long. At the time of our visit it was in excellent repair, but the eastern span, destroyed by the Mahometans during the rebellion, had only recently been rebuilt. At the time of Baber's visit it was 'in a dangerous state of dilapidation.' The stream was running rapidly below the eastern span, but the western was quite dry.

We halted at Lu-Chiang-Pa, a little village about a quarter of a mile beyond the end of the bridge, and here at 10.30 a.m. the thermometer marked 80° Fahr. Baber noticed the sultriness of this place, for on April 29 his thermometer registered 96° Fahr.

On leaving this we went straight up the mountain by a very fairly-paved, but exceedingly steep, road to Ho-Mu-Shu, 2,800 feet above the river, and 5,486 above the sea.

Our lodging was a shed made of split bamboos, over which mud had been thrown in some places to fill up the interstices, and so exclude the wind and rain sufficiently to enable people to smoke their opium pipes, without having their lamps put out by either one element or the other. There was a good thatched roof, and the hut was divided into three compartments by partitions of split bamboos, reaching not more than half-way up. My pony was lodged in one, we occupied the other two, and as it was easy to see through the partitions, I could watch over my animal whilst sitting writing. The weather, fortunately, was fine, sunshiny and without wind, so we did not find our airy apartment in any way uncomfortable.

The little village of Ho-Mu-Shu was more than crowded when we arrived. We occupied as much room as twenty or thirty Chinamen, and it appeared as if our fellow-travellers were obliged to take it in turns to go to bed, and cook their provisions for the morrow; for we were kept awake all night by their lively conversations and culinary operations; everything that was going on being seen and heard quite plainly through the wicker-work partitions of the rooms. The other guests were mostly candidates on their way to the examinations at Yung-Ch'ang, but besides these there were a good many traders and private travellers.

The landlord of the inn was a very greedy person. I wanted some cash, and told Chin-Tai to change some of my silver, which was the best to be found in China, in fact, nearly pure, whilst the stuff generally in use about this part of the country was dirty rubbish, full of alloy and impurities. At the same time, on account of the travelling officials passing backwards and forwards, who generally use tolerably pure silver, the people here were not one whit behind the rest of the Chinese in the remarkable power of judging silver by its appearance, and it is really wonderful how a Chinaman, out of a number of pieces of all shapes and sizes, will pick out a good or a bad bit without a moment's hesitation.

The landlord, however, thinking to make a few extra cash, declared that although my silver might be very good, yet that, as he only understood the particular kind in general use here, he could only give me 1,500 cash for a tael. We were, however, equal to the emergency, and produced some very inferior stuff; but our landlord now, with bold effrontery, said he had never seen silver so worthless, and could not think of accepting it at any price. Our negotiations then dropped through for a time, but when we were leaving we paid our bill in copper cash borrowed from one of the chair-coolies. We gave the landlord about twice as much as he deserved, but still he was not satisfied, and so we expressed our sorrow that all our cash was used up, and that it was no use offering him silver, as he had already pronounced all ours to be worthless.

October 16.—The road from Ho-Mu-Shu was a vexatious one for man and beast. We continued the ascent of the valley we had followed from the Lu-Chiang, and having at length attained a point on the summit of the great range that divides the Lu and Lung rivers, instead of at once commencing the descent into the plain on the other side, or following a contour along the face of the hill, the road ran for a long distance along the crest of the ridge, now going up four hundred feet, then down by a paved road in ruins; next there would be a short ascent by a kind of staircase, followed by a descending zigzag—all quite unnecessary if the smallest engineering skill had been employed.

On the road we met a train of two hundred mules bringing cotton to Yung-Ch'ang, but most of the animals were returning unladen. There were on the road great numbers of men on their way to the military examinations, all gaily dressed as well as their ponies, whose headstalls and breast-straps were neatly covered with red cloth; their saddles and saddle-cloths were also gaily decked with bits of red here and there; and parties of them winding about amongst trees and rocks formed many a pretty picture. Upwards of one thousand present themselves for examination every year from the T'êng-Yüeh district; if they pass they gain a certain social position in their town or village, and are eligible to serve in the capacity of petty municipal officers. Few of them have any idea of becoming soldiers, but pass the examination for the sake of the importance they thereby obtain. It seems at first somewhat inconsistent that the Chinese, who usually hold all military officers more or less in contempt, should offer advantages to the men who pass military examinations, which are tests of physical strength only. But it is the old custom handed down from generations. In days of yore and of much hard fighting, when the sword, bow and arrow, and the spear were the ordinary weapons, it required stout, skilful, lusty fellows to wield them well; so the Government established these athletic sports, as they might well be called, at which the prizes were social positions amongst the people, and were well worth striving for. So everywhere military exercises became common, nearly everyone practised them, and thus the State had always ready-made soldiers that they could call on when required. The old custom still survives, though the reason, in its full force, no longer exists.

The manes were hogged and the tails cropped of a great many of the ponies these men were riding; but there were none of the docked tails mentioned by Marco Polo.[1]

As we ascended from Ho-Mu-Shu there was many a charming view down the well-wooded valleys that ran down from the ridge; and the Lu-Chiang was nearly always visible below us. Early in the morning a vapour was hanging over the river and the ground immediately on its banks; as the morning advanced this gradually rose, but low clouds seemed to shroud the valley all day, while we, high up above it, were enjoying a pure, delicious mountain air.

There were some wild raspberries growing amongst the woods, but the leaf of the bush was, as usual, quite unlike its European congeners, and almost exactly like that of a currant; and had it not been for the fruit, which was very good, this plant would certainly have been noted as a wild currant.

From the final summit, 2,643 feet above Ho-Mu-Shu, and 8,129 feet above the sea, a tolerable road took us through a thick wood to Tai-P'ing-P'u (Peaceful Village), where we stopped for breakfast. As we descended we rode through a very thick wood, and could see little until we reached the limit of the trees, and then the Lung-Chiang was winding below; not in a plain, nor yet in a gorge, as nearly all the rivers we had seen or crossed hitherto had been, but the mountains bounding it on both sides ended in

[1] *Marco Polo*, Bk. ii. chap. xlix.

very gentle slopes that ran down to the water. These were well cultivated with rice, but there was again a great extent of ground where cultivation had been discontinued.

The Lung-Chiang [2] is crossed by another very good iron chain suspension bridge in one span of about fifty yards, supported on eleven chains below, with two more above. Both this river and the Lu-Chiang were very low, but when they are full a vast body of water must flow down them; the Lung-Chiang is the more rapid of the two, but its bed is much higher above the sea level—four thousand five hundred and two feet.

The road from the bridge ascended gently through rice-fields to Kan-Lan-Chan, where we were saluted for the third time by three soldiers in charge of a Pah-Tsung, who sent and apologised for not heading his army, as he did not expect us so soon, and was not dressed. The number of times that we were saluted during the day must have been a serious expense to his Celestial Majesty, who pays for the powder. At Tai-P'ing-P'u I was some distance ahead, and as I approached the village three soldiers fired off a musket apiece, went down on their knees to the 'Imperial Commissioners,' as they were pleased to call us, and repeated the formula usual on these occasions:

'Welcome, Great Excellency! The men of Tai-P'ing-P'u have come out to salute you.'

This they did in a droning, chanting way that sounded like the 'responses' in a church where the parson is short of a congregation.

Mesny arrived about half an hour afterwards,

[2] Lung-Chiang is the Chinese name of the Irawadi's tributary, called by the Burmese Shwé-li.— *Y.*

but his salute was reduced by one gun, for one of the dirty old matchlocks spluttered for a minute or so, like an indifferent squib of amateur manufacture, and gradually burnt itself out without any report. But in these parts a few guns more or less in a salute are not of much moment.

In the afternoon some soldiers at a village, and some more at the Lung-Chiang Bridge, burnt gunpowder for us. One lot were rather put out because a little boy who had brought up a matchlock from the last place arrived too late to have it loaded ready, and there was an awkward pause between the second and third guns. As far as I was concerned, the dignity of the thing was quite spoiled by the behaviour of the pony I was riding, who always shied away from the soldiers at the critical moment. My other mischievous grey, too, would insist on contributing to my discomfiture by intruding himself between me and the army; rushing up against me, and knocking me completely out of time, or breaking through the ranks (a single rank of three), with a snort and a toss of the heels. The white pony was always in mischief, if he could find any to get into. If he could leave the road and wander away into the forest, he would; especially if he saw another horse or two likely to follow him. Nothing pleased him better than to jump violently into a mud-hole just when someone was in a position to be splashed all over. If he saw the pony I was riding balancing itself on some narrow or slippery stone, where there was barely room for one foot, that was the moment of all others that he chose for running me down and knocking me, with perhaps a drop of a couple of feet or so, into the bog; and the pranks that pony played were worthy of an English schoolboy.

A man came up in the evening with a 'celt,' which may or may not have been genuine. It looked to me quite new, though we wondered who the antiquarians could be who made it worth while for anyone in these parts to fabricate relics of the stone age;[3] but since a certain occasion on which I was astonished by the offer of sham Roman coins near Damghan in Persia, I have always been prepared to find false antiquities in the most unlikely spots.

October 17.—The road was very good for a change, and there was a generally easy ascent amongst undulating hills with but little wood on them. Now and then we passed through a deep lane cut in the soft sandstone, the banks at each side covered with ferns, grass, creepers, and shrubs or small trees, that brought to mind many a lane in Surrey or Kent. Then the road would emerge into a downy country, where in a hollow the margin of some small pond would be lined with rushes, reeds, and ferns, now turning yellow and red—the very place for a duck, if the whole country had not been disturbed by a train of eighty or a hundred mules laden with salt, which had passed just before, on their way, like ourselves, to T'êng-Yüeh.

We wound along, now up and now down, but steadily rising, till we reached Ch'in-T'sai-T'ang, where soldiers turned out as usual to salute. I was walking on ahead, alone, but just at this moment Chung-Erh galloped up and passed me, anxious to

---

[3] Baber explains the discovery of a copper knife and a 'celt' at the fair in Ta-Li-Fu. He says, 'The knife is undoubtedly genuine; the celt —called locally, and, indeed, all the world over, "thunder stone" (lei-ta-shih)—bears traces of sharpening on the axe-edge, and is well adapted for use; but as these objects are now employed as charms, on account of their supposed supernatural origin and properties, and as there is a brisk demand for them, it is difficult to satisfy oneself of their authenticity.'— (*F. O. Report, China*, No. 3, 1878.)

be ready to receive me in proper style at the door of the hut. The soldiers never for a moment imagining that either of their excellencies would be on foot, mistook Chung-Erh for one of them, and as he passed bent one knee, and, much to our diversion, gravely informed him that ' the men of Ch'in-Ts'ai-T'ang had come out to salute him.'

We had already mounted about two thousand feet from Kan-Lan-Chan, and another five hundred feet brought us to the summit of Mount Urh-T'ai-P'o, whence we descended two thousand feet by a very fair road to the city of T'êng-Yüeh, or Momein, situated at the head waters of the Ta-Ying river, in a perfectly flat and treeless plain, some five miles broad and long. This was entirely covered with rice-fields, where the crop was being harvested, and was bounded on all sides by uncultivated grassy slopes, from which every trace of trees had disappeared.

We found roomy quarters in an excellent inn outside the south gate.

Two soldiers had escorted us from Kan-Lan-Chan, who on arrival at this place received the usual gift; but presently Chin-Tai came in, saying that there had been four men with us; no one had seen the other two, but their word was taken, and a present was made to them; soon afterwards the number increased to seven, and as some guns had been exploded in our honour just outside the city, the extra three were after some deliberation duly veiled; but when someone else came in to say that the number had been truly ten, our incredulity surpassed our generosity, and we declined to remunerate these buckram men.

October 18.—The Chen-Tai paid us a visit in the morning, apparently for the purpose of frightening us.

CH. IX. THINGS MANAGED BETTER IN SSŬ-CH'UAN.

He told us that in the year 1876 the king of Burmah had asked the Chinese Government to send some troops across the frontier, and put down some tribes who were giving trouble. Our visitor had been in command, and had not succeeded in his mission without much hard fighting and a great deal of sickness; for the campaign had been carried out during the unhealthy season. The king of Burmah had paid the whole expenses of the expedition, and had asked for and obtained the loan of three hundred soldiers after the main body had returned to China.

The Chen-Tai told us that this body of men had just been disbanded, on the demand of the British Government[4]; that they were roving over the country in lawless bands, that travelling was very dangerous, and that he could not be responsible for our safety unless we would give him time to recall these men, and get them out of the way. He also said that the governor-general of Yün-Nan intended to raise three million taels to work the mines in the province, under the superintendence of Europeans.

The mines of Yün-Nan no doubt are exceedingly rich; but before they can be made to pay, communications must be improved, and the country better governed. It struck us very forcibly that the government of Ssŭ-Ch'uan was far better than that of Yün-Nan. In Ssŭ-Ch'uan the officials were invariably more than attentive, and it was easy to see that their orders were promptly and efficiently carried out. The difference was apparent the very day we crossed the boundary into Yün-Nan. The Margary proclamation, which had been universally posted in Ssŭ-Ch'uan, was rarely seen; and although the officials

[4] The British Government, of course, had had nothing to do with the matter.

were almost always civil and polite, there was a marked difference in our treatment in the two provinces. It must be said that the higher magistrates seemed to pay us most attention, but their orders were not carried out by the petty officials with the alacrity and regularity always observed in Ssŭ-Ch'uan. It happened on more than one occasion that despatches sent on from the prefectoral city before our departure did not reach their destination until after our arrival; and although these are trivial matters they serve to compare the government of the two provinces.

On the whole, there can be little doubt that the central Government of Peking wields a potent sway even in these distant provinces; it is due to the Chi-Fu Convention that Englishmen may travel in comfort throughout the vast empire; and this one fact alone will stamp the term of office of Sir Thomas Wade as one memorable in the annals of our dealings with the Chinese Government; and it is to be hoped and expected that it will do much to bring about that intercourse with foreigners which is the one and only means by which cordial and comprehensible relations can be established between the Chinese and European nations.

October 19.—Chin-Tai told us in the morning that he had found some muleteers who were willing to take us to Chiu-Ch'êng (Old Town), for seven and a half tenths of a tael per mule, and wanted to know if we would agree to that price. Although it was much too high, we readily consented, and bade him go and strike the bargain and have the paper written; but when the muleteers unexpectedly found their offer accepted without any attempt at reduction, they naturally regretted that they had not asked a larger sum at first, and casting about for an excuse said that when they had made their offer they had not understood that they were to provide any riding animals, and under these circumstances they must have eight-tenths of a tael. 'Anything for a quiet life,' we said, as the nearness to our journey's end was making us reckless; but a soothing cigar had scarcely been lighted when Chin-Tai returned afresh, and said that although the muleteers had come to terms about the riding animals, yet they would not let us have two riding saddles unless we paid eight and a half tenths of a tael for every animal.

The hatred of being imposed upon, which is innate in the human breast, now began to assert itself; and we told Chin-Tai that we would not give that sum; but he had not reached the door when we repented of our rashness, and inwardly ashamed of our weakness consented to the amount demanded; but when they declared that they had now changed their minds, and would not go for less than a tael, our spirit of independence made us say that we had now changed our minds, and would not pay more than the seven and a half tenths of a tael originally offered. They refused this, and went away, no doubt intending to reopen negotiations later; but in the meanwhile the price was sufficiently good to induce another set of muleteers to accept the offer, and the paper was at length written and delivered.

In the afternoon we discovered that the civil official of Man-Yün was also staying in the hotel; he paid us a visit, and told us the Chen-Tai's story with considerable variations. He said that thirty soldiers, not three hundred, had been lent to the king of Burmah, but that the officer in command was of so bad a character that the king had disbanded the company; that the officer had been disgraced by the Chinese Government, that he now did not dare to return to Chinese territory, and was roving the country, committing depredations, and robbing whomsoever fell in his way.

The Man-Yün magistrate and another military officer came again in the evening to endeavour to induce us to wait a few days; and now the former said that the disbanded soldiers numbered one hundred, that they were very dangerous, that in any case we should be compelled to wait at Man-Yün until he could join us, and that we had much better remain at T'êng-Yüeh-T'ing, where the quarters were comfortable. To all these blandishments we lent a deaf ear; and, ultimately, the magistrate sent his steward and a lot of people with us, amongst them two Cantonese, to help us on the road, and placed his residence at Man-Yün at our disposal.

October 20.—There are two roads from T'êng-Yüeh-T'ing to Hsiao-Ho-Ti, both roads cut across the spur, round which the river of T'êng-Yüeh, or the Ta-Ying-Ho, makes a long sweep.

The road led us over grassy hills almost without a tree, though there were patches of scrub here and there, amongst which Chin-Tai discovered a pheasant.

We met little traffic, and passed scarcely a house. The road wound about, sometimes up and sometimes down, but it was unpaved, quite firm, and fairly level, until we commenced our descent into the valley of the Ta-Ying. The little village of Hsiao-Ho-Ti lay beside the stream fifteen hundred feet below the crest of the spur, and here we found a shanty which did duty for a restaurant, and was the finest place for the servants that we had been in for a long time. The number of dainties quite brought to mind a Ssŭ-

Ch'uan tea-house; and the face of my Ma-Fu, as he made short work of innumerable dishes of pork, onions, chilies, bean-curds, and good bowls of rice, was a sight well worth paying for.

The road beyond Hsiao-Ho-Ti was excellent, winding along the edge of the perfectly flat plain, which was entirely devoted to rice cultivation, although there was a considerable area of uncultivated ground. There was a fair sprinkling of villages in the plain, with a few trees about them, and one or two ruins. It was bounded by gently sloping grassy hills, on which there was little wood, and where there was still a large extent of disused terraces, but where the traces of the Mahometan rebellion were fast disappearing.

The road was good and level, but the inhabitants have a most eccentric custom of using it not only as an aqueduct but as a reservoir.

The numerous streams that flow out from the mountains are turned on to the road wherever it is hollowed out between banks; little dams about a foot or eighteen inches high are made across the track to keep in the water; and thus the adjacent rice-fields can be flooded when required.

We met thirty loads of cotton, but, besides these, there were few people about; we saw some men thrashing with flails made of bamboo, one in each hand, but everything was still thoroughly Chinese, and there were no signs of the manners and customs of the Burmese, or of the wild mountain tribes between Man-Yün and Bhamo, except the turbans of the women, which were built up like towers on their heads.

We passed through the walled town of Nan-Tien, and about two miles further found the house of the native chief, or T'u-Sze, in the small village of Che-Tao-Ch'êng. There appeared not to be enough soldiers to fire a salute with matchlocks, but, instead of this, three iron guns about eight inches long, and with a calibre of about an inch, were planted upright in the ground, and were touched off by a man with a bit of lighted paper at the end of a bamboo, quite in the style of a professed pyrotechnist; and what they wanted in dignity was made up for by the loudness of the report.

The native chief has the rank of a Yu-Chi, and wears the clear blue button, as the English always call it, though a more inappropriate term could hardly have been devised. The French call it *globule*, just what it is: a globule a little more than an inch in diameter, which is worn on the official hat. Strangely enough, there is no regulation size for this, though for almost every part of a Chinaman's dress there are stern rules and regulations; and every man, official or non-official, must shave his head and wear a plait, for if he leaves his hair it is a sign that he is a rebel.

The 'Ugly Chief of Homely Virtues,' who entertained Margary, had died of grief for the loss of all his fortune during the rebellion; the boy of whom Margary speaks was not yet of age, and the honours of the house were done by an old relative holding the Chinese rank of Pa-Tsung.

The circular about us sent from T'êng-Yüeh-T'ing five days before, arrived about an hour after us; but it was not wanted, as the family of the chief were quite ready to dispense their hospitality without it. The way in which these despatches used always to arrive just in time to be too late, was both amusing and instructive; and I thought with the 'Sentimental

Traveller' that 'they manage these things better in' Ssŭ-Ch'uan.

The mother of the former chief, and the grandmother of the children (two sons, the eldest fifteen, and three girls), looked after the house, and invited us into her rooms after dinner to drink tea.

She wore a white jacket, with sleeves turned up, and a good deal of embroidery in gold on the cuffs ; this was fastened at the throat by a brooch with twelve (or fifteen, I am not certain which) different coloured stones, set in three rows, like the pictures of the breast-plate of the Jewish high priests. Silver bracelets adorned her wrists, and she wore white trousers with some red stripes ; but the room was so dark it was impossible to make out the details of her costume. A majestic turban rose to a height of eighteen inches above her head, and bulged out about half-way up, as though swelling with honest pride at its exalted position.

She introduced her two sons, and told us that their territory stretched for a length of thirty miles by the river-side, and extended back from it for a distance of sixty miles.

She evidently entertained a sincere regard for Margary, and told us that he had given a sword and a microscope to the late chief, but she did not mention the 'fine pair of scissors' which he gave to the 'amiable spouse.' What has become of her we did not learn. The language of these people is alphabetic with nineteen letters, and they write as we do from left to right.

October 21.—As we rode out of the gates, where a couple of Burmese priests were standing about in their yellow garments—the first signs of a change of country—three more terrific explosions startled our animals. The old Pa-Tsung had promised us sixty soldiers as a guard of honour, but only two sorry-looking fellows turned up. The morning was fine, but it came on to rain at about nine o'clock, and rained all day; a perfect deluge falling in the afternoon.

As we advanced the scenery changed; hitherto the hills had been grass-covered, now trees appeared, many of them of a kind not seen before, and the vegetation was almost tropical in appearance; creepers with huge leaves trailing up the trees, plantains growing here and there, and an occasional banyan, all indicated a change of climate.

The road again was bad: sometimes at the river level, and strewn with huge stones, then it went up a ravine and down again, here it was feet deep in mud, or as before turned into a series of reservoirs. We found some sheds of matting by the road-side, where we sat down to discuss our sandwiches; but presently some people from a neighbouring village appeared and took possession. These proved to be the proprietors; they had brought with them great baskets of hot cooked rice, and some of the little dishes that always accompany it, and soon they were doing a fine trade. Their shanties are midway between Chin-Ch'êng and Nan-Tien, and there is hardly any other halting-place on the road. People going backwards and forwards generally start much about the same time, so the owners of these huts know pretty well when to expect the 'up and down trains,' and come from their village to feed the hungry passengers. The natives here are very fond of tattooing their legs with all sorts of figures, and they wear on each leg, just below the knee, a number of very fine rings of rattan cane painted black. They chew a mixture of lime and very coarse tobacco, as well as betel-nut. In this they are unlike the Chinese, who never chew tobacco or lime, though they sometimes make use of the betel.

We stayed here about an hour, to let somebody get ahead; both my Peking boys were now in a deadly fright that we, and they with us, would share Margary's fate; so not only would they not go in front, but they always remained at a safe distance behind, looking particularly mean; but once they found themselves inside the walls of a house, all the old northern bluster came out, and they were tremendous fire-eaters, especially Chin-Tai, who was the greater coward of the two. On the road we met a chair, in which the mother of Li-Sieh-Tai was travelling; she was supposed to be a sister of the king of Burmah, though this was disputed by most of the Chinese. Some of the coolies and people with her were heard to say:

'What! Here they are again; are they not frightened yet?'

These sort of remarks from people about had been rather frequent lately, and they had not contributed to raise the courage of my two servants.

We had made up our minds that, as we wanted fresh mules the next day, our best plan would be to go to the house of the native chief at Kan-^{Ng}ai. This name includes two towns, Chiu-Chêng, or 'Old Town,' on the eastern side of the river, and on the direct road from T'êng-Yüeh to Man-Yün, and Hsin-Ch'êng, or 'New Town,' which being on the western or right bank is considerably off the highway. The chief used to live at Chiu-Ch'êng, but probably to be out of the way of the Chinese officials continually passing and repassing, and whose exactions were almost insupportable, he moved to Hsin-Ch'êng.

Our head muleteer, however, had ideas of his own, and saying to the two servants sent with us that a friend of the innkeeper at T'êng-Yüeh had prepared a house for us at Chiu-Ch'êng,[5] he took us there without asking our opinion. Here, in a deluge of rain, we were told that it was at the inn in the market that preparations had been made for us; thereupon we rode back, and at a most miserable straw shed, with no rooms in it, the people refused us admittance. Seeing the state of affairs, we told the two Cantonese servants to take us to Hsin-Ch'êng, whither we had intended to have gone at first. The head muleteer, whose mules were some distance behind, grumbled, and said that his animals could not reach Hsin-Ch'êng, and that he had not agreed to take us there. Nevertheless we set our ponies into a gallop, to show that we were in earnest, and made our way to the river-banks, passing on our road under a magnificent grove of banyan trees.

It was now astounding to note the manner in which this river (Ta-Ying-Ho or Ta-Ho) had grown since T'êng-Yüeh, where it was but a stream a few yards wide. Here the bed was nearly a mile across; at this season it was not, of course, full, but we were seven minutes fording the main branch, the water being above the horse's belly, and flowing two to three miles an hour. Besides this there were four other channels twenty yards wide, and several smaller ones, and I began to understand, what had hitherto appeared almost incredible, that the sources of the majestic Irawadi might be as far south as they are represented on all maps.

We were not expected by the native chief, whose

[5] Kan-Ngai is the Muangla of Anderson, but he coming from Burma appears always to have learnt the native names, whilst we coming from China were never told anything but the Chinese names.

rank is that of Sieh-Tai, and the customary explosion of three guns was omitted. This chief is probably the lad of fifteen visited by Sladen in 1868,[6] and seemed better off than he of Che-Tao-Ch'êng. He was rebuilding and making a fine place of his house, which had been destroyed during the rebellion. His reception-room was large and lofty, and divided into three compartments by magnificent screens of teak, about twelve feet high. He gave us the centre for our eating, drinking, and writing, and one of the rooms at the side to sleep in. The servants were lodged in another building beyond.

The mules came up almost immediately, and we were able to put an end politely to the ceremonious interview that always takes place immediately on arrival at these places; for no matter how wet, tired, dirty, hungry, or thirsty one may be, the reception must always proceed in a regular and formal manner.

October 22.—After the heavy rain of the previous day, all nature was fresh and green; there was a delicious feeling in the air, and the sun was shining in a clear sky, as a salute of three guns announced our departure, with an escort that reached the respectable number of thirty.

The direct road to Chan-Ta[7] is on the right bank of the river, but a portion of it was so bad and muddy that it was deemed advisable to cross and recross the stream. The Taping river joins the Ta-Ho or Ta-Ying from the right about a couple of miles below Muang-La, and they form one really fine river, which we crossed in the dug-out trunk of a tree, our animals swimming over. This was an amusing performance, chiefly on account of my grey,

[6] *Mandalay and Momein,* by Dr. Anderson, p. 175.
[7] *Sanda* of Anderson and others.—*Y.*

who was a pony of remarkable force of character; the other animals always submitted to the authority he established over them, and here the half-dozen, with whom he had only been acquainted a couple of hours, obeyed him with the utmost docility. Being driven into the river, he led as long as he could find the bottom, but, unaccustomed to swimming, he did not like the deep water, and as soon as he was nearly out of his depth he wandered about in a purposeless manner, leading the rest to follow his example. Our canoe then made for the group; we drove them on, and they swam until they reached a shoal, where the grey, always ready for mischief, anchored himself, the other beasts doing the same with admirable regularity. And now all the persuasive eloquence of the Ma-Fu on the opposite bank was powerless to induce them to move, until they were driven on by another canoe that passed that way. Once started there was no other halting-place until they reached the land; and here it was really delightful to see how thoroughly they all enjoyed a good roll in a fine bed of bright, clean sand, which seemed as if it had been laid there expressly for that purpose. For the last month my grey had been covered with mud, which had never been removed, and now that at length his coat was really clean, it was scarcely possible to recognise him.

This business had been a somewhat lengthy one, and although Chin-Tai and Chung-Erh had taken very good care to start long after everybody else, they had by this time overtaken us, and might have been observed standing about on the opposite bank in a nonchalant sort of way, trying to look as if they had no connection whatever with anyone else.

As we proceeded, fresh sights continually presented themselves; here there were some ricks of unthrashed rice, in shape like English hayricks, and standing in fields of rice-stubble; and, notwithstanding their un-English surroundings, they could not fail to bring recollections of many an English country scene. In a hamlet hard by we saw the first Burmese pagoda, with a high steeple; and the huge leaf of the plantain and the delicate bamboo sheltered the mud walls and thatched roofs of every village. Yellow wild-ducks, that apparently knew no fear of man, paddled lazily in the broad reaches of the river, but these, they say, are sacred; the Lamas, or Phoongyees, as the priests ought now to be called, throwing over them their protecting ægis of sanctity, chiefly on account of their colour.

A good and level road took us to the boundary of the Kan-^{Ng}ai or Muang-La chiefship, and here we halted under a magnificent banyan tree, where two old women kept a couple of refreshment stalls, and were selling quinces cut up into slices, and rice blancmange. Here we dismissed our soldiers, and others took their place, or were supposed to do so, for we saw nothing of them until we were leaving Chan-Ta, when they punctually came for the usual present.

Leaving the river, we ascended a spur, the end of a ridge dividing the Taping from a small tributary called by Anderson the Nam-Sanda,[8] and from the crest, about two hundred feet above the river, there was a glorious view to the south-west, over the magnificent valley. It was a lovely scene: the plain was covered with rice-fields, the crop now nearly, or quite ripe, and as yellow as a September corn-field at home; dotted over it were numerous villages all enclosed,

[8] *I.e.* the 'Sanda (or Chan-Ta) River.'—*Y.*

and the houses nearly hidden, by fine bamboo or banyan trees. Here and there would be a noble old banyan placed by nature on the summit of some grassy knoll that rose up from the midst of the golden meadows; in other places these trees might be standing up amongst the rice on an artificial mound, and often some young sapling just planted would be protected by a fence of split bamboo. On both sides rose a fine range of mountains, their slopes diversified with woods, patches of cultivation, and stretches of fine grass; and winding through the plain, the fine river rolled smoothly down to join the Irawadi.

An excellent road took us to Chan-Ta, the residence of another Shan chief, who holds the Chinese rank of Yu-Chi. This was probably the little boy who became the adopted heir of Sladen in 1868, for he was now fifteen years of age, his affairs being conducted by a relative. He seemed miserably poor, and was dreadfully ill at ease during the ceremonious reception. He wanted to learn English, and asked us whether anyone would teach him if he went to Rangoon. He talked Chinese very slightly, and complained that it was a very difficult language. The whole house was half tumbling down, and very dirty—a remarkable contrast to the 'handsome structure of blue gneiss on a large and handsome scale' described by Anderson.[9] We had the usual suite of a building to ourselves: one room in the centre quite open in front, and another at each side; but the whole of it would have gone into the central part of our quarters at Kan-Ngai. There was a wretched old table with one leg missing, and the other three tied up with bamboo strips. The beds

[9] *Mandalay and Momein*, p. 169.

were made of doors on trestles, and everything betokened poverty and ruin. The people complained bitterly of Chinese oppression; they said that nothing was left to them, even their very tables and chairs being taken if they had any. One man standing about had been to Rangoon; he loudly praised the English rule, declared that their own government was abominable, and, though he did not say so, evidently wished from the bottom of his heart that England would walk in and annex this country. We were not saluted on arrival, but the affairs of this chief seemed so badly managed that it was no matter for surprise; a fowl and a duck were brought in and flung under a bench; we afterwards learnt that these were meant as a present, but we had no idea of it at the time, as the presentation of a gift of this kind ought to be, and almost invariably is, attended with considerable ceremony.

The conduct of the two Cantonese men who were with us was abominable. The rank of the native chief was higher than that of their master, but amongst the Chinese that counts for nothing, a Chinese coolie thinking himself as good as, if not better than, the highest native chief. Directly we arrived the Cantonese wanted their opium and a place to smoke it in; they called for this and for that, and spoke to the people as if they were so much dirt. One place was too draughty for the lamps, another not comfortable, and they grumbled and cursed and made themselves generally disagreeable. In the hearing of all they told us in a loud voice that these natives did not understand common politeness, that no guns had been fired for us on our arrival, and that it would be a good thing when they were all killed. The native chiefs must put up with all this, as they dare not say a word even to the servants of a Chinese official. The exactions they have to support, too, are terrible. Chinese officials passing and repassing take lodging, food, coolies, horses, and everything without payment, and grind down the people till they can scarcely live. Some time previously, in the territory of Kan-^{Ng}ai, some of the wild mountaineers, subject to no one, made a descent, attacked a party of traders, and stole some bales of cotton. The chief was called on to pay not only the full value, but double the amount over again in squeezes to the various Chinese officials, though he was quite powerless to have prevented the attack. No wonder all these people who live so close to our good rule wish that we would come and govern their country.

October 23.—This was a day on which a great festival of some sort was held at Chan-Ta, and nearly two hundred retainers had been brought into the house to accompany the chief to a temple. The people are not much earlier in their habits here than generally in Yün-Nan. I enjoyed a quiet hour's writing before anyone else was astir, and then watched the people get up one by one and perform their scanty ablutions in the courtyard, after rubbing the sleep out of their eyes.

As the Shan chief went off to his devotions he passed through a double line of men, who were attired in most picturesque costumes. All were armed with swords or guns, some had both, and after the sober dresses of the Chinese, the contrast of the brilliant colours in which these people love to deck themselves was very striking. The Chinese almost invariably wear the dark blue cotton in winter, and in summer they dress in white; the Tibetans, too, indulge in little bright colouring, for the clothes of the Lamas are but a dull red; but here, all of a sudden, there were people wearing green, red, yellow, or purple cummerbunds and turbans.

The Shan chief rode in his chair, carried by some very ragged and clumsy chair-coolies; his official red umbrella, seal of office, and diploma, all done up in red or yellow, went before him. There were two or three big muskets, like punt-guns, carried by two men apiece, the rest of the retainers being armed with matchlocks or old percussion guns. The chief was away at this business a little more than an hour, and as soon as he returned the people began to stir themselves about getting mules, and as all the men and women for miles round came in to pay a state visit to the chief, everyone was very glad to get us out of the way, and as much haste as possible was made to find us what we wanted.

The consumption of pork in the house on this festive occasion was enormous; half the pigs in the village must have been killed for the purpose. Every two or three minutes a man passed through the door leading to the private apartments, carrying a huge lump of pork. Up to this point, as far as eating is concerned, the people had been exactly like the Chinese: at all the little stalls, under the trees, the usual Chinese dishes had been invariably found, and here the regular Chinese love of pork was most evident. As we started, and rode off through the village, where numbers of a small, but particularly repulsive-looking, breed of pigs, with unusually long snouts, were wallowing in the mire, and where, as a contrast, there were some very handsome ducks and geese, we met all the people dressed in their best clothes coming in, and really it was a very pretty sight. The women mostly wore tight black cotton garments, which were folded many times round their hips, giving to this part of the body the appearance of great breadth. Some, instead of black bodies to their dresses, wore them of blue, green, or almost any bright colour except red, and some wore white. The people looked very much cleaner than the Chinese generally do, their white clothing, whether on men or women, always being clean and fresh. Their sleeves were generally ornamented with red cuffs. They wore loose black trousers reaching a little below the knee, the rest of their legs and feet being quite bare. Round their waists there were brilliant cummerbunds, mostly of cotton, but some of silk. These were of every hue, red being the favourite tint, and there was a bunch of bits of cloth of all sorts of bright colours, like a large tassel, tucked in behind. Their turbans, swelling as they rose high above their heads, were black, and decorated with pins, from which hung large ornaments of beads, with very large and bright-coloured tassels, generally red. A narrow slip of black cloth formed a necktie, and was fastened at the throat with a large brooch of silver, sometimes set with fifteen stones in three horizontal rows.

Round their necks they wore two or three heavy silver hoops, eighteen inches in diameter; earrings, with bright red tassels, played against their cheeks; their wrists were weighted with three or four massive silver bracelets, and their fingers were tricked with a quantity of heavy silver rings, set with stones of a very inferior description. The ears of some of the women were pierced with holes about half an inch in diameter, in which silver tubes, two or three inches long, were inserted; and a bunch of the delicate black rings of rattan cane encircled the legs of all.

They were very fond of flowers, nearly all having a brilliant yellow flower in their turban, or somewhere else about their dress.

It was very amusing, too, to see that at least half the men wore buttons from England, made in imitation of half-rupees, with the head of her Most Gracious Majesty embossed upon them. We met also a few women from the wild mountain tribes. They were dressed quite differently, with bare heads, and their hair cut in a horizontal fringe across the forehead, and with a skirt to their dresses, embroidered in front;[1] and here and there a good many Lamas or Phoongyees stood lazily about in bright yellow dresses and flat yellow turbans, their lips and mouths all red with the betel-nut that they chew. It was altogether very interesting watching these people, and the first hour of the journey passed very quickly.

Our road generally led through rice-fields. Most of the rice was now cut, and the fields were quite dry; but a good deal was still standing, and the horses we were riding could hardly be prevented enjoying an occasional mouthful of this delicious food, for the path was but a track, with the crops growing close on both sides. In some cases a little fence of split bamboo was erected at the edge; and every now and then, where the road became wider, running between banks or hedges of cactus, there would be fences across the track with gates, the first gates I had seen since leaving Europe. Now there would be a little undulating stretch of beautiful turf; and at another time we rode for nearly a quarter of a mile under a fine grove of banyan trees. Here, under a gigantic banyan, would be an old man or woman seated with a little refreshment stall, where a picturesque group

[1] These must have been Kakyens (or Kach'yens). — Y.

of people, horses, or mules would be collected, resting and taking a dish of rice, blancmange, pickled quince, or a piece from a gigantic cucumber, the size and shape of a melon.

Presently up came an old man, riding a fine chestnut pony; he smiled when he saw us, made a European salute, and, very pleased, stopped to say a word or two. He had lived twelve years in Ava, and loudly sang the praises of the English, who, he said, had treated him, though only a poor trader, like a prince. He wanted us to buy his pony. He said that our honourable countrymen always liked to buy good horses, and his was just the thing to suit them. We did not make the purchase, but wishing him good-bye, we rode on. It was very pleasant to find that those who had been amongst the English in Burmah were always glad to see us, and spoke of our people and our rule as so good and just. Here the villages were almost hidden by very fine trees and bamboos, but I never saw a bamboo of the extraordinary dimensions of which I have heard. All the way from Ch'êng-Tu I examined every bamboo grove that I passed, and I never saw one more than six inches in diameter.

The road was generally very good and level, about a mile or so from the river, but now and then coming close down by the edge, where we could see people fishing or poling about in their dug-out canoes. Great numbers of white paddy-birds flapped about; there were a few cormorants, and a yellow wild duck or two; the magpie was as much a part of the landscape as ever, and in the banyan trees a kind of black and white chattering bird was generally in flocks of ten or a dozen. The day was very fine, the temperature just pleasant, and the ride would have been perfect but for the unpleasant habit the people have of purposely keeping their roads under water. Once we came to a drop of about two feet into a bog, where one of our ponies literally sank up to his nose in the mud, and it was all the poor beast could do to extricate himself.

To our great surprise Chin-Tai galloped on ahead with one of the Cantonese, and we wondered what had caused this sudden access of courage ; the natural suggestion would have been ' cash,' and so it was.

Shortly before we arrived at the market-place of T'ai-P'ing-Chieh (or Karahokah of Anderson), he returned, and, scarcely intelligible with rage, poured out a torrent of words, explaining as well as his excitement would permit how, as his pony was unable to travel fast, he had said to the Cantonese :

' Dear sir, would you be so good as to go on first, and kindly find a house for their excellencies to breakfast in ; and if, honourable sir, you could make it convenient to command rice for the little ones, I should esteem it a very high favour.'

' Whereat,' said Chin-Tai, ' the Cantonese began to curse and swear, and said that he was no servant of the foreigners, and would do nothing for them.' Such in effect was Chin-Tai's tale, and now the Cantonese, who had by this time rejoined us, gave us his version.

He said that no sooner was he a long way ahead with Chin-Tai, than the latter had accused him of extracting eight taels from the native chief to pay for our horses and mules, and that Chin-Tai had demanded half of this sum, which existed only in the imagination of our follower ; and that he had said to Chin-Tai :

' Dear Chin-Tai, you are quite mistaken, for I

have received nothing. I am but a poor Cantonese, and really have no money, while you come from the noble city of Peking; if I had a few cash, I would willingly share it with so honourable a person, but I have nothing, really nothing.'

'Then,' said the Cantonese, 'Chin-Tai drew his sword and beat me twice, and as I was unwilling to be on anything but the most friendly terms with your excellencies' servants, rather than defend myself I ran away.'

That both tales were a string of lies went without saying; for if King David had only lived a little further east, his verdict, delivered as he confesses in haste, might safely have been pronounced in his moments of leisure after the most mature deliberation. Not that the Chinese are worse than other Eastern nations, in fact they are not so bad as many. A Chinaman will always tell the truth for choice, if there is no conflicting interest; but it would be of course too much to expect that he would sacrifice either his pocket or his convenience to the exigencies of veracity; on the other hand, I have noticed that some Orientals will always lie merely for the pleasure of doing so.

We poured very cold water on the complaints of both the disputants with most discriminating impartiality, and so contrived to extinguish the flames of their wrath.

When we arrived at Tai-P'ing-Chieh, which consisted of one very broad street between low huts of bamboo wicker-work, splashed with mud, with thatched roofs, Chin-Tai proposed one house and the Cantonese another. Anxious to retain the credit we had acquired for holding the scales of justice even-handed, we went first to the house of the Cantonese

CH. IX.   CHINESE FRONTIER-TOWN ('MANWYNE').   383

selection, and then finding no rice cooked, moved across the road to Chin-Tai's choice, thus hurting the feelings of neither party.

The weather was hot, and the room was small, but it was soon densely packed with inquisitive Chinese, who settled themselves down comfortably to enjoy the show, until we expressed our regret that we could not invite all of them to breakfast, for what was one bowl of rice amongst so many? This shamed the greater part of them into a retreat, and we were allowed to finish our meal in peace.

We had a pleasant afternoon's march through the same magnificent and fertile valley; the trees, with which all the villages were surrounded, giving the plain the appearance of being well wooded. The ground was nearly covered with yellow rice, with here and there a small patch of beans, cotton, tobacco, or cabbages; and we arrived at Man-Yün,[2] the frontier town of China, at about 6 o'clock in the evening.

We were conducted to the residence of the civil magistrate, of which that officer had spoken in such unctuous terms at T'êng-Yüeh. He, however, had no house, but lived at the back of a temple, the eaves of which projected about nine feet; the space underneath, for a length of twenty feet, had been walled in by a straw mat, and divided into two compartments by another. One of the rooms so formed was the house of the Chinese magistrate; the other was the mansion of one of his subordinates; and an open cesspit was just outside. The poor fellow in giving us his house had certainly done the best he could for us, and as it is never wise to be critical with regard

[2] This is Manwyne (Manwain), known by the treacherous murder of Augustus Margary there.—*Y.*

to a gift horse, we settled ourselves down as well as circumstances would admit.

October 24.—The march to Ma-Mou or Sicaw was a difficult one, and long and frequent were the legends told us of the fearsome nature of the path itself, and the savage conduct of the 'wildmen,' as the Chinese called the mountaineer inhabitants of the border land between Cathay and Amien.

It is customary for travellers to pay tribute to the heads of all the places passed through; if this is not done they have a pleasing habit of cutting down trees and putting them in the way; then the traveller must make a detour to some other village, where he may find more trees across the road, if he has not been robbed before arrival. In this way the journey, if performed at all, naturally occupies some days, but sometimes traders will band themselves together, to the number of seventy or eighty, and pass through in one march, regardless of the 'wildmen.' There was no native chief here; he being dead, a woman, his widow, reigned in his stead. She was a stout little woman of already fifty summers at the time of Sladen's visit, and ten years had probably not added to her activity; but we did not see her. Her affairs were conducted by some deputy, and were, as a consequence, all more or less 'east and west'; but he promised to find us mules and coolies, and a 'wildman' to take a letter to Bhamo.

We had already sent a letter from Ta-Li-Fu, which the Tao-Tai had informed us would travel at the rate of fifty miles a day; at T'êng-Yüeh I had written another, and had entrusted it to the officials, but it had been returned to me the same evening with the excuse that it had been opened by someone in mistake; and although it is probable that this was true I did not deem it worth while to make another attempt; but Li-Sieh-Tai, who called on us, told us that my letter despatched from Ta-Li-Fu on September 30, had only reached Man-Yün on October 20, and as the 'wildmen' demanded 5*l*. for taking a letter to Sicaw, and there seemed much uncertainty of its getting beyond that place, we abandoned the idea.

Li rather made light of the difficulties of the road, but said he did not think we could reach Sicaw in one day.

We naturally looked with peculiar interest at this man, whose career had been so remarkable, and on whom so much suspicion hung with regard to the deplorable death of our countryman Mr. Margary.

A Burmese officer called on us, who astonished us by shaking hands in European fashion; he wore a bright yellow embroidered silk handkerchief on his head, and a Chinese jacket, with the regulation five buttons, and lined with fur, though the thermometer was between 70° and 80°. A long piece of silk, about a yard broad, striped yellow, green, red, and white—yellow being the predominating colour—was wound round his waist, forming a skirt; and the end, folded three or four times into a sash, hung down in front. His legs were bare, and his feet were encased in a pair of wooden sandals turned up in front. He was in some way connected with the place, held a Chinese official rank, and talked Chinese very well.

Our meals at Man-Yün put me in mind of the Zoological Gardens; we used to take them in the chief part of the temple, which was open to the front, except for some large wooden gates with vertical bars about nine feet high. Here the inquisitive crowd used to collect and stare through at us. It only wanted a placard outside—' Animals fed at 11 and 7 '—to make the resemblance complete.

The Chinese civil magistrate, whose house we were occupying, arrived in the evening; but he would not let us turn out, and he found a small garret adjacent. He told us that he would make all the arrangements with the chiefs of the districts, and that we should find twenty native soldiers sufficient as an escort.

He advised us to take some opium as a present for the heads of the villages. He added that the mountaineers had a superstition that if people rode through their villages ill-luck would follow, and he counselled us to dismount and walk through them. We asked if the officers of the British force that marched through dismounted at the villages. He said he thought not, but that they were a strong body, and could do as they pleased. The number of disbanded soldiers had again risen, and according to the latest intelligence there were three bands, of two hundred or three hundred each; and instead of being between Man-Yün and T'êng-Yüeh, it was now stated that they were at Ma-Mou, or between that place and Man-Yün.

October 25, 26.—A steady rain kept us indoors during our stay at Man-Yün, but we managed to visit the market between the showers. Some of the Pa-I people were seen about. The customs of the Pa-I in South-Eastern Yün-Nan, as related by Garnier, seem similar to those of all the tribes in this district, especially the delight in silver ornaments; but none of the dresses in Garnier's picture are much like those of the natives here. The Pa-I women in Man-Yün were certainly very good-looking as compared with the Chinese.[3]

There were quantities of English goods in the market—needles, buttons, balls of thread, and English

---

[3] Pa-I is the Chinese name of a Shan race widely diffused in Yün-Nan, or rather is the synonym of *Shan*. See *Marco Polo*, 2nd ed., ii. 51.—*Y.*

cotton—and a long train of two hundred or three hundred mules came in from Bhamo laden with salt from England. The caravan had been attacked on the road, and had lost twenty mules. The salt reaches Nan-Tien, although it has no business to go even as far as that, for Nan-Tien, though under a native chief, is ruled by, and is a part of China, where salt is a Government monopoly, and where the importation is forbidden by law.

As far as T'êng-Yueh, we passed trains of salt going the same way as ourselves, and beyond Nan-Tien we saw it coming up from the other direction.

At the time of our visit to Man-Yün there was a head-man with one of the most villainous faces that it had ever been my lot to see, but he appeared all-powerful, and even the Chinese magistrate seemed more or less in his hands. He seemed to have had a guilty conscience about something, for when the Grosvenor expedition was here he cleared out and ran away. The Chinese magistrate was, of course, determined to make as much out of us as possible. He averred that he had no authority whatever over the people between Man-Yün and Bhamo; but although we completely failed to get mules without his assistance, directly he made sure that he could gain a large profit the mules were arranged for. He professed to be very much annoyed when the people asked us five taels per animal for the journey, and assured us that one tael was quite enough. He, however, made arrangements with a chief to conduct us for 2·2 taels. He told us that we must pay a further sum of ten taels to this chief as a kind of tribute, and also give him one hundred taels weight of opium to distribute along the road. He said that we ought to pay the money and opium through him, and he wanted us to give the whole in advance. This we refused, but paid him half the opium, the whole of the ten taels tribute, and half the mule hire.

This sum of 2·2 taels was very high, but Sladen's expedition, Margary, and subsequently the Grosvenor expedition, all paid more, and it was consequently very difficult to make arrangements even at this rate; but we were determined, as far as possible, to consider those who might follow us; otherwise, as it was the last stage, we would willingly have paid whatever was asked, to avoid the haggling.

Whilst at Man-Yün we received a warm letter of welcome from the late Mr. T. T. Cooper, and after all our wanderings it was a pleasant thing to feel ourselves once again so nearly under the shadow of the British flag.

Before leaving Man-Yün I instituted a gun-bearer, for during the journey to Ma-Mou, we should be more or less liable to an attack of some sort, and the coolie was given strict injunctions never, under any circumstances, to leave my stirrup.

October 29.—The muleteers kept us a long time waiting, so we started in advance, and sat down under a banyan-tree until our caravan should catch us up; the air was pleasant, and we were well amused watching the people pass; the men—even the agriculturists, without a single exception, were armed with swords, and sometimes with guns as well, and were tattooed from the waist to the knee. This tattooing is commenced at the age of puberty, and it must be a long time before it is complete, for no man could stand the pain and inflammation on so large a surface of the body at one time.

The wild Kakyen women from the mountains were coming in to Man-Yün, all with their hair cut in a fringe across the forehead.

When at last we moved off together we were an imposing force, with twenty native soldiers carrying swords and guns.

Just beyond a stream we came to a hot spring. We asked the chief if this was the scene of poor Margary's death. No, he said, but just by the edge of the water where we had crossed it.

Standing thus at the scene of his cruel murder, I could not but feel what a loss the country had sustained in that brilliant young officer, who, through sickness and the difficulties attending a pioneer in new and untravelled districts, had carried out with singular tact the delicate duties entrusted to him, and may, in the words of Dr. Anderson, be said ' to have bequeathed it as a public duty, made more imperative by its being the most fitting tribute to his worth, —to establish in those border-lands the right of Englishmen to travel unmolested.'

' The name of AUGUSTUS RAYMOND MARGARY will be most fitly honoured by a party of his countrymen formally asserting the right to traverse, in honour and safety, the route between Burmah and China which he was the first Englishman to explore, and which should be maintained as his most durable monument.'

It was our fortune to be the humble instruments of thus honouring his name, but any feeling of gratification was lost in the thoughts of the rueful scene that had been enacted on that fatal shore. We had claimed the legacy bequeathed by him, but it was in sorrow that I felt that we had redeemed the right his life had purchased. For a moment I thought of sketching a spot which will ever be a hallowed one to Englishmen; but it might have raised suspicions in the superstitious minds of our companions; and long after such a paltry record would have perished his name will stand bright and clear in the recollection of his regretful countrymen. I uncovered my head as the only tribute of respect that I could pay to the memory of one who will ever be dear to the hearts not only of those who knew him, but of all who value the noble qualities of uprightness, courage, and determination.

There are three roads between Man-Yün and Ma-Mou,[4] and the one we followed does not keep close to the river, but winds about amongst spurs thrown out from a high wooded range of mountains that bounds the valley, and separates it from that of the Nampoung river. There is but little cultivation, the country being entirely inhabited by Kakyens, who mostly live in small huts by themselves, though at about every ten or twelve miles there is a collection of perhaps half a dozen forming a village. These solitary huts generally have no walls, but simply consist of a gabled roof of thatch supported on bamboo stakes, with a raised floor, rather higher than the lower edge of the roof, underneath a portion of it. The floor is made of thin strips of split bamboo, and the supports, like almost everything in these parts, are of bamboo. The thatch is made from the long grass that grows to a height of seven or eight feet, and through which the narrow track, which cannot be called a road, passes.

There were no rocky places nor steep gradients, the great difficulties we had to contend with being the frequent bogs, one of which was so deep that we were obliged to cut branches of trees and grass, and

---

[4] Ma-Mou is apparently 'Old Bhamo' of our maps, at which (according to Dalrymple) the East India Company had a factory in the middle of the seventeenth century.— Y.

make a path before the animals could cross. We passed through a regular jungle of thorns, very long grass, and trees, but as yet did not enter the forest of magnificent trees of which I had heard so much. The country is very undulating, and admirably adapted for robbers' purposes : even a couple of men, hidden away amongst the grass on the top of a hill, could easily throw a caravan into confusion ; and our chief showed us a place where, as the grass was much trampled down some yards off the track, he considered there must have been a robbery during the last day or two.

We had been informed that our chief was going to conduct us to his house, a march of only ten miles ; but after having ridden about seven miles, in answer to my inquiries he said that he had taken a different road, as there were a good many troublesome people on the other, and that now we had come about half-way to our halting place.

As it was now about two o'clock, I determined to eat my breakfast without dismounting, and soon afterwards became so absorbed in the interesting occupation of peeling a hard-boiled egg, that I failed to notice a group of some twenty or thirty people in a clearing at a little distance.

The sound of a shot caused me to look up ; but it did not strike me as anything more serious than a man frightening birds, until Mesny called out—

'Won't you load your rifle ? They are firing at us !'

The *bolt* that I made of that egg would even have astonished '*Pip*,' as I sprang down and clapped a couple of cartridges into a heavy double express. The bullet had struck a bamboo just in front of the chief who was riding first.

And now how our old friend Marco would have revelled in the telling of how the mules turned tail and fled, and nothing on earth would have induced them to turn. How off they sped with such a noise and uproar that you would have trowed the world was coming to an end! And how, too, they plunged into the wood, and rushed this way and that, dashing their burdens against the trees, bursting their harness, and smashing and destroying everything that was on them! How the battle raged furiously; how you might see swashing blows dealt and taken! How the din and uproar were so great, from this side and from that, that God might have thundered and no man would have heard it!

The necessities of a truthful tale, however, compel me to admit that the above animated description, adapted from that of the battle of Vochan, is in no way applicable to the attack at Pung-Shi. No one seemed either excited or alarmed; the animals, when they were stopped, began quietly to nibble the grass; even the Peking braves shared the general apathy, and scarcely turned their heads. The native chief put a fresh quid of betel into his mouth, as he assured us that it was nothing, and begged us to move on. Not another shot was fired; and the scene was far more ludicrous than thrilling, as one of our party with an old sixteen-bore muzzle-loader, the best lock of which had a useful knack of tumbling off at critical moments, and which was charged with No. 7 shot, stood at the *ready* behind a hedge so thick that he could not have seen the whole Russian army if it had been at the other side.

The excitement soon 'dwindled to a calm,' and we quietly marched away from our assailants, who were some of the people living in these solitary huts, and who, notwithstanding the patch of rice with which they surround their dwellings, are more robbers than agriculturists. If they see a small train of twenty or thirty animals they fire a shot, when, if the travellers are Chinese, they generally take fright, stop, or run away. The wild-man then takes tribute or helps himself, seldom killing anybody. In this case the assailants, in all probability, had not the faintest idea that there were foreigners, and when they saw that we were prepared to fight they made no further attempt to interfere with us.

We passed on, and presently came to the outskirts of a small village, where all sorts of wonderful things had been put up to frighten away the spirits. Two posts were driven into the ground, sloping at an angle of about 60°, on which curious cabalistic signs were painted in black and white; little square or triangular platforms were erected on bamboo stakes for the spirits to sit on; these were decorated with dried branches of leaves and tufts of grass; and there were long rows of bamboo stakes, to each of which a bit of small bamboo, about a foot long, was fastened. The history and meaning of each and all we could not learn. The people would only tell us that they were a protection against the spirits, or 'Nats.'

Immediately after this we arrived at the village of Pung-Shi,[5] consisting of about half a dozen bamboo huts. These are all exactly alike. A level platform is first cut out on the slope of the hill, leaving a steep bank on the upper side, against which the hut is erected, the thatched roof coming down to the top of the bank, which thus forms a sort of wall. Three feet above the level platform there is a flooring, extend-

---

[5] Pung-Shi is the 'Ponsee' of Anderson and Sladen.—*Y.*

ing over about half the covered area, the other half having no floor. The upper portion is divided into compartments by bamboo matting; the flooring is of split bamboo, supported on bamboos resting on piles, and it is reached by a sloping log of wood, in

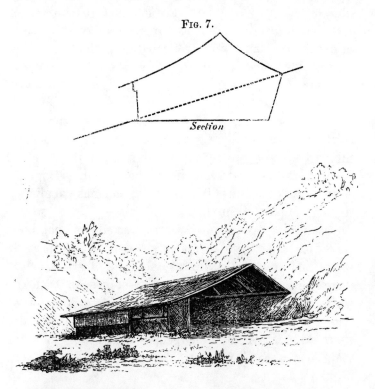

Fig. 7.

which there may, or may not, be notches for the feet. The gable at the upper end is closed by matting, with a door leading out. This is a private door for the use of the family only; the other gable is either half or altogether open.

We were shown into the largest compartment, in the middle of which some sticks were burning on some earth that had been plastered over the flooring to make a hearth.

On entering I saw a nice little square wicker shelf in the corner, the very place for my hat, I thought, and put it down there. Straightway a man leapt up from the ground on which he was seated, and with anxiety pictured on his face snatched it away. At the same moment Chung-Erh deposited my saddle-bags underneath this shelf, but hardly were they there when another of the men hastily removed them. This was the spirit's corner; for in every house there is a portion set apart for the spirit, so that he may not intrude himself elsewhere, and if people put anything or sit down in the spirit's corner the consequences that ensue are terrible.

We had a long conversation with this chief, who told us that he had not received one cash of the ten taels paid for safe conduct, nor one little piece of the opium, all of which had been retained by the Chinese magistrate, who probably divided the spoil with the head-man of the villainous face.

The natives have an apparatus by which they strike a light by compressed air. The apparatus consists of a wooden cylinder, two and a half inches long by three quarters of an inch diameter. This is closed at one end; the bore being about the size of a stout quill pen, an air-tight piston fits into this with a large flat knob at the top. The other end of the piston is slightly hollowed out, and a very small piece of tinder is placed in the cup thus formed. The cylinder is held in one hand, the piston inserted, and pushed about half-way down; a very sharp blow is then delivered with the palm of the hand on to the top of the knob; the hand must at the same time close on the knob, and instantly withdraw the piston, when the tinder will be found alight. The compression of the air produces heat enough to light the tinder; but this will go out again unless the piston is withdrawn very sharply. I tried a great many times, but covered myself with confusion in fruitless efforts to get a light, for the natives themselves never miss it. Altogether, however, I thought that Bryant & May were preferable, whose matches are sold at Man-Yün for twenty-five cash a box (less than a penny farthing), though the lowness of the price seems incredible. We dined off some beef of the buffalo. When it first came in hot, the odour seemed strangely familiar, and suddenly the dining hall of the Royal Military Academy flashed upon me. I again saw it as it used to be, the tables and forms, with many long-forgotten details. For there was a peculiar smell appertaining to the beef supplied at this institution that I never met with before or since. It is curious how a smell will sometimes call to memory scenes of long ago. This buffalo beef was exceedingly coarse, but it was eatable, and not particularly tough. I tried to believe that the animal had not died a natural death, but wisely asked no questions on the subject.

October 30.—The muleteers would not start before daylight, but left soon after six; they all took cold rice with them wrapped up in plantain leaves; everything betokened a long march, and we thought we should sleep at Ma-Mou that night.

It was a lovely morning, and soon we plunged into a forest of mighty teak trees with a dense undergrowth of long grass, brambles, and bushes, the large forest trees growing widely apart. It was magnificent forest scenery, and might well have originated some of the wildest fancies of Gustave Doré; creepers growing to a huge size and twisting round the limb of some tree like a gigantic python, the resemblance being all the more complete, as the creeper in its growth gradually crushes the life out of the limb that has supported it. The dead limb then rots away, and the cruel creeper, like some monstrous corkscrew, stretched across the path, supports itself with difficulty for awhile, and then shares the fate of its victim. Sometimes after reaching the top of a tree a creeper drops down to the ground, so perpendicularly, and so straight, that it is difficult to believe that it is not a stout rope suspended from a branch. One very remarkable fact about these creepers is that they all train, without exception, from right to left (against the hands of a watch). Then there are trees, with weird-looking roots above the ground, grasping an unyielding rock, that fancy might conjure into the form of some antediluvian cuttlefish which, in its dying agony, was clutching at and striving to crush the rock. Butterflies of marvellously brilliant and varied hues flutter about amongst the glossy fronds of great tree-ferns, and bamboos of a length almost incredible shoot up, till, often unable to bear their own weight, they fall across the road. The bamboos do not attain any great thickness, the largest I measured being five and a half inches, and the largest I saw being certainly not more than seven inches in diameter; but their height is extraordinary, as is the number of them; they grow in clumps of twenty or thirty together, and as the road is traversed, there are always two if not three of these fine groups in sight. They are used for nearly every purpose, even that of water buckets; lengths of about three feet are taken for this, and in the houses there are always some half-dozen of them in a corner.

We had rather a tiring march, a great deal of up-and-down hill over a somewhat indifferent road, rocky and very steep, but the mules kept up a steady pace until mid-day, when after fording the Nampoung river we came to a little opening. Here the packs were taken off, and the animals let loose to graze and roll themselves after their six hours' march. All the people had brought cold rice with them, and even the chief himself sat down to his cold meal. This looked as if they were determined to push on till night, and, as some rice had also been put up for the animals, we thought the halt would not be long. These people partook very contentedly of their uninteresting food; no Chinaman would eat cold rice unless he were driven to very hard straits, for he would at least pour some hot or warm water over it, and my boys even preferred cold potatoes, some of which they fortunately had with them. Potatoes grow in the hills here, so that all the way from Ta-Chien-Lu this valuable root is found, and, notwithstanding the contempt in which the Chinese hold it, the culture has spread with wonderful rapidity.

A wood fire is easily lighted, and so tea was ordered; Chin-Tai then came up, and asked if we would like some poached eggs. We were hungry, a portion of a stale loaf, and hard-boiled eggs, one of which was bad, was not a tempting meal even in the forest. Chin-Tai's proposal was most seductive; everything appeared handy, still we did not quite like the idea of commencing cooking operations; but he who hesitates is lost. We hesitated, finally acquiesced, and, as the sequel shows, were lost!

The chief and all the muleteers, though they had clearly made up their minds to start soon, had done so sorely against their will, and only on account of the tremendous presents that had been promised should a Ma-Mou roof shelter us this night. When they saw us making cooking preparations the temptation was too strong : numerous fires were lighted, men sent off to cut grass for the animals, and the unthrashed rice prepared for them was reserved for another occasion. The despatch of the grass-cutters we did not notice until it was too late ; but after our breakfast and a cigar, when we mooted the question of moving, the chief quietly replied we would move to-morrow morning, and reach Ma-Mou in plenty of time. Threats, persuasions, and offers of egregious reward were alike useless, the chief and the muleteers sat stolidly smoking or cooking their rice, and simply took not the slightest notice of anything we said. We determined, however, to go off, hoping the rest would follow ; and ordering the Ma-Fu to get the ponies we packed up the few things we had out, amongst others the thermometer. This instrument had much exercised the chief, and he asked if it was the machine by which we found out whether there were thieves on the road. 'No,' we answered. 'We can't show you that affair—this is to see whether it is hot or cold.' 'Why you needn't trouble yourself to do that, you have only to ask me and I can tell you—without using anything like that,' was the rather obvious retort.

The people were all very lazy, and even the Ma-Fu, who was generally most active and willing, seemed to share the general lethargy. Mesny's boy had gone off to cut wood to make a bed with, and Chin-Tai and Chung-Erh would not catch their animals, and expected the muleteers to do it for them, who looked on with a grin. Seeing we were in earnest, the muleteers gave our riding animals some rice ; and cowardice prevailing over laziness, my two boys, all at once becoming very humble, captured their erring beasts, and saddled them without more ado. The chief, though clearly more or less uneasy in his mind, made no motion of stirring, and we started alone, our party being Mesny, Chin-Tai, my Ma-Fu, my gun coolie, Mesny's boy, and myself. Chung-Erh was not quite ready, but he followed as fast as he could, fear now keeping him close to our heels.

Of course none of us knew anything of the road, but we recklessly plunged again into the forest, trusting to good luck and the chances of war. We had not gone far before a guide sent by the chief overtook us, for, having undertaken to conduct us in safety, he did not like the idea of our wandering about by ourselves in the dense forest. We rode on steadily, and as fast as we could, but after a little more than an hour our guide remarked—

'It will be dark before we get to Ma-Mou.'

'Will it?' we answered; 'then we shall not get to Ma-Mou before dark.'

This reply, though it ought to have satisfied anyone, did not seem to please him, for immediately afterwards he stopped at a stream to drink, fell behind us, and at a village a hundred yards further on he disappeared, and we saw him no more.

In about half an hour we came to a bifurcation of the trail, a halt was called to consider the momentous question, and we decided to wait six minutes for the guide. That period having expired, the gun coolie, who coming out quite in a new light displayed the instinct of a Mohican chief, now examined one path, then the other, and gave it as a deliberate opinion that the left hand or upper road was the right one. We consulted a little, and all voting for the motion of the gun coolie we again went off.

Mesny was hopefully of the opinion that '*Tout chemin mène à Rome.*' I could only give a doubtful assent to this pleasing theory. We were riding through a very narrow track, our faces being continually brushed by the grass or leaves, when, all of a sudden, I felt as if there were a necklace of thorns on me. I fancied that a bramble stretching across the road must have caught me, and thinking that I should get sadly torn I tried to bring my pony up abruptly; this animal, however, accustomed to follow in a string, stop if the beast in front of him stops, and go on when his leader moves, would not come to a halt as quickly as I wished; and then I found that though there was no feeling of scratching, the pain was becoming every second more intense, as if my necklace of thorns was being gradually tightened. In a very few moments I could bear it no longer, and nearly frantic with wonder as to what had occurred, and sharp, stinging pain, I shouted out to Mesny in front. He looked back and called out—'Get down, get down! you are covered with ants.' I jumped off the pony, and found that thousands of huge red ants nearly a quarter of an inch long were in my hair, under my shirt, all over my clothes, and viciously biting with one accord; I was simply covered with them. I took off my clothes, and though the Ma-Fu and both my boys came to my aid, it was a long time before I was altogether free.

Mesny had gone on a few yards with the gun coolie, and now I suddenly heard his voice in loud altercation with someone. I was behind a hedge, was completely hidden from everything in front, and could not see what was going on. I hurried up, and saw Mesny pointing his revolver at a man who, at that moment, disappeared into a hut about twenty yards off the road, and the gun coolie squatting down and struggling with the buckle of my rifle-case.

Chinamen are never able to manage a buckle; indeed, I have often thought a buckle almost as good as a padlock. I quickly extracted and loaded my rifle, and asked Mesny what was up.

'Oh, he says, we may not ride past his house, but must walk.'

'Does he?' I answered; and we both jumped quickly on our ponies, and, revolver in hand, rode on till the bushes and trees hid him from view.

Mesny now told me that as he went on, the gun coolie just in front sat down to brush off the ants, some of which had attacked him also, when a man armed with a sword ran towards Mesny, who, he thought, was alone, as owing to the bushes he could see no one else, and whom he evidently mistook for a Chinaman.

The native called out that no one was to ride past. Mesny shouted to the coolie to bring the foreign gun, but the latter was so busy with the ants that he did not understand what was happening. The man then went into his hut, and came back with a gun, followed by another man, or woman, unarmed; he knelt down, and resting his gun against a post, aimed at Mesny, who, pointing his revolver at the native, called out—

'If you shoot, you are a dead man.'

'Oh,' he cried, 'I'm not shooting your way.'

'I don't care where you shoot; if you fire, I shall hit you,' said Mesny; whereat the man put away his gun, again said that no one must ride past, but retired into his hut.

This was the work of a few seconds, and it was

at this point that I came up. Nothing more came of this adventure, but we gave orders to our people not to make any noise when there were any huts about, for we did not want any more of these occurrences.

This is the way these natives always treat passers-by; the Chinese are afraid, they give way, and then, when the wild-man has gained the first point and made them walk, he calls his neighbours, and all demand blackmail—tobacco, opium, and silver, so that the wretched trader has altogether a mean time of it. But these people, like all bullies, are regular cowards, and the smallest show of resistance brings them down from their tall trees.

We rode on, and as it began to get dark we wondered where we should pass the night. Though there was no moon, the stars were very brilliant, but here and there not a ray of light could penetrate the thick and heavy foliage. At last the darkness became so pitchy that we might have been knocked out of our saddles by some overhanging bough before we could have seen it, and not deeming it safe to ride we dismounted. We stepped into a regular quagmire, and up to our knees in mud groped and stumbled about amongst great rocks and stones for fifty yards or so, until the trees being a little more open it was again safe to mount. We saw a hut or two, but, doubtful as to what sort of a reception we should meet with, we passed them by; but, at last, getting very uneasy as to whether or not we were on the right track, we called out to a man sitting over a fire. He would give us, however, no reply of any sort, and we passed on. At seven o'clock we came across a deserted hut; here we halted, lit a fire, and began to prepare some torches; but while this was going on we held a consultation. We did not even know whether we were on the right road, we could see nothing; we did not know how far Ma-Mou might be, nor what sort of a place it was; it might have walls and gates, and if so the latter might be shut if we ever arrived there. From what we had seen of the plain before it became dark it looked as if there was a great deal of water about; here was, at least, a shelter for the night. It was true that Ma-Mou, if we had kept the right road, could not be far off, but, even if thus far we had not gone wrong, we might possibly stray in the next few miles. At length the final verdict was given to stop, and then everyone was more or less glad. I had some bread still in my saddle-bags, my servants had the remains of a few potatoes, and a kind of root like a turnip was discovered by some eyes that could see in the dark. There was also a pinch of tea and a small brass wash-hand-basin to make it in; so after a meal that was rather simple than plentiful, I smoked one cigar, wrapped myself in a blanket, and defied the mosquitoes until dawn.

The mosquitoes annoyed Mesny sadly, but they let me alone somehow, although I dreamt I was under mosquito-curtains engaged in hunting some dozen of venomous brutes the size of dragon-flies; but it was after all only a dream, and I managed to get through a considerable number of hours of sleep before the first grey streak of dawn.

There had, as usual, been a very heavy dew, and riding through the jungle we were soon all wet enough to welcome the sun when at last he topped the hills, and began to dry us as well as the bushes. There had been nothing for us to eat or drink before starting, but the thoughts that now we really were at the end of our land journey, quite drove away unpleasant feelings of any kind. A very short ride brought us to a village at the foot of the hills, but before reaching it we passed a stream over some planks which would most effectually have stopped us in the dark. At the village, the promise of a rupee procured us a guide, who, soon after turning abruptly to the left, showed us that even if we had passed the bridge we should most certainly and inevitably have taken the wrong road at this point, and it was comforting to think that we had after all done right in stopping.

Presently we met a man carrying a couple of mallards and a double barrelled-breechloader; while I was wondering who he might be he made a military salute, and in a language of which no one of us understood one single word explained that a letter had arrived somewhere; he then turned back with us. We now dismissed the guide with a rupee borrowed from the gun coolie, and, much elated, continued our ride, discussing meanwhile our new companion, who continually shouted to an invisible person somewhere away in the jungle on our right. We, however, could make nothing of him, and soon Ma-Mou was pointed out; we were taken straight to a house, where a Bengali met us, and, making a salute, told us that Cooper had sent him, his boat, and a number of men to meet us, and that he had been waiting for us eight days. This my almost forgotten Hindustani enabled me to gather, but I must confess in shame that I was not long in discovering that he spoke English very much better than I did his language. A house had been prepared for us; here was a pile of newspapers, a letter from Cooper, two huge boxes of eatables and drinkables, pipes, tobacco, cigars, candles, candlesticks, matches, and everything one thoughtful and experienced traveller could send to comfort the heart and mind of another.

Though the news was not very recent it was the latest Cooper had, and no one who has not gone through the anxiety that I had felt during the last two months can form the slightest conception of the feeling that came over me—one that I utterly failed to comprehend even myself, a feeling of peace, ease, and contentment quite indescribable, and so apparent that Mesny told me afterwards that I appeared to be quite a different person. Chin-Tai was set to work at once to cook rice for our unfortunate and starving crew. Much to our delight the mules arrived before the rice-eating was finished. A bath, clean clothes, and breakfast, was followed by one of Cooper's Havannah cigars, and I exclaimed with Pangloss, 'All is for the best in the best of worlds possible.'

The corporal, for such he was who spoke the English, told us that the steamer would not be at Bhamo for two or three days, and would not start for two or three more, so we determined to take it easy, remain at Ma-Mou till the next morning, and then drop quietly down the river. The corporal added that a letter that we had sent from Pung-Shi arrived the previous afternoon, and that he had found a means to forward it to Cooper.

The house we were in was of bamboo, raised about five feet above the ground on piles of teak; the roof was of thatch, and the walls and floor of split bamboos, and here we looked on to the river of T'êng-Yüeh, now grown to a fine large rolling stream.

In the afternoon the corporal came to tell us that the Burmese officer had arranged that 'Burmese woman should make ball this evening,' if we liked, by which we eventually understood that a dance had been arranged outside our house, for our edification after dinner.

The dance eventually took place by the light of a quantity of crude petroleum in a large broken earthen jar, in which there was a long branch of a live tree. There were three women performers, their hair done neatly and quietly up in a knot on their heads; they had small earrings, a white jacket, and a red skirt reaching down to the ankles ; each had a yellow silk handkerchief laid over one shoulder, but which was continually used during the dance, and they all wore heavy silver bracelets. There were four or five men, one or two naked to the waist, and with broad cloths of different colours and patterns wound round their hips. Music and drumming was kept up all the time ; first the women danced for half an hour a very slow quiet measure, simply moving their feet and hands very gently, and with infinitely more grace than the hideous, impossible, and unnatural postures that our most admired European opera-dancers fling themselves into. Then it was the turn of the men, and after some time one who must have been the recognised favourite of the troupe appeared, for his entrance was greeted with a burst of laughter, though there did not appear to be anything particularly amusing about it. That he did the low comedy business was subsequently quite clear, by the continual laughter that was showered on his words and actions ; for besides dancing there was a sort of play, in which all the men and women joined, and in which the picking of leaves off the branch of the tree seemed to take an important part. What it was all about I never had the faintest idea, and had it not been for Cooper's packet of cigars, which I finished, I might have wearied of the performance ; but sitting outside the house on a large platform in the lovely cool starlight night, and looking down on the play by the fitful light of the petroleum, which sometimes flared up and at others nearly died away, casting wonderful lights and shadows over the performers, and a great number of people collected to see the theatricals, some leaning against the posts of the next house, some lying or sitting on the ground, others on great logs of wood, but all in graceful attitudes, there was a pleasant feeling of ease and comfort which lasted as long as the cigars. When these were done, I gave the corporal some silver for the actors, and retired to bed.

November 1.—Cooper had sent up his boat for us, a fast and comfortable one, with a little covered-in house at the stern, where we could just lie down or sit. Another native boat was hired for luggage, servants, and the horse, and we started easily at nine o'clock.

It was rather warm on the way down; the river winds between low and uninteresting banks of high grass jungle with trees; hills are away in the background; a porpoise rolls here and there, kingfishers dart about amongst the reeds, huge pelicans drop into the water with a splash that seems to threaten destruction to their breast-bones; a few native boats are passed made on Dicey's plan—two dug-out canoes, separated by three or four feet, with a deck across, and a house at the stern; single canoes also are paddled about; but the scenery did not vary sufficiently to raise excitement, and after reading the last word of the last newspaper I slept during the greater part of the journey.

The T'êng-Yüeh river enters the Irawadi about a mile above Bhamo, and in the Irawadi I was much disappointed. I had expected an immense and gigantic river, like the Yang-Tzŭ at Chin-Kiang. It is true that the river even now is very wide—a quarter of a mile perhaps—but it is shallow, and the current is slow. Now that I had seen the marvellous way in which the T'êng-Yüeh river increases in a few miles, had also seen how much less water passes Bhamo than I had thought, and when I considered the almost continual rain that falls over the basin of this river, the Irawadi ceased to be a mystery.

We reached Bhamo before three, the distance being about twenty miles, and here was Cooper to shake our hands. Oh, the pleasure of a hearty British shake of the hand! who shall measure it after the everlasting ceremonies of the Chinese. I know of strange people in England who object to shaking hands as an ungraceful, rough, and barbarous salutation. It may be so, but I know that that one hearty grasp meant more than ten thousand Chinese flowery expressions about my honourable self, and did me more good and put me in better spirits than anything that had happened for many a long and weary day. The attempt to convey an idea of the kindness Cooper showered upon me would be vain, nor can I describe the delightful feeling of being once again in a clean house, where I could walk without tucking my trousers inside my socks, of seeing a damask tablecloth, and the thousand and one things never noticed at home, but which seem such luxuries after long separation. We had something to eat immediately, and to drink. We passed the afternoon and evening in talking over our travels, and went to bed early.

We were generously and hospitably entertained by the late Mr. T. T. Cooper until November 6, when pack-animals were left behind, and we were swiftly borne down the broad bosom of the Irawadi towards home and civilisation.

How quickly those first hours of idleness sped by, though the past had already begun to assume the misty outlines of a dream when the day of our departure arrived; one warm shake of the hand, and, as we stepped into the boat and left our host alone to his solitary life with cheery wishes that we soon might meet again, it was well that the future was hidden from us.

Death, alas! has made sad havoc amongst those to whose kindness in distant lands I owe so much, and there is something of irony in the fate that permitted Mr. Cooper in his desperate attempt to pass from China to Burmah, through the rebel camp during the Mahometan insurrection, to live so long with his life in his hands, and to escape to tell the tale, and that yet, whilst in comparative security, and with the British flag floating above him, gave him over to the bullet of an assassin.

The boat pushed off into the swirling stream; in a few minutes we stood on board the 'Ta-Pa-Ing,' the last rope was cast loose, and as Bhamo disappeared from our view, a veil seemed as it were to pass over the recollections of the old travelling life, and it almost appeared as if a new phase of existence had been entered on as we swept past the wooded shores, with here and there a town, where high-sterned boats would be drawn up in tiers, ard where, in the early morning, we might hear the musical swell and fall of the phoongyee's bell.[6]

[6] In Burma the priests, or phoongyees, go round from house to house collecting rice from the well-disposed; as they walk slowly round the village or town they strike from time to time a silver-toned piece of gun-metal.

CH. IX. BHAMO TO DOVER; THE JOURNEY ENDED.

On by New Mandalay, with its temples, whose gilded roofs bring to mind the gold and silver towers of Mien;[7] by Prome, and at last to Rangoon. A few days' steaming on a sea of glass, and the City of Palaces was reached; a few weeks amongst many Indian friends, and everywhere the kindness and hospitality surrounding me helped to banish from my recollection the fatigues and discomforts of travel; never can I sufficiently thank those amongst whom I passed those first few weeks of civilisation, and of enjoyment such as I can hardly hope for again.

Westward again; and it was with mingled feelings that my glances first lighted on the European Continent; soon the white cliffs of Dover rose on the horizon, and after twenty months of travel I was home at last.

[7] *Marco Polo,* vol. ii. chap. liv.

APPENDICES.

APPENDIX A.

CAPTAIN GILL'S ITINERARIES.

I. Northern Journey in Province Pe-Chi-Li.

Date		Altitude above Sea-level in Feet.	Miles between	Days' Journey in Miles
1876				
Sept. 20	Tien-Tsin	—	—	
,, ,,	Yang-Tsun	—	18	
,, ,,	Nan-Tsai-Tsan	—	7	
,, ,,	Ho-Se-Wu	—	11	36
,, 21	Chang-Chia-Wan	—	21	
,, ,,	Peking	80	18	39
,, 25	Tung-Chou	—	—	13
,, 26	Yen-Ch'iao	—	—	7
,, 27	Tsao-Lin	—	14	
,, ,,	Pang-Chün	—	11	25
,, 28	Chi-Chou	—	$8\frac{1}{2}$	
,, ,,	Shih-Men	—	18	$26\frac{1}{2}$
,, 29	Ma-Lan-Yu	—	5	
,, ,,	Tsun-Hua-Chou	—	11	16
,, 30	Sha-Ho-Chiao	—	—	22
Oct. 1	Nan-Yang-Cheng	—	$3\frac{1}{2}$	
,, ,,	Saddle	600	$4\frac{1}{2}$	
,, ,,	Hsi-Feng-K'ou, Gate in Great Wall	—	$1\frac{1}{2}$	
,, ,,	Po-Lo-Tai	—	$6\frac{1}{2}$	16
,, 2	Kuan-Ching	1,060	16	
,, ,,	Leng-Ling-Tzǔ	1,500	$18\frac{1}{2}$	$34\frac{1}{2}$
,, 3	Pa-K'ou - Ying or Ping - Chuan-Chou	1,640	—	$25\frac{1}{2}$

APPENDIX A.

Date			Altitude above Sea-level in Feet	Miles between	Days' Journey in Miles
1876					
Oct.	4	Pei-Kung	1,560	—	30
,,	6	Ta-Tzŭ-K'ou	1,445	—	24
,,	7	Chi-Chien-Fang . . .	—	12	
,,	,,	Ha-Go-Ta	—	5½	
,,	,,	San-Tai	955	3	20½
,,	8	Ku-Ch'iao-Tzŭ	1,530	—	23
,,	9	Pass, Ta-Liang-Tzŭ . . .	2,010	15	
,,	,,	Kang-K'ou	1,180	6	21
,,	10	Pass, Huang-Shih-Ling . .	1,690	1½	
,,	,,	Pass, Tsai-Chia-Ling . .	1,690	9	
,,	,,	Pass, Ku-Ling	1,660	12	
,,	,,	San-Cha	—	2½	25
,,	11	I-Yuan-K'ou, Gate in Great Wall	—	5	
,,	,,	Shih-Men-Chai	—	7½	
,,	,,	Shan-Hai-Kuan . . .	—	11	23½
,,	13	Ning-Hai	—	2	
,,	,,	Niu-Tou-Yai	—	28	30
,,	14	P'u-Ho-Ying	—	14	
,,	,,	Sha-Ho	—	4	18
,,	15	T'uan-Lin	—	5½	
,,	,,	River Lan	—	17	
,,	,,	Lê-Ting-Hsien	—	8	30½
,,	16	Tang-Chia-Ho	—	—	9
,,	17	Lao-Mu-K'ou	—	8	
,,	,,	Ma-T'ou-Ying	—	18½	26½
,,	18	Ho-Chuang	—	—	10
,,	19	Shuang-Tou	—	16	
,,	,,	Kai-Ping	—	14½	30½
,,	20	Han-Ch'êng	—	11½	
,,	,,	Fêng-T'ai	—	18	29½
,,	21	Lu-T'ai	—	—	20
,,	22	Han-K'ou	—	6	
,,	,,	Pei-T'ang	—	10½	16½
,,	23	Huai-Tien	—	13½	
,,	,,	Fêng-T'ai-Tzŭ	—	8½	
,,	,,	Urh-Chuang	—	8	30
,,	25	Pei-Tsai-T'sun	—	—	19½
,,	26	Tsin-Huang-Chen . . .	—	10½	
,,	,,	Huang-Chuang	—	22	32½
,,	27	Lin-Ting-Chen	—	11	
,,	,,	Fêng-Tai	—	16	
,,	,,	Wo-La-Ku	—	8½	
,,	,,	Chang-Yai-Chuang . . .	—	9	
,,	27	Hsin-An-Chen	—	5½	50

APPENDIX A.   415

Date		Altitude above Sea-level in Feet	Miles between	Days' Journey in Miles
1876				
Oct. 28	Lin-Ting-Chen	—	9	
,, ,,	Pao-Ti-Hsien	—	14½	23½
,, 29	River Chiang-Kan	—	10	
,, ,,	Hsiang-Ho-Hsien	—	11	
,, ,,	Tai-Tzŭ-Fu	—	7½	28½
,, 30	Tung-Chou	—	15	
,, ,,	Peking	80	13	28
Nov. 9	Chang-Chia-Wan	—	18	
,, ,,	Ho-Se-Wu	—	21	39
,, 10	Nan-Tsai-Tsan	—	11	
,, ,,	Yang-Tsun	—	7	
,, ,,	Tien-Tsin	—	18	36
	Total			935

II. VOYAGE ON THE YANG-TZŬ-CHIANG FROM SHANGHAI TO HANKOW IN STEAM VESSELS.

                                                              Miles
1877, January 24 to 30 680

III. BOAT JOURNEY FROM HANKOW TO CH'UNG-CH'ING.

Date		Altitude above Sea-level in Feet	Miles between	Days' Journey in Miles
1877				
Feb. 8	Hankow	—	—	
,, ,,	Han-Yang	—	1	
,, ,,	Chuan-K'ou	—	7	8
,, 9	Chin-K'ou	—	—	10
,, 10	Teng-Chia-K'ou	—	12	
,, ,,	At moorings	—	9	21
,, 11	Pai-Chou-Ssŭ	—	—	12
,, 12	Hua-K'ou	—	—	15
,, 13	Lao-Chu-T'a	—	—	15
,, 14	Lu-Chi-K'ou	—	13	
,, ,,	Mao-P'u	—	9	22
,, 15	Hsin-Ti	—	6	
,, ,,	Lo-Shan	—	11	17
,, 16	Pai-Lo-Ssŭ	—	—	13½

APPENDIX A.

Date		Altitude above Sea-level in Feet	Miles between	Days' Journey in Miles
1877				
Feb. 17	Entrance to Tung-Ting Lake	—	3	
,, ,,	In Huc Reach	—	7½	10½
,, 18	Chih-Pa-K'ou	—	—	15
,, 19	Commencement of Camel Reach	—	—	12
,, 20	Shang-Chê-Wan	—	—	23
,, 21	At moorings in Brine Bend	—	—	5
,, 22	Chien-Li-Hsien-Ma-Tou	—	—	10
,, 24	Adam's Point	—	—	6½
,, 25	Farmer Point	—	9	
,, ,,	Tiao-Kuan	—	14½	23½
,, 26	Parson's Point	—	25½	
,, ,,	How-Ko-Lao	—	3½	29
,, 27	Skipper Point (Shi-Show-Hsien)	—	—	12
,, 28	Huo-Hsüeh	—	—	25
March 1	Sha-Shih	—	—	31
,, 2	Entrance of Tai-Ping Canal	—	9½	
,, ,,	Chiang-K'ou	—	20	
,, ,,	At moorings in Boone Reach	—	3	32½
,, 3	Tung-Shih	—	—	7½
,, 4	Grant Point	—	11	
,, ,,	Chi-Chiang	—	7	
,, ,,	I-Tu	—	13	31
,, 5	Tiger Teeth gorge	—	11	
,, ,,	I-Ch'ang	—	12	23
,, 9	P'ing-Shan-Pa	—	—	15
,, 10	Huang-Ling-Miao	—	—	6½
,, 11	Ta-Tung rapid	—	—	7½
,, 12	Ch'ing-Tan rapid	—	—	7
,, 13	Upper end of Mi-Tsang gorge	—	—	3½
,, 14	Kuei-Chou	—	5	
,, ,,	Niu-K'ou	—	9½	14½
,, 15	Pa-Tung-Hsien	—	—	6
,, 16	Nan-Mu-Yuan	—	—	11½
,, 17	Ch'ing-She-K'ou	610	—	9
,, 19	Tiao-Shih	—	—	7
,, 20	Wu-Shan	—	8	
,, ,,	Near entrance to Kuei-Fu gorge	—	14	22
,, 21	Kuei-Chou-Fu	—	—	8½
,, 23	At moorings	—	—	6
,, 24	At moorings 1 mile below Miao-Chi	—	—	17
,, 25	Yün-Yang	—	10	
,, ,,	At moorings	—	4	14
,, 26	Situ-Chiang	—	9	
,, ,,	A moorings	—	10	19

APPENDIX A. 417

Date		Altitude above Sea-level in Feet	Miles between	Days' Journey in Miles
1877	Brought forward			
Mar. 27	Wan	—	—	10
,, 28	Ta-Ch'i-K'ou	—	—	18½
,, 29	Shih-Pao-Chai	—	—	16½
,, 30	Shang-Kuan-Chi	—	10½	
,, ,,	At moorings	—	8½	19
,, 31	Chung	—	6	
,, ,,	Wu-Yang	—	6	
,, ,,	Yang-Tu-Chi	—	12½	24½
April 1	Pa-La-Kiang	—	12½	
,, ,,	At moorings	—	7½	20
,, 2	Feng-Tu	—	7	
,, ,,	Ma-Chin-Tzŭ	—	10½	17½
,, 3	Lan-Tu	—	6½	
,, ,,	St. George's Island	—	8	14½
,, 4	Fu-Chou	—	14	
,, ,,	Li-Tu	—	8	22
,, 5	Shih-Kia-To	—	15½	
,, ,,	At moorings	—	6½	22
,, 6	Chung-Chou	—	4	
,, ,,	Lo-Chi	730	14	
,, ,,	Hu-Tung	—	12½	30½
,, 7	Iron gorge	—	14	
,, ,,	Ta-Fo-Ssŭ	—	7	21
,, 8	Ch'ung-Ch'ing	1,050	—	4
	Total			844

IV. Ch'ung-Ch'ing to Ch'êng-Tu.

Date		Altitude above Sea-level in Feet	Miles between	Days' Journey in Miles
1877				
Apr. 26	Pai-Shih-Yi-Ch'ang	—	—	16
,, 27	Ma-Fang-Chou	—	—	20
,, 28	Yung-Ch'uan-Hsien	—	11½	
,, ,,	T'ai-P'ing-Chên	—	10	21½
,, 29	Fêng-Ngan-P'u	—	9½	
,, ,,	Jung-Ch'ang-Hsien	—	5½	15
,, 30	Lung-Ch'ang-Hsien	—	—	23
May 1	Wang-Chia-Ch'ang	—	12	

VOL. II. E E

418   APPENDIX A.

Date		Altitude above Sea-level in Feet	Miles between	Days' Journey in Miles
1877				
May 1	Niu-Fu-Tu	—	$6\frac{1}{4}$	$18\frac{1}{4}$
„ 2	Hsien-Tang	—	$9\frac{1}{2}$	
„ „	Tzŭ-Liu-Ching	—	7	$16\frac{1}{2}$
„ 4	Wei-Yuan-Hsien	—	—	$14\frac{1}{2}$
„ 5	Tzŭ-Chou	—	—	$20\frac{1}{2}$
„ 6	Nan-Ching-Yi	—	—	$22\frac{3}{4}$
„ 7	Yang-Hsien	—	7	
„ „	Yang-Chia-Kai	—	13	20
„ 8	Chien-Chou	—	$7\frac{1}{2}$	
„ „	Shih-Ch'iao	—	$2\frac{1}{2}$	
„ „	Ch'a-Tien-Tzŭ	—	12	22
„ 9	Ch'êng-Tu	1,504	—	16
	Total			246

V. Ch'êng-Tu to Sung-P'an-T'ing, and Return.

Date		Altitude above Sea-level in Feet	Miles between	Days' Journey in Miles
1877				
May 18	Ch'êng-Tu	1,504	—	
„ „	Pi-Hsien	1,766	—	$11\frac{1}{4}$
„ 19	Ngan-Te-P'u	1,776	6	
„ „	Kuan-Hsien	2,347	$12\frac{1}{2}$	$18\frac{1}{2}$
„ 21	Yu-Ch'i	2,670	$9\frac{1}{2}$	
„ „	Yin-Hsiu-Wan	3,187	$5\frac{3}{4}$	$15\frac{1}{4}$
„ 22	Hsin-Wên-P'ing	3,241	7	
„ „	T'ao-Kuan	3,623	$8\frac{1}{2}$	$15\frac{1}{2}$
„ 23	Wên-Ch'uan-Hsien	3,899	$9\frac{1}{2}$	
„ „	Pan-Ch'iao	4,275	$5\frac{1}{4}$	15
„ 24	Hsin-P'u-Kuan	4,398	$6\frac{1}{4}$	
„ „	Ku-Ch'eng	4,888	8	$14\frac{1}{4}$
„ 25	Kan-Ch'i	4,821	$6\frac{1}{2}$	
„ „	Li-Fan-Fu	5,312	$4\frac{3}{4}$	$11\frac{1}{4}$
„ 26	Return to Hsin-P'u-Kuan	—	—	$19\frac{1}{4}$
„ 27	Wên-Chêng	4,670	—	$7\frac{1}{2}$
„ 28	Pai-Shui-Chai	4,717	6	
„ „	Mao-Chou	4,996	10	16
„ 30	Wei-Mên-Kuan	5,123	$4\frac{1}{2}$	

APPENDIX A.

Date		Altitude above Sea-level in Feet	Miles between	Days' Journey in Miles
1877				
May 30	Ch'a-Erh-^Ngai	5,423	$6\frac{1}{2}$	11
„ 31	Ch'ang-Ning-P'u	—	$3\frac{1}{2}$	
„ „	Confluence of the Lu-Hua-Ho with the Sung-Fan-Ho	—	$1\frac{1}{2}$	
„ „	Mu-Su-P'u	5,344	$1\frac{1}{4}$	
„ „	Ta-Ting	5,798	$5\frac{1}{2}$	$11\frac{3}{4}$
June 1	Shui-Kou-Tzŭ	6,940	3	
„ „	Tieh-Chi-Ying	7,837	$5\frac{1}{4}$	
„ „	Sha-Wan	7,017	$2\frac{3}{4}$	11
„ 2	P'ing-Ting-Kuan	7,436	7	
„ „	Chên-Fan-Pao	—	4	
„ „	Chêng-P'ing-Kuan	7,807	$1\frac{1}{2}$	$12\frac{1}{2}$
„ 3	Pin-Fan-Ying	—	$6\frac{3}{4}$	
„ „	Chên-Chiang-Kuan	8,159	$1\frac{1}{4}$	
„ „	Lung-Tan-P'u	8,729	$10\frac{1}{4}$	$18\frac{1}{4}$
„ 4	^Ngan-Hua-Kuan	9,032	8	
„ „	Sung-P'an-T'ing	9,470	$9\frac{3}{4}$	$17\frac{3}{4}$
„ 6	Lamassery	—	4	
„ „	Hsüeh-Lan-Kuan	10,881	$4\frac{3}{4}$	
„ „	Fêng-Tung-Kuan	11,884	$2\frac{1}{4}$	11
„ 7	Hsüeh-Shan, summit of pass	13,184	$2\frac{1}{2}$	
„ „	Hung-^Ngai-Kuan or Sung-^Ngai-P'u	10,529	$6\frac{1}{2}$	
„ „	Chêng-Yuan	9,021	10	19
„ 8	Yueh-Erh-^Ngai	7,874	$5\frac{1}{2}$	
„ „	Shih-Chia-P'u	5,995	$8\frac{3}{4}$	$14\frac{1}{4}$
„ 9	Hsiao-Ho-Ying	5,297	$8\frac{1}{2}$	
„ „	Yeh-T'ang	4,339	$8\frac{1}{4}$	$16\frac{3}{4}$
„ 10	Shui-Ching-P'u	3,962	$5\frac{1}{4}$	
„ „	Shui-Chin-Chan	3,675	$6\frac{1}{4}$	$11\frac{1}{2}$
„ 11	Ko-Ta-Pa	3,391	—	$8\frac{3}{4}$
„ 12	T'i-Tzu-Yi	3,283	4	
„ „	Tieh-Lung-Kuan	3,150	$8\frac{1}{2}$	
„ „	Lung-An-Fu	3,094	$5\frac{1}{3}$	18
„ 13	Ku-Ch'êng	2,890	$8\frac{1}{2}$	
„ „	Kuang-Yi	2,811	$6\frac{1}{4}$	$14\frac{3}{4}$
„ 14	Hei-Shui-Kou	2,700	$6\frac{1}{2}$	
„ „	Chiu-Chou	2,560	10	$16\frac{1}{2}$
„ 15	Hsiang-^Ngai-Pa	2,412	$9\frac{1}{2}$	
„ „	P'ing-I-P'u	2,464	$7\frac{1}{4}$	$16\frac{3}{4}$
„ 16	Chiang-Yu-Hsien	2,211	12	
„ „	Chung-Pa-Ch'ang	2,064	$10\frac{1}{4}$	$22\frac{1}{4}$
„ 17	Chang-Ming-Hsien	1,949	$3\frac{1}{2}$	

APPENDIX A.

Date		Altitude above Sea-level in Feet	Miles between	Days' Journey in Miles
1877				
June 17	Mien-Chou	1,918	$19\frac{3}{4}$	$23\frac{1}{4}$
,, 18	Chao-Chiao-P'u	—	6	
,, ,,	Hsin-P'u	2,078	$3\frac{1}{4}$	
,, ,,	Bridge across the Lo-Chiang-Ta-Ho	—	$9\frac{1}{4}$	
,, ,,	Lo-Chiang-Hsien	2,033	2	$20\frac{1}{2}$
,, 19	Pai-Ma-Kuan, or Pass of the White Horse	2,232	$2\frac{1}{2}$	
,, ,,	Tê-Yang-Hsien	1,983	$11\frac{1}{4}$	
,, ,,	Han-Chou	—	12	$25\frac{3}{4}$
,, 20	P'i-T'iao-Ch'ang	—	$6\frac{1}{4}$	
,, ,,	Hsin-Tu-Hsien	—	$5\frac{1}{4}$	
,, ,,	Ch'êng-Tu	1,504	$10\frac{1}{2}$	22
	Total			487

VI. Ch'êng-Tu to Bhamo.

Date		Altitude above Sea-level in Feet	Miles between	Days' Journey in Miles
1877				
July 10	Ch'êng-Tu	1,504	—	
,, ,,	Tsu-Ch'iao	1,575	$3\frac{1}{2}$	
,, ,,	Shuang-Liu	1,647	4	$7\frac{1}{2}$
,, 11	Kao-Ch'iao (bridge over river)	—	$3\frac{3}{4}$	
,, ,,	Chan-To-P'u	—	$2\frac{1}{4}$	
,, ,,	Hua-Ch'iao-Tsu	1,532	$3\frac{3}{4}$	
,, ,,	Hsin-Chin-Hsien	1,595	$3\frac{1}{4}$	13
,, 12	Yang-Chia-Ch'ang	1,585	6	
,, ,,	Ch'iung-Chou	1,637	$12\frac{1}{2}$	$18\frac{1}{2}$
,, 13	Bridge over the Nan-Ho	—	1	
,, ,,	Ta-T'ang-P'u	1,681	$6\frac{3}{4}$	
,, ,,	Pai-Chang-Yi	1,920	$6\frac{3}{4}$	
,, 14	Ming-Shang-Hsien	1,660	$10\frac{3}{4}$	$14\frac{1}{2}$
,, ,,	Chin-Chi-Kuan, summit of pass	2,036	4	
,, ,,	Yao-Ch'iao	—	$1\frac{3}{4}$	
,, ,,	Tung-Tze-Yuen	—	$1\frac{1}{2}$	
,, ,,	Ya-Chou-Fu	1,671	$3\frac{1}{2}$	$21\frac{1}{2}$
,, 15	Tzŭ-Shih-Li	2,004	$6\frac{1}{2}$	
,, ,,	Kuan-Yin-P'u	2,484	$3\frac{1}{4}$	$9\frac{3}{4}$
,, 16	Fei-Lung-Kuan, summit of pass	3,583	$1\frac{3}{4}$	

APPENDIX A. 421

Date		Altitude above Sea-level in Feet	Miles between	Days' Journey in Miles
1877				
July 16	Kao-Ch'iao	—	$2\frac{3}{4}$	
,, ,,	Shih-Chia-Ch'iao	2,190	$1\frac{1}{2}$	
,, ,,	Hsin-Tien-Chan	—	$1\frac{3}{4}$	
,, ,,	Yung-Ching-Hsien	2,299	$4\frac{3}{4}$	$12\frac{1}{2}$
,, 17	Ching-K'ou-Chan	2,670	7	
,, ,,	Huang-Ni-P'u	3,725	4	11
,, 18	Hsiao-Kuan	4,809	2	
,, ,,	Ta-Kuan	5,754	$2\frac{1}{2}$	
,, ,,	T'ai-Hsiang-Ling-Kuan, summit of pass	9,366	6	
,, ,,	Ch'ing-Ch'i-Hsien	5,478	$4\frac{1}{2}$	15
,, 19	Fu-Hsing-Ch'ang	3,873	—	$7\frac{1}{2}$
,, 20	Pan-Chiu-Ngai	4,279	$4\frac{1}{2}$	
,, ,,	I-T'ou-Ch'ang	4,882	$5\frac{1}{2}$	10
,, 21	Kao-Ch'iao	6,002	$3\frac{1}{2}$	
,, ,,	San-Ch'iao-Ch'êng	5,830	2	
,, ,,	Wu-Yai-Ling or Fei-Yueh-Ling, summit of pass	9,022	$5\frac{1}{2}$	
,, ,,	Hua-Lin-P'ing	7,073	$1\frac{1}{4}$	$12\frac{1}{4}$
,, 22	Lêng-Chi	4,633	—	$5\frac{1}{4}$
,, 23	Lu-Ting-Ch'iao	4,640	—	11
,, 24	Hsiao-P'êng-Pa	4,653	6	
,, ,,	Wa-Ssŭ-Kou	4,933	$6\frac{1}{2}$	$12\frac{1}{2}$
,, 25	Liu-Yang	6,570	6	
,, ,,	Ta-Chien-Lu	8,346	$5\frac{1}{4}$	$11\frac{1}{4}$
Aug. 7	Cheh-Toh	10,838	—	$7\frac{1}{4}$
,, 8	Cheh-Toh-Shan, summit of pass	14,515	$9\frac{1}{4}$	
,, ,,	Ti-Zu or Hsin-Tien-Chan	13,335	3	
,, ,,	Nah-Shi	—	$4\frac{1}{2}$	
,, ,,	An-Niang or Ngan-Niang-Pa	12,413	$4\frac{3}{4}$	$21\frac{1}{2}$
,, 9	Tung-Che-Ka	—	$4\frac{1}{2}$	
,, ,,	Ngoloh or Tung-Golo	12,027	7	$11\frac{1}{2}$
,, 11	La-Tza or Shan-Kên-Tzŭ	13,040	$5\frac{1}{4}$	
,, ,,	Ka-Ji-La or Ko-Urh-Shi-Shan, summit of pass	14,454	1	
,, ,,	Do-Kû-La-Tza, summit of pass	14,597	2	
,, ,,	Wu-Rum-Shih or Wu Ru-Chung-Ku	12,048	$5\frac{3}{4}$	14
,, 12	Ker-Rim-Bu or Pa-Ko-Lo	10,435	10	
,, ,,	Ho-K'ou or Nia-Chŭ-Ka	9,222	$8\frac{3}{4}$	$18\frac{3}{4}$
,, 13	Ma-Geh-Chung	11,971	—	7
,, 14	Ra-Ma-La, first summit	14,915	5	
,, ,,	La-Ni-Ba (hollow between mountains)	14,335	$\frac{1}{2}$	

APPENDIX A.

Date			Altitude above Sea-level in Feet	Miles between	Days' Journey in Miles
1877					
Aug. 14		Ra-Ma-La, second summit	15,110	2	
,,	,,	Mu-Lung-Gung or P'u-Lang-Kung	—	3½	
,,	,,	Lit'ang-Ngoloh or Shih-Wolo	12,451	3½	14½
,,	16	Niu-Chang	13,900	3	
,,	,,	Tang-Gola, summit of pass	14,109	½	
,,	,,	Zou-Gunda	13,235	1¼	
,,	,,	Cha-Ma-Ra-Doñ	—	2¼	
,,	,,	Deh-Re-La, summit of pass	14,584	3½	
,,	,,	Wang-Gi-La, summit of pass	15,558	6¼	
,,	,,	Ho-Chŭ-Ka	13,250	6¼	23
,,	17	Shie-Gi-La, summit of pass	14,425	8	
,,	,,	Lit'ang	13,280	5	13
,,	19	Che-Zom-Ka (bridge over Li-Chŭ river	—	5½	
,,	,,	Jiom-Bu-T'ang	14,718	6½	12
,,	20	Nga-Ra-La-Ka, summit of pass	15,753	4½	
,,	,,	Lake Cho-Din	—	1½	
,,	,,	Dzong-Da	14,896	2¼	
,,	,,	Source of Ye-Chŭ river	—	3¾	
,,	,,	Ma-Dung-La-Tza	—	2¾	
,,	,,	La-Ma-Ya or Ra-Nung	12,826	5¼	20
,,	21	Ye-La-Ka, summit of pass	14,246	1½	
,,	,,	Bridge over Dzeh-Dzang-Chŭ river	13,162	1¾	
,,	,,	Cha-Chŭ-Ka (hot springs)	—	¼	
,,	,,	Mang-Ga-La, summit of pass	13,412	1¼	
,,	,,	La-Ka-Ndo	—	2	
,,	,,	Nen-Da	13,133	3¾	10½
,,	22	Yunnan-Chiao	—	5	
,,	,,	Ra-Ti or San-Pa	13,794	6	11
,,	23	Rung-Se-La or San-Pa-Shan, summit of pass	15,796	5¾	
,,	,,	Ta-Shiu or Ta-So	13,347	10¼	16
,,	24	J'rah-La-Ka or Ta-So-Shan, summit of pass	16,568	5¼	
,,	,,	Pun-Jang-Mu or Pung-Cha-Mu	13,158	8	13¼
,,	25	Ba-Jung-Shih or Hsiao-Pa-Chung	10,691	6	
,,	,,	Bat'ang or Ba	8,546	6½	12½
,,	29	Ch'a-Shu-Shan or Cha-Keu, summit of pass	9,388	5	
,,	,,	Niu-Ku	—	3	
,,	,,	Leh or Choui-Mao-Kiu, monastery	—	¾	
,,	,,	Chu-Ba-Lang or Chru-Ba-Long	8,165	10	18¾

Date		Altitude above Sea-level in Feet	Miles between	Days' Journey in Miles
1877				
Aug. 30	Ferry over Chin-Sha-Chiang .	—	3	
,, ,,	Gue-Ra	8,660	4¾	
,, ,,	Kong-Tze-La-Ka, summit of pass	11,972	7	
,, ,,	Kong-Tze-Ka	11,675	¾	15½
,, 31	Mûng-M'heh or Chung-Mong-Li	12,189	6¼	
,, ,,	Jang-Ba or Pa-Mu-T'ang . .	12,793	3¾	
,, ,,	Kia-Ne-Tyin	13,135	9	19
Sept. 1	Dzung-Ngyu	10,792	—	10¼
,, 2	Boah-Tsa	—	5¾	
,, ,,	Nieh-Ma-Sa	10,868	15¾	21½
,, 3	Ma-Ra	11,505	2½	
,, ,,	Tsa-Leh	12,690	6	8½
,, 4	Tsa-Leh-La-Ka, summit of pass .	15,788	6½	
,, ,,	Jieh-Kang-Sung-Doh. . .	—	1¾	
,, ,,	Lûng-Zûng-Nang . . .	12,684	4¼	12½
,, 5	Dong	9,000	15	
,, ,,	Jo-La-Ka, summit of pass . .	12,389	6½	
,, ,,	A-Tun-Tzŭ or N'geu . . .	11,029	1	22½
,, 9	Jing-Go-La, summit of pass .	12,300	3¼	
,, ,,	Pung-Gien-Tiyin (guard hut) .	—	5¾	
,, ,,	Pa-Ma-La, summit of pass. .	14,307	2½	
,, ,,	Mien-Chu-La, summit of pass .	14,227	3½	
,, ,,	Shwo-La, summit of pass . .	14,307	2½	
,, ,,	Deung-Do-Lin-Ssŭ or Tung-Chu-Ling	9,335	12	29½
,, 11	Sha-Lu	9,287	—	13
,, 13	Jing-Go-La, summit of pass .	13,699	6½	
,, ,,	Ka-Ri	9,610	6½	13
,, 14	Bridge over Chiu-Chŭ river .	—	4½	
,, ,,	Shieh-Zong	—	5½	
,, ,,	N'doh-Sung	7,417	2½	12½
,, 15	Sa-Ka-Tying	7,075	10½	
,, ,,	Bridge over Kung-Chŭ river .	—	4¼	
,, ,,	Ron-Sha	6,916	2¾	17½
,, 16	Rûng-Geh-La-Ka, summit of pass	12,134	7½	
,, ,,	Shio-Gung or Hsiao-La-Pu .	—	5¾	
,, ,,	Ta-Chio or La-Pu . . .	6,777	2	15¼
,, 17	Jie-Bu-Ti or Chi-Dzung . .	6,621	11	
,, ,,	Lu-Jiong or Wai-Ta-Chen . .	6,647	3½	14½
,, 18	Mu-Khun-Do or Hsiao-Ken-To	—	10½	
,, ,,	Ku-Deu or Chi-Tien . . .	6,200	9¼	19¾
,, 20	Pai-Fên-Ch'iang	—	10½	
,, ,,	Ch'iao-T'ou	—	5¾	
,, ,,	Tz'ŭ-Kua	6,645	4¼	20½

Date		Altitude above Sea-level in Feet	Miles between	Days' Journey in Miles
1877				
Sept. 21	San-Hsien-Ku	6,390	10¼	
,, ,,	Shih-Ku	5,952	10¼	20½
,, 22	Chin-Ku-P'u, summit of pass .	8,391	5½	
,, ,,	Chiu-Ho	7,565	10¼	15¾
,, 23	Chien-Ch'uan-Chou . . .	7,489	—	8
,, 24	I-Yang-T'ang	8,681	10	
,, ,,	T'ai-P'ing-Chên.	—	6½	
,, ,,	Niu-Chieh	7,113	3½	20
,, 25	Lang-Ch'iung-Hsien . . .	6,970	—	10
,, 26	Yu-So	6,758	9½	
,, ,,	Têng-Ch'uan-Chou . .	6,573	2½	12
,, 27	Shang-Kuan	—	4½	
,, ,,	Wan-Ch'iao	—	10	
,, ,,	Ta-Li-Fu	6,666	7½	22
Oct. 4	Chua-Yao	—	3¼	
,, ,,	Hsia-Kuan	6,544	4	7¼
,, 5	T'ang-Tzŭ-P'u	—	2¼	
,, ,,	Shih-Ch'uan-Shao . . .	—	3¼	
,, ,,	Mao-T'sao-T'ang . . .	—	1¾	
,, ,,	Hsiao-Ho-Chiang . . .	—	2	
,, ,,	Ho-Chiang-P'u	5,196	1	10¼
,, 6	Chi-I-P'u	—	1¾	
,, ,,	Chin-Niu-Tun	—	1¾	
,, ,,	Ma-Ch'ang	—	2½	
,, ,,	Yang-Pi	5,299	3½	
,, ,,	Pei-Mên-P'u	—	2¼	
,, ,,	Ch'ing-Shui-Shao, summit of pass	8,233	2	
,, ,,	Ch'ing-Shui-Shao, village . .	—	3	
,, ,,	T'ai-P'ing-P'u	6,624	2½	19¼
,, 7	Tou-Po'-Shao	—	2	
,, ,,	Niu-P'ing-P'u	5,783	2¾	
,, ,,	Bridge over the Shun-Pi river .	5,238	2	
,, ,,	Huang-Lien-P'u	5,420	2½	9¼
,, 8	Chiao-Kou-Shan	—	2	
,, ,,	Pai-T'u-P'u	—	2½	
,, ,,	Wan-Sung-An	—	2	
,, ,,	T'ien-Ching-P'u.	8,148	1¾	
,, ,,	Sung-Shao	—	1¾	
,, ,,	Mei-Hua-P'u	—	2	
,, ,,	P'ing-Man-Shao	—	1¾	
,, ,,	Hei-Yu-Kuan	—	2¾	
,, ,,	Ch'ü Tung	5,555	3	19½
,, 9	T'ieh-Ch'ang	—	4	

APPENDIX A.

Date		Altitude above Sea-level in Feet	Miles between	Days' Journey in Miles
1877 Oct. 9	Hsiao-Hua-Ch'iao	—	$1\frac{1}{2}$	
,, ,,	Ta-Hua-Ch'iao	—	1	
,, ,,	T'ien-Ching-P'u or Hua-Ch'iao range, summit	8,229	2	
,, ,,	Yung-Kuo-Ssü-T'ang	—	$\frac{1}{2}$	
,, ,,	Sha-Yang or Sha-Mu-Ho	5,145	$4\frac{1}{4}$	$13\frac{1}{4}$
,, 10	Yung-Fêng-Chuang	—	$2\frac{1}{2}$	
,, ,,	Summit of ridge	5,432	$\frac{1}{2}$	
,, ,,	Lan-Ts'ang-Chiang or Mekong river	3,953	$\frac{1}{2}$	
,, ,,	P'ing-P'o	—	1	
,, ,,	Shui-Chai	6,569	$1\frac{3}{4}$	
,, ,,	Ta-Li-Shao	7,412	$2\frac{1}{2}$	$8\frac{3}{4}$
,, 11	T'ien-Ching-P'u, summit	7,795	$1\frac{1}{2}$	
,, ,,	Kuan-P'o	—	$2\frac{1}{4}$	
,, ,,	Pan-Ch'iao	5,692	$4\frac{1}{4}$	
,, ,,	Yung-Ch'ang	5,645	$4\frac{3}{4}$	$12\frac{3}{4}$
,, 14	Wo-Shih-Wo	—	$3\frac{1}{2}$	
,, ,,	Lêng-Shui-Ching	—	4	
,, ,,	Summit of ridge	7,733	1	
,, ,,	Hun-Shui-T'ang	5,874	1	
,, ,,	P'u-P'iao	4,711	4	
,, ,,	Fang-Ma-Ch'ang	4,910	$4\frac{3}{4}$	$18\frac{1}{4}$
,, 15	Ta-Pan-Ching	—	$2\frac{1}{4}$	
,, ,,	Lu-Chiang or Salwen river	2,620	$7\frac{1}{2}$	
,, ,,	Lu-Chiang-Pa	—	$\frac{1}{2}$	
,, ,,	Ho-Mu-Shu	5,486	$3\frac{3}{4}$	$13\frac{3}{4}$
,, 16	Hsiang-P'o	6,924	$2\frac{1}{2}$	
,, ,,	Summit, Kao-Li-Kung range	8,129	2	
,, ,,	Tai-P'ing-P'u	7,538	2	
,, ,,	Ta-Li-Shu	—	3	
,, ,,	Lung-Chiang or Shuay-Li river	4,502	2	
,, ,,	Kan-Lan-Chan	5,029	$1\frac{1}{2}$	13
,, 17	Kan-Lu-Ssŭ	—	$3\frac{1}{2}$	
,, ,,	Ch'in-Ts'ai-T'ang	7,082	$2\frac{1}{2}$	
,, ,,	Urh-T'ai-P'o, summit	7,500	$1\frac{3}{4}$	
,, ,,	Li-Chia-P'u	—	3	
,, ,,	T'êng-Yüeh-T'ing or Momein	5,489	2	$12\frac{3}{4}$
,, 20	Shuayduay	—	$5\frac{1}{2}$	
,, ,,	Hsiao-Ho-Ti	4,178	$6\frac{3}{4}$	
,, ,,	Nan-Tien	—	8	
,, ,,	Che-Tao-Ch'êng	3,625	2	$22\frac{1}{4}$
,, 21	Mawphoo	—	$8\frac{3}{4}$	

Date		Altitude above Sea-level in Feet	Miles between	Days' Journey in Miles
1877				
Oct. 21	Nahlow	—	$4\tfrac{3}{4}$	
,, ,,	Kan-^{Ng}ai or Muangla	2,957	4	$17\tfrac{1}{2}$
,, 22	Namon	—	$8\tfrac{1}{2}$	
,, ,,	Chan-Ta or Sanda	2,945	5	$13\tfrac{1}{2}$
,, 23	T'ai-P'ing-Chieh or Karahokah	—	8	
,, ,,	Man-Yün or Manwyne	2,647	10	18
,, 29	Pung-Shi or Ponsee	3,584	—	$9\tfrac{1}{4}$
,, 30	Lakong	—	6	
,, ,,	Confluence of the Nampoung river with the Taping	1,198	$8\tfrac{1}{2}$	
,, ,,	Ponline	—	5	
,, ,,	Near Ma-Mou	—	14	$33\tfrac{1}{2}$
,, 31	Ma-Mou or Sicaw	462	—	6
Nov. 1	Haelone	—	9	
,, ,,	Shuaykeenah	—	$8\tfrac{1}{2}$	
,, ,,	Bhamo	430	$3\tfrac{1}{2}$	21
	Total			1,110

APPENDIX B.

LIST OF THE MOST IMPORTANT BOTANICAL SPECIMENS BROUGHT HOME AND NAMED IN THE BOTANICAL DEPARTMENT OF THE BRITISH MUSEUM.

Gathered at Hsin-Tien-Chan:

Delphinum grandiflorum: Fisch. Plant used for killing lice. See vol. ii., p. 126.

Stellera—conf. S. concinna: Edge. Plant used for making parchment. See vol. ii., p. 126.

Gathered at La-Ni-Ba on Mount Ra-Ma-La:

Spenceria Ramalana. A new genus of Rosaceæ. Described by Mr. Trimen, M.B., F.L.S., in the 'Journal of Botany,' No. 196, with a plate.

Saxifraga Gillii: Trimen. A new species.
Primula Gillii: Britten. A new species.
Pedicularis Ramalana: Britten. A new species.

APPENDIX C.

CHINESE OFFICIAL RANK.

THE following table of the rank of Chinese officials is partly deduced from the 'Desultory Notes' of T. T. Meadows, partly from information furnished me by Mr. Mesny, and from other sources; but the difficulties of acquiring this kind of information are such that I can hardly hope that the table even now is altogether free from error.

The civil and military officials have been placed side by side to give some idea of relative rank. It must not, however, be assumed that the table can be taken as an absolute Table of Precedence.

The officials of different ranks are chiefly to be distinguished by the balls or globules worn on their official hats. These are often styled '*buttons*' by the English. They are, in fact, spherical in shape, and generally about an inch in diameter. As they button nothing, the term does not seem altogether appropriate.

The ball, however, is very deceptive. High officials are frequently deprived of their proper one, while low officials constantly are entitled to wear, and do wear, a ball of a higher rank than their own.

It is almost impossible to translate the military titles, or give the comparative European and Chinese ranks, for circumstances are so different, and duties so unlike those of a European officer, that it would be little better than misleading to attempt it.

A European Consul ranks with a Tao-T'ai, and a Tao-T'ai ranks about with a Tu-Ssŭ, and that is all that can be said absolutely with regard to relative rank.

The Chinese, moreover, generally hold military officials in contempt, and the command of an army in time of war is frequently given to a civilian; for (the Chinese argue) as the military officials have not passed the literary examinations, the literati, whose talents have been tested, and who at all events may have read about war, are more fit to make war than an uneducated military official.

CIVIL OFFICIALS.	MILITARY OFFICIALS.
According to some authorities the four Cabinet Ministers and members of the great council of the nation are the only Civil Officials of the first class first rank. But the Heads or Presidents, 'Shang-Shu,' of the six boards, 'Liu-Pu,' and the	CHIANG-CHÜN. An official of the first class first rank, and has generally some civil rank in addition to his military rank. He is nearly always at the capital of the province. He wears a carbuncle in

CIVIL—cont.

Governors-General of provinces, Tsung-Tu, are generally considered as belonging to the first rank of the first class. Meadows, in his 'Desultory Notes,' puts them in the first class, but does not say to which rank they belong. Archdeacon Grey, who appears to have studied the subject, and Mr. Mesny put them in the second class. The Tsung-Tu are the highest officials in the provinces. They are the direct representatives of the Emperor, and have the power of life and death.

The proper 'ball' for the first class first rank is a carbuncle in court, and a coral in ordinary dress.

LIU-PU-SHANG-SHU.

The Heads of boards, are of the first or second rank of the first class. They wear the ball of the first rank as above, or else a coral ball in court dress, and a red ivory or porcelain ball in ordinary dress.

TSUNG-TU OR CHIH-T'AI.

These are the Governors-General of provinces. They sometimes have two provinces under them, and sometimes only one.

They are generally members of the Board of War, and such are superior to the T'i-Tu. They reside at the capital of the province. These are sometimes styled Viceroys by Europeans.

FU-T'AI OR HSUN-FU.

These are the Governors of provinces under the Chih-T'ai. When a Chih-T'ai governs two provinces, the Chih-T'ai would

MILITARY—cont.

court dress, and a coral ball in ordinary dress.

TU-T'UNG.

These have almost equal rank with the Chiang-Chün. They are generally in command of Tatar troops at the capitals of provinces. They generally have some civil rank in addition to their military rank, and wear a carbuncle ball in court, and a coral ball in ordinary dress.

T'I-TU.
Second Class, First Rank.

He is the chief military official in a province where there are no Tatar troops. If afloat

APPENDIX C. 429

CIVIL—cont.

have two Fu-T'ai under him, residing at the capitals of the two provinces.

He is of the second class first rank. He may be a member of the Board of War, and as such would rank with a T'i-Tu. He is entitled to a red ball.

FAN-T'AI OR PU-CH'ÊNG-SSŬ.
Second Class, Second Rank.

Superintendent of the finances of a province. Resides at the capital of the province. He wears a red ball.

NIEH-T'AI OR NGAN-CH'A-SSŬ.
Third Class, First Rank.

Provincial Judge residing at the capital of a province. Clear light blue ball.

YEN-YÜN-SSŬ.
Third Class, Second Rank.

He is the Provincial Salt Commissioner. Wears a clear blue ball rather darker than that of a Nieh-T'ai.

LIANG-T'OU OR LIANG-T'AI.
Fourth Class, First Rank.

Provincial Grain Collector. Wears the same ball as the Yen-Yün-Ssŭ, or else an opaque blue ball.

MILITARY—cont.

his duties are those of an Admiral, for the Chinese make little distinction between the services. He wears a coral ball in court dress, and a red ivory or porcelain ball in ordinary dress.

CHÊN-T'AI.
Second Class, First Rank.

He has command of Chinese troops only. He is in command of a district. There may be as many as six in one province. He wears a coral ball in court dress, and a red ivory or porcelain ball in ordinary dress.

HSIEH-T'AI.
Second Class, Second Rank.

There are usually several of these under a Chên-T'ai. He wears a red ball.

TSAN-CHIANG.
Third Class, First Rank.

Wears a clear light blue ball.

FU-CHIANG.
Third Class, Second Rank.

He wears a clear blue ball rather darker in colour than that of a Tsan-Chiang.

YU-CHI.
An Officer of the Third Class, Second Rank.

And almost the same thing as the Fu-Chiang. Wears a clear blue ball a little darker than that of a Tsan-Chiang.

CIVIL—*cont.*
TAO-T'AI.
Fourth Class, First Rank.

District Intendent of a circuit; he has generally under him three principal towns of the rank of Fu.
He lives at the chief of these, and wears a dark opaque blue ball. An English Consul ranks with a Tao-T'ai.

CHIH-FU, GENERALLY CALLED A FU.
Fourth Class, Second Rank.

Prefect of a department, and wears a dark opaque blue ball.

CHIH-LI-CHIH-CHOU.
Fifth Class, First Rank.

Prefect of an inferior department, and wears a clear crystal ball.

CHIH-LI-TUNG-CHIH, OTHERWISE CALLED T'ING.
An Officer of the Fifth Class.

He is an independent Sub-Prefect of inferior departments. These inferior departments are generally not larger than the sub-divisions of the larger departments called Fu and Chih-Li-Chou; but independent sub-prefects are stationed in them on account of circumstances which make the administration of affairs unusually difficult.
He wears a clear crystal ball.

TUNG-CHIH.
Fifth Class.

Sub-Prefect of department. Wears a clear crystal ball.

CHIH-CHOU, GENERALLY CALLED A CHOU.
Fifth Class.

He is a district magistrate, and wears a clear crystal ball.

MILITARY—*cont.*
TU-SSŬ.
Fourth Class, First Rank.

Wears a dark opaque blue ball.

SHOU-PEI.
An Officer of the Fifth Class,

And wears a clear crystal ball.

CHIEN-TSUNG.
Fifth Class.

Wears a clear crystal ball.

APPENDIX C. 431

CIVIL—*cont.*
TUNG-PAN.
Sixth Class.

Deputy Sub-Prefect of department. Wears a white porcelain ball.

CHIH-HSIEN, GENERALLY CALLED A HSIEN.
Seventh Class.

He is a District Magistrate, and wears a gilt ball.

HSIEN-CHENG.
Eighth Class.

Assistant District Magistrate, and wears a gilt ball with flowers in relief.

CHU-PU.
Ninth Class.

Township Magistrate. Wears a gilt ball.

HSUN-CHIEN.
Ninth Class.

Also a Township Magistrate, and wears a gilt ball.

MILITARY—*cont.*
PA-TSUNG.
Sixth Class.

Wears a white porcelain ball.

WAI-WEI.
Eighth Class.

Wears a gilt ball.

NGÊ-WAI.
Ninth Class

Wears a gilt ball.

MA-PING.

A horse soldier.

PU-PING.

A foot soldier.

INDEX.

ABE

ABEL-RÉMUSAT, i. 125
    Accident to ship, how repaired, i. 254, 255
Accounts, eccentric system of Chinese, ii. 192
Adam's Point on the Yang-Tzŭ-Chiang, ii. 416
Agricultural and Horticultural Society of India, extract from Journal of, ii. 62-64
Agriculture, Chinese, i. 113, 114
Alarm, a false, ii. 209, 210
Ambassadors, rank of Chinese, ii. 44; the Lama, 85
Amien of Marco Polo, ii. 343, 384
Anderson, Dr., ii. 304n, 344, 370n, 371n, 373, 374, 381, 389
ANIMALS AND BIRDS.
    Bear, i. 348, ii. 71, 173, 181
    Boar, wild, i. 169, 338
    Buffalo, i. 414, ii. 266, 324
    Cormorant, ii. 380
    Deer, i. 164, 227, 338, 348, 370, 377, 389, ii. 127, 149
    Dog, Tibetan, i. 382, ii. 151, 247
    Donkey, i. 63, 64
    Ducks, wild, ii. 380
    Geese, wild, i. 192, ii. 106
    Goat, wild, i. 377, ii. 127
    Hare, i. 377, ii. 127
    Horse, i. 26, 34, 46, 85, ii. 108, 114, 115, 133, 358
    Jay, ii. 142
    Magpie, i. 55, ii. 380
    Monkey, ii. 173, 181, 217
    Mule, i. 23, 24, 46, 47, ii. 133, 314
    Musk-deer, i. 348, 377, ii. 149
    Oxen, wild, ii. 220
    Panther, ii. 136, 173, 324
    Parrot, ii. 52, 135, 140, 180, 264
    Partridge, i. 384
    Pelican, ii. 294
    Pheasant, i. 164, 227, 367, 370, 377, 381, 389, ii. 142, 322

VOL. II.

BAM

    Pigeon, description of, i. 373
    Sheep, wild, i. 377, ii. 127
    Squirrel, ii. 140
    Tiger, i. 6, 7, ii. 173
    Wolf, i. 79, ii. 172, 173
    Yak, i. 369, 370, 381, ii. 81, 99, 103, 104, 121, 122, 133, 143, 149, 194
An-Niang, ii. 127, 421
Antiquities of Ch'êng-Tu, ii. 17-21
Ants, perils from, ii. 401
April, 1877, i. 253-286, ii. 417, 418
Aqueduct at Kuei-Chou, i. 238, 239
Arches, triumphal, i. 290, 291, 414
Army, character of Chinese, i. 153, 154
Art, realistic, i. 291
Asian problem, Chinese a factor in, i. 145
Asiatic Society, the Royal, i. 129
'As others see us,' translation of Chinese poem, i. 265
Aspect of mountains, sublime, ii. 173
Atenze of Cooper, (A-Tun-Tzŭ), ii. 230
A-Tun-Tzŭ, ii. 8, 423; position of, 206; arrival at, 230; evil character of, 232; form of houses in, 233
August, 1876, i. 5, 6; 1877, ii. 99-216, 421-423
'Ava,' steamer, i. 4, 10, ii. 313

BA, (a Tibetan encampment) ii. 151; Ba or Bat'ang, ii. 422
Baber, Mr., i. 159, 211, 212, 278, ii. 304n, 319n, 326, 327, 344, 345, 359n; arrives at Ch'ung-Ch'ing, i. 260
Baggage arrangements, i. 277
Ba-Jung-Shih, ii. 182, 422
Balista, i. 148
Ballet, a Burmese (Po-é), ii. 406
Bamboos, the young shoots eaten, i.

F F

434   INDEX.

BAN

337; the first seen, ii. 227; the first large one, 297; size of, 380, 397; uses of, 397
Banduk, i. 148
Bandukdar, i. 148
Banking transactions, ii. 26, 28-31
Bank-notes, early use of, in China, i. 148
Barber's shops, i. 415
Barrow, Mr., remark by, on Chinese industry, i. 247
Bat'ang, i. 165, ii. 422; start for, 8; account given of road to, 89; welcome on arrival at, 183; Huc's derivation of name, 186; plain of, 187; river of, 188; government of, 192; pleasant stay at, 199; departure from, 206; descend river of, 208
Beads used for payments, ii. 77
Beans, how prepared by Chinese, i. 68
Bees, i. 231, 381
Bell, the great, at Peking compared with other large bells, i. 131, 132
Bendocquedar, i. 148
Bhamo, ii. 426; arrival at, 409; departure from, 410
Biet, Monsieur, now Bishop of Ta-Chien-Lu, ii. 206
BIRDS, see ANIMALS
Black Man-Tzŭ, (Ju-Kan), i. 365
Blakiston, i. 194, 256
Boah-Tsa, ii. 423
Boat engaged for navigating Upper Yang-Tzŭ, description of, i. 180, 181; boats on the Pei-Ho, i. 21
Boccaccio on plague at Florence, 1348, ii. 346
Bogle, Mr., i. 268
Bonze, ii. 189
Boone Reach on the Yang-Tzŭ-Chiang, ii. 416
Botanical specimens, list of, ii. 426
Bowls, the wooden (Pu-Ku), of Tibetans, ii. 113; superstition concerning, 233
Boxes, how packed for travelling, i. 159, 160n
Brassey, Mrs., i. 126
Bread, little used by Chinese, i. 48; a treat of, ii. 320
Breakfast, invitations to, ii. 53; a delightful *al fresco*, 124
Brick Tea, manufacture of, i. 176-178
Bridge, a suspension, i. 335, 391; a haul, 359, 360; a chain, 397; a stone, 399; covered bridge of Marco Polo, 418; a handsome, ii. 42; a suspension, 55, 69, 317,

CHA

320, 330, 352; a bridge swept away, 66; covered bridge of Marco Polo not to be found, 9, 289
Brine-bend in the Yang-Tzŭ-Chiang, ii. 416
Brine-wells of Tzŭ-Liu-Ching, i. 298-302
Brodie, Mr. R. C., ii. 64
Buddha, i. 130
Building, peculiarities of Chinese, i. 392; Tibetan mode of, ii. 167, 168
Bullies, perils from, ii. 401
Bunds, i. 402
Burmah, road from Ta-Li-Fu to, ii. 207; capital of (Pagan), 345
Bushell, Dr., i. 171
Buttered-tea churn, the, ii. 125
Butterflies, Chinese belief concerning, ii. 13; brilliant, 397
Button, the Chinese, ii. 366, 427
Byzantium, Imperial, i. 143

CACHAR MODUN, i. 111
Calico, ii. 330
Caliphate, the, i. 143
Camel Reach in the Yang-Tzŭ-Chiang, i. 188, ii. 416
Canal, the Grand, of China, description of, i. 166-168
Candles, Chinese, i. 149
Carajan of Marco Polo (Ta-Li-Fu), ii. 299
Carles, Mr., i. 14, 24, 26, 46
Carts of Northern China, i. 22, 23, 46
Cash, rate of exchange of, i. 270
Caspian Sea, the, i. 139, 143
Cathay, a cycle of, i. 139, 144n, ii. 384
Cattle plague at Ngoloh, ii. 132
Cave habitations, i. 230, 231; a miracle cave, 389
Cen-Chien, the 'Sun-Kiouen' of De Mailla, ii. 19
Ceremonies, Chinese, ii. 200, 233, 234
Cha-Chŭo-Ka, ii. 165, 422
Cha-Chŭo-T'ing, i. 354
Ch'a-Erh-Ngai, i. 364, ii. 419
Cha-Keu, ii. 208, 422
Cha-Ma-Ra-Doñ, ii. 150, 422
Chang, ten Chinese feet, ii. 44n
Chang-Chia-Wan, i. 41, ii. 413, 415
Chang-Fi, ii. 18
Chang-Gan, i. 142
Chang-Ming-Hsien, i. 404, ii. 419
Ch'ang-Ning-P'u, ii. 419
Chang-Shien-Chung, a brigand, ii. 6
Chang-Shou-Pao (Long-lived Gem), a Ma-Fu, ii. 103, 141, 153, 179

INDEX. 435

CHA

Chang-Tu-Sze, ii. 157
Chang-Yai-Chuang, i. 119, ii. 414
Chan-Ta (Sanda of Anderson), ii. 371, 426
Chan-To-P'u, ii. 420; arrival at, 374
Chao-Chiao-P'u, ii. 420
Chao-Ho-Ching, ii. 319; possibly Chao-Chou, 319*n*
Chao Ta-Laoye,' ii. 183; visit from, 184; hospitality of, 200; accident to, 216; farewell visit to, 233; absence of, regretted, 243
Chao-Tsiao, ii. 19
Charm boxes, ii. 112, 113
Ch'a-Shu-Shan, ii. 422
Ch'a-Tien-Tzŭ, i. 314, ii. 418
Chauveau, Monseigneur, Bishop of Ta-Chien-Lu, i. 268, ii. 51, 75, 78, 79; kindly aid from, 93, 102, 109; noble character of, 111, 112
Cheh-Toh, the ' Jeddo,' of Cooper, ii. 119, 120, 421
Cheh-Toh-Shan, ii. 122, 421
Chên-Chiang-Kuan, ii. 419
Chên-Fan-P'ao, ii. 419
Chêng-P'ing-Kuan, ii. 419
Ch'êng-Tu-Fu, ii. 418, 420; plain of, i. 314; arrival at, 316; take house at, 320, 321; departure from, for Li-Fan-Fu, 325; return to, 420; described by Marco Polo, ii. 1, 2; by Padre Martin Martini, 3–5: description of modern city, 4–6; probable changes at, 5, 6; historical changes at, 6, 7; wal's of, 13; ' Literary Book Hall Temple ' of, 13–17; antiquities of, 17–21; preparing to start from, 31, 32; departure from, for Ta-Chien-Lu, 33; concluding notice of,—extracts from Ritter and Martini, 34–36
Ch'êng-Yuan, i. 386, ii. 419
Chên-T'ai, military official, i. 316 ii. 22, 429; visit from, i. 270, 271
Che-Tao-Ch'êng, ii. 425
Che-Zom-Ka, ii. 159, 422
Chiamdo, road to, ii. 123
Chiang, meanings of, i. 148
Chiang-Chün, military official, ii. 427
Chiang-Kan river, ii. 415
Chiang-K'ou, ii. 416
Chiang-Pei-Ling, persecution of Christians at, i. 267
Chiang-Yu-Hsien, i. 402, ii. 419
Chiao-Kou-Shan, ii. 424
Ch'iao-T'ou, ii. 271, 423
Chi-Chiang, ii. 416
Chi-Chien-Fang, i. 78, ii. 414

CH'I

Chi-Chou, i. 59, ii. 413
Chi-Dzûng (Jie-Bu-Ti), ii. 264, 423
Chien-Chou, i. 312, ii. 418
Chien-Ch'uan-Chou, ii. 424; arrival at, 279; fair words of the Chou of, 281; departure from, 282; plain of, 283
Chien-Li-Hsien-Ma-Tou, i. 189, ii. 416
Chien-Lung, emperor, i. 74
Chien-Tsung, military official, ii. 247, 430
Chi-Fu, i. 10, 13, 15, 16, 18, 26, 138; convention of, 14, 178; good effects of convention, ii. 362
Chih-Chou, civil official, ii. 430
Chih-Fu, civil official, ii. 97, 430
Chih-Hsien, civil official, ii. 431
' Chih-Li,' American steamer, i. 15; on mud bank, 21, 25; afloat again, 31
Chih-Li-Chih-Chou, civil official, ii. 430
Chih-Li-Tung-Chih, civil official, ii. 430
Chih-Pa-K'ou, ii. 416
Chih-T'ai, civil official, ii. 428
Chi-I-P'u, ii. 424
Chiká, i. 360*n*
China, travelling in, i. 24; ponies of, 24; climate of North of, in October, 80
Chin-Chi-Kuan, ii. 420
Chinese, gambling of, i. 28, 80; courage of, 30, 153; curiosity of, 64, 101; dirty habits of, 91; farming, 114; cradle of the Chinese nation, 139; absence of imagination and originality among, 146, 147; industry of, 246-248; their care of the dead, 249; commercial probity of, 275; their want of ideality, 291; relentless advance of, 342, 343; superstition of, 131, 181, 204, 229, 250, 266, 272, 343, 344, 358, 380, 382, 389, ii. 13, 38, 56; their dislike of light, i. 388; moderate drinking of, 400; religion of, ii. 10; honesty of, 33, 270; their ideas of foreigners, 40; contrast between, and Tibetans, 89–91; civility of Chinese officers, 95; Chinese etiquette, 200, 234; rank of Chinese officials, 427
Ch'ing-Ch'i-Hsien, ii. 58, 59, 421
Ching-K'ou-Chan, ii. 53, 421
Ch'ing-She-K'ou, i. 229, 416
Ch'ing-Shui-Shao, ii. 317, 424

F F 2

CH'I

Ch'ing-Tan, arrival at, i. 215; rapid of, 216, 219, ii. 416
Chin-Kiang, i. 166, ii. 19
Chin-King-Sze, ii. 13
Chin-K'ou, ii. 415
Chin-Ku-P'u, Mt., ii. 275, 424
Chin-Niu-Tun, ii. 424
Chin-Sha-Chiang (River of Golden Sand), i. 165, 256, ii. 122, 154; approach to, 180; Bat'ang river joins it, 188; no road down valley of, 207; valley of, 209; cross and quit it, 211, 423; in view again, 246; river regained, 264; breadth of, 265; swollen by rains, 274; final departure from, 274
Chin-Tai, a servant, engagement of, i. 26; discovers a weakness for cookery, 41; wrangles with innkeeper, 112; in good spirits, 135; instructed in bread-making, 171; consults doctors, 271; tales of, 382; tempers of, ii. 87; his method of packing, 121
Ch'in-Ts'ai-T'ang, ii. 359, 425
Chin-Yen-Ch'iao, a bridge, ii. 9
Ch'i-Shan, i. 128
Chi-Tien (Ku-Deu), ii. 269, 423
Chiu-Ch'êng (Old Town), ii. 362
Chiu-Chou, i. 398, ii. 419
Chiu-Chŭ river, ii. 253, 258, 423
Chiu-Ho, ii. 277, 424
Ch'iung-Chou, ii. 41, 420; unfinished buildings at, 42
Chi-Zom-Ka, a bridge, ii. 159, 422
Cho, or Thso, note on, by Colonel Yule, ii. 163
Cho-Din (The Sea), Lake, ii. 163, 422
Chou dynasty, i. 128
Chou, civil official, i. 316, 361, ii. 280, 430
Choui-Mao-Kiu, ii. 422
Christians, persecution of, at Chiang-Pei-Ling, i. 267
Chru-Ba-Long, ii. 422
Chuan-Kou, i. 182, ii. 415
Chua-Yao, ii, 424
Chu-Ba-Lang, ii. 210, 422
Chu-Ko-Liang (Kung-Ming-Sien-Sen), ii. 18, 19, 20
Chung, ii. 417
Ch'ung-Chiang, valley of the, i. 313
Ch'ung-Ching, ii. 417; hiring boats for, i. 176; arrival at, 260; price of provisions at, 270; departure from, 278
Chung-Chou, i. 258, ii. 417
Chung-Erh, a servant, i. 171; escapes

CUR

an accident, ii. 228; a march stolen on, 306
Chung-Mong-Li, ii. 423
Chung-Pa-Ch'ang, ii. 419
Chu-Pu, civil official, ii. 431
Chŭ-Sung-Dho (meeting of three waters), the 'Jessundee' of Cooper, ii. 219
Ch'ü-Tung, ii, 324, 325, 327, 424
Clothing of people in summer, i. 410; of Tinc-Chais, 416
Coal districts, at Kai-Ping, i. 103; on the Yang-Tzŭ-Chiang, 222
Coffins, Chinese, i. 167
Cold, preparations to meet, ii. 91, 92
Compass affected by rocks, ii. 177
Compliments, exchange of, ii. 97, 98, 234
Compradore, a, i. 155
Compressed air used to produce fire, ii. 395
'Comptes Rendus,' the, ii. 63
Concert, a Chinese, i. 240
Confucius, doctrine of, i. 142, 149
Consulate at Ch'ung-Ch'ing proclaimed, i. 269
Coolies, elaborate contract made with, i. 275; hiring of, ii. 32; habits of, 32, 33; quarrel among, 40; loads of, 43, 47; tax on, 71; gambling of, 80
Cooper, the late Mr., ii. 51, 54; relics of, 100; 'Jeddo' of, 119, 122; Vermilion river of, 218; 'Jessundee' of, 219; 'Atenze' of, 230; kept in ignorance of road, 242; description of the Lan-Ts'ang-Chiang, by, 330; letter from, 388; messenger arrives from, 405; sends boat for us, 408; meeting with, 409; are entertained by, 409, 410; death of, 410
Copper mines at Huang-Ni-P'u, ii. 55
Corea, i. 80
Costume of self, i. 36
Cotton manufacture, preparation of warp, ii. 297
Couriers, Imperial, ii. 230
Cow's Mouth, cave of the, i. 224
Cradle of the Chinese nation, i. 139
Criticism, personal, ii. 39
Crystal Village, the, (Shiu-Ching-P'u), i. 392
Curiosity, frantic, at Kuan-Hsien, i. 332, 333
Currency, ii. 77, 78
Curtains, travelling, of Chinese, ii. 8, 9

INDEX. 437

DAL

DALAI-LAMA, the, ii. 82, 83
 Davenport, Mr., ii. 319n
Dead, Chinese care of, i. 249
Decoration, domestic, ii. 254, 255
Deh-Re-La, ii. 151, 422
Desflêches, Monseigneur, Bishop of Ch'ung-Ch'ing, i. 261, 267, 317, ii. 64
Desgodins, Monsieur, letter received from, ii. 175; kindness of, 183, 184; a visit to, 185; engages us two Ma-Fus, 203; is a delightful companion, 206
Deung-Do-Lin-Ssŭ, ii. 423; road from, to Tz'ŭ-Kua, 207; analysis of water at, 232n; arrival at, 241
Dinner, a Chinese, i. 125–127; ii. 23–26, 200, 201
Disappointments of travellers, i. 369, 370
Do-Kû-La-Tza, ii. 421
Dong, gorge of (Cooper's 'Duncanson Gorge'), ii. 227; village of, 229, 423
Douglas, Mr., on the birthplace of the Chinese nation, i. 139; his translation of inscription, ii. 11, 12
Dragon-bridge of Cooper, the, i. 290
Dragon-hill Mountains, i. 60
Drinking, moderate, of Chinese, i. 400
Drought, prayers during, i. 131, ii. 27, 38; threatenings of, 27; at Yün-Nan-Fu, 302
Drums at Peking, i. 127, 128
Duck-gun, a Chinese, i. 190
Duncanson Gorge of Cooper, the (Dong), ii. 227
Dyer, Mr. Bernard, his analysis of water of Deung-Do-Lin, ii. 232n
Dzeh-Dzang-Chŭ, river, ii. 165, 422
Dzong-Dä (Dry Sea), ii. 163, 422
Dzung-Ngyu (Dzongun), ii. 217, 423

EDUCATION, defects of Chinese, i. 307
Eggs, preserved ducks', i. 400
'Encyclopædia Britannica,' i. 139, 141n
Erh-Hai, lake (Ta-Li), ii. 295, 300
Etiquette, Chinese, ii. 200, 234
Europeans, Chinese dread of, i. 155
Examinations, public, i. 133, 149, 316; military, ii. 297, 355
Exchange, troubles of, i. 272, 273; rate of at Ta-Chien-Lu, ii. 91

FANG-MA-CH'ANG, ii. 345, 425; arrival at, 349
Fans, Chinese use of, i. 280

FLO

Fan-T'ai, civil official, ii. 22, 429
Farewell ceremonies, ii. 233, 234
Farmer Point, on the Yang-Tzŭ-Chiang, ii. 416
Fasts, ii. 27, 38, 302
February, 1877, i. 179-192, ii. 415, 416
Fei-Lung-Kuan (Flying Dragon Pass), ii. 49, 420
Fei-Yueh-Hsien, ii. 66
Fei-Yueh-Ling (Fly beyond Range), ii. 65, 67, 420
Female decorations, ii. 145, 148
Fêng-Hsiang-Fu, i. 128
Fêng-Ngan-P'u, ii. 417
Fêng-T'ai, i. 100, 118, ii. 414
Fêng-T'ai-Tzŭ, i. 107, ii. 414; ferry over canal at, i. 107, 108
Fêng-Tu, ii. 417
Fêng-Tung-Kuan, (Wind Cave Pass,) i. 381-383, ii. 419
Fields-within-Fields,' General, visit from, i. 270, 271
Fires in towns, superstitions as to, ii. 38
Fire-wells at Tzŭ-Liu-Ching, i. 299
Fish, of Western China, i. 347; Tibetan aversion to, ii. 138
Fleming, i. 85
Fleuve-Bleu, Le, French appellation for the Yang-Tzŭ-Chiang, i. 166
FLOWERS AND PLANTS—
 Arnica, ii. 133
 Azalea, i. 386
 Buttercup, ii. 122, 127, 133, 143
 Castor-oil plant, i. 37
 Chia-Chu-T'ao (Rhododendron), ii. 174n
 China Aster, ii. 220
 Cotton-plant, i. 37
 Creepers, i. 256; flowering, 338; gigantic, ii. 396
 Crocus, purple, i. 367
 Delphinum Grandiflorum (possessing the property of destroying lice), ii. 126, 426
 Ferns, i. 206, 230, 256, 335, 338, 386, 388, ii. 359; maiden-hair, i. 230; tree, ii. 397
 Flowers, bright yellow, in great masses, ii. 143, 212, 213
 Flowers, wild, i. 372, 381, 385, 392, ii. 57, 122, 126, 127, 132, 133, 143, 160, 167, 169, 192, 225
 Geranium, i. 4
 Indigo, ii. 297
 Iris, i, 367
 Jasmine, i. 230, 338
 Lily, i. 392
 Lotus, i. 380, ii. 17

438   INDEX.

FLOWERS AND PLANTS—continued.
Mint, ii. 182
Mistletoe, i. 78
Moss, i. 256 ; long, hanging from trees, ii. 134, 238
Pedicularis Ramalana, ii. 426
Peony, i. 386
Plant, with leaves growing downwards, ii. 162
Poppy (Opium), i. 230, 231, 252, 254, 279, 282, 293, 314, 326, 392
Primula, pink, i. 230; *Gillii*, ii.426
Rape, i. 197, 244, 245, 246
Rhododendron (called Ta-Ma by the Tibetans), ii. 54, 174, 180, 224, 238, 250, 263
Rhubarb, i. 379, ii. 35, 151
Rose, i. 372, ii. 65
Safflower, i. 311
Saxifraga Gillii, ii. 426
Shrub like the laurel, with a white flower like the wild rose, ii. 134
Spenceria Ramalana, ii. 426
Stellera-conf. S. concinna (plant used for making parchment), ii. 426
Sunflower, ii. 51, 260, 266, 273, 293
Ta-Ma (Rhododendron), ii. 54, 174, 180, 224, 238, 250, 263
Tzh-Tzh-Hua (tree with a flower like the Gardenia), ii. 54
Violet, i. 197
Flowery Bridge, the lesser, (Hsiao-Hua-Ch'iao), ii. 327 ; the greater (Ta-Hua-Ch'iao), 327
Foreigners in Chinese eyes, i. 73
Foreign names, Chinese rendering of, ii. 179, 208
Fruit trees, rarely noticed, i. 63 ; absence of, 411

FRUITS AND VEGETABLES—
Apples, i. 28, 62, 247, ii. 48
Apricots, i. 396
Artichoke, Jerusalem, ii. 342
Blackberries, i. 398
Brinjalls, i. 62, 405, 412, ii. 342
Cabbages, i. 108, 205, 252, 270, 405, ii. 156, 251, 253
Cherries, ii. 135, 140
Chestnuts, i. 62, ii. 263, 273, 323
Chilies, ii. 320, 342
Cucumbers, i. 57, 396, 405, 412, ii. 37, 51, 380
Currants, ii. 122, 140, 221, 250, 356
Date, Chinese, i. 37
Date-plum(*DiospyrosKaki*),ii.246n
Fruits, i. 28, 36, 62
Gooseberries, ii. 122, 221
Gourd, i. 65

Grapes, i. 28, 62, 66, 96
Greengages, i. 411, ii. 48
Lemon, ii. 315
Loquot (*Eriobotrya Japonica*, Pi-Pa in Chinese), i. 338, 396
Melons, i. 208, 406, ii. 37 ; fruit like a melon, ii. 315
Mushrooms, ii. 129, 218
Onions, i. 142
Peaches, i. 28, 62, 247, 411, ii. 48, 140, 184, 246, 260
Pears, i. 28, 62, 247, ii. 273, 284, 306, 317, 335, 342
Persimmon (*Diospyros Virginiana*), ii. 246, 314, 342
Pi-Pa or Loquot, i. 338, 396
Plums, i. 247, ii. 48
Pomegranates, ii. 273, 306
Potatoes, i. 58, 65, 345, 378, ii. 73, 156, 282, 342
Pumpkins, i. 57, 62, ii. 263, 266
Quinces, ii. 373, 380
Raspberries, wild, ii. 356
Strawberries, i. 282, 391, ii. 250
Turnips, i. 405, ii. 156, 266
Walnuts, i. 62, 399, ii. 260, 273, 312, 314, 317
Water-melons, i. 51
Yams, i. 20, 62, 246
Fu, civil official, i. 252, 316, ii. 80, 255, 280, 430
Fu-Chiang, military official, ii. 429
Fu-Chou, i. 254, 256, ii. 417
Fu-Hsing-Ch'ang, ii. 59, 60, 421
Fu-Ma, the, ii. 81
Fu-Piau(P'u-P'iao of Baber ?), ii. 345
Fu-T'ai, civil official, i. 316, ii. 428
Fu-Tu-Sze, ii. 157

GABET, ii. 171
Gala day, a Chinese, i. 399
Galle, i. 5
Garnier, Lieut., ii. 135n, 296n, 386
Geese, wild, on the Yang-Tzŭ, i. 192; used as house guards, ii. 106
Gelatine, the foundation of every Chinese delicacy, i. 126
Ghora-Walla, i. 157
Gobi, i. 84
Gôitre, prevalence of, ii. 232
Gold, washing for, i. 82, 392, 393, ii. 149 ; from Lit'ang, 93
Golden Sand, river of, (Chin-Sha-Chiang), i. 165, 256, ii. 122, 154 ; approach to, 180 ; Bat'ang river joins, 188 ; no road down valley of, 207 ; valley of, 209 ; cross, and quit it, 211, 423 ; in view again,

INDEX.

246; river regained, 264; breadth of, 265; river swollen by rains, 274; finally left, 274
Gombo-Kung-Ka mountain, ii. 166
Gordon, Colonel, i. 154
Gorges of the Yang-Tzŭ entered, i. 206; gorge of Mi-Tsang, 220; of Wu-Shan, 227, 228, 229, 232; a luxuriant gorge, 387, 388; a desperate gorge, ii. 219, 220; gorge of Dong, 227
Government, the local, of Ta-Chien-Lu, ii. 80, 81
Grant Point on the Yang-Tzŭ-Chiang, ii. 416
Grapes, preserved Chinese, i. 28
Grass, a beautiful, ii. 315, 316
Graves of aborigines, ii. 348
Great Wall of China, first seen, i. 61; a pass in, 67; impressions of, 84; date of erection, 142; the builder of, ii. 17
Grosvenor expedition, the, ii. 343, 387
Gue-Ra, ii. 211, 423
Guide, description of our, ii. 178, 179
Gunpowder not invented by Chinese, i. 147, 148

HAELONE, ii. 426
Ha-Go-Ta, i. 79, ii. 414
Han dynasty, ii. 218
Han-Ch'êng, i. 100, ii. 414
Han-Chou, i. 417, ii. 420
Han-K'ou, ii. 414
'Hankow,' steamer, i. 172
Han-Kow, i. 13, ii. 415; correct spelling of, i. 175n; arrival at, 175; factory at, 176
Han-Yang, i. 179, ii. 415
Hats, official, i. 415, 416
Haul Bridge, a, i. 359, 360
Hei-Shui-Kou, ii. 419
Hei-Yu-Kuan, ii. 424
Hiang-Kouan of Garnier (Shang-Kuan), ii. 296n
Hill district of Upper Yang-Tzŭ entered, i. 196
History, characteristics of Chinese, i. 141
Hiung-Nu, i. 142, 143
Ho-Chiang-P'u, ii. 314, 424
Ho-Chuang, i. 96, ii. 414
Ho-Chŭ-Ka, ii. 153, 422
Hogg's Gorge, ii. 330
Ho-Hia, of Blakiston, i. 194
Ho-K'ou, (River mouth), ii. 77, 157, 421

Ho-Mu-Shu, ii. 353, 425; greedy host at, 353, 354
Honan, i. 128
Hong-Kong, i. 9, 10
Horse-dealing, i. 26, 34, 46, ii. 114, 115; search for a horse, ii. 108
Ho-Se-Wu, i. 41, 135, ii. 413, 415
Hostility of the Lamas, ii. 86
Hot spring at Hsin-Tien-Chan, ii. 124
Hou-Pei, i. 229
House, endeavours to take a, i. 320, 321; form of houses in A-Tun-Tzŭ, ii. 233; at Ka-Ri, 252; at Sha-Yang, 328
House-fittings, Chinese, i. 99
How-Ko-Lao, ii. 416
Hsia-Ken-To, (Mu-Khun-Do), ii. 266, 423
Hsia-Kuan, ii. 424; temple on road to, 311; situation of, 311; market day at, 312
Hsiang-Ho-Hsien, ii. 415
Hsiang-Ngai-Pa, i. 399, ii. 419
Hsiang-P'o, ii. 425
Hsiao-Ho (Little River), frequent occurrence of term, i. 166
Hsiao-Ho-Chiang, ii. 424
Hsiao-Ho-Ti, road to, ii. 364; village of, 364, 425
Hsiao-Ho-Ying, (Small river camp), i. 390, ii. 419
Hsiao-Hua-Ch'iao, (The Lesser Flowery Bridge), ii. 327, 425
Hsiao-Kuan (Little Pass), ii. 56, 421
Hsiao-La-Pu(Shio-Gung), ii. 259, 423
Hsiao-Pa-Chung, ii. 182, 422
Hsiao-P'êng-Pa, ii. 72, 421
Hsieh-T'ai, military official, i. 316, 351, ii. 81, 185, 196, 429
Hsien, civil official, i. 316, 340, ii. 326, 431
Hsien-Cheng, civil official, ii. 431
Hsien-Tang, i. 295, ii. 418
Hsi-Fêng-K'ou (Gate in Great Wall), i. 67, ii. 413
Hsin, river (Hsi-Ho?), i. 339
Hsin-An-Chen, i. 118, 119, ii. 415
Hsin-Ch'êng (New Town), ii. 369
Hsin-Chin-Hsien, ii. 40, 41, 420
Hsin-P'u, i. 407, ii. 420
Hsin-P'u-Kuan, i. 344, ii. 418
Hsin-Ti, i. 187, ii. 415
Hsin-Tien-Chan, (New Inn Stage), ii. 50, 421; another, 123, 421; hot spring at, 124; botanical specimens collected at, 426
Hsin-Tu-Hsien, i. 418, ii. 420
Hsin-Wên-P'ing, i. 338, ii. 418
Hsuan, visits of, i. 267

HSÜ

Hsüeh-Lan-Kuan, ii. 419
Hsüeh-Lung-Shan (Snow Dragon Mountain), i. 356
Hsüeh-Shan, (Snow Mountain), alleged terrors of, i. 382, 383; summit of, 384, ii. 419
Hsun-Chien, civil official, ii. 431
Hsun-Fu, civil official, ii. 428
Hua-Ch'iao, ii. 327, 425
Hua-Ch'iao-Tsu, ii. 420
Huai-Tien, i. 107, ii. 414
Hua-K'ou, ii. 415
Hua-Ling-P'ing, ii. 65, 421
Huang-Chuang, i. 116, ii. 414
Huang-Fu, illness of, ii. 134, 147, 157; recovery of, 168, 180, 235; reports accident to himself, 241; returns home, 309
Huang-Lien-P'u, ii. 320, 424
Huang-Ling-Miao, i. 208, ii. 416
Huang-Ni-P'u, ii. 54, 421; copper and iron mines at, 55
Huang-Shih-Ling, ii. 414
Huc, ii. 140, 156, 171, 174; remarks by, on water-melons, i. 51-53; on a Chinese dinner, 125, 126
Huc Reach on the Yang-Tzŭ-Chiang, ii. 416
Hui-Hui (Mahometans), rebellion of, ii. 303
Hung-Ngai-Kuan (Red Rock Pass) i. 385, ii. 419
Hung-Pai, ii. 257
Hun-Shui-T'ang (Troubled Water Station) ii. 348, 425
Huo-Hsüeh (Ho-Hia), i. 194, ii. 416
Hu-Tung, ii. 417

I-CH'ANG, ii. 416; Consul to, i. 178; arrival at, 199; disturbances at, 199-202; departure from, 205
Ili or Kuldja, i. 24n
Imam-Reza, shrine of the, ii. 85
Imperial tombs, i. 60
Industry of Chinese, i. 246-248
Inn, a Chinese, i. 38, 39; inns of Ssŭ-Chuan, 285, 297; a dirty, 398
Insect, the Wax, treatment of, ii. 62, 63
Interpreter engaged, an, ii. 93, 94
I-Ran, or I-Jen (the Man-Tzŭ), i. 355, 356, 359
Irawadi, first sight of, ii. 409
Iron Gorge in the Yang-Tzŭ-Chiang, ii. 417
Iron mines at Huang-Ni-P'u, ii. 55
Irrigation, a method of, i. 411, 412
1 T'ou-Ch'ang, ii. 421

KAN

I-T'u, i. 197, ii. 416
I-Yang-T'ang, ii. 284, 424
I-Yuan-Kou, i. 84, ii. 414

JANG-BA (Pa-Mu-T'ang) ii. 214, 423
January, 1877, i. 172-175, ii. 415
Ja-Ra (King of Mountains), ii. 133
Jarlung (White River), the cradle of Tibetan monarchy, ii. 137n
Jaxartes river, i. 143
Jeddo of Cooper, ii. 119, 122
Jehol, i. 74
Jen-ma, ii. 104
Jervois, Sir William, i. 5, 6
Jessundee of Cooper (Chŭ-Sung-Dho), ii. 219
Jewellery, largely worn by Tibetans, ii. 79, 145; abundance of, at Yung-Ch'ang, 341
Jie-Bu-Ti (Chi-Dzûng), ii. 264, 423
Jieh-Kang-Sung-Doh, ii. 225, 423
Jing-Go-La, ii. 237, 250; two mountains of same name, 250, 423
Jinnyrickshaw (a public conveyance in Shanghai), i. 10; accident to, 11, 12; derivation of word, 163n
Jiom-Bu-T'ang (the Flat Plain), ii. 160, 422
Johnson, Mr., i. 211
Jo-La-Ka, ii. 229, 423
Joss-house, derivation of term, i. 278
J'ra-La-Ka, ii. 177, 422
Ju-Kan (Black Man-Tzŭ), i. 365
July, 1876, i. 4; 1877, ii. 26-99, 420, 421
June, 1876, i. 2; 1877, i. 366-418, ii. 7-26, 419, 420
Jung-Ch'ang-Hsien, i. 286, 287, ii. 417

K'AI-FUNG-FU, i. 128
Kai-Ping, i. 97, 98, ii. 414; coal found at, i. 103
Ka-Ji-La, ii. 133, 421
Kakyen huts, ii. 394
Kan-Ch'i, ii. 418
Kang, a, described, i. 39; mismanaged, 135, 136
Kang-K'ou, i. 81, ii. 414
Kang-Shi, or Khang-Hi, emperor ii. 13, 14
Kan-Lan-Chan, ii. 357, 425
Kan-Lu-Ssu, ii. 425
Kan-Ngai (Chiu-Ch'êng and Hsin-Ch'êng), the 'Muangla' of Anderson, ii. 369, 370n, 426

INDEX

KAO

Kao-Ch'iao, ii. 420, 421
Kao-Li-Kung, ii. 425
Karahokah of Anderson (T'ai-P'ing-Chieh), ii. 381, 426
Ka-Ri, ii. 251, 423; houses of, 252
Kashgar, i. 143; proposed journey to, given up, ii. 7, 8
Ker-Rim-Bu (the Octagon Tower), ii. 136, 421
' Kestrel,' H.M.S., at Hankow, i. 178; visit to, at Tung-Shih, 194, 195
Khatas (scarves of felicity), ii. 105, 107
Khitan, (Liao Tartars), i. 128, 143, 144
Khoten, i. 143
Kia-Ching, or Kia-King, fifth emperor of the present Chinese dynasty, ii. 14
Kia-Ne-Tyin, ii. 215, 423; house at, 216
Kiang-Ka ('Vermilion River' of Cooper), ii. 218
Kian-Suy, ii. 2
Kia-Ting-Fu, ii. 61, 67
Ki-Chan (Keshen), ii. 172
Kin (Niuchih Tartars), i. 129, 143, 144
King, Mr., consul at I'Ch'ang, i. 178
King of mountains (Ja-Ra), ii. 133
Kirin (Central Manchuria), i. 80
Kiun-Liang-Fu, military provision storekeeper, ii. 77, 80, 97
Ko-Erh-Shi-Shan, ii. 133, 421
Koko-Nor, district of, i. 331, 379
Kong-Tze-Ka, ii. 212, 423
Ko-Ta-Pa, ii. 419
K'ou-Lung-Shan, a mountain with a natural tunnel, i. 79
Kuan-Ching, i. 69, ii. 413
Kuang-Yi, i. 397, ii. 419
Kuan-Hsien, ii. 418; old gates of, i. 330; temple at, 332; curiosity of people of, 332
Kuan-P'o, ii. 425
Kuan-Yin-P'u, ii. 420; road from, 48
Kuan-Yu, ii. 18
Kublai Kaan, i. 144; sporting grounds of, 98, 111
Ku-Ch'êng, i. 347, ii. 418; return to, i. 357; another, 397, ii. 419
Ku-Ch'iao-Tzŭ, ii. 414
Ku-Deu (Chi-Tien), the ' Ku-Tung ' of Baber (?), ii. 269, 423
Kuei-Chou, i. 220, ii. 416
Kuei-Chou-Fu, i. 233, 235, ii. 416; salt manufacture at, i. 236, 237; aqueduct of, 238, 239
Kuldja (Ili), i. 24n, 153

LA-N

Ku-Ling Pass, the, i. 82, ii. 414
Kung, Prince of, a visit to, i. 51
Kung-Chŭ, ii. 262n, 423
Kung-Kuan, i. 312; a wretched one, ii. 153, 154, 179
Kung-Ming-Sien-Sen (Chu-Ko-Liang), ii. 18
Kung-Nung, ii. 76
Ku-Tung of Baber, ii. 269
Kwei-Chou province, i. 140
Kwei-Yang-Fu, ii. 7

LA (Tibetan for ' Pass '), ii. 132
Ladder Village, the (T'i-Tzŭ-Yi), i. 393
Ladrone Islands, i. 9
Lady Skipper, the, i. 179; history of, 223, 224; the last of, 263, 264
Lai river, the, crossing of, i. 66
La-Ka-Ndo, ii. 166, 422
Lakong, ii. 426
Lamas, the, violent hatred of, to foreigners, i. 268; a Si-Fan Lama, 377; Lamas of Tibet, ii. 82–86; hostility of, 86, 197, 198; Lama, Tibetan word for monk, 189; Lamas a curse to the country, 191; oppression of, 214; a magnificent Lama, 230; description of Lamas at Ku-Tung, 270; Lamas called Phoongyees, 373
Lamassery, a Si-Fan, i. 380; a visit to, ii. 103–106; Lamassery at Lit'ang, 156; Lamassery at Bat'-ang, 189; great number of, and how filled, 189–192; Lamassery of Deung-Do-Lin-Tz'u, 242; origin of word, 242; Chinese name for, 242
Lama-Tz'u, Chinese name for Lamassery, ii. 242
La-Ma-Ya, ii. 165, 422
Lamp, a Chinese, description of, i. 49
Lan, river, ii. 414
Landlord, the Christian, i. 297
Lang-Ch'iung-Hsien, plain of, ii. 283, 285, 294; position of, 289; road to, 290; arrival at, 290, 424; dilapidated state of, 291; civility of a landlord at, 291, 292; sulphur spring near, 293; departure from, 294
Language, difficulties of the Chinese, i. 212; the Si-Fan, 378; an alphabetic, ii. 367
Lan-Ho, the, i. 94
La-Ni-Ba (Hollow between Two Mountains), ii. 143, 421; botanical specimens collected at, 426

INDEX

LAN

Lan-Ts'ang-Chiang or Mekong River, ii. 274, 425; water-parting between, and Chin-Sha-Chiang, 224, 239, 275, 276; no bifurcation of, 316; crossing of, 330
Lan-Tu, ii. 417
Lao-Chu-T'a, ii. 415
Lao-Mu-K'ou, i. 95, 96, ii. 414
La-Pu (Ta-Chio), ii. 259, 423
La-so, a term of respect, ii. 121
Lassa, expected arrival of English and Russian missions at, i. 268; journey to, abandoned, ii. 8; pilgrims on the way to, 161; armed opposition on road to, 215
La-Tza (Root of the Mountain), ii. 132, 421
Legation, the British, in Tien-Tsin, i. 21; in Peking, 43
Leguilcher, Monsieur, the Provicaire, ii. 301, 303; a visit from, 305 bid farewell to, 311
Leh, ii. 422
Le-Ka-Ndo, ii. 166, 422
Lekin taxes, the, ii. 231; in Li-Kiang, 278
Lêng-Chi, ii. 67, 68, 421
Leng-Ling-Tzŭ, ii. 413
Lêng-Shui-Ching, ii. 425
Lê-Ting-Hsien, i. 94, ii. 414
Letters of credit, use of, i. 32
Liang-T'ai, or Liang-T'ou, civil official, ii. 157, 183, 429
Liao Tartars (Khitan), i. 128
Liau-Tung, i. 80
Li-Chia-P'u, ii. 425
Li-Chŭ, ii. 154, 159, 422
Liebig's Meat Extract, value of, i. 50
Li-Fan-Fu, i. 319, ii. 418; arrival at, i. 349; wonders of, 351-353; description of, 353
Life-boats on the Yang-Tzŭ-Chiang, i. 213, 214
Light, Chinese dislike of, i. 388; light from pine-splinters, ii. 138
Li-Hung-Chang, Chinese minister, i. 14, 15; departure of, 16
Li-Kiang-Fu, ii. 159
Lily Pool, i. 238, 239
Limestone scenery, ii. 53
Lin-Ting-Chen, i. 116-119, ii. 414, 415
Li-Sieh-Tai, the mother of, travelling, i. 369; visit from, 385
'Li-Ta,' translations of, ii. 244
Lit'ang, ii. 422; jurisdiction of, 81; the town in sight, 156; Lamassery at, 156; alleged meaning of name, 156

MA-F

Lit'ang-Ngoloh, ii. 145-149, 422
Littré, on derivation of word 'Pagoda,' i. 157
Li-Tu, i. 256, 257, ii. 417
Liu-Liu (The Willow), a servant, i. 273; behaviour of, 281; accident to, 389
Liu-Pi, emperor, history of, ii. 18 20; drives back barbarians, 76
Liu-Pu, Chinese boards of administration, ii. 427
Liu-Pu-Shang-Shu, civil officials, ii. 428
Liu-Yang, ii. 421
Living water, Chinese, ii. 112
Lo-Chi, ii. 417
Lo-Chiang-Hsien, i. 407, ii. 420
Lo-Chiang-Ta-Ho, ii. 420
Loh-Chung-Wan-Tun, a Hsieh-Tai, visit from, ii. 185; account of, 196
Lo-Shan, ii. 415
Lou, a watch-tower, i. 157
Lu, river (also Tung-Ho, or Ta-Ho), ii. 67
Lu-Chiang, or Salwen River, ii. 425; unhealthiness of valley of, 350, 351
Lu-Chiang-Pa, ii. 352, 425; sultry climate of, 352
Lu-Chi-K'ou, ii. 415
Lu-Hua-Ho, i. 365, ii. 419
Lu-Jiong-La-Ka, Mt., ii. 265, 423
Lung-An-Fu, i. 395, ii. 419
Lung-Ch'ang-Hsien, ii. 417; seen in the distance, i. 290; arrival at, 292; departure from, 293
Lung-Chiang (Shuay-Li, or Shwé-Li River), ii. 344, 357, 425
Lung-Tan-P'u, i. 373, ii. 419
Lung-Zung-Nang, ii. 225, 423
Lu-T'ai, i. 102, ii. 414
Lu-Ting-Ch'iao, ii. 421; suspension bridge of, 69; road from, 72
Lying and truth-telling, i. 239
Ly-Kouo-Ngan (Pacificator of Kingdoms), ii. 171

MACARONI-making, i. 339, 340; rice macaroni, ii. 334
Macartney, Earl, i. 74
Ma-Ch'ang, ii. 424
Ma-Chin-Tzu, ii. 417
Ma-Dung-La-Tza, ii. 422
Ma-Fang-Chou, ii. 417
Ma-Fu, the (horse-boy), i. 34; dress of, 35; loses road, 40; comes to grief, 59; leads ponies, 76; his method of tying up horses, 112, 113; his vagaries, 115; the term

INDEX.   443

MAH

untranslatable, 157; the new Ma-Fu, ii. 100; dress of, 226; another new Ma-Fu, 310
Mah-Geh-Chung, ii. 140, 421
Mahometans (Hui-Hui), rebellion of, ii. 303
Maize-loaves, i. 372
Major and Smith, of Hankow, their agent at Kuei-Chou-Fu, and his torrent of speech, i. 234, 235
Ma-Kian-Dzung, ii. 140
Malacca, Straits of, i. 5
Ma-Lan-Chen, i. 61
Ma-Lan-Yu, i. 60, ii. 413
Ma-Mou (Sicaw), ii. 384, 426; old Bhamo, 390n; journey to, 396-405
Manchus, the, i. 12, 144, ii. 3
Mandarin, derivation of word, i. 158n
Man-Ga-La, ii. 165, 422
Mani inscriptions, ii. 84
Manning, Mr., i. 268
Man-Tzŭ, the (Barbarians), tradition concerning, i. 323; first village of, 341; search for a village of, 354, 355; black and white, 365
Manwyne (Man-Yün), ii. 344, 426
Man-Yün (Manwyne), ii. 344, 426
Manzi of Marco Polo, i. 144, ii. 1
Mao-Chou, i. 360, ii. 418
Mao-Niu, ii. 104
Mao-P'u, ii. 415
Mao-T'sao-T'ang, ii. 424
Ma-Pen, a, ii. 212; house of, 214
Ma-Ping, a horse-soldier, ii. 431
Ma-Ra, ii. 221, 423
March, 1877, i. 194-252, ii. 416, 417
Marco Polo, i. 61n., 115, 144, 148, 418, ii. 356, 411n; on Kublai Kaan's sporting grounds, i. 98, 111; his description of Peking, 132; of Ch'êng-Tu, ii. 1; covered bridge of, not to be found, 9, 289; 'Carajan' of, 299; his description of lake of Yün-Nan, 300; his mention of gold found in neighbourhood of Ta-Li-Fu, 302; his description of Yün-Nan salt, 312; 'Vochan' of, 335; his mention of large pears, 342; discussion of itinerary of, from Ta-Li-Fu, 342, 343; his account of valley of the Lu-Chiang, 350
Margary, Mr., i. 266; the Margary proclamation, 349, 350; is taken care of, ii. 259; favourable impression made by, 331, 367; scene of his murder, 389

MOS

Market of I-Ch'ang, peculiarity of, i. 205
Martini, Padre Martin. His description of Ch'êng-Tu, ii. 3, 4, 5, 35, 36
Ma-T'ou-Ying, i. 96, ii. 414
'Mauvais Pas,' a, i. 394
Mawphoo, ii. 425
May, 1877, i. 293, 339, ii. 418, 419
Mayers, Mr., i. 147
M'beh, ii. 257
McCarthay, Mr., illness of servants of, ii. 321
Mê, a Chinese Christian, i. 263; troubles of, 266
Meadows, T. T., 'Desultory Notes' of, ii. 427, 428
Medicine Mountains, ii. 123, 152; ascent of one (Ra-Ma-La), 142
Mei-Hua-P'u, ii. 424
Mekong River, the (Lan-Ts'ang-Chiang), ii. 274, 425; water-parting between, and Chin-Sha-Chiang, 224, 239, 275, 276; no bifurcation of, 316; crossing of, 330
'Mélanges Posthumes' of Abel-Rémusat, i. 125
Mesny Mr., i. 318; ii. 401, 402, 404, 427, 428; arrives at Ch'êng-Tu, 7
Miao-Chi, rapid, and river-wall of, i. 242, ii. 416
Miau-Tzŭ of Kwei-Chou, aborigines of China, i. 140
Mien, King of, ii. 335n.; province of, 343, 411
Mien-Chou, i. 405, ii. 420
Mien-Chu-La, ii. 239, 423
Min River, i. 334, 339
Min-Ching, ii. 334
Ming dynasty, i. 239, ii. 6, 13
Ming-Shan-Hsien, ii. 420; road to, 45
Miracle cave, a, i. 389
Mi-Tsang gorge on the Yang-Tzŭ-Chiang, i. 220, ii. 416
Mixt Court of Shanghai, i. 170
M'jeh (Tibetan for rice), ii. 257
Momein (T'êng-Yüeh), ii. 360, 425
Monastery of Wen-Shu-Yüan, ii. 13-17
Money current, at Shanghai, i. 32; between Peking and Tien-Tsin, 33; at Peking, 44
Mongolia, i. 24, 142
Mongols, the, i. 143, 144
Moquoor of Cooper (Mu-Kwa), ii. 259
Morra, a Chinese game, ii, 203
Morrison, Mr., i. 147n
Moscow, Great bell of, i. 131

INDEX

Mosquito-nets, ii. 8
Mou-Pin, disturbances of, i. 324
Muangla of Anderson (Kan-Ngai), ii. 370*n*, 426
Mu-Kwa (Moquoor of Cooper), native chief, ii. 259
Mu-Khun-Do (Hsia-Ken-To), ii. 266, 423
Mu-Lung-Gung, ii. 144, 422; name mispronounced, 179
Mûng-M'heh, ii. 423
Mu-Su people, the (Moosoo), ii. 265, 269, 270
Mu-Su-P'u, i. 364, ii. 419

NA, meaning of, ii. 239
Nahlow, ii. 426
Nah-Shi, ii. 421
Namon, ii. 426
Nampoung River, ii. 426
Nam-Sanda of Anderson, ii. 373
Nan-Ching, commonly called 'Nanking,' i. 42*n*
Nan-Ching-Yi, i. 308, ii. 418
Nan-Ho (Southern river), ii. 42, 420
Nan-Mu-Yuan, i. 227, ii. 416
Nan-Tien, ii. 345, 365, 425
Nan-Tsai-Tsan, ii. 413, 415
Nan-Yang-Chen, i. 66, ii. 413
Nats (spirits), ii. 393
Nawa, a Si-Fan Lama, i. 377
N'doh-Sung, ii. 254, 423
Nen-Chū river, ii. 166
Nen-Da (the Sacred Mountain), ii. 166; glorious view of, 167, 169; village, 169, 422
Nepal, Chinese name for, ii. 103
Nescradin, ii. 336
New Mandalay, ii. 411
Newton, i. 145
New-year festival, a, i. 185, 186
Ngan-Ch'a-Ssū, civil official, ii. 429
Ngan-Hua-Kuan, i. 374, ii. 419
Ngan-Niang-Pa, ii. 127, 421
Ngan-Te-P'u, ii. 418
Nga-Ra-La-Ka, ii. 162, 422
N'geu-La-Ka (Pai-Na-Shān), ii. 239, 423
Ngê-Wai, military official, ii. 431
Ngoloh, ii, 129, 421; a halt at, 130; cattle plague at, 132
Nia-Chū, ii. 135
Nia-Chū-Ka, ii. 137, 421
Nia-Rung, or Nia-Jung, ii. 143
Nieh-Ma-Sa, ii. 220, 423
Nieh-T'ai, civil official, ii. 429
Nine Nails Mountain, the, i. 362

Ning-Hai, i. 88; ii. 414; desecration of, i. 88, 89
Ning-Yuan-Fu, ii. 61
Niu-Chang (cattle feeding-ground), ii. 149, 422
Niu-Chieh, ii. 285; rudeness of officer at, 285; his servility, 286; wretchedness of people of, 287; market day at, 288; position of, 288, 424
Niuchih Tartars (Kin), i. 129
Niu-Fu-Tu, i. 294, ii. 418
Niu-Kan Gorge, scenery of, i. 211
Niu-K'ou, i. 224, ii. 188, 416
Niu-K'u, ii. 209, 422
Niu-P'ing P'u, ii. 424
Niu-T'ou-Yai, i. 91, ii. 414
Niu-Wa, ii. 346
Nomekhan, the, ii. 215
Nomenclature, difficulties of Chinese, i. 211, 212
November, 1876, i. 127–138, ii. 415; 1877, ii. 408–411, 426

OCTOBER, 1876, i. 66–22, ii. 413–415; 1877 ii. 310–408, 424–426
Odoric, Friar, i. 144, 165
Offerings, sham, i. 279
Om Mani Pemi Hom,' the universal Buddhist prayer, ii. 79, 83, 103, 165
Opium, smoked by coolies, i. 194; its effect on bees, 231; increased cultivation of, 282; smoking of, prevalent in Ssū-Ch'uan, ii. 45; and Yün-Nan, 235; proverb concerning, 235: laziness of opium-smokers, 281, 293
Ornaments of people of Ta-Chien-Lu, ii. 79; of Tibetan women, 145, 148
Orthography of Chinese names, i. 212
Owen, Professor, i. 145
Oxus river, the, i. 139

PAGODA, derivation of word, i. 157, 158
Pa-I people, ii. 386
Pai (a hundred), ii. 44*n* (White), 239
Pai-Chang-Yi, ii. 44, 420
Pai-Chou Ssū, ii. 415
Paidza, a grain somewhat resembling rice, ii. 283
Pai-Fên-Ch'iang, ii. 271, 423
Pai-La, the white wax of Ssū-Ch'u-an, ii. 61

INDEX. 445

PAI

Pai-Lo-Ssŭ, ii. 416
Pai-Ma-Kuan (Pass of the White Horse), i. 410, ii. 420
Pai-Na-Shăn (N'geu-La-Ka), ii. 239
Pai-Shi-Yi-Ch'ang, i. 279, ii. 417
Pai-Shui-Chai, ii. 418
Pai-T'u-P'u, ii. 424
Pai-Yen-Ching, ii. 319n, 335
Pa-Kou-Lou, ii. 136, 421
Pa-K'ou-Ying (Ping-Chuan-Chou), i. 71, ii. 413
Palaces, city of, arrival at, ii. 411
Pa-La-Kiang, ii. 417
Palisade barrier, the, non-existent, i. 80
Pa-Ma-La, ii. 238, 423
Pa-Mu-T'ang (Jang-Ba), ii. 212, 423
Pan-Ch'iao (in Ssŭ-Ch'uan), i. 341. ii. 418; (in Yün-Nan), native description of road to, 333; town of, 334, 425; river of, 335
Pan-Chiu-Ngai, ii. 61, 64, 421
Pang-Chün, i. 58, ii. 413
Panthays, or Panthés, the so-called, ii. 303n
P'ao, various meanings of, i. 148
Pao-Ho, i. 69
Pao Ta-Laoye, ii. 81; courtesies of, 97; visit from, 101; bids us farewell, 112; excellent character of, 163
Pao-Ti, i. 119
Pao-Ti-Hsien, ii. 415
'Pao-Ting,' steamer, i. 137; makes better passage than 'Chih-Li,' i. 138
Parsimony, Chinese, i. 285, 286
Parson's Point, on the Yang-Tzŭ-Chiang, i. 190, ii. 416
Passports, a name necessary for, i. 55, 56
Patience exercised, ii. 236
Pa-Tsung, military official, ii. 255, 431
Pa-Tung-Hsien, i. 226, ii. 416
Pau, a cake of tea, ii. 47
Peat-streams, i. 348
Pe-Chi-Li, province, i. 4, 15, ii. 169; plain of, i. 98; gulf of, 17, 102, 138
Peh-Ma, an interpreter, history of, ii. 94; character of, 95; illness of, 180; returns home, 186
Peh-T'ang, i. 98
Pei-Ching, correct spelling of Peking, i. 42n
Pei-Ho, i. 18, 114, 121; navigation of, 18, 20; crossing of, 56, 123

POP

Pei-Kung, ii. 414
Pei-Mên-P'u, ii. 424
Pei-T'ang, i. 100, 102, 103, ii. 414; commandant of, i. 104; ferry at, 105
Pei-Tsai-T'sun, i. 114, ii. 414
Peking, i. 2, 14, 15, 21, 22, 23, 24, 26, 41, ii. 413, 415; arrival at, i. 42; fair in, 42; 'Temple of Heaven' at, 43; the city left behind, 54; return to, 123; fuel used in, 123; Confucian temple at, 127; summer palace at, 129; bell temple at, 131; second departure from, 134, 141
Peun-Kop-Pa, ii. 198
Phoongyees (Lamas), ii. 373; bell of, 410
Photography under difficulties, i. 271, 272
Picnic, invitation to, ii. 21-26
Pieman, wares of Chinese, i. 93, 94
Pien-Ching, i. 128
Pigeon-English used at Shanghai, i. 155-157
Pigeon-poultice, a, ii. 139
Pi-Hsien, i. 327, ii. 28, 418; politeness of official at, 328
Pinchon, Monseigneur, Bishop of Ch'êng-Tu, visit from, i. 319
Pin-Fan-Ying, ii. 419
P'ing-I-P'u, i. 400, ii. 419
P'ing-Man-Shao, ii. 424
P'ing-P'o, ii. 425
P'ing-Shan-Pa, ii. 416
P'ing-Ting-Kuan, i. 371, ii. 419
'Pioneer of Commerce,' the, ii. 51
Pipes, Chinese, i. 81; a coolie's pipe, 374
Pi-Pon-Tzŭ, Chinese name for Nepal, ii. 103
'Pirate,' the, of Woo-Sung, i. 153
P'i-T'iao-Ch'ang, ii. 420
Plague, a recent, ii. 345; the plague at Florence in 1348, and of London, 346
PLANTS, see FLOWERS and TREES
Po-Lo-Tai, i. 67, ii. 413
Pong-Tung, or Pong-Chou, grave of, i. 408; account of, 408, 409; death of, 410
Ponline, ii. 426
Ponsee, ii. 426
Pony, accident to Chung-Erh's, ii. 228; eccentricities of, 358
Population, mistakes concerning. i. 72, 73; density of, on rice-plains, ii. 276, 277; diminution of, in Yün-Nan, 315, 319, 322

446　INDEX.

POR

Porpoises in the Yang-Tzŭ-Chiang, i. 192
Porridge, buckwheat, ii. 332
Postal system of Chinese merchants, i. 261
Posts, Imperial, ii. 130
Potato, rapid spread of, i. 345
Potosi, altitude of, ii. 156*n*
Poung-Dze-Lan (Pong-Sera), wrongly placed on maps, ii. 246
Prayer-cylinder, the, or Prayer-wheel, ii. 83, 84, 114; endeavours to buy a, 233
Prejevalsky, ii. 151
Presents of local delicacies, ii. 306
Probity, commercial, of Chinese, i. 275
Prome, ii. 411
Property, transfers of, i. 309
Provisions, price of, at Peking, i. 124; at Ch'ung-Ch'ing, 270; at Ch'êng-Tu, 325; at Ta-Chien-Lu, ii. 91
Provôt, Monsieur, French missionary at Ch'ung-Ch'ing, i. 261; visit from, 267
Pu-Ch'êng-Ssŭ, civil official, ii. 429
Pu-Erh tea, how made, ii. 267
P'u-Ho-Ying, i. 92, ii. 414
Pu-Ku, a wooden cup, ii. 113
Pu-Lang-Kung, ii. 144, 422
Pung-Cha-Mu, or Pun-Jang-Mu, ii. 179, 422
Pung-Gien-Tiyin, ii. 423
Pung-Shi, ii. 426; attack at, 392
Pun-Jang-Mu, ii. 122, 175, 422
P'u-P'iao of Baber (Fu-Piau?), ii. 345, 425
Pu-Ping, a foot soldier, ii. 431

RAIN, prayers for and against, and superstitions about, i. 131, ii. 27, 38, 301, 302
Ra-Ma-La, ii. 143, 421; ascent of, 142; botanical specimens collected on, 426
Rangoon, ii. 411
Ra-Nung, ii. 422
Rapids, passage of Ta-Tung, i. 209; on Yang-Tzŭ-Chiang, 209, 210; dangers of, 213, 214; passage of Ch'ing-Tan, 216-219; passage of Wu-Shan, 232, 233
Rasselas, the happy valley of, probable position of, ii. 152
Ra-Ti, or San-Pa, the 'Tsanba' of Cooper, ii. 170, 422
Raw hide, manifold uses of, ii. 176
Red basin of Ssŭ-Ch'uan, the, i. 308

SAN

Red Rock Pass (Hung-Ngai-Kuan), i. 385
Regiment, a Chinese, i. 305
Renou, Monsieur, makes bargain with a Lama, ii. 242
Restaurant, characteristics of a Chinese, i. 288, 289
Rice, not sole food of Chinese, i. 82; as food, 193; miraculous supply of, 250; used to repair vessels, 255; culture of, 280; rice cakes, ii. 264; rice macaroni, 334
Richthofen, Baron von, i. 2, 3, 24*n*, 140, 301, 308, ii. 157, 271, 296; on the insect tree, 63
Ritter, ii. 4*n*, 5; his description of Ch'eng-Tu, 34-36; of Ch'ing-Ch'i, 58; of Fei-Yueh-Ling, 66
River, a divided, i. 334; direction of the great rivers of Western China, ii. 228
RIVER OF GOLDEN SAND, *see* Chin-Sha-Chiang
Road and rivers, relation between, ii. 206, 207
Roads, when repaired, ii. 60; fragment of an ancient, 74
Roadside pictures, i. 412-414
Rock, a curious, i. 249
Rock-discs, oracular, i. 229
Ron-Sha, hospitable welcome at, ii. 258, 423
Roofs, Chinese fondness for, i. 283
Ru-Ching, mountain, ii. 119
Rûng-Geh-La-Ka, ii. 259, 423
Rung-Se-La, ii. 173, 422
Rupee, the Indian, current at Ta-Chien-Lu, ii. 77

SACRED Temple Slope Mountain, i. 359
Saigon, i. 6, 7; unhealthiness of, 8; post-office at, 8
Sailors, good qualities of Chinese, i. 17, 19
Sa-Ka-Tying, ii. 257, 423
Salt, Chinese, i. 50; manufacture of, 92, 236, 237; Marco Polo's cakes of, ii. 312, 334
Salt-wells, of Tzŭ-Liu-Ching, description of, i. 298-302; of Chao-Ho-Ching, ii. 319*n*
Salwen river, the (Lu-Chiang), ii. 344, 345, 425
San-Cha, i. 83, ii. 414
San-Chiao-Ch'êng, ii. 421; road beyond, 65
Sanda, of Anderson (Chan-Ta), ii. 371, 426

INDEX. 447

SAN

Sand-ridge, a miraculous, i. 344
San-Ho, the, crossing of, i. 58
San-Hsien-Ku, ii. 273, 424
San-Pa (the three plains), the 'Tsanba' of Cooper, ii. 170, 422
San-Pa-Shan, ii. 172, 422
San-Tai, i. 79, ii. 414
School, a village, i. 306
Sculls, use of the great, on Chinese junks, i. 258, 259
September, 1876, i. 10-62, ii. 413; 1877, ii. 216-310, 423, 424
Serai, Arabic word, ii. 242
Servants, the, threaten desertion, ii. 88; discharge of, at Bat'ang, 185
Sewage, utilisation of, in Chinese farming, i. 114n
Sha-Ho, i. 93, ii.414
Sha-Ho-Chiao, ii. 413
Sha-Lu, ii. 247, 423
Sha-Mu-Ho, ii. 425
Shan lady, costume of a, ii. 367
Shang-Chê-Wan, i. 188, ii. 416
Shanghai, i. 10, 11, 12, 13, 14, 25, 33, 123, 138, 152; in winter, 155; theatre of, 160-163; sport at, 164-169; mixt court of, 170; second departure from, 172
Shang-Kuan (the upper barrier), the 'Hiang-Kouan' of Garnier, ii. 296, 424
Shang-Kuan-Chi, ii. 417
Shang-Shu, civil officials, ii. 427, 428
Shan-Hai-Kuan, i. 54, 84, 86, 134, ii. 169, 414; derivation of name, i. 86
Shan-Kên-Tzŭ (Root of the Mountain), ii. 132, 421
Shan-Si, province, Chinese settlement in, i. 139, 140
Shan-Tung, i. 13, 141
Sha-Shih, i. 194, ii. 416
Sha-Wan (the sandy hollow), i. 368, ii. 419
Sha-Yang, plain of, ii. 328; houses at, 328, 425
Shen-Si, province, i. 128, 142; road to, 398
Shie-Gi-La, mountain, ii. 155, 422
Shieh-Zong, ii. 253, 423
Shih-Chia-Ch'iao, ii. 421
Shih-Ch'iao, i. 312, ii. 418; inn at, 313
Shih-Chia-P'u, 388, ii. 419
Shih-Ch'uan-P'u or Shih-Ch'uan-Shao, ii. 313, 424
Shih-Kia-To, ii. 417
Shih-Ku (Stone-drum Town), ii. 273, 424

SNO

Shih-Men, i. 60, ii. 413
Shih-Men-Chai, ii. 414
Shih-Pan-Fang (Stone-slab-house Mt.), i. 371, 372
Shih-Pao-Chai (Stone-jewel Fort), curious rock of, i. 249, ii. 417
Shih-Wolo, ii. 422
Shin-Ka, ii. 140
Shio-Gung (Hsiao-La-Pu), ii. 259, 423
Shio-Gung-Chŭ, ii. 259, 262, 264
Shi-Show-Hsien, ii. 416
Shoes, Mongol, i. 75
Shou, a Tu-Ssŭ, ii. 183; hospitality of, 201-203
Shou-Pei, military official, ii. 157, 430
Shrine of the Imam Reza, ii. 85
Shuang-Liu, ii. 420; the march to, 37; departure from, 38
Shuang-Pao (Double Gem), a Ma-Fu, ii. 102, 153
Shuang-Tou, ii. 414
Shuayduay, ii. 425
Shuaykeenah, ii. 426
Shuay-Li river (Lung-Chiang), ii. 344, 425
Shui-Chai, ii. 425
Shui-Ching-Chan, i. 392, ii. 419
Shui-Ching-P'u (Crystal Village), i. 392, ii. 419
Shui-Kou-Tzŭ, i. 367, ii. 419
Shui-Yin-Ssu, temple of, ii. 331
Shun-Pi, river, ii. 318, 320, 424
Shu-Wang (King of Shu), grave of a concubine of, ii. 17
Shwé-Li (Shuay-Li) ii. 344, 357n
Shwo-La, ii. 239, 423
Siau-Chiang, ii. 416
Sicaw (Ma-Mou), ii. 384, 426
Si-Fan, the, i. 375, 385
Silk, manufacture of, i. 69
Silver, used as money, i. 33, 45; weighing of, 45, 46, ii. 28, 31
Sindafu, of Marco Polo, ii. i.
Si-Ngan-Fu, i. 24n, ii. 10, 171
Singapore, arrival at, i. 5; botanical gardens at, 6
Skipper Point on the Yang-Tzŭ-Chiang, ii. 416
Sladen, ii. 371, 384
Slavery in Tibet, ii. 194
Small-pox, fatality of, among Tibetans, ii. 115
Snow-Dragon Mountain (Hsüeh-Lung-Shan), i. 356
Snow Mountain, the (Hsüeh-Shan), alleged terrors of, i. 382, 383; summit of, 384

SNO

Snowy Mountains around Ta-Li-Fu, ii. 301
Soil, a rich piece of, i. 244
Soldiers, dress of Chinese, i. 329
Soongaris, i. 80
Southern River (Nan-Ho), ii. 42
Sport at Shanghai, i. 164-169
Sporting grounds of Kublai Kaan, i. 98, 111
Spot, (Mr. Carles' dog), occasions much speculation, i. 58, 59; taken for a lion, 78, 79
Spring, a hot, ii. 124
'Squeeze, the,' memorial of the system of, ii. 42
Ssŭ-Ch'uan, (Province of Four Streams), i. 229, ii. 33; good roads of, i. 284; good manners of people, 286; character of people of, ii. 70, 71
Ssŭ-Ma-Ch'ih, i. 128
Steamers on the Yang-Tzŭ-Chiang, i. 170, 171; how built, 172
St. George's Island in the Yang-Tzŭ-Chiang, ii. 417; accident to junk at, i. 254, 255
St. Jaques' Point, i. 7
Stone, with inscription, ii. 10; translation, 11, 12; a stone called 'Tooth of Heaven,' 12, 13
Stone-Slab-House Mt., the (Shih-Pan-Fang), i. 371, 372
Stores taken by Captn. Gill from Shanghai to Ch'ung-Ch'ing, i. 159n
Stormonth, on derivation of word 'Pagoda,' i. 157
Suliman, Sultan (Tu-Wên-Hsiu), ii. 303
Sulphur-baths, hot, ii. 159; hot sulphur spring at Hsin-Tien-Chan, 124; sulphur-spring near Lung-Ch'ien-Hsien, 293
Summer palace at Peking, a visit to, i. 129
Su-Mu (White Man-Tzŭ), i. 365
Sun, nephew of the Fu of Wei-Si, ii. 259, 260, 261, 265
Sunflower seeds used instead of water-melon seeds, ii. 51, 260, 273, 293
Sung dynasty, i. 128, ii. 13
Sung-Fan-Ho, ii. 419
Sung-Ngai-P'u, ii. 419
Sung-P'an-T'ing, a military station, i. 324, ii. 419; description of, i. 375; limestone valley of, ii. 53
Sung-Shao, ii. 424
Sun-Kiouen of De Mailla, ii. 19n

T'AN

Surong Mountains, ii, 155
Suspension bridge, a rope, i. 335; an iron, 391

TA, a peculiar kind of tower, i. 157
Table-manners and customs, ii. 23-26
Ta-Cha-Ho, i. 211
Ta-Chien-Lu, bishop of, i. 268; river of, ii. 67; approach to, 73; arrival at, 75, 421; meaning of name, 76; the boundary of China, 76; situation of, 78; local government of, 80, 81; departure from, 117, 118
Ta-Ch'i-K'ou, ii. 417
Ta-Chio, (La-Pu), ii. 259, 423
Tael, value of, i. 69, 272
Ta-Fo-Ssŭ, ii. 417
Ta-Ho, (Great River), frequent occurrence of the term, i. 185, 186; boat descent of, 401
Ta-Hua-Ch'iao, (Greater Flowery Bridge), i. 328, 425
T'ai-Hsiang-Ling-Kuan (Pass of Great Minister's Range), ii. 56, 57, 421
Tai-Ping canal, ii. 416
T'ai-P'ing-Chên, i. 283, ii. 417, 424
T'ai-P'ing-Chieh, the 'Karahokah' of Anderson, ii. 381, 426
Tai-P'ing-P'u (Peaceful Village), ii. 318, 356, 424, 425
Tai-Tzŭ-Fu, i. 121, ii. 415
T'ai-Wang, i. 128
Taking it easy, ii. 244, 245
Ta-Kuan (Great Pass), ii. 57, 421
Taku Bar, i. 18, 138
Ta-Laoye (Great Excellency), ii. 97n, 183
Ta-Li, lake of (Erh-Hai), ii. 295, 300, 311
Ta-Liang-Tzŭ, ii. 414
Ta-Li-Fu, road from, to Burmah, ii. 207, 313; road to, 296; ruined condition of the town and country round, 296-303; arrival at, 298; 'Carajan' of Marco Polo, 299; position of, 299, 424; plain of 300, 302; marble quarries of, 302; visiting at, 307, 308
Ta-Li-Shao, ii. 333, 425
Ta-Li-Shu, ii. 425
Tan-Fu, i. 128
Tang (a military post), ii. 150
Tang-Chia-Ho, i. 95, ii. 414
T'ang dynasty, i. 128
Tang-Go-La Mt., ii. 149, 422
T'ang-Tzŭ-P'u, ii. 424

INDEX. 449

TAN

Tang-Ye, ii. 182
T'ao-Kuan, ii. 418
Tao-T'ai, civil official, i. 263, ii. 301, 307, 308, 430
'Ta-Pa-Ing,' boat on the Irawadi, ii. 410
Ta-Pan-Ching, ii. 425
Taping river, ii. 371, 426
Ta-Shiu, or Ta-So, ii. 175, 422; name mispronounced by Chinese, 179
Ta-So-Shan, ii. 177, 422
Ta-T'ang-P'u, ii. 44, 420
Ta-Ting, i. 366, ii. 419
Tattooing, ii. 388
Ta-Tung rapid, i. 209, ii. 416
Ta-Tzŭ-K'ou, i. 77, ii. 414
Taxes, Chinese, ii. 193, 231, 278
Ta-Ying river, ii. 360; increased size of, 370
Tea, scarcity of, in North China, i. 50; delicious, 374; Chinese method of making, 162, ii. 267; the kind sent to Tibet, 47; used for making payments, 77, 78, 107; buttered tea, 110
Tea-oil tree, Cooper's mistake about, ii. 54
Teashop, a way-side, i. 280; characteristics of, 287, 288
Teen Shan mountains, i. 139
Telescope, the, causes amusement, i. 198
Temple of Heaven at Peking, i. 43
Teng-Chia-K'ou, i. 183, ii. 415
Têng-Ch'uan-Chou, ii. 424; road and distance to, 290
T'êng-Yüeh (Momein), ii. 336, 425; arrival at, 360; river of, 408
Terrors of the hills, alleged, i. 382, ii. 56
Tê-Yang-Hsien, i. 415, ii. 420; official civilities at, i. 416
Thang dynasty, i. 143
Theatre, the Chinese, at Shanghai, i. 160-163
Three Brothers confederacy, the, ii. 18
Three Kingdoms, the, i. 143, ii. 19
Threshing-ground, a, i. 38; threshing in Mongolia, 70
Thrift, excess of, i. 315
Thsin, Princes of, i. 142
Thunderstorm, a grand, i. 284
Tiao-Kuan, ii. 416
Tiao-Shih, i. 232, ii. 416
'Tib', a dog, purchase of, i. 171; his sense of smell, 173; goes to dinner, 241; left behind, ii. 114
Tibet, Lamas of, ii. 82-86; gradual

TRE

depopulation of, 195, 221; boundary of, 215
Tibetans, great quantity of jewellery worn by, ii. 79, 145; independence of, 87; chief food of, 89; contrast between Tibetans and Chinese, 89-91; houses of, 126; hospitality of, 127, 128; groups of, 127, 128; horse-shoeing of, 136; their aversion to fish, 138; female decorations, 145; their mode of building, 167, 168; independent, 181; some customs of, 195; have no dread of mountain passes, 225; their superstition concerning their wooden bowls, 233
T'ieh-Ch'ang (ironworks), ii. 326, 424
Tieh-Chi-Ying, i. 367, ii. 419
Tieh-Lung-Kuan, ii. 419
T'ien-Ching-P'u, Mt. ii. 321, 326, 327, 425; village of, 323, 424
Tien-Tsin, i. 13, 15, 18, 20, 21, ii. 413, 415; hotels of, i. 27, 31; scenes on wharf at, 28; building at, 30; our guns at custom-house of, 33, 34; departure from, 36-38; supplies from, 110; river of, in winter, 134; second arrival at, 137; cheap living at, 137; second departure from, 137
Tiger's-Tooth rock, the, i. 204
Tiger Teeth Gorge on the Yang-Tzŭ-Chiang, ii. 416
Tinc-Chais, i. 322; dress of, 416; officiousness of, ii. 49
T'ing, civil official, i. 358, 361, ii. 430
Ting-Ko, ii. 118, 123
Ti-T'ai, the, of Ta-Li-Fu (General Yang), ii. 308
Ti-T'u, military official, ii. 428
T'i-Tzŭ-Yi (the ladder village), i. 393, ii. 419
Ti-Zu, ii. 124, 421
Tombs of the Manchu Emperors, i. 60
Tou-P'o-Shao, ii. 318, 424
Tracking on the Yang-Tzŭ, vicissitudes of, i. 189, 190, 222
Transport arrangements, ii. 81, 82
Travelling by torchlight, ii. 241
Trees cut down to place in the way of travellers, ii. 384
TREES—
Acacia, i. 61, ii. 48
Acacia Rugata (the soap tree) i. 364
Apricot, i. 225, 227, 230, 330, 338, 342, 348, ii. 67, 135

VOL. II.   G G

450 INDEX.

TRE

TREES—*continued*
Artemisia, fragrant, i. 106
Ash, i. 338, 397, 413, ii. 48
Bamboo, i. 167, 168, 196, 208, 225, 244, 251, 252, 278, 290, 325, 337, 386, 388, 391, 413, ii. 21, 37, 48, 55, 67, 227, 297, 314, 373, 380, 390, 391, 393, 394, 397
Banyan, i. 244, 249, 278, ii. 48, 368, 374, 379, 388
Barberry, i. 338, 366, 372, 392, ii. 122, 135, 140, 167, 253, 257
Beech, i. 330, 338
Birch, ii. 135
Bramble, i. 230, 338
Buckthorn (*Rhamnus Theezans*, the Chinese Date Tree—Tsao) i. 37
Cactus, ii. 379
Cassia (Kwei-Hua), ii. 263
Cedar, i. 196
Ch'a-Yo (Tea-oil Tree), ii. 54
Cherry, ii. 135, 140, 237
Chestnut, i. 65, 67, ii. 257, 258, 263, 271, 322, 323
Chien-a-ragi, ii. 217
Ch'ung-Shu (a tree whose leaves are used for making a decoction instead of tea), i. 253
Cinnamon, ii. 303
Currant, ii. 122, 134, 135, 140, 167, 172, 180, 182, 221, 224, 245, 250, 356
Diospyros Kaki (the Date-plum Tree) ii. 246n
Diospyros Virginiana (the Persimmon Tree) ii. 246n
Elm, i. 78
Eriobotrya Japonica, or Loquot (Pi-Pa), i. 216, 338
Fir, i. 196, 208, 227, 252, 290, 293, 348, 367, 372, ii. 48, 131, 134, 228, 257
Gooseberry, ii. 122, 131, 132, 134, 135, 167, 172, 180, 182, 212, 221, 224, 245
Hawthorn, i. 62
Hazel, ii. 250
Hibiscus, a species of (Shwui-Kin), ii. 63
Holly, i. 391, ii. 134, 174, 258; dwarf, 132
Holly-leaved Oak, ii. 149, 174, 180, 212, 253, 256, 257, 263, 265, 268
Insect Tree, ii. 61
Juniper, ii. 185, 213, 217
Kwei-Hua (Cassia) ii. 263
Laburnum, i. 44
Lemon, wild, ii. 315

TRE

Ligustrum Glabrum (Tong-Tsing), ii. 63
Ligustrum Lucidum, ii. 62
Mulberry, i. 397, ii. 188
Niu-Tching (*Rhus Succedanea*), ii. 63
Oak, ii. 142, 180, 182, 212, 215, 220, 221, 224, 225, 227, 228, 237, 238, 245, 247, 250, 253, 257, 258, 271, 322, 323; with leaves like holly, 149, 174, 180, 212, 253, 256, 257, 263, 265, 268; on the leaves of which silkworms are fed (*Quercus obovata*), i, 69; water oak, ii. 48
Orange, i. 208, 216, 227, 293, 330, ii. 48, 264
Peach, i. 230, 338, 342, ii. 135, 140, 219, 246, 253, 257
Pear, ii. 246, 257
Persimmon (*Diospyros Virginiana*) ii. 246, 257, 314
Pine, i. 206, 225, 316, 366, 369, 372, 384, ii. 43, 68, 132, 133, 134, 135, 140, 142, 149, 160, 164, 167, 169, 173, 174, 178, 180, 182, 212, 213, 215, 217, 220, 221, 224, 225, 227, 237, 238, 239, 246, 247, 250, 256, 258, 263, 265, 268, 269, 274, 276, 279, 318, 322
Pi-Pa or Loquot (*Eriobotrya Japonica*), i. 216, 338
Plane, i. 4
Plantain, ii. 368, 373
Plum, ii. 67, 135, 182, 253, 257
Pomegranate, i. 278, 290, 392, ii. 257, 263
Poplar, i. 65, 121, ii. 212, 221, 224, 250, 253
Prickly Pear, ii, 295, 296, 322
Quercus obovata, i. 69
Raspberry, ii. 356
Rhamnus Theezans (the Chinese Date Tree—Tsao—a kind of Buckthorn or Jujube), i. 37
Rhus Succedanea, ii. 63
Shwui-Kin (a species of Hibiscus), ii. 63
Soap Tree (*Acacia Rugata*), i. 364
Tea Tree, ii. 208
Teak, ii. 396
Tea-oil Tree (Ch'a-Yo), ii. 54
Tong-Tsing, (*Ligustrum Glabrum*), ii. 63
Tree, whose flowers are used for making a decoction instead of tea, (Ch'iung-Shu), i. 253
Tree, with a flower like the Gardenia, (Tzh-Tzh-Hua), ii. 54

INDEX. 451

TRE

Tree, something like an orange with strongly-scented blossoms (the Insect Tree), ii. 61
Tree, something like a Willow (the Wax Tree), ii. 61
Tree, with a leaf like the Plane, ii. 135
Tree, in appearance like the Walnut, ii. 140
Tree, with thorns, and leaves like the English oak, ii. 250
Tree-fern, ii. 397
Trees growing like one another, ii. 323
Tsao (*Rhamnus Theezans*, the Chinese Date Tree), i. 37
Tung-oil Tree, i. 304, 396, ii. 48
Tzh-Tzh-Hua, ii. 54
Vines, i. 63, 399, ii. 41, 46, 188
Walnut, i. 65, 67, 338, 342, 348, ii. 48, 65, 67, 69, 74, 135, 219, 229, 246, 247, 253, 257, 262, 263, 269, 271, 314
Water-oak, ii. 48
Wax Tree, ii. 61
White-wood Tree, ii. 67
Willow, i. 1, 38, 61, 65, 67, 70, 75, 119, 121, 185, 192, 227, 290, 325, 330, 348, ii. 134, 167; weeping, 132, 134, 219
Yew, i. 44, 67, 249, 278, 330, ii. 122, 169, 176, 213, 217; dwarf, 172
Triumphal Arches, i. 290, 291, 414
Tsai-Chia-Ling pass, ii. 414
Tsa-Leh (Tsali), ii. 221; arrival at, 222; altitude of, 223, 423
Tsa-Leh-La-Ka, ii. 207, 223, 224, 423
Tsanba (oatmeal porridge), ii. 89; a supper of, 119
Tsanba of Cooper, ii. 170
Tsan-Chiang, military official, ii. 429
Ts'ao-Hsieh-Ping (straw-sandal flat), a tea-house, ii. 58
Tsao-Lin, i. 58, ii. 413
T'sao-T'sao, (Chao-Tsiao) ii. 19*n*
Tsao-Yun (Tze-Lung), ii. 19, 20
Tsin-Huang-Chen, ii. 414
T'sin - Shih - Hwang - Ti, emperor, builder of Great Wall, ii. 17
Tsu-Ch'iao, ii. 420
Tsung-Tu or Chih-T'ai, civil official, ii. 428
Tsun-Hua-Chou, i. 62, ii. 413
T'uan-Lin, i. 93, ii. 414
Tu-Fu, poet, ii. 21
Tung-Che-Ka, ii. 421
Tung-Chih, civil official, ii. 430
Tung-Chou, i. 21, 22, ii. 413, 415; arrival at, i. 54; return to, 122

WAD

Tung-Chu-Ling, ii. 423
Tung-Golo (Tung-olo of Cooper), ii. 129, 421
Tung-Kwan, i. 139
Tung-Pan, civil official, ii. 431
Tung-Shih, i. 194, ii. 416
Tung-Ting lake, pass entrance to, i. 188, ii. 416
Tung-Tze-Yuen, ii. 420
'Tun-Sin,' steamer, i. 174, 175
Turfan, i. 143
Turkai, ii. 222
Turquoise beads used as money, ii. 77, 78, 107
Tu-Ssŭ, military official, ii. 183, 430
Tu-Sze, a Tibetan chief, ii. 366
Tu-T'ung, military official, ii. 428
Tu-Wên-Hsiu (Sultan Suliman), ii. 303
Tzu-Chou, i. 304, ii. 418
Tz'ŭ-Kua, ii. 423; road to, 207; arrival at, 272
Tzŭ-Liu-Ching, i. 293, 295, ii. 418; inn at, i. 296; salt-wells of, 298-302; fire-wells of, 299
Tzŭ-Shih-Li, ii. 420

'UGLY Chief of Homely Virtues,' the, ii. 366
Ulo, the, a gigantic scull, i. 179
Urh-Chuang, i. 107, ii. 414; arrival at, i. 110, 111
Urh-T'ai-P'o, ii. 344, 360, 425
Urumchi, ii. 7
Usuri (Russian territory), i. 80

VANDALISM, an act of, i. 89
VEGETABLES. See FRUITS
Vegetables, compressed, value of, i. 50
Vegetation, luxuriant, i. 251
Vermilion river, the (Kiang-Ka), ii. 218
'Vigilant,' the, despatch boat, i. 14, 15, 16, 26
Village school, breakfast in a, i. 305
Villages, picturesque, i. 57; the first aboriginal village, ii. 72
Vines, absence of, i. 399
Visiting cards, Chinese, i. 325
Visits, official, i. 262, 263; visits of ceremony, ii. 233, 234
Vochaṇ of Marco Polo, ii. 335, 342, 344

WADE, SIR THOMAS, i. 14, 15, 21, ii. 362; leaves Peking, i. 127; his system of orthography, 212, 213

G G 2

WAI

Wai-Ta-Chen, ii. 423
Wai-Wei, military official, ii. 431
Wall, the Great, of China, first seen, i. 61; a pass in, 67; impressions of, 84; date of erection, 142; builder of, ii. 17
Wan, ii. 417
Wan-Ch'iao, ii. 424
Wang, a Chien-Tsung, ii. 247
Wang-Chia-Ch'ang, i. 293, ii. 418
Wang-Gi-La, ii. 152, 422
Wang-Tua-Shan, i. 129
Wan-Li-Ch'iao, ii. 18
Wan-Sung-An, ii. 424
Wa-Ssŭ-Kou, ii. 73, 421; valley beyond, 73
Water of Deung-Do-Lin, analysis of, ii. 232n
Water-melon seeds, extraordinary predilection of Chinese for, i. 51–53
Water-mills, i. 337
Water-power, a simple application of, ii. 275
Water-wheels, i. 401
Wax insect, the, treatment of, ii. 62, 63
Wax, white, of Ssŭ-Ch'uan (Pai-La), ii. 61
Wei, or Goei, dynasty, ii. 19n
Wei (Siberian nomads), i. 143; a Fan-Tai, ii. 22
Wei-Mên-Kuan, i. 363, ii. 418
Wei-Si, rapacity of the Fu of, ii. 255
Wei-Yuan-Hsien, i. 304, ii. 418
Wên-Chêng, i. 359, ii. 418
Wên-Ch'uan-Hsien, i. 340, ii. 418
Wên-Shu-Yüan (Literary Book Hall Temple of Ch'êng-Tu), ii. 13–17
Wet season, a, ii. 301
White Cloud Mountain, i. 358
White Horse, the Pass of the (Pai-Ma-Kuan), i. 410
White Man-Tzŭ (Su-Mu), i. 365
Williams, ii. 342
Wind-Cave Pass (Fêng-Tung-Kuan), i. 381–383, ii. 419
Wind-points, Chinese view of, i. 186
Wo-La-Ku, i. 118, ii. 414
Wonders, local, i. 351, 352, 358
Woo-Sung railway, attacks on the, i. 152, 153
Wo-Shih-Wo, ii. 425
Wrangles, Chinese, i. 112
Wu-Ho-Tz'ŭ Temple (Military Marquis' Memorial Chapel), ii. 6, 18, 20, 21
Wu-Ngo-Nü, enchantress of Mount Wu, i. 227n
Wu-Ru-Chung-Ku, ii. 421

YUL

Wu-Rum-Shih, ii. 130, 134, 421
Wu-San-Kwei (Pacificator of the West), account of, ii. 169
Wu-Shan, gorge, i. 227; aspect of, 228, 229; town of, 232, ii. 416
Wu-Yai-Ling (Range without a Fork), ii. 65, 421
Wu-Yang, i. 252, ii. 417
Wylie, Mr., ii. 9

YA-CHOU, ii. 46, 47, 420
Ya-Lung-Chiang, ii. 122, 135, 137; ferry across, 139
Yamên, the, description of, ii. 97
Yang, ii. 22; meanings of word, 311
Yang-Chia-Ch'ang, ii. 420
Yang-Chia-Kai, i. 310, ii. 418
Yang, General, ii. 308
Yang-Hsien, i. 310, ii. 418
Yang-Pi, ii. 424; official at, 315; river of, not a bifurcation of the Mekong, 316
Yang-Shan, i. 81
Yang-T'ai, i. 227n
Yang-Tsun, i. 38, 137, ii. 413, 415
Yang-Tu-Chi, ii. 417
Yang-Tzŭ-Chiang (ocean river), i. 10, 13, 143; various names of, 165; size of, at Hankow, 175, 176; boat engaged for navigating Upper, 180, 181; tracking on, 189, 190, 222; porpoises in, 192; change of scenery on, 206, 207; the gorges entered, 207; mode of navigating the gorges, 206, 207, 222; passage of Ta-Tung rapid, 209, 210; the river called 'Ta-Cha-Ho,' 211; life-boats on, 213, 214; Ch'ing-Tan rapid, 216–219; coal districts entered, 222; Wu-Shan rapid, 232, 233; distances on, ii. 415–417
Yao-Ch'iao, ii. 420
Ye-Chŭ river, ii. 422
Yeh, a silk merchant, i. 320, ii. 22; commissioner, i. 400n
Yeh-T'ang, i. 391, ii. 419
Ye-La-Ka, ii. 165, 422
Yellow river, i. 139
Yen-Ch'iao, i. 57, ii. 413
Yen-Yün-Ssŭ, civil official, ii. 429
Yin-Hsiu-Wan, i. 337, ii. 418
Yu-Chi, military official, ii. 366, 429
Yu-Ch'i, i. 336, ii. 418
Yueh-Erh-Ngai, i. 387, ii. 419
Yule, Colonel, i. 2, 148; note by, on word 'Pagoda,' 158; on word 'Mandarin,' 158n; notes by, throughout both volumes

INDEX.

YU-L

Yu-Lung-Chou, ii. 334
Yung-Ch'ang, plain of, ii. 333;
  city of, 336, 425; Scene at inn of,
  339, 340; abundance of jewels at,
  341; market of, 342
Yung-Ch'ing-Ho, ii. 55
Yung-Ching-Hsien, ii. 51, 52, 53, 421
Yung-Ch'uan-Hsien, i. 282, ii. 417;
  politeness of people of, i. 282
Yung-Fêng-Chuang, ii. 425
Yung-Kuo-Ssü-T'ang, ii. 425
Yung-P'ing, plain of, ii. 324; river, 327
Yung-P'ing-Hsien, ii. 324; destruction of, 325

ZU-G

Yün-Nan, prov., roads to, ii. 49, 60,
  137, 207; devastated country, 314
Yün-Nan-Ch'iao (Bridge of Yün-Nan), ii. 169, 422
Yün-Nan-Fu, drought at, ii. 302
Yün-Yang, i. 243, ii. 416
Yu-So, ii. 424
Yvo, emperor, ii. 3

ZARDANDAN of Marco Polo, ii. 342
'Zing-Nan-Zing,' steamer, i. 13
Zu-Gunda, ii. 150, 422

THE END.

LONDON: PRINTED BY
SPOTTISWOODE AND CO., NEW-STREET SQUARE
AND PARLIAMENT STREET